APERIODIC CRYSTALS

INTERNATIONAL UNION OF CRYSTALLOGRAPHY BOOK SERIES

IUCr BOOK SERIES COMMITTEE

Ch. Baerlocher, *Switzerland*
G. Chapuis, *Switzerland*
P. Colman, *Australia*
J. R. Helliwell, *UK*
K.A. Kantardjieff, *USA*
T. Mak, *China*
P. Müller, *USA*
Y. Ohashi, *Japan*
A. Pietraszko, *Poland*
D. Viterbo (*Chairman*), *Italy*

IUCr Monographs on Crystallography
1. *Accurate molecular structures*
   A. Domenicano, I. Hargittai, editors
2. *P.P. Ewald and his dynamical theory of X-ray diffraction*
   D.W.J. Cruickshank, H.J. Juretschke, N. Kato, editors
3. *Electron diffraction techniques, Vol. 1*
   J.M. Cowley, editor
4. *Electron diffraction techniques, Vol. 2*
   J.M. Cowley, editor
5. *The Rietveld method*
   R.A. Young, editor
6. *Introduction to crystallographic statistics*
   U. Shmueli, G.H. Weiss
7. *Crystallographic instrumentation*
   L.A. Aslanov, G.V. Fetisov, J.A.K. Howard
8. *Direct phasing in crystallography*
   C. Giacovazzo
9. *The weak hydrogen bond*
   G.R. Desiraju, T. Steiner
10. *Defect and microstructure analysis by diffraction*
    R.L. Snyder, J. Fiala and H.J. Bunge
11. *Dynamical theory of X-ray diffraction*
    A. Authier

12 *The chemical bond in inorganic chemistry*
   I.D. Brown
13 *Structure determination from powder diffraction data*
   W.I.F. David, K. Shankland, L.B. McCusker, Ch. Baerlocher, editors
14 *Polymorphism in molecular crystals*
   J. Bernstein
15 *Crystallography of modular materials*
   G. Ferraris, E. Makovicky, S. Merlino
16 *Diffuse X-ray scattering and models of disorder*
   T.R. Welberry
17 *Crystallography of the polymethylene chain: an inquiry into the structure of waxes*
   D.L. Dorset
18 *Crystalline molecular complexes and compounds: structure and principles*
   F. H. Herbstein
19 *Molecular aggregation: structure analysis and molecular simulation of crystals and liquids*
   A. Gavezzotti
20 *Aperiodic crystals: from modulated phases to quasicrystals*
   T. Janssen, G. Chapuis, M. de Boissieu
21 *Incommensurate crystallography*
   S. van Smaalen
22 *Structural crystallography of inorganic oxysalts*
   S.V. Krivovichev
23 *The nature of the hydrogen bond: outline of a comprehensive hydrogen bond theory*
   G. Gilli, P. Gilli
24 *Macromolecular crystallization and crystal perfection*
   N.E. Chayen, J.R. Helliwell, E.H.Snell
25 *Neutron protein crystallography: hydrogen, protons, and hydration in bio-macromolecules*
   N. Niimura, A. Podjarny
26 *Intermetallics: structures, properties, and statistics*
   W. Steurer, J. Dshemuchadse
27 *The chemical bond in inorganic chemistry: The bond valence model*, second edition
   I.D. Brown
28 *Aperiodic crystals: From modulated phases to quasicrystals: Structure and properties*, second edition
   T. Janssen, G. Chapuis, M. de Boissieu

IUCr Texts on Crystallography
 1 *The solid state*
   A. Guinier, R. Julien
 4 *X-ray charge densities and chemical bonding*
   P. Coppens

8 *Crystal structure refinement: a crystallographer's guide to SHELXL*
   P. Müller, editor
9 *Theories and techniques of crystal structure determination*
   U. Shmueli
10 *Advanced structural inorganic chemistry*
   Wai-Kee Li, Gong-Du Zhou, Thomas Mak
11 *Diffuse scattering and defect structure simulations: a cook book using the program DISCUS*
   R. B. Neder, T. Proffen
13 *Crystal structure analysis: principles and practice, second edition*
   W. Clegg, editor
14 *Crystal structure analysis: a primer, third edition*
   J.P. Glusker, K.N. Trueblood
15 *Fundamentals of crystallography, third edition*
   C. Giacovazzo, editor
16 *Electron crystallography: electron microscopy and electron diffraction*
   X. Zou, S. Hovmöller, P. Oleynikov
17 *Symmetry in crystallography: understanding the International Tables*
   P.G. Radaelli
18 *Symmetry relationships between crystal structures: applications of crystallographic group theory in crystal chemistry*
   U. Müller
19 *Small angle X-ray and neutron scattering from solutions of biological macromolecules*
   D.I. Svergun, M.H.J. Koch, P.A. Timmins, R.P. May
20 *Phasing in crystallography: a modern perspective*
   C. Giacovazzo
21 *The basics of crystallography and diffraction, fourth edition*
   C. Hammond

Aperiodic Crystals

From Modulated Phases to Quasicrystals: Structure and Properties

2nd edition

Ted Janssen
Institute of Theoretical Physics, University of Nijmegen

Gervais Chapuis
École Polytechnique Fédérale de Lausanne

Marc de Boissieu
CNRS and Université Grenoble Alpes

Great Clarendon Street, Oxford, OX2 6DP,
United Kingdom

Oxford University Press is a department of the University of Oxford.
It furthers the University's objective of excellence in research, scholarship,
and education by publishing worldwide. Oxford is a registered trade mark of
Oxford University Press in the UK and in certain other countries

© Ted Janssen, Gervais Chapuis, and Marc de Boissieu 2018

The moral rights of the authors have been asserted

First Edition published in 2007
Second Edition published in 2018

All rights reserved. No part of this publication may be reproduced, stored in
a retrieval system, or transmitted, in any form or by any means, without the
prior permission in writing of Oxford University Press, or as expressly permitted
by law, by licence or under terms agreed with the appropriate reprographics
rights organization. Enquiries concerning reproduction outside the scope of the
above should be sent to the Rights Department, Oxford University Press, at the
address above

You must not circulate this work in any other form
and you must impose this same condition on any acquirer

Published in the United States of America by Oxford University Press
198 Madison Avenue, New York, NY 10016, United States of America

British Library Cataloguing in Publication Data
Data available

Library of Congress Control Number: 2018933084

ISBN 978–0–19–882444–2

DOI: 10.1093/oso/9780198824442.001.0001

Links to third party websites are provided by Oxford in good faith and
for information only. Oxford disclaims any responsibility for the materials
contained in any third party website referenced in this work.

Preface

In the eighteenth and nineteenth centuries the idea developed that, on the atomic level, crystals are constructed by regularly spaced unit cells. The mathematical theory for this idea was created by Bravais, Schoenflies, and Fedorov. After the discovery of X-rays in 1912 one really could prove this underlying periodic structure. Since then the prevailing idea has been that the ground state of matter at zero temperature is lattice periodic with a lattice constant of the order of a nanometre. Although there was actually no theoretical proof for this idea, it was generally accepted. In that view disordered systems, like glasses, form at most metastable configurations.

In the 1960s, several materials became known where besides the reflections at the points of a reciprocal lattice, there are also sharp peaks which do not fit into this scheme. The first to describe this situation was Pim de Wolff from Delft, who realized that the satellite peaks observed in anhydrous γ-Na_2CO_3 are not defects, but belong really to the structure, which is in this case an aperiodic modulated crystal phase. Although at very low temperature, the ground state is again periodic, the intermediate state between the periodic high-temperature and low-temperature phases is thermodynamically stable and aperiodic, but not disordered! Later several such materials were found. In fact, a substantial part of the minerals of the earth's crust turn out to be aperiodic. However, the term 'aperiodic' is not the proper characterization. As can be seen from the sharpness of the diffraction spots, these materials are just as well ordered as the usual crystals. One must agree that these materials should be called crystals as well. They have sharp diffraction peaks and flat facets. Nevertheless, these materials were not generally considered to be interesting. This changed with the discovery in 1982 of quasicrystals by Shechtman. These materials were intriguing because they show sharp diffraction spots and a rotation symmetry which is not crystallographic, in the sense that the symmetry is not compatible with lattice periodicity. This shows that the new materials are well ordered, but aperiodic. In fact, they are quasiperiodic in the mathematical sense, and, probably for that reason, they are called quasicrystals. After some years it turned out that the non-crystallographic symmetry is not an essential property. In that sense quasicrystals are not really different from modulated phases. They are all quasiperiodic crystals.

Since the discovery approximately 40 years have passed, and aperiodic crystals have been generally accepted as an interesting class of materials. This has brought about a change in well-accepted ideas about the structure of matter. Much work has been done on the determination of their structure and of their physical properties. Because of the aperiodicity the usual techniques, very often based on the presence of a Brillouin zone, cannot be used. New techniques had to be developed. The new materials have properties that are quite different from those of the usual lattice periodic crystals. This has led to new developments in chemistry, physics, crystallography, and even mathematics.

The field has now become mature, although there are still many open questions. A number of books on modulated structures, and, in particular, on quasicrystals have appeared in the meantime. However, the common property of quasiperiodicity means that many of the techniques can be used on the whole family of quasiperiodic materials. Therefore, it seemed a good idea to us to consider quasiperiodic crystals from a unified point of view, with applications in modulated phases and quasicrystals. Actually, it is very difficult to make a real distinction between these. It is easy to find materials where it is difficult to tell whether they belong to one class or to the other.

The book is intended for materials scientists, physicists, and crystallographers. We do not assume previous knowledge of aperiodic crystals. We only assume a basic knowledge of solid state physics and crystallography. We have tried to avoid a heavy mathematical formulation, but on the other hand we want to be sufficiently precise. Proofs are not always given. Instead, there are many references to the literature, and in a number of cases we give a more rigorous discussion of some point in a paragraph one can skip if one wants. These parts are indicated by ♯ at the beginning and ♮ at the end. Sections intended to give some more fundamental background, and which could be skipped, are also indicated by the sign ♯.

One of us (T.J.) wants to thank the Tohoku University in Sendai, in particular Professor An-Pang Tsai, for the hospitality during a visit, when part of this book was written. We thank many people for interesting discussions, and want to mention Alla Arakcheeva, Michael Baake, Esther Bélin-Ferré, S.I. Ben-Abraham, Roland Currat, Jean-Marie Dubois, Michal Dušek, Sonia Francoual, Franz Gähler, Denis Gratias, Mark Hollingsworth, Yoshihiro Ishibashi, Aloysio Janner, Ronan McGrath, Marek Mihalkovič, Lukas Palatinus, Manuel Perez-Mato, Václav Petříček, Penelope Schobinger-Papamantellos, Pat Thiel, Andreas Schönleber, Bertrand Toudic, Hans-Rainer Trebin, Sander van Smaalen, and Akiji Yamamoto.

Preface to the Second Edition

Since the first edition of this book in 2007, the field of aperiodic crystals has developed considerably. One has found new materials, and new structures. Progress has been made in the structure determination, in the interpretation and understanding of the structural characteristics and in the calculation of electrons and phonons. Therefore, we decided to update the book. New developments discussed in this edition include natural quasicrystals, incommensurate magnetic and multiferroic structures, and photonic and mesoscopic quasicrystals. The size of the book does not permit us to go into detail for these topics. So, the reviews are rather condensed and meant as a reference. Also, many new structures have been determined and new ways to find phononic and electronic properties have been developed. Furthermore, we have added a number of exercises to give the reader an opportunity to check his understanding of the material.

There are several excellent books on aperiodic crystals. We mention here van Smaalen (2012) and Steurer and Deloudi (2009). The first handles modulated phases and incommensurate composites, and the second, quasicrystals. Both deal with the crystallography of these compounds, not with their physical properties. A general introduction on quasicrystals, including their physical properties, is given in Janot (2012). Physical properties of aperiodic crystals are discussed in, among others, Blinc and Levanyuk (1986a, 1986b) for incommensurate phases and Stadnik (1999) and Trebin (2003) for quasicrystals.

Shortly after sending the new version of this book to the editor, we were deeply saddened to learn that our friend and coauthor Ted Janssen had passed away after a very sudden illness. He was still working on the book a few days before leaving us. Both remaining authors will dearly miss the frequent interactions with a very dear friend and great physicist, always willing to give us some professional and competent advices. We will also miss his presence in conferences where he always surprised us by his broad knowledge not only of the field of aperiodic crystals but also of the vast field of solid state physics.

Contents

Glossary		xvii
1	**Introduction**	**1**
1.1	Periodic crystals	1
	1.1.1 History	1
	1.1.2 Description	2
	1.1.3 The role of space group symmetry in the structure determination	6
	1.1.4 Symmetry and physical properties	7
	1.1.5 Examples	9
	1.1.6 Conclusion	10
1.2	Aperiodic crystals	10
	1.2.1 History	10
	1.2.2 Classes and examples	16
	1.2.3 Modulated phases	17
	1.2.4 Incommensurate composites	22
	1.2.5 Quasicrystals	25
	1.2.6 Morphology	30
1.3	Summary	31
2	**Description and symmetry of aperiodic crystals**	**32**
2.1	Aperiodic and quasiperiodic functions	32
2.2	Quasiperiodic structures	35
	2.2.1 Modulated phases	35
	2.2.2 Composites	40
	2.2.3 Quasicrystals	43
	2.2.4 Natural aperiodic crystals	46
	2.2.5 ♯ Electromagnetic crystals in space-time	47
2.3	Description in superspace	49
	2.3.1 Embedding	49
	2.3.2 Modulated phases	60
	2.3.3 Incommensurate composites	64
	2.3.4 Quasicrystals	71
	2.3.5 The classification into three (or four) types is not unique!	82
2.4	Symmetry	83
	2.4.1 Point group symmetry of diffraction patterns	83
	2.4.2 Superspace groups	86
	2.4.3 Examples	88

xii   *Contents*

	2.4.4	Approximants	90
	2.4.5	Superspace groups for commensurate phases	91
	2.4.6	Consequences of superspace group symmetry	93
2.5	Scaling symmetries		96
2.6	Alternative descriptions		100
2.7	Magnetic symmetry of quasiperiodic systems		104
	2.7.1	Magnetic systems and time-reversal symmetry	104
	2.7.2	Magnetic point groups	105
	2.7.3	Magnetic space groups	105
	2.7.4	The magnetic groups for quasiperiodic crystals	106
2.8	Summary		109

3 Tilings: mathematical models for quasicrystals — 111

3.1	Model sets		111
3.2	Introduction to tilings		113
3.3	Substitutional chains		117
	3.3.1	Substitutions with an alphabet	117
	3.3.2	Embedding of substitutional chains	118
	3.3.3	Tilings by substitution	119
3.4	Aperiodic tilings		122
	3.4.1	Construction of aperiodic tilings	122
	3.4.2	Embedding of tilings	127
	3.4.3	Symmetry of tilings	135
3.5	Approximants		140
3.6	Coverings		143
3.7	Random tilings		144
3.8	Summary		146

4 Structure — 148

4.1	Diffraction		148
	4.1.1	Diffraction from periodic and aperiodic crystals	148
	4.1.2	Indexing the diffraction pattern	162
	4.1.3	♯ Mathematical questions	165
4.2	Diffraction techniques		167
	4.2.1	X-ray area detectors	168
	4.2.2	Neutron area detectors	169
	4.2.3	Measurement techniques	170
4.3	Determination of modulated phases and composites		171
	4.3.1	Introduction	171
	4.3.2	The structure factor of incommensurate structures	173
	4.3.3	Possible expressions of the modulation functions	174
	4.3.4	Additional expressions of modulation functions	176
	4.3.5	Practical aspects of structure determination and refinement	177
	4.3.6	Ab initio methods	184

		4.3.7 Relation between harmonics and satellite orders	187
		4.3.8 Composite structures	189
		4.3.9 Commensurately modulated structures	191
	4.4	Typical examples of modulated phases and composites	193
		4.4.1 Introduction	193
		4.4.2 The modulated phases of Na_2CO_3	193
		4.4.3 The composite structure of $La_2Co_{1.7}$	198
		4.4.4 Alkane–urea compounds	202
		4.4.5 Aperiodicity in the structures of elements	203
		4.4.6 *p-Chlorobenzamide*	208
		4.4.7 Modular structures	211
		4.4.8 Superspace and crystal chemistry	215
		4.4.9 Conclusion	217
		4.4.10 Structure determination of quasicrystals	218
		4.4.11 A simple one-dimensional quasiperiodic model	218
		4.4.12 Structure determination of a one-dimensional quasicrystal	222
		4.4.13 Structure determination of icosahedral and decagonal phases	232
	4.5	Examples of quasicrystal structures	233
		4.5.1 Introduction	233
		4.5.2 Structure of the i-AlPdMn phase	234
		4.5.3 Atomic structure of the CdYb icosahedral phase	269
		4.5.4 Structure of the AlNiCo decagonal phase	282
		4.5.5 Dodecagonal quasicrystals	290
		4.5.6 Reversible phase transitions	293
	4.6	Diffraction by an imperfect crystal	296
		4.6.1 Diffuse scattering when an average lattice exists	298
		4.6.2 Diffuse scattering when there is no average lattice	301
5	**Physical properties**		305
	5.1	Introduction	305
	5.2	Tensorial properties	306
	5.3	Hydrodynamics of aperiodic crystals	312
		5.3.1 Hydrodynamic theory of fluids and periodic crystals	312
		5.3.2 Hydrodynamic theory of aperiodic crystals and phason modes	314
	5.4	Phonons and phasons: Theory	316
		5.4.1 Introduction	316
		5.4.2 Simple models	320
		5.4.3 Eigenvectors and spectrum	348
		5.4.4 Damping	349
		5.4.5 Calculation of phonons for real incommensurate phases	350
		5.4.6 Dynamics of quasicrystals	350
		5.4.7 Diffuse scattering and Debye–Waller factors	351
	5.5	Non-linear excitations	353
	5.6	Electrons	357

	5.6.1	Introduction	357
	5.6.2	Simple models	358
	5.6.3	Electrical conductivity	363
	5.6.4	Realistic potentials	363
	5.6.5	Quantum criticality in a magnetic quasicrystal	364
5.7		Summary of the theoretical situation	364
5.8		Phonons: Experimental findings	365
	5.8.1	Scattering	365
	5.8.2	Modulated phases and composites	367
	5.8.3	Quasicrystals	368
5.9		Phasons: Experiment	378
	5.9.1	Introduction	378
	5.9.2	Phason modes in modulated crystals	378
	5.9.3	Diffuse scattering and phason modes in icosahedral quasicrystals	383
	5.9.4	Phason modes in the i-AlPdMn icosahedral quasicrystal	386
	5.9.5	Phason modes in other quasicrystals	398
5.10		Summary of the experimental findings	404

6 Origin and stability — 405

6.1	Introduction	405
6.2	The Landau theory of phase transitions	406
6.3	Semi-microscopic models	412
	6.3.1 Substrate models	412
	6.3.2 Spin models	416
	6.3.3 Models with continuous degrees of freedom	418
	6.3.4 Specific models for incommensurate phases	424
6.4	Composites	437
6.5	Quasicrystals	437
6.6	Electronic instabilities	438
	6.6.1 Charge-density and spin-density systems	438
	6.6.2 Hume–Rothery compounds	440
6.7	Growth of quasicrystals	441
6.8	Summary	442

7 Other topics — 444

7.1	Morphology of aperiodic crystals	444
	7.1.1 The puzzling habit of the mineral calaverite	444
	7.1.2 The morphology of the TMA Zn phases	448
	7.1.3 The morphology of icosahedral and decagonal quasicrystals	450
7.2	Surfaces	453
	7.2.1 Introduction	453
	7.2.2 Structure of surfaces of aperiodic crystals	455
	7.2.3 Generalization of the morphological laws	458
	7.2.4 Physical properties of quasicrystalline surfaces	460

7.3	Magnetic quasiperiodic systems	462
7.4	Incommensurate multiferroics	466
7.5	Aperiodic photonic crystals	470
7.6	Mesoscopic quasicrystals	471
7.7	Defects	472

Appendix A Space groups in arbitrary dimensions   475

A.1	Crystallographic operations in n dimensions	475
A.2	Lattices	476
A.3	Crystal classes	478
A.4	Space groups	479
A.5	Classification	482
A.6	Space groups for aperiodic crystals	485
A.7	Notation	486
	A.7.1 Superspace groups for incommensurate phases	486
	A.7.2 General space groups	488
A.8	Equivalence of (super)space groups	489
A.9	Extinction rules	490
A.10	Tables	491
	A.10.1 Introduction	491
	A.10.2 Tables for irreducible representations of point groups: Point groups of 5-, 8-, 10-, 12-fold, or icosahedral symmetry	492
	A.10.3 Examples of superspace groups for modulated phases	493
	A.10.4 Superspace groups for quasiperiodic structures with 5-, 8-, 10-, 12-fold, or icosahedral symmetry	496

Appendix B Exercises: Solutions   499

References   503

Index   529

Glossary

⋆	Complex conjugation OR convolution: $(f * g)(x) = \int_{-\infty}^{\infty} f(y)g(x-y)dy$
$\mathbf{a}_i$	Basis of 3D direct lattice
$\mathbf{a}_{si} = (\mathbf{a}_i, \mathbf{a}_{Ii})$	Basis of a lattice in superspace
$\mathbf{a}_i^*$ with $(\mathbf{a}_i.\mathbf{a}_j^* = \delta_{ij})$	Basis of a reciprocal lattice
$\mathbf{a}^*, \mathbf{b}^*, \mathbf{c}^*$	Basis of a 3D reciprocal lattice
$(\mathbf{a}_i^*, \mathbf{b}_i^*)$	Basis of reciprocal lattice as vectors in reciprocal superspace
BCCD	Betaine calcium chloride dihydrate
CCD	Charge-coupled device
d	Dimension of the internal space
D	Dimension of the physical space
D_μ	Irreducible representation μ
$\mathbf{e}_{\mathbf{q}\nu j}$	Vibration of atom j in the mode $\mathbf{q}, \nu$
$f_j(\mathbf{H})$	Atomic scattering factor
$\hat{f}(\mathbf{k})$	Fourier transform of $f(\mathbf{r})$ $\left(= \frac{1}{2\pi} \int \hat{f}(\mathbf{k}) \exp(i\,\mathbf{k}.\mathbf{r})\right)$
$F(\mathbf{H})$	Structure factor $\left(= \frac{1}{N} \sum_i^N \exp(-2\pi i\,\mathbf{H}.\mathbf{r}_i)\right)$
Frac(x)	Frac: value of x minus the largest integer smaller than x
g	Metric tensor
$\mathbf{H} = \sum_{j=1}^n h_j \mathbf{a}_j^*$	Vector from Fourier module, uniquely associated with a $\mathbf{H}_s$
$\mathbf{H} = \mathbf{K} + \mathbf{q}$	Vector from Fourier module for a modulated phase
$\mathbf{H}_s$	Vector of the nD reciprocal lattice: $\mathbf{H}_s = (\mathbf{H}, \mathbf{H}_I)$
IDOS	Integrated density of states
$I(\mathbf{H}) = |F(\mathbf{H})|^2$	Scattering intensity
$\mathbf{k}$	Wave vectors (phonons, electrons, etc.)
$\mathbf{k}_s = (\mathbf{k}, \mathbf{k}_I)$	Reciprocal superspace vector
$\mathbf{K}$	Reciprocal lattice vector 3D
LEED	Low-energy electron diffraction
M*	Fourier module (the projection of Σ^* on V_E)
n	Dimension superspace
N	Number of unit cells in a sample
P	Participation ratio
PBZ	Pseudo-Brillouin zone
PMT	Photomultiplier tube
PSL	Photo-stimulated luminescence
$\mathbf{q} = \mathbf{k_f} - \mathbf{k_i}$	Scattering vectors
$\mathbf{q}$	Modulation vector
$Q_{\mathbf{k}\nu}$	Normal coordinate of a mode characterized by $\mathbf{k}, \nu$
$\{R|\mathbf{t}\}$	Rigid motion, i.e. non-homogeneous Euclidean transformation
$S(\mathbf{q}, \omega)$	Dynamical structure factor, scattering function (Eqs pp. 332, 366)
$\mathbf{S}(\mathbf{r})$	Spin wave

STM	Scanning tunnelling microscopy
$\mathbf{u}_j(x)$	Modulation function for particle j
V_E or $E_\parallel$ or E_{par}	Physical (external, or parallel) space
V_s or E_{sup}	Superspace (or hyperspace)
V_I or $E_\perp$ or E_{perp}	Internal (or perpendicular) space
VDOS	Vibrational density of states
Z_{ij}^ν	Decomposition of the reciprocal lattices of a composite (Eqs 2.12 and 2.13)
$\Gamma_E, \Gamma_I, \Gamma_M$	Parts of the integral matrix $\Gamma(R)$ (Eq. 1.15)
Φ	Golden mean $(\sqrt{5}-1)/2 \approx 0.618$
$\Phi(p)$	Euler function (p. 136)
λ	Wave length ($=1/k$)
Λ	Lattice in physical space
Λ^*	Reciprocal lattice in physical space
$\pi = \pi_E$	Projections $V_s \to V_E$
π_I	Projection $V_s \to V_I$
$\rho(\mathbf{r})$	Density
$\hat{\rho}(\mathbf{k})$	Fourier transform of $\rho(\mathbf{r})$
σ_{ij}	Components of the modulation vector (Eq. 2.9)
Σ	Lattice in superspace
Σ^*	Reciprocal lattice in superspace
τ	Inverse golden mean $(\sqrt{5}+1)/2 = 1/\Phi = 1 + \Phi \approx 1.618$
Ω	Atomic surface (also called window or acceptance domain for quasicrystals)

1
Introduction

1.1 Periodic crystals

1.1.1 History

People have always been intrigued by the beautiful shapes of crystals. The symmetrical objects with their flat, shiny surfaces and beautiful optical properties have been used for decoration since ancient times. Scientific interest in them also started very early. Johannes Kepler was fascinated by the beauty of snow crystals. In 1611 he wrote his *On the six-cornered snowflake* in which he tried to explain the six-fold symmetry of snow crystals on the basis of a close-packing of spheres. René Descartes studied snowflakes in Amsterdam and wrote his observations of their morphology in February 1634. Robert Hooke showed detailed drawings of the many different shapes of snowflakes in his book *Micrographia* (1665). So, two important ingredients of crystallography, symmetry and morphology, were studied in detail in the seventeenth century already. During the French Revolution, Adrien-Marie Legendre wrote his *Eléments de géométrie* (1794), in which he gave a precise definition of symmetrical polyhedra. Shortly afterwards a new step forward was formulated in the *Traité de minéralogie* by René Just Haüy. He described the construction of a regularly shaped crystal out of smaller symmetrical units (cubes, octahedra, and tetrahedra). He omitted icosahedra, 'because in crystals no five-fold symmetries occur'. From his composition procedure follows the law of rational indices, which states that the planes of crystal surfaces intersect properly chosen axes in points with integer coordinates.

The macroscopic symmetry is based on the symmetry of the microscopic periodic arrangement of the unit building blocks. Lattices and their symmetries were studied for the first time by Moritz Ludwig Frankenheim in 1835. Previously, he and Johann Hessel had derived the 32 crystal classes. In 1856 Frankenheimer showed the 14 lattice classes in three dimensions. Auguste Bravais then derived the Bravais classes on the basis of purely geometrical reasoning.

The next step was the derivation of the full Euclidean symmetry of periodic patterns in three dimensions, the space groups. The mineralogist Evgraph Fedorov and the mathematician Arthur Schoenflies gave, largely independently, the full classification of the 230 space groups in 1890. An interesting account of the history of symmetry in crystallography can be found in Burckhardt (1988).

The macroscopic consequences of the microscopic symmetry could be checked on the morphology, in particular the law of rational indices. The proof that the microscopic structure of an ideal crystal has space group symmetry had to wait till the discovery of X-rays by Max von Laue in 1912. This started a new era of crystallography. Structure determination was then possible, based on the work of the Braggs, and it was highly successful. This led to the idea that the ground state of matter was an arrangement of atoms and molecules with lattice periodicity, which means with space group symmetry. In this view, which prevailed for decades, each ideal crystal could be described by one of the 230 space groups. At non-zero temperature there are always defects, and metastable glassy matter may exist, but in principle the ground state is lattice periodic.

This view changed after 1960 when states of matter were discovered which reasonably should be classified as crystals, but which lacked the lattice periodicity. Before starting in that direction, the main topic of this book, we shall briefly review the role symmetry plays in conventional, lattice periodic crystals.

1.1.2 Description

Two ingredients of the description of (lattice periodic) crystals are space group symmetry, and morphology. Macroscopically, the point group symmetry can be seen in the morphology. A crystal is invariant under certain rotations, reflections, and combinations of these. The group of all these operations is the *point group*. At the atomic level, accessible via neutron and X-ray diffraction and via high-resolution electron microscopy, one may distinguish the lattice periodicity. Then *orthogonal transformations* (proper and improper rotations as the combinations of a rotation and a reflection are called) leave the framework of the translations invariant, but, to keep the positions of the atoms invariant, they sometimes have to be complemented by a translation. These combinations leave the distances the same.

A *space group* is a group of rigid, distance-preserving transformations, also called *Euclidean transformations*. Its elements are pairs $\{R|\mathbf{t}\}$ of an orthogonal transformation R and a translation $\mathbf{t}$ acting on a position vector $\mathbf{r}$ as $\{R|\mathbf{t}\}\mathbf{r} = R\mathbf{r}+\mathbf{t}$. After the choice of an origin and a basis, the transformation is given by matrices:

$$\mathbf{r} = \begin{pmatrix} x \\ y \\ z \end{pmatrix} \rightarrow \begin{pmatrix} R_{11} & R_{12} & R_{13} \\ R_{21} & R_{22} & R_{23} \\ R_{31} & R_{32} & R_{33} \end{pmatrix} \begin{pmatrix} x \\ y \\ z \end{pmatrix} + \begin{pmatrix} t_1 \\ t_2 \\ t_3 \end{pmatrix} = R\mathbf{r} + \mathbf{t}. \quad (1.1)$$

Such transformations are physically important, because they leave the physics the same, as the distances between the particles do not change.

A group of Euclidean transformations is a *space group* if its *translation subgroup*, all the translations in the group, is generated by three independent basis vectors. Each element of the translation group can be written as

$$\mathbf{a} = \sum_{j=1}^{3} n_j \mathbf{a}_j, \quad \text{(integers } n_j\text{)}. \tag{1.2}$$

The set of all orthogonal transformations R in the space group elements forms the point group, which is isomorphic to the quotient of the space group and its translation group: $K = G/A$, where G is the space group, A its translation group, and K its point group. The translation parts of space group elements may belong to the translation group or not. If $\mathbf{t}$ in $g = \{R|\mathbf{t}\}$ belongs to the translation group, it is called a *primitive translation*. Otherwise, it is a *non-primitive translation*. In three dimensions, space group elements with non-primitive translations are transformations with either a screw axis or a glide plane. Notice that the translation part **a** depends on the origin of the rotation. Therefore, a non-primitive translation may disappear as such, when the origin is shifted. If there is an origin such that all translation parts of a space group become primitive translations, then the group is called a *symmorphic space group*. Otherwise it is a *non-symmorphic space group*. An example of a non-symmorphic group in two dimensions is given in Fig. 1.1.

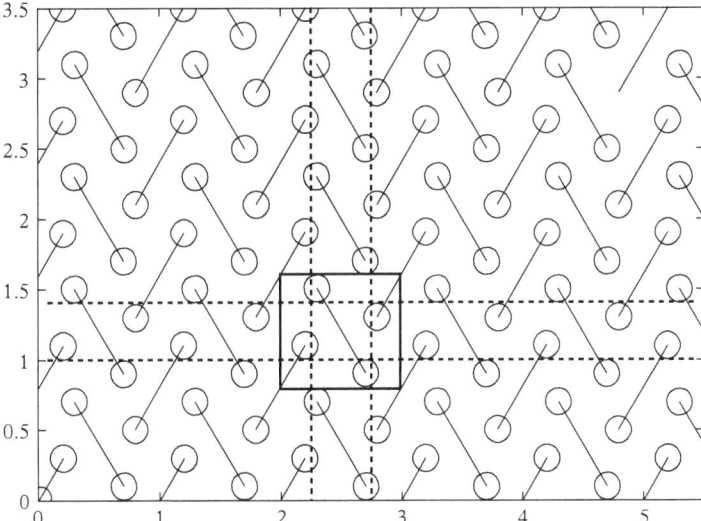

Figure 1.1 *A two-dimensional periodic array. The rectangle is the unit cell. It contains two dumbbell-shaped molecules. The array is left invariant by the translations along the edges of the unit cell, and by a rotation of π around a corner. Moreover, there are two glide operations: mirrors along a line followed by a translation along the line. The operations are $\{1|m_1\mathbf{a}_1 + m_2\mathbf{a}_2\}$, $\{2|0\}$, $\{m_x|(\mathbf{a}_1 + \mathbf{a}_2)/2\}$, and $\{m_y|(\mathbf{a}_1 + \mathbf{a}_2)/2\}$. The symbol for this space group is p2gg. Its point group is 2mm, which has four elements. Indicated are a unit cell and four glide lines.*

The lattice of a space group is invariant under the point group. This means that on a lattice basis the point group transformations R are represented by integer matrices M defined by

$$R\mathbf{a}_i = \sum_{j=1}^{3} M(R)_{ji}\mathbf{a}_j \quad (i=1,2,3).$$

On the other hand, on an orthogonal basis the matrices corresponding to R are orthogonal matrices, which can be written on a properly chosen basis as

$$M'(R) = \pm \begin{pmatrix} \cos(\phi) & -\sin(\phi) & 0 \\ \sin(\phi) & \cos(\phi) & 0 \\ 0 & 0 & 1 \end{pmatrix}.$$

Under a basis transformation the trace of a matrix does not change. Therefore, the traces of $M(R)$ and $M'(R)$ are the same. Now, $M(R)$ is an integer matrix, and consequently $1+2\cos(\phi)$ is an integer, which is only possible if $\phi = \pi$, $2\pi/3, \pi/2, \pi/3$, or 0. If R is a rotation, it is a two-, three-, four-, or six-fold rotation, or the identity. This strongly restricts the number of possible rotations in a point group of a lattice. The condition that $1+2\cos(\phi)$ is an integer is called 'the crystallographic condition'.

If a system is left invariant by a lattice of translations, there is a finite region in space such that every point in space can be reached by a lattice translation from a uniquely determined point in this region. The structure is a repetition of a basic unit, called the *unit cell*. A crystal is described by the contents of its unit cell. The positions of the atoms in the unit cell are $\mathbf{r}_j$ ($j = 1, \ldots, s$). Each position has a site symmetry, the group that leaves the point fixed modulo lattice vectors. This site symmetry characterizes the different positions, the *Wyckoff positions*. Once a unit cell is chosen, every point in space corresponds to a uniquely determined point in the unit cell, from which it can be reached by lattice translations. However, the unit cell is itself not uniquely determined. A convenient choice is the parallelepiped spanned by the basis vectors. With respect to this basis, the coordinates of the unit cell range from 0 to 1. However, another choice is the same unit cell shifted, such that the coordinates run from $-\frac{1}{2}$ to $+\frac{1}{2}$. A choice that takes the symmetry better into account is the *Wigner–Seitz cell* defined as the set of points closer to a chosen lattice point than to any other lattice point. Such cells all have the same shape and size, and they have the full point group symmetry of the lattice.

The notion of space groups may also be used for other dimensions. For two dimensions, these are called *plane groups*. The crystallographic condition is the same and the only non-primitive translations occur in transformations with a glide mirror. These transformations consist of a mirror operation together with a translation along the mirror line. For higher dimensions, the notion of space groups may also be extended, as we shall do further on in this book. Then the groups are called space groups, in general. These are briefly discussed in Appendix A.

The facets of a crystal are parallel to planes spanned by lattice translations, called *net planes*. Net planes in a crystal may be characterized by their normals. It is convenient to choose these in a particular way, as reciprocal lattice vectors. These are introduced as follows. The density of a crystal may be decomposed into a Fourier series, a sum of plane waves with wave vectors **K**. Because the crystal is invariant under translations **a** of the lattice, this must hold also for the constituting plane waves. Therefore, $\exp(2\pi i \mathbf{K}.\mathbf{r})$ and $\exp(2\pi i \mathbf{K}.(\mathbf{r}+\mathbf{a}))$ must be the same for all **a** from the lattice. The wave vectors **K** for which all products **K.a** are integers, or in other words for which **K.a** = 0 (mod 1), form a lattice, called the *reciprocal lattice*. (Notice that crystallographers usually write plane waves as $\exp(2\pi i \mathbf{k}.\mathbf{r})$, whereas physicists prefer the expression $\exp(i\mathbf{k}.\mathbf{r})$. The wave length λ is $1/|\mathbf{k}|$ for the crystallographic convention.) For a basis $\mathbf{a}_j$ ($j=1,2,3$) of the translation lattice, the basis of the reciprocal lattice is given by the *reciprocal basis* $\mathbf{a}_i^*$ satisfying the relations

$$\mathbf{a}_j.\mathbf{a}_i^* = \delta_{ij}, \tag{1.3}$$

which in three-dimensional space is equivalent to

$$\mathbf{a}_1^* = \frac{\mathbf{a}_2 \times \mathbf{a}_3}{\mathbf{a}_1.(\mathbf{a}_2 \times \mathbf{a}_3)} \quad \text{(and cyclic)}. \tag{1.4}$$

The vectors **K** of the reciprocal lattice, as defined here, satisfy **K.a** = 0 (mod 1) for every lattice vector **a**, because $\left(\sum_i h_i \mathbf{a}_i^*\right).\left(\sum_j n_j \mathbf{a}_j\right) = \sum_{ij} h_i n_j \delta_{ij} = 0$ (mod 1), since h_i and n_j are integers. The space of all wave vectors is called *reciprocal space*.

♯ In principle, the *reciprocal space* V^* is the space of all linear functions on a linear space V. If there is a metric, both spaces are identical because, for every linear function $f(\mathbf{r})$ on V, there is a wave vector **k** such that $f(\mathbf{r}) = \mathbf{k}.\mathbf{r}$. Therefore, $\mathbf{a}_i$ and $\mathbf{a}_j^*$ span the same space (as shown by Eq. 1.4), but, in general, the lattices spanned by them are not the same.♮

The Wigner–Seitz cell of the reciprocal lattice is called the *Brillouin zone*. An example in two dimensions is given in Fig. 1.2. In three dimensions the vectors of the reciprocal lattice are often written as $\mathbf{K} = h\mathbf{a}^* + k\mathbf{b}^* + \ell\mathbf{c}^*$. The basis vectors of the lattice also form a basis for the space. Therefore, any point in space **r** has three coordinates with respect to this basis:

$$\mathbf{r} = \sum_{i=1}^{3} \xi_i \mathbf{a}_i, \quad \text{with } \xi_i = \mathbf{r}.\mathbf{a}_i^*.$$

In the unit cell the values of ξ_i are between 0 and 1. They are called *lattice coordinates*. The notation of a point in terms of these is $\mathbf{r} = [\xi_1, \xi_2, \xi_3]$.

A net plane is then given by **K.a** = n, with constant integer n, for given **K** and all **a** in the plane. Each net plane is characterized by the smallest perpendicular vector **K** from the reciprocal lattice. The distance between two adjacent parallel net planes is given by the inverse of the length of **K**. Then, in principle, the morphologically most important net planes are those with the smallest $|\mathbf{K}|$, but that is not always the case, because of 'extinction rules' we shall discuss later.

6 Introduction

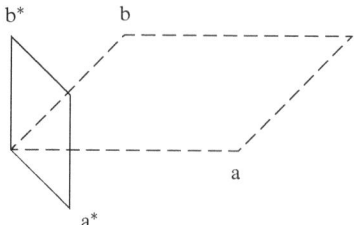

Figure 1.2 *The relation between the basis* **a, b** *and the reciprocal basis* **a★, b★**.

1.1.3 The role of space group symmetry in the structure determination

The space group symmetry is very helpful in the structure determination. Suppose the crystal has a density function $\rho(\mathbf{r})$ and space group G. If $\{R|\mathbf{t}\}$ is an element of G, the following relation holds.

$$\rho(\mathbf{r}) = \rho(R\mathbf{r} + \mathbf{t}). \tag{1.5}$$

The Fourier transform of this relation gives

$$\hat{\rho}(\mathbf{k}) = \exp(2\pi i R\mathbf{k}.\mathbf{t})\,\hat{\rho}(R\mathbf{k}). \tag{1.6}$$

For a vector $\mathbf{k}$ satisfying $R\mathbf{k} = \mathbf{k}$ the relation (1.6) implies that either $\mathbf{k}.\mathbf{t} = 0$ (mod 1) or $\hat{\rho}(\mathbf{k}) = 0$. This is a very important relation, because the Fourier transform of the density, or the potential, plays an important role in scattering experiments. In such an experiment, a neutron or X-ray beam with wave vector $\mathbf{k}_i$ is scattered by the sample into beams with wave vector $\mathbf{k}_f$. The scattering vector is $\mathbf{k} = \mathbf{k}_f - \mathbf{k}_i$. Quantum mechanics teaches us that the scattering is determined by the matrix element of the scattering potential between incoming and outgoing states. In this case we have

$$\langle \mathbf{k}_f | V(\mathbf{r}) | \mathbf{k}_i \rangle \sim \int \rho(\mathbf{r}) \exp(2\pi i \mathbf{k}.\mathbf{r}) d\mathbf{r} = \hat{\rho}(\mathbf{k}).$$

The relation of the Fourier transforms leads to extinction rules $\hat{\rho}(\mathbf{k}) = 0$ for $\mathbf{k} = R\mathbf{k}$ in the case that (in a non-symmorphic space group with a screw axis or a glide plane) $\mathbf{k}.\mathbf{t} \neq 0$ (mod 1), which is a great help in the structure determination.

The main steps in the structure determination are as follows. The latter is based on the measured intensities of the Bragg peaks:

$$I(\mathbf{K}) \sim \left|\sum_j \exp(2\pi i \mathbf{K}.\mathbf{r_j})\right|^2, \tag{1.7}$$

where the sum is over all particles in the unit cell. The sum is called the structure factor. The positions of the Bragg peaks are $\mathbf{K} = h\,\mathbf{a}^* + k\,\mathbf{b}^* + \ell\,\mathbf{c}^*$. A basis for the direct lattice is then formed by three vectors $\mathbf{a}, \mathbf{b}, \mathbf{c}$. Together with the composition this gives the number of structure units in the unit cell.

The symmetry of the diffraction pattern contains at least the point group of the space group. Assuming a point group, all possible space groups then can be found in the *International Tables for Crystallography*, and these space groups are distinguished by the extinction rules. The difference between space groups with the same point group is the difference in the presence of screw axes and glide planes. These are associated with well-defined extinction rules.

The next step is the choice of a model. For that purpose, the positions $\mathbf{r}_j$ of the particles in the unit cell are expressed in free parameters, compatible with the space group. Because the space group links positions of atoms, the number of parameters is, generally, much less than three times the number of particles in the unit cell. The parameters are varied to yield a minimum in the difference between calculated and observed intensities, which gives the final answer. That is a very short account of crystal structure determination. It is just a reminder. For details we refer to introductory texts on crystallography.

There are a number of problems. A very difficult one is the phase problem. The intensities are measured experimentally, but these are connected to the absolute value of the structure factor. This means that the structure is not yet uniquely determined. Several methods have been developed (such as direct methods, the *charge-flipping* method, and the *maximum entropy* method) to help to solve this problem. Another problem is the fact that different models may give comparable differences between calculated and measured intensities, because these differences are never zero.

The structure determination has become a powerful technique, applicable to very large unit cells, with many atoms. Its principal basis is the knowledge of the space groups.

1.1.4 Symmetry and physical properties

Another important role of the space group symmetry is the possibility it gives to characterize electrons and lattice vibrations. From the theory of applications of group theory in physics, it is known that elementary excitations and electron states may be described using *representations of the symmetry group*. A matrix representation of a group G is a mapping from the group to a group of non-singular matrices $D(G)$ such that $D(g_1)D(g_2) = D(g_1 g_2)$ for every two elements g_1 and g_2 of G. For translations this means that for two translations, $\mathbf{a}$ and $\mathbf{b}$, the relation $D(\mathbf{a})D(\mathbf{b}) = D(\mathbf{a} + \mathbf{b})$ holds. The relevant representations of the translation group of the space group G are characterized by a vector $\mathbf{k}$ such that

$$D_\mathbf{k}(\mathbf{a}) = \exp(-2\pi i\, \mathbf{k}.\mathbf{a}).$$

These are one-dimensional representations. Two representations with $\mathbf{k}_1$ and $\mathbf{k}_2$ are the same if the two vectors differ by a reciprocal lattice vector. Therefore, the irreducible representations of the translation subgroup are characterized by vectors in the unit cell of the reciprocal lattice. Choose as unit cell the set of vectors $\mathbf{k}$ closer to the origin than to any other reciprocal lattice vector. This unit cell is called the *Brillouin zone*. So, states in a crystal are characterized by a vector from the Brillouin zone.

A consequence of the lattice translation symmetry is the fact that wave functions in the system may be written as *Bloch functions*. If $\Psi(\mathbf{r})$ is an energy eigenfunction, it may be written as

$$\Psi(\mathbf{r}) = \exp(2\pi i \mathbf{k}.\mathbf{r}) U_\mathbf{k}(\mathbf{r}), \tag{1.8}$$

where the function $U(\mathbf{r})$ is lattice periodic, that is $U(\mathbf{r}+\mathbf{a}) = U(\mathbf{r})$ for all lattice vectors $\mathbf{a}$. Indeed, the translation of such a function gives $\Psi(\mathbf{r}-\mathbf{a}) = \exp(-2\pi i \mathbf{k}.\mathbf{a})\Psi(\mathbf{r})$. Using Bloch functions reduces the determination of the electron eigenvalues and eigenfunctions to the unit cell of the lattice. Instead of a problem in the infinite crystal, only the problem in the finite unit cell has to be solved. Similarly, the lattice vibration problem with an infinite number of degrees of freedom can be reduced to the degrees of freedom inside the unit cell, and this is three times the number of particles in the unit cell. Another consequence of the periodicity is that the density, the square of the absolute value of the wave function, is the same in every unit cell, because $\rho(\mathbf{r}) = |\Psi(\mathbf{r})|^2 = |U(\mathbf{r})|^2$ and this is invariant under any lattice translation $\mathbf{a}$. In an ideal crystal with lattice periodicity there are no localized states.

Representations of the space group are generally more than one-dimensional. These are now characterized by stars of vectors in the Brillouin zone, where the star of a vector $\mathbf{k}$ is the set of vectors $R\mathbf{k}$ (modulo reciprocal lattice vectors) for all elements R of the point group. An irreducible representation of the space group is now characterized by a star of a vector $\mathbf{k}$ and an irreducible representation of the group of $\mathbf{k}$, which is the subgroup of the point group consisting of all elements leaving $\mathbf{k}$ invariant (modulo the reciprocal lattice vectors). An irreducible representation is one that cannot be written as the sum of two or more representations. We shall come back to this point. For the moment it is sufficient to say that phonons and electrons in lattice periodic crystals are characterized by a star in the Brillouin zone, and an *irreducible representation* of a subgroup of the point group, the group of $\mathbf{k}$.

The eigenvalues of the Hamilton operator for a particle in a lattice periodic crystal depend smoothly on the characterizing wave vector $\mathbf{k}$. The consequence is that the density of states consists of bands.

All these properties rely on the lattice periodicity of the crystal. That is why the symmetry is so important.

1.1.5 Examples

1.1.5.1 NaCl

The crystal structure of NaCl has space group symmetry $Fm\bar{3}m$ with a cubic lattice, invariant under the cubic point group $m\bar{3}m$. The lattice is spanned by three mutually perpendicular vectors of equal length, and there are four formula units per primitive unit cell, Na at positions $(0,0,0)$, $(\frac{1}{2},\frac{1}{2},0)$, $(\frac{1}{2},0,\frac{1}{2})$, $(0,\frac{1}{2},\frac{1}{2})$ and Cl at positions $(\frac{1}{2},0,0)$, $(0,\frac{1}{2},0)$, $(0,0,\frac{1}{2})$, $(\frac{1}{2},\frac{1}{2},\frac{1}{2})$. They have as site symmetry the full point group. Because the space group does not have screw axes or glide planes, it is symmorphic, and there are no systematic extinctions, except the centring condition $(h_i + h_j = \text{even})$.

Phonons and electrons are characterized by the wave vector **k** in the Brillouin zone. If $(a,0,0)$, $(0,a,0)$, and $(0,0,a)$ are the basis vectors of the lattice, the vectors $(a^{-1},0,0)$, $(0,a^{-1},0)$, and $(0,0,a^{-1})$ span the reciprocal lattice. The vectors in the Brillouin zone are (k_1, k_2, k_3) with $-1/(2a) \leq k_i \leq 1/(2a)$. Actually, the representations are given by stars of **k**, where **k** belongs to the asymmetric unit cell, the tetrahedron with four corners $(0,0,0)$, $(1/(2a),0,0)$, $(1/(2a),1/(2a),0)$, and $(1/(2a),1/(2a),1/(2a))$.

1.1.5.2 Long-period compounds

The unit cell of a lattice periodic crystal may become very big. Consider as an example the long-periodic structures of some alloys. In the so-called Nowotny phases or chimney-ladder structures, the composition strongly influences the periodicity. They are alloys of the type TA_x, where T = Ti, Zr, V, Mo, Cr, Mn, or Rh, and A = Si, Ge, or Sn. There is a pseudo-unit cell for the two subsystems: with lattice constants a_T and c_T for the T subsystem, and a_A and c_A for the A subsystem with $a_T = a_A$ and $c_T = xc_A$. For $Mn_{11}Si_{19}$ the unit cell has a lattice constant $11c_{Mn} = 19c_{Si}$. The positions in the 11 pseudo-unit cells of Mn are modulated, as are the positions of Si (Fig. 1.3).

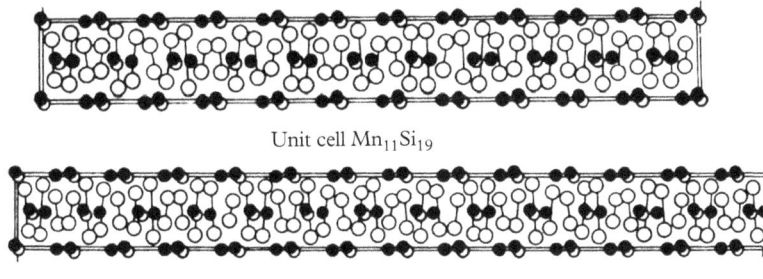

Unit cell $Mn_{11}Si_{19}$

Half unit cell $Mn_{27}Si_{47}$

Figure 1.3 *Nowotny phases* $Mn_{11}Si_{19}$ *and* $Mn_{27}Si_{47}$ *(After (De Ridder et al., 1976)).*

1.1.5.3 The monster

Unit cells may be very large. An example is the Samson β-phase. β-Mg_2Al_3 has 1168 atoms in a unit cell of volume 22 526 $Å^3$.

1.1.5.4 Tilings

The theory of *plane groups* is very similar to that of space groups. They are subgroups of the two-dimensional Euclidean group with a lattice group spanned by two independent vectors. There are 17 different plane groups, and many of them can be found as symmetry group of a decoration in the Alhambra in Granada. These are coverings of the walls with a specific symmetry. In mathematics such coverings have been studied under the name of tilings. Tilings are coverings of the plane by means of copies of a limited number of tiles, such that no overlaps and no gaps occur. Nice examples of structures with plane group symmetry are various tilings created by the Dutch artist Maurits Escher. In his periodic patterns one easily recognizes the unit cell, and the asymmetric unit cell, a part from the unit cell such that no two points of it can be transformed into each other by an element of the plane group.

In science the plane groups are the symmetry groups of crystalline surfaces. An ideal crystal has facets with a characteristic plane group symmetry which depends on the orientation of the facet.

1.1.6 Conclusion

For conventional crystals with lattice periodicity, the symmetry is very important. Both structure determination and the study of physical properties are based on this periodicity which allows the study to be simplified to the unit cell or the Brillouin zone.

1.2 Aperiodic crystals

1.2.1 History

The conviction that the ground state of matter is always an ideal crystal with three-dimensional space group symmetry, that is with lattice periodicity, started to weaken after 1960. Then researchers started to pay more attention to scattering besides the Bragg peaks of the reciprocal lattice. Mathematical models for lattice periodic crystals show sharp Bragg peaks only. Real crystals always have defects, and are always of finite size. The consequence is broadening of the Bragg peaks, additional diffuse scattering, and other scattering phenomena. The occurrence of additional spots ('Gittergeister' or lattice ghosts) as a consequence of a periodic variation, for example in the crystal growth, was well known. These effects were considered as perturbations of the ideal periodic structure.

It was also known that *helical magnetism* sometimes occurs. Because the pitch of the helix is determined by the interaction in the magnetic system, and the coupling with the crystal structure is weak, a difference between the period of the magnetism and that of the crystal structure may easily occur. An example is thulium which shows magnetic satellite reflections below approximately 56 K (Koehler et al., 1962). These satellites were interpreted as due to a modulated anti-ferromagnetic structure with a period 7/2. The spin is a periodic function $S(\mathbf{r})=f(\mathbf{q}.\mathbf{r})$, where f has period 1. In neutron scattering, diffraction peaks were found at positions

$$\mathbf{H} = h\mathbf{a}^* + k\mathbf{b}^* + \ell\mathbf{c}^* + m\mathbf{q}, \quad \mathbf{q} = \alpha \mathbf{c}^*$$

in the hexagonal close-packed basis with unique axis $\mathbf{c}$. For $T > 56$ K there are no satellites, for 40 K $< T <$ 56 K there are satellites with $m=1$, and for $T < 40$ K higher-order satellites ($m > 1$) occur. The value of α is (approximately) 2/7. The appearance of satellites with $m > 1$ can be interpreted as an amplitude increase of the originally sinusoidal function f. Later it was found that the value of α is temperature dependent. It increases from 0.27 at 56 K to 2/7 at approximately 35 K (Brun et al., 1970).

By the modulation the periodicity in the c-direction is broken. The structure is periodic with periodicity 7 if $\alpha = 2/7$, but it is no longer periodic if α is an irrational number. That this may happen is suggested by the temperature dependence of $\mathbf{q}$. This point was only noticed later, when more materials of this kind were found. In the beginning the system was considered as periodic with a large seven-fold unit cell.

Other long-period structures were known in alloys. In the beginning these were considered to be metastable. An early example is a Cu-Ni-Fe alloy in a metastable region of the phase diagram, described as a modulated structure (Daniel and Lipson, 1943, 1944; Hargreaves, 1951). Another example is CuAu. In phase CuAu I it has a face-centred orthorhombic unit cell with Cu in (0, 0, 0) and $\left(\frac{1}{2},\frac{1}{2},0\right)$ positions and Au in the $\left(\frac{1}{2},0,\frac{1}{2}\right)$ and $\left(0,\frac{1}{2},\frac{1}{2}\right)$ positions. In CuAu II there are domains. In one domain the positions of Cu and Au are the same as in phase I, in the other domain the positions are exchanged. The domain walls are periodic and perpendicular to the c-axis. The domains are called anti-phase domains. (Sato and Toth, 1961; Jehanno and Perio, 1962; van Landuyt et al., 1974; Yamamoto, 1982a). If the domain walls are five lattice constants apart, this results in a ten-fold superstructure. The longer period in the c-direction is reflected in the diffraction pattern. The Bragg peaks are at positions $\mathbf{H} = h\,\mathbf{a}^* + k\,\mathbf{b}^* + \ell\mathbf{c}^* + m\,\mathbf{q}$, with $\mathbf{q} = \mathbf{c}^*/(2d)$, where d is the distance between the domain walls. Again, it was assumed that $d = 5$, but later it was discovered that the domain walls are five or six lattice constants apart and that the average d is actually an irrational number (Sato and Toth, 1961). This implies that the periods of the domain walls and of the orthorhombic lattice do not have a common multiple, and that the structure is not lattice periodic in the c-direction. See Fig. 1.4.

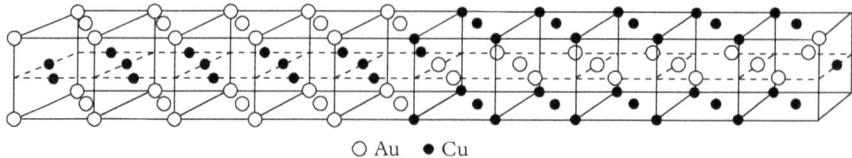

Figure 1.4 *Phase II of CuAu. The atoms are arranged on the positions of a tetragonal lattice. In a unit cell Cu is either at* (000) *and at* $\left(\frac{1}{2}\frac{1}{2}0\right)$ *or at* $\left(0\frac{1}{2}\frac{1}{2}\right)$ *and* $\left(\frac{1}{2},0\frac{1}{2}\right)$. *At domain boundaries, with incommensurate average distance, the positions of Cu and Au are interchanged.*

Satellites were also observed in $NaNO_2$ (Tanisaki, 1961; Gesi, 1965). In a tiny temperature interval just above the ferroelectric transition temperature X-ray diffuse scattering experiments revealed the existence of a sinusoidally varying dipole ordering along the a-axis with a period of approximately eight lattice constants. Above the phase transition, the orientation of the NO_2 molecules is disordered. Below the phase transition, the probability of finding one of two orientations of the molecules varies periodically. It is a *order–disorder transition*. There is a clear sign of a phase transition in the specific heat measurements at the temperature where the sinusoidal structure disappears. The new phase was called the sinusoidal phase. Because the period of the sinusoidal modulation varies approximately from ten to eight lattice constants, as function of temperature, the structure is aperiodic in the a-direction.

Evidence for aperiodic structures in other structures started to grow. *Pim de Wolff* and collaborators found an anomaly in the crystal structure of Na_2CO_3 (Brouns *et al.*, 1964). Indexing the powder diffraction pattern turned out to be very hard, until they introduced satellites for the analysis of an intermediate phase. At room temperature the compound crystallizes in a C-centred monoclinic lattice with four structural units per unit cell, with additional satellites. In what they called the γ-phase, at lower temperature, each 'normal' reflection is surrounded by up to six satellites at

$$\mathbf{H} = h\,\mathbf{a}^* + k\,\mathbf{b}^* + \ell\mathbf{c}^* + m\,\mathbf{q}, \tag{1.9}$$

where $\mathbf{q} = \alpha \mathbf{a}^* + \gamma \mathbf{c}^*$ ($\alpha \approx 0.182, \gamma \approx 0.318$). For increasing temperature the values of α and γ change (Dubbeldam and de Wolff, 1969). They coined the name *incommensurate modulated phase*, and gave an interpretation in terms of a displacement wave. This contrasts with a substitution modulation discussed in Jamieson *et al.* (1969), where the probability of finding a certain species of atom in a given position in the unit cell varies in space. The variation in position for γ-Na_2CO_3 was called a *displacive modulation*. The picture is that such modulated structures can be viewed as a lattice periodic 'basic structure' and a modulation. If the modulation period λ does not fit to the lattice periodicity of the basic structure, the resulting compound is aperiodic (cf. Fig. 1.5).

In 1972 de Wolff came up with an entirely new description of the modulated phases (de Wolff, 1974). It was clear that incommensurate modulated phases,

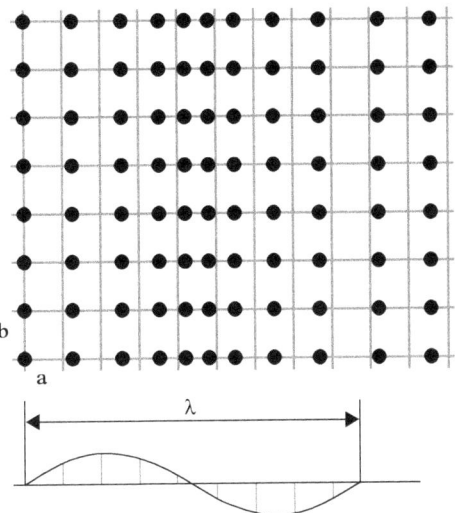

Figure 1.5 *For an incommensurate displacively modulated crystal, the atoms deviate in a periodic fashion from the positions of a conventional periodic crystal.*

not having three-dimensional lattice periodicity, could not be characterized by one of the 230 space groups. He noticed, however, that the positions of Eq. 1.9 could be seen as the projection of a reciprocal lattice in four dimensions on the three-dimensional physical space. He succeeded in showing that there is a four-dimensional structure with a four-dimensional space group symmetry such that the intersection of this structure with the three-dimensional space gives exactly the structure of the incommensurate modulated phase. We shall come back to this in Chapter 2. De Wolff now showed that the structure of γ-Na_2CO_3 can be characterized by the four-dimensional space group. The four-dimensional space is an abstract space, and not the four-dimensional space-time. Actually the coordinates are x, y, z and the position of the modulation wave with respect to the origin. Four-dimensional space groups for space-time had been studied by Janner and Janssen for the description of properties of electromagnetic fields (Janssen *et al.*, 1969; Janssen, 1969). (See Section 2.2.5.) It turned out that abstractly the space groups studied by them were exactly the same space groups as needed to describe incommensurate modulated phases with a single modulation wave vector **q**.

A long series of such modulated crystal phases was subsequently discovered. Nowadays hundreds of compounds with a modulated structure are known. It was clear that the traditional concept of a crystal as solid-state phase with lattice periodicity had to be revised.

Another class of materials with aperiodic structure were the *incommensurate composites*. The compound $Hg_{3-\delta}AsF_6$ consists of mercury chains inside a host lattice of AsF$_6$ octahedra (Brown *et al.*, 1974; Cutforth *et al.*, 1976; Schultz *et al.*, 1978). Above 120 K there are diffuse sheets in the diffraction pattern, below that temperature the mercury orders, but with a lattice constant that is incommensurate with that of the host lattice. Both the AsF$_6$ subsystem and the mercury subsystem show up in the diffraction pattern with Bragg peaks. Actually, there are two Hg subsystems, one with chains parallel to the a-direction, and one parallel to the b-direction. So, in principle, there are nine basis vectors of three reciprocal lattices, but these are not all independent. All diffraction spots can be obtained as linear combinations of four reciprocal space vectors. None of these four is a linear integer combination of the others. Consequently, the system is aperiodic. Another example of such a structure is TTF$_7$I$_{5-\delta}$, where the iodine is a subsystem inside the host lattice of tetrathiafulvalene molecules (Johnson and Watson, 1976). In both crystals the subsystems are not stable by themselves. The subsystems need each other. Moreover, they are not lattice periodic, but modulated. The modulation function of one subsystem contains wave vectors from the reciprocal lattices of the other subsystems, because the modulation is caused by their mutual interaction. In these two systems one may distinguish a host and one or more guest systems. That is not always the case. For example, the misfit structure (BiSe)$_{1.09}$TaSe$_2$ consists of alternating layers with incommensurate lattice constants. Here, it is not possible to distinguish a '*host*' from a '*guest*'.

In general, the spots of the diffraction pattern of an incommensurate composite are located on linear combinations of vectors from the reciprocal lattice vectors of the basic structures of the subsystems. If there are s subsystems, numbered by v, there are $3+d$ ($\leq 3s$) basis vectors such that all the spots are at positions which are linear combinations of these basis vectors. They are

$$\mathbf{H} = \sum_{v=1}^{s}\sum_{i=1}^{3} h_{vi}\mathbf{e}_{vi} = \sum_{j=1}^{3+d} h_j \mathbf{a}_j^* \quad \text{(integers } h_j\text{).} \quad (1.10)$$

Other compounds with similar properties were inclusion compounds, monolayers adsorbed on a crystal substrate, and intercalation compounds.

A very exciting new development was the discovery of *quasicrystals* by Shechtman in 1982, published in Shechtman *et al.* (1984). In the alloy AlMn, produced by rapid quenching from the melt, a diffraction pattern was found with five-fold symmetry. In fact, the diffraction pattern had the symmetry of an icosahedron, a point group that is not compatible with three-dimensional periodicity. The discovery raised much interest. The authors were able to show that the symmetry was a real property of the solid, and not the consequence of, for example, twinning. The diffraction spots were located at the positions

$$\mathbf{H} = \sum_{j=1}^{6} h_j \mathbf{a}_j^*, \quad (1.11)$$

and the spots are as narrow as those of high-quality conventional crystals. Other symmetry groups were also found, such as decagonal (ten-fold rotation symmetry) and dodecagonal (twelve-fold symmetry) types. The discovery led to a concentrated activity and a new field of crystallography.

The reason why it took more than 2 years between the sensational discovery and the publication lies probably in the unfortunate choice of the journal where Shechtman and Blech wanted to publish their results. Probably, the referees were not aware of the mathematical work on aperiodic tilings. However, several crystallographers had played already with the idea that structures like 3D aperiodic tilings could exist in nature (Mackay (1981): 'it (the Penrose tiling) gives an example of a pattern of the type which might well be encountered but which might go unrecognized if unexpected.') But they did not know where to look, and it was *Dan Shechtman* who discovered the material where this was realized. Only when the paper was rewritten and pointed out the role of aperiodicity, it could be published in *Physics Review Letters*. Another point was that the famous crystallographer *Linus Pauling* thought that the material was just periodic, but with twinning. It took several years before he gave up. As a third reason it has been suggested that the acceptance was delayed by the fact that the discovery was made with electrons, and not with X-rays. This seems unlikely, because there were several highly esteemed labs working with electrons, especially in works on alloys. For structure determination, there is a difference, but not for the diffraction experiments. Finally, everyone was convinced of the reality of quasicrystals, and Shechtman received in 2011 the *Nobel Prize in Chemistry* for his discovery.

Most people agreed that the aperiodic crystals with sharp diffraction peaks had to be considered as crystals as well. Therefore, incommensurate modulated phases were also called incommensurate crystal phases, implying that these are also crystals. After the discovery of quasicrystals people felt that a formal definition was needed. Therefore, after a couple of years the International Union of Crystallography proposed a new *definition of 'crystal'*. According to this definition a material is a crystal if it has 'essentially' a sharp diffraction pattern. The word 'essentially' means that most of the intensity of the diffraction is concentrated in relatively sharp 'Bragg' peaks, besides the always present diffuse scattering. Moreover, the vectors of the sharp spots should not be restricted to a subspace of the (usually three-dimensional) physical space. In the cases considered in this section, the positions of the diffraction peaks could be given as

$$\mathbf{H} = \sum_{j=1}^{n} h_j \mathbf{a}_j^* \quad (n \geq 3). \tag{1.12}$$

Here the vectors $\mathbf{a}_j^*$ span the physical space. Structures with this property are *quasiperiodic* according to the mathematical definition. The number n of basis vectors is the *rank*. The conventional crystals then are a special, though very large, class for which $n=3$. Besides, one can have, in principle, structures with sharp diffraction peaks that are not quasiperiodic. (For example, the so-called limit periodic structures.) To stress the difference with conventional crystals, one calls the quasiperiodic crystals with rank higher than three '*aperiodic crystals*'. The quasicrystals are a special case of aperiodic crystal. Usually, an aperiodic crystal with diffraction symmetry that is incompatible with 3D lattice periodicity is called a *quasicrystal*.

1.2.2 Classes and examples

As mentioned in Section 1.2.1, crystal structures were found with a diffraction pattern that excludes lattice periodicity. The diffraction spots are often very narrow, but of course no mathematical delta peaks. For real crystals, the peaks are always broadened. Their position may be given by

$$\mathbf{H} = \sum_{j=1}^{n} h_j \mathbf{a}_j^* \quad \text{(integers } h_j\text{)} \tag{1.13}$$

for some integer n. The number n is the minimum for which the positions can be described with integer coefficients h_j. All vectors $\mathbf{H}$ of this form (Eq. 1.13) form the *Fourier module*, which is said to be of *rank n*. The Fourier module is the generalization of the reciprocal lattice to the case of aperiodic crystals.

Experimentally, this creates obviously a problem. Choosing a basis $\mathbf{a}^*, \mathbf{b}^*, \mathbf{c}^*$ of the space, the vector $\mathbf{H}$ may be written as $\alpha\mathbf{a}^* + \beta\mathbf{b}^* + \gamma\mathbf{c}^*$, and because of the finite resolution it is always possible to choose the basis in such a way that α, β, and γ are fractions. However, generally, the integers in the approximate fraction L/N will be large and will vary wildly if external parameters, like the temperature, change. In contrast to that the indices $h_1, \ldots, h_n$ will remain constant, and the temperature dependence will only be in the basis vectors $\mathbf{a}_i^*$. Besides, the diffraction peaks will have a finite width, but this is also the case for conventional crystals. Finally, if $n > 3$, the positions of the Fourier module form lines, planes, or volumes of which every point is arbitrarily close to a vector of the module. These are called dense sets. Take as a simple example a case with $n = 4$, and basis vectors $(1, 0, 0), (0, 1, 0), (0, 0, 1)$, and $(0, 0, \sqrt{2})$. Then every point $(0, 0, z)$ on the z-axis is arbitrarily close to a vector $(0, 0, n + m\sqrt{2})$, because $\sqrt{2}$ is an irrational number. If one takes all the limit points, one gets a line. The diffraction pattern of a single domain will, however, show discrete peaks above a certain threshold. We assume that these observable peaks have a minimal distance. In the case of powders the diffraction will not show peaks, but discrete rings.

Generally the diffraction pattern has a certain symmetry, the *Laue symmetry*. It is a point group with elements R such that each vector $\mathbf{H}$ of the Fourier module is transformed to $R\mathbf{H}$ in the module with the same diffraction intensity. One can describe, in the usual way, such an element by an integer matrix $\Gamma(R)$:

$$R\mathbf{a}_j^* = \sum_{k=1}^{n} \Gamma(R)_{jk} \mathbf{a}_k^* \quad (j = 1, \ldots, n). \tag{1.14}$$

The matrix $\Gamma(R)$ is an n-dimensional matrix with integer entries. Because of the discreteness of the peaks the matrices form a finite group. Generally, however, this is not a crystallographic point group, at least not in three dimensions. However, we shall make use of it later on.

For incommensurate modulated phase, the main peaks are transformed onto main peaks by R. Therefore, the matrix $\Gamma(R)$ has the form

$$\Gamma(R) = \begin{pmatrix} \Gamma_E(R) & 0 \\ \Gamma_M(R) & \Gamma_I(R) \end{pmatrix}, \tag{1.15}$$

where the matrices $\Gamma_E(R)$, $\Gamma_I(R)$, and $\Gamma_M(R)$ are $D \times D$, $d \times d$, and $d \times D$, resp,. (The dimension of the physical space V_E is D and that of the internal space V_I is d, with $D + d = n$).

Nowadays very many quasiperiodic materials are known. Roughly speaking, they fall into one or more of three classes: incommensurate modulated crystal phases, incommensurate composites, and quasicrystals. As we shall see, this is not a strict classification. Several materials may be seen as belonging to more than one class. However, we start with a discussion of the various classes.

1.2.3 Modulated phases

A *modulated phase* consists of a lattice periodic *basic structure* and a periodic modulation. A simple example is a basic structure with lattice spanned by **a**, **b**, and **c** and one particle per unit cell, with a displacement of the particle at lattice site **n** given by $\mathbf{A}\cos(2\pi\mathbf{q}.\mathbf{n})$. Here the displacement is parallel to **A**. The periodicity is along **q** with period $1/q$. If for none of the lattice vectors **n** ($\neq 0$) the scalar product **q.n** is an integer, the structure including the displacements is no longer periodic.

The particle at position **n** in the basic structure is displaced to $\mathbf{r_n}$ and the density is given by the function $\rho(\mathbf{r}) = \sum_\mathbf{n} \delta(\mathbf{r} - \mathbf{r_n})$. The Fourier transform of this function is the structure factor. The Fourier coefficients may be written as:

$$F(\mathbf{H}) = \sum_\mathbf{n} \exp(2\pi i \mathbf{H}.(\mathbf{n} + \mathbf{A}\cos(2\pi\mathbf{q}.\mathbf{n}))).$$

It is a set of delta peaks at positions $h\mathbf{a}^* + k\mathbf{b}^* + \ell\mathbf{c}^* + m\mathbf{q}$. Because $\exp(iz\cos\theta)$ has periodicity 2π it may be developed in a simple Fourier series. The coefficients are Bessel functions:

$$\exp(iz\cos\theta) = \sum_{m=-\infty}^{\infty} i^m \mathcal{J}_m(z)\exp(im\theta). \tag{1.16}$$

(This is the so-called Jacobi–Anger relation.) Together one gets

$$F(\mathbf{H}) = \sum_{\mathbf{K},m} \delta(\mathbf{H} - \mathbf{K} - m\mathbf{q}) i^{-m} \mathcal{J}_m(2\pi\mathbf{H}.\mathbf{A}),$$

18 Introduction

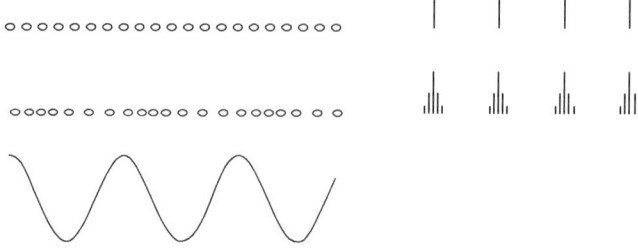

Figure 1.6 *A modulated chain and its diffraction. Left: (1) upper line unmodulated chain, (2) middle modulated chain, with (3) bottom modulation function. Right: first line: diffraction of the periodic chain (1), second line: the modulation (2) is seen from the satellites appearing besides the original main reflections.*

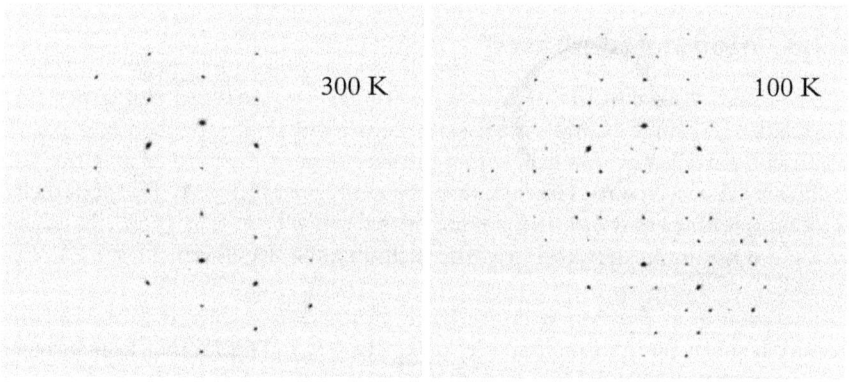

Figure 1.7 *Left: Diffraction pattern of bis(4-chlorophenyl)sulfone (BCPS) at 300 K, where only main reflections occur. Right: The same at 100 K, in the modulated phase, when additional satellites appear. (Courtesy: Bertrand Toudic, Rennes)*

where the summation **K** runs over the reciprocal lattice of the basic structure. A simple one-dimensional example with rank two is given in Fig. 1.6

This is the structure factor of a quasiperiodic function of rank 4, provided **q** is not a rational combination of the reciprocal lattice basis vectors. One may write

$$\mathbf{q} = \alpha \mathbf{a}^* + \beta \mathbf{b}^* + \gamma \mathbf{c}^*. \tag{1.17}$$

The rank of the Fourier module is four if at least one of the coefficients α, β, or γ is irrational. Generally, incommensurate modulated phases occur in a certain temperature range (Fig. 1.9). At the phase transition, satellites appear (see Fig. 1.7). In the range where incommensurate modulated phases occur, phase transitions between different incommensurate phases may be observed. An example is betaine calcium chloride dihydrate (BCCD), where there are many phases with different modulation wave vectors, or biphenyl, where a rank 5 modulated phase changes to a rank 4 phase. In one phase one has modulation vectors $(\alpha, \pm\beta, 0)$ and in the other $(\alpha, 0, 0)$. This is called a *partial lock-in transition*. For lower temperatures the modulation wave vector often becomes commensurate. This is the *lock-in transition*.

If the modulation wave vector **q** is incommensurate, the phase of the modulation becomes unimportant. The origin of the modulation wave may be anywhere in the crystal. This implies, as we shall see, under certain conditions, the existence of a free sliding motion of the modulation function with respect to the lattice, next to the zero frequency motion described by a rigid motion of the whole crystal. So, in this case, as a consequence of a dynamical symmetry, there is an additional zero frequency phonon mode.

If there are more particles in the unit cell a displacive modulation with one period gives particles at positions $\mathbf{r}_{\mathbf{n}j}$, where **n** is a lattice vector and $\mathbf{r}_j$ the position of particle j in the unit cell. The positions are

$$\mathbf{r}_{\mathbf{n}j} = \mathbf{n} + \mathbf{r}_j + \mathbf{u}_j(\mathbf{q}.\mathbf{n}), \tag{1.18}$$

where vectors $\mathbf{u}_j$ are periodic in their arguments with period 1. The diffraction pattern again has peaks at the positions of the Fourier module

$$\mathbf{H} = h\mathbf{a}^* + k\mathbf{b}^* + \ell\mathbf{c}^* + m\mathbf{q}. \tag{1.19}$$

The functions $\mathbf{u}_j$ are called the *modulation functions*, the vector **q** is the *modulation vector*. A priori, the modulation functions are not necessarily smooth. We shall see that the existence of discontinuities poses problems for the structure determination, and for the interpretation.

In the diffraction pattern peaks exist which remain when the modulation vanishes. These are the main reflections. The other peaks are called *satellites*. If there is one wave vector, the satellites are in rows around the main reflections (Fig. 1.7, (Etrillard *et al.*, 1993)). Their intensity is not always smaller than that of the *main reflections*. When the modulated phase originates at a second-order phase transition, the intensity near the transition temperature will be small.

Assigning a satellite to a main reflection is not unique. Neither, of course, is the choice of the basis of the reciprocal lattice of the basic structure. For example, if the satellite position is given by $\mathbf{q} = \gamma \mathbf{c}^*$ from the origin, the satellite reflection may be considered to be a satellite of $\mathbf{c}^*$ with modulation wave vector $(1-\gamma)\mathbf{c}^*$ as well.

One of the components of the modulation wave vector must be irrational to have an aperiodic structure. The other components may well be rational. For example, a modulation wave vector in an orthorhombic structure might be $\left(0, \frac{1}{2}, \gamma\right)$. These are the coefficients with respect to the basis $\mathbf{a}^*, \mathbf{b}^*, \mathbf{c}^*$. Introducing a new basis of the reciprocal space $\mathbf{a}^*$, $\mathbf{b}'^* = \frac{1}{2}\mathbf{b}^*$, and $\mathbf{c}^*$, and considering the satellite as belonging to $\mathbf{b}'^*$, the coordinates become $(0, 0, \gamma)$. This leads to extinction rules similar to those for centred lattices in usual crystallography.

Another type of modulation is created by a composition wave. In the example of CuAuII the roles of Cu and Au are exchanged at the domain boundaries. This means that the position 1 in the orthorhombic unit cell is taken by either Cu or Au, and position 2 by the other species. This may be described as a position-dependent probability of finding Cu at position 1: $p(\mathbf{n})$. Then the probability of finding Au at this position is $1 - p(\mathbf{n})$. In modulated phases this probability or composition wave is periodic with a period which is commensurate or incommensurate. When it is incommensurate the structure is aperiodic. (However, when it is commensurate, the structure is, strictly speaking, also aperiodic, because the function p only indicates a probability. There is lattice periodicity in the mean.) Again, the probability may be decomposed into a Fourier series.

$$p_j(\mathbf{n}) = \sum_{\mathbf{k}=h_1\mathbf{q}_1+h_2\mathbf{q}_2+\ldots} \hat{p}(\mathbf{k})_j \exp(2\pi i \mathbf{k}.\mathbf{n}). \qquad (1.20)$$

Satellites were studied for determining the ordering of atoms, e.g. for $Nd_2(Mo)_4)_3$, as early as 1969 (Jamieson et al., 1969). In general, composition modulation is accompanied by a displacive modulation. There is an interaction between the two order parameters.

The *basic structure* is not a uniquely defined object. It is chosen conveniently in order to have a relatively small modulation in the positions or in the occupation. Even the notion of main reflection is not unique in general. However, very often a three-dimensional lattice to which most of the strong peaks belong stands out. This is the reciprocal lattice of the basic structure. Using this, one may define an *average structure*. It is the inverse Fourier transform of the main reflections. If $\rho(\mathbf{r})$ is the density of the crystal, and $\hat{\rho}(\mathbf{k})$ its Fourier transform, the latter has peaks at the Fourier module $\mathbf{H} = \sum_{j=1}^{n} h_j \mathbf{a}_j^*$. The main reflections are at positions $\sum_{j=1}^{3} h_j \mathbf{a}_j^*$. The average structure then is

$$\rho_{av}(\mathbf{r}) = \sum_{\mathbf{H}} \hat{\rho}(\mathbf{H}) \exp(2\pi i \mathbf{H}.\mathbf{r}); \quad \mathbf{H} = h_1\mathbf{a}_1^* + h_2\mathbf{a}_2^* + h_3\mathbf{a}_3^*.$$

This function has lattice periodicity with a lattice for which the reciprocal lattice is spanned by $\mathbf{a}_1^*, \mathbf{a}_2^*, \mathbf{a}_3^*$. It gives the distribution of the density modulo the lattice of the basic structure. For atoms at positions $\mathbf{n} + \mathbf{r}_j + \mathbf{u}_j(\mathbf{n})$, these are the positions $\mathbf{r}_j + \mathbf{u}_j(\mathbf{n}) \pmod{\mathbf{n}}$. Consider, for example, a one-dimensional chain with a sinusoidal modulation: the positions are $na + A\sin(2\pi qna)$. Then the points in the unit cell are $u = A\sin(z)$ with $-\pi < z < \pi$ and the distribution in the unit cell, proportional to $1/A\cos(z))$ which is given by

$$X(u) = \frac{1}{\pi}\left(\sqrt{A^2 - u^2}\right)^{-1},$$

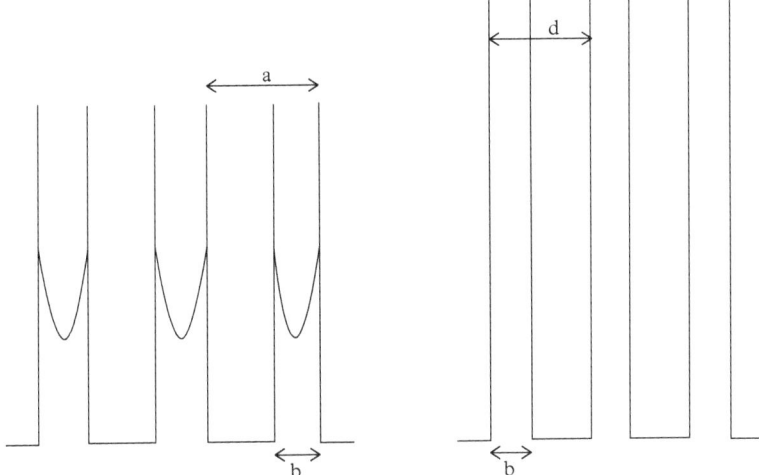

Figure 1.8 *Average structure. Left: sinusoidal modulation with basic structure lattice parameter a, b = twice the amplitude A. Right: Fibonacci chain: d = 3Φ–1 (average lattice parameter), b = 1 – Φ.*

and the average structure is this function periodically repeated with period a. Notice that in this case the maximal probability occurs for $u = \pm A$. This is close to a two-site model, where the atom is either at position $u = A$ in the unit cell, or at $u = -A$. (See Fig. 1.8.)

Incommensurate modulated phases turn out to occur very frequently, in all types of materials: in ionic and molecular crystals, in alloys, in high-T_c superconductors and ferroelectrics, even in elements under pressure. Usually, they occur in certain regions of the $p - T$-composition phase diagram (See Fig. 1.9.). The phase transition from periodic to aperiodic is the *incommensurate phase transition*, the transition at lower temperature is the *lock-in transition*. The incommensurate phases are also important in total quantity. A substantial part of the earth's crust consists of incommensurate minerals, like feldspar. Usually the amplitudes are small, but sometimes these are as big as 1 Å.

The origin of the occurrence of such phases also varies widely. A common term is 'frustration' or 'competition', in the sense that two or more mechanisms favour a certain periodicity, whereas these periodic structures are mutually incompatible. It may be the optimal packing of rigid molecules in A_2BX_4 compounds, or the ordering of vacancies. Sometimes the reason lies in the interaction between structure and electrons. Here the Peierls or the Hume–Rothery mechanism is at work. In certain circumstances, a deformation of the basic structure may lower electron energy levels. Then the balance between the elastic energy of the deformation and the gain in electron energy determines the ground state.

Sterical hindrance, coupling to the electron system and the tendency towards an energetically favourable packing of rigid elements lead, in many cases, to an incommensurate phase, originating from an instability in the structure. A soft

22 Introduction

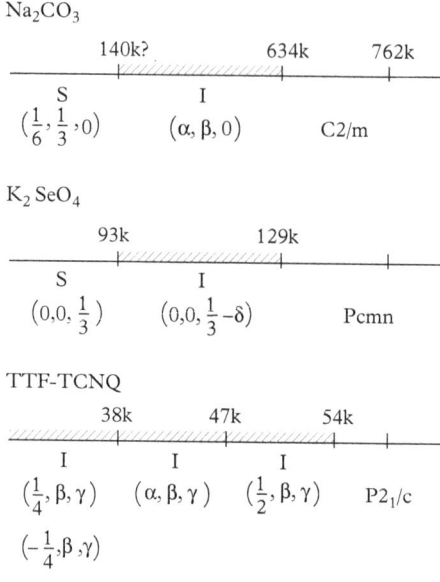

Figure 1.9 *Incommensurate phases often occur in an intermediate temperature range: three examples of phase diagrams. For high temperatures, there is a conventional periodic crystal structure; for low temperatures, a superstructure; and, in the intermediate range, an incommensurate phase (hatched bars).*

phonon mode, with a frequency tending to zero, may introduce such an instability at a point in the Brillouin zone that does not correspond to a superstructure.

1.2.4 Incommensurate composites

Inter-growth compounds, also called composites, need another basic structure for the description (Janner and Janssen, 1980b; van Smaalen, 1991). Let us consider a hypothetical structure with space group symmetry, in the channels of which another compound is found with an own space group symmetry. Very often, the materials do not exist as such, but together they form a stable compound if one takes into account the mutual interaction which causes modulation of the constituting components (Fig. 1.10). The compound may then be considered as consisting of two subsystems, each with a modulated structure.

Suppose the two subsystems have basic structures with reciprocal lattices Λ_ν^* ($\nu = 1, 2$). The atoms of subsystem 1 will undergo a modulation with a periodicity of subsystem 2, and vice versa. The atoms of system 1 will diffract with diffraction vectors

$$h\mathbf{a}_{11}^* + k\mathbf{a}_{12}^* + \ell\mathbf{a}_{13}^* + m\mathbf{q}_1$$

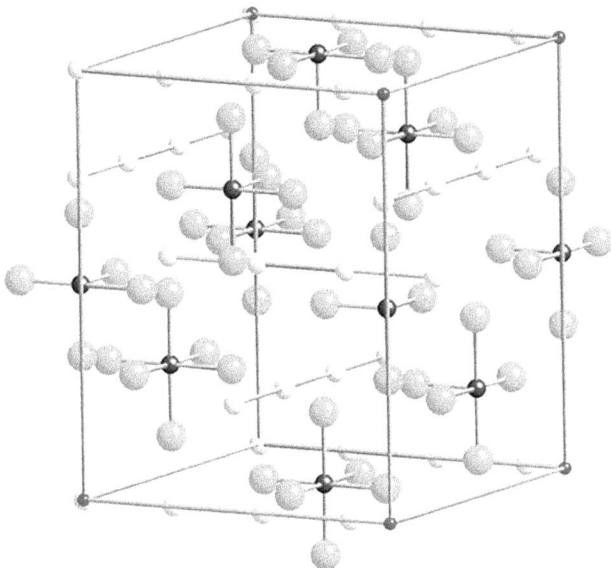

Figure 1.10 *The composite $Hg_{3-\delta}AsF_6$ has a host system consisting of AsF_6 octahedra. In the channels parallel to the x- and the y-axis there are chains of Hg atoms with a distance incommensurate to the lattice parameter of the host lattice. It is an incommensurate composite.*

and those of subsystem 2 with diffraction vectors

$$h\mathbf{a}_{21}^* + k\mathbf{a}_{22}^* + \ell\mathbf{a}_{23}^* + m\mathbf{q}_2$$

with $\mathbf{q}_1$ belonging to the reciprocal lattice Λ_2^* of subsystem 2 and vice versa. Therefore, all diffraction peaks are at positions which are linear combinations of the six basis vectors. Generally, these six vectors $\mathbf{a}_{\nu i}^*$ ($\nu = 1, 2$, $i = 1, 2, 3$) are not independent. One may choose a basis of rank $n \leq 6$ vectors $\mathbf{b}_j^*$ such that each diffraction wave vector belongs to the Fourier module generated by these n vectors. One has

$$\mathbf{a}_{\nu i}^* = \sum_{j=1}^{n} Z_{ij}^{\nu} \mathbf{b}_j^* \quad (i=1,2,3, \ \nu=1,2). \tag{1.21}$$

For a simple model with two types of chain, the positions of the atoms are (Fig. 1.11)

$$x_{n\ell} = x_0 + na + f(na), \ y_{n\ell} = \ell c$$
$$X_{m\ell} = X_0 + mb + g(mb), \ y_{m\ell} = (\ell + 1/2)c,$$

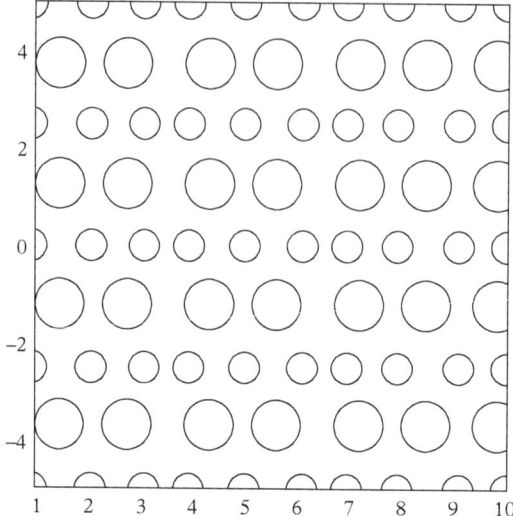

Figure 1.11 *A two-dimensional model for an incommensurate composite with* $b/a = \sqrt{2}$.

with $f(x) = f(x+b)$, and $g(y) = g(y+a)$. The Fourier module of this system consists of $\mathbf{k} = (i/a + j/b, k/c)$. The matrices Z are

$$Z^1 = \begin{pmatrix} 1 & 0 & 0 \\ 0 & 0 & 1 \\ 0 & 1 & 0 \end{pmatrix}, \quad Z^2 = \begin{pmatrix} 0 & 1 & 0 \\ 0 & 0 & 1 \\ 1 & 0 & 0 \end{pmatrix}.$$

If a/b is irrational, the structure is aperiodic, and the rank of the Fourier module is $n = 3$. Similar to the case of incommensurate modulated phases one has two dynamical symmetries. For the first one there is a common translation δ for both chains ($x_{n,\ell} \to x_{n,\ell} + \delta$, $X_{m,\ell} \to X_{m,\ell} + \delta$). The potential energy is invariant under this operation. It corresponds again to conservation of total momentum. The other dynamical symmetry will be discussed later. It corresponds to the relative shift of the two subsystems ($x_{n,\ell} \to x_{n,\ell} + \delta$, $X_{m,\ell} \to X_{m,\ell} - \delta$).

The mechanisms for the origin of incommensurate composites are again of several types. A compound may form a metastable structure with large channels, which may be stabilized by introducing another material inside the channels. If the lattice constants are mutually incommensurate, the result is aperiodic. The interaction may be weak, as in several inclusion compounds with organic molecules as guests, or strong, as is the case in misfit layered structures, such as PbS-TiS. Usually the composites are non-stoichiometric. A monolayer of material A on a substrate of material B, just as argon on graphite, is in principle aperiodic, because of the interaction. The system consisting of the monolayer and the top layer of the substrate forms an incommensurate composite.

1.2.5 Quasicrystals

The discovery of an alloy with icosahedral symmetry by *Dan Shechtman* in 1982 (see Fig. 1.12) meant a further breakthrough for aperiodic crystals. In rapidly quenched AlMn a diffraction pattern was observed with fairly narrow peaks at positions

$$\mathbf{H} = \sum_{i=1}^{6} h_i \mathbf{a}_i^*,$$

where the six basis vectors point to the six pairs of parallel facets of a dodecahedron. The intensity distribution had the symmetry of an icosahedron, a point group not allowing lattice periodicity in three-dimensional space. The coherent domains were at that time still very small (of the order of 50 μm).

The discovery led at first to a controversy between scientists who believed that what was observed was due to twinning, and those who thought that the material at that time was not ideal, but that an ideal aperiodic but well-ordered phase would be possible. Aperiodicity was already known from incommensurate phases. Acceptance of the idea of aperiodicity in crystals was further stimulated by other

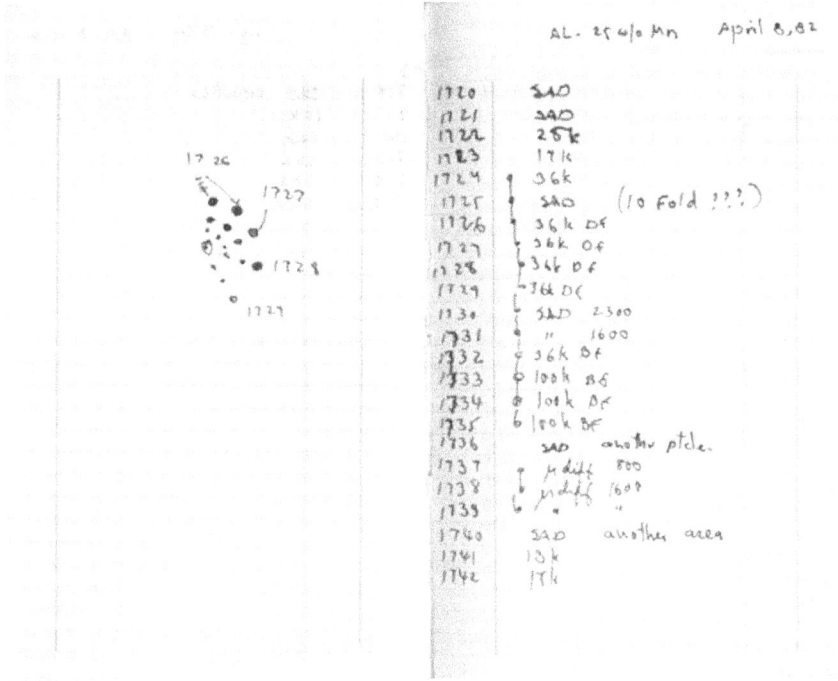

Figure 1.12 *Shechtman's notebook, in which the discovery of the icosahedral phase was written down: April 1982. (Courtesy: Dan Shechtman)*

results. In the theory of tilings, the discovery of an aperiodic tiling of the plane by Penrose showed that such well-ordered systems were mathematically possible (Penrose, 1979). An experimental proof was given by Mackay (1982), who constructed an optical diffraction pattern for a Penrose tiling. In the diffraction pattern consisting of fairly sharp peaks the five-fold symmetry is clearly visible. Locally five-fold symmetric units in Frank–Kasper phases and icosahedral clusters as, for example, the Mackay clusters were known. In computer simulations, icosahedral orientation ordering had been found by Levine, Steinhardt, and Nelson. A paper by Levine and Steinhardt (Levine and Steinhardt, 1984) on quasiperiodically ordered structures with crystallographically forbidden symmetry (at least forbidden for three-dimensional periodic structures) appeared in the same issue as where the discovery by Shechtman and collaborators was announced. The notion of quasiperiodicity was known, and in the latter paper the term 'quasicrystal' was coined. Perhaps this was not a very good name, because it suggests something that looks almost as a crystal but isn't and, moreover, if it means quasiperiodic, it implies that all lattice periodic crystals are quasicrystals because, according to the mathematical definition, a periodic function is also quasiperiodic. At first, the term was used mainly for alloys with the 'non-crystallographic' symmetries. We shall discuss this issue later. Earlier, Mackay had already introduced the term 'quasi-lattice', for what here is called Fourier module, in Mackay (1982).

The main difference between quasicrystals and incommensurate modulated phases is that in the latter it is usually possible to recognize a lattice of the main reflections and satellites, whereas in the former the basis vectors of the Fourier module are on the same footing. A possible basis for the Fourier module of an icosahedral quasicrystal is given by the six vectors (Fig. 1.13)

$$\mathbf{a}_i^* = \frac{1}{a}(1, \tau, 0), \quad \frac{1}{a}(-1, \tau, 0), \quad \frac{1}{a}(0, 1, \tau), \qquad (1.22)$$
$$\frac{1}{a}(\tau, 0, 1), \quad \frac{1}{a}(\tau, 0, -1), \quad \frac{1}{a}(0, 1, -\tau),$$

with $\tau = (\sqrt{5}+1)/2$ and $i=1,\ldots,6$. With respect to this basis, a five-fold and a three-fold rotation are given by the matrices

$$A = \begin{pmatrix} 1 & 0 & 0 & 0 & 0 & 0 \\ 0 & 0 & 1 & 0 & 0 & 0 \\ 0 & 0 & 0 & 1 & 0 & 0 \\ 0 & 0 & 0 & 0 & 1 & 0 \\ 0 & 0 & 0 & 0 & 0 & 1 \\ 0 & 1 & 0 & 0 & 0 & 0 \end{pmatrix}, \quad B = \begin{pmatrix} 0 & 0 & 0 & 0 & 0 & 1 \\ 1 & 0 & 0 & 0 & 0 & 0 \\ 0 & 0 & 0 & 0 & 1 & 0 \\ 0 & 0 & -1 & 0 & 0 & 0 \\ 0 & 0 & 0 & -1 & 0 & 0 \\ 0 & 1 & 0 & 0 & 0 & 0 \end{pmatrix}.$$

These two matrices generate, together with the total inversion (the opposite of the identity), a point group with 120 elements, the symmetry group of an icosahedron (or a dodecahedron). This group contains 60 rotations: 6 five-fold axes (giving 24

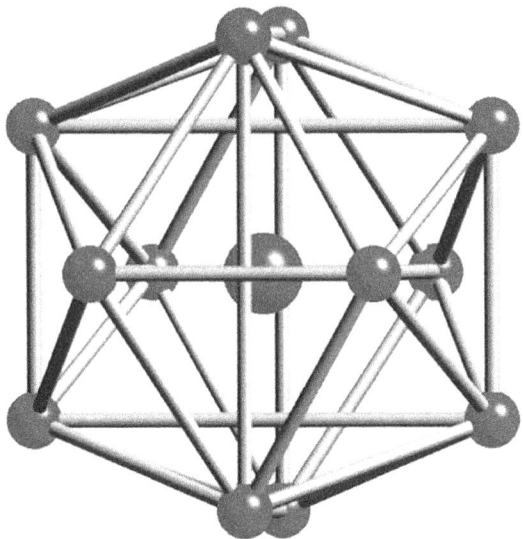

Figure 1.13 *The six basis vectors of the icosahedral Fourier module (1,...,6) and their inverses form the vertices of an icosahedron. The 12 vectors point to the faces of a dodecahedron. They form three mutually perpendicular rectangles with edge ratio* $\tau(\pm 1, \pm 2;\ \pm 3, \pm 6;\ \pm 4, \pm 5)$.

rotations), 10 three-fold axes (for 20 rotations), 15 two-fold axes (for 15 rotations), and the identity.

A simple one-dimensional model for a quasicrystal is the *Fibonacci chain*. This chain with the golden ratio $\Phi = \tau - 1$ in the Fourier module is obtained by putting atoms at positions

$$x_n = n(3\Phi - 1) + (\Phi - 1)\text{Frac}(n\Phi),$$

where the fractional part $\text{Frac}(x)$ is x minus the largest integer smaller than x. The constructed chain has intervals ordered aperiodically with lengths 1 or Φ. This is so, because $\text{Frac}((n+1)\Phi)$ differs from $\text{Frac}(n\phi)$ by Φ or $1 - \Phi$. The diffraction pattern has peaks at the positions of the Fourier module of rank 2 with generators ('basis vectors') $1/(2-\Phi)$ and $\Phi/(2-\Phi)$, because the periodicity of the basic lattice is $3\Phi - 1$ and that of the modulation function is $1/\Phi$. The diffraction peaks are at the positions

$$h_1/(2-\Phi) + h_2\Phi/(2-\Phi).$$

The modulation function $(\Phi - 1)\text{Frac}(x/\Phi)$ is special in the sense that it is a discontinuous function which is linear on intervals. On the other hand, the resulting chain consists of a sequence of two types of intervals (one-dimensional tiles), and in that respect it resembles the structure of a quasicrystal.

♯ Fibonacci numbers were introduced by Fibonacci in his study of the following problem. One starts with a pair of rabbits. After a month they produce a new pair. After another month all pairs produce a new pair. The rabbits do not die and grow up to fertility in one month. Then the question is 'How many rabbits are there after n months?' This number is the n-th *Fibonacci number* F_n, which satisfies the recurrence relation $F_{n+1} = F_n + F_{n-1}$. Starting with one pair of baby rabbits, one has in the following months 1, 2, 3, 5, 8, etc., pairs of rabbits. Each pair S of babies gives rise in a month to a pair of adults L, and a pair of adults L is replaced by a pair of babies S and a pair of adults. This is a so-called substitution rule. The ratio F_{n+1}/F_n tends to $\tau = (\sqrt{5} + 1)/2$ when n tends to infinity. This follows from $x_{n+1} = F_{n+1}/F_n = (F_n + F_{n-1})/F_n = 1 + 1/x_n$. Therefore, the limit x satisfies $x = 1 + 1/x$, for which the positive solution is τ. Its inverse is Φ, which satisfies $x : (1-x) = 1 : x$. This is the golden mean. Because $\tau^2 = 1 + \tau$ and $\tau^{-1} = \tau - 1 = \Phi$ every polynomial in τ can be written as $n + m\tau$ with integer values for n and m. For example, $\tau^3 = 1 + 2\tau$, and $\tau^4 = 2 + 3\tau$. The golden mean pops up in many places in art (Leonardo da Vinci and Dürer) and science (phyllotaxis).♯

When quasicrystals were discovered, perfectly ordered structures with aperiodic order were known not only from modulated phases, but also from mathematics. R. Penrose had found a tiling of the plane, without voids or overlaps, by copies of two tiles, such that the structure is quasiperiodic. Moreover, the Fourier transform has five-fold symmetry. So the Penrose tiling was quickly mentioned as a two-dimensional model for a quasicrystal.

One version of the *Penrose tiling* consists of rhombi, with an angle of 36° or 72°, the skinny and the fat rhombi. All have the same edge lengths. If certain matching rules, which we shall discuss later on, are obeyed the result is an aperiodic tiling. The Fourier transform of a function that is a sum of delta functions on the vertices of the tiling has sharp peaks at the positions of the Fourier module which is of rank 4: every spot is at a linear combination of the vectors $(\cos(2\pi j/5), \sin(2\pi j/5))/a$,

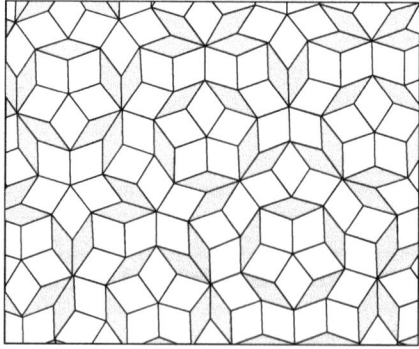

Figure 1.14 *The Penrose tiling with fat and skinny rhombi.*

where $j = 1, \ldots, 5$. The Fourier transform has five-fold symmetry, as was already shown by Mackay using optical Fourier transform. If the tiles are decorated by putting 'atoms' in each tile, this is an example of a perfectly ordered not periodic but quasiperiodic structure (Fig. 1.14).

Soon after the discovery of quasicrystals their relation with the Penrose tiling was seen. A first type of model was a three-dimensional tiling of the space by means of rhombohedra. Using two types of rhombohedra, a three-dimensional analogue of the Penrose tiling can be constructed. The first attempts were made using a decoration of such a tiling.

The structure determination of quasicrystals turned out to be very hard, but there is a clue because some of the quasicrystals have related compounds with slightly different composition and with locally similar structure. These so-called *approximants* are lattice periodic, and for them the usual techniques could be used. A knowledge of the structure of approximants was very useful for constructing models of quasicrystals.

Another way to the structure determination was the construction of models using *clusters*. It was observed that locally one may see the structure as a piling of (overlapping) clusters, just as was already known before for alloys. As building blocks one may consider clusters connected by *glue atoms*. Three types of clusters have, for example, been considered, the so-called *Bergman, Mackay,* and *Tsai clusters* (Fig. 1.15). The first one consists of 12 atoms at the vertices of an icosahedron inside a double sized icosahedron decorated at the vertices (12 atoms) and mid-edges (30 atoms). The second one has a similar central icosahedron, now inside a larger triacontahedron with atoms at its vertices (32 atoms). These were known before the discovery of quasicrystals. The third cluster was found

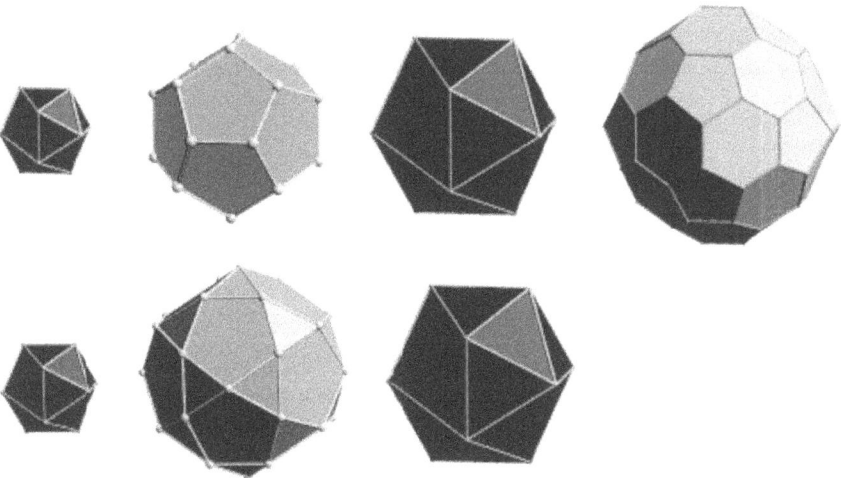

Figure 1.15 *Decomposition of the successive shells which constitute the Bergman cluster (top) and the Mackay cluster (bottom) (Courtesy: Cesar Pay Gomez)*

30 *Introduction*

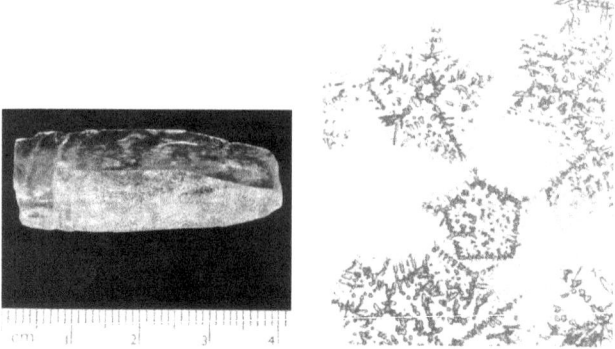

Figure 1.16 *Left: A crystal of Rb_2ZnBr_4. The largest facet is connected with the wave vector of a satellite for this modulated phase. (Courtesy: Theo Rasing, Nijmegen) Right: Five-fold symmetry in microcrystallites of AlMn.*

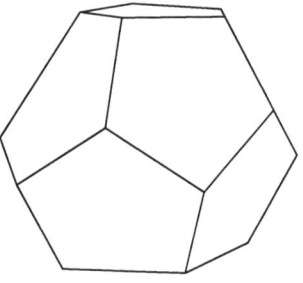

Figure 1.17 *Specimen of icosahedral HoMgZn. Quasicrystals may have the shape of a dodecahedron. Six Miller indices are needed in this case. (Photograph courtesy: I. R. Fisher (Stanford) and P. C. Canfield (Ames Laboratory))*

afterwards. Note that the triacontahedron can be obtained as the union of an icosahedron and a dodecahedron. These clusters with icosahedral symmetry are quasiperiodically packed.

1.2.6 Morphology

Just as lattice periodic crystals show flat facets, with an orientation connected to the reciprocal lattice, quasiperiodic crystals have facets connected to the Fourier module. This means that one needs more Miller indices than the dimension of the physical space. Each facet is connected to a vector of the Fourier module, which means that one needs n indices (cf. Figs 1.16 and 1.17). The morphology of aperiodic crystals will be discussed in Section 7.1. The morphology of quasicrystals in particular has been discussed in Janssen *et al.* (1989).

1.3 Summary

It has become clear that the ground state of solid matter is not necessarily lattice periodic. This new paradigm has the consequence that a crystal should be defined in another way than just by requiring lattice periodicity. Another consequence is that a new approach is needed for those crystals that are not lattice periodic. The usual techniques for studying physical properties and for describing the structure are heavily based on the assumption of lattice periodicity.

Perfect quasiperiodic crystals have, just as periodic crystals, sharp Bragg peaks, on the positions of a Fourier module. When the rank of the quasiperiodic structure is equal to the dimension of the space, the Fourier module is identical to the reciprocal lattice. A quasiperiodic crystal with a rank higher than the dimension of the physical space is called an aperiodic crystal.

Among the aperiodic crystals there are classes, such as modulated phases, incommensurate composites, and quasicrystals, but these classes are not mutually exclusive.

The morphology of aperiodic crystals may be described by generalized Miller indices. One needs as many such indices as the rank of the Fourier module.

In the following chapters we shall have a closer look at the structure and the physical and mathematical properties of aperiodic crystals.

....

EXERCISES

1.1. In a crystal with space group P4/mmm a soft mode develops with wave vectors in the star of the wave vector $\mathbf{q} = \alpha \mathbf{a}^* + \beta \mathbf{b}^* + \gamma \mathbf{c}^*$, i.e. all the vectors obtained from this vector under the operation of the point group. If the frequency of these modes goes to zero, a modulated structure is formed. Determine the rank of the Fourier module of this structure for arbitrary values of α, β, γ.

1.2. Calculate the structure factor F(H) for a one-dimensional modulated chain with lattice constant equal to 1. It has a displacive sinusoidal modulation $u(x)$ with amplitude a and irrational wave vector q. Calculate the structure factor for some vectors from the Fourier module.

2
Description and symmetry of aperiodic crystals

2.1 Aperiodic and quasiperiodic functions

The notion of regular, but aperiodic motion is well known in dynamical systems. The orbital periods of the earth around the sun and the moon around the earth are incommensurate. This is an important fact for the stability of a dynamical system. Because of the incommensurability there is no common period, which means that the motion is aperiodic. However, it is a very regular motion, which is called quasiperiodic. In mathematics, such quasiperiodicity has been studied already before the twentieth century (Bohl 1893). Let us start with periodic functions.

A periodic function on the line is a function that is invariant under a translation a:

$$f(x) = f(x+a)$$

from which it follows that $f(x+na) = f(x)$ for all integers n. Such a function has a Fourier transform consisting of delta peaks at $k^* = h/a$, that is on the reciprocal lattice.

$$f(x) = \sum_{m=-\infty}^{\infty} c_m \exp(2\pi i m x/a), \quad c_m = \hat{f}(m/a).$$

In d dimensions the translations form a lattice with elements

$$\mathbf{a} = \sum_{j=1}^{d} n_j \mathbf{a}_j,$$

and the *reciprocal lattice* is a d-dimensional lattice with basis vectors $\mathbf{a}_i^*$, such that $\mathbf{a}_i \cdot \mathbf{a}_j^* = \delta_{ij}$ for $i, j = 1, \ldots, d$.

Aperiodic Crystals: From Modulated Phases to Quasicrystals: Structure and Properties.
Second Edition. Ted Janssen, Gervais Chapuis, and Marc de Boissieu.
© Ted Janssen, Gervais Chapuis, and Marc de Boissieu 2018. Published in 2018 by Oxford University Press.
DOI: 10.1093/oso/9780198824442.001.0001

$$f(\mathbf{r}) = \sum_{h_1 h_2 \ldots h_d} c(\mathbf{H}) \exp(2\pi i \mathbf{H}.\mathbf{r}), \quad \mathbf{H} = \sum_{j=1}^{d} h_j \mathbf{a}_j^*.$$

The function $g(x) = \cos(2\pi x/a) + \cos(2\pi x/b)$ has a simple Fourier decomposition with wave vectors $\pm 1/a$ and $\pm 1/b$. If a/b is a rational number it is a periodic function with as period the smallest common multiple of a and b. However, when a/b is irrational, the function is aperiodic (it has no periodicity), but it is by no means a chaotic function. It is an example of a quasiperiodic function. To obtain a general quasiperiodic function, we can do the following.

Consider a function with n variables, which is periodic in each of these variables. Then $f(\mathbf{r}) = f(x_1, x_2, \ldots, x_n)$ satisfies

$$f(x_1, x_2, \ldots, x_n) = f(x_1 + a_1, x_2, \ldots, x_n), \quad \text{etc.}$$

One may construct a function on the line from this by taking a line $\mathbf{r} = x\mathbf{R}$, where $\mathbf{R}$ is a fixed vector in n-dimensional space. Then the function

$$g(x) = f(x\mathbf{R}) \qquad (2.1)$$

has Fourier components $c(m_1, m_1, \ldots)$ and a Fourier decomposition

$$g(x) = f(x\mathbf{R}) = \sum_{m_1, m_2 \ldots} c(m_1, \ldots, m_n) \exp(2\pi i (\mathbf{R}.\mathbf{H})x) = \sum_q c(q) \exp(2\pi i q x),$$

with
$$q = \sum_{j=1}^{n} h_j \mathbf{R}.\mathbf{a}_j^* = \sum_{j=1}^{n} h_j q_j. \qquad (2.2)$$

So, the Fourier decomposition of the function $g(x)$ has wave numbers that are linear combinations of n quantities q_j. For $g(x) = \cos(2\pi x/a) + \cos(2\pi x/b)$ the value of n is 2 (provided a/b is irrational).

Analogously a three-dimensional function may be constructed. Consider a function given by its discrete Fourier decomposition

$$f(\mathbf{r}) = \sum_{\mathbf{H} \in M^*} \hat{f}(\mathbf{H}) \exp(2\pi i \mathbf{H}.\mathbf{r}), \qquad (2.3)$$

with M^* defined as the set of vectors

$$\mathbf{H} = \sum_{j=1}^{n} h_j \mathbf{q}_j. \qquad (2.4)$$

If the minimal number of basis vectors $\mathbf{q}_j$ is n, and n is larger than 3, the function $f(\mathbf{r})$ has no lattice periodicity (periodicity in three directions), it is aperiodic. However, that is not a good characterization, because a fully random function is also aperiodic. A function with a Fourier decomposition consisting of delta peaks on positions given by Eq. 2.4 is called *quasiperiodic*. The set M^* is called its Fourier module, and the (minimal) number of basis vectors n its *rank*. The difference $n-3$ between the rank and the dimension of the space is sometimes called the *co-dimension*. For the special case that $n=3$ (or co-dimension zero), the structure is lattice periodic if the basis vectors $\mathbf{q}_j$ are linearly independent, that is, span the whole space, and not a plane or a line only. Then the Fourier module is just the reciprocal lattice.

♯ A property of a quasiperiodic function is the following. Roughly speaking, there are 'enough' translations which map the function to a function that differs arbitrarily little from the original one. More precisely, for each value $\epsilon > 0$, there is a '*relatively dense*' set of translations $\mathbf{a}$ such that $|f(\mathbf{r}+\mathbf{a}) - f(\mathbf{r})| < \epsilon$ for all $\mathbf{r}$. 'Relatively dense' means that there are for each ϵ values R_1 and R_2 such that every sphere of radius R_2 contains at least one $\mathbf{a}$ satisfying the condition, and in every sphere with radius R_1 around any translation $\mathbf{a}$ satisfying the condition, there is no other translation but $\mathbf{a}$.

A function satisfying this condition is called *almost periodic*. The theory of almost periodic functions was developed by H. Bohr (cf. (Bohr, 1952)). Every quasiperiodic function is almost periodic. And every periodic function is quasiperiodic. ♮

A simple example of a quasiperiodic function is $f(x) = \cos(2\pi x) + \cos(2\pi \alpha x)$ if α is irrational. Then the Fourier module consists of $(m + n\alpha)$ and a basis of the module is 1 and α. Its rank is 2. The function is almost periodic because there are integers n and m such that for the translation $a = (n + m\alpha)$ the shifted function differs arbitrarily little from the function itself:

$$|f(x+a) - f(x)| < \epsilon.$$

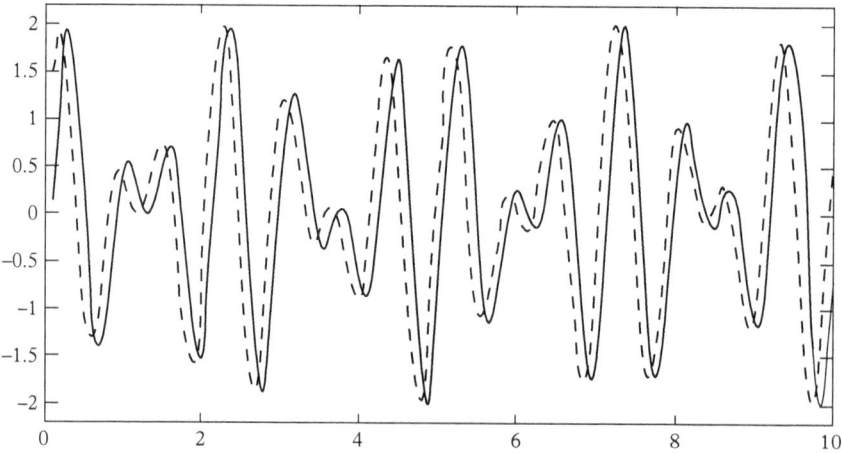

Figure 2.1 *The function* $\cos(2\pi x) + \cos(2\pi x\sqrt{2})$ *and its translate over* 29 *lattice constants (offset by 0.1 for clarity).*

An example is given in Fig. 2.1. The force applied to the earth by sun and moon would be a function of this type.

2.2 Quasiperiodic structures

2.2.1 Modulated phases

Incommensurate modulated phases are aperiodic, but well-ordered, structures. For a modulated phase there exists a *basic structure* with lattice periodicity, hence with space group symmetry. This hypothetical structure is used as basis for the description of the real structure. It is a hypothetical structure, because it does not exist as such at the same temperature. The positions of the atoms are given by their displacements from the basic structure positions. Suppose the position of the j-th atom in the unit cell at $\mathbf{n}$ of the basic structure is $\mathbf{n}+\mathbf{r}_j$ with $j=1,\ldots,s$. Then the positions in the modulated phase are $\mathbf{n}+\mathbf{r}_j+\mathbf{u}_j(\mathbf{n}+\mathbf{r}_j)$. Suppose further that the modulation functions $\mathbf{u}_j$ are periodic. If this periodicity is a translation vector of the basic structure, then the modulated phase is also lattice periodic. The modulation is called commensurate, and the structure is a *superstructure* for the basic structure. However, when the periodicity $\mathbf{a}$ has irrational components with respect to the lattice of the basic structure, the modulated structure is still perfectly ordered, but no longer lattice periodic. If the modulation is a plane wave with wave vector $\mathbf{q}$

Figure 2.2 *A two-dimensional system with a one-dimensional displacive modulation. The wave vector is along the horizontal axis. The vertical lines give the period of the modulation. The modulation here is a rotation in the plane.*

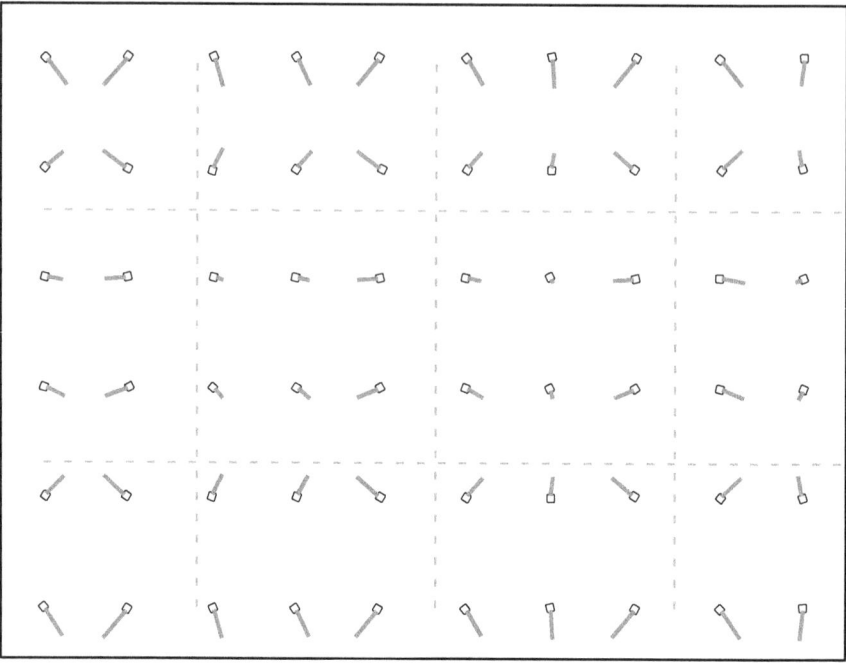

Figure 2.3 *A two-dimensional system with a two-dimensional displacive modulation. The points indicate the basic structure, the lines the displacements. Horizontal lines give the period of the modulation in the **b**-direction, vertical lines that in the **a**-direction.*

this is called the modulation wave vector. In the Fourier transform then appear the multiples of **q**. In general, there may be modulation wave vectors in several directions.

In Figs 2.2, and 2.3 two examples are given for a two-dimensional crystal with an incommensurate modulation. In the first case, the rank is 3: the modulation has one basic period. In the second case, there are two basic vectors. In the first example, the modulation is displacive, but the displacement consists of a rotation of a rigid building block.

For a three-dimensional structure with reciprocal lattice basis $\mathbf{a}^*$, $\mathbf{b}^*$, and $\mathbf{c}^*$, and modulation wave vector $\mathbf{q}$, the wave vectors of the Fourier components are of the form $h\mathbf{a}^* + k\mathbf{b}^* + \ell\mathbf{c}^* + m\mathbf{q}$. Assume that the atoms are point particles. Then the density function is a sum of delta peaks at the atomic positions, and is given by

$$\rho(\mathbf{r}) = \sum_{\mathbf{n}j} A_j \delta\left(\mathbf{r} - \mathbf{n} - \mathbf{r_j} - \mathbf{u_j}(\mathbf{n})\right).$$

Because $\mathbf{a}^*$, $\mathbf{b}^*$, and $\mathbf{c}^*$ already span the space, the modulation wave vector $\mathbf{q}$ may be expressed in terms of them:

$$\mathbf{q} = \alpha \mathbf{a}^* + \beta \mathbf{b}^* + \gamma \mathbf{c}^*. \tag{2.5}$$

The Fourier transform of the density function is

$$\hat{\rho}(\mathbf{H}) = \sum_{\mathbf{n}j} A_j \exp\left(-2\pi i \mathbf{H}.(\mathbf{n} + \mathbf{r}_j + \mathbf{u}_j(\mathbf{n}))\right), \tag{2.6}$$

and is a sum of delta peaks on the positions of the Fourier module spanned by $\mathbf{a}^*$, $\mathbf{b}^*$, $\mathbf{c}^*$, and $\mathbf{q}$.

A more general modulation has more than one modulation vector. The modulation function may have more periodicities. In that case, the position of the atom displaced from the site $\mathbf{n} + \mathbf{r}_j$ in the basis structure is given by

$$\mathbf{r}_{\mathbf{n}j} = \mathbf{n} + \mathbf{r}_j + \mathbf{u}_j(\mathbf{n}), \quad \mathbf{u}_j(\mathbf{n}) = \sum_{\mathbf{H}=\sum_i h_i \mathbf{q}_i} \hat{\mathbf{u}}_j(\mathbf{H}) \exp(2\pi i \mathbf{H}.\mathbf{n}). \tag{2.7}$$

An example of a helicoidal structure is an orthorhombic basic structure with a modulation function

$$u_x(\mathbf{n}) = a\cos 2\pi \mathbf{q}.\mathbf{n}, \quad u_y(\mathbf{n}) = a\sin 2\pi \mathbf{q}.\mathbf{n}, \quad \mathbf{q} = (0,0,q).$$

When there are more basic vectors $\mathbf{q}_j$, the rank of the Fourier module is, generally, greater than 4. A lattice of satellites may be observed in the diffraction pattern. The Fourier module is spanned by a basis of the reciprocal lattice and the basic satellites $\mathbf{q}_j$.

♯ If the functions $\mathbf{u}_j$ have several Fourier components expressible in a number of basic modulation vectors as $\mathbf{q} = \sum_m h_m \mathbf{q}_m$, the expression for the Fourier transform $\hat{\rho}$ becomes a sum of Bragg peaks at positions belonging to the Fourier module spanned by the basis vectors $\mathbf{a}_i^*$ of the reciprocal lattice of the basic structure and the vectors $\mathbf{q}$ in the Fourier transform of the periodic modulation. Therefore,

$$\mathbf{H} = \sum_{i=1}^{3} h_i \mathbf{a}_i^* + \sum_{j=1}^{d} h_{3+j} \mathbf{q}_j. \quad ♭ \tag{2.8}$$

Because we want to deal also with one- or two-dimensional aperiodic crystals, we consider a more general case. We denote the dimension of the physical space by D. If the two sets of basis vectors (of reciprocal lattice and of the modulation) are independent for integer coefficients, the structure is quasiperiodic of rank $n = D + d$. The basis of the reciprocal lattice of the basic structure is already a basis of the physical space. Therefore, the modulation wave vectors may also be expressed with them, but the coordinates will generally be irrational numbers. For the basic satellites, one has

$$\mathbf{q}_i = \sum_{j=1}^{D} \sigma_{ij} \mathbf{a}_j^* \quad (i = 1, \ldots, d). \tag{2.9}$$

Compare this with the expression $\mathbf{q} = \alpha \mathbf{a}^* + \beta \mathbf{b}^* + \gamma \mathbf{c}^*$ for the case $D = 3$, $d = 1$.

♯ A small complication may occur. Suppose that $D = 3$, that the reciprocal basis vectors are $\mathbf{a}^*$, $\mathbf{b}^*$, and $\mathbf{c}^*$, and the basis vectors for the modulation are $\frac{1}{2}\mathbf{a}^*$ and $\gamma \mathbf{b}^*$ (with γ irrational). A new basis for the basic structure may be chosen with $\frac{1}{2}\mathbf{a}^*$ instead of $\mathbf{a}^*$. Then the basis vectors for the Fourier module are independent for integer coefficients, and the rank of the module is 4. Notice that the new basis structure has a unit cell which is the double of that of the old cell. This, however, is not a problem, because the basic structure is not unique.♯

As an example, consider K_2SeO_4 in the temperature range from 93 to 128 K. The basic structure is orthorhombic, with space group *Pnma*. There are four formula units per unit cell: four almost undeformed tetrahedra SeO_4, and eight K atoms. The modulation wave vector is $\mathbf{q} = \frac{1}{3}(1 - \delta)\mathbf{a}^*$, where δ is temperature dependent: it goes to zero at 93 K, where the incommensurate modulation becomes commensurate. Above 128 K the modulation disappears and the structure has space group symmetry *Pnma*. This is a typical situation. Above the *phase transition* temperature T_i the structure has lattice periodicity; below T_c the symmetry is lower than in the high-temperature phase (see Fig. 1.9). It is a superstructure or a structure with the same unit cell, but lower point group symmetry. In the temperature interval between T_c and T_i the structure is incommensurately modulated.

In later chapters we shall come back to the question of what the origin could be for the incommensurability.

The diffraction pattern of an incommensurately modulated phase has a Fourier module with points arbitrarily close to each other. The symmetry group of the structure is therefore not a three-dimensional space group. The point group symmetry of the diffraction pattern may be described on the basis of the Fourier module. For example, γ-Na_2CO_3 has a monoclinic basic structure with reciprocal lattice basis $\mathbf{a}^*$, $\mathbf{b}^*$, $\mathbf{c}^*$ and satellites with a basis vector $\mathbf{q} = \alpha \mathbf{a}^* + \gamma \mathbf{c}^*$. The intensities have point group symmetry $2/m$. The two generators, the two-fold rotation R and the mirror m, transform the basis vectors of the Fourier module as follows.

$$\begin{array}{ll} R\mathbf{a}^* = -\mathbf{a}^* & m\mathbf{a}^* = +\mathbf{a}^* \\ R\mathbf{b}^* = +\mathbf{b}^* & m\mathbf{b}^* = -\mathbf{b}^* \\ R\mathbf{c}^* = -\mathbf{c}^* & m\mathbf{c}^* = +\mathbf{c}^* \\ R\mathbf{q} = -\mathbf{q} & m\mathbf{q} = +\mathbf{q} \end{array}$$

Hence the two operations are described by the 4×4 matrices

$$D(R) = \begin{pmatrix} -1 & 0 & 0 & 0 \\ 0 & 1 & 0 & 0 \\ 0 & 0 & -1 & 0 \\ 0 & 0 & 0 & -1 \end{pmatrix}, \quad D(m) = \begin{pmatrix} 1 & 0 & 0 & 0 \\ 0 & -1 & 0 & 0 \\ 0 & 0 & 1 & 0 \\ 0 & 0 & 0 & 1 \end{pmatrix}.$$

The two matrices generate a group of four elements consisting of 4 × 4 matrices. This point group characterizes the point group of the diffraction pattern. In this case, the point group is a normal three-dimensional point group, but presented by four-dimensional matrices. This suggests considering a four-dimensional space, as we shall do further on.

The modulation considered above consists of a displacement wave. It is called a *displacive modulation*. Another type of modulation is the *density or occupation modulation*. Here the position $\mathbf{r}_j$ in the unit cell $\mathbf{n}$ is occupied by an atom of certain species with a certain probability, which is position dependent (Fig. 2.4). This could be an alternation between the atomic species A and B. Then the probabilities are $p_1(\mathbf{r})$ and $p_2(\mathbf{r}) = 1 - p_1(\mathbf{r})$. The periodic probability function can be sinusoidal, but also alternate between 0 and 1. For example, in CuAuII the roles of the Cu and Au positions are exchanged with a period of approximately 11 lattice constants in one direction. The term species may also refer to the orientation of a molecule, as in NaNO$_2$. Then the difference between displacive and occupational modulation is no longer clear-cut. The density function of the occupation modulated crystal is

$$\rho(\mathbf{r}) = \sum_{\mathbf{n}j} \left(A_{j1} p_{j1}(\mathbf{r}) + A_{j2} p_{j2}(\mathbf{r}) \right) \delta(\mathbf{r} - \mathbf{n} - \mathbf{r}_j). \quad (2.10)$$

When the functions $p(\mathbf{r})$ have Fourier components with wave vectors $\mathbf{q}$, the Fourier transform of $\rho(\mathbf{r})$ has a Fourier module with main reflections and satellites.

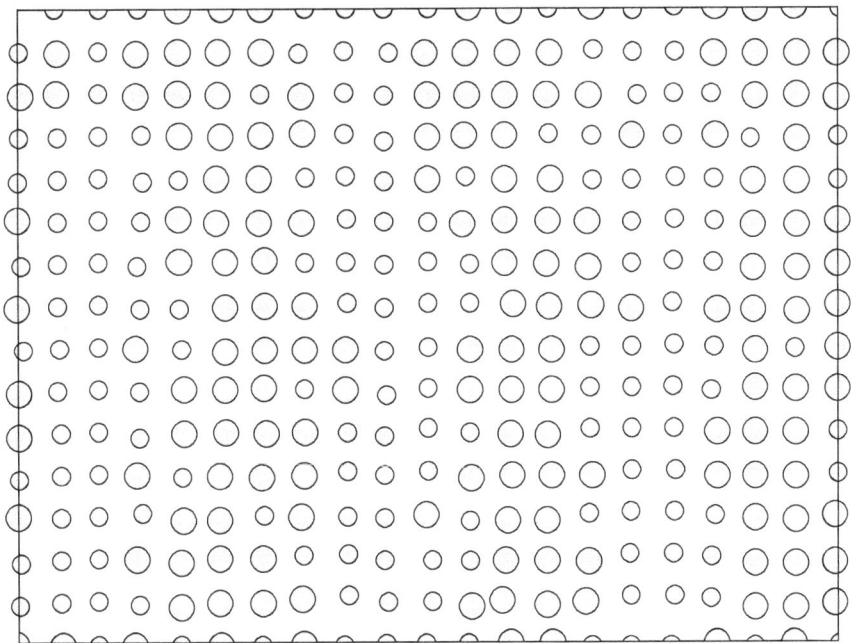

Figure 2.4 *A two-dimensional occupation modulated phase with two types of atoms.*

Because an aperiodic crystal does not have three-dimensional space group symmetry, the description of such a structure is not the conventional one. One might give the basic structure and specify the modulations. A more convenient way will be treated further on.

Displacements in a crystal may be described in terms of the normal modes of the crystal, characterized by the *irreducible representations of the space group*. This gives a way to describe the modulation functions. An irreducible representation of a space group is characterized by a star of wave vectors in the Brillouin zone (a complete set of wave vectors mutually transformable by the elements of the point group), and an irreducible representation of the so-called group of $\mathbf{k}$, which consists of all space group elements $\{R|\mathbf{t}\}$ such that $R\mathbf{k} = \mathbf{k}$ (up to a reciprocal lattice vector). A displacement field $\mathbf{u_n}$ can be written as

$$\mathbf{u}_{\mathbf{n}j} = \sum_{\mathbf{k}\in B.Z.} \sum_{\nu} Q_{\mathbf{k}\nu} \epsilon(\mathbf{k}\nu|j) \exp(2\pi i \mathbf{k}.\mathbf{n}) + c.\,c. \qquad (2.11)$$

Here *c. c.* means complex conjugated, and ν indicates an irreducible representation of the point group of the 'group of $\mathbf{k}$', j the various particles in the unit cell. For a modulation function with one wave vector, the wave vectors in the decomposition are the higher harmonics of the modulation wave vector $\mathbf{q}$. Hence only vectors $\mathbf{k}$ in the Brillouin zone occur which satisfy $\mathbf{k} = m\mathbf{q}$ (mod reciprocal lattice vectors). The modulation function, and in particular its symmetry, is characterized by the irreducible representation with a label ν. For example, the basic structure of γ-Na_2CO_3 has monoclinic symmetry with point group 2/m. The wave vector of the modulation is $\mathbf{q} = \alpha\mathbf{a}^* + \gamma\mathbf{c}^*$. The group of this $\mathbf{q}$ is the space group Pm, because the two-fold rotation inverts $\mathbf{q}$. The point group of this group Pm has two elements, and two irreducible representations: one for which the vector ϵ remains constant (sign +1), and one for which this vector gets a minus sign under the mirror m. Therefore, there are two different types of modulation possible: one with a vector ϵ (the polarization) in the mirror plane (sign is +1) and one with polarization perpendicular to the plane (sign is −1). In the early days of incommensurate phases the structure was often described by an irreducible representation of the space group of the basic structure. This was usually the representation associated with the soft mode leading to the incommensurate instability.

2.2.2 Composites

Composite structures consist of two or more structural *subsystems* which are modulated phases. If the basis structures for the subsystems are mutually incommensurate, one calls it an *incommensurate composite*. Usually the subsystems are not stable as such. They obtain their stability from the interaction with the other subsystems. The interaction between the subsystems causes the modulations.

We number the subsystems with the letter ν. Each subsystem has a basis structure with lattice Λ_ν and reciprocal lattice Λ_ν^*. For three dimensions, a subsystem has a basis $\mathbf{a}_\nu$, $\mathbf{b}_\nu$, and $\mathbf{c}_\nu$, and it has modulation functions $\mathbf{f}_{\nu j}(\mathbf{r})$. The modulation wave vectors belong to the module spanned by all other reciprocal lattices, if the modulation is due to the interaction between the subsystems. The Fourier module spanned by all reciprocal lattice vectors $\mathbf{a}^*_{\nu i}$ and possibly additional modulation wave vectors has a basis $\mathbf{a}^*_j$. Usually, the modulation wave vectors belong to the span of the basis vectors of the reciprocal lattice of the various subsystems. Then there are no 'additional' modulation vectors, but it is possible to have other mechanisms, e.g. interaction with the electron system, such that not all spots belong to the span of the reciprocal lattice vectors of the various subsystems. For simplicity, we shall disregard this possibility.

Then the reciprocal lattice of subsystem ν may be expressed in the basis vectors of the Fourier module, with integer coefficients.

$$\mathbf{a}^*_{\nu i} = \sum_{k=1}^{n} Z^\nu_{ik} \mathbf{a}^*_k, \qquad (2.12)$$

if $\mathbf{a}^*_k$ are the basis vectors of the total Fourier module. The integer matrices Z^ν are $D \times n$ matrices.

The satellites for the system ν can also be expressed in terms of the basis of the Fourier module:

$$\mathbf{q}_{\nu j} = \mathbf{a}^*_{\nu D+j} = \sum_{k=1}^{n} Z^\nu_{D+j\,k} \mathbf{a}^*_k. \qquad (2.13)$$

Together with the first D lines of Z^ν these form $n \times n$ matrices which have an inverse.

Again, if there are commensurate modulation vectors, a new basis structure has to be found for the subsystem. Finally, notice that, also in this case, the last d basis vectors of the Fourier module may be expressed in terms of the first D vectors:

$$\mathbf{a}^*_{D+j} = \sum_{k=1}^{D} \sigma_{jk} \mathbf{a}^*_k. \qquad (2.14)$$

That is useful if one has a clear distinction between subsystems, one as the *host*, and the other(s) as the *guest(s)*.

Each subsystem is a modulated crystal phase with positions

$$\mathbf{n}_\nu + \mathbf{r}_{\nu j} + \sum_{\mathbf{H} \in M^*} \mathbf{u}_{\nu j}(\mathbf{H}) \exp\left(2\pi i \mathbf{H}.(\mathbf{n}_\nu + \mathbf{r}_{\nu j})\right). \qquad (2.15)$$

The description of these positions is as for modulated phases. Again this could be based on the basic structures of the subsystems, supplemented by information about the modulation.

Consider as an example the compound $Hg_{3-\delta}AsF_6$ (fool's gold). The AsF_6 octahedra form a rather rigid host lattice with tetragonal symmetry. It contains channels in which Hg is present. The Hg atoms form the two guest systems. Above $T_c = 120$ K the diffuse streaks in the diffraction pattern show that the mercury is disordered. Below 120 K the mercury becomes ordered. Because the lattice constants of the Hg chains are incommensurate with respect to the host lattice, the crystal is an incommensurate composite. There are two orientations for the channels of the Hg chains. Therefore, one may distinguish three subsystems: the host structure consisting of the AsF$_6$ octahedra, the Hg chains in the **a**-direction, and the Hg chains in the **b**-direction. The host lattice is body-centred tetragonal. Its lattice is spanned by three vectors, which with respect to the primitive tetragonal basis (three mutually perpendicular vectors, two of which are of equal length) can be written as

$$\mathbf{a}_{11} = (-1/2, 1/2, 1/2), \quad \mathbf{a}_{12} = (1/2, -1/2, 1/2), \quad \mathbf{a}_{13} = (1/2, 1/2, -1/2),$$

with reciprocal basis

$$\mathbf{a}_{11}^* = (0, 1, 1), \quad \mathbf{a}_{12}^* = (1, 0, 1), \quad \mathbf{a}_{13}^* = (1, 1, 0).$$

The system of Hg chains in the **a**-direction has reciprocal basis

$$\mathbf{a}_{21}^* = (3 - \delta, -1 - \delta, 0), \quad \mathbf{a}_{22}^* = (0, 1, 1), \quad \mathbf{a}_{23}(0, 1, -1),$$

and the system parallel to the **b**-direction has reciprocal basis

$$\mathbf{a}_{31}^* = (-1 - \delta, 3 - \delta, 0), \quad (1, 0, 1), \quad (1, 0, -1).$$

Then lattice Λ_2 is A-centred monoclinic, and Λ_3 is B-centred monoclinic. All the basis vectors may be written as linear combinations of four reciprocal space vectors:

$$\mathbf{a}_1^* = (0, 1, 1), \quad \mathbf{a}_2^* = (1, 0, 1), \quad \mathbf{a}_3^* = (1, 1, 0), \quad \mathbf{a}_4^* = (-\delta, -\delta, 0).$$

Then the matrices Z^ν become

$$Z^1 = \begin{pmatrix} 1 & 0 & 0 & 0 \\ 0 & 1 & 0 & 0 \\ 0 & 0 & 1 & 0 \\ 0 & 0 & 0 & 1 \end{pmatrix}, \quad Z^2 = \begin{pmatrix} -2 & 2 & 1 & 1 \\ 1 & 0 & 0 & 0 \\ 0 & -1 & 1 & 0 \\ 0 & 1 & 0 & 0 \end{pmatrix},$$

$$Z^3 = \begin{pmatrix} 2 & -2 & 1 & 1 \\ 0 & 1 & 0 & 0 \\ -1 & 0 & 1 & 0 \\ 1 & 0 & 0 & 0 \end{pmatrix}.$$

The fourth row of each matrix gives the satellite, which in all cases is incommensurate with respect to the vectors in the first three rows. Using real coefficients, instead of integer ones, the fourth basis vector $\mathbf{a}_4^*$ of the Fourier module may be expressed in terms of the first three:

$$\mathbf{a}_4^* = -\delta \mathbf{a}_3^* \rightarrow \sigma_{1j} = (0,0,-\delta).$$

2.2.3 Quasicrystals

From the diffraction pattern of quasicrystals with icosahedral symmetry it is clear that such a compound cannot have lattice periodicity, because that is incompatible with the five-fold rotation axes. (This is the 'crystallographic condition') The description of such a structure can not use the same approach as that for modulated phases and composites, because, generally, there is no obvious basis structure. Several quasicrystalline phases, however, may be described using a three-dimensional tiling of the space with building blocks of two or more types. The atomic structure may then be given by describing the positions of atoms within these building blocks.

The most striking symmetry aspect of quasicrystals at their discovery was the *non-crystallographic rotation symmetry* (see Fig. 2.5). One may ask the question whether this is an essential condition for a quasicrystal. If one looks at the diffraction spots of an icosahedral quasicrystal, a basis for the Fourier module needs six basis vectors (see Section 1.2.6). In these vectors the inverse golden mean $\tau = (\sqrt{5}+1)/2$ appears in the coordinates. If one changes the value of these coordinates which have the value τ for the quasicrystal by a small amount, to another irrational value, the structure is still quasiperiodic, but the five-fold symmetry is lost. We shall see (Section 3.4.3) that it has tetrahedral symmetry, which is perfectly compatible with lattice periodicity, but in this case the structure is aperiodic. The structure does not have to change very much. So it is a matter of taste whether one should also call the deformed structure a quasicrystal or not. Usually, the term '*quasicrystal*' is reserved for quasiperiodic structures with rotation symmetry that is incompatible with lattice periodicity in three dimensions.

Another argument that supports the idea that non-crystallographic symmetry is not a necessary condition for quasicrystals is the fact that they can often be described as decorated tilings. In one dimension such a tiling could be the Fibonacci chain. In one dimension, however, there are no non-crystallographic rotations. Because many of the properties of quasiperiodic tilings may already be found in the Fibonacci chain, it is an argument to consider a structure based on this chain also as a quasicrystal. However, in the strict sense a quasicrystal is a quasiperiodic compound with non-crystallographic point group symmetry.

This deviates slightly from the definition given in the paper (Levine *et al.*, 1985) where the term 'quasicrystal' was introduced. There the definition is given as a structure that (a) is discrete, (b) has a self-similar lattice (that holds also for lattice periodic crystals), (c) has long-range bond orientational order, and (d) has quasiperiodic order with k linearly independent (incommensurate) lattice

44 *Description and symmetry of aperiodic crystals*

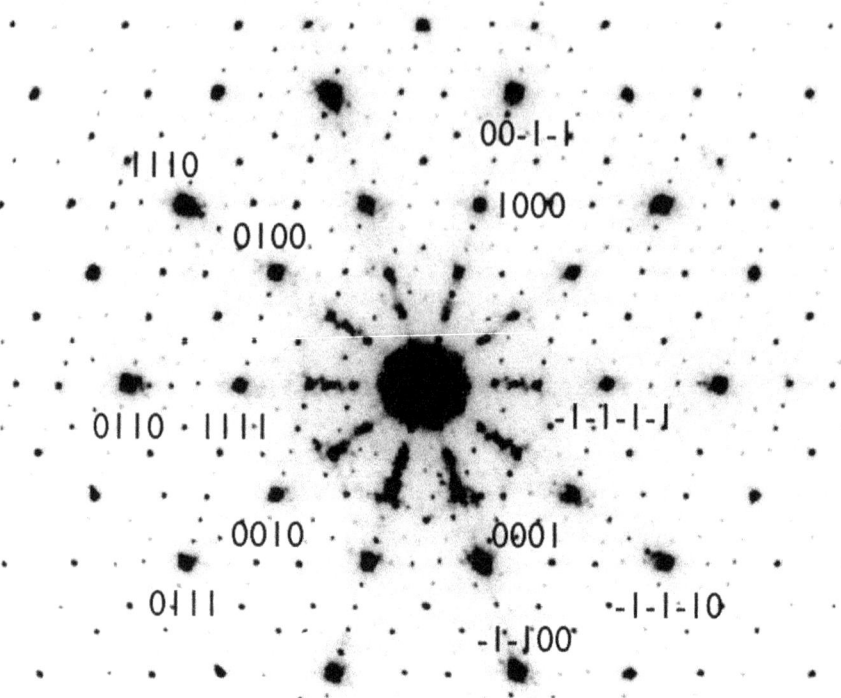

Figure 2.5 *The diffraction pattern of a decagonal phase, with ten-fold symmetry. Indexing this plane perpendicular to the ten-fold axis needs four integers.*

spacings. (It was remarked already there that lattice periodic systems are also quasiperiodic: $k=3$.)

The Fibonacci chain is a sequence of long and short intervals, denoted by L and S, respectively. Giving atomic positions in each of these two intervals leads to a quasicrystalline phase.

The Penrose tiling consists of tiles of either of two types: rhombi with angles of $36°$ and $72°$. The positions of these rhombi need an algorithm for their description. This will be discussed in Chapter 3. Inside the rhombi atomic positions may be given with respect to the tile edges.

A three-dimensional quasiperiodic space filling structure with icosahedral symmetry may be constructed in analogy to the Penrose tiling. It consists of copies of two types of rhombohedra: a prolate one and an oblate one with a volume ratio of τ. For such a structure, a decoration may be imagined, which could be considered as a model for a quasicrystal. There are, however, also quasicrystals which cannot be regarded as a decorated tiling. The question how to describe such a structure is not easy to answer. One answer will be given in the next section.

A quasicrystal may often be described in terms of clusters. When an intermetallic liquid cools down above the melting temperature such clusters are often present. In terms of close packing, an atom may be surrounded by at most 12 atoms

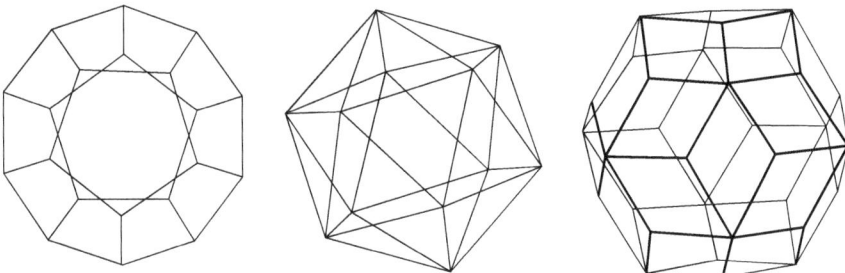

Figure 2.6 *Three regular polytopes playing a role in quasicrystals with icosahedral symmetry: a dodecahedron, an icosahedron, and a triacontahedron.*

of the same kind. This gives a dodecahedral surrounding. Such a structure may not be continued in a periodic way with the same symmetry. However, in quasicrystals the structure becomes quasiperiodic and in the structure one may distinguish clusters with (broken) dodecahedral or icosahedral symmetry. Generally, these clusters overlap. The building blocks, however, are the regular polytopes with a symmetry which forbids that the unit cell has this shape. In quasicrystals with five-fold symmetry the clusters are dodecahedra, icosahedra, and triacontahedra, or generalizations like Bergmann and Mackay clusters. In these clusters, the golden mean $\Phi = (\sqrt{5}-1)/2$ plays an important role. It is the same number as occurs in the Fibonacci chain. This gives a hint why in a quasicrystalline structure Fibonacci sequences occur frequently (see Fig. 2.6).

After the discovery of quasicrystals with icosahedral symmetry, other compounds were found with symmetry groups with ten-, eight-, and twelve-fold rotation symmetry. These compounds, called decagonal, octagonal, and dodecagonal quasicrystals, respectively, consist of quasiperiodic puckered layers periodically stacked along the axis perpendicular to the planes. However, also here there are clusters that connect the consecutive layers. These quasicrystals have rank 5: four basis vectors are needed for the diffraction spots in the plane, and a fifth one for the third direction. With respect to a properly chosen basis, the rotations are given by the matrices R_{10}, R_8, and R_{12}, for the decagonal, octagonal, and dodecagonal cases, respectively:

$$R_{10} = \begin{pmatrix} 0 & 0 & 0 & -1 & 0 \\ 1 & 1 & 1 & 1 & 0 \\ -1 & 0 & 0 & 0 & 0 \\ 0 & -1 & 0 & 0 & 0 \\ 0 & 0 & 0 & 0 & 1 \end{pmatrix}; \quad R_8 = \begin{pmatrix} 0 & 0 & 1 & 0 & 0 \\ 0 & 0 & 0 & 1 & 0 \\ 0 & 1 & 0 & 0 & 0 \\ -1 & 0 & 0 & 0 & 0 \\ 0 & 0 & 0 & 0 & 1 \end{pmatrix};$$

$$R_{12} = \begin{pmatrix} 0 & 0 & 1 & 0 & 0 \\ 0 & 0 & 0 & 1 & 0 \\ 0 & 1 & 0 & 0 & 0 \\ -1 & 1 & 0 & 0 & 0 \\ 0 & 0 & 0 & 0 & 1 \end{pmatrix}.$$

These quasicrystals are sometimes called two-dimensional quasicrystals, because the diffraction spots are arranged in two-dimensional layers: the aperiodicity occurs in planes. In this terminology, icosahedral quasicrystals are three-dimensional. When the spots are arranged along lines, as, for example, for a periodic arrangement of Fibonacci chains, the quasicrystal is called one-dimensional. However, in all three cases the reciprocal basis vectors span a three-dimensional space.

2.2.4 Natural aperiodic crystals

Aperiodic crystals are quite abundant in nature. Many minerals belong to the class of incommensurately modulated crystals. One has estimated that about 10 per cent of the earth crust consists of modulated structures. One of the earliest materials for which the law of rational indices did not seem to be satisfied was calaverite, which shows a morphology that puzzled crystallographers for a long time, till one discovered that this mineral is incommensurately modulated, and the facets can be indexed by four instead of three indices. Other *minerals* that show satellite peaks are feldspars (Kalning *et al.*, 1997), melilite (Bindi *et al.*, 2001), mullite (Angel *et al.*, 1991), pyrrhotite (Li and Franzen, 1996), quartz (Dolino *et al.*, 2005), sylvanite (Krutzen and Inglesfield, 1990), and trydimite (Pryde and Dove, 1998), to mention a few. Some of these phases have a very large range in temperature, while for others the range is small; e.g. the incommensurate phase of quartz exists in an interval of only approximately 1 K.

For many years the available quasicrystals were grown in the laboratory. But in 2009 it was found that also quasicrystals occur in nature. It was also in a mineralogical collection that a *natural quasicrystal* was found. Luca Bindi made this discovery in the collection of the Museo di Storio Naturale of the Università degli Studi di Firenze. It was a sample that came from the Koryak Mountains in Siberia (Bindi *et al.*, 2009) and its composition was one often produced in the laboratory, AlCuFe ($Al_{63}Cu_{24}Fe_{13}$). The sample has icosahedral symmetry and contains grains of micrometre size. Later, other samples were found in that region. One of them, found in the same meteorite, has decagonal symmetry and composition $Al_{71}Ni_{24}Fe_5$ (Bindi *et al.*, 2015). Another, again with icosahedral symmetry, has composition $Al_{62}Cu_{31}Fe_7$ (Bindi *et al.*, 2016).

Very interesting results gave the analysis of the isotope composition of the samples. This composition differs from what is normally found on earth. The origin of this sample must be sought in the interstellar space. Evidence for that statement is given in Bindi *et al.* (2012). The grains are embedded in stishovite, a mineral that is formed at high pressures. However, one expects that somewhere on earth the formation conditions for quasicrystals have been satisfied as well. But up to now, no natural quasicrystals with a terrestrial origin have been found. Of course, other minerals of extraterrestrial origin were already known from the moon expeditions Jagodzinski and Korekawa (1972). The authors of the latter article had already done pioneering work on the interpretation of satellites in minerals Jagodzinski and Korekawa (1965); Korekawa (1967).

2.2.5 ♯ Electromagnetic crystals in space-time

The way to describe the symmetry of aperiodic crystals of rank 4 has much in common with that for the symmetry groups of electromagnetic fields. These can be described in four-dimensional space-time. Space-time is a space with a metric tensor of Lorentz type, which we take in the form

$$g = \begin{pmatrix} -1 & 0 & 0 & 0 \\ 0 & 1 & 0 & 0 \\ 0 & 0 & 1 & 0 \\ 0 & 0 & 0 & 1 \end{pmatrix},$$

where $x_0 = ct$ (c is the speed of light), and $x_1 = x, x_2 = y, x_3 = z$. This means that the square of the norm of a vector (x, y, z, t) is given by $x^2 + y^2 + x^2 - c^2 t^2$.

An electromagnetic field is described by a four-dimensional potential A with components $A_\mu(x)$. The corresponding field tensor is $F_{\mu\nu} = \partial A_\mu / \partial x_\nu - \partial A_\nu / \partial x_\mu$:

$$F_{\mu\nu} = \begin{pmatrix} 0 & E_1 & E_2 & E_3 \\ -E_1 & 0 & H_3 & -H_2 \\ -E_2 & -H_3 & 0 & H_1 \\ -E_3 & H_2 & -H_1 & 0 \end{pmatrix}. \quad (2.16)$$

A Poincaré transformation (a pair of a Lorentz transformation and a translation in space-time) $g = (L|v)$ leaves the field invariant if

$$L.F.L^T(g^{-1}x) = F(x). \quad (2.17)$$

This does not imply that the potential A is left invariant, but there is always a gauge transformation with gauge function $\chi(x)$ such that

$$L.A(g^{-1}x) = A(x) + \partial \chi, \quad \text{or} \quad L_\nu^\mu A_\mu(g^{-1}x) = A_\nu(x) + \partial \chi / \partial x_\nu. \quad (2.18)$$

This gauge transformation is called a *compensating gauge transformation*. The compensating gauge transformation depends on the symmetry element g. If one considers a constant and homogeneous field, which is invariant under any space-time translation, the potential cannot be constant, because the field would be zero in that case. The potential transforms into a potential that is obtained from the original one by a gauge transformation.

The symmetry group of the potential consists of pairs (χ_g, g) of a Poincaré transformation g and a compensating gauge transformation χ_g. The gauge transformations form an Abelian subgroup of the symmetry group, and the structure of the latter is comparable to that of a crystallographic space group, which has as elements pairs of a orthogonal transformation and a translation (Janner and Janssen, 1971).

As an example, we consider a linearly polarized plane wave with wave vector q along the x_2 axis, electric field along the x_3 axis and magnetic field in the x_1 direction: $E_3 = H_1 = A\cos(2\pi q.x)$ with $q^\mu = (\omega, 0, \omega, 0)$. This field has as symmetry group a group of Poincaré transformations with translations a such that $q^\mu a_\mu = 0$ (mod 1), and a point group which consists of a Lie group with generators

$$L_1(t) = \begin{pmatrix} 1+\frac{1}{2}t^2 & \sigma & -\frac{1}{2}t^2 & 0 \\ t & 1 & -t & 0 \\ \frac{1}{2}t^2 & t & 1-\frac{1}{2}t^2 & 0 \\ 0 & 0 & 0 & 1 \end{pmatrix}; \quad L_2(t) = \begin{pmatrix} 1+\frac{1}{2}t^2 & 0 & -\frac{1}{2}t^2 & t \\ 0 & 1 & 0 & 0 \\ \frac{1}{2}t^2 & 0 & 1-\frac{1}{2}t^2 & \rho \\ t & 0 & -t & 1 \end{pmatrix}$$

and a discrete subgroup generated by the transformations

$$2_y = \begin{pmatrix} 1 & 0 & 0 & 0 \\ 0 & -1 & 0 & 0 \\ 0 & 0 & 1 & 0 \\ 0 & 0 & 0 & -1 \end{pmatrix}, \quad 2'_x = \begin{pmatrix} -1 & 0 & 0 & 0 \\ 0 & 1 & 0 & 0 \\ 0 & 0 & -1 & 0 \\ 0 & 0 & 0 & -1 \end{pmatrix},$$

$$m_x = \begin{pmatrix} 1 & 0 & 0 & 0 \\ 0 & -1 & 0 & 0 \\ 0 & 0 & 1 & 0 \\ 0 & 0 & 0 & 1 \end{pmatrix}.$$

Elements λ of the Lie algebra satisfy $\lambda g + g\lambda^T = 0$ and $\lambda F + F\lambda^T = 0$. Then the generators of the Lie group are found by exponentiation $L(t) = \exp(\lambda t)$. The transformation 2_y changes the sign of the field tensor, which has to be compensated for by a translation b with $q.b = 1/2$ (mod 1) (Janner and Ascher, 1970; Janner and Janssen, 1972). The discrete point group is the group mm'm. If there are four independent wave vectors, there is a lattice of symmetry translations in space-time. The point group depends on the polarizations of the plane waves. What is left is generally a point group in four dimensions consisting of three-dimensional transformations or three-dimensional transformations combined with time reversal. Such a point group is a subgroup of the product of the orthogonal group in physical space and the time-reversal group. When the translations in the symmetry group form a lattice the group is a four-dimensional space group with special property that it belongs to the intersection of the Poincaré and the inhomogeneous Galilei group.

In general, the restriction of a lattice periodic field A to a three-dimensional hyperplane, such as the space $t =$ constant, is not periodic, but quasiperiodic. The quasiperiodic three-dimensional field may be embedded in space-time such that it becomes periodic. The symmetry group of such a field has as elements space-time translations and Lorentz transformations. If the potential is quasiperiodic, the translations form a four-dimensional lattice, and the point group is a discrete subgroup of the Lorentz group. The group is called a *space-time group*.

When the point group is a subgroup of the direct product of the orthogonal group in three dimensions, and the group generated by the time-reversal operator, it is called a *generalized magnetic space-time group*. The list of such generalized magnetic space-time groups turned out to be identical to that of (3 + 1)-dimensional superspace groups used for incommensurate crystal phases.

Quasiperiodic crystals may conveniently be described by means of so-called superspace (or hyperspace) groups. Superspace groups are groups of transformations in an abstract space, of which the physical space is a subspace. In other words, the superspace is an extension of physical space with new degrees of freedom. In the case of space–time-dependent electromagnetic fields, this is an extension by the timeline. In other cases in the field of elementary particles, the physical space is extended by a space of internal degrees of freedom, such as the isospin. For quasiperiodic crystals these internal degrees of freedom have the character of a phase of the modulation function with respect to the lattice, or the relative position of subsystems in the case of composites. These crystals may conveniently be described by means of these superspace (or hyperspace) groups.

2.3 Description in superspace

2.3.1 Embedding

A very convenient way to present aperiodic crystals is suggested by the way one may obtain a quasiperiodic function of one variable from a multi-periodic function in more variables, as discussed in Section 2.1. Going in the other direction it is possible to embed a quasiperiodic function into a higher-dimensional space, that is, to describe the function as a periodic one in an abstract space with more dimensions, restricted to the physical space. The additional variables have, generally, a simple physical interpretation, such as the phase of a modulation function or the relative positions of subsystems in a composite.

This idea gets support from the following observation. Suppose that we have a quasiperiodic covering of the plane with eight-fold symmetry, called an octagonal tiling. If we put scatterers at the vertices one may calculate the diffraction pattern, which is given in Fig. 2.7. The spots display, of course, the eight-fold symmetry, but they do not belong to a two-dimensional reciprocal lattice. On the contrary, the spots are all over the place, although their intensities are in most cases very low. The position of every spot, however, may be reached by a linear combination of four basis vectors, indicated in the figure. Now consider the 16 points with coordinates 0 or 1. They form a regular pattern which is the projection of a four-dimensional cube on the two-dimensional plane. If one takes this seriously, each diffraction spot corresponds to a vertex of the four-dimensional reciprocal lattice for which the hypercube is the unit cell (Fig. 2.7). Going one step further, consider a periodic function in four dimensions for which the Fourier component at a certain vertex of the reciprocal lattice is exactly the Fourier component of the two-dimensional

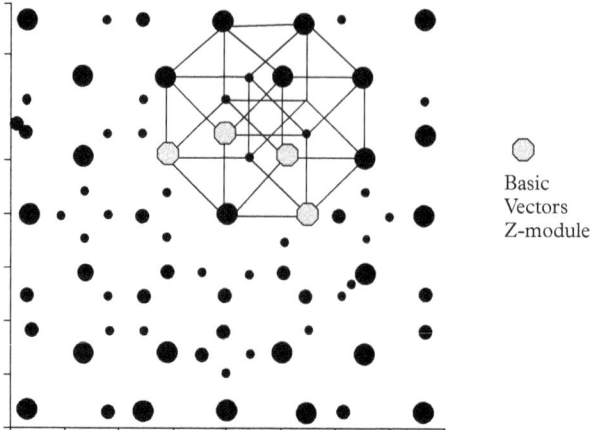

Figure 2.7 *Diffraction pattern of an octagonal tiling. Only spots with an intensity above a certain threshold are shown. The basis vectors of the Fourier module are projections of four-dimensional basis vectors of a reciprocal lattice; the projection of the unit cell is indicated.*

structure at the corresponding diffraction spot. Then there is a simple relation between the aperiodic structure in two dimensions and an abstract periodic function in four dimensions. We shall see that the correspondence between the two has a simple interpretation.

The idea of using a higher-dimensional space for simplifying the presentation is simple. Suppose we get the function of Fig. 2.1. How can we decide whether it is a random function, and, if this is not the case, how can we present it in a easy way? We could take its Fourier transform, which has just four components: for the vectors $q = \pm 1$ and $\pm \sqrt{2}$. For each $\hat{f}(q) = \frac{1}{2}$. The fact that the Fourier transform consists of delta peaks means that the system is perfectly ordered. Now define $y = x\sqrt{2}$ and consider the function in two variables $f(x,y) = \cos(2\pi x) + \cos(2\pi y)$. It is periodic in both x and y. In the unit cell the points on the original line now trace a path $(x, y = x\sqrt{2})$ (mod 1) (see Fig. 2.8). But if we consider the square as the unit cell of a square lattice, the path is just the straight line $y = x\sqrt{2}$. This line ('the physical space' V_E) crosses a periodic pattern and the value of the aperiodic function in x is just the value of the periodic function in the point $(x, x\sqrt{2})$. The lattice has a reciprocal lattice, and the projection of this reciprocal lattice on the crossing line is the Fourier module, spanned by 1 and $\sqrt{2}$.

For a crystal with density given by the quasiperiodic function $f(x)$, the Fourier transform consists of a series of delta peaks. Suppose that the Fourier transform is just of rank 2. Then the one-variable function with delta peaks gives points in a two-dimensional unit cell which give lines in the usual case. Consider a modulated chain with particles at positions $x_n = na + \Delta \cos(2\pi \alpha na)$. If α is irrational, this is

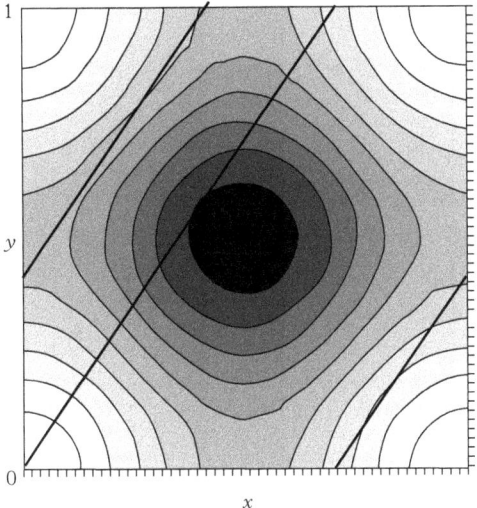

Figure 2.8 *Contour lines of the function* cos $(2\pi x) + \cos(2\pi y)$ *in a unit cell. The embedded quasiperiodic function is determined by the restriction of the periodic function to the line* $y = x\sqrt{2}$ *('the physical space'). At the end, the line* $y = x\sqrt{2}$ *(mod 1) fills the whole unit cell.*

aperiodic. The chain is aperiodic, but well ordered: the position of the n-th particle is precisely defined. This is in agreement with the fact that the Fourier transform consists of delta peaks at positions $(h_1 + \alpha h_2)/a$, where h_1 and h_2 are integers. These points are projections of two-dimensional vectors (q_1, q_2) with $q_1 = (h_1 + \alpha h_2)/a$ and $q_2 = h_2/a$. These form a two-dimensional reciprocal lattice with basis vectors $(1/a, 0)$ and $(\alpha/a, 1)$. The direct lattice is spanned by $(a, -\alpha)$ and $(0, 1)$. Each point of the one-dimensional chain corresponds with a point in the two-dimensional unit cell (see Fig. 2.9, where a number of points are given). When one takes the infinite number of points, these fill the line $x_1 = \Delta \cos(2\pi x_2)$ in the unit cell. This is the embedding in two-dimensional space. In general, the embedding is in n-dimensional space, further on denoted as nD space.

The construction of the lattice and its reciprocal lattice in the higher-dimensional space goes as follows. Suppose a quasiperiodic crystal has a structure factor consisting of delta peaks. Two examples are given in Fig. 2.10. The first is the spectrum of a sinusoidally modulated crystal, the second that of the Fibonacci chain. The first shows main reflections and satellites. Starting from the origin the intensity of the satellites increases, that of the main reflections decreases. For the Fibonacci chain it is more difficult to see main reflections, because, considered as modulated phase, it has a large amplitude. Both spectra are of rank 2, and the

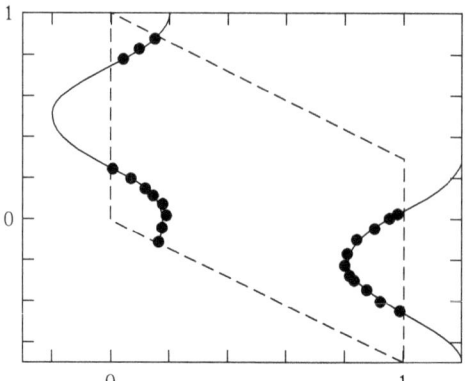

Figure 2.9 *The embedding of the chain* $na + \Delta \cos(2\pi na)$. *The n-th particle corresponds via the two-dimensional translations to a point in the unit cell in 2D (dotted line). All the points fall on a cosinusoidal line, with the atomic surface inside the unit cell. The graph shows 21 such points but in reality, the curves are densely occupied by such points.*

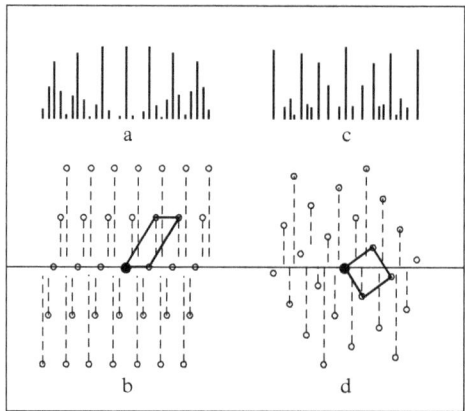

Figure 2.10 (*a*) *The spectrum of a sinusoidally modulated chain.* (*b*) *The points of the spectrum are the projection of 2D reciprocal lattice vectors.* (*c*) *The spectrum of the Fibonacci chain.* (*d*) *The reciprocal lattice vectors in 2D project on the points of the spectrum. The unit cells of the 2D reciprocal lattices are indicated.*

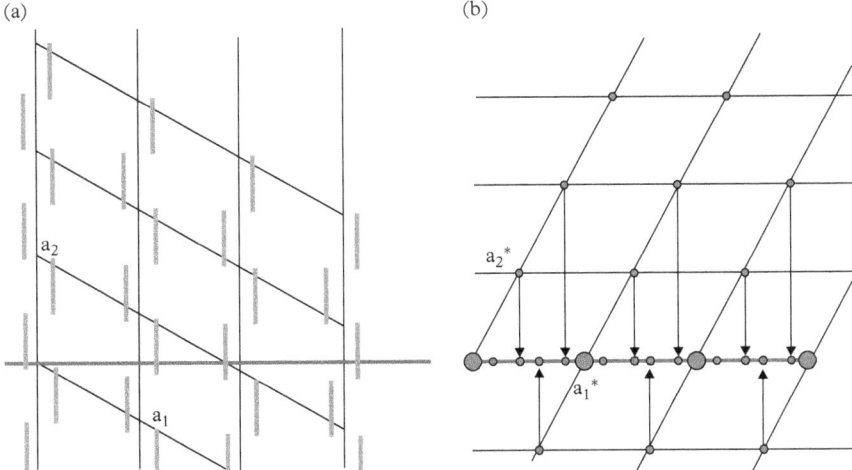

Figure 2.11 (a) *Embedding of a linear chain with discontinuous modulation function.* (b) *The reciprocal lattice in two-dimensional superspace. The projection of the reciprocal lattice vectors on physical space gives the positions of the Fourier module.*

basis vectors may be considered as projections of 2D (two-dimensional) reciprocal lattice vectors. Every spot in the diffraction is the projection of a reciprocal lattice vector (see Fig. 2.12). The intensity of the spots depends strongly on the distance from reciprocal lattice vector to the physical space (on which the reciprocal lattice is projected). Roughly speaking, the intensity drops with the distance. This is clearly visible for the Fibonacci chain. Its reciprocal lattice vectors have a high intensity if it comes close to the physical space. These vectors are special combinations of the basis vectors.

To illustrate the concepts we consider a modulated chain with a stepwise constant modulation function. The basis structure has the lattice constant a, and the displacement is b if na (mod b) is between 0 and 1/2, and $-b$ otherwise. The atomic surfaces run parallel to V_I and form a two-dimensional array, and the projections of the reciprocal lattice points are the positions of the Fourier module. See Fig. 2.11.

Let us formulate this situation in a more general way. Consider a density function, say of electrons in a solid, denoted by $\rho(\mathbf{r})$, which is quasiperiodic. Then the Fourier components belong to a Fourier module M^* with basis $\mathbf{a}_1^*, \mathbf{a}_2^*, \ldots$. Therefore,

$$\rho(\mathbf{r}) = \sum_{\mathbf{H}} \hat{\rho}(\mathbf{H}) \exp(2\pi i \mathbf{H}.\mathbf{r}), \quad \mathbf{H} = \sum_{i=1}^{n} h_i \mathbf{a}_i^*. \tag{2.19}$$

We denote the dimension of the space by D. Then the n D-dimensional basis vectors of the Fourier module are the projection of vectors of a reciprocal lattice in

54 Description and symmetry of aperiodic crystals

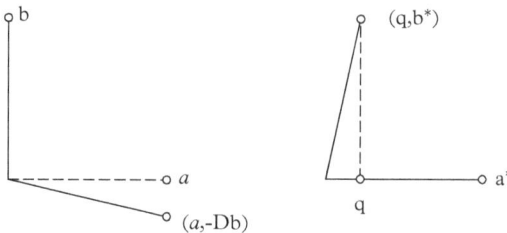

Figure 2.12 *The Fourier module basis as projection of a higher-dimensional reciprocal lattice basis and the basis for the corresponding direct lattice. Left: The basis of the basic structure* (**a**) *and of the two-dimensional lattice* Σ: (**a**, −D**b**) *and* (0, b), *with* Db = qa. *Right: Basis vectors* **a*** *and* **q** *of the Fourier module are projections of two-dimensional basis vectors of a reciprocal lattice.*

n dimensions. The vector **H** corresponds to an n-dimensional vector $\mathbf{H_s} = (\mathbf{H}, \mathbf{H}_I)$ (Fig. 2.12). The projection of the vector $\mathbf{H}_s$ is the vector $\mathbf{H} = \pi \mathbf{H}_s$. We shall come back to the question, how to determine the additional components which make up the vector $\mathbf{H}_I$, which is of dimension $d = n - D$. In particular, the basis vectors of the Fourier module correspond to the basis vectors of the reciprocal lattice, which we denote by Σ^*. The Fourier module M^* is the projection of the reciprocal lattice Σ^*: $M^* = \pi \Sigma^*$. The basis of Σ^* is given by

$$\mathbf{a}^*_{si} = (\mathbf{a}^*_i, \mathbf{b}^*_i) \quad (i = 1, \ldots, n). \tag{2.20}$$

Here $\mathbf{a}^*_i$ is a D-dimensional vector in physical space, and $\mathbf{b}^*_i$ a d-dimensional vector in internal space. If $\mathbf{e}_k$ ($k = 1, \ldots, D$) are orthogonal vectors in physical space, and $\mathbf{e}_k$ ($k = D+1, \ldots, n$) in internal space, then

$$\mathbf{a}^*_i = \sum_{k=1}^{D} \alpha_{ki} \mathbf{e}_k, \quad \mathbf{b}^*_i = \sum_{k=D+1}^{n} \beta_{ki} \mathbf{e}_k.$$

Next we take the Fourier decomposition of the density function $\rho(\mathbf{r})$ and construct from it a function in the n-dimensional space.

$$\rho_s(\mathbf{r}_s) = \rho_s(\mathbf{r}, \mathbf{r}_I) = \sum_{\mathbf{H}_s} \hat{\rho}(\pi \mathbf{H}_s) \exp(2\pi i(\mathbf{H}.\mathbf{r} + \mathbf{H}_I.\mathbf{r}_I)). \tag{2.21}$$

The sum runs over all vectors of the reciprocal lattice in n dimensions, and because every reciprocal lattice vector corresponds to a vector of the Fourier module, the Fourier components $\hat{\rho}(\pi \mathbf{H}_s) = \hat{\rho}(\mathbf{H})$ are known.

To the basis of the reciprocal lattice Σ^* corresponds the basis $\mathbf{a}_{si}$ determined by $\mathbf{a}_{si}.\mathbf{a}^*_{sj} = \delta_{ij}$, which spans the n-dimensional lattice Σ. For each vector $\mathbf{a}_s$ of this lattice, the function $\rho(\mathbf{r_s})$ is invariant. The vector has components **a** and $\mathbf{a_I}$, and

$$\mathbf{H}.\mathbf{a} + \mathbf{H_I}.\mathbf{a_I} = 0 \pmod 1.$$

So, finally we have a lattice periodic function in n dimensions for which the Fourier components correspond in a unique way to the Fourier components of the aperiodic function $\rho(\mathbf{r})$. Moreover, as is clear from Eq. 2.21, the restriction of the function ρ_s to the physical space ($\mathbf{r_I} = 0$) is just the density function $\rho(\mathbf{r})$:

$$\rho_s(\mathbf{r}, 0) = \rho(\mathbf{r}). \tag{2.22}$$

Therefore, we have embedded the aperiodic function into a higher-dimensional space where it corresponds to a periodic function. Because the Fourier components are identical, the information in both descriptions is the same. The advantage of the embedding is that now the crystallographic techniques can again be applied.

For the construction of a basis for the reciprocal lattice, we have used the equation $\mathbf{a}_{si}.\mathbf{a}_{sj}^* = \delta_{ij}$. This means that we assume that there is a metric tensor in the superspace. However, we know that the length scale in V_I is arbitrary, because only the projection on V_E is used. But we can introduce a distance in superspace by $|\mathbf{r}_s|^2 = \mathbf{r}^2 + \mathbf{r}_I^2$. Introducing a metric tensor in this way means that the direct superspace V_s and the *reciprocal superspace* V_s^* can be identified. Of course, the direct lattice Σ and the reciprocal lattice Σ^* are not the same, in general.

♯ The relation between intersection in direct space and projection in reciprocal space is quite general. Consider a simple example. Take a lattice periodic function in two variables $f(x, y)$. Its Fourier decomposition consists of delta peaks.

$$f(x, y) = \sum_{(k_1, k_2)} \hat{f}(k_1, k_2) \exp(-2\pi i(k_1 x + k_2 y)).$$

The restriction of this function to the line $y = 0$ gives $f(x, 0)$ for which the Fourier transform is equal to

$$\hat{f}(k) = \int dx f(x, 0) \exp(2\pi i k x) = \sum_{k_1, k_2} \hat{f}(k_1, k_2), \tag{2.23}$$

and because there is exactly one two-dimensional vector with projection k, the values are equal: $\hat{f}(k) = \hat{f}(k, k_2)$. One can formulate this as follows: the Fourier transform of the restriction $f(x, 0)$ to the line $y = 0$ is the projection of the Fourier transform of $f(x, y)$ on the line $y = 0$.

To every vector of the reciprocal lattice Σ^* corresponds a vector of the Fourier module, given by the projection π. Because of the incommensurability, two different vectors of Σ^* give different vectors of M^*. This means that there is an inverse map: to each vector of the module M^*, there is a unique vector of Σ^*, which in turn has a unique projection on the additional d-dimensional space:

$$\mathbf{a}_i^* \rightarrow \mathbf{a}_{si}^* = (\mathbf{a}_i^*, \mathbf{b}_i^*) \rightarrow \mathbf{b}_i^*. \tag{2.24}$$

This map is called the *star map*: $\mathbf{H}_I = {}^*\mathbf{H}$ for every $\mathbf{H}$ from the Fourier module.♮

The space in which the structure is located is the D-dimensional space V_E, and is called *physical space*, or *external space*. The additional d-dimensional space is denoted by V_I and is called *internal space*. The sum of the two spaces is the n-dimensional space V_s ($n = D + d$) and is called *superspace* (or sometimes hyperspace). For quasicrystals one uses mostly another terminology: *parallel space* E_{par} for the physical space, and *perpendicular space* E_{perp} for the internal space.

A point in superspace is the sum of a vector in V_E and one in V_I: $\mathbf{r} = (\mathbf{r}_E, \mathbf{r}_I)$ or $\mathbf{r} = (\mathbf{r}_E, 0) + (0, \mathbf{r}_I)$. It may be expressed in the n-dimensional basic lattice vectors: $\mathbf{r} = \sum_j x_j \mathbf{e}_j$. But it may also be expressed in a basis of space of which D vectors lie in V_E and $d = (n - D)$ in V_I. Such a basis is called a *split basis* (see Fig. 2.12). With respect to a split basis the coordinates of lattice vectors are, generally, not integers.

♯ The basic lattice vectors may be expressed in terms of a basis of which D vectors lie in V_E and d vectors in V_I. For example, one may choose an orthonormal basis in both subspaces. The coordinates with respect to this split basis are given by

$$\begin{pmatrix} y_1 \\ y_2 \\ \vdots \\ y_n \end{pmatrix} = \begin{pmatrix} e_{11} & e_{12} & \cdots & e_{1n} \\ e_{21} & e_{22} & \cdots & e_{2n} \\ \vdots & & \cdots & \vdots \\ e_{n1} & e_{n2} & \cdots & e_{nn} \end{pmatrix} \begin{pmatrix} x_1 \\ x_2 \\ \vdots \\ x_n \end{pmatrix} = M \begin{pmatrix} x_1 \\ x_2 \\ \vdots \\ x_n \end{pmatrix}.$$

Here e_{ij} is the j-th component of the i-th basis vector. The first D components are the components in V_E and the last $n - D$ components those in V_I. The coordinates in V_E and V_I are expressed with the n-dimensional matrices P_E and P_I:

$$P_E = \begin{pmatrix} 1_D & 0 \\ 0 & 0 \end{pmatrix}, \quad P_I = \begin{pmatrix} 0 & 0 \\ 0 & 1_d \end{pmatrix},$$

with 1_p the unit matrix in p dimensions. In matrix form:

$$\mathbf{r}_E = M^* P_E M \mathbf{r} = \Pi_E \mathbf{r}, \quad \mathbf{r}_I = M^* P_I M \mathbf{r} = \Pi_I \mathbf{r}. \tag{2.25}$$

The matrices M and M^* have the elements

$$M_{ij} = e_{ji}, \quad M_{ij}^* = e_{ij}^*.$$

If the basis in superspace is orthonormal, M is orthogonal and $M^* = M$ is the transpose of the inverse of M. The matrices Π_E and Π_I are *projectors*: they satisfy

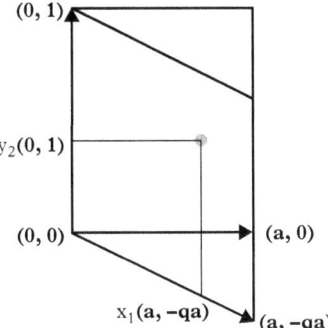

Figure 2.13 *A point in superspace has coordinates x_i with respect to a lattice basis, and coordinates y_j with respect to a split basis. The components $\mathbf{r}_E$ and $\mathbf{r}_I$ of $\mathbf{r}$ may be obtained by projection. In this case in a $(1+1)$-dimensional superspace, the lattice basis is $(a, -qa)$ and $(0, 1)$, the split basis is $(a, 0)$ and $(0, 1)$, and the relation between the coordinates is given by $y_1 = x_1, y_2 = -x_1 qa + x_2$.*

$$\Pi_E^2 = \Pi_E, \quad \Pi_I^2 = \Pi_I, \quad \Pi_E \Pi_I = \Pi_I \Pi_E = 0, \quad \Pi_E + \Pi_I = 1.$$

For example, consider the *Fibonacci chain*. It has rank 2 and the dimension of the superspace is also 2. The length scale in V_I is arbitrary, but it is custom to choose a scale such that the unit cell becomes a square. (Although that does not have any physical significance.) When the lengths of the intervals are a and $a\Phi = a/\tau$, the basis vectors are conveniently chosen as $a(1, -\Phi)$ and $a(\Phi, 1)$. Then the basis vectors of the reciprocal lattice are $(1, -\Phi)/(a(2-\Phi))$ and $(\Phi, 1)/(a(2-\Phi))$. For $a=1$, one then has

$$\Pi_E = \frac{1}{2-\Phi} \begin{pmatrix} 1 & -\Phi \\ \Phi & 1 \end{pmatrix} \begin{pmatrix} 1 & 0 \\ 0 & 0 \end{pmatrix} \begin{pmatrix} 1 & \Phi \\ -\Phi & 1 \end{pmatrix} = \frac{1}{2-\Phi} \begin{pmatrix} 1 & \Phi \\ \Phi & 1-\Phi \end{pmatrix}$$

and

$$\Pi_I = \frac{1}{2-\Phi} \begin{pmatrix} 1 & -\Phi \\ \Phi & 1 \end{pmatrix} \begin{pmatrix} 0 & 0 \\ 0 & 1 \end{pmatrix} \begin{pmatrix} 1 & \Phi \\ -\Phi & 1 \end{pmatrix} = \frac{1}{2-\Phi} \begin{pmatrix} 1-\Phi & -\Phi \\ -\Phi & 1 \end{pmatrix}$$

It is easy to verify that these matrices are projection operators. For example, the basis vector $\begin{pmatrix} 1 \\ -\Phi \end{pmatrix}$ has a projection to $\begin{pmatrix} 1 \\ 0 \end{pmatrix}$ under Π_E and to $\begin{pmatrix} 0 \\ -\Phi \end{pmatrix}$ under Π_I.

For the six-dimensional icosahedral case, the projectors are:

$$\Pi_E = \frac{1}{4\tau - 2}\begin{pmatrix} 2\tau-1 & 1 & 1 & 1 & 1 & 1 \\ 1 & 2\tau-1 & 1 & -1 & -1 & 1 \\ 1 & 1 & 2\tau-1 & 1 & -1 & -1 \\ 1 & -1 & 1 & 2\tau-1 & 1 & -1 \\ 1 & -1 & -1 & 1 & 2\tau-1 & 1 \\ 1 & 1 & -1 & -1 & 1 & 2\tau-1 \end{pmatrix},$$

$$\Pi_I = \frac{1}{4\tau - 2}\begin{pmatrix} 2\tau-1 & -1 & -1 & -1 & -1 & -1 \\ -1 & 2\tau-1 & -1 & 1 & 1 & -1 \\ -1 & -1 & 2\tau-1 & -1 & 1 & 1 \\ -1 & 1 & -1 & 2\tau-1 & -1 & 1 \\ -1 & 1 & 1 & -1 & 2\tau-1 & -1 \\ -1 & -1 & 1 & 1 & -1 & 2\tau-1 \end{pmatrix}.$$

The projectors for incommensurate modulated phases are similar. Consider a modulated crystal with a one-dimensional modulation with modulation vector q. The basis in superspace is given by $(\mathbf{a}_i, -\mathbf{q}.\mathbf{a}_i)$ and $(0, 1)$ (with $i = 1, 2, 3$). A point in superspace is given by

$$\left(\sum_{i=1}^{3} x_i \mathbf{a}_i, -\sum_{i=1}^{3} x_i \mathbf{q}.\mathbf{a}_i + x_4\right).$$

The projections on physical and internal space are $\left(\sum_{i=1}^{3} x_i \mathbf{a}_i, 0\right)$ and $\left(0, -\sum_{i=1}^{3} x_i \mathbf{q}.\mathbf{a}_i + x_4\right)$, respectively. The coordinates with respect to the basis of these two points are x_i ($i = 1, 2, 3$) and $\sum_i x_i \mathbf{q}.\mathbf{a}_i$ for the first, and 0 ($i = 1, 2, 3$) and $-\sum_i x_i \mathbf{q}.\mathbf{a}_i + x_4$ for the second. Then $\mathbf{r}_E = \Pi_E \mathbf{r}$ and $\mathbf{r}_I = \Pi_I \mathbf{r}$ with

$$\Pi_E = \begin{pmatrix} 1 & 0 \\ \mathbf{q}.\mathbf{a}_i & 0 \end{pmatrix}, \quad \Pi_I = \begin{pmatrix} 0 & 0 \\ -\mathbf{q}.\mathbf{a}_i & 1 \end{pmatrix}.$$

It is easily seen that these are two projectors adding up to the identity. ▫

If the aperiodic structure consists of point atoms, the density function $\rho(\mathbf{r})$ consists of delta peaks: $\rho(\mathbf{r}) = \sum_i \delta(\mathbf{r} - \mathbf{r}_i)$. To each atom in physical space correspond an infinite number of translationally equivalent points in superspace, and one of them belongs to the n-dimensional unit cell. The relation between the positions in physical space and those in the n-dimensional unit cell is as follows. The atoms located in V_E have n-dimensional positions $(\mathbf{r}_i, 0)$. The coordinates of this point with respect to the n-dimensional basis are x_{ik} ($k = 1, \ldots, n$) given by the scalar product of the n-dimensional position vector and the k-th basis vector of the reciprocal lattice:

$$x_{ik} = \mathbf{a}_{sk}^*.(\mathbf{r}_i, 0) = \mathbf{a}_k^*.\mathbf{r}_i.$$

The point in the unit cell to which the point $(\mathbf{r}_i, 0)$ corresponds has coordinates $\mathbf{a}_k^* \cdot \mathbf{r}_i$ (mod 1). All the points together form a countable set: the number of points is the number of particles. For quasiperiodic structures taking all limit points gives the so-called *atomic surfaces*. These are not necessarily continuous.

♯ The choice of the internal components $\mathbf{b}_j$ so far is still free. There are, however, restrictions, if there exist symmetry relations between the vectors of the Fourier module. Suppose that the module is left invariant under a rotation R in the physical space, and that the rotation transforms diffraction spots to spots of the same intensity. Intensities generally fall off if the length of the internal component corresponding to a vector of the Fourier module increases. That means that the diffraction spots above a certain threshold are discrete. That implies that the symmetry rotation R is of finite order. Applying the rotation to a basis vector of the Fourier module transforms it to a vector of the module, which is a integer combination of the basis vectors. This means that

$$R\mathbf{a}_i^* = \sum_{j=1}^{n} \Gamma^*(R)_{ij} \mathbf{a}_j^*, \tag{2.26}$$

for integer coefficients $\Gamma^*(R)_{ij}$. These form an n-dimensional matrix. For every rotation, proper or improper, which is a symmetry operation, there is such a matrix and these form a group. It is a finite group because of discreteness. A well-known theorem of group theory states that such a group of matrices can be brought into the form of a sum of orthogonal matrices:

$$\Gamma^*(R) \sim \begin{pmatrix} R_E & 0 \\ 0 & R_I \end{pmatrix},$$

If the dimension of the physical space is D and of the internal space $d = n-D$, then R_E^* is D-dimensional and R_I^* d-dimensional. Therefore, it is a pair of orthogonal transformations: one in physical space and one in internal space. It acts on the reciprocal lattice Σ^* with basis vectors $\mathbf{a}_{si}^*$ having components $\mathbf{a}_i^*$ in V_E and $\mathbf{b}_i^*$ in V_I.

$$R\mathbf{a}_{si}^* = \left(R_E \mathbf{a}_i^*, R_I \mathbf{b}_i^*\right) = \left(\sum_{j=1}^{D} D_E^*(R)_{ij} \mathbf{a}_j^*, \sum_{k=1}^{d} D_I^*(R)_{jk} \mathbf{b}_k^* \right). \tag{2.27}$$

Take as an example a five-fold rotation on a Fourier module of rank 4. The integer matrix on a lattice basis is equivalent (via a basis transformation) with an orthogonal matrix on a split orthonormal basis:

$$\begin{pmatrix} 0 & 1 & 0 & 0 \\ 0 & 0 & 1 & 0 \\ 0 & 0 & 0 & 1 \\ -1 & -1 & -1 & -1 \end{pmatrix} \sim \begin{pmatrix} \cos(2\pi/5) & -\sin(2\pi/5) & 0 & 0 \\ \sin(2\pi/5) & \cos(2\pi/5) & 0 & 0 \\ 0 & 0 & \cos(6\pi/5) & -\sin(6\pi/5) \\ 0 & 0 & \sin(6\pi/5) & \cos(6\pi/5) \end{pmatrix}$$

It is a rotation of $2\pi/5$ in a two-dimensional physical space, combined with a rotation of $6\pi/5$ in internal space. Starting from one basis vector one may obtain a basis by applying the four-dimensional rotation to get the other three. In physical space the angle between two consecutive basis vectors is $2\pi/5$, but in internal space it is $6\pi/5$. Therefore, there are strong restrictions on the internal components $\mathbf{b}_j^*$. ♮

2.3.2 Modulated phases

As a first example of the embedding, we consider a three-dimensional incommensurate modulated phase with a single modulation vector, and we consider a displacive modulation. This was the first type of quasiperiodic crystals for which the embedding in a higher-dimensional space was developed (de Wolff, 1974; de Wolff et al., 1981; Janner and Janssen, 1977, 1980a). The diffraction spots are at the positions

$$\mathbf{H} = h\mathbf{a}^* + k\mathbf{b}^* + \ell\mathbf{c}^* + m\mathbf{q}, \qquad (2.28)$$

where the first three vectors are the three basis vectors of the reciprocal lattice of the basic structure, and $\mathbf{q}$ is the modulation vector. The latter can be written with respect to the basis vectors of Λ^*:

$$\mathbf{q} = \alpha\mathbf{a}^* + \beta\mathbf{b}^* + \gamma\mathbf{c}^*. \qquad (2.29)$$

To have an incommensurate structure, at least one of the coordinates α, β, γ has to be irrational. The three coordinates are usually chosen between 0 and 1.

The basic structure positions of the atoms are given by a lattice vector $\mathbf{n}$ and a position $\mathbf{r}_j$ in the unit cell. There are s atoms in the unit cell. The atoms are displaced by the modulation, which is periodic with wave vectors which are multiples of the modulation wave vector $\mathbf{q}$. Then the positions are given by

$$\mathbf{r}_{\mathbf{n}j} = \mathbf{n} + \mathbf{r}_j + \sum_m \mathbf{u}_j(m\mathbf{q}) \exp(2\pi i m\, \mathbf{q}.(\mathbf{n}+\mathbf{r}_j)). \qquad (2.30)$$

Following the general prescription, the basis of the Fourier module is the projection of a four-dimensional basis. Without loss of generality we take three of these basis vectors in the physical space, and the fourth has components $\mathbf{q}$ and q_I. The point atoms correspond each to an infinite number of points in superspace falling on lines

$$(\mathbf{r}_{\mathbf{n}j}(t), t) = \left(\mathbf{n} + \mathbf{r}_j + \sum_m \mathbf{u}_j(m\mathbf{q}) \exp(2\pi i(m\mathbf{q}(\mathbf{n}+\mathbf{r}_j) + mq_I t), t\right). \qquad (2.31)$$

Here t runs from $-\infty$ to $+\infty$. The points form a countable, dense set, but in this case we get continuous lines if all limit points are added. The embedded structure

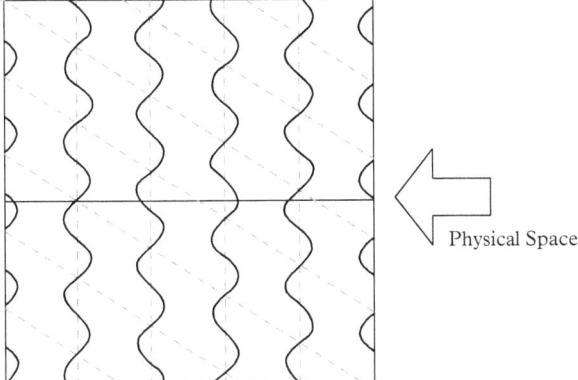

Modulated phase as intersection in 4D space

Figure 2.14 *Embedding of a displacively modulated phase. Solid lines: atomic surfaces. Dashed lines: lattice of translations. Arrow: physical space. The 3D modulated crystal phase is the intersection of the physical space with a 4D pattern of lines with 4D lattice periodicity.*

has lattice periodicity (Fig. 2.14). The additional variable t has the meaning of the phase of the modulation function in the origin of the lattice.

♯ This can be seen as follows. Take a lattice vector (**a**, b) from the four-dimensional lattice Σ. Then the lines are displaced to

$$\mathbf{n} + \mathbf{a} + \mathbf{r_j} + \sum_m \mathbf{u}_j(m\mathbf{q}) \exp(2\pi i \left(m\mathbf{q}(\mathbf{n} + \mathbf{r_j}) + mq_I(t - b) \right))$$

$$= \mathbf{n'} + \mathbf{r_j} + \sum_m \mathbf{u}_j(m\mathbf{q}) \exp(2\pi i (m\mathbf{q}(\mathbf{n'} + \mathbf{r_j}) + mq_I t - \mathbf{q}.\mathbf{a} - q_I.b)),$$

which is the same set of lines because $\mathbf{q}.\mathbf{a} + q_I.b = 0$ (mod 1), since $(\mathbf{q}, q_I)$ belongs to the lattice Σ^* and (**a**, b) to the lattice Σ. This shows that the array of lines in superspace is periodic with lattice Σ. ♮

More generally, the Fourier module of a quasiperiodic structure is spanned by the vectors $\mathbf{a}_i^*$ ($i = 1, \ldots, n$). For convenience, we choose the basis vectors of the reciprocal lattice for the (periodic) basic structure as the first D vectors. The d vectors that complete the basis for the Fourier module after the choice of the basis for the reciprocal lattice of the basic structure may be expressed in terms of the latter:

$$\mathbf{a}_{D+j}^* = \sum_{i=1}^{D} \sigma_{ji} \mathbf{a}_i^*. \quad (2.32)$$

A convenient embedding is determined by the embedding of the basis of the Fourier module as follows:

$$\mathbf{a}_{si}^* = (\mathbf{a}_i^*, 0) \quad (i = 1, \ldots, D), \qquad (2.33)$$
$$\mathbf{a}_{sD+j}^* = (\mathbf{a}_{D+j}^*, \mathbf{b}_j^*) \quad (j = 1, \ldots, d).$$

As a consequence, the basis for the lattice Σ then is given by

$$\mathbf{a}_{si} = \left(\mathbf{a}_i, -\sum_{j=1}^{d} \sigma_{ji} \mathbf{b}_j\right) \quad (i = 1, \ldots, D), \qquad (2.34)$$
$$\mathbf{a}_{sD+j} = (0, \mathbf{b}_j) \quad (j = 1, \ldots, d),$$

where the d vectors $\mathbf{b}_j$, which span the internal space, are related to the vectors $\mathbf{b}_j^*$ by $\mathbf{b}_i . \mathbf{b}_j^* = \delta_{ij}$.

The embedding of the incommensurate phase into the superspace now gives the arrays

$$(\mathbf{r}_{\mathbf{n}j}(t), \mathbf{t}) = \left(\mathbf{n} + \mathbf{r}_j + \sum_{\mathbf{q}} \mathbf{u}_j(\mathbf{q}) \exp\left(2\pi i(\mathbf{q}.\mathbf{n} + \mathbf{q}.\mathbf{r}_j) + \mathbf{q}_I.\mathbf{t}\right), \mathbf{t}\right). \qquad (2.35)$$

The vector components $\mathbf{t}$ in internal space are now d-dimensional. Therefore, the array consists of d-dimensional surfaces in superspace. The surfaces intersect the physical space in points, the positions of point atoms. Therefore, the d-dimensional surfaces are called 'atomic surfaces'. When the dimension of the internal space is 1, the surfaces become lines. In the same way as above, it is shown that the array of parallel atomic surfaces is lattice periodic with lattice Σ. The corresponding vectors of the reciprocal lattice Σ^* are $(\mathbf{k}, \mathbf{k}_I)$. The vectors $\mathbf{k}$ are the positions of the Fourier module M^*.

The embedding of Eq. 2.35 gives a mapping from V_E to V_I in the following way. Any translation $\mathbf{t}$ in physical space may be expressed in the basis vectors $\mathbf{a}_i$ of the basic structure. These are components of basis vectors $\mathbf{a}_{si}$ in superspace, with well-defined internal components. So,

$$\mathbf{t} = \sum_{i=1}^{D} x_i \mathbf{a}_i \quad \rightarrow \quad \sum_{i=1}^{D} x_i \mathbf{a}_{si} \quad \rightarrow \quad -\sum_{i=1}^{D} x_i \sum_{j=1}^{d} \sigma_{ji} \mathbf{b}_j = -\Delta \mathbf{t},$$

which is an element of V_I. Therefore, there is a map Δ which maps a vector $\mathbf{t}$ in physical space to a vector $-\Delta \mathbf{t}$ in internal space. Compare this to the star map introduced earlier.

The embedding of an *occupation* or *composition modulated crystal* follows the same lines. The atomic surfaces are now flat and parallel to internal space. On the atomic surfaces probability functions are defined, one for each atomic species. If an atomic surface intersects the physical space, the intersection point gives the atomic position, and the value of the probability function at the intersection point gives the probability of finding a specific species at that position. The embedding of displacive and occupation modulated crystals is illustrated in Fig. 2.15. In general,

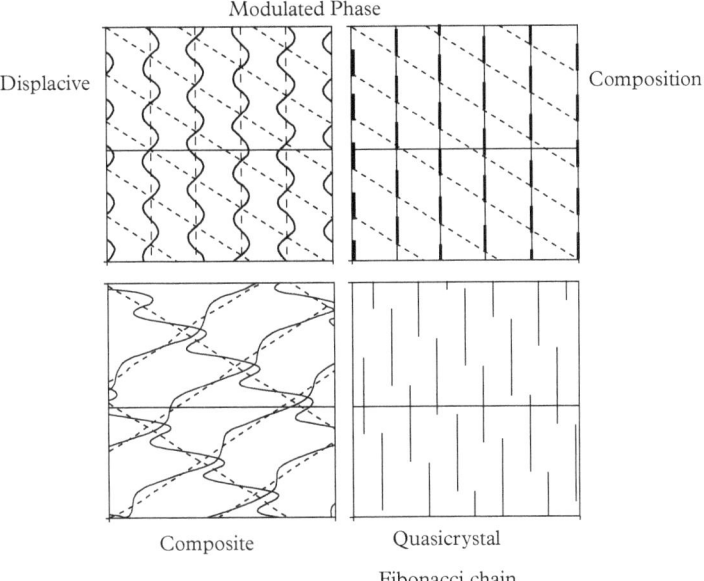

Figure 2.15 *Embeddings of four types of quasiperiodic structures: displacive and composition modulated (thick lines for atoms A with probability p_A and for atoms B with probability $1 - p_A$, and thin lines for atoms A with probability $1 - p_B$, and for atoms B with probability p_B), composite, and quasicrystalline.*

an incommensurate composition modulated crystal phase is also displacively modulated. In that case, the atomic surfaces on which the probability functions are defined are no longer flat, but curved as in the displacively modulated case.

A special case of a composition modulated phase is the following. For this, the probability of finding a particle in a certain position is given by 1 or 0. It is a way to characterize aperiodic ordering. Atoms of types A and B, where B may also represent a vacancy, ordered in a quasiperiodic way, may be embedded in superspace, giving rise to atomic surfaces which are flat and parallel to the internal space. On this surface, the probability of finding an atom A is given by a periodic step function (sometimes called a '*crenel function*') with values 0 and 1. If the value is 1, then the position is taken by an atom A, otherwise by an atom B (see Fig. 2.16) or by a vacancy. We shall come back to this point in Chapter 4.

There is a constraint on the regions representing A's. Because the system is quasiperiodic, the projection of the lattice for the embedding is dense in internal space. Each realization of the crystal for a value of the internal coordinate r_I is identical, up to a translation, to one for a value $r_I + a_I$, if **a** is a lattice vector. This means that an arbitrarily chosen shift of the internal coordinate must give a crystal with the same density. If one requires that this holds locally, then an arbitrarily small shift which induces a jump of an atom from one position to another must keep the local density the same. If the intersection of the physical space leaves a

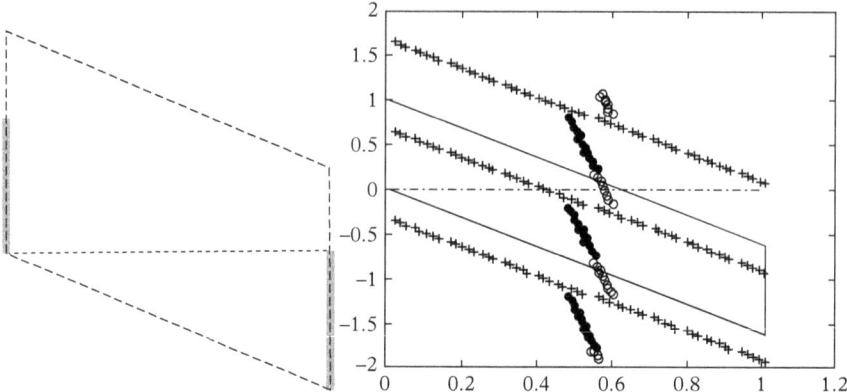

Figure 2.16 *Left: The closeness condition. The atomic surface to the right ends where, in projection, the left atomic surface starts. Right: Embedding of an incommensurate composite with two types of atoms in the first subsystem (steep slope, light and dark grey points), and one type in the second subsystem (with gentle slope). The solid line outlines the unit cell.*

region of the atomic surface representing particles A, it should enter somewhere else in the neighbourhood a region of another atomic surface also representing particles A. This means that (if the internal space is one-dimensional) the end of the projection of a region with probability 1 for one atomic surface should coincide with the beginning of a similar region on a neighbouring atomic surface. This is called the *closeness condition*. If the dimension of internal space is higher, the projections of the regions should touch (along a line if the dimension is two). Consider an example in two-dimensional superspace. The lattice has basis vectors $(a, -qa)$ and $(0, 1)$. Suppose the atomic surface in the origin is the interval of length 1, with atoms of type A on the interval $[0, x]$ and of type B on $[x, 1]$. Then the atomic surface at a is the interval $[-qa, 1 - qa]$. The closeness condition then is $-qa + x = 0$, which gives a relation between the composition x and the modulation wave vector q.

The fact that an incommensurate modulated phase has a basic structure means that the main reflections have to be mapped on a main reflection under a symmetry operation. This means that the point group of the basic structure has to obey the *crystallographic condition*: only rotations of order 1, 2, 3, 4, and 6 are allowed.

2.3.3 Incommensurate composites

Incommensurate composites are also quasiperiodic and aperiodic. According to the general scheme, they can also be embedded into a higher-dimensional superspace. The details differ slightly from those for modulated phases. The embedding procedure was developed in Janner and Janssen (1980*b*), and later extended in van Smaalen (1991).

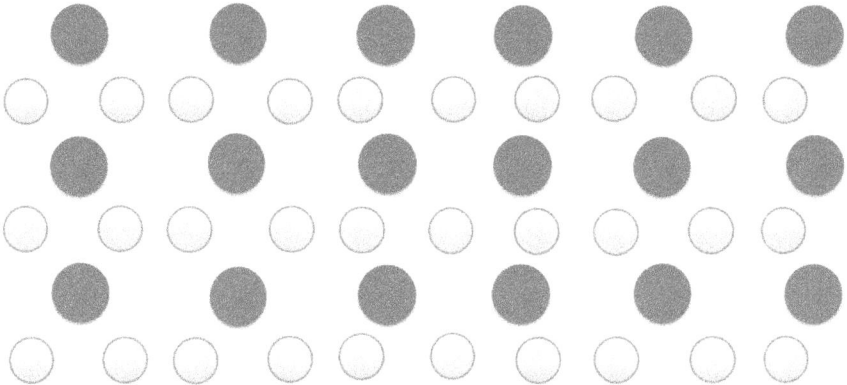

Figure 2.17 *Two-dimensional incommensurate composite. The positions of the first subsystem are modulated by the other, and vice versa.*

We consider first a simple two-dimensional structure with two subsystems (Fig. 2.17). The basic structure of the first system is rectangular with a lattice with basis vectors $(a, 0)$ and $(0, c)$, the second is also rectangular with basis $(b, 0)$ and $(0, c)$. We suppose that a/b is irrational.

The first subsystem is modulated by the interaction with the second subsystem, and the modulation function is periodic in the x-direction with periodicity b, whereas the modulation of the second has periodicity a. The two-dimensional positions of the atoms are

$$(na + f(na), \ell c), \quad \left(mb + g(mb), \left(k + \frac{1}{2}\right)c\right)$$

with

$$n, m, k, \ell \text{ integers}, f(x + b) = f(x), g(y + a) = g(y).$$

The Fourier module of the structure has basis

$$(1/a, 0), \ (1/b, 0), \ (0, 1/c).$$

A convenient choice for the embedding into the three-dimensional superspace would be

$$\mathbf{a}^*_{s1} = (1/a, 0, 0), \ \mathbf{a}^*_{s2} = (1/b, 0, 1), \ \mathbf{a}^*_{s3} = (0, 1/c, 0).$$

The lattice Σ then has the basis

$$\mathbf{a}_{s1} = (a, 0, -a/b), \ \mathbf{a}_{s2} = (0, 0, 1), \ \mathbf{a}_{s3} = (0, c, 0).$$

The basic structure of subsystem 1 has positions $(na, 0)$, which correspond to points in the unit cell of Σ with coordinates $x_1 = n \equiv 0 \pmod{1}$, $x_2 \equiv na/b \pmod{1}$, and $x_3 \equiv 0$. The values of x_2 cover the whole interval from 0 to 1. The points $(mb, 0)$ of the second subsystem correspond to points with coordinates $x_1 = mb/a \pmod{1}$, $x_2 = 0$, and $x_3 = \frac{1}{2}$. Therefore, the embedding of the unmodulated structure is given by $(na, \ell c, t)$ and $(mb - bt, (\ell + \frac{1}{2})c, t)$. Here t is the continuous variable, corresponding to the phase of the modulation in the preceding section. The full (modulated) composite then has an embedding

$$(na + f(na + bt), \ell c, t), \quad \left(mb - bt + g(mb - bt), \left(\ell + \frac{1}{2}\right)c, t\right)$$

It is easily seen that this pattern in three dimensions is left invariant by the three-dimensional lattice spanned by

$$\mathbf{a}_{s1} = (a, 0, -a/b), \quad \mathbf{a}_{s2} = (0, 0, 1), \quad \mathbf{a}_{s3} = (0, c, 0)$$

(cf. Fig. 2.18). In the case of a sinusoidal modulation function, the atomic surfaces extend to infinity. For more complicated cases, the atomic surfaces may become discontinuous (see Fig. 2.19).

♯ Clearly this embedding is not symmetric in the two subsystems. One could also choose the basis vectors of the reciprocal lattice of the second system to embed them in physical space. And in the most general way the embedding is given by

$$(na - z_1 t + f(na + \Delta t), \ell c, t), \quad (mb - z_2 t + g(mb + \Delta t), \ell' c, t),$$

where $\Delta = z_2 - z_1$. The lattice Σ in superspace is generated by

$$\mathbf{a}_1 = (z_2 a/\Delta, 0, -a/\Delta), \quad \mathbf{a}_2 = (-z_1 b/\Delta, 0, b/\Delta), \quad \mathbf{a}_3 = (0, c, 0). \tag{2.36}$$

The reciprocal basis is

$$\mathbf{a}_1^* = (1/a, 0, z_1/a), \quad \mathbf{a}_2^* = (1/b, 0, z_2/b), \quad \mathbf{a}_3^*(0, 1/c, 0). \tag{2.37}$$

Invariance under $\mathbf{a}_1$ follows from

$$(na - z_1 t + z_2 a/\Delta + f(na + \Delta t), \ell c, t - a/\Delta) \equiv (n'a - z_1 t + f(n'a + \Delta t), \ell c, t),$$
$$(mb - z_1 t + z_2 a/\Delta + g(mb - \Delta t), \ell c, t - a/\Delta) \equiv (mb - z_2 t + g(mb - \Delta t)),$$

and under $\mathbf{a}_2$ from

$$(na - z_1 a - z_1 b/\Delta + f(na + \Delta t, \ell c, t + b/\Delta) \equiv (na - z_1 t + f(na + \Delta t), \ell c, t),$$
$$(mb - z_2 t - z_1 b/\Delta + g(mb - \Delta t), \ell c, t + b/\Delta) \equiv (m'b - z_2 t + g(m' - \Delta t), \ell c, t).$$

Description in superspace 67

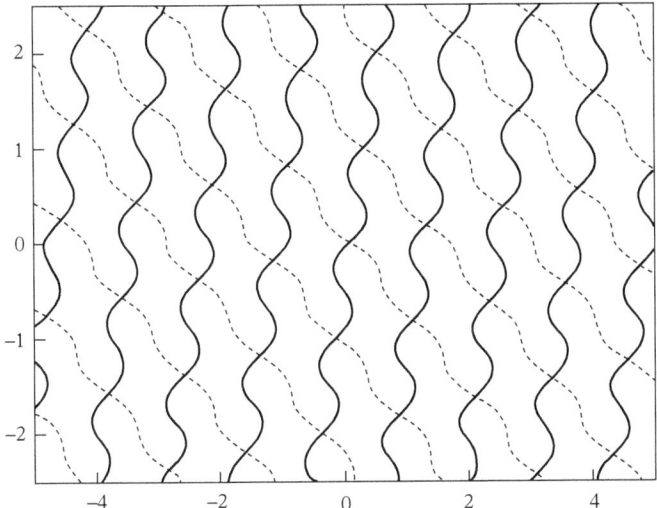

Figure 2.18 *The embedding of an incommensurate composite. The modulation functions are smooth, and the atomic surfaces continue to infinity. Notice that the solid and dashed lines are not in the same plane, and do not intersect.*

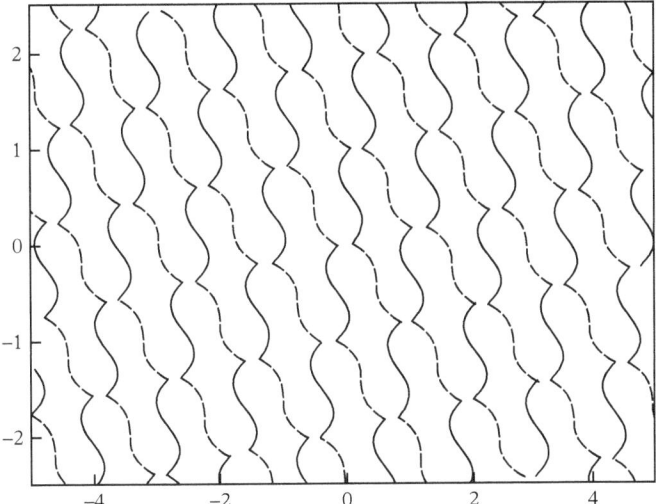

Figure 2.19 *The embedding of an incommensurate composite. The modulation functions are discontinuous. The atomic surfaces are finite and disjoint. (Dashed and solid lines not in the same plane.)*

The intersection of the three-dimensional periodic structure with the two-dimensional physical space remains the same when z_1 and z_2 change, and this holds also for the projection of the three-dimensional Σ^* on the physical space, which is the Fourier module M^*. Changing the internal coordinate t moves both subspaces with 'speeds' z_1 and z_2. In the former convention $z_1 = 0$. ▯

The procedure can be generalized to the situation of three-dimensional systems with several subsystems. For this we have to formalize what has been done in the first part of this section. We suppose that there is a D-dimensional crystal consisting of N subsystems, each of which is a modulated crystal. We number the subsystems with ν. The basic structure of the subsystem ν has a lattice Λ_ν with basis vectors $\mathbf{a}_{\nu i}$ ($i = 1, \ldots, D$). Its reciprocal lattice Λ_ν^* has basis vectors $\mathbf{a}_{\nu i}^*$. As always, one has the relation $\mathbf{a}_{\nu i} \cdot \mathbf{a}_{\nu j}^* = \delta_{ij}$. The wave vectors of the composite structure are all the vectors of the reciprocal lattices, and the wave vectors of the modulations. However, the latter are just combinations of the set of all reciprocal basis vectors, if we assume that the origin of the modulation is the interaction with the other subsystems. Therefore, the Fourier module of the composite is the span (all linear combinations with integer coefficients) of all $\mathbf{a}_{\nu i}^*$ ($i = 1, \ldots, D$; $\nu = 1, \ldots, N$).

$$M^* = \mathrm{Span}\left(\mathbf{a}_{\nu i}^*\right) = \mathrm{Span}\left(\mathbf{a}_k^*\right) \quad (i = 1, \ldots, D; \ \nu = 1, \ldots, N),$$

where the n vectors $\mathbf{a}_k^*$ are the basis of the Fourier module. Each reciprocal basis vector of a subsystem belongs to the Fourier module. Therefore, there are (integer) coefficients Z_{ik}^ν such that

$$\mathbf{a}_{\nu i}^* = \sum_{k=1}^n Z_{ik}^\nu \mathbf{a}_k^*; \tag{2.38}$$

D of the basis vectors of M^* already span the physical space. Call them $\mathbf{a}_k^*$ with $k = 1, \ldots, D$. The other $n - D$ vectors then are linear combinations of them with, generally, irrational coefficients:

$$\mathbf{a}_{D+j}^* = \sum_{i=1}^D \sigma_{ji} \mathbf{a}_i^*. \tag{2.39}$$

The embedding can then be performed in the same way as for modulated structures. The module M^* is considered to be the projection of an n-dimensional reciprocal lattice Σ^* with basis

$$\begin{aligned} \mathbf{a}_{sk}^* &= \left(\mathbf{a}_k^*, 0\right) \quad (k = 1, \ldots, D) \\ &= (\mathbf{a}_{D+j}^*, \mathbf{b}_j^*) \quad (j = 1, \ldots, d = n - D). \end{aligned} \tag{2.40}$$

The d vectors $\mathbf{b}_j^*$ form a basis for the internal space, because of the irrationality of the system. The lattice Σ is then spanned by the basis vectors

Description in superspace

$$\mathbf{a}_{sk} = \left(\mathbf{a}_k, \sum_{j=1}^{d} \sigma_{jk}\mathbf{b}_j\right) \quad (k = 1, \ldots, D) \tag{2.41}$$
$$= (0, \mathbf{b}_j) \quad (k = D + j),$$

where the d vectors $\mathbf{b}_j$ span the lattice for which the vectors $\mathbf{b}_j^*$ form the reciprocal basis. They again form a basis for the internal space V_I.

To find the corresponding embedding of the unmodulated atom positions we associate a shift of the ν-th subsystem in physical space with a change of the internal coordinate. For each subsystem we define a map from internal to physical space as follows:

$$\pi_\nu \mathbf{b}_j = \sum_{k=1}^{D} Z_{k\ D+j}^\nu \mathbf{a}_{\nu k}. \tag{2.42}$$

Because the vectors $\mathbf{b}_j$ form a basis of the internal space, any shift in internal space is a linear combination of these vectors: $\mathbf{t} = \sum_{j=1}^{d} z_j \mathbf{b}_j$. Consequently, a shift in internal space implies a change of position

$$\pi_\nu \mathbf{t} = \pi_\nu \sum_{j=1}^{d} z_j \mathbf{b}_j = \sum_{j=1}^{d} \sum_{k=1}^{3} z_j Z_{k\ D+j}^\nu \mathbf{a}_{\nu k}$$

of the ν-th subsystem. The unmodulated positions of the atoms in subsystem ν are $\mathbf{n}_\nu + \mathbf{r}_{\nu j}$, where $\mathbf{n}_\nu$ is a lattice vector in the lattice Λ_ν and $\mathbf{r}_{\nu j}$ the position of the j-th atom in its unit cell. Then the statement is that the embedding

$$(\mathbf{n}_\nu + \mathbf{r}_{\nu j} - \pi_\nu \mathbf{t},\ \mathbf{t}) \tag{2.43}$$

is invariant under the n-dimensional translation lattice Σ. It is an embedding because for $\mathbf{t} = 0$ one gets the physical structure back.

♯ The proof of the statement goes as follows:
(1) Under a translation $\mathbf{a}_{sk}$ ($k > D$) the embedding of subsystem ν transforms to $(\mathbf{n}_\nu - \pi_\nu \mathbf{t}, \mathbf{t} + \mathbf{b}_j)$. Substitution of $\mathbf{t}'$ for $\mathbf{t} + \mathbf{b}_j$ gives $\left(\mathbf{n}_\nu + \sum_{k=1}^{D} Z_{k\ D+j}^\nu \mathbf{a}_{\nu k} - \pi_\nu \mathbf{t}', \mathbf{t}'\right)$ and this can be written as $(\mathbf{n}_\nu' - \pi_\nu \mathbf{t}, \mathbf{t})$, because the coefficients $Z_{k\ D+j}^\nu$ are integers.
(2) Under a translation $\mathbf{a}_{si}$ ($i = 1, \ldots, D$)) it transforms to $(\mathbf{n}_\nu - \pi_\nu \mathbf{t} + \mathbf{a}_i, \mathbf{t} - \sum_{j=1}^{d} \sigma_{ji} \mathbf{b}_j)$. By denoting the internal component by $\mathbf{t}'$ this becomes $(\mathbf{n}_\nu - \pi_\nu \mathbf{t} + \mathbf{v}, \mathbf{t})$ with $\mathbf{v} = \mathbf{a}_i - \sum_{j=1}^{d} \pi_\nu \sigma_{ji} \mathbf{b}_j$. Now $\mathbf{v}$ is a vector of Λ_ν as its scalar product with any basis vector $\mathbf{a}_{\nu m}^*$ is an integer:

$$\mathbf{v}.\mathbf{a}_{\nu m}^* = \left(\mathbf{a}_i - \sum_{j=1}^{d}\sum_{k=1}^{D} \sigma_{ji} Z_{k\ D+j}^\nu \mathbf{a}_{\nu k}\right).\mathbf{a}_{\nu m}^* = \mathbf{a}_i.\mathbf{a}_{\nu m}^* - \sum_{j=1}^{d} \sigma_{ji} Z_{m\ D+j}^\nu = Z_{mi}^\nu.$$

Here, the relation $\mathbf{a}_i.\mathbf{a}^*_{D+j} = \sigma_{ji}$ has been used. Because $\mathbf{v}$ belongs to Λ_ν the transform can be written as $(\mathbf{n}'_\nu - \pi_\nu \mathbf{t}, \mathbf{t})$. This proves the invariance of the embedding under the n-dimensional lattice Σ. ♮

Finally, the modulated structures may be embedded. For subsystem ν the modulation is a quasiperiodic function with wave vectors in the Fourier module M^*. The positions in physical space are

$$\mathbf{n}_\nu + \mathbf{r}_{\nu j} + \sum_{\mathbf{H} \in M^*} \mathbf{u}_{\nu j}(\mathbf{H}) \exp(2\pi i \mathbf{H}.(\mathbf{n}_\nu + \mathbf{r}_{\nu j})). \qquad (2.44)$$

The wave vectors $\mathbf{H}$, belonging to M^*, are projections of $(\mathbf{H}, \mathbf{H}_I)$ of the n-dimensional reciprocal lattice Σ^*. And the embedding becomes

$$(\mathbf{n}_\nu + \mathbf{r}_{\nu j} - \pi_\nu \mathbf{t} + \sum_{\mathbf{H} \in M^*} \mathbf{u}_{\nu j}(\mathbf{H}) \exp\left(2\pi i \mathbf{H}.(\mathbf{n}_\nu + \mathbf{r}_{\nu j} - \pi_\nu \mathbf{t}) + \mathbf{H}_I \mathbf{t}\right), \mathbf{t}). \qquad (2.45)$$

Also this array of atomic surfaces is invariant under Σ. The proof is given below. A first, important conclusion is that the embedded structure has as symmetry group an n-dimensional space group.

♯ The embedded modulated structure transforms under translations from Σ as follows. Under $\mathbf{a}_{sk} = (\mathbf{a}_E, \mathbf{a}_I)$ one gets

$$(\mathbf{n}_\nu + \mathbf{a}_E + \mathbf{r}_{\nu j} - \pi_\nu \mathbf{t}$$
$$+ \sum_{\mathbf{H} \in M^*} \mathbf{u}_{\nu j}(\mathbf{H}) \exp(2\pi i \mathbf{H}.(\mathbf{n}_\nu + \mathbf{r}_{\nu j} - \pi_\nu \mathbf{t}) + \mathbf{H}_I \mathbf{t}), \mathbf{t} + \mathbf{a}_I)$$
$$\equiv (\mathbf{n}_\nu + \mathbf{a}_E + \mathbf{r}_{\nu j} - \pi_\nu \mathbf{t} - \mathbf{a}_I$$
$$+ \sum_{\mathbf{H} \in M^*} \mathbf{u}_{\nu j}(\mathbf{H}) \exp(2\pi i \mathbf{H}.(\mathbf{n}_\nu + \mathbf{r}_{\nu j} - \pi_\nu \mathbf{t} - \mathbf{a}_I) + \mathbf{H}_I \mathbf{t} - \mathbf{a}_I), \mathbf{t})$$
$$= (\mathbf{n}'_\nu + \mathbf{r}_{\nu j} - \pi_\nu \mathbf{t}$$
$$+ \sum_{\mathbf{H} \in M^*} \mathbf{u}_{\nu j}(\mathbf{H}) \exp(2\pi i \mathbf{H}.(\mathbf{n}'_\nu + \mathbf{r}_{\nu j} - \pi_\nu \mathbf{t}) + \mathbf{H}_I \mathbf{t} - iw), \mathbf{t}),$$

with $w = \mathbf{H}.\mathbf{a}_E + \mathbf{H}_I.\mathbf{a}_I$.

Because $(\mathbf{H}, \mathbf{H}_I)$ belongs to Σ^* and $(\mathbf{a}_E, \mathbf{a}_I)$ to Σ, their scalar product is 0 (mod 1), which means that one may put w equal to 0. Thus, the transformed structure is the same as the original one. ♮.

As pointed out in Section 2.3.3, the embedding is not symmetric in the subsystems. Like there, we could consider a more general embedding as well. Here we stay with the simpler formulation, which has become more or less standard.

Notice that the atomic surfaces of modulated structures and incommensurate composites are not necessarily smooth or connected. When the subsystems are very weakly coupled, the modulation will be small, and the atomic surfaces smooth. However, for strong coupling the modulation function may become discontinuous and the atomic surfaces split up in disconnected parts. We shall come back to this mechanism later.

In Perez-Mato et al. (1999) the *closeness condition* was used to describe a family of incommensurate composites. An example of an embedding of such a compound, with disconnected atomic surfaces, is given in Fig. 2.16.

A difference between the symmetries of modulated phases and composites lies in the fact that the *crystallographic condition* is not necessarily fulfilled. The reason is that atoms of one subsystem may be mapped by a symmetry operation onto another subsystem. Then main reflections from one subsystem may be mapped on main reflections to another subsystem. Therefore, the crystallographic condition breaks down. An example of a composite where one subsystem is mapped by a symmetry operation to another subsystem is that of the compound $Hg_{3-\delta}AsF_6$. But this system has only symmetries that are allowed in three-dimensional crystals. A fictitious system is the following. It consists of modulated layers perpendicular to the z-axis. The layers of the first subsystem alternate with those of the second. The basic structure in the layer is tetragonal, and the basis vectors of the system of odd layers point in the x- and y-directions, but the squares in the even layers are rotated over 45° with respect to those in the odd layers. This is an aperiodic composite of rank 5. A basis of the Fourier module consists of the vectors $\mathbf{a}^*$ (along x), $\mathbf{b}^*$ (along y), the two $(\mathbf{a}^* \pm \mathbf{b}^*)/\sqrt{2}$, and $\mathbf{c}^*$. The point group may be 8/mmm, depending on the modulation caused by the interaction between the two subsystems. In this case, the crystallographic condition is not satisfied.

2.3.4 Quasicrystals

For quasicrystals we start exemplifying the embedding on a simple one-dimensional model, the *Fibonacci chain*, an aperiodic chain with intervals of length 1 and $\Phi = (\sqrt{5}-1)/2 = 1/\tau$. There are several ways to construct this chain, and one of them is to give explicitly the positions x_n of the vertices.

$$x_n = x_0 + na + f(qna + \phi), \qquad (2.46)$$

with $a = 3\Phi - 1$, $q = \Phi/a$, $f(x) = (\Phi - 1)\text{Frac}(x)$, where $\text{Frac}(x)$ is x minus the largest integer smaller than x. From the expression it is seen that the chain can be considered as a modulated phase with a discontinuous modulation function with period one. The values of $x_n - x_{n-1}$ are $3\Phi - 1 + (\Phi - 1)\Phi = 1$ and $3\Phi - 1 + (\Phi - 1)(\Phi - 1) = \Phi$. Thus, the chain consists of intervals of length 1 and Φ. It should be noted that the length scale in V_I is still arbitrary. For the choice made here one obtains a two-dimensional lattice that is square, but this does not mean that the embedded structure has square symmetry!

The embedding follows from the Fourier module, spanned by $1/(3\Phi - 1)$ and $\Phi/(3\Phi - 1)$. An arbitrary point of the Fourier module is $(h+k\Phi)/(3\Phi - 1)$. Then the reciprocal basis in superspace consists of

$$\mathbf{a}_1^* = (1/(3\Phi - 1), 0), \quad \mathbf{a}_2^* = (\Phi/(3\Phi - 1), 1), \tag{2.47}$$

and the basis for the lattice Σ consists of

$$\mathbf{a}_1 = (3\Phi - 1, -\Phi), \quad \mathbf{a}_2 = (0, 1). \tag{2.48}$$

The modulation function is piecewise linear. This means that the atomic surfaces are disjoint straight intervals (Fig. 2.20).

The same points of the Fourier module are obtained as projections of the vertices of a square lattice with two-dimensional basis vectors

$$\mathbf{a}_1^* = \frac{1}{2-\Phi}(1, -\Phi), \quad \mathbf{a}_2^* = \frac{1}{2-\Phi}(\Phi, 1) \tag{2.49}$$

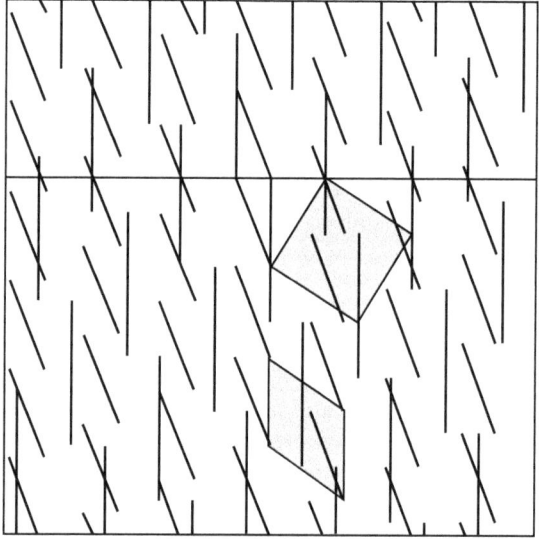

Figure 2.20 *Two embeddings of the quasiperiodic Fibonacci chain superimposed: one as a modulated chain (oblique atomic surfaces), and one with atomic surfaces parallel to the internal space. They give the same chain, the intersection of the vertical and oblique lines, respectively, with the physical space (horizontal line). Examples of unit cells for the two lattices are also indicated. The length scale in V_I is arbitrary, but usually one takes a value such that the 2D unit cell is a square. However, this has no physical consequences!*

for which the direct lattice has the basis

$$\mathbf{a}_1 = (1, -\Phi), \quad \mathbf{a}_2 = (\Phi, 1). \tag{2.50}$$

If one attaches vertical line segments (atomic surfaces) of length $1 + \Phi$ to the vertices, the intersection points with V_E are the end points of the intervals of a Fibonacci chain of lengths 1 and Φ, i.e. exactly the same points as given by Eq. 2.46. Then the vertices of the chain correspond to the intersection of an atomic surface of length $1 + \Phi$ with V_E.

A difficult point for aperiodic tilings and quasicrystals is connected to the scale symmetry, which will be treated later on (Section 2.5). The diffraction spots are dense and this makes it not so easy to choose a basis for the Fourier module. We illustrate this again with the Fibonacci chain. Suppose that we have chosen a basis for the Fourier module and constructed the reciprocal basis as in Eq. 2.49. Then the corresponding basis in superspace is given by Eq. 2.50. However, one may choose another basis for the Fourier module. Take, for example, $\mathbf{b}_1^* = (\Phi, 1)/(2-\Phi) = \mathbf{a}_2^*$ and $\mathbf{b}_2^* = (1-\Phi, -1-\Phi)/(2-\Phi) = \mathbf{a}_1^* - \mathbf{a}_2^*$. This is an invertible basis transformation. The corresponding unit cell in direct superspace is spanned by $\mathbf{b}_1 = (1+\Phi, 1-\Phi)$ and $\mathbf{b}_2 = (1, -\Phi)$. This unit cell is longer (by a factor τ) along V_E, but shorter by a factor Φ along V_I, and the basis vectors of the Fourier module are a factor Φ smaller, but they span the same module. The number of atomic surfaces (1, in this case) does not change, of course, because the unit cell has the same volume (Fig. 2.22). Since the scale in V_I is still free, one can multiply the internal coordinate by a factor c. Choosing this factor such that $c^2 = (1+\Phi)/2\Phi - 1)$, the basis vectors $\mathbf{b}_1$ and $\mathbf{b}_2$ transform to vectors that span a square. Then one has the same picture as before, but with a larger 2D unit cell. In the example of Fig. 2.22, one takes $c = 2 + \Phi$, which gives a 2D unit cell with bases $(1+\Phi, 1)$ and $(1, -1-\Phi)$. Then the extension along V_E is a factor $\tau = 1 + \Phi$

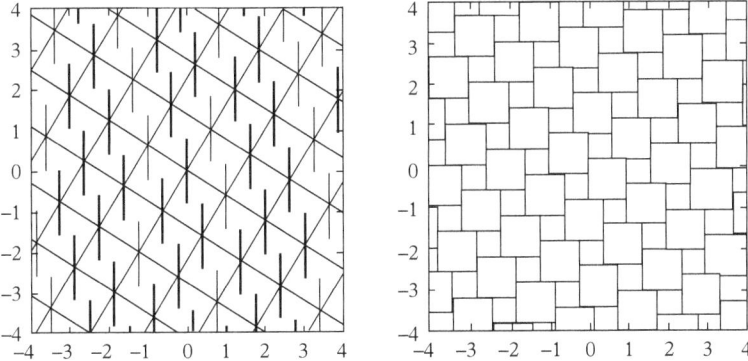

Figure 2.21 *The unit cell of the embedding with a square lattice of the Fibonacci chain is not unique. Examples are the canted square unit cell and the unit cell consisting of two square blocks: a small one and a bigger one.*

74 *Description and symmetry of aperiodic crystals*

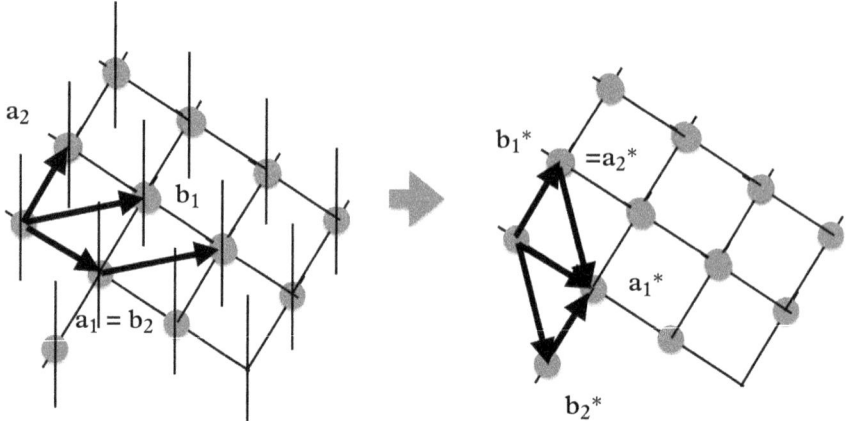

Figure 2.22 *Two bases for the embedding of the Fibonacci chain. Left: Direct space, the first basis* $\mathbf{a}_1$ *and* $\mathbf{a}_2$, *and the second basis* $\mathbf{b}_1$ *and* $\mathbf{b}_2$. *Right: corresponding bases for the reciprocal lattice. Vertical lines are the atomic surfaces.*

larger, but the Fourier module (in V_E) remains the same. In the same spirit, for an icosahedral quasicrystal, one usually takes a six-dimensional hypercube for the elementary cell. Also here, this choice does not have a physical meaning. There is no hypercubic symmetry in this case.

It should be noted that the atomic surfaces are not necessarily line intervals, or simple two- or three-dimensional objects. For several tilings the atomic surface is a fractal. An example is given in Fig. 3.4. However, there are at the moment no examples of real quasicrystals having a fractal atomic surface, although they may be quite complicated (see for example, the modelling of icosahedral quasicrystals in Chapter 4). For displacively modulated aperiodic crystals, the atomic surfaces are not necessarily parallel to V_I. Their projection on V_E usually is not a point. The standard view on the atomic surfaces of quasicrystals, however, is that their projection on V_E is a point. This is, however, not necessarily so. If the atomic surface of a quasicrystal is not flat, it has a modulation function defined on the atomic surfaces. The rank of the structure does not change if the modulation of the atomic surfaces is the same on all surfaces.

As we have seen in the section on composites, the embedding is not unique. A general strain may be applied which leaves the intersection of physical space and embedded structure the same. The strained lattice has basis vectors

$$(-z_1 b/(z_2 - z_1), b/(z_2 - z_1)) \text{ and } (z_2 a/(z_2 - z_1), -a/(z_2 - z_1)). \quad (2.51)$$

For $a = 3\Phi - 1$, $b = (3\Phi - 1)/\Phi$, $z_1 = 0$, $z_2 = b$ the given basis is recovered. Another convenient embedding has a basis obtained for $a = b = 2 - \Phi$, $z_1 = -\Phi$, and $z_2 = 1$:

$$\mathbf{a}'_1 = (1, -\Phi), \quad \mathbf{a}'_2 = (\Phi, 1), \tag{2.52}$$

with reciprocal basis $(1, -\Phi)/(2 - \Phi)$ and $(\Phi, 1)/(2 - \Phi)$. This basis spans a square lattice and the strain is chosen such that the atomic surfaces are parallel to the internal space. Notice that the choice of a square unit cell in the 2D superspace is arbitrary. The corresponding Fourier module is the same as that for the earlier embedding, because $(n + m\Phi)/(2 - \Phi) = (h + k\Phi)/(3\Phi - 1)$ for $m = h$ and $n = h + k$. The new atomic surfaces are again straight, and have a length equal to $1 + \Phi$, which is the length of the projection of the unit cell on the internal space. This gives a second approach to the embedding. Choose the middle of the atomic surface at a vertex of the lattice. It will give an intersection with the physical space if the absolute value of the internal coordinate is less than $(1 + \Phi)/2$. Therefore, if we take a strip of width $1 + \Phi$ and parallel to the physical space, every lattice vertex inside the strip carries an atomic surface that intersects the physical space (cf. Figs 2.19, 2.21). Consequently, all these lattice vertices correspond to a vertex of the chain in the physical space. This is the essence of the method of *cut-and-project*: select all points inside the strip and project them on the physical space. The projection is a perpendicular projection, because the atomic surfaces are parallel to the internal space. In general, atomic surfaces are not necessarily flat, and then the atomic positions are obtained as intersections, not as projections.

Consider the embedding of the Fibonacci chain as given by Eq. 2.52. The atomic surface is the projection of the unit cell on the internal space, that is, a line interval of length τ. The atomic surface with internal coordinate of its centre equal to x_I^o intersects the physical space if

$$x_I^o - \frac{\tau}{2} < 0 < x_I^o + \frac{\tau}{2}.$$

Its intersection with the physical space is x_E. Now, the next point in the chain is $x_E + 1$ if the atomic surface with internal coordinate $x_I^o - \Phi$ intersects the physical space; otherwise, the next point is at $x_E + \Phi$. Hence

$$x_I^o - \tau/2 - \Phi < 0 < x_I^o + \tau/2 - \Phi \rightarrow x'_E = x_E + 1$$
$$x_I^o - \tau/2 + 1 < 0 < x_I^o + \tau/2 + 1 \rightarrow x'_E = x_E + \Phi.$$

This gives a very natural choice for the unit cell. It consists of two blocks. One block has width Φ and lies right of the upper part, with length Φ, of the atomic surface, the other has width 1 and lies right of the lower part, with length 1, of the atomic surface. The two blocks together have an area of $2 - \Phi$, which is the area of the square with edges Eq. 2.52 (Figs 2.19 and 2.21). The same considerations may be made for the left neighbours. Then the coordination of a point in the quasicrystal is determined by its position on the atomic surface. For the Fibonacci chain, a vertex of the tiling produced by an intersection of the physical space with the atomic surface at a distance y from the lower end of the atomic surface lies

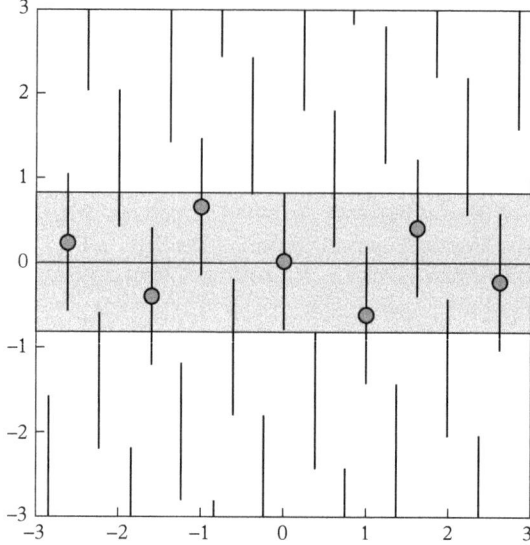

Figure 2.23 *The vertices of the Fibonacci chain may be obtained as the intersection of the atomic surfaces with the physical space ($y = 0$) or by projection of the lattice vertices inside the shaded strip on V_E.*

between a short and a long interval if $0 < y < \Phi$, between two short intervals if $\Phi < y < 1$, and between a long and a short if $1 < y < 1 + \Phi$ (Fig. 2.23).

A decoration of a long interval gives atomic positions in the interval of length 1. In the embedding these give vertical lines with length 1 in the block with area 1. The decoration of the short interval gives vertical lines of length Φ in the block with area $1 - \Phi = \Phi^2$. On top of that, the positions may be modulated (Fig. 2.24). Supposing that the Fourier module is that of the modulated structure, the unit cell remains the same. The vertical lines are now replaced by wavy lines. The intersection of these lines with the physical space gives the atomic positions of the modulated quasicrystal. Then the positions of the atoms are different in different tiles. This means that the structure is not a *decorated tiling*. A decorated tiling is one where the positions of the atoms in tiles of the same type (e.g. L or S for the Fibonacci chain) are the same. The modulation may also be occupational. Then there is a function on the atomic surface which gives the probability of finding a certain species in the intersection point. This determines the chemical order. If part of the atomic surface consists of aluminium and part of manganese, the Al and Mn probabilities are 0 or 1 for either of the elements.

This procedure may be generalized to higher-dimensional structures. Consider, instead of the square planar lattice above, a hypercubic lattice in four dimensions. The x_1-x_2-plane is considered to be the physical space and the x_3-x_4-plane the

Description in superspace 77

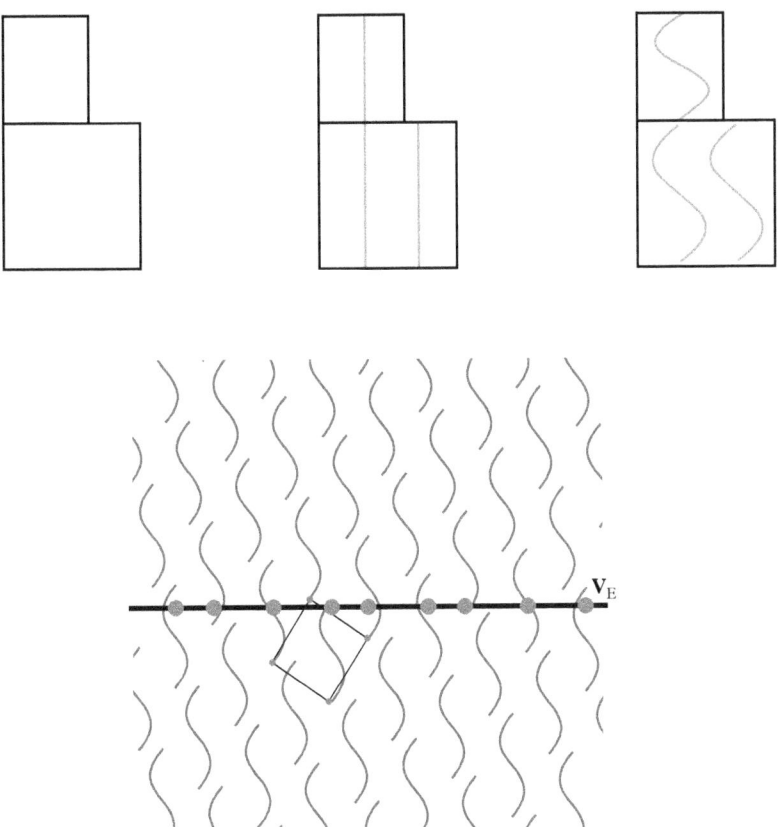

Figure 2.24 *Above:* (1) *The unit cell of the embedded Fibonacci chain consists of two blocks.* (2) *A simple decoration of the tiles gives straight atomic surfaces.* (3) *A modulation is represented by atomic surfaces which are not straight.*
Below: Embedding of a modulated Fibonacci chain with the same Fourier module. The unmodulated chain is given by the dots. The small square is the unit cell in superspace. If the modulation is incommensurate, the Fourier module changes its rank and the dimension of superspace increases.

internal space. Choose eight vectors in the physical space which form a regular star, with an angle of $\pi/4$ between two neighbours. Only four of these vectors are independent if one allows integer coefficients only. Take for these the vectors $\mathbf{e}_j = (\cos((j-1)\pi/4), \sin((j-1)\pi/4))$. A cubic (reciprocal) lattice is now spanned by the four four-dimensional vectors

$$\mathbf{e}^*_{s1} = (\mathbf{e}_1, \mathbf{e}_1),\ \mathbf{e}^*_{s2} = (\mathbf{e}_2, \mathbf{e}_4),\ \mathbf{e}^*_{s3} = (\mathbf{e}_3, -\mathbf{e}_3),\ \mathbf{e}^*_{s4} = (\mathbf{e}_4, \mathbf{e}_2).$$

The projection of this four-dimensional lattice on the physical space is a Fourier module of rank 4. The direct lattice is the same as the reciprocal lattice, up to a factor 2:

$$\mathbf{e}_{sj} = \mathbf{e}_{sj}^*/2.$$

The projection of the Wigner–Seitz unit cell, the hypercube spanned by the four basis vectors, on internal space is a regular octagon, with eight-fold rotation symmetry. If one now attaches a copy of this octagon in every vertex of the lattice in superspace, it can be considered as the embedding of an aperiodic, quasiperiodic structure. Its Fourier module is of rank 4 and is generated by the four vectors $\mathbf{e}_i$. The positions in physical space are the intersections of the octagonal atomic surfaces with the physical space. These positions belong to the set generated by $\mathbf{e}_j/2$, also a module of rank 4 sometimes called the Bravais module. The vertices are connected by edges $\pm\mathbf{e}_i/2$. This leads to a tiling of the plane by means of two types of rhombi: with an angle $\pi/2$ (a square) or an angle $\pi/4$. In the same way as for the Fibonacci chain, we arrive at a quasiperiodic structure with two types of tile.

The main problem is, of course, the determination of the atomic positions, or 'Where are the atoms?' Bak (1986). One has tried to construct models for quasicrystals as decorated tilings, but this approach meets some problems, like unphysical distances. It could help to consider more general descriptions, where the atomic surfaces are not parallel to the internal space. As an example, in Fig. 2.25 one compares a Penrose tiling with two types of atoms, first at positions of the vertices, then after relaxation, using a non-central symmetric interaction between the atoms. As a consequence the positions change, which corresponds to a deformation of the atomic surfaces with displacements in V_E.

The atomic surfaces contain more information than just the vertex positions. An atomic surface at $(\mathbf{r}_E, \mathbf{r}_I)$ gives a vertex of the tiling at $\mathbf{r}_E$ if the physical space

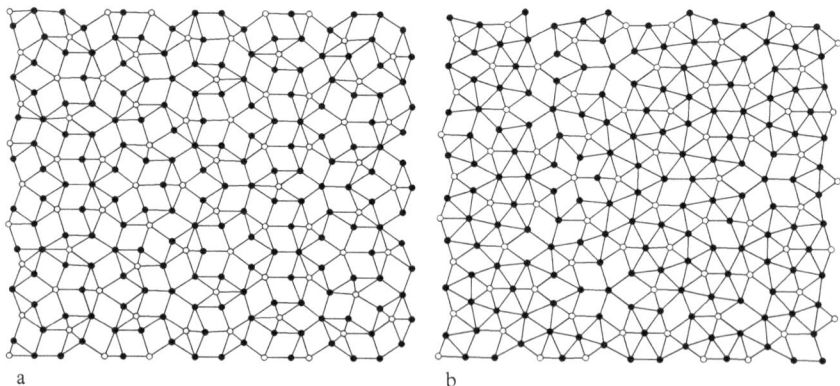

Figure 2.25 (a) *Penrose tiling with two types of atoms at the vertices.* (b) *Relaxed positions using a non-central symmetric potential. The tiles themselves are deformed, but the Fourier module remains the same.*

intersects the atomic surface: $\mathbf{r}_I \in \Omega$. The point $\mathbf{r}_I$ of Ω may contain information about the vertex. When the tiling is a model for a quasicrystal, an atomic species A may be assigned to the point $\mathbf{r}_I$ of the atomic surface. That implies that an atom of type A is located at vertex $\mathbf{r}_E$. Thus, chemical order means a distribution function on the atomic surface: it is divided into domains corresponding to one atomic species.

The coordination may also be found via the atomic surface. Consider as simple example, the Fibonacci chain, obtained from a two-dimensional array with lattice generated by $(1, -\Phi)$ and $(\Phi, 1)$, and with atomic surfaces of length $1 + \Phi$, the interval from $-\Phi$ to 1. If V_E intersects the atomic surface at (r_E, r_I), then there is a vertex at r_E. A lattice translation $(1, -\Phi)$ brings the atomic surface to the next unit cell. Its projection shifts over $-\Phi$. It will again intersect V_E if $r_I + \Phi < 1$: $-\Phi < r_I < 1 - \Phi$. The atomic surface shifted over $(\Phi, 1)$ intersects V_E if $r_I > 1 - \Phi$. Therefore, points with $-\Phi < r_I < 1 - \Phi$ correspond to vertices r_E which are left end points of intervals of length 1, and points with $1 - \Phi < r_I < 1$ correspond to left end points of intervals of length Φ. The projection of the lattice on V_I is uniform, the density of points is everywhere the same. This implies that $1/(1+\Phi)$ of the vertices are left points of long intervals, and $\Phi/(1+\Phi)$ of the vertices are left points of a short interval. The average length, therefore, is equal to

$$\frac{1}{1+\Phi} + \Phi \frac{\Phi}{1+\Phi} = 3\Phi - 1 \approx 0.85.$$

The ratio between long and short intervals is the ratio between the corresponding parts of the atomic surface, that is, $1/\Phi = \tau$.

An analogous argument for the octagonal tiling gives the neighbours of a vertex. A vertex is an intersection of the octagonal atomic surface with V_E. The intersection point is $\mathbf{r}_I$. If a translation $(\mathbf{e}_{iE}, \mathbf{e}_{iI})$ is such that $\mathbf{r}_I - \mathbf{e}_{iI}$ belongs also to the atomic surface, then there is also a vertex of the tiling at $\mathbf{r}_E + \mathbf{e}_{iE}$. In this way it is possible to determine the neighbours of a vertex. The atomic surface for the octagonal tiling may be subdivided in domains with a certain coordination.

A special property of quasiperiodic tilings is the *local isomorphism*. Consider a Fibonacci chain, and select a patch of vertices. The corresponding lattice points are (r_E, r_I) with $-\Phi < r_I < 1$. Because the patch is finite there is a maximum and a minimum value of r_I: $-\Phi < Min < Max < 1$. Select a translation vector $\mathbf{n}$ with $|n_I| < Min + \Phi$ and $|n_I| < 1 - Max$. Then the shifted patch of atomic surfaces will all intersect V_E. Therefore, the new vertices are just shifted by a translation n_E. There are an infinite number of such translations. However, the translations will become larger and larger when the size of the patch grows, because then $|1 - Max|$ and $|Min + \Phi|$ will tend to zero. This is the property of local isomorphism: every finite patch of vertices will occur infinitely often, just shifted by a translation.

A third property of quasiperiodic tilings is the *scaling symmetry*. We illustrate this again on the Fibonacci chain. The point $n + m\Phi$ is a vertex of the chain if $-n\Phi + m$ belongs to the atomic surface, that is, if $-\Phi < -n\Phi + m < 1$. Now multiply the

Figure 2.26 *Scaling the vertices of the Ammann–Beenker tiling, one obtains a subset of the vertices, which form a similar tiling (thick lines).*

positions by a factor $\tau = 1 + \Phi$. This gives a point $(1 + \Phi)(n + m\Phi) = n' + m'\Phi$ with $n' = n + m, m' = n$. This is again a vertex of the chain, because

$$-n'\Phi + m' = -(n+m)\Phi + n = -\Phi(-n\Phi + m).$$

The internal coordinate is multiplied by a factor $-\Phi$ which is smaller than unity in absolute value. Therefore, the atomic surface at the new vertex lies closer to V_E than the original, and therefore intersects V_E also. According to this property, multiplying all the vertex positions by a factor τ maps vertices on vertices. Similar properties occur for general quasiperiodic tilings. For example, the octagonal tiling is mapped into itself by multiplication by a factor $1 + \sqrt{2}$ (cf. Fig. 2.26).

The description of real quasicrystals goes along the lines explained here for tilings, which serve as models for quasicrystals. From a basis of the Fourier module one constructs a basis for the n-dimensional lattice (where n is the rank of the module), taking into account the symmetry relations. This, generally, leaves at least one degree of freedom, the scale in internal space. Then the description of the unit cell consists of a description of positions and shapes of the atomic surfaces, which are finite, but may have a complicated structure. The chemical composition of the quasicrystals may be reflected in the composition regions of the atomic surfaces. Too short distances are avoided, and local jumps made possible by the *closeness condition*: the projection of the border of an atomic surface on the internal space should coincide with the border of another atomic surface in the neighbourhood. This keeps the chemical composition and density locally constant, when the position of the physical space in the superspace changes.

For an icosahedral quasicrystal, the basis of the Fourier module, given in Eq. 1.23, may be lifted to a basis of a reciprocal lattice in six dimensions. One may choose the length scale in V_I such that the direct six-dimensional lattice has as its basis

$$\mathbf{a}_1 = b(1, \tau, 0, 1, \tau, 0); \quad \mathbf{a}_2 = b(-1, \tau, 0, 1, -\tau, 0) \quad (2.53)$$
$$\mathbf{a}_3 = b(0, 1, \tau, -\tau, 0, 1); \quad \mathbf{a}_4 = b(\tau, 0, 1, 0, -1, -\tau)$$
$$\mathbf{a}_5 = b(\tau, 0, -1, 0, -1, \tau); \quad \mathbf{a}_6 = b(0, 1, -\tau, -\tau, 0, -1),$$

with $b = a/(4+2\tau)$. This is a basis for a hypercubic lattice. The projection of the unit cell is a triacontahedron. If one uses that as atomic surface in a lattice point one gets an icosahedral three-dimensional tiling. In real icosahedral quasicrystals, the structure has to be given by a specification of the atomic surfaces in the unit cell. Each atomic surface is a three-dimensional object. The choice of the length scale in V_I made here is arbitrary. It is the hypercubic lattice for the *3D standard icosahedral tiling*, also called the *icosahedral Ammann tiling*. It gives a *hypercubic unit cell*, for which one may determine the lattice constant. But, this embedding has no physical signification. And the atomic surfaces remain the same up to a scaling factor.

From the embedding it is simple to derive the density of the particles in physical space and the chemical composition of the quasicrystal. Suppose there are s atomic surfaces in the nD unit cell, each with a projection of volume Ω_μ on the internal space. (Here, and further on, nD means n-dimensional.) The sum Ω of all these atomic surfaces is minimally the volume of the projection of the unit cell, because there may be overlap. Denote by V the total volume of the unit cell in n dimensions. Because the physical space intersects the atomic surfaces in points which are equivalent with points in the unit cell, and these points fill the atomic surfaces uniformly, the average volume per particle is V/Ω, and the *particle density* is $\rho = \Omega/V$. The chemical composition of the aperiodic crystal is $A_x B_y \ldots C_z$. To the element A correspond regions on the atomic surfaces. Denote the volume of atomic surface μ occupied by atoms A by $V_{A\mu}$. The total volume in the unit cell is $V_A = \sum_\mu V_{A\mu}$. The percentage x corresponding to the A content of the crystal then is equal to V_A/Ω.

The notion of approximant is important, particularly for describing quasicrystals. An approximant tiling is a periodic tiling which is locally identical with a quasiperiodic tiling. Take as an example the classical Fibonacci chain. It is quasiperiodic, and may be obtained from a two-dimensional periodic array of atomic surfaces. The lattice is generated by the vectors $(1, -\Phi)$ and $(\Phi, 1)$. Because $-n\Phi + m$ is never equal to zero (provided $n, m \neq 0, 0$), because of the irrationality of Φ, there is no lattice vector which leaves the physical space invariant. Such a lattice vector necessarily has an internal component equal to zero. The situation would be different if the internal component of the first basis vector would be rational. Replace the irrational Φ by a $p/q \approx \Phi$. Then the internal component of a lattice vector would be $-np/q + m$. This equals zero if $np = mq$. A basic solution of this equation is $m = p, n = q$. The corresponding lattice vector is $(q + p\Phi, 0)$. A sequence which converges to Φ is the sequence $p/q = F_n/F_{n-1}$, where F_n are the Fibonacci numbers. For more accurate approximations, the numbers p and q grow. Therefore, the unit cell of the periodic approximant increases in length, and tends to infinity. The building blocks, the intervals of length 1 and Φ, remain the same, and the sequence of intervals inside the unit cell is equal to sequences inside the aperiodic Fibonacci chain. Take, for example, $p/q = 3/5$. Starting from an atomic surface in the origin with $r_I = 0$, the atomic surfaces which intersect are at positions $r_I = 0.6, -0.4, 0.2, -0.8, -0.2, 0.4, -0.6$, and 0. The intervals form a series LSLSLLSL inside the unit cell, which also occurs in the Fibonacci chain. There are five long and three short intervals. In the same way, approximants in higher dimensions are constructed by replacing the irrational numbers in the internal coordinates of the basis vectors by rational numbers that approximate the irrational numbers. The transition from the lattice of the aperiodic chain to the periodic one is an affine transformation. It is a strain of the lattice.

2.3.5 The classification into three (or four) types is not unique!

In the preceding sections, we have seen three types of aperiodic structures, modulated phases, composites, and quasicrystals. Is this a rigorous classification? Modulated phases are defined as structures not too far from a periodic structure. Positions and chemical species may differ from the basic structure. Especially in the superspace formulation it is easily seen that there is space for ambiguity. Look at Fig. 2.19. In this case with disconnected atomic surfaces, one might associate one subsystem with the lines of one type (solid or dashed), and the other subsystem with the lines of the other type. However, if one follows one type across the gap far from the other type, and if one follows the other type after the gap where the subsystems practically cross, one obtains an array of parallel structures, composed of two types of atoms, with discontinuous atomic surfaces. Because these structures are all parallel they form a modulated phase, in this case a combination of displacive and composition modulation.

We have seen another ambiguity in the case of the Fibonacci chain. Its embedding may be considered as a modulated phase with a discontinuous modulation function, but it may also be considered as a quasicrystal, because it is a one-dimensional model for a quasiperiodic tiling.

In addition, the difference between displacive modulation and occupation modulation becomes vague in certain borderline cases. The long-period alloy CuAuII may be considered as an occupation modulated structure, where the positions are taken by either Cu or Au. Its structure shows domain walls separating regions with an interchange of Cu and Au. This may be achieved by a displacement, albeit over a relatively large distance. But in this sense the modulation is displacive.

We have, in principle, a clear distinction between periodic and aperiodic crystals. So we could consider all mentioned types as sharing the property of quasiperiodicity. But it is worthwhile to keep in mind that another assignment of the type of quasiperiodicity could be advantageous. In practice, very often the assignment of a type is natural.

However, even the distinction between periodic and aperiodic crystals should be considered with care. Of course, mathematically there is no problem. A number is rational or irrational. But, there are families of compounds or phase diagrams with incommensurate phases, where the modulation wave vector changes, but sometimes remains constant in a domain with some width. We shall see examples of such phases where the commensurate phases fit in very well with the incommensurate phases, and can be described using one superspace group. Then the commensurate phases may be described in exactly the same manner as the incommensurate phases.

2.4 Symmetry

2.4.1 Point group symmetry of diffraction patterns

Because the diffraction spots for an aperiodic crystal above a certain threshold form a discrete set there is a finite group of rotations, mirrors, and roto-inversions leaving the pattern invariant, taking into account also the intensities. Suppose R is such a transformation; then

$$R\mathbf{a}_i^* = \sum_{j=1}^{n} \Gamma^*(R)_{ij} \mathbf{a}_j^*. \tag{2.54}$$

Here the matrix elements $\Gamma^*(R)_{ij}$ are all integers. The group of all orthogonal transformations R leaving the diffraction pattern invariant form a finite point group K. The matrices corresponding to all these elements R form a group of n-dimensional integer matrices $\Gamma^*(K)$. Because a vector $\mathbf{a}_{si}^*$ of a reciprocal lattice Σ^* in n-dimensional superspace corresponds to a vector $\mathbf{a}_i^*$ of the Fourier module, these matrices also describe transformations of Σ^*.

Group theory tells us that, for such a finite group of matrices, there is a basis such that every matrix is in block diagonal form:

$$\Gamma^*(K) \sim S(K), \text{ with } S(R) = \begin{pmatrix} S_1(R) & 0 & 0 \\ 0 & \ddots & 0 \\ 0 & 0 & S_{\max}(R) \end{pmatrix}, \tag{2.55}$$

for every operation R from the group K. A block of matrices $S_i(K)$ is irreducible, which means that it is not further decomposable. And the blocks may be chosen to be in the form of orthogonal matrices. The vectors in the physical components are transformed among themselves. So, the physical space is an invariant space, and one or more of the blocks $S_i(K)$ make up a D-dimensional representation. The blocks R_E in physical space are just D-dimensional orthogonal transformations. The element R_E has to be supplemented by an orthogonal transformation R_I in internal space. Therefore, the matrices $\Gamma^*(R)$ correspond to pairs (R_E, R_I) of a D-dimensional orthogonal transformation in physical space and a d-dimensional orthogonal transformation in internal space. All these pairs form the n-dimensional point group K. The D-dimensional components R_E form a point group K_E and the d-dimensional components R_I the point group K_I. These point groups K_E and K_I are not necessarily crystallographic: there is not always a D-dimensional lattice invariant under K_E or a d-dimensional lattice invariant under K_I. This implies that the group of integer matrices $\Gamma^*(R)$ can generally not be brought into a block diagonal form with integer components.

Consider as an example the diffraction pattern of K_2SeO_4. It has a Fourier module of rank 4 with orthorhombic symmetry. The basis vectors of the module are

$$\mathbf{a}^*, \mathbf{b}^*, \mathbf{c}^*, \mathbf{q} = \alpha \mathbf{a}^*.$$

Symmetry transformations are the mirrors m_x, m_y, and m_z, which give the three matrices $\Gamma^*(m_i)$:

$$\begin{pmatrix} -1 & 0 & 0 & 0 \\ 0 & 1 & 0 & 0 \\ 0 & 0 & 1 & 0 \\ 0 & 0 & 0 & -1 \end{pmatrix}, \begin{pmatrix} 1 & 0 & 0 & 0 \\ 0 & -1 & 0 & 0 \\ 0 & 0 & 1 & 0 \\ 0 & 0 & 0 & 1 \end{pmatrix}, \begin{pmatrix} 1 & 0 & 0 & 0 \\ 0 & 1 & 0 & 0 \\ 0 & 0 & -1 & 0 \\ 0 & 0 & 0 & 1 \end{pmatrix}.$$

These matrices are already in block diagonal form, the blocks being one-dimensional. The two point groups K_E and K_I are the three-dimensional point group mmm, and the one-dimensional point group $\bar{1}$. Both are crystallographic.

A second example is the diffraction pattern of the octagonal tiling. It has a symmetry group with two generating elements which give for the Fourier module of rank four the matrices

$$\Gamma^*(R_{\pi/4}) = \begin{pmatrix} 0 & 1 & 0 & 0 \\ 0 & 0 & 1 & 0 \\ 0 & 0 & 0 & 1 \\ -1 & 0 & 0 & 0 \end{pmatrix}, \Gamma^*(m_x) = \begin{pmatrix} -1 & 0 & 0 & 0 \\ 0 & 0 & 0 & 1 \\ 0 & 0 & 1 & 0 \\ 0 & 1 & 0 & 0 \end{pmatrix}.$$

The eigenvalues of the first are $\exp(\pm \pi i/4)$ and $\exp(\pm 3\pi i/4)$, a rotation over $\pi/4$ in the physical space, and a rotation over $3\pi/4$ in internal space. Therefore, the groups K_E and K_I are both the non-crystallographic two-dimensional point group

8mm, but there is no split basis giving integer matrices. If a matrix in block diagonal form is wanted, a split basis must be chosen, and then the blocks are non-integer:

$$R_{\pi/4} = \begin{pmatrix} \sqrt{\frac{1}{2}} & \sqrt{\frac{1}{2}} & 0 & 0 \\ -\sqrt{\frac{1}{2}} & \sqrt{\frac{1}{2}} & 0 & 0 \\ 0 & 0 & -\sqrt{\frac{1}{2}} & \sqrt{\frac{1}{2}} \\ 0 & 0 & \sqrt{\frac{1}{2}} & -\sqrt{\frac{1}{2}} \end{pmatrix}, \quad m_x = \begin{pmatrix} -1 & 0 & 0 & 0 \\ 0 & 1 & 0 & 0 \\ 0 & 0 & -1 & 0 \\ 0 & 0 & 0 & 1 \end{pmatrix}.$$

The first two coordinates are Cartesian coordinates in physical space, and the last two are in internal space. There is no split basis such that the matrices are integer. This shows that the point group occurring in the description of an aperiodic crystal is not always a combination of a D-dimensional and a d-dimensional crystallographic point group.

For incommensurate modulated phases there is a special situation. The basis vectors of the Fourier module may be distinguished in main reflections $\mathbf{a}_i^*$ ($i = 1, \ldots, D$) and satellites $\mathbf{a}_{D+j}^*$ ($j = 1, \ldots, d$). The main reflections are transformed among themselves:

$$R\mathbf{a}_i^* = \sum_{j=1}^{D} \Gamma_E^*(R)_{ij} \mathbf{a}_j^* \quad (i = 1, \ldots, D). \tag{2.56}$$

The satellites can be expressed in main reflections with the matrix σ. Then

$$R\mathbf{a}_{D+j}^* = \sum_{k=1}^{D} \sigma_{jk} R\mathbf{a}_k^* = \sum_{k=1}^{D} \sum_{m=1}^{D} \sigma_{jk} \Gamma_E^*(R)_{km} \mathbf{a}_m^*. \tag{2.57}$$

Because

$$R\mathbf{a}_{D+j}^* = \sum_{k=1}^{D} \Gamma_M^*(R)_{jk} \mathbf{a}_k^* + \sum_{m=1}^{d} \Gamma_I^*(R)_{jm} \mathbf{a}_{D+m}^*,$$

for integer matrices Γ_M^* and Γ_I^*, the matrix $\Gamma^*(R)$ may be written as

$$\Gamma^*(R) = \begin{pmatrix} \Gamma_E^*(R) & 0 \\ \Gamma_M^*(R) & \Gamma_I^*(R) \end{pmatrix}, \tag{2.58}$$

with a relation between the three constituting matrices:

$$\Gamma_M^*(R) = \sigma \Gamma_E^*(R) - \Gamma_I^*(R)\sigma. \tag{2.59}$$

This relation puts restrictions on both the internal rotation R_I and the basic satellites, given by the matrix σ. Because $\Gamma_M^*(R)$ is an integer matrix the matrix elements of $\sigma\Gamma_E^*(R) - \Gamma_I^*(R)\sigma$ should be 0 (mod 1). Take as an example the point group of the modulated phase K_2SeO_4. The group K_E is mmm. Because the internal space is one-dimensional, the element R_I is $\epsilon = \pm 1$, and σ is the modulation vector $(\alpha, \beta, \gamma) = \mathbf{q}$. Then for $R_E = m_x$ the relation $\sigma\Gamma_E^*(m_x) - \epsilon\sigma \equiv 0$ (mod 1) gives $\epsilon = -1$. Given the pair (R_E, ϵ) the modulation vector has to satisfy $(0, 2\beta, 2\gamma) \equiv 0$, which means that β and γ are either 0 or $\frac{1}{2}$, but cannot be irrational.

Because there is a simple relation between a lattice and its reciprocal lattice, the action of the point group on the two is also related in a simple way. If a point group transformation R acts on the Fourier module basis by the n-dimensional matrix $\Gamma^*(R)$, it acts with the same matrix on the reciprocal lattice Σ^*. Then the action on the direct lattice Σ is given by

$$R\mathbf{a}_{si} = \sum_{j=1}^{n} \Gamma(R)_{ji}\mathbf{a}_{sj}, \quad \text{with } \Gamma(R) = \Gamma^*(R)^{-1}. \tag{2.60}$$

2.4.2 Superspace groups

Space groups in n dimensions are groups of distance preserving transformations. These may be written as a combination of an orthogonal transformation, leaving a point invariant, and a translation. The pure translations in a space group form an n-dimensional lattice. For the embedding of a quasiperiodic crystal, the orthogonal transformations are pairs of orthogonal transformations, one in physical space, and one in internal space. Also the translations may be written as a sum of a translation in each of these two spaces. Therefore, one may write

$$g = \{R|\mathbf{t}\} = \{(R_E, R_I) \mid (\mathbf{t}_E, \mathbf{t}_I)\}, \tag{2.61}$$

where $(R_E, \mathbf{t}_E)$ is a distance preserving transformation in physical space, and $(R_I, \mathbf{t}_I)$ one in the internal space.

The space groups in n dimensions are relevant for the description of the symmetry of quasiperiodic structures with rank n if their point groups are reducible into a D-dimensional and a d-dimensional component, where D is the dimension of the physical space. Such groups are called superspace groups. Space groups in n dimensions with point groups which mix the D-dimensional and the d-dimensional spaces are not relevant, because such symmetry operations are only present for special ratios between the length scale in physical space and that in internal space. This ratio can be changed for quasiperiodic structures, because the Fourier module is the projection of the reciprocal lattice, and this does not depend on the length scale ratio. Therefore, not all n-dimensional space groups are superspace groups.

A superspace group symmetry element is a transformation of the superspace which leaves the embedded density function the same.

$$\rho_s(\mathbf{r}_s) = \rho_s(\{R|\mathbf{t}\}\mathbf{r}_s). \tag{2.62}$$

In the two components this becomes

$$\rho_s(\mathbf{r}_E, \mathbf{r}_I) = \rho_s(R_E\mathbf{r}_E + \mathbf{t}_E, R_I\mathbf{r}_I + \mathbf{t}_I).$$

In particular, for modulated structures this comes down to

$$(R_E\mathbf{r}_{\mathbf{n}j}(\mathbf{t}) + \mathbf{t}_E,\ R_I\mathbf{t} + \mathbf{t}_I) = (\mathbf{r}_{\mathbf{n}'j'}(\mathbf{t}'),\ \mathbf{t}'), \tag{2.63}$$

for a permutation $\mathbf{n},j,\mathbf{t} \to \mathbf{n}',j',\mathbf{t}'$.

Just as in three-dimensional crystallography the translation parts $\mathbf{t}$ may be decomposed into an intrinsic and a non-intrinsic part. The latter disappears when a proper origin is chosen, the former is independent of the origin. The intrinsic part of the translation $\mathbf{t}$ accompanying the orthogonal transformation R is

$$t^i = \frac{1}{N}\sum_{n=1}^{N} R^n \mathbf{t} = \left(\frac{1}{N}\sum_n R_E^n \mathbf{t}_E,\ \frac{1}{N}\sum_n R_I^n \mathbf{t}_I\right).$$

For a three-dimensional rotation this is the part along the rotation axis. Any orthogonal transformation in three dimensions may be decomposed into the sum of a two-dimensional rotation (which may be absent) and 1 till 3 values ± 1. In n dimensions an orthogonal transformation may decomposed into a sum of two-dimensional rotations, and a number (from 0 to n) of values ± 1. The intrinsic translations are along the axis where the values are $+1$. For each element of a superspace group one has to indicate the orthogonal transformation (R_E, R_I) and the intrinsic part of the translation $\mathbf{t}$.

In the case of an embedding with the first D basis vectors of the reciprocal lattice in the physical space one may go further. Then there exists a mapping Δ from the physical space to the internal space. Therefore, for the translation part of the superspace group element $\{(R_E, R_I)|(\mathbf{t}_E, \mathbf{t}_I)\}$, a part of $\mathbf{t}_I$ is due to the internal space component that goes with $\mathbf{t}_E$. So, we can write

$$\mathbf{t}_I = -\Delta \mathbf{t}_E + \mathbf{t}_I^o.$$

The mapping Δ is defined by means of the matrix σ, the matrix of the d fundamental satellites. This σ satisfies the relation

$$\Gamma_M^*(R) = \sigma \Gamma_E^*(R) - \Gamma_I^*(R)\sigma \equiv 0 \ (\mathrm{mod}\ 1).$$

A special solution is given by

$$\sigma^i \Gamma_E^*(R) - \Gamma_I^*(R)\sigma^i = 0.$$

Then σ may be decomposed into the sum of σ^i and σ^r. The latter matrix is rational, and the first one is irrational. A second mapping from V_E to V_I is given by

$$\Delta^i \mathbf{a}_k = \sum_{j=1}^{d} \sigma_{jk}^i \mathbf{b}_j. \tag{2.64}$$

It can be shown that the component $\mathbf{t}_I + \Delta^i \mathbf{t}_E$ has rational components with respect to the lattice basis $\mathbf{b}_j$ ($j = 1, \ldots, d$) in internal space. This may be used for the notation.

A simple situation occurs for $D=3$ and $d=1$, a one-dimensional modulation. Then σ is the row vector with the three components of the modulation wave vector $\mathbf{q}$. In this case $\Delta \mathbf{t}_E = \mathbf{q}.\mathbf{t}_E$. The rational intrinsic component in this case is $\mathbf{t}_I + \mathbf{q}^i.\mathbf{t}_E$, where $\mathbf{q}^i$ is the solution of $R_E \mathbf{q}^i = \epsilon \mathbf{q}^i$. ($\epsilon = R_I = \pm 1$). The rational component $\mathbf{t}_I + \mathbf{q}^i + \mathbf{q}^*.\mathbf{t}_E$ is a fraction such that N times this fraction is an integer. This leads to a simple notation for the superspace group elements. The components R_E of the orthogonal transformations form a point group K_E, the components R_I a point group K_I. The symbol for the D-dimensional group K_E is followed by the symbol for the d-dimensional point group in parentheses. Often a non-standard setting is needed, because the convention prescribes that there are equally many symbols for both groups and they should match. For example, if K_E is the group 2/m, and K_E the one-dimensional group $\bar{1}$, the symbol for the superspace point group is either $2/m(1\bar{1})$ (if $\mathbf{q}$ is along the rotation axis) or $2/m(\bar{1}1)$ (if $\mathbf{q}$ is in the mirror plane). The translation components $\mathbf{t}_E$ are indicated in the same way as in the three-dimensional case. For example $2_1/m$ means that the two-fold rotation has an intrinsic translation of half a lattice vector along the rotation axis. Taking for the unique axis the $\mathbf{a}$-axis, $\mathbf{t}_E = \frac{1}{2}\mathbf{a}$. Then the translation $\mathbf{t}_I$ consists of $-\Delta \mathbf{t}_E = -\mathbf{q}.\mathbf{a}/2$. and the part $\mathbf{t}_I^o$, which is either 0 or $\frac{1}{2}$. If it is 0, a 0 is placed behind the parentheses, if it is $\frac{1}{2}$, then an s is placed. Then there are superspace groups like $2_1/m(1\bar{1})00$, $2_1/m(\bar{1}1)00$, and $2_1/m(\bar{1}1)0s$.

2.4.3 Examples

Here three examples are given of space groups for incommensurate phases, and one for a quasicrystal. This may help to clarify the rather abstract treatment of the theory.

First we look at a modulated phase with basic structure $P2$ and modulation wave vector along the unique axis: $\mathbf{q} = \gamma \mathbf{c}^*$. Because the generator $R = 2$ (a two-fold rotation) leaves $\mathbf{q}$ invariant, the corresponding generator for the 4D group is (2, 1). The matrix σ is $(0, 0, \gamma)$, and consequently the matrix Γ_I is given by

$$R\sigma - \sigma = (0, 0, 0).$$

This fixes the point group $\Gamma(K)$. There is a superspace group which associates a translation $(0, 0, 0, \frac{1}{2})$ to the generator of the point group. The corresponding space group element is

$$\left(\begin{pmatrix} -1 & 0 & 0 & 0 \\ 0 & -1 & 0 & 0 \\ 0 & 0 & 1 & 0 \\ 0 & 0 & 0 & 1 \end{pmatrix} \begin{pmatrix} 0 \\ 0 \\ 0 \\ \frac{1}{2} \end{pmatrix} \right),$$

or $(x_1, x_2, x_3, x_4) \rightarrow (-x_1, -x_2, x_3, x_4 + \frac{1}{2})$. The basis of the Fourier module may be lifted to a basis for the 4D reciprocal lattice consisting of $(\mathbf{a}^*, 0)$, $(\mathbf{b}^*, 0)$, $(\mathbf{c}^*, 0)$, and $(\gamma \mathbf{c}^*, 1)$. In split coordinates a general point then transforms according to

$$(x_1\mathbf{a} + x_2\mathbf{b} + x_3\mathbf{c}, -\gamma x_3 + x_4) \rightarrow \left(-x_1\mathbf{a} - x_2\mathbf{b} + x_3\mathbf{c}, -\gamma x_3 + x_4 + \frac{1}{2}\right).$$

The translation $(0, 0, 0, \frac{1}{2})$ is intrinsic. It cannot be transformed by choosing another origin. A value $\frac{1}{2}$ is indicated by the letter s. Therefore, the symbol for this superspace group is $P2(00\gamma)s$.

A second example is a group for a modulated structure having a basic structure with space group B2/m, and modulation wave vector along the unique axis. The basis of the 4D lattice becomes $(\mathbf{a}, 0)$, $(\mathbf{b}, 0)$, $(\mathbf{c}, -\gamma)$, $(0, 1)$, and the centring vector is $(\mathbf{a}+\mathbf{c}/2, -\gamma/2)$. The matrices $\Gamma_M(R)$ are 0 for both generators (2 and m). The point groups generators with their corresponding translation parts are

$$\left(\begin{pmatrix} -1 & 0 & 0 & 0 \\ 0 & -1 & 0 & 0 \\ 0 & 0 & 1 & 0 \\ 0 & 0 & 0 & 1 \end{pmatrix} \begin{pmatrix} 0 \\ 0 \\ 0 \\ \frac{1}{2} \end{pmatrix} \right), \quad \left(\begin{pmatrix} 1 & 0 & 0 & 0 \\ 0 & 1 & 0 & 0 \\ 0 & 0 & -1 & 0 \\ 0 & 0 & 0 & -1 \end{pmatrix} \begin{pmatrix} 0 \\ 0 \\ 0 \\ \frac{1}{2} \end{pmatrix} \right).$$

The translation of the first transformation is intrinsic, but that of the second may be transformed away. Therefore, the symbol is $B2/m(00\gamma)s0$.

For the next example we take the 3D space group P4/n as the symmetry group of the basic structure. The modulation wave vector is $(\mathbf{a}^* + \mathbf{b}^*)/2 + \gamma \mathbf{c}^*$. The generators of the point group are 4 and m. The rotation 4 leaves $\mathbf{q}$ invariant, and m inverts it. Therefore, the generators of the 4D point group are $(4, 1)$ and $(m, -1)$. The corresponding matrices Γ_M are not zero in this case. They lead to the 4D matrices

$$\begin{pmatrix} 0 & -1 & 0 & 0 \\ 1 & 0 & 0 & 0 \\ 0 & 0 & 1 & 0 \\ 0 & -1 & 0 & 1 \end{pmatrix} \text{ and } \begin{pmatrix} 1 & 0 & 0 & 0 \\ 0 & 1 & 0 & 0 \\ 0 & 0 & -1 & 0 \\ 1 & 1 & 0 & -1 \end{pmatrix}.$$

One space group with this point group has, associated to the generators, the translations $(\frac{1}{2},\frac{1}{2},0,\frac{1}{4})$ and $(\frac{1}{2},\frac{1}{2},0,0)$. For points given with respect to a split basis by $(x_1\mathbf{a}+x_2\mathbf{b}+x_3\mathbf{c},-(x_1+x_2)/2-\gamma x_3+x_4)$, the transformation is to points $((\frac{1}{2}-x_2)\mathbf{a}+(\frac{1}{2}+x_1)\mathbf{b}+x_3\mathbf{c},-(x_1+x_2)/2-\gamma x_3+x_4+\frac{1}{4})$ and $((x_1+\frac{1}{2})\mathbf{a}+(x_2+\frac{1}{2})\mathbf{b}-x_3\mathbf{c},(x_1+x_2)/2+\gamma x_3-x_4)$, respectively. By a change of origin the translation of the first generator may be reduced to its intrinsic part, which is $(0, 0, 0, \frac{1}{4})$. The translation $\mathbf{a}_4/4$ is indicated with the letter q, the translation $(\frac{1}{2},\frac{1}{2}, 0, 0)$ by n. Finally, the symbol for the group becomes P4/n$(00\gamma)q0$.

The fourth example has five-fold symmetry. The Fourier module is of rank 5, and has four basis vectors in the x,y-plane and one along the z-axis. It may be lifted to a reciprocal lattice basis in five dimensions, for which the basis of the direct lattice is given with respect to a split orthonormal basis by $a(\cos(2\pi j/5)$, $\sin(2\pi j/5), \cos(4\pi j/5), \sin(4\pi j/5),0)$ with $j = 1,\ldots,4$, and $(0, 0, 0, 0, c)$. The point group is 5m with generators

$$A = \begin{pmatrix} 0 & 0 & 0 & -1 & 0 \\ 1 & 0 & 0 & -1 & 0 \\ 0 & 1 & 0 & -1 & 0 \\ 0 & 0 & 1 & -1 & 0 \\ 0 & 0 & 0 & 0 & 1 \end{pmatrix}, \quad B = \begin{pmatrix} 0 & 0 & 0 & 1 & 0 \\ 0 & 0 & 1 & 0 & 0 \\ 0 & 1 & 0 & 0 & 0 \\ 1 & 0 & 0 & 0 & 0 \\ 0 & 0 & 0 & 0 & 1 \end{pmatrix}.$$

The symmorphic superspace group with this point group transforms the point in superspace with lattice coordinates $(x_1, x_2, x_3, x_4, x_5)$ to $(-x_4, x_1 - x_4, x_2 - x_4, x_3 - x_4, x_5)$ under A and to $((x_4, x_3, x_2, x_1, x_5)$ under B. But in coordinates x_i with respect to the split basis the point transforms to

$$(x_1 \cos(2\pi/5) - x_2 \sin(2\pi/5), x_1 \sin(2\pi/5) + x_2 \cos(2\pi/5),$$
$$x_3 \cos(4\pi/5) - x_4 \sin(4\pi/5), x_3 \sin(4\pi/5) + x_4 \cos(4\pi/5), x_5).$$

The matrix A can be decomposed into a rotation over $2\pi/5$ and one over $4\pi/5$, and B into the direct sum of two mirrors (which is a two-fold rotation). Therefore, the symbol for the group is P5m(5^2m). For convenience this is often written as P5m. This is not ambiguous because in three dimensions the point group 5m does not occur as crystallographic group.

2.4.4 Approximants

Many incommensurate phases have a wave vector which changes with temperature. For specific temperatures, this wave vector may become commensurate, which means that the lattice Σ in superspace is deformed in such a way that the intersection with the D-dimensional physical space becomes a lattice. The deformed structure, which is lattice periodic, then has locally a structure that closely resembles that of the incommensurate phase. It is called an *approximant*. Such approximants also play a role in quasicrystals. Very often, a quasicrystal is formed within a certain composition range. The ground state of the system with slightly different composition often has a very similar structure locally to that of the quasicrystal, but may be lattice periodic. In general, an *approximant* is a lattice periodic structure that locally is very similar to that of an aperiodic crystal.

The intersection of an nD lattice of an aperiodic crystal structure with the physical space does not contain a D-dimensional lattice. The projection on the internal space is dense. So, arbitrarily close to V_E there are lattice vectors of Σ. By a small strain one may bring three independent vectors of Σ into physical space, and from the quasiperiodic structure one gets a lattice periodic structure in V_E.

For example, the embedding of the Fibonacci chain has a lattice with basis vectors $a(1, -\Phi)$ and $a(\Phi, 1)$. The irrational number Φ may be approximated by the fraction L/N. If one replaces the Φ in the first basic vector by L/N, this may be achieved by a small strain applied to the two-dimensional structure. The new lattice has lattice points $(n + m\Phi, -nL/N + m)$. The lattice points lie in V_E if $nL = mN$. These points are spanned by the vector $(N + L\Phi, 0)$. The approximant will have N long intervals and L short intervals in the unit cell of length $N + L\Phi$. The configuration of the unit cell of the approximant also occurs an infinite number of times in the Fibonacci chain, but then aperiodically.

A displacive modulated structure with wave vector $\mathbf{q}$ close to $\mathbf{c}^*/3$ will, for a long interval, be similar to a period 3 superstructure, which is an approximant. However, the phase of the incommensurate will change slowly. If the modulation function is smooth, this change will be continuous. When it is piecewise constant, the phase will jump at regular distances, at the so-called discommensurations. Between the discommensurations (or domain walls) the structure of the incommensurate crystal will be as a piece of the approximant.

2.4.5 Superspace groups for commensurate phases

For an aperiodic crystal the lattice vectors in the physical space do not form a D-dimensional lattice. Otherwise it would not be aperiodic. But, because the projection of the n-dimensional lattice on the internal space is dense, there are lattice points arbitrarily close to the physical space. A small strain in the n-dimensional structure may bring a full lattice in the physical space, and the structure then becomes lattice periodic, and has D-dimensional space group symmetry.

The number of superspace groups in n dimensions is finite, but strain is a continuous quantity. Therefore, strain will change the symmetry group discontinuously, but locally a small strain will keep the structure very similar. When will a superspace group be the symmetry group of a lattice periodic structure, and how can one determine the resulting D-dimensional space group?

Let us take, as an example, the 4D superspace group Pnma(α00)$\bar{1}$s1. The point group is generated by (m$_x$, $\bar{1}$), (m$_y$,1), and (m$_z$,1), and the modulation wave vector is (α00). The basis vectors of the lattice Σ are (**a**, $-\alpha$), (**b**, 0), (**c**, 0), and (**0**, 1). If α is rational, say equal to the fraction L/N, there are three basis vectors in V_E: N**a**, **b**, and **c**. They span a lattice Λ. Then the subgroup of the superspace group leaving V_E invariant is a three-dimensional space group with lattice Λ. What is its point group? In other words, which elements R,**v** of the superspace group leave V_E invariant?

The mirror m$_z$ has internal component 1, and non-primitive translation **a**/2 (in V_E). To this we can add a 4D lattice translation, for which we take n(**a**, $-\alpha$) + m(**0**, 1) for values of n and m we are going to determine. The point (x**a** + y**b** + z**c**, t) in superspace is transformed by this combination to

$$\left(\left(x + \frac{1}{2} + n\right)\mathbf{a} + y\mathbf{b} - z\mathbf{c}, t - n\alpha - \alpha/2 + m\right).$$

The superspace group element leaves the physical space at t invariant, if there are values of n and m such that $-n\alpha - \alpha/2 + m = 0$. If $\alpha = L/N$, then this gives the equation

$$-2nL - L + 2mN = 0.$$

This has only integer solutions for n and m if the greatest common divisor of $2L$ and $2N$ divides L. This only happens if L is even and N odd. For example, if $\alpha = 2/3$, then $n = m = 1$ is a solution.

The mirror m$_x$ has internal component -1, and non-primitive translation $\frac{1}{2}$(**a** + **b** + **c**) in V_E. Combined with a 4D lattice translation n(**a**, $-\alpha$) + m(**0**, 1), it transforms (x**a** + y**b** + z**c**, t) to

$$(-x + n + 1/2)\mathbf{a} + (y + 1/2)\mathbf{b} + (z + 1/2)\mathbf{c}, -t - 1/2\alpha - n\alpha + m).$$

This leaves the hyperplane t = constant invariant if n and m satisfy the equation

$$2t = m - 1/2\alpha - n\alpha \quad \rightarrow \quad 2Nm - L - 2Ln = 4Nt.$$

Now the existence of a solution depends not only on the values of L and N, but also on the value of t. The 3D space group may depend on the precise position of the physical hyperplane with respect to the origin. There is a solution if L and N are both odd, provided $4Nt$ is an odd integer, if L/N is even/odd and $4Nt$ is an even integer, and if L/N is odd/even, provided $4Nt$ is an odd integer.

In this way one may determine whether there are lattice translations which, in combination with the 4D superspace group generators, leave the physical space t = constant invariant. One obtains the following results:

α	Phase $4Nt$	3D Space group
Odd/odd	Even	$P2_1/n11$
	Odd	$P2_12_12_1$
	Non-integer	$P2_111$
Odd/even	Even	$P12_1/a1$
	Odd	$Pna2_1$
	Non-integer	$P1a1$
Even/odd	Even	$P112_1/a$
	Odd	$Pn2_1a$
	Non-integer	$P11a$

This shows that several different space groups have to be used when one wants to describe the symmetry group of the rational cases of the modulation. All these 3D space groups follow from one 4D superspace group. The various symmetry elements of the superspace group cannot be retained simultaneously if one considers various commensurate approximants. In this way the superspace group gives a unified view on the whole family of closely related structures. And the 3D space groups of the cases where the modulation wave vector reaches a commensurate value are fully determined by the single superspace group, the fraction L/N, and the phase of the section. For example, the phase diagram of several A_2BX_4 compounds, as function of temperature and composition, may be described using one single superspace group, in fact the group $Pnma(\alpha 00)0s0$ we have used here as example (Janssen, 1986, 1992). The latter completely determines the 3D space groups for the cases for which the modulation is commensurate.

2.4.6 Consequences of superspace group symmetry

An n-dimensional density is invariant under an element $g = \{R|\mathbf{t}\}$ of a superspace group if Eq. 2.62 is satisfied. For the Fourier transform there is an equivalent relation:

$$\hat{\rho}_s(\mathbf{H}_s) = \exp(-2\pi i \mathbf{H}_s.\mathbf{t})\hat{\rho}_s(R_s^{-1}\mathbf{H_s}), \qquad (2.65)$$

or, in components

$$\hat{\rho}(\mathbf{H}) = \exp(-2\pi i \mathbf{H}_I.\mathbf{t}_I)\exp(-2\pi i \mathbf{H}.\mathbf{t}_E)\,\hat{\rho}(R_E^{-1}\mathbf{H}). \qquad (2.66)$$

If $\mathbf{t}_I = 0$, this is just the condition for the Fourier components of a function invariant under the Euclidean transformation $\{R|\mathbf{t}_E\}$. The internal degrees of freedom allow a generalized type of symmetry. In fact, this can be called a gauge transformation. (Think of the change of a quantum-mechanical wave function under a change of electromagnetic gauge.) Because it compensates for an incorrect phase of the Fourier component, it is called a *compensating gauge transformation* (cf. Section 2.2.5).

♮ The electromagnetic fields **E** and **H** can be derived from the electromagnetic scalar potential Φ and the vector potential **A** with

$$\mathbf{E} = -\mathrm{Grad}\Phi - \frac{1}{c}\partial\mathbf{A}/\partial t, \quad \mathbf{H} = \mathrm{Rot}\mathbf{A}.$$

The potentials are not unique. The same fields are obtained from potentials

$$\mathbf{A} + \mathrm{Grad}\chi, \quad \Phi - \frac{1}{c}\partial\chi/\partial t$$

for a so-called gauge function $\chi(\mathbf{r}, t)$. The potentials, and not the fields, appear in the Hamiltonian. This is reflected in the fact that the wave functions $\psi(\mathbf{r},t)$ change when the potentials are changed via a gauge transformation according to

$$\psi(\mathbf{r}, t) \rightarrow \exp(i\chi(\mathbf{r}, t))\psi(\mathbf{r}, t).$$

For a homogeneous field, there is translation invariance, but this invariance does not hold for the potentials. However, if the potentials change by a translation such that the fields are the same, the shifted potential is related to the original one via a gauge transformation. The same holds for Euclidean symmetries of fields. They transform the corresponding potentials to others related by a gauge transformation to the original potentials. This associates a gauge function $\chi_{\{R|\mathbf{a}\}}(\mathbf{r}, t)$ with each symmetry transformation. The gauge function of the product of two symmetry operators is related to the gauge functions of the two in the same way as non-primitive translations associated with elements of the point group. So, there is a direct correspondence between non-primitive translations and gauge transformations (Janner and Janssen, 1971). ♮

From Eq. 2.65 follow very important rules, which are very useful in structure determination. Consider a symmetry element $\{R|\mathbf{t}\}$ of the structure, and take a wave vector $\mathbf{H}_s$ which is left invariant. This is the same as the condition that $R_E\mathbf{H} = \mathbf{H}$. From Eq. 2.65 then follows that either $\mathbf{H}.\mathbf{t} = 0$ (mod 1) or $\hat{\rho}(\mathbf{H}) = 0$. Taking an element R_E such that its **t** satisfies $\mathbf{H}.\mathbf{t} \neq 0$ (mod 1), leads to the conclusion that $\hat{\rho}(\mathbf{H}) = 0$. This is an *extinction rule*. When can we use this rule? The wave vector **H** and the (non-primitive) translation **t** may be expanded in terms of the reciprocal and direct bases:

$$\mathbf{H} = \sum_{i=1}^{n} h_i \mathbf{a}_i^*, \quad \mathbf{t} = \sum_{j=1}^{n} x_j \mathbf{a}_j.$$

Then

$$\hat{\rho}(\mathbf{H}) = 0, \text{ if } R_E\mathbf{H} = \mathbf{H} \text{ and } \sum_{i=1}^{n} h_i x_i \neq 0 \pmod{1}.$$

Actually only the intrinsic part, which does not change if the origin changes, plays a role. Under an origin shift in superspace equal to $\mathbf{v}_s$ the translation part $\mathbf{t}$ changes to $\mathbf{t}' = \mathbf{t} + (1 - R_s)\mathbf{v}_s$, for which one has

$$\mathbf{H}_s.\mathbf{t}' = \mathbf{H}_s\mathbf{t} + (1 - R_s)\mathbf{v}_s.\mathbf{h}_s = \mathbf{H}.\mathbf{t}.$$

The intrinsic parts always have fractions as components.

Let us consider the superspace group Pnma$(00\gamma)0s0$. Its generators lead to extinction rules given in the following table. For example, the Fourier module vector $[H_1, H_2, H_3, H_4]$ is left invariant by m_x if $H_1 = 0$. The non-primitive translation is $(\frac{1}{2}, \frac{1}{2}, \frac{1}{2}, 0)$. This leads to the reflection condition $H_2/2 + H_3/3 = 0$ (mod 1), or $H_2 + H_3 =$ even. The other conditions follow in a similar way.

Point group element	Translation part	Intrinsic part	Reflection condition	For
$(m_x, 1)$	$(\frac{1}{2}, \frac{1}{2}, \frac{1}{2}, 0)$	$(0, \frac{1}{2}, \frac{1}{2}, 0)$	$H_2 + H_3 = 2n$	$[0, H_2, H_3, H_4]$
$(m_y, 1)$	$(0, \frac{1}{2}, 0, \frac{1}{2})$	$(0, 0, 0, \frac{1}{2})$	$H_4 = 2n$	$[H_1, 0, H_3, H_4]$
$(m_z, -1)$	$(\frac{1}{2}, 0, \frac{1}{2}, \frac{1}{2})$	$(\frac{1}{2}, 0, 0, 0)$	$H_1 = 2n$	$[H_1, H_2, 0, 0]$

These extinction rules are *general extinction rules*. There are other extinction rules stemming from the fact that the lattice basis is not always the most convenient choice. Consider the situation where the modulation wave vector(s) contain(s) fractional components. For example, for a structure with Pnmn as space group symmetry of the basic structure, and modulation wave vector $\mathbf{b}^*/2 + \gamma\mathbf{c}^*$ a basis for the Fourier module is

$$\mathbf{a}^*, \quad \mathbf{b}^*, \quad \mathbf{c}^*, \quad \mathbf{b}^*/2 + \gamma\mathbf{c}^*.$$

It may be lifted to a reciprocal basis in superspace

$$(\mathbf{a}^*, 0), \quad (\mathbf{b}^*, 0), \quad (\mathbf{c}^*, 0), \quad (1/2\mathbf{b}^* + \gamma\mathbf{c}^*, 1).$$

If one chooses a basis $\mathbf{a}^*, \frac{1}{2}\mathbf{b}^*, \gamma\mathbf{c}^*$, then the Fourier module is a sublattice of this: there are points on the module created by the latter basis which do not occur in the Fourier module. In other words, they have intensity zero. If the coordinates with

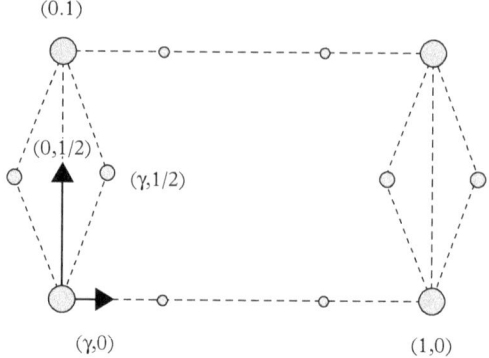

Figure 2.27 The basis of the Fourier module is given by $(1,0)$, $(0,1)$, and $(\gamma, \frac{1}{2})$. A conventional basis would be $(1,0)$, $(\gamma, 0)$, and $(0, \frac{1}{2})$.

respect to the first basis are h_i ($i = 1, \ldots, 4$), and those with respect to the latter H_i, then

$$h_1 = H_1, \quad h_2 = (H_2 - H_4)/2, \quad h_3 = H_3, \quad h_4 = H_4$$

(cf. Fig. 2.27). Because all coefficients h_i should be integer, one has the condition $H_2 - H_4$ (or $H_2 + H_4$) must be even. These extinction rules for the indices H_i are called the *centring extinctions*, because they stem from using a supermodule for the description of the Fourier components. The unit cell in superspace is larger than that which is strictly required. The benefit is in the more symmetric unit cell, just as in 3D crystallography.

2.5 Scaling symmetries

The property that there is a factor λ such that, multiplying all distances by this factor, one obtains a subset of the original set is called scaling. For lattice periodic crystals there is scaling for integer values of λ. Such *scaling symmetries* also occur for quasicrystals and tilings, and there the scaling factors may be non-trivial. They are also present in their Fourier modules.

Consider as an example the Fourier module of a Fibonacci chain. The peaks are found at positions $(h_1 + h_2 \Phi)/a$. When these positions are multiplied with τ their positions become $(h_1 \tau + h_2)/a$, which belong to the Fourier module as well: they are $(H_1 + H_2 \Phi)/a$ with $H_1 = h_1 + h_2$ and $H_2 = h_1$. The corresponding reciprocal lattice vectors are mapped to each other by the basis transformation

$$S = \begin{pmatrix} 1 & 1 \\ 1 & 0 \end{pmatrix}.$$

This has eigenvalues τ and $-\Phi$, and eigenspaces V_E and V_I, respectively. The same transformation in direct space gives the relations

$$\mathbf{a}_1 + \mathbf{a}_2 = (1,-\Phi) + (\Phi,1) = (\tau a_{1E}, -\Phi a_{1I}), \quad \mathbf{a}_1 = (1,-\Phi) = (\tau a_{2E}, -\Phi a_{2I})$$

The physical space is dilated by a factor τ, and the internal space contracted by a factor $1/\tau$. The first is called *inflation*, and the second, *deflation*. In reciprocal space all combinations $h_1 + h_2\Phi$ correspond to a diffraction peak. This peak corresponds to a vertex $(h_1\mathbf{a}_1 + h_2\mathbf{a}_2)/a$ of the reciprocal lattice Σ^*. The matrices S and S^{-1} map this reciprocal lattice on itself. Because, in principle, every reciprocal lattice vector corresponds to a vector in the Fourier module, the latter is invariant under both S and S^{-1}.

The situation in physical space is different from that in reciprocal space. Only atomic surfaces at vertices close to V_E intersect the latter. Acting with S on a vertex of Σ that gives an intersection brings it closer to the physical space by a factor τ, and the new vertex will again give an intersection. The chain is mapped by the factor τ in physical space on itself. For the matrix S^{-1} this does not hold. By its action a vertex giving rise to an intersection will move away from the physical space, and eventually its atomic surface no longer intersects. In direct space, therefore, the group generated by S only gives a semi-group of transformations in physical space: there is no inverse. Moreover, the transformations are not real symmetry transformations if the set is not a point set. If there are atoms with a certain size at the vertices, these are transformed to 'atoms' of a different, rescaled, size.

For the embedding of the Fibonacci chain every basis transformation leaving physical and internal space invariant leads to a scale transformation. All the matrices corresponding to such basis transformations are of the form

$$S = \begin{pmatrix} m & n \\ n & m-n \end{pmatrix} \quad \text{with } m^2 - mn - n^2 = \pm 1.$$

They are all powers of S up to a sign. For higher-dimensional tilings, the symmetry group plays a role in determining the scale transformations.

The scale invariance is a problem in the structure determination. Consider, as an example, the Fibonacci chain again. The vectors of the Fourier module may be written as $h_1 b + h_2 b\Phi$. Choosing a square lattice in reciprocal space, its basis consists of $b(1,-\Phi)$ and $b(\Phi,1)$. Then the lattice in direct space has the basis vectors $a(1,-\Phi)$ and $a(\Phi,1)$, with $a = 1/b(2-\Phi)$. This means that the intervals of the chain have length $L = a$ and $S = a\Phi$. However, another basis of the Fourier module consists of $b\Phi$ and $b\Phi^2 = b(1-\Phi)$. Its embedding in two dimensions leads to another basis in direct space, and the lengths of the intervals are $L' = L/\Phi$ and $S' = S/\Phi$, i.e. a factor of $\tau = 1 + \Phi$ larger. That means that for the construction of a model there is more room for atoms.

For another example of *scaling symmetry*, consider the standard octagonal tiling. It has rank 4, and its point symmetry group is generated by

$$A = \begin{pmatrix} 0 & 0 & 0 & -1 \\ 1 & 0 & 0 & 0 \\ 0 & 1 & 0 & 0 \\ 0 & 0 & 1 & 0 \end{pmatrix}, \quad B = \begin{pmatrix} 0 & 0 & 0 & 1 \\ 0 & 0 & 1 & 0 \\ 0 & 1 & 0 & 0 \\ 1 & 0 & 0 & 0 \end{pmatrix}.$$

The transformation A is equivalent with the sum of two rotations: one in physical space over an angle $\pi/4$ and one in internal space over an angle $3\pi/4$. Together with the two mirrors in both spaces, corresponding to B, they span a two-dimensional non-crystallographic group 8mm in each of the spaces. These two two-dimensional groups are two non-equivalent representations of the group 8mm. They are different because the element in physical space corresponding to A has trace $\sqrt{2}$ and that in internal space has trace $-\sqrt{2}$. A scale transformation is a multiplication with λ in one space and by $\pm\lambda^{-1}$ in the other. Therefore, it commutes with all elements of the point group. In group-theoretical terms it belongs to the centralizer of the group. (The *centralizer* of a group G of n-dimensional matrices is the group of matrices S such that $SAS^{-1} = A$ for all elements A of G. The *normalizer* of G is the set of all matrices S such that $SGS^{-1} = G$.) According to a well-known theorem of group theory, a matrix commuting with all elements of a representation consisting of two non-equivalent components acts as a multiple of the identity in each of the two representation spaces. An element of the centralizer of the group generated by A and B is

$$T = \begin{pmatrix} 1 & 1 & 0 & -1 \\ 1 & 1 & 1 & 0 \\ 0 & 1 & 1 & 1 \\ -1 & 0 & 1 & 1 \end{pmatrix},$$

which has two eigenvalues: $1 \pm \sqrt{2}$. The transformation gives a scaling in physical space with a factor $1+\sqrt{2}$, and a contraction in internal space with a factor $\sqrt{2}-1$. If T is the transformation in reciprocal space, then T^{-1} is the transformation in direct space. The transformations T^{-n} form a semi-group for the vertices of the tiling. This means that, by inflating the octagonal tiling by a factor of $1+\sqrt{2}$, one gets a tiling that, after a shift, corresponds with a subset of the original tiling (see Fig. 2.26).

♯ The generators A and B leave a four-dimensional lattice invariant with the metric tensor

$$g = \begin{pmatrix} b & a & 0 & -a \\ a & b & a & 0 \\ 0 & a & b & a \\ -a & 0 & a & b \end{pmatrix} \tag{2.67}$$

with arbitrary parameters a and b. The eigenvalues of g are $b \pm a\sqrt{2}$. This implies that the metric is not always positive definite, but, depending on a and b, can be an *indefinite metric*, just like the metric for the Minkowski space. In that space the space-time groups (see Section 2.2.5) are defined. ♮

For the primitive 6D lattice of the icosahedral quasicrystals, the point group is $5\bar{3}m(5^2\bar{3}m)$, or shorter $5\bar{3}m$. It may be generated by two orthogonal transformations: one a combination of a five-fold rotation in V_E and one in V_E, and the other a six-fold roto-inversion in V_E and one in V_I. Their matrices are, with respect to a basis of the 6D lattice,

$$A = 5(5^2) = \begin{pmatrix} 1 & 0 & 0 & 0 & 0 & 0 \\ 0 & 0 & 1 & 0 & 0 & 0 \\ 0 & 0 & 0 & 1 & 0 & 0 \\ 0 & 0 & 0 & 0 & 1 & 0 \\ 0 & 0 & 0 & 0 & 0 & 1 \\ 0 & 1 & 0 & 0 & 0 & 0 \end{pmatrix}, \quad B = \bar{3}(\bar{3}) = \begin{pmatrix} 0 & 0 & 0 & 0 & 0 & -1 \\ -1 & 0 & 0 & 0 & 0 & 0 \\ 0 & 0 & 0 & 0 & -1 & 0 \\ 0 & 0 & 1 & 0 & 0 & 0 \\ 0 & 0 & 0 & 1 & 0 & 0 \\ 0 & -1 & 0 & 0 & 0 & 0 \end{pmatrix}. \quad (2.68)$$

An element of the centralizer is

$$S = \begin{pmatrix} 2 & 1 & 1 & 1 & 1 & 1 \\ 1 & 2 & 1 & -1 & -1 & 1 \\ 1 & 1 & 2 & 1 & -1 & -1 \\ 1 & -1 & 1 & 2 & 1 & -1 \\ 1 & -1 & -1 & 1 & 2 & 1 \\ 1 & 1 & -1 & -1 & 1 & 2 \end{pmatrix}.$$

This element S has three eigenvalues $2 + \sqrt{5}$ and three eigenvalues $2 - \sqrt{5}$. The two eigenspaces are V_E and V_I, respectively. Using the same argument as in the octagonal case, there is a scaling factor $2+\sqrt{5} = \tau^3$ in V_E. Multiplying the resulting tiling with this factor gives a scaled version of the same tiling. In the Fourier module it means that the latter is invariant under multiplication with τ^{3n} for arbitrary n. The factor τ does not leave the Fourier module invariant. For example, take the module vector $(1, \tau, 0)/a$. The vector $\tau(1, \tau, 0)/a$ has non-integer coefficients $[\frac{1}{2}, \frac{1}{2}, \frac{1}{2}, \frac{1}{2}, \frac{1}{2}, \frac{1}{2}]$ with respect to the module basis in Eq. 1.23. Therefore, it does not belong to the Fourier module.

There are two other icosahedral 6D lattices, obtained from this one by centring: a face-centred and a body-centred icosahedral lattice, respectively. A vector in V_E belongs to the projection of the 6D reciprocal lattice of the body-centred lattice if it belongs to the Fourier module of the primitive lattice, and if its coefficients satisfy $\sum_i h_i =$ even. For example $h_i = 1$ for all i. The projection is the vector $(2\tau, 2 + 2\tau, 0)$ which has rational module coefficients [111111], whereas τ times this vector has coefficients [311111] and also satisfies the condition $\sum_i h_i =$ even. This holds for all basis vectors, which shows that the Fourier module for the body-centred icosahedral lattice is invariant under the factor τ. This holds for the face-centred icosahedral lattice as well. For the two centred reciprocal lattices the Fourier modules are invariant with a scale factor τ.

The scaling is apparent in the diffraction pattern as well. Consider the diffraction pattern for the decagonal phase (Fig. 2.5). The basis vectors [1000], [0100], [0010], and [0001] multiplied by a factor τ go to spots of the Fourier module, respectively [00–1–1], [1110], [0111], and [–1–100]. For the module, this is a basis transformation with as matrix an integer matrix S of determinant 1, and eigenvalues τ and $-\Phi$ (each double). Each point of the Fourier module is mapped on another point of the module by S (and by S^{-1}).

♯ Systems that are invariant under space group transformations and scale transformations have as symmetry group a *scale space group* Janssen (1991). In this case the generalized point group is generated by a point group and its centralizer or normalizer. For the Ammann–Beenker tiling the generators of this generalized point group L are

$$A = \begin{pmatrix} 0 & 0 & 0 & -1 \\ 1 & 0 & 0 & 0 \\ 0 & 1 & 0 & 0 \\ 0 & 0 & 1 & 0 \end{pmatrix}, \quad B = \begin{pmatrix} 0 & 0 & 0 & 1 \\ 0 & 0 & 1 & 0 \\ 0 & 1 & 0 & 0 \\ 1 & 0 & 0 & 0 \end{pmatrix}, \quad C = \begin{pmatrix} 1 & 1 & 0 & -1 \\ 1 & 1 & 1 & 0 \\ 0 & 1 & 1 & 1 \\ -1 & 0 & 1 & 1 \end{pmatrix}.$$

The point group K is generated by A and B, and the generalized point group L by A,B. and C. The scale-space group G is an extension of L with the translation group U. U is an invariant subgroup of G and G/U ≡ L. For the Amman–Beenker tiling L has the defining relations

$$A^8 = B^2 = (AB)^2 = ACA^{-1}C^{-1} = BCB^{-1}C^{-1} = E \tag{2.69}$$

and there are four non-equivalent extensions:

$$\text{Group 1}: \{A|0000\}, \{B|0000\}, \{C|0000\}$$

$$\text{Group 2}: \{A|0000\}, \left\{B\Big|\frac{1}{2}\frac{1}{2}\frac{1}{2}\frac{1}{2}\right\}, \{C|0000\}$$

$$\text{Group 3}: \{A|0000\}, \{B|0000\}, \left\{C\Big|\frac{1}{2}\frac{1}{2}\frac{1}{2}\frac{1}{2}\right\}$$

$$\text{Group 4}: \{A|0000\}, \left\{B\Big|\frac{1}{2}\frac{1}{2}\frac{1}{2}\frac{1}{2}\right\}, \left\{C\Big|\frac{1}{2}\frac{1}{2}\frac{1}{2}\frac{1}{2}\right\}.$$

The generators $A = (A_E, A_I)$ and $B = (B_E, B_I)$ of the point group act separately in V_E and V_I as orthogonal transformations. The generator C gives an expansion with factor $1 + \sqrt{2}$ in V_E and a contraction with factor $1 - \sqrt{2}$ in V_I. With these four scale-spacegroups one may construct scale invariant octagonal tilings. ♯

2.6 Alternative descriptions

Displacements from a structure with symmetry may always be described using the representations of the symmetry group. Consider as an example a two-dimensional molecule with an equilateral triangle as equilibrium shape. The symmetry group of the molecule is the dihedral group 3m. This group has six elements, and three irreducible representations, with the character table

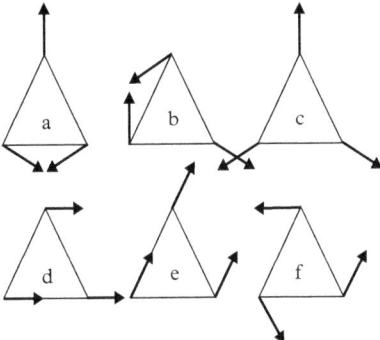

Figure 2.28 *The six modes for a triangular molecule: two translation modes (d, e: representation 3), one rotation mode (f: representation 2), two shear modes (a, b: representation 3), and one breather mode (c: representation 1).*

	E	A, A²	B, AB, A²B
D_1	1	1	1
D_2	1	1	−1
D_3	2	−1	0

The number of degrees of freedom is six: twice the number of particles. Each particle can move in the *x*- and *y*-directions. The displacement patterns are the linear sum of six basic displacement patterns (Fig. 2.28). Two of them correspond to a rigid translation, and one to a rigid rotation. In these three cases the mutual distances do not change. There are two shearing displacements, and because of the symmetry the two basis displacement modes have the same energy. Finally, there is a breather mode. In the last four modes the centre of mass is invariant. Rotation over $2\pi/3$ of the first mode gives the second one, and rotation of the second one gives a sum of the two. Rotation leaves the third and sixth invariant, transforms the fourth to the fifth, and transforms the fifth to a combination of the fourth and the fifth. The mirror operation m_x leaves the sixth invariant, gives a minus sign for the third, and transforms the first and second to a combination of these two, and analogously for the fourth and fifth. The sixth belongs to the trivial representation of the symmetry group, the fourth to the fully anti-symmetric representation, and the pairs 4, 5 and 1, 2 belong to two-dimensional *irreducible representations*.

102   Description and symmetry of aperiodic crystals

The same is true for space groups. Any displacement of a lattice periodic structure can be decomposed into components belonging to the irreducible representations of the space group.

The irreducible representations of the translation group are one-dimensional, because the group is Abelian. The irreducible representations of a space group are characterized by a vector $\mathbf{k}$ in the Brillouin zone and an irreducible representation of the stabilizer point group of $\mathbf{k}$. Then the phase factor $\exp(2\pi i\mathbf{k}.\mathbf{a})$ corresponds to the translation $\{E|\mathbf{a}\}$.

For a vector $\mathbf{k}$ in the Brillouin zone, the group of $\mathbf{k}$ is the subgroup of the space group for which the homogeneous terms leave $\mathbf{k}$ invariant up to a reciprocal lattice vector. This is a space group, with the same translation group. An element $\{R|\mathbf{t}\}$ of the group of $\mathbf{k}$ is represented in an irreducible representation as $\exp(2\pi i\mathbf{k}.\mathbf{t})\Gamma_\nu(R)$. The d_ν-dimensional matrices $\Gamma_\nu(R)$ form an irreducible representation of the point group of the group of $\mathbf{k}$. The subindex ν labels the various irreducible representations.

Consider, as an example, the space group Pnma. Its point group is the orthorhombic group mmm. The generator m_x has as translation the vector $\left(\frac{1}{2}, \frac{1}{2}, \frac{1}{2}\right)$, the generator m_y has $\left(0, \frac{1}{2}, 0\right)$, and the element m_z has the translation $\left(\frac{1}{2}, 0, \frac{1}{2}\right)$. The group of the wave vector $\alpha \mathbf{a}^*$ is the group with point group 2mm. It has four irreducible representations, labelled by ν. Under the generators of the group of $\mathbf{q}$ the displacements transform according to a multiplication with

$$\chi_\nu(m_y), \text{ and } \exp(\pi i\alpha)\chi_\nu(m_z).$$

The elements χ_ν are ± 1, depending on the representation. This fixes the polarization: the direction in which the atom is displaced. If $\chi_\nu(m_y) = -1$, it means that the displacement is inverted by the mirror operation. This implies that the displacement is in the y-direction.

Bienenstock and Ewald formulated the theory of symmetry of crystals in Fourier space, instead of direct space. The formulations in direct and Fourier space are dual: periodicity in direct space implies a Fourier transform $F(h,k,l)$ consisting of delta peaks on a reciprocal lattice, the points of which are indexed by integers h, k, and l. And the Fourier transform of a real function on direct space is, in general, complex. Symmetry operations are then linear transformations which leave the absolute value of the structure factor invariant. They may shift the phase of the structure factor. This may be expressed in the following form:

$$\begin{pmatrix} h' \\ k' \\ l' \\ \phi' \end{pmatrix} = A \begin{pmatrix} h \\ k \\ l \\ \phi \end{pmatrix} + \begin{pmatrix} 0 \\ 0 \\ 0 \\ \Delta\phi \end{pmatrix}, \text{ with } A = \begin{pmatrix} R_{11} & R_{12} & R_{13} & 0 \\ R_{21} & R_{22} & R_{23} & 0 \\ R_{31} & R_{32} & R_{33} & 0 \\ a_1 & a_2 & a_3 & 1 \end{pmatrix}, \quad (2.70)$$

with the R_{ij} forming a point group element, and the a_i a vector determining the gauge transformation in the space of phases ϕ. Conditions on the symmetry operations stem from the fact that the phase shift should be zero if the space transformation is the identity. Groups may be defined by their defining relations, expressions in the generators which equal the identity operation. This means that with each defining relation there is a restriction on the phase shifts. Using these restrictions one may determine the possible combinations of a Euclidean transformation and a transformation of the complex plane. This gives an alternative way for determining the space groups in two and three dimensions. Mathematically, this approach is equivalent with that in direct space.

Mermin and collaborators used these ideas for the formulation of the symmetry of aperiodic crystals in an attempt to avoid a higher-dimensional space. They considered functions in reciprocal space, with symmetry operations which may change the phase of the complex functions. For the definition of equivalence, they introduced the physically interesting notion of indistinguishability (Mermin, 1992): two functions $f_1(\mathbf{k})$ and $f_2(\mathbf{k})$ are *indistinguishable* if all their n-point correlation functions are the same. This means that for all m-tuples of reciprocal space vectors $\mathbf{k}_1, \ldots, \mathbf{k}_m$ with $\sum_j \mathbf{k}_j = 0$ the relation

$$f_1(\mathbf{k}_1)f_1(\mathbf{k}_2)\ldots f_1(\mathbf{k}_m) = f_2(\mathbf{k}_1)f_2(\mathbf{k}_2)\ldots f_2(\mathbf{k}_m)$$

holds. This is equivalent to the requirement

$$f_1(\mathbf{k}) = \exp(2\pi i \chi(\mathbf{k}))f_2(\mathbf{k}),$$

for all vectors $\mathbf{k}$ from the Fourier module. The function $\chi(\mathbf{k})$ is the gauge function in the Bienenstock–Ewald formulation, which we had seen earlier in Section 2.4.6: $\chi(\mathbf{k}) = -\mathbf{k}_I.\mathbf{t}_I$. It is a linear function on the Fourier module modulo integers: $\chi(\mathbf{k}_1 + \mathbf{k}_2) = \chi(\mathbf{k}_1) + \chi(\mathbf{k}_2)$. A symmetry operation transforms a function to a function that is indistinguishable from the first one. That means R is an orthogonal symmetry transformation if

$$f(R\mathbf{k}) = \exp(2\pi i \chi_R(\mathbf{k}))f(\mathbf{k}), \tag{2.71}$$

with a *gauge function* χ_R. The gauge functions obey rules which are similar to the rules for non-primitive translations:

$$\chi_{RS}(\mathbf{k}) = \chi_R(S\mathbf{k}) + \chi_S(\mathbf{k}) \mod \mathbb{Z},$$

because the internal components $\mathbf{t}_I$ satisfy these relations. These can be used to determine the possible combinations of a point group element, a translation, and a phase transformation, which make up the symmetry group of the system. In mathematical terms, it is the determination of the first cohomology group. Finally, one has to define an equivalence relation for these groups, and eliminate groups to find the non-equivalent representatives of the equivalence classes. The formulation in Fourier space has the advantage that the functions are defined in physical space. But, one has to introduce the space of phases, which is equivalent to the internal space. As shown earlier, the Fourier space expressions follow directly from the formulation in superspace. However, in superspace there is a one-to-one correspondence between the vectors of the Fourier module (in physical space) and the vectors of the n-dimensional reciprocal lattice, which makes it possible to stay purely in physical space.

2.7 Magnetic symmetry of quasiperiodic systems

2.7.1 Magnetic systems and time-reversal symmetry

The crystallographic operations that are important for physics are Euclidean transformations, because these leave physical laws, like the Schrödinger equation, invariant. If relativistic effects are important, one should consider Lorentz and Poincaré transformations. However, even in non-relativistic physics other transformations may be relevant. An important example is time-reversal symmetry. This operation reverses magnetic fields and flips spins. Therefore, its action on a physical system is, in general, not trivial. Consider, for example, an antiferromagnetic chain, with regularly spaced spins alternately pointing up and down. If we call the distance between the spins a, the space symmetry is the translation over $2a$. However, a translation over a combined with time reversal leaves the spin configuration invariant as well.

This leads to magnetic symmetry groups, consisting of pairs of a Euclidean transformation and either the identity or the time-reversal operator. Abstractly, these groups are also known as black-and-white groups. Then, instead of up and down spins, one has black and white points, and the symmetry operators considered are pairs of a Euclidean transformation and a permutation of the two colours. This is readily generalized to cases with more colours. This leads to the study of colour groups. Here, however, we shall limit ourselves to the case of two colours, and we shall use the language of magnetic groups.

A magnetic group is a subgroup of the direct product of the Euclidean group (in n dimensions) and a group with two elements, the identity E and the time-reversal T. The latter group is isomorphic to the group C_2.

A magnetic group is a subdirect product of a group of space transformations G and C_2. (A *subdirect product* of groups G_1 and G_2 is a subgroup of the direct product $G_1 \times G_2$ such that for its elements (g_1, g_2) the elements g_1 form group G_1, and the elements g_2, the group G_2.) One may distinguish three types of such groups. The first is that of non-magnetic groups, simply a subgroup G of the Euclidean group. The second type is that of groups where T is an element. Then the group is the direct product of G and the group C_2. These are the trivial magnetic groups. Finally, to the third type belong those groups for which there are elements which are the product of a Euclidean transformation and T, but where T itself is not an element. These are the non-trivial magnetic groups. There are a subgroup H in G (of index 2) and an element $g' = gT$ (the product of a Euclidean transformation and T) such that $G = H + g'H$. We call all magnetic and non-magnetic groups which are the subdirect product of a Euclidean group G and the identity or the group C_2 the family of G. G is also called the parent group of the family. The latter two types are subdirect products of a group G and C_2. In one dimension the family of the Euclidean point group $\bar{1}$ consists of the non-trivial group of order 2 with generator $\bar{1}$, the trivial magnetic group of order 4 generated by $\bar{1}$ and T, and the non-trivial group of order 2 generated by the product of $\bar{1}$ and T. The conventional notation indicates elements that are products of a Euclidean transformation and T by a prime. Then the notation for the three groups is: $\bar{1}, \bar{1}1'$, and $\bar{1}'$. Notice that a spin configuration is never invariant under a trivial magnetic group, because T does not leave any spin invariant.

For each Euclidean group G there are a non-magnetic group and a trivial magnetic group. A non-trivial magnetic group has a subgroup of index 2 consisting of all unprimed elements. These form a subgroup because the product of two unprimed elements is unprimed. Therefore, a Euclidean group admits only a non-trivial magnetic group if it has a subgroup of index 2. If G has a subgroup H of index 2, that is, with half the number of elements, then a non-trivial group is obtained by giving the elements in G, but not in H, a prime, and the elements of H no prime. In this way one may construct all magnetic groups in the family of the Euclidean group G.

2.7.2 Magnetic point groups

The magnetic point groups may be constructed from the non-magnetic point groups. For a point group K there are always the groups K and $K \times C_2$. If K is of odd order there are no non-trivial groups, but when K is of even order, there is a non-trivial magnetic group associated with each subgroup of index 2. Two magnetic groups belong to the same geometric crystal class if the parent groups belong to the same geometric crystal class via a transformation that maps primed elements on primed elements. For example, in two dimensions the parent group mm2 has non-trivial magnetic groups in its family: m'm2', mm'2', and m'm'2. The first two belong to the same geometric crystal class because they are connected by an interchange of the x- and y-axes.

2.7.3 Magnetic space groups

A space group G has a translation group A such that G/A is isomorphic with the point group K. To find the magnetic groups in its family one may apply the general procedure. There is always a trivial magnetic group $G \times T$ denoted by $G1'$. For the non-trivial magnetic groups one may distinguish two cases. The group A may contain a subgroup of index 2, A_0, with non-primed elements. A non-trivial group of the family of A then has elements $a = (a, 1)$ if a belongs to A_0 and $a' = (a, T)$ if a does not belong to A_0. Then for any $R \in K$ there is always a non-primed element in its coset. Therefore, we can take the point group elements without prime. The parent group and the magnetic group have the same point group. They are '*klassengleich*' (class equivalent). The magnetic group is an extension of a non-magnetic point group by a non-trivial magnetic translation group. The other possibility is that A has only non-primed elements. Then the elements $\{R|\mathbf{t}_R\}$ in one coset (for fixed R) are all primed or all non-primed. The prime then can be used as label for the point group elements. To have a magnetic group there is a subgroup K_0 of the point group K (of index two) of non-primed elements. Then the magnetic space group is an extension of a non-trivial magnetic point group with a non-magnetic translation group: the magnetic space group has the same translations as the parent group. This is called '*translationengleich*' (translation equivalent).

The notation for the non-trivial magnetic space groups with a non-magnetic translation group is simple: in the symbol for the parent group one attaches a prime to each symbol for a point group generator which corresponds to a primed operation. For example, in two dimensions the two non-trivial translation equivalent magnetic space groups in the family of Pmm2 are Pm'm2' and Pm'm'2. For the class equivalent groups (the point group is the same as that of the parent group) one may indicate a translation that is combined with time reversal. This is a primed translation. For the example Pmm2 this is $P_a$mm2, which means that the translation group has three generators: $\{E|\mathbf{a}\}'$ (with prime), $\{E|\mathbf{b}\}$, and $\{E|\mathbf{c}\}$ (without prime).

The equivalence of magnetic space groups is formulated in the same way as for the point groups. Notice that the non-trivial magnetic space groups in the family of a space group G are isomorphic to G. Two non-magnetic space groups are considered to be equivalent if they are isomorphic. Then two non-trivial magnetic space groups are equivalent, if they are isomorphic via an isomorphism that maps primed elements on primed elements. Correspondingly, the finer equivalence relation by an affine conjugation with determinant +1 may also be formulated for magnetic groups.

2.7.4 The magnetic groups for quasiperiodic crystals

The discussion of magnetic symmetry for point and space groups may directly be repeated for quasiperiodic crystals, because they also are described with point and space groups, now in more dimensions. The procedure for obtaining the magnetic family of an n-dimensional point or space group is identical to that discussed in the preceding sections. And the notation is also straightforward. Below we give some examples. Other cases have been discussed in Janner and Janssen (1980a) for modulated phases and in Niizeki (1990a,b) in the context of transformations in quasicrystals. An extensive list for a selected choice of groups for quasiperiodic systems in physical spaces of two and three dimensions can be found in Lifshitz (1997). This paper discusses the more general notion of colour group as well.

The action of a magnetic space group element $g = \{R|\mathbf{v}\}T^n$ ($n = 0, 1$) on a function $f(\mathbf{r}, t)$ is given by

$$T_g f(\mathbf{r}, t) = f\left(R^{-1}(\mathbf{r} - \mathbf{v}), (-1)^n t\right). \tag{2.72}$$

This equation holds, independent of the dimension. The position $\mathbf{r}$ may be a position in superspace. However, for a quantum-mechanical wave function this action becomes

$$T_g \psi(\mathbf{r}) = \begin{cases} \psi(R^{-1}(\mathbf{r} - \mathbf{v})) & \text{if } g \text{ is not primed,} \\ \psi^*(R^{-1}(\mathbf{r} - \mathbf{v})) & \text{if } g \text{ is primed.} \end{cases} \tag{2.73}$$

For spin-$\frac{1}{2}$ particles the action is

$$T_g \psi(\mathbf{r}) = u(R)\psi(R^{-1}(\mathbf{r} - \mathbf{v}))$$

for unprimed g and

$$T_g \psi(\mathbf{r}) = u(R)\sigma_y \psi^*(R^{-1}(\mathbf{r} - \mathbf{v}))$$

for primed g. Here, σ_y is the Pauli spin matrix. The time-reversal operator is an anti-unitary operator. If the wave function is not a scalar, but a spinor, an additional spin operator has to be introduced. So a spin function transforms according to

$$T_g \mathbf{S}(\mathbf{r}, t) = \text{Det}(R) R \mathbf{S}(R^{-1}(\mathbf{r} - \mathbf{v}), (-1)^n t).$$

The determinant $\text{Det}(R)$ appears because the spin is a pseudo-vector.

For a quasiperiodic spin system, the atomic surfaces are decorated with a spin distribution. The atomic surfaces are invariant under the parent space group. The spin decoration must be invariant under the trivial or non-trivial magnetic space group. Consider the standard octagonal tiling. The atomic surfaces are regular octagons. If these are divided into eight sectors, with alternating spin direction, the point group is $8'mm'$. The corresponding quasiperiodic spin configuration is shown in Fig. 2.29. Its superspace group is $P8'mm'(8^3mm)$. Notice that the point group elements in the point group symbol are $(8, 8^3)'$, $m_x(m_x)$, and their product. It is sufficient to attach the prime to the external component, because there is the operator T combined with the n-dimensional operation in case of a prime. Another spin configuration, with up spin in a central octagon of half the area and down spins in the remaining part, has non-magnetic symmetry $P8mm(8^3mm)$. The third example is the octagonal spin configuration in Fig. 2.30. It has a non-magnetic point group and a non-trivial magnetic translation group. Its space group is $P_a 8mm(8^3mm)$, because the primed translation $\mathbf{a}_1 = (1, 0, 1, 0)$ is a representative of the primed coset. In this case, all even vertices of the Bravais module have spin up, and the odd vertices have spin down.

108 *Description and symmetry of aperiodic crystals*

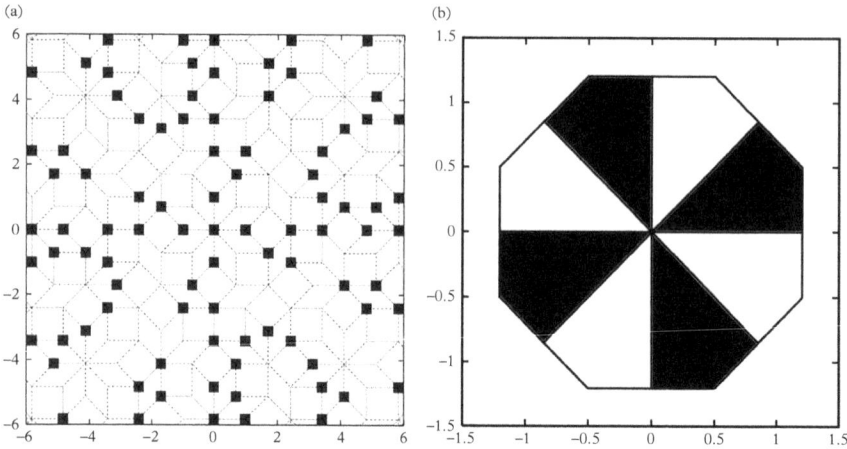

Figure 2.29 (a) *A spin configuration on the standard octagonal tiling with magnetic space group* $P8'mm(8^3mm)$. (b) *The corresponding atomic surface (dark: spin up, light: spin down).*

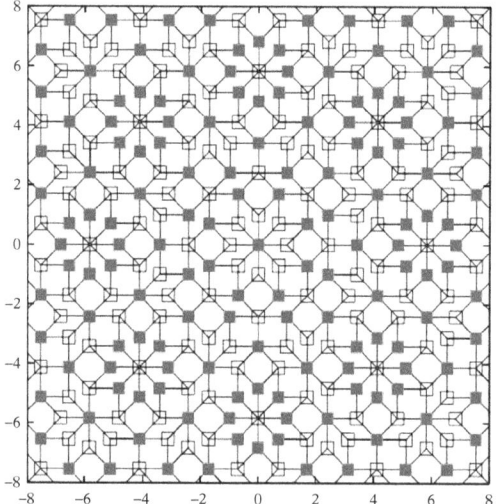

Figure 2.30 *A spin configuration on the standard octagonal tiling with magnetic space group* P_a8mm (8^3mm).

The same groups may be used to describe chemical ordering. In the example of the spin-decorated octagonal atomic surfaces, one may consider two chemical species instead of two spin directions. The operation then is not time reversal, but permutation of colours. The groups to describe these configurations are, nevertheless, isomorphic.

States in a lattice periodic crystal are characterized by a wave vector **k** in the Brillouin zone. In an aperiodic crystal this **k** is no longer a good quantum number. However, the functions may still be described in reciprocal space. The transform of a reciprocal space function under the element $g = \{R|\mathbf{v}\}T^n$ ($n = 0, 1$) of a magnetic group is given by

$$T_g \hat{f}(\mathbf{k}, t) = \exp(2\pi i \mathbf{k}.\mathbf{v}_E) \exp(2\pi i \mathbf{k}_I.\mathbf{v}_I) \hat{f}(R^{-1}\mathbf{k}, (-1)^n t); \quad (2.74)$$

T_g is a symmetry operator if $T_g \hat{f}(\mathbf{k}, t) = \hat{f}(\mathbf{k}, t)$. For a spin distribution the transform reads

$$T_g \hat{\mathbf{S}}(\mathbf{k}, t) = \text{Det}(R) \exp(2\pi i \mathbf{k}.\mathbf{v}_E) \exp(2\pi i \mathbf{k}_I.\mathbf{v}_I) R \hat{\mathbf{S}}(R^{-1}\mathbf{k}, (-1)^n t).$$

2.8 Summary

A convenient way to describe the structure and symmetry of quasiperiodic crystals is by embedding them into a higher-dimensional 'superspace' as lattice periodic structures. Their intersection with the physical space gives the real structure, shifting the physical space parallel to the original one gives another possible realization of the crystal with the same energy.

The Fourier transform of the quasiperiodic structure, and its diffraction pattern, are projections of the corresponding quantities in higher dimensions.

The appearance of lattice periodicity means that the usual crystallographic techniques for the description of structures apply again, although in higher-dimensional space.

The symmetry groups of quasiperiodic structures are superspace groups, higher-dimensional space groups for which the point group can be decomposed into a component in physical space and one in the additional, internal space.

The structure determination reduces to the determination of number and positions of atomic surfaces in the higher-dimensional unit cell, and that of the shape of the atomic surfaces. The number of parameters needed for the description of the shape is, in principle, large. That is the reason why structure determination is still hard. The characterization of the electron density differs only from the periodic case in that the dimension of the unit cell is larger.

....

EXERCISES

2.1. (a) Consider a one-dimensional chain of particles. The basic structure is lattice periodic and has lattice constant a. The chain is modulated with a modulation function, that is equal to $0.1a$ for atoms between nb and $(n+0.5)b$ and $-0.1a$ for atoms between $(n + 0.5)b$ and $(n + 1)b$ (for integer n and irrational b/a. Calculate the structure factor $F(\mathbf{H})$ for this chain within the superspace formalism.

(b) Calculate $F(\mathbf{H})$ also for the commensurate values $b/a = \frac{1}{2}$ and $b/a = 2$.

(c) Draw the two-dimensional embedding for $b/a = \sqrt{2}$.

2.2. (a) A two-dimensional commensurate magnetic structure has as symmetry elements the (anti-)translations $(a,0)'$ and the elements $\{2|0\}$ and $\{m_x|\,0\}'$. Give the symbol for this group.

(b) Give generators for the magnetic superspace groups Pmmm$'(\alpha 00)$ and P$_d$mmm$(\alpha 00)$.

2.3. Consider a two-dimensional square lattice. Inside the unit cell, there are five atoms at the vertices of a regular pentagon. Calculate the structure factor. Is the five-fold symmetry of the filling of the unit cell visible in the diffraction pattern?

2.4. Calculate the structure factor $F(\mathbf{H})$ for a one-dimensional aperiodic chain of rank 2. The lattice in superspace is spanned by the vectors $(1, -\Phi)$ and $(\Phi, 1)$ (the same as for the Fibonacci chain $\Phi = (\sqrt{5} - 1)/2$). This is a modulated chain (a modulated quasicrystal) with atomic surface that is a sinusoidal displacement of the points of the lines of length $(1 + \Phi)$ with displacement along V_E of the form $A \sin(2\pi \Phi r_I)$. Calculate in the superspace the structure factor $F(\mathbf{H}) = F(\mathbf{H}_s)$.

2.5. A two-dimensional aperiodic composite has two subsystems. The basic structure of the first is primitive rectangular with lattice constants a and b, the second is centred rectangular with lattice constants c and $2b$. 1) Determine the Fourier module. 2) Determine the superspace group, first taking the pmm as host structure, then with the cmm structure as host. 3) Show that these superspace groups are non-equivalent, when equivalence is defined as for incommensurate modulated phases.

3
Tilings: mathematical models for quasicrystals

3.1 Model sets

To study the general properties and characteristics of solids one needs a general, mathematical framework. Models for crystals are supposed to carry the essential features, and should be simple enough for general statements.

The solid state consists of atoms, which in the simplest models are represented by points in the physical space. Real crystals have finite size, and the finiteness has important consequences. This is, in particular, true when the size becomes small, as in nanostructures. Here we shall be concerned mainly with bulk properties. For that reason, the models are chosen to have an infinite size. An infinite point set has to satisfy two more requirements. There should be a minimal distance between the points, because atoms cannot come arbitrarily close to each other, and there should be no arbitrarily large gaps. With this requirement one excludes, for example, self-similar fractals. Mathematically, such point sets are called *Delaunay sets* (or Delone sets). Considering infinite model sets has advantages and disadvantages. Infinite systems may have convergence problems when these do not occur for finite systems. And real crystals have a finite size. On the other hand, convergence problems may point to the fact that properties of finite size samples depend critically on their size.

A point set is periodic if there is a translation **a** such that all positions $\mathbf{r}_n + \mathbf{a}$ belong to the set if the $\mathbf{r}_n$ belong to the set. Then multiples of **a** and the sum of two translation vectors **a** and **b** are also translation vectors under which the set is invariant. If the d-dimensional translation vectors form a lattice, that is, the set of translations is spanned by d linearly independent vectors, then the point set is lattice periodic. For that case, the group of all translations in d-dimensional space, modulo the lattice, is a compact set: it may be identified with the unit cell. If there is not such a lattice, then the system is called aperiodic.

One may consider different types of atoms, but to start with we only consider systems of points, representing identical atoms. Then the Voronoi construction partitions the space into non-overlapping regions: the region around a point **r** consisting of all points that are closer to **r** than to any other point of the set is

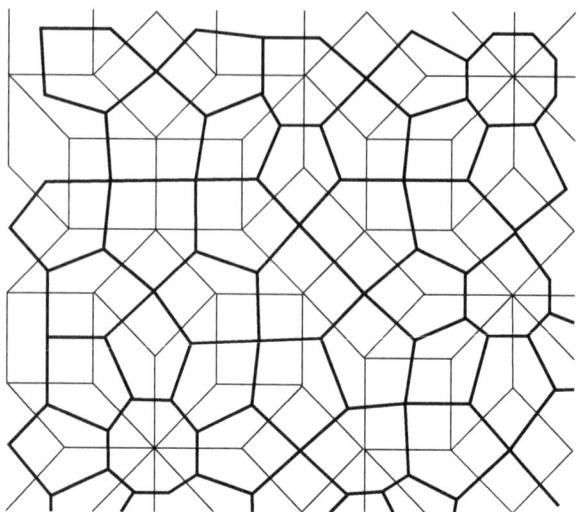

Figure 3.1 *A point set consisting of the vertices of an octagonal tiling, and its Voronoi cells.*

called a *Dirichlet domain* or *Voronoi cell*. If the set is lattice periodic, it corresponds to the *Wigner–Seitz cell*. Any point in space belongs to one cell, or is on the border of a cell. Connecting the centres of neighbouring Voronoi cells gives a second partitioning of the space. It is the dual partitioning, and its cells are called *Delaunay cells* (Fig. 3.1). For example, for a rectangular two-dimensional lattice with points on the vertices both the Voronoi and the Delaunay cells are rectangles (of equal size). For the two-dimensional hexagonal lattice the vertices form a set with a regular hexagon as Voronoi cell and a rhombus with angle $\pi/6$ as Delaunay cell. The Dirichlet domains for a lattice periodic point set are all the same, but for an incommensurate crystal structure they may be all different. If there are several types of atoms, one may consider the Dirichlet domains for one type. Then the other atoms decorate these cells. This becomes especially important if there is a finite number of different types of Voronoi cells, where two cells are of the same type if they have the same shape and the same decoration, just as for the unit cells of a lattice periodic structure. The atomic content of each unit cell is the same in that case.

Lattice periodic crystals also have another special property. The set of all difference vectors $\mathbf{x} - \mathbf{y}$ between two points of the lattice is the lattice itself. The difference set in general does not have this property. The vertices of a tiling give a difference set that is still discrete, that is, around each point there is a sphere without another point. This is no longer true for incommensurate modulated phases, where the difference set is, generally, no longer discrete. A point set which has a discrete difference set is called a *model set*.

Yves Meyer, awarded the Abel Prize 2017, introduced what is called *Meyer sets*. Consider the space $R^D + R^d = R^n$, the sum of two real vector spaces. We consider projections from R^n to R^D (π_E) and from R^n to R^d (π_I). Then consider a compact set Ω in R^d and consider all points x of a lattice Σ in R^n. Then the collection of all points $\pi_E x$ for which $\pi_I x \in \Omega$ forms a Meyer set. (There are some additional conditions such as the condition that both $\pi_I \Sigma$ and $\pi_E \Sigma$ are dense.) Meyer sets are model sets. We have adapted our notation such that it is clear that this is very close to the way the atom positions in an aperiodic crystal are obtained.

For a deeper view into the mathematical properties of quasiperiodic structures we refer to (Baake and Moody, 2000), (Moody, 1997), (Kellendonk *et al.*, 2015) and (Grimm and Baake, 2013). Among the model sets, for periodic and aperiodic crystals, tilings are especially useful. We start with a discussion of these structures.

3.2 Introduction to tilings

A crystal with lattice periodicity can be seen as a filling of space with identical unit blocks, the *unit cells*. The discovery of quasiperiodic crystals has shown that the solid state cannot always be described with such a tiling of the space. Every unit cell of the basic structure becomes different when the modulation is present. Then, the notion of a space filling with identical tiles ceases to be applicable. The discovery of quasicrystals then showed that there are condensed matter systems that, in a first approximation, may be described as a space filling with copies of two or more types of tiles. This gives a connection with a mathematical notion for which a large literature exists, that of *tilings*. We shall start with a brief introduction to the fundamentals of the theory of tilings.

A tiling of the plane is a covering, without gaps or overlaps, of the plane with a family of sets called tiles. Analogously a tiling of the space is a filling of the space with copies of a set of closed sets, also called tiles. The family should be finite or countable. Here we shall require that the tiles are two- or three-dimensional objects, topologically equivalent to a disc, or a sphere, respectively. This means that we can deform the tiles without cutting or gluing, such that they become a disc or a sphere. This excludes lower-dimensional protuberances on or holes in the tiles. The boundaries do not obey special rules, although we shall limit ourselves to polygons and polyhedra. A polygon is a planar object delimited by straight line intervals, the edges, which connect at the corners. A polyhedron is a three-dimensional volume delimited by plane surfaces, the facets. The straight boundaries of the faces are the *edges*, and in their turn the edges connect corners or *vertices*.

Among the tiles there may be congruent ones. So the tiles are copies of a set of *prototiles*. This set of non-congruent prototiles may be infinite, as for a general incommensurate displacively modulated crystal, or finite. If all the tiles are congruent, then the tiling is called *monohedral*. If the set of non-congruent tiles has k elements, then the tiling is *k-hedral*. A lattice periodic crystal gives a monohedral tiling. In that case the tiles are the unit cells or the asymmetric fundamental domains of the space group of the crystal. A general tiling is obtained if one takes the *Dirichlet domains* of the set of points, since this gives a gapless covering with non-overlapping units.

Two tilings are isomorphic if there is a Euclidean transformation which maps every vertex on a vertex, every edge on an edge, etc. The tilings then are the same, up to a rigid motion. The symmetry of a tiling is the group of Euclidean transformations that leave the tiling invariant. For a lattice periodic crystal, that is a crystallographic space group.

Tilings may have the property of local isomorphism. Then every patch (a simply connected domain) of the tiling is found elsewhere in the tiling infinitely many times. In general, the distances between the occurrences of the patches increase strongly when the patch size grows. The Fibonacci chain is an example of an aperiodic chain with the local isomorphism property. Each string of L's and S's that occurs somewhere, occurs infinitely many times in the chain. Two tilings are *locally isomorphic* if every patch in one tiling occurs infinitely many times in the other with a minimum frequency. This means that for a radius R_1 there is a radius R_2 such that every patch with radius R_1 occurs at least once in every sphere with radius R_2. This gives one possible equivalence relation between tilings. Another equivalence relation between tilings is the *mutual local derivability* (MLD). Two tilings are MLD related if by a local shift of the vertices of one tiling one obtains the vertices of the other. A third equivalence relation is connected with symmetry. Two tilings are *symmetry equivalent* if there is a Euclidean transformation such that this transformation applied to one tiling gives a tiling that is locally isomorphic with the second tiling.

We have already seen the simplest aperiodic tiling, the one-dimensional *Fibonacci chain*, consisting of a quasiperiodic string of intervals, of length 1 and Φ. Its vertices are at positions

$$x_n = x_0 + n(3\Phi - 1) + (\Phi - 1)\operatorname{Frac}(n\Phi + \psi),$$

where ψ is a phase, x_0 the origin, and $\operatorname{Frac}(x)$ the value $x \pmod 1$. The intervals have either length 1 or length Φ, and form a quasiperiodic array of rank 2.

There are many ways to construct aperiodic tilings. In this chapter we want to give a brief review of the various techniques. These include substitutions, inflation and deflation methods, matching rules, the cut-and-project method, the grid method, and other techniques implicitly or explicitly using the superspace.

It is simple to construct quasiperiodic tilings in two dimensions. Consider a square lattice. In each square unit cell one draws one of the two diagonals. Which one depends on whether the value of a lattice periodic function on the plane,

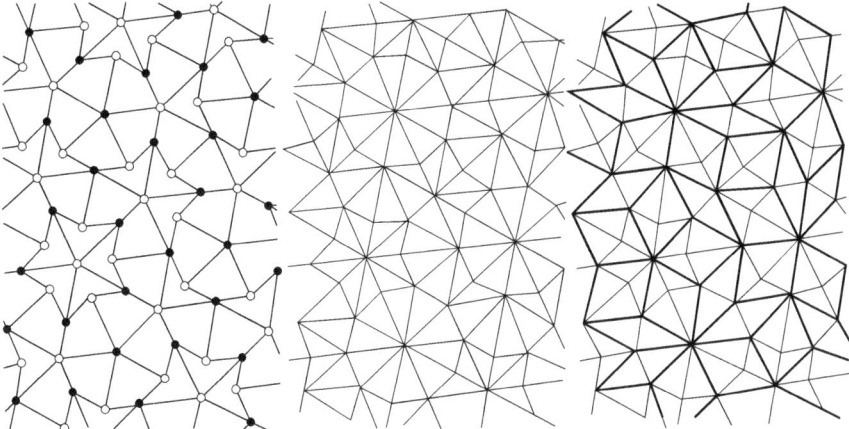

Figure 3.2 *The Penrose kite-and-dart tiling, the associated Robinson tiling, and the rhombic tiling (thick lines).*

taken at the centre of the square, is positive or negative. If the periodicity lattice is incommensurate with the square lattice, the structure is quasiperiodic. The tiling is monohedral and quasiperiodic, but with the same tiles a completely periodic tiling or a completely random tiling may be obtained.

The problem of finding tiles such that they enforce aperiodicity is more difficult. In 1966 Berger discovered such a set of tiles. However, this set has 20,426 elements. Later the number of tiles that is required could drastically be reduced. In 1971 *Robinson* found a set of six tiles producing an aperiodic tiling. His prototiles are modifications of squares, with notches and protuberances added. Then, in 1974, *Penrose* discovered a series of such sets. His first set still has six tiles, but the other two need only two tiles (see, e.g., (Grünbaum and Shephard, 1987)). One set consists of two tetragons, one in the form of a dart (two angles equal to 36°, one to 72°, and one to 216°), and one in the form of a kite (a tetragon with three 72° angles and one 144° angle). The other set consists of two rhombi: a 'fat' one with a 72° angle, and a 'skinny' one with a 36° angle (cf. Fig. 3.2).

At a vertex of the tiling a number of edges meet. Because all angles are multiples of 36°, the vertices may be written as

$$\mathbf{r_n} = \sum_{j=0}^{4} n_j \mathbf{e}_j, \quad \text{with} \quad \mathbf{e}_j = (\cos(2\pi j/5), \sin(2\pi j/5)).$$

The vertices belong to a vector module with five-fold symmetry, the Bravais module. The resulting tiling has a Fourier transform with five-fold symmetry. A *Bravais module*, in general, is a set of points in direct space, which can be written as

$$\mathbf{v} = \sum_{i=1}^{n} n_i \mathbf{e}_i.$$

In contrast to the Robinson tiling the *Penrose tiling* does not have tiles enforcing quasiperiodicity by their shape alone. However, if one decorates the tiles in a certain way, and requires that the decoration of the tiles coincides on the edges, one has matching rules. For example, the edges may be decorated with arrows of two types. Then coinciding edges should have the same type and direction of the arrows. For some tilings (such as the Penrose tiling) matching rules have been found such that a tiling with the decorated tiles satisfying the *matching rules* are necessarily aperiodic.

In the case of the rhombic Penrose pattern, matching rules can be formulated using a marking of the edges. Each tile has single or double arrows along the edges, tiles of the same type have identical arrowing. Neighbouring tiles have edges in common. The matching rules require that coinciding edges have the same type and direction of the arrows. This enforces quasiperiodicity of the tiling. Matching rules that enforce a single local isomorphism class are called *perfect matching rules*. An example is the case of the Penrose tiling (cf. Fig. 3.3).

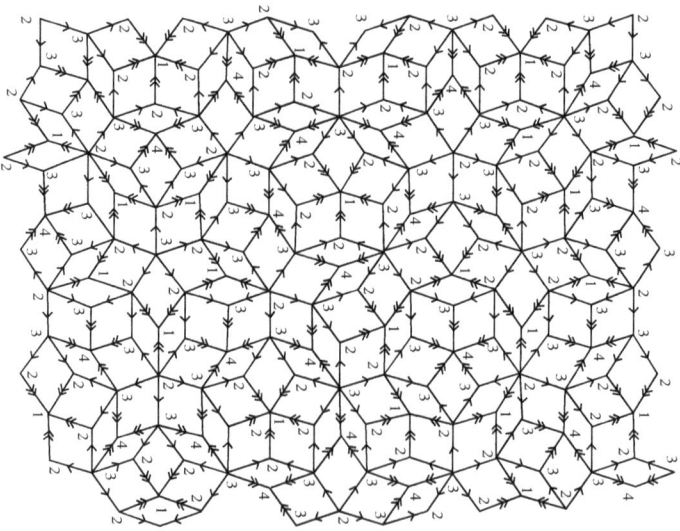

Figure 3.3 *The rhombic Penrose pattern, marked with arrows. The matching rules require that tiles touch via edges with the same marking. The vertices are labelled with the integers 1, 2, 3, and 4. Going from one vertex to the next via a basis vector* $\mathbf{e}_i$ $(i = 1, \ldots, 5)$ *this integer increases by 1, and via* $-\mathbf{e}_i$ *decreases by 1. The numbers remain nevertheless limited to the four values 1, 2, 3, and 4.*

3.3 Substitutional chains

3.3.1 Substitutions with an alphabet

There are several methods of constructing aperiodic tilings. One of them is by *substitution*. We take again as example the Fibonacci chain. One starts with a seed, consisting of a word in the letters S and L. Not every seed is allowed, but one might start with the letter S. Then every letter S is substituted by a letter L, and every letter L by a pair LS. This procedure is iterated. The limit is a semi-infinite sequence of S's and L's. Then for every letter L an interval of length 1 is taken on the real line, and for every S an interval of length Φ. This results in the *Fibonacci chain*. Only a word of length ℓ that occurs somewhere in the sequence constructed before is allowed as a seed. If one starts with a seed of length 3 or more, and one keeps one letter always at position 0, then an aperiodic chain is obtained which is infinite in both directions. For example, from the seed LSL follows

$$\ldots LSLLSLSLLSLSLLSLLSLSLLSLSL \ldots$$

The number of letters in the n-th step is the Fibonacci number F_n, which satisfies $F_{n+1} = F_n + F_{n-1}$. The procedure was proposed by the Italian Fibonacci from Pisa. It results in a chain of infinite length. In the construction, the number of letters L and S increases. If N_L is the number of L's and N_S the number of S's, the numbers after the n-th iteration become

$$N_L^n = N_S^{n-1} + N_L^{n-1}, \quad N_S^n = N_L^{n-1}.$$

Starting from the letter S, that is, $N_L^1 = 0$, $N_S^1 = 1$, the ratio of the numbers N_L and N_S becomes

$$\lim_{n \to \infty} \frac{N_L^n}{N_S^n} = \frac{N_S^{n-1} + N_L^{n-1}}{N_L^{n-1}} = \frac{N_S^{n-1}}{N_L^{n-1}} + 1. \quad (3.1)$$

In the limit this gives the equation $x = 1/x + 1$ with solutions $x = (1 \pm \sqrt{5})/2 = \tau$ or $-\Phi$. The ratio of the number of tiles of both types is then equal to τ, the inverse golden mean. The values of x are the eigenvalues of a matrix S, the substitution matrix, defined by the relation

$$\begin{pmatrix} N_L^n \\ N_S^n \end{pmatrix} = S \begin{pmatrix} N_L^{n-1} \\ N_S^{n-1} \end{pmatrix}, \quad \text{with} \quad S = \begin{pmatrix} 1 & 1 \\ 1 & 0 \end{pmatrix}.$$

If $N_L^n = a$ and $N_S^n = b$, then (a, b) goes to $(a+b, a)$ and in turn to $(2a+b, a+b)$. The total number satisfies $N_{n+1} = 3a + 2b = N_n + N_{n-1} = (2a+b) + (a+b)$. If $a = 0$ and $b = 1$, the total number $N_n = F_n$, where the Fibonacci numbers F_n satisfy $F_{n+1} = F_n + F_{n-1}$ with $F_1 = F_2 = 1$.

The procedure can be generalized by taking an alphabet of p letters $A_1,\ldots,A_p$ and expressing substitution rules: A_i is replaced by the 'word' $W_i(A_1,\ldots,A_p)$. The i-th row in the substitution matrix is given by the number of A_js ($j=1,\ldots,p$) in the word W_i. The matrix has non-negative integer entries, and p eigenvalues λ_i. Notice that the matrix does not depend on the precise form of the substitutions, only on the numbers of letters of type j in the substitution of A_i. The resulting sequence is quasiperiodic if all eigenvalues are between 0 and 1, in absolute value, except one which is larger than 1. This is the *Pisot condition*. The highest eigenvalue, in absolute value, is a *Pisot number*, or Pisot–Vijayaraghavan number, if it is the only eigenvalue larger than unity in absolute value. This and many other properties of substitution rules are treated in Queffélec (1987). Note that the order of the letters in the words W_i does not play a role for the substitution matrix S, and therefore for the question of whether the resulting chain is quasiperiodic or not. The atomic surfaces, however, for which the role is described in Section 3.4, do depend on the letter order, and may be fractal (cf. Fig. 3.4). Such fractal atomic surfaces (or acceptance domains) were also found for the dodecagonal square-triangle tiling in Baake et al. (1992).

A sequence that has been used in the study of quasicrystals is the *octonacci chain* or *Pell chain*. It is created with the two-letter substitution rule S → L, L → LLS. The eigenvalues are $1 \pm \sqrt{2}$. Such substitution chains have been used as well for the construction of 2D aperiodic tilings. A double-Fibonacci chain is formed by the points (x_n, y_m) in the plane, where x_n and y_n are the points of a Fibonacci chain. A so-called labyrinth lattice is formed by the points (x_n, x_m) ($n+m$ even), where points x_n form an octonacci chain.

3.3.2 Embedding of substitutional chains

Consider a substitution rule based on two letters, A and B, with substitutions $W_1(A,B)$ and $W_2(A,B)$, which are words in the letters. We iterate the substitution. The results after the n-th iteration are $A_n = W_1(A_{n+1}, B_{n+1})$ and $B_n = W_2(A_{n-1}, B_{n-1})$ with $A_0 = A$ and $B_0 = B$. Consider then a two-dimensional square lattice, with basis vectors $\mathbf{e}_1$ and $\mathbf{e}_2$. The word A_n corresponds to a finite path on this lattice. Start at the origin, follow $\mathbf{e}_1$ if the first letter is an A, and $\mathbf{e}_2$ if it is a B, and repeat this. In the limit of n tending to infinity, this gives a semi-infinite path. The average slope (if it exists) is determined by the substitution matrix. In a substitution every A is replaced by S_{11} As, and S_{21} Bs, and every B by S_{21} As, and S_{22} Bs. A vector $(N_A, N_B)_n$ corresponding to N_A As and N_B Bs in the n-th step gives $(S_{11}N_A + S_{12}N_B)$ As and $(S_{21}N_A + S_{22}N_B)$ Bs for the next step. These are the elements of the column vector which is the substitution matrix multiplied by the column vector $(N_A, N_B)_n^T$. If the substitution has the Pisot property, one of the eigenvalues is larger than unity. The eigenvector of this eigenvalue has elements for which the ratio is the ratio of the numbers of As and Bs in the limit, and determines the average slope of the path on the lattice. The eigenspace can then be considered as the subspace V_E and the other eigenspace is V_I. Projection of the path on V_I

gives the atomic surface. Placing a copy of this projected set at each vertex of the lattice gives the embedded structure, corresponding to a sequence of long and short intervals.

The fluctuations around the average slope can also be calculated and these turn out to depend on the precise substitution, on the words W_1 and W_2, and not only on the matrix S. Therefore, the atomic surface depends specifically on the substitution rule.

For the *Fibonacci chain*, the substitution matrix is given by

$$\begin{pmatrix} 1 & 1 \\ 1 & 0 \end{pmatrix},$$

for which the Pisot eigenvalue is τ with eigenvector $(1, \Phi)$. The projection of the two edges on this eigenspace have lengths $\cos(\alpha)$ and $\sin(\alpha)$, where $\tan(\alpha) = \Phi$. Therefore, the length ratio $\ell_A/\ell_B = \tau$.

Also the substitution rule for the octonacci chain has a Pisot eigenvalue. As given before, it has the binary rule $W_1(A, B) = AAB$, $W_2(A, B) = A$. Its substitution matrix

$$S = \begin{pmatrix} 2 & 1 \\ 1 & 0 \end{pmatrix}$$

has eigenvalues $1 \pm \sqrt{2}$. The Pisot eigenvalue is $1 + \sqrt{2}$. The intervals have length 1 and $1 + \sqrt{2}$, respectively, and the two-dimensional lattice has basis vectors $(1, -1 - \sqrt{2})$ and $(1 + \sqrt{2}, 1)$. The ratio $\sqrt{2} - 1$ is called the silver mean.

The substitution $W_1(A, B) = ABBA$, $W_2(A, B) = AA$ has substitution matrix with eigenvalues twice as big as those of the Fibonacci case. Then both eigenvalues are larger than 1 in absolute value, which means that it is a non-Pisot substitution, and that the resulting chain is not quasiperiodic. In the embedding, a lattice point will move away from the internal space, and from the physical space as well.

The embedding of a substitutional chain, and its corresponding atomic surface, are not determined by the substitution matrix only. Chains with the same matrix share the Pisot property, if one has it, but the atomic surfaces may be different. The square of the Fibonacci substitution is $W_1(A, B) = AAB$, $W_2(A, B) = AB$ and leads to the same atomic surface as the Fibonacci chain, but $W_1(A, B) = BAA$, $W_2(A, B) = AB$ has the same matrix and a very different atomic surface. It is a fractal. This shows immediately that an atomic surface is not necessarily a simple object. For example, the three-letter substitution $W_1(A, B, C) = B$, $W_2(A, B, C) = C$, $W_3(A, B, C) = AC$ gives a quasiperiodic chain but its *atomic surface* is a fractal in the two-dimensional internal space. It is shown in Fig. 3.4.

3.3.3 Tilings by substitution

The substitution technique may also be applied to sets of tiles. Consider first again the Fibonacci chain. One may obtain it starting from two intervals, of lengths

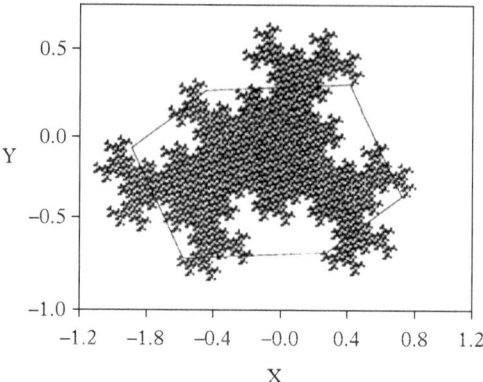

Figure 3.4 *The atomic surface for the substitution $A \to B, B \to C, C \to AC$ is a fractal. The thin straight lines indicate the projection of the three-dimensional unit cell in superspace. It has the same area as the projection of the atomic surface.*

Φ and unity, respectively. There are two substitution rules. The short interval is replaced by a long interval and the long interval by the sum of a long and a short interval. The substitution matrix in this case is given by

$$M = \begin{pmatrix} 1 & 0 \\ 1 & 1 \end{pmatrix},$$

with eigenvalues $-\Phi$ and $1 + \Phi$. This has the Pisot property, and the resulting chain is quasiperiodic, as we know already. The technique consists of two steps. In the first step the length is decreased by a factor (*deflation*), and in the second step the original length scale is recovered (*inflation*).

♯ Similar constructions are possible for higher-dimensional tiles as well. Consider as example two tiles of the shape of an isosceles triangle, with unit length for the equal edges, and with top angle equal to $\pi/10$ or $3\pi/10$ (Fig. 3.5). There are several ways to partition the tiles such that the pieces are of the same shape as the original ones, but scaled down with a factor. One way is the following. If A, B, C are the corners of the acute triangle, and F, G, H those of the obtuse triangle, the substitution rules are as follows. Partition the acute triangle into three triangular pieces with corners $A' = A, B' = C, C' = \Phi C + \Phi^2 B$, $A'' = \Phi C + \Phi^2 B, B'' = B, C'' = \Phi A + \Phi^2 B$, and $F' = C, G' = A, H' = \Phi C + \Phi^2 B$, and the obtuse triangle into two triangles with corners $A' = G, B' = F, C' = \Phi H + \Phi^2 F$, and $F' = H, G' = \Phi H + \Phi^2 F, H' = G$. Then the substitution matrix is

$$M = \begin{pmatrix} 2 & 1 \\ 1 & 1 \end{pmatrix},$$

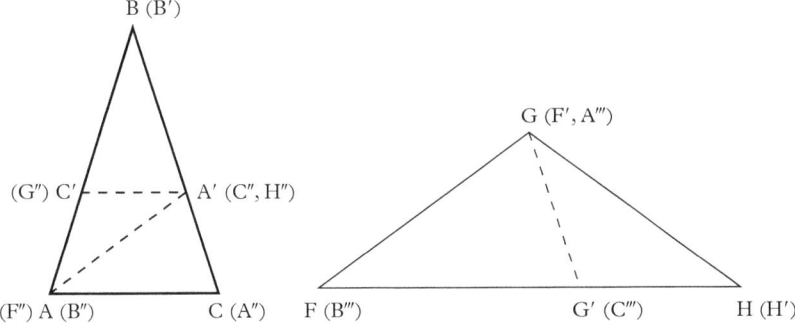

Figure 3.5 *An example of a substitution rule for tilings: the deflation step;* $ABC = A'B'C' + A''B''C'' + F'G''H''$, $FGH = F'G'H' + A'''B'''C'''$

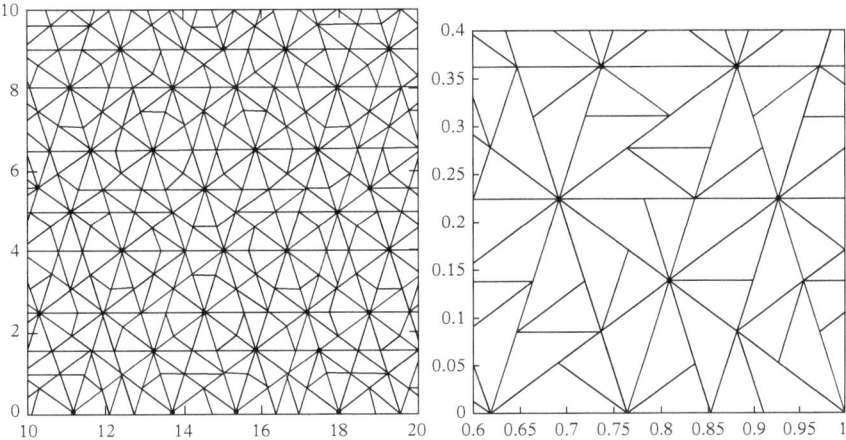

Figure 3.6 *Left: The triangle tiling which follows from the substitution rules in Fig. 3.5. Right: A non-edge-to-edge tiling with five-fold symmetry.*

with eigenvalues Φ^2 and Φ^{-2}. This is again a Pisot substitution. After each partitioning the length scale is multiplied by $\Phi^{-1} = 1 + \Phi$, in agreement with the eigenvalues of the substitution matrix. (The square stems from the dimensionality 2.) The resulting structure is a quasiperiodic tiling, with triangles (Fig. 3.6). In this case the tiling is edge to edge. That means that no edges end on another edge apart from the beginning or the end. This tiling is sometimes called the *Tübingen Triangle Tiling*, because it has been studied in detail in the group in Tübingen (P. Kramer and co-workers). Another partitioning leads to the Robinson triangular tiling, which is a tiling also obtained from the Penrose kite-and-dart tiling by cutting the kites and darts into isosceles triangles. This is also an edge-to-edge tiling. Another partitioning yields a non-edge-to-edge tiling (see Fig. 3.6).

3.4 Aperiodic tilings

3.4.1 Construction of aperiodic tilings

3.4.1.1 The section method

A quasiperiodic tiling may also be obtained using the technique of superspace, as discussed in previous chapters. In a space of dimension n, equal to the rank of the Fourier module, one distinguishes a subspace, called the physical space, and an additional subspace, called the internal space. In the n-dimensional space there is a lattice, with a reciprocal lattice for which the projection on the physical space is the Fourier module. In the vertices of the lattice flat atomic surfaces parallel to the internal space are attached. If the atomic surfaces are the projection of the unit cell of the lattice, the intersections of the atomic surfaces with the physical space are the vertices of a tiling in the physical space.

We summarize here the procedure, and exemplify it again with the Fibonacci chain. The lattice in this case is two-dimensional, and has basis vectors $(1, -\Phi)$ and $(\Phi, 1)$, written with respect to a split basis, with one vector in V_E and one in V_I. The atomic surfaces are straight line intervals of length $\tau = 1 + \Phi$. These intersect the physical space at the points

$$x_n = x_0 + n(3\Phi - 1) + (\Phi - 1)\,\text{Frac}(n\Phi + \psi),$$

the positions of the Fibonacci chain (Eq. 2.46).

3.4.1.2 The grid method

The grid method was developed by de Bruijn for the construction of a Penrose tiling (de Bruijn, 1981).

The grid method may be derived from the section method if the atomic surfaces are flat, parallel to the internal space and congruent to the projection of the unit cell in superspace. Suppose the n-dimensional lattice has the translation vectors $\mathbf{e}_i$ ($i = 1, \ldots, n$). Then the vertices of the lattice are given by

$$\mathbf{r_n} = \sum_{j=1}^{n} (n_j + x_j) \mathbf{e}_j, \tag{3.2}$$

where $\sum_j x_j \mathbf{e}_j$ is the position of the origin of the lattice. The atomic surfaces are the projection of the unit cell attached to the vertices $\mathbf{r}_n$. If the atomic surface intersects the D-dimensional physical space, the intersection point is a vertex of the tiling. The atomic surface at $\mathbf{r}_n$ intersects the physical space if and only if the unit cell $\mathbf{n}$ intersects V_E. Consider the net plane of the lattice with constant n_i for some value of i. This net plane intersects V_E in points given by

$$\mathbf{r} = (n_i + x_i)\mathbf{e}_i + \sum_{k \neq i} x_k \mathbf{e}_k \quad \text{with } \pi_I \mathbf{r} = 0. \tag{3.3}$$

This gives a line if $D = 2$ and a plane if $D = 3$. In general, it is a manifold of dimension $D - 1$. These lines or planes are denoted by L_i. The intersection of D such manifolds is a point. We suppose here that in this point there is not another manifold. This means that every intersection point is the intersection of exactly D manifolds L_i. This is called a non-singular intersection point. The intersection point has $d = n - D$ other lattice coordinates, which are not integer. If the intersection point is $\mathbf{p}$ (in the physical space), the coordinates x_i with $i = i_1, \ldots, i_D$ are $n_i + x_i$, and the others are given by the scalar product of $(\mathbf{r}, 0)$ with $\mathbf{e}_j^*$ ($j \neq i_1, \ldots, i_D$), that is,

$$\mathbf{p} \cdot \mathbf{e}_{jE}^* = n_j + x_j. \tag{3.4}$$

This means that the unit cell with lattice coordinates $n_1, \ldots, n_D$ intersects the physical space, and consequently that there is a tiling vertex at

$$\pi_E \sum_{j=1}^{n} (n_j + x_j)\mathbf{e}_{jE}.$$

Moreover, because the coordinates n_j with $j = i_1, \ldots, i_d$ change by unity at the manifolds L_j, this means that there are also tiling vertices at positions $\sum_j (m_j + x_j)\mathbf{e}_{jE}$ with $m_j = n_j$ or $= n_j - 1$ for all $j = i_1, \ldots, i_D$. This gives a rhombus or rhombohedron with origin at x_j (Eq. 3.4).

As a first example consider the Fibonacci chain (Fig. 3.7). The two basis vectors have components in physical and internal space given by

$$\mathbf{e}_1 = (1, -\Phi), \quad \mathbf{e}_2 = (\Phi, 1).$$

The reciprocal basis is given by $\mathbf{e}_i^* = \mathbf{e}_i/(2 - \Phi)$. The net planes are given by

$$(n_1 + x_1)\mathbf{e}_1 + t\mathbf{e}_2, \quad \text{and} \quad (n_2 + x_2)\mathbf{e}_2 + t\mathbf{e}_1.$$

These intersect the physical space in points $L_1 = (n_1 + x_1)(2 - \Phi)$ and $L_2 = (n_2 + x_2)(1 + 2\Phi)$, respectively. These form two periodic chains with lattice constants $(2 - \Phi)$ and $(1 + 2\Phi)$, respectively. We assume that no points coincide. A point of the first chain has coordinates $n_1 + x_1$ and $L_1\Phi/(2 - \Phi) = n_2 + x_2$ (n_2 is an integer). This point gives vertices at $x = (n_1 + x_1) + (n_2 + x_2)\Phi$ and $y = x - 1$. A point of the second chain has coordinates $n_1 + x_1 = L_2/(2 - \Phi)$ and $n_2 + x_2$, and gives vertices at $x = (n_1 + x_1) + (n_2 + x_2)\Phi$ and $y = x - \Phi$. The vertices are

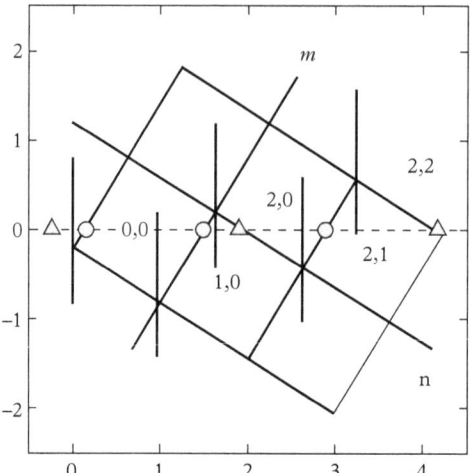

Figure 3.7 *The grid method for the Fibonacci chain. The circles and triangles in the physical space are the points L_i. The interval between two subsequent points belongs to a unit cell in which the atomic surface intersects the physical space. It is given by the coordinates i, j. The lines m and n are $(n_1 + x_1)\mathbf{e}_1 + t\mathbf{e}_2$ and $(n_2 + x_2)\mathbf{e}_2 + t\mathbf{e}_1$.*

the endpoints of the tiles that cover the real line without overlap or gap. The 'grid lines' are in this case the points L_1 and L_2.

As a second example consider a tiling with eight-fold symmetry, called the Ammann–Beenker tiling (Fig. 3.8). The four-dimensional superspace has as basis the vectors $\mathbf{e}_j = (\cos(\phi_j), \sin(\phi_j), \cos(3\phi_j), \sin(3\phi_j))$, ($\phi_j = (j-1)\pi/4$, $j = 1, \ldots, 4$). The reciprocal lattice has basis $\mathbf{e}_j^* = \mathbf{e}_j/2$. For each net plane with fixed n_i (for some i) there is a line L_i which is the intersection of the net plane with the physical space. For each pair i, j from 1, 2, 3, 4 there is an intersection point of the lines L_i and L_j. The lines L_i are parallel, have a mutual distance of 2, and are perpendicular to one of the four vectors $\mathbf{e}_{jE}$. If $\mathbf{p}$ is the intersection point the two coordinates x_k with $k \neq i, j$ are given by the scalar product

$$x_k = \mathbf{p} \cdot \mathbf{e}_{kE},$$

and the unit cell that intersects the physical space is given by n_i, n_j, and the two remaining integers n_k ($k \neq i, j$) such that n_k is the largest integer smaller than x_k. The intersection point $\mathbf{p}$ then determines a vertex of the tiling

$$\mathbf{v} = \sum_{j=1}^{4} (n_j + x_j) \mathbf{e}_{jE}.$$

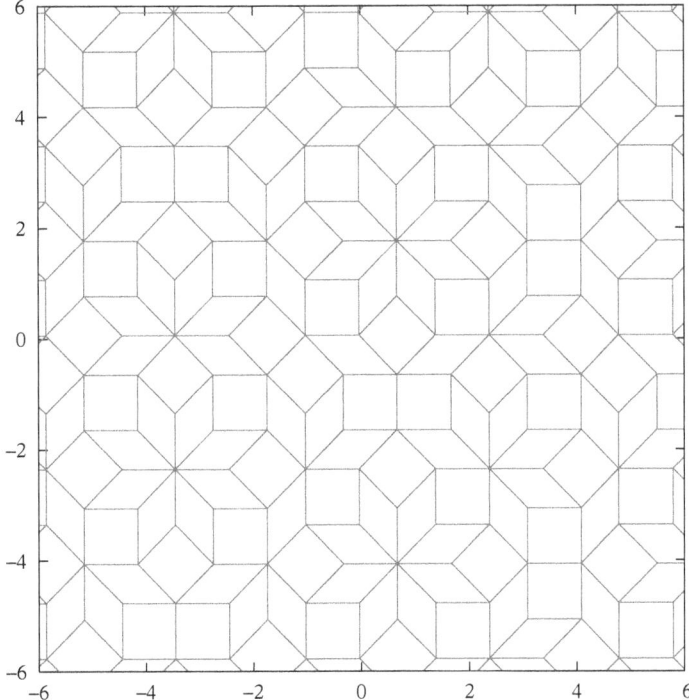

Figure 3.8 *The octagonal (Ammann–Beenker) tiling obtained by the grid method.*

Besides this vertex there are vertices at $\mathbf{v} + \mathbf{e}_{iE}$, $\mathbf{v} + \mathbf{e}_{jE}$ and $\mathbf{v} + \mathbf{e}_{iE} + \mathbf{e}_{jE}$. The four vertices are the vertices of a tile of the tiling, and form a rhombus. There is also another '*Ammann–Beenker tiling*', one with tiles in the shape of a rhomb and an equilateral triangle. It has also octagonal symmetry.

We are now in a position to see the connection between the section and the grid method. The grid lines are the intersections of net planes with the physical space. The intersection point of D of the grid lines corresponds to an atomic surface. Expressing the physical coordinates of this atomic surface in terms of the quantities n_i and x_i gives the position of the corresponding vertex of the tiling. The set of grid lines partitions the space into cells, and each cell corresponds to a unit cell in superspace which intersects the physical space. Singular intersections lead to some more technicalities, which will be neglected here. We can point out that the grid method corresponds to the construction by means of the section method, provided the atomic surfaces are flat and congruent to the projection of the unit cell.

3.4.1.3 The cut-and-project method

Another way to construct aperiodic tilings is inspired by the derivation of the Penrose tiling from a five-dimensional space, as given by de Bruijn. It was developed further in Duneau and Katz (1985). Consider an n-dimensional lattice, and choose a unit cell. Often a (hyper)cubic lattice is taken, but that is not the most general case. The choice of the unit cell is also of importance. For the (hyper)cubic lattice it is convenient to choose the Wigner–Seitz cell, which is then a hypercube, or its dual, but that is congruent with the first in this case. Then choose a D-dimensional hyperplane with an irrational slope with respect to the lattice. The slope determines the incommensurability, in the sense that the latter follows from the rank of the module which is the projection of the reciprocal lattice (also hypercubic) on the irrational hyperplane. We then project the unit cell on a plane that is orthogonal to the chosen irrational plane. If one drags the projection of the unit cell along a hyperplane which is parallel to the chosen irrational hyperplane, this defines an n-dimensional strip which is the product of the projected unit cell and a D-dimensional plane. Now every lattice point is projected parallel to the perpendicular space onto the irrational hyperplane. This then gives the vertices of a tiling.

As an example we consider again the standard example of the Fibonacci chain. Consider a two-dimensional square lattice with basis vectors $\mathbf{a}_1$ and $\mathbf{a}_2$, both of length a. Take a straight line $x(\mathbf{a}_1 + \Phi\mathbf{a}_2)$. (The dimension D is equal to 1 here.) The perpendicular space is given by $y(-\Phi\mathbf{a}_1 + \mathbf{a}_2)$. The projection of the Wigner–Seitz cell is an interval of length L. The projected unit cell, called *window*, *acceptance domain*, or *atomic surface*, dragged along a line parallel to the chosen line, yields a corridor of width L parallel to the 'physical space' V_E (cf. Fig. 3.9). All points inside the strip are projected on V_E. They have distances $a\cos(\alpha)$ and $a\sin(\alpha)$, with $\tan(\alpha) = \Phi$. The ratio of the distances is τ, and the sequence of intervals is that of a Fibonacci chain. A more subtle distinction between atomic surface and acceptance domain is the following. The *atomic surfaces* are the d-dimensional objects in n-dimensional superspace for which the intersections with the physical space give the atomic positions. The *acceptance domain* then is the product of the projection of an atomic surface on the internal space V_I and V_E (Fig. 3.9).

This can be understood starting from a Fibonacci series of letters A and B. If one traces a path on the lattice, by going along $\mathbf{a}_1$ if the letter is A and along $\mathbf{a}_2$ otherwise, the path is never further away from the line with slope Φ than half the length of the projected unit cell.

This procedure is applicable for any n-dimensional lattice, for which a D-dimensional irrational hyperplane is chosen. The elements of the procedure remind very much that of the section method. If there is one atomic surface per unit cell, and the atomic surfaces are flat and perpendicular to the physical space, the two methods are equivalent (cf. Fig. 3.9). A tiling obtained with a single atomic surface in the unit cell (the projection of this cell on the internal space) is called a *canonical tiling*.

A method to construct three-dimensional tilings with icosahedral symmetry was also developed in Kramer (1982) and Kramer and Neri (1984). In the first paper it was shown that such a 3D tiling could be obtained using seven 'elementary cells'.

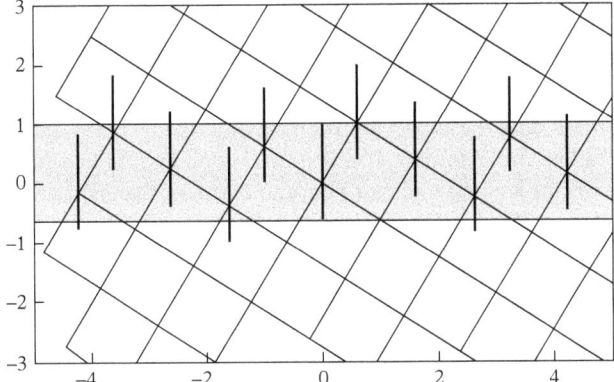

Figure 3.9 *The cut-and-project method, applied to the Fibonacci chain. The lattice points in the strip carry atomic surfaces intersecting the physical space. Their intersections with the physical space are vertices of the tiling.*

In the second paper, the construction is based on the Lie algebra A(5) and can be seen as a generalization of the method used in de Bruijn (1981).

3.4.2 Embedding of tilings

There are several ways to arrive at a tiling using the section method.

In the first place one may look for tilings with a certain D-dimensional point group symmetry. This implies that the Fourier module of the diffraction pattern is invariant under this point group. Then start with an initial vector in reciprocal space. Under the action of the point group the initial vector gives rise to a set of vectors which is invariant under the point group. This set now generates a Fourier module of rank n. Of course, n depends on the initial vector. We suppose that the vectors span the whole physical D-dimensional space. One may now classify the stars of vectors by the rank n. For example, if $D = 2$ and the point group the octagonal group 8mm, with mirrors along x- and y-axes and along the diagonals, then the rank will be 4 if the initial vector lies along a mirror line, and otherwise it will be 8. When the rank is n, the Fourier module is the projection on the physical space of an n-dimensional reciprocal lattice Σ^*. For the octagonal example of rank 4, this is unique up to a scale factor in the internal space. Then the direct lattice Σ is determined. The vertices of a tiling then may be obtained by attaching an atomic surface to each node of Σ, parallel to the internal space. A convenient choice of atomic surface is the projection of a unit cell on V_I. Now, a unit cell is not uniquely determined. Therefore, again a choice has to be made. A convenient choice is that of the Wigner–Seitz cell of the lattice or its dual. These are the Voronoi and the Delaunay (or Delone) cells, respectively. Because the atomic surfaces are parallel, the intersection points with V_I form a discrete set, with a minimal distance. It is even 'relatively dense': there are no arbitrarily small nor arbitrarily large holes. Taking these points as vertices yields a tiling. The number of different tiles depends on the symmetry, and the embedding.

As an example take the two-dimensional point group 8mm, generated by an eight-fold rotation and a mirror. It is a non-crystallographic point group in the plane. Take a vector **k** along a mirror line, and consider the eight vectors obtained from the starting vector by the action of the point group. They span a module of rank 4, because a rotation over π gives each vector a minus sign, and the vectors of such a pair are dependent. For convenience we take vectors of unit length. Then a basis for the module is given by the four vectors $\mathbf{a}^*_{j+1} = (\cos(\pi j/4), \sin(\pi j/4))$ with $j = 0, \ldots, 3$. The action of the point group is given by the action of the eight-fold rotation R_1 and the mirror R_2. In matrix form, where $R\mathbf{a}^*_j = \sum_{m=1}^{4} \Gamma(R)^*_{mj} \mathbf{a}^*_m$, one gets

$$\Gamma^*(R_1) = \begin{pmatrix} 0 & 0 & 0 & -1 \\ 1 & 0 & 0 & 0 \\ 0 & 1 & 0 & 0 \\ 0 & 0 & 1 & 0 \end{pmatrix}; \quad \Gamma^*(R_2) = \begin{pmatrix} -1 & 0 & 0 & 0 \\ 0 & 0 & 0 & 1 \\ 0 & 0 & 1 & 0 \\ 0 & 1 & 0 & 0 \end{pmatrix}. \quad (3.5)$$

These matrices generate a group of 16 elements which form a representation of the point group. There are two irreducible, invariant subspaces for this group of matrices, both two-dimensional. Using standard techniques from group representation theory one may transform the two generators into matrices which are the direct sum of the two components:

$$\Gamma^*(R_1) \to \begin{pmatrix} \cos(\pi/4) & -\sin(\pi/4) & 0 & 0 \\ \sin(\pi/4) & \cos(\pi/4) & 0 & 0 \\ 0 & 0 & \cos(3\pi/4) & -\sin(3\pi/4) \\ 0 & 0 & \sin(3\pi/4) & \cos(3\pi/4) \end{pmatrix},$$

$$\Gamma^*(R_2) \to \begin{pmatrix} -1 & 0 & 0 & 0 \\ 0 & 1 & 0 & 0 \\ 0 & 0 & -1 & 0 \\ 0 & 0 & 0 & 1 \end{pmatrix}.$$

In other words, the first generator acts as a rotation over $\pi/4$ in one subspace (identified with the physical space V_E), and as a rotation over $3\pi/4$ in the other (V_I). Basis vectors of the sum space with respect to two orthonormal vectors in V_E and two orthonormal vectors in V_I are

$$\mathbf{a}^*_1 = (1, 0, 1, 0), \quad \mathbf{a}^*_2 = \frac{1}{2}(\sqrt{2}, \sqrt{2}, -\sqrt{2}, \sqrt{2}),$$

$$\mathbf{a}^*_3 = (0, 1, 0, -1), \quad \mathbf{a}^*_4 = \frac{1}{2}(-\sqrt{2}, \sqrt{2}, \sqrt{2}, \sqrt{2}).$$

These four vectors are of equal length and mutually perpendicular. Therefore, they span a hypercubic lattice in four dimensions. The corresponding direct lattice Σ is spanned by four vectors $\mathbf{a}_i = \frac{1}{2}\mathbf{a}_i^*$. If one takes as unit cell the Wigner–Seitz cell of the lattice, its projection on V_I is a regular octagon. When one attaches such an octagon to each node of the lattice, parallel to V_I, then the intersections with V_E form the vertices of the Ammann–Beenker tiling, with squares and 45° rhombi as tiles.

Closely related is the procedure where one starts with a tiling. Suppose we can calculate the diffraction pattern for this tiling. If the tiling is quasiperiodic, the (sharp) spots of the diffraction pattern fall on a Fourier module of rank n. In the same way as in the previous case, the Fourier module determines a reciprocal lattice Σ^* and a direct lattice Σ. The vertices of the tiling are now points in the subspace V_E of the n-dimensional embedding space. Each vertex corresponds to a point in the unit cell of Σ. Its coordinates are the fractional parts of the scalar product of the vertex vector $(\mathbf{r}, 0)$ and the corresponding basis vector of Σ^*. The closure of this set in the unit cell forms an atomic surface, or a set of atomic surfaces. As an example, take our standard example, the Fibonacci chain. Its positions are given by a formula. The diffraction spots are at positions $(m + n\Phi)/(2 - \Phi)$, which are projections of a two-dimensional reciprocal lattice. The vertices correspond each to a point in the two-dimensional unit cell, and form a line interval parallel to V_I if all limit points are taken. This then is the atomic surface.

A third method starts with an n-dimensional lattice Σ. Suppose the holohedry is the group H. If the dimension of the physical space is D, then H may occur only as symmetry group for an embedded structure if it leaves a D-dimensional space invariant which can then be identified with physical space. In general, this is not the case, and then one has to look for a subgroup G of H which satisfies this condition. As an example we consider a lattice in four dimensions, which in Janssen et al. (1999) is the lattice of the family 21. It has a holohedry of order 240, but this is an irreducible group: there is no invariant, proper subspace. A subgroup G which leaves a two-dimensional subspace invariant is the group $10mm(10^3mm)$, generated by a ten-fold rotation and a two-fold rotation. The whole four-dimensional space is the sum of two invariant subspaces, which are orthogonal to each other. The ten-fold rotation acts as a rotation over $\pi/5$ in one invariant space, and as a rotation $2\pi/5$ in the orthogonal complement. The two-fold rotation acts as a mirror in each of the two subspaces. (For an explanation of the symbols for higher-dimensional point and space groups, see Appendix A.)

For the invariant D-dimensional subspace there is a projection operator π_E, and a second operator π_I such that the sum is the unit operator. These are the two projection operators on the two complementary subspaces, each carrying an irreducible representation, or a sum of irreducible representations, of the group G. The next step is to choose a unit cell for the lattice and project it on the second space via the operator π_I. This projection is the atomic surface, a copy of which is attached to each node of the lattice, parallel to the second space (V_I).

Then the section method gives rise to a tiling of the d-dimensional space (V_E). A convenient choice for the unit cell is the Wigner–Seitz cell. Because this contains all points closer to a given lattice point than to any other, it is invariant under the point group of the lattice. Now we use the fact that the point group of the embedding is reducible, it leaves both the physical space and the internal space invariant. Therefore, the projection of the Wigner–Seitz cell on the internal space is invariant under the internal point group K_I. The corresponding atomic surface, therefore, has the full point group symmetry.

♯ In the example of the group $G = 10\text{mm}(10^3\text{mm})$ consider the lattice of the family 21. It is characterized by the metric tensor

$$g = \begin{pmatrix} 1 & \frac{1}{2} & \frac{1}{2} & \frac{1}{2} \\ \frac{1}{2} & 1 & \frac{1}{2} & \frac{1}{2} \\ \frac{1}{2} & \frac{1}{2} & 1 & \frac{1}{2} \\ \frac{1}{2} & \frac{1}{2} & \frac{1}{2} & 1 \end{pmatrix}. \tag{3.6}$$

The *metric tensor* corresponding to a basis choice $\mathbf{a}_i$ ($i=1,\ldots,n$) of an n-dimensional lattice is a tensor with as element g_{ij} the scalar product of the i-th and j-th basis vector. Therefore, it gives the lengths and the mutual angles of the basis vectors. As said before, the holohedry of this lattice is of order 120. The group $10\text{mm}(10^3\text{mm})$ is the holohedry of a lattice with more than one free parameter, but for convenience the more symmetric lattice may be chosen. The symmetry of the structure we are going to construct will indeed have only symmetry G because of the filling of the unit cell. A realization of the lattice is given by four basis vectors. They are given here in orthogonal components, two in V_E and two in V_I. A possible choice is

$$\mathbf{a}_1 = (d_1, d_2, d_3, d_4)$$
$$\mathbf{a}_2 = (d_3, d_4, d_1, -d_2)$$
$$\mathbf{a}_3 = (d_3, -d_4, d_1, d_2)$$
$$\mathbf{a}_4 = (d_1, -d_2, d_3, -d_4),$$

with $d_1 = \cos(3\pi/5)$, $d_2 = \sin(3\pi/5)/\sqrt{5}$, $d_3 = \cos(4\pi/5)$, and $d_4 = \sin(4\pi/5)/\sqrt{5}$. However, this basis is not the basis with shortest vectors, and the projection of the parallelepiped spanned by the vectors is not the projection of the Wigner–Seitz cell. Instead choose a new basis with basis vectors

$$\mathbf{b}_1 = \mathbf{a}_2 - \mathbf{a}_1, \quad \mathbf{b}_2 = \mathbf{a}_2, \quad \mathbf{b}_3 = \mathbf{a}_3, \quad \mathbf{b}_4 = \mathbf{a}_3 - \mathbf{a}_4. \tag{3.7}$$

Now the lengths of the basis vectors are equal. The corresponding tiling is given in Fig. 3.10. The tiling has five-fold symmetry (even ten-fold). As tiles one recognizes a star, a boat, a pentagon, and an acute rhombus. Its Fourier module is given by the projection of the four-dimensional reciprocal lattice Σ^*. Choosing as basis for

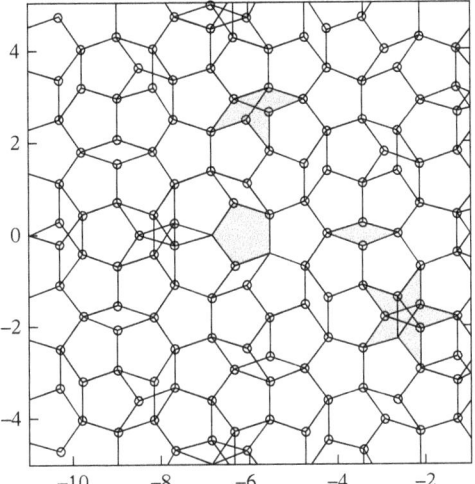

Figure 3.10 *A two-dimensional tiling following from a lattice belonging to the four-dimensional family 21 (Janssen et al., 1999). The tiles are pentagons, stars, narrow rhombi, and boats. Examples of the four types are given in grey.*

Σ^* the reciprocal basis corresponding to the metric tensor g (Eq. 3.6) the basis for the Fourier module is

$$\frac{2}{\sqrt{5}}(\cos(2\pi/5), \sin(2\pi/5)), \quad \frac{2}{\sqrt{5}}(\cos(4\pi/5), \sin(4\pi/5)),$$

$$\frac{2}{\sqrt{5}}(\cos(4\pi/5), -\sin(4\pi/5)), \quad \frac{2}{\sqrt{5}}(\cos(2\pi/5), -\sin(2\pi/5)),$$

and this is the same as the module of the Penrose tiling. This is the reason we took the reciprocal basis corresponding to g, and not the basis which is reciprocal to the basis of Eq. 3.7. The new reciprocal basis shows more clearly the rotation symmetry.

In the superspace, *i.e.* the sum of V_E and V_I, the embedded tiling consists of atomic surfaces parallel to V_I. If there is one atomic surface per unit cell the points of the atomic surface in the cell $\mathbf{n}$ are given by $\mathbf{n} + \Omega$. Its projection on V_I is $\mathbf{n}_I + \Omega$. The atomic surface gives a vertex if the physical space, which has internal component 0, intersects it, that is, when $0 \in \mathbf{n}_I + \Omega$ or $-\mathbf{n}_I \in \Omega$. This gives a way to represent a tiling on Ω. If $\mathbf{n}_E$ is a vertex, it means that the corresponding $\mathbf{n}_I = {}^\star\mathbf{n}_E = \pi_E\mathbf{n}$ satisfies $-\mathbf{n}_I \in \Omega$. A neighbour $\mathbf{n}_E \pm \mathbf{e}_i$ is also a vertex if $-(\mathbf{n}_I \pm {}^\star\mathbf{e})_i \in \Omega$. The vector ${}^\star\mathbf{e}_i$ is the internal component of the n-dimensional basis vector $(\mathbf{e}_i, {}^\star\mathbf{e}_i)$. So, all neighbours of $\mathbf{n}$ are found as the vertices $\mathbf{n} \pm \mathbf{e}_i$ for which

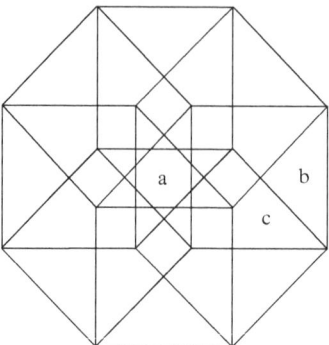

Figure 3.11 *The atomic surface of the octagonal tiling, partitioned into regions corresponding to the coordination. If the projection of a vertex falls into region a, then the tiling vertex has eight neighbours, at positions $\pm \mathbf{e}_i$. When it falls into region b, it is surrounded by three vertices differing from the tiling vertex by $\mathbf{e}_2$, $-\mathbf{e}_1$, and $-\mathbf{e}_4$. In the region c it has four neighbours at $-\mathbf{e}_1$, $\mathbf{e}_2$, $-\mathbf{e}_3$, and $-\mathbf{e}_4$. To each region corresponds a specific surrounding.*

also $-(\mathbf{n}_I \pm {}^\star\mathbf{e}_i)$ belongs to Ω. So Ω may be partitioned into sectors, one for each configuration around a point of the sector. As an example, in Fig. 3.11 the octagonal atomic surface is partitioned into sectors, each corresponding to a certain coordination. ♮

♯ Another example of a tiling corresponding to a higher-dimensional lattice is a *dodecagonal tiling* corresponding to a lattice of family 16 in Table 3 in Janssen et al. (1999) The Fourier module is of rank 4, and a basis may be chosen as

$$\mathbf{a}_1^* = (1,0), \quad \mathbf{a}_2^* = \left(1/2\sqrt{3}, 1/2\right), \quad \mathbf{a}_3^* = \left(1/2, 1/2\sqrt{3}\right), \quad \mathbf{a}_4^* = (0,1), \qquad (3.8)$$

up to a factor. The point group is denoted by $12m(12^5m)$ and has 24 elements. Two generators with respect to this basis are given by the matrices

$$\Gamma^*(R_1) = \begin{pmatrix} 0 & 0 & 0 & -1 \\ 1 & 0 & 0 & 0 \\ 0 & 1 & 0 & 0 \\ 0 & 0 & 1 & 0 \end{pmatrix}, \quad \Gamma^*(R_2) = \begin{pmatrix} -1 & 0 & -1 & 0 \\ 0 & -1 & 0 & 0 \\ 0 & 0 & 1 & 0 \\ 0 & 1 & 0 & 1 \end{pmatrix}.$$

The four-dimensional integer representation is irreducible when rational representations are considered but, when real numbers are used, it has two non-equivalent irreducible components. Both are two-dimensional. One is carried by V_E and the other by V_I. The pairs (R, R_I) are determined as follows. The rotation R_1 over 30° in V_E corresponds to a rotation over 150° in V_I, and with the mirror R_2 corresponds to a mirror in V_I. A split basis for the embedding is

$$\mathbf{a}_{s1} = (\mathbf{a}_1, \mathbf{a}_1), \quad \mathbf{a}_{s2} = (\mathbf{a}_2, \mathbf{a}_6), \quad \mathbf{a}_{s3} = (\mathbf{a}_3, -\mathbf{a}_5), \quad \mathbf{a}_{s4} = (\mathbf{a}_4, \mathbf{a}_4),$$

where the components of the two-dimensional vector $\mathbf{a}_m$ are given by $\cos(2\pi(m-1)/12)$ and $\sin(2\pi(m-1)/12)$. The scale in the internal space has been chosen as unity. Then the metric tensor becomes

$$g = \begin{pmatrix} 2 & 0 & 1 & 0 \\ 0 & 2 & 0 & 1 \\ 1 & 0 & 2 & 0 \\ 0 & 1 & 0 & 2 \end{pmatrix},$$

which is the metric tensor of family 16 with a special choice for the two parameters ($e = 0$). It has holohedry $12m(12^5m)$ of order 24. The special choice of scale in V_I would give a higher symmetry for the lattice, but this is broken by the orientation of the atomic surfaces. The intersection of the Wigner–Seitz cell with the planes 1–3 and 2–4 are regular hexagons, rotated over 30° with respect to each other. Therefore, the projection of the Wigner–Seitz cell is a regular dodecagon, which can be used as atomic surface for the construction of the dodecagonal tiling. It is a tiling with squares, equilateral triangles, and acute rhombi with an angle of 30°. It is closely related to the *dodecagonal square-triangle tiling*, a tiling with squares and equilateral triangles only (Baake et al., 2002). ▯

Finally, we have a look at the *Penrose tiling* and its embedding. The Fourier module has rank 4. As we shall see in Section 3.5, this is connected with the fact that the Euler function for argument 5 has value 4. With a given starting vector along a mirror of the point group 5m, its star has five prongs. However, these are not independent, because their sum is 0. Therefore, the minimal embedding dimension is 4. However, let us start with a basis consisting of five vectors. Its point group, 5m, has as generators a rotation over 72° and a mirror. With respect to this basis the generators are

$$\Gamma^*(R_1) = \begin{pmatrix} 0 & 1 & 0 & 0 & 0 \\ 0 & 0 & 1 & 0 & 0 \\ 0 & 0 & 0 & 1 & 0 \\ 0 & 0 & 0 & 0 & 1 \\ 1 & 0 & 0 & 0 & 0 \end{pmatrix}, \quad \Gamma^*(R_2) = \begin{pmatrix} 0 & 0 & 0 & 0 & 1 \\ 0 & 0 & 0 & 1 & 0 \\ 0 & 0 & 1 & 0 & 0 \\ 0 & 1 & 0 & 0 & 0 \\ 1 & 0 & 0 & 0 & 0 \end{pmatrix}.$$

This five-dimensional integer representation, has (as real representation) three irreducible components, two two-dimensional, and one one-dimensional. The rotation R_1 over 72° corresponds with a three-dimensional rotation over 144° in the three-dimensional internal space, and the reflection R_2 corresponds to a reflection through the rotation axis in V_I. A split basis for the five-dimensional space is given by

$$\mathbf{a}_{sj} = (\mathbf{e}_j, c\mathbf{e}_{2j}, d), \quad \text{with} \quad \mathbf{e}_j = (\cos(2\pi(j-1)/5), \sin(2\pi(j-1)/5)) \quad (j=1,\ldots,5)$$

The five vectors generate a lattice with metric tensor

$$g = \begin{pmatrix} A & B & C & C & B \\ B & A & B & C & C \\ C & B & A & B & C \\ C & C & B & A & B \\ B & C & C & B & A \end{pmatrix}, \qquad (3.9)$$

for $A = 1 + c^2 + d^2$, $B = \Phi/2 - c^2(1 + \Phi)/2 + d^2$, and $C = -(1 + \Phi)/2 + c^2\Phi/2 + d^2$. The numbers c and d are arbitrary parameters. Choosing $c = 1$ and $d = \frac{1}{2}\sqrt{2}$, the basis spans a hypercubic lattice in five dimensions ($A = 5/2, B = C = 0$). The general lattice belongs to the five-dimensional family 20, with three parameters and holohedry $10m(10^3 m) \perp m$ of order 40. For the special choice of parameters, for which the lattice is hypercubic, the projection of the Wigner–Seitz cell on the three-dimensional internal space is a rhombic icosahedron. For general values of the parameters the icosahedron is elongated along the third axis.

As noticed before, 5 is not the minimal embedding dimension for the Penrose tiling, because the five basis vectors of the Fourier module sum up to 0. The rank of the Fourier module is 4, and therefore, an embedding in four dimensions is possible. The Fourier module is already generated by the four vectors $\frac{2}{5}\mathbf{e}_j$ with $j = 2, \ldots, 5$. The symmetry group generated by R_1 and R_2 is then represented by the four-dimensional matrices

$$\Gamma^*(R_1) = \begin{pmatrix} 0 & 0 & 0 & -1 \\ 1 & 0 & 0 & -1 \\ 0 & 1 & 0 & -1 \\ 0 & 0 & 1 & -1 \end{pmatrix}, \quad \Gamma^*(R_2) = \begin{pmatrix} 0 & 0 & 0 & 1 \\ 0 & 0 & 1 & 0 \\ 0 & 1 & 0 & 0 \\ 1 & 0 & 0 & 0 \end{pmatrix}.$$

This representation of the point group 5m has two irreducible components, both two-dimensional. Bases which carry the representation, and the representation in direct space are:

$$\mathbf{b}^*_{si} = \frac{2}{5}(\mathbf{e}_i, \mathbf{e}_{2i}), \quad \mathbf{b}_i = (\mathbf{e}_i - \mathbf{e}_0, \mathbf{e}_{2i} - \mathbf{e}_0), \quad i = 1, \ldots, 4.$$

The embedding in four dimensions may be obtained from that in five dimensions by projecting the five-dimensional structure parallel to the fifth axis. A lattice point of the five-dimensional lattice has fifth coordinate equal to $m = \sum_{i=1}^{5} n_i$. The points in the hyperplane m (mod 5) are projected on the five points along the diagonal of the four-dimensional unit cell. Because the projection is in V_I, an atomic surface giving a vertex in V_E corresponds to one of the five atomic surfaces in the four-dimensional unit cell. For the Penrose tiling the atomic surface with $m = 0$ (mod 5) is a point, those for values 2 and 5 are pentagons, and those for values 3 and 4 are also pentagons, but larger. Therefore, using the four-dimensional description, the Penrose tiling has four atomic surfaces per unit cell. Non-equivalent five-fold symmetry tilings are obtained if the atomic surfaces are situated not on multiples of 1/5 along the diagonal.

3.4.3 Symmetry of tilings

♯ A tiling is invariant under a Euclidean transformation if every vertex is transformed into a vertex by the transformation. An aperiodic tiling does not have lattice translation symmetry, and, generally, there is no rotation invariance either. The symmetry concept has to be generalized in this case, just as was done for general quasiperiodic systems.

Let us first look at the Fourier module of the tiling. Suppose it is a module of rank n in a d-dimensional space ($n \geq d$). Then the d-dimensional orthogonal transformation (proper or improper rotation) R is a symmetry operation if every vector $\mathbf{k}$ of the module is transformed to a module vector with the same intensity: $I(R\mathbf{k}) = I(\mathbf{k})$. Because the peaks with an intensity above a certain threshold form a discrete set, in each sphere around the origin there is only a finite number of peaks above the threshold. That implies that the symmetry operations form a finite d-dimensional point group. The physical space is one-, two-, or three-dimensional. Therefore, all the relevant point groups are known. In one dimension there is only one, in two dimensions there are the cyclic groups of order m, and the dihedral groups of order $2m$, and in three dimensions there are the infinite series of cyclic groups, dihedral groups, their products with the total inversion, and the tetrahedral, cubic, and icosahedral groups.

The modules can be distinguished in the cyclic modules, which are spanned by the transforms of one vector, and the non-cyclic modules. To start with, we consider cyclic modules only. The rank of the module is determined by the orientation of a starting vector with respect to the axes of the point group. For example, if the point group is 4mm, then the rank is 2 if the starting vector lies along one of the mirror lines, and otherwise it is 4. If the point group is mmm, and the starting vector lies along a mirror line, it generates a rank 1 module only, and the rank of the Fourier module is 4.

If the starting vector $\mathbf{k}$ under the point group K gives a set which generates a module of rank n, and the module has basis $\mathbf{a}_i^*$ ($i = 1, \ldots, n$), then the action of an element R is given by the matrix $\Gamma^*(R)$ by

$$Ra_i^* = \sum_{j=1}^{n} \Gamma^*(R)_{ij} a_j^*, \quad (i = 1, \ldots, n).$$

The matrices give an integer representation of the point group K (the entries of the matrices are integers). Another, non-essential, assumption is that the representation is irreducible as a real representation. As a complex representation it might be reducible. The so-called crystallographic condition, which requires that the orders of the elements are 1, 2, 3, 4, or 6, does not apply here. The condition that the n-dimensional integer representation is an irreducible representation of the point group does put restrictions, on the embedding dimension n.

A first example is the octagonal group 8mm as symmetry group for a two-dimensional tiling. The group 8mm has seven irreducible, real representations, four of dimension 1, and three of dimension 2. The character table is given by

	E	A, A^5	A^2, A^6	A^3, A^5	A^4	$B \ldots$	$AB, \ldots$
D_1	1	1	1	1	1	1	1
D_2	1	1	1	1	1	-1	-1
D_3	1	-1	1	-1	1	1	-1
D_4	1	-1	1	-1	1	-1	1
D_5	2	$\sqrt{2}$	0	$-\sqrt{2}$	-2	0	0
D_6	2	0	-2	0	2	0	0
D_7	2	$-\sqrt{2}$	0	$\sqrt{2}$	-2	0	0

For a starting vector along the x-axis, the generated module has rank 4, and basis vectors on which the two generating elements A and B act according to the matrices

$$\Gamma^*(A) = \begin{pmatrix} 0 & 1 & 0 & 0 \\ 0 & 0 & 1 & 0 \\ 0 & 0 & 0 & 1 \\ -1 & 0 & 0 & 0 \end{pmatrix}, \quad \Gamma^*(B) = \begin{pmatrix} -1 & 0 & 0 & 0 \\ 0 & 0 & 0 & 1 \\ 0 & 0 & 1 & 0 \\ 0 & 1 & 0 & 0 \end{pmatrix}.$$

The representation generated by these matrices is four-dimensional and reducible as a real representation (the sum of the irreducible representations D_5 and D_7), but irreducible as integer representation. This is clear from the character table, because for each irreducible component there are non-integer characters. So, the representation space can be decomposed into the sum of two real spaces (V_E and V_I), but the invariant reciprocal lattice does not have a split basis.

The situation can be formulated in a general statement: the minimal embedding dimension for a p-fold rotation is the value of the *Euler function* $\Phi(p)$, defined as the number of integers between 0 and p which are co-prime with p. For $p = 8$ this $\Phi(8)$ is equal to 4: 1, 3, 5, and 7 are co-prime with 8, but 2, 4, and 6 are not.

The reciprocal lattice in n-dimensional space corresponding to a Fourier module with symmetry K is characterized by its metric tensor, for which the n^2 elements are the scalar products of the basis vectors: $g^*_{ij} = \mathbf{a}^*_i.\mathbf{a}^*_j$. This reciprocal lattice is invariant under the n-dimensional point group $\Gamma(K)^*$ if for every generator the relation

$$\Gamma(A)^*.g^*.\text{Transpose}(\Gamma(A)^*) = g^*$$

holds. This puts symmetry restrictions on the metric tensor. Usually a lattice with higher symmetry is taken. For example, for the standard octagonal tiling one often uses a hypercubic lattice, although the symmetry restriction allows more parameters. The freedom in the length scale in internal space is then used to make the lattice more symmetric for convenience. From the reciprocal lattice Σ^* follows directly the (direct) lattice Σ. A convenient choice of the unit cell yields, after projection onto the space V_I, the most symmetric atomic surface.

Attaching the atomic surfaces to the nodes of the lattice gives a periodic array in n dimensions. The space group of this array may be considered to be the symmetry group of the tiling, in the sense discussed in Chapter 2.

In three dimensions the icosahedral point group does not have an integer representation. From the character table of the group (see Table A.7 in Appendix A) follows that the three-dimensional irreducible representation corresponding with the action of the group in physical space has to be combined with the other non-equivalent irreducible representation. The dimension of the embedding, therefore, is 6. A *basis* for the *six-dimensional lattice* follows from the same principles as before. A possible choice is

$$\mathbf{a}_{s1} = (1, \tau, 0, c, c\tau, 0); \quad \mathbf{a}_{s2} = (-1, \tau, 0, c, -c\tau, 0)$$
$$\mathbf{a}_{s3} = (0, 1, \tau, -c\tau, 0, -c); \quad \mathbf{a}_{s4} = (\tau, 0, 1, 0, -c, c\tau)$$
$$\mathbf{a}_{s5} = (\tau, 0, -1, 0, -c, -c\tau); \quad \mathbf{a}_{s6} = (0.1. - \tau, -c\tau, 0, c),$$

where c is an arbitrary constant. The corresponding metric tensor is

$$g = \begin{pmatrix} A & B & B & B & B & B \\ B & A & B & -B & -B & B \\ B & B & A & B & -B & -B \\ B & -B & B & A & B & -B \\ B & -B & -B & -B & B & A \end{pmatrix},$$

with $A=(2+\tau)(1+c^2)$, $B=\tau(1-c^2)$. For the choice $c=1$ the metric tensor is diagonal, and the lattice is hypercubic. For arbitrary values of c it is the six-dimensional icosahedral lattice with point group $5\bar{3}m(5^2\bar{3}m)$. For $c=1$ the reciprocal lattice is the same, up to a factor. Then the Fourier module is generated by the six three-dimensional vectors $(1, \tau, 0)/(4+2\tau)$, $(-1, \tau, 0)/(4+2\tau)$,

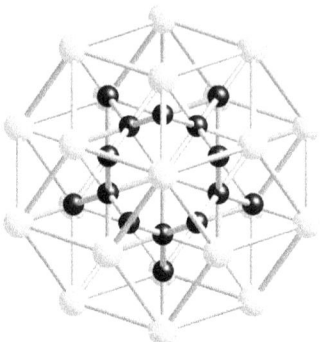

Figure 3.12 *The atomic surface of the three-dimensional icosahedral tiling: a rhombic triacontahedron. It is composed of 20 rhombohedra, the vertices of which are on the surface (light spheres) or in the interior (dark spheres). For clarity we show only the vertices and edges. The atomic surface is the full interior of the triacontahedron.*

$(0, 1, \tau)/(4+2\tau)$, $(\tau, 0, 1)/(4+2\tau)$, $(\tau, 0, -1)/(4+2\tau)$, and $(0, 1, -\tau)/(4+2\tau)$, pointing to 6 (not parallel) of the 12 facets of a dodecahedron (see Fig. 1.13). Note that the dodecahedron has the same symmetry as the icosahedron.

The point group of the holohedry is the icosahedral group, generated by two elements, working on the reciprocal basis according to the matrices

$$\Gamma^*(R_1) = \begin{pmatrix} 1 & 0 & 0 & 0 & 0 & 0 \\ 0 & 0 & 1 & 0 & 0 & 0 \\ 0 & 0 & 0 & 1 & 0 & 0 \\ 0 & 0 & 0 & 0 & 1 & 0 \\ 0 & 0 & 0 & 0 & 0 & 1 \\ 0 & 1 & 0 & 0 & 0 & 0 \end{pmatrix}, \quad \Gamma^*(R_2) = -\begin{pmatrix} 0 & 0 & 0 & 0 & 0 & 1 \\ 1 & 0 & 0 & 0 & 0 & 0 \\ 0 & 0 & 0 & 0 & 1 & 0 \\ 0 & 0 & -1 & 0 & 0 & 0 \\ 0 & 0 & 0 & -1 & 0 & 0 \\ 0 & 1 & 0 & 0 & 0 & 0 \end{pmatrix}.$$

If one takes for the atomic surface the projection of the Wigner–Seitz cell on V_I, then the atomic surface is a rhombic triacontahedron (Fig. 3.12). Placing copies of these on the vertices of the lattice, the intersections with V_E give the vertices of a three-dimensional tiling with icosahedral symmetry. The tiles are copies of two prototiles, rhombohedra with a volume ratio of τ. This is a kind of a three-dimensional version of the Penrose tiling, and is called sometimes the *three-dimensional Penrose Tiling* (3DPT), although Penrose did not introduce it.

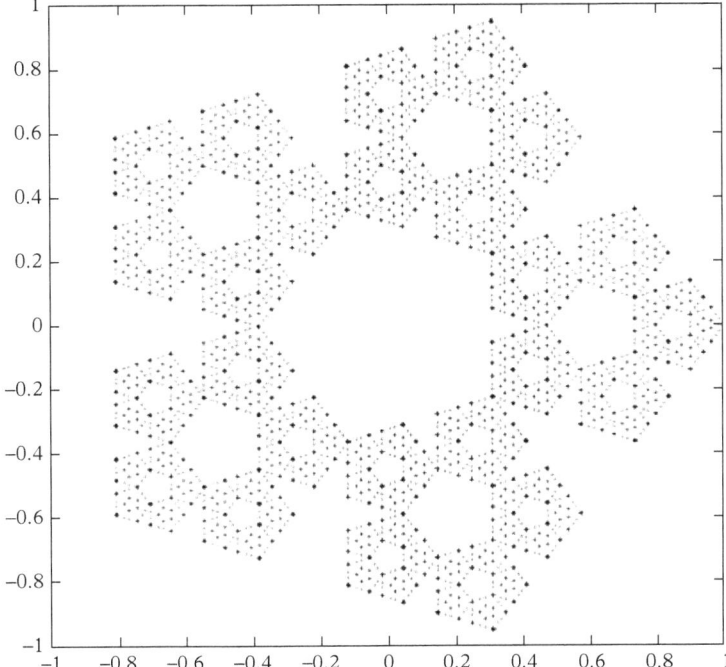

Figure 3.13 *An example of a set of points with five-fold symmetry but not relatively dense. The points are given for S_4.*

This tiling was invented by Robert Ammann, though he did not publish this result. Therefore, a better name is *icosahedral Ammann tiling*, not to be confused with the two-dimensional Ammann–Beenker tiling.

For the choice $c = 1$ the lattice is hypercubic in six dimensions. The holohedry of such a lattice is of much higher order (46,080). It is irreducible. The symmetry will be broken in the embedding by the atomic surfaces. The atomic surface is the projection of the Wigner–Seitz cell, which is a rhombic triacontahedron, a regular polyhedron with 30 faces. The polyhedron may be decomposed into ten oblate and ten prolate rhombohedra. The unit cell of the lattice may be chosen as the Wigner–Seitz cell, or as the sum of the product of the 20 rhombohedra in V_I with their dual in V_E, where the dual of a rhombohedron with edges $\mathbf{e}_i$, $\mathbf{e}_j$, and $\mathbf{e}_k$ with i, j, k a triplet from the six values $1, \ldots, 6$, is the rhombohedron with as edges the complementary triple. (For 1, 3, 5 this is 2, 4, 6.) The volume of the prolate rhombohedron is $2 + 2\tau$, and that of the oblate one 2τ. (These are the determinants of the matrices with as rows the coordinates of the three edge vectors.) The ratio of the two is τ.

As is known from crystallography in six dimensions, there are three Bravais lattices with the icosahedral holohedry in six dimensions: the primitive, the face-centred, and the body-centred lattices. These have 1, 16, and 2 points per unit cell, respectively.

A collection of points with N-fold rotation symmetry that is not quasiperiodic is the following. Consider in the complex plane the orbit of a point $z = 1$ under the group of N-fold rotations. We consider

$$S_1 = \{\text{all } \exp(2\pi i m/N)\} \quad (m = 1, \ldots, N), \tag{3.10}$$

$$S_{n+1} = \lambda z + (1-\lambda)z', \quad z, z' \text{ in } S_n, \tag{3.11}$$

$$S = \lim_{n \to \infty} S_n. \tag{3.12}$$

At each step the points are multiplied with a factor λ such that the minimum distance is kept constant. This collection S of discrete points has N-fold rotation symmetry but is not *relatively dense*, because there are holes of arbitrary large size. See Fig. 3.13. It has scaling symmetry with scale factor λ.

3.5 Approximants

For aperiodic tilings the lattice of the embedding has a projection on V_I which is dense. Every point in V_I is arbitrarily close to the projection of a lattice point. (We deal here with the minimal embedding. The statement is, for example, not true for the five-dimensional embedding of the Penrose tiling.) Quasiperiodic crystals often have approximants, periodic structures close to the aperiodic ones. In the embedding these may be obtained by a strain of the lattice such that a D-dimensional lattice of points is brought into the physical space. A simple way is to consider the internal components of the basis vectors. If one replaces irrational values by rational approximants, the projection of the lattice on V_I is no longer dense, but discrete. In Fig. 3.14 the change of basis is illustrated for the example of the Fibonacci chain, when the value Φ in internal components is replaced by 2/3. This produces a strain in superspace where vertices are moved in the vertical direction.

By the strain the symmetry of the lattice will change. In general, the point group symmetry of the lattice will be lowered to a subgroup. This subgroup can be determined using the metric tensor for the lattice.

As an example, consider the standard octagonal tiling. Its lattice has four basis vectors:

$$\mathbf{a}_{s1} = (a, 0, b, 0), \quad \mathbf{a}_{s2} = \frac{1}{2}\left(a\sqrt{2}, a\sqrt{2}, -b\sqrt{2}, b\sqrt{2}\right),$$

$$\mathbf{a}_{s3} = (0, a, 0, -b), \quad \mathbf{a}_{s4} = \frac{1}{2}\left(-a\sqrt{2}, a\sqrt{2}, b\sqrt{2}, b\sqrt{2}\right).$$

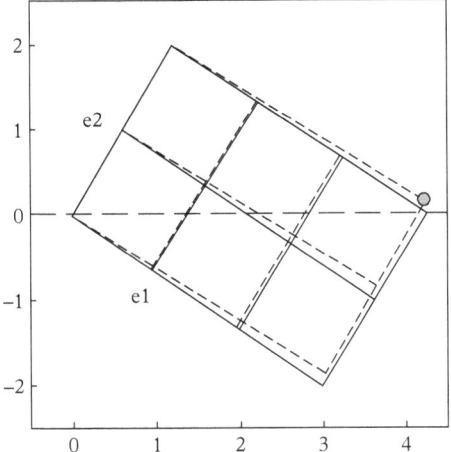

Figure 3.14 *A part of the lattice in superspace for the Fibonacci chain, with (solid lines) and without (dashed lines) phason strain. The strain is obtained by replacing Φ by 2/3 in the internal components of the vertices.*

The metric tensor for this lattice is given by

$$g = \begin{pmatrix} A & B & 0 & -B \\ B & A & B & 0 \\ 0 & B & A & B \\ -B & 0 & B & A \end{pmatrix},$$

with $A = a^2+b^2$ and $B = (a^2-b^2)/\sqrt{2}$. The point group of this lattice is $8m(8^3m)$, and becomes hypercubic if $a = b$. If one keeps the irrational factors $\frac{1}{2}\sqrt{2}$ in the first two coordinates, but replaces them by a fraction L/N in the last two, then there is a lattice in physical space V_E. If N is odd, the lattice translations in V_E are generated by the basis vectors $(L+\frac{1}{2}N\sqrt{2}, L+\frac{1}{2}N\sqrt{2})$ and $(L+\frac{1}{2}N\sqrt{2}, -(L+\frac{1}{2}N\sqrt{2}))$, which span a square lattice. If N is even, there are centring vectors $(0, L+\frac{1}{2}N\sqrt{2})$ and $(L+\frac{1}{2}N\sqrt{2}, 0)$. The centred lattice is again square. The point group, therefore is broken from $8m(8^3m)$ to $4m(4m)$. This is in agreement with the holohedry of a lattice with the metric tensor for a new basis ($e_{s1} = a_{s1}$, $e_{s2} = a_{s3}$, $e_{s3} = (a_{s2} - a_{s4})/\sqrt{2}$, and $e_{s4} = (a_{s2} + a_{s4})/\sqrt{2}$) On this basis the metric tensor becomes

$$g = \begin{pmatrix} A & 0 & B & 0 \\ 0 & A & 0 & B \\ B & 0 & C & 0 \\ 0 & B & 0 & C \end{pmatrix},$$

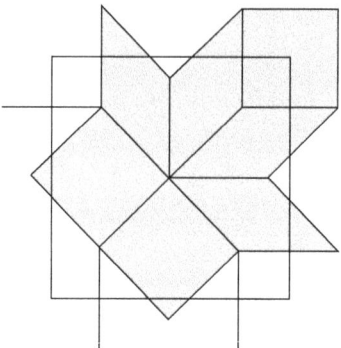

Figure 3.15 *The unit cell of a low approximant for the octagonal tiling, with seven vertices per unit cell.*

with $A = a^2 + b^2$, $B = a^2 + b^2(L/N)\sqrt{2}$, and $C = a^2 + 2b^2(L/N)^2$. This is the metric tensor of the four-dimensional family 10. It is clear that only for $L/N = \frac{1}{2}\sqrt{2}$ (which is impossible if L/N is a fraction) the lattice has the higher symmetry group. Also for a replacement of $\frac{1}{2}\sqrt{2}$ by an irrational number which is not $\sqrt{2}$, the symmetry is broken. In this case the resulting tiling is not periodic, but has a simple two-dimensional crystallographic point group. An aperiodic tiling does not have necessarily a 'forbidden' symmetry group. Notice that the shape of the tiles does not change going from the octagonal tiling to the square tiling. Both have tiles in the form of squares and 45° rhombi. This is because we have not changed the parallel components of the four-dimensional basis vectors, but only the internal components (cf. Fig. 3.15).

The approximants to the icosahedral tiling in three dimensions are quite similar. If one replaces the irrational number τ in the internal components of the basis vectors of the six-dimensional lattice by a rational number L/N, then the tiling becomes periodic, because the vectors

$$N(\mathbf{a}_{s1} - \mathbf{a}_{s2}) + L(\mathbf{a}_{s2} + \mathbf{a}_{s5}), \; L(\mathbf{a}_{s1} + \mathbf{a}_{s2}) + N(\mathbf{a}_{s3} + \mathbf{a}_{s6}), \; L(\mathbf{a}_{s3} - \mathbf{a}_{s6}) + N(\mathbf{a}_{s4} - \mathbf{a}_{s5})$$

have components zero in internal space, and therefore span a lattice in the three-dimensional space V_E. These three vectors of length $2(L^2 + N^2)/N$ span a cubic lattice. The holohedry of the cubic lattice ($m\bar{3}m$) is not a subgroup of the icosahedral group, but the tetrahedral group $m\bar{3}$ is one. Therefore, the point symmetry of the tiling is broken from the icosahedral group to the tetrahedral group. The space group of the periodic tiling then is $Pm\bar{3}$. A general strain of the lattice which corresponds to the replacement of τ by a different value (rational or irrational) leads to a periodic or aperiodic tiling with tetrahedral symmetry.

When L and N are both odd, the vector

$$\frac{1}{2}(N+L)(\mathbf{a}_{s1}+\mathbf{a}_{s3}+\mathbf{a}_{s4}) - \frac{1}{2}(L-N)(\mathbf{a}_{s2}+\mathbf{a}_{s5}-\mathbf{a}_{s6})$$

also belongs to V_E. This gives a centring translation. The periodic tiling in this case is a body-centred cubic lattice, and the arithmetic point group is $Im\bar{3}$.

The approximant one obtains by replacing τ in the basis vectors for the lattice of the embedding of the icosahedral tiling by a fraction L/N stems from a six-dimensional lattice with metric tensor

$$g = \begin{pmatrix} A & B & C & C & C & C \\ B & A & C & -C & -C & C \\ C & C & A & C & -C & -B \\ C & -C & C & A & B & -C \\ C & -C & -C & B & A & C \\ C & C & -B & -C & C & A \end{pmatrix},$$

with $A = (2+\tau)a^2 + (1 + L^2/N^2)c^2$, $B = \tau a^2 + (1 - L^2/N^2)c^2$, and $C = \tau a^2 - (L^2/N^2)c^2$. For L/N tending to τ, B tends to C, and we get the icosahedral six-dimensional lattice with holohedry $5\bar{3}m(5^2\bar{3}m)$. Because L/N is different from τ, the holohedry lowers to $m\bar{3}(m\bar{3})$. This is a centring of the lattice with holohedry $m\bar{3}m(m\bar{3}m)$. It has three free parameters instead of the two parameters for the icosahedral six-dimensional lattice. An icosahedral quasicrystal to which one applies a strain in six dimensions such that the lattice becomes the above mentioned lattice with L/N instead of τ, becomes lattice periodic in physical space. It is called an L/N-approximant. One may consider a series of such approximants with 1/1, 2/1, 3/2, 5/3, etc.

3.6 Coverings

The aperiodic tilings considered in the previous sections need at least two different tiles. Sometimes an aperiodic tiling has three or more different tiles. If an aperiodic structure is obtained by decorating a tiling, this means that we need to decorate at least two tiles. And if we would like to consider such decorated tiles as (pseudo-)unit cells, this means that there are at least two 'unit cells'. Conway proved (see (Grünbaum and Shephard, 1987)) that a Penrose tiling may completely be covered by decagons, which then have overlap. Intuitively one may already have this impression by looking at the set of vertices of a tiling. One may see a pattern of points appear everywhere in a tiling, but these patterns have overlap. Burkov (Burkov, 1991) has proposed a description of decagonal quasicrystals in terms of overlapping decorated decagons. Gummelt (Gummelt, 1996) could prove that one may obtain the Penrose tiling from a covering of the plane by decagons with well-chosen overlap rules. These rules enforce the quasiperiodicity.

144 *Tilings: mathematical models for quasicrystals*

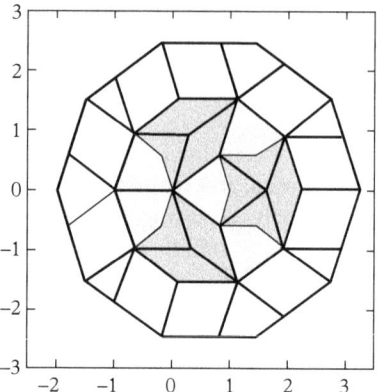

Figure 3.16 *A decagonal patch of a Penrose pattern with the marked unit cluster. (cluster: light, marks: darker)*

Steinhardt and Jeong (Jeong and Steinhardt, 1997; Steinhardt *et al.*, 1998) gave another proof and showed that the overlap rules correspond to conditions for a most dense covering by the decagons. These results are very interesting from the mathematical point of view, but physically it is also very interesting, because then one has only to consider one pseudo-unit cell (Fig. 3.16). It is also related to an old tradition in material science of describing structures in terms of clusters. In the case of quasicrystals, one has used overlapping clusters. Not all atoms belong to such a cluster. In general. there are *glue atoms*.

3.7 Random tilings

Tilings and other model sets are used as models for quasicrystals. The resulting structures studied in this chapter are ideal, quasiperiodic structures. Like models for lattice periodic crystals, these are unrealistic in the sense that real structures always have defects, already because of thermodynamical reasons. For quasicrystals one might argue that defects are intrinsic to the structure. There has been a long discussion whether to see quasicrystals as realizations of ideal structures, but with defects, or as intrinsically defective structures. The starting point for models in the latter view are random tilings.

Randomness introduced in tilings may give an approach to this question. Consider a simple one-dimensional model, our standard example of the Fibonacci chain. A perfect chain may be embedded in two dimensions. The vertices are intersections of the atomic surfaces with the physical space. If such an intersection lies close to the end of the atomic surface, a small shift in the position of this atomic surface can eliminate the intersection, but because the projection of this atomic surface touches the projection of another atomic surface at another position (closeness condition), the shift will create a new intersection somewhere else. Since

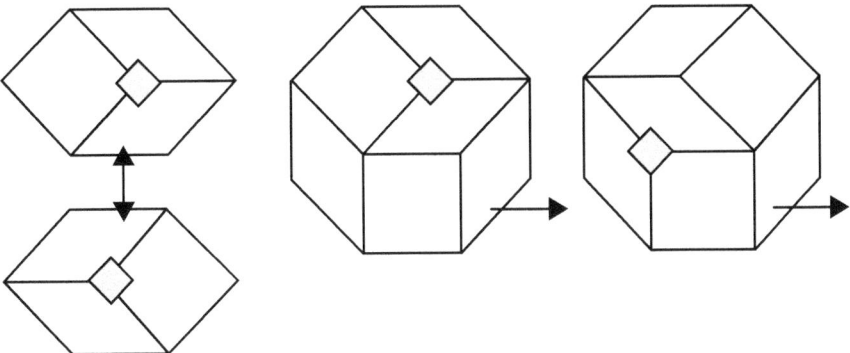

Figure 3.17 *A phason jump (left), and a diffusing phason jump (right). The jumping sites are indicated. The jump goes from the upper left to the lower left configuration and back. The jump in the middle situation creates a different configuration which gives the possibility for a jump elsewhere, indicated by the mark in the situation on the right.*

the end sections of the atomic surfaces correspond to SL or LS pairs of intervals, the shift of the atomic surface means the change from an LS pair to an SL pair, or vice versa. This is a *phason jump*.

In a two-dimensional octagonal tiling occur hexagons, formed by two acute rhombi and a square. The vertex inside the hexagon is the intersection with an octagonal atomic surface, and the intersection point lies close to the border of the atomic surface. Again, a shift of the atomic surface will annihilate the vertex and create another one such that the hexagon still has one square and two rhombi, but in another configuration (Fig. 3.17). The phason jump has changed the environment of the jumping position, and this might have created a new hexagon, where a phason jump may occur. In this way the phason jump may diffuse (Fig. 3.17). After one or many phason jumps the tiling is no longer ideally quasiperiodic, but it has some disorder. The disorder is related to the long-wavelength phason fluctuations discussed in Chapter 6.

The disorder increases the entropy of the system. This means that for temperatures different from zero there is competition between the internal energy and the entropy. Therefore, the stability of quasicrystals can be considered from the point of view of minimization of the internal energy, or that of the free energy

$$F = U - ST,$$

where U is the internal energy, and S the entropy. The internal energy is determined by a Hamiltonian. If this Hamiltonian favours quasiperiodic tilings, the first term in F favours a strict quasiperiodic ground state. The term S can be increased by introducing disorder via phason flips. Another starting point is that one starts, for example, with a set of tiles with interactions. Then the ground state is an average, which might be a quasiperiodic structure, but there are intrinsically deviations from the quasiperiodic structure.

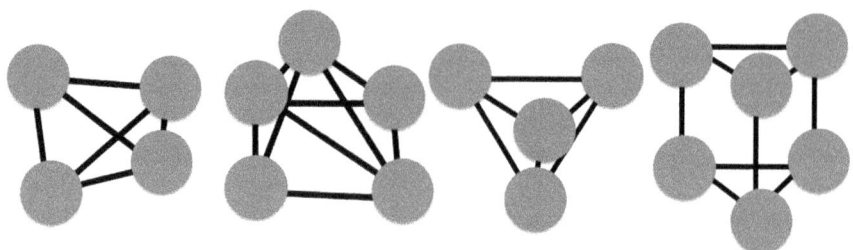

Figure 3.18 *The four canonical cells as used by Mihalkovič and Widom. After Fig. 2 of (Mihailkovič and Widom 2006).*

Random tilings have been studied extensively, both numerically and analytically. Henley (1991a) has introduced the concept of *canonical cell tiling*. He considers a small set (4) of cluster structures connected by packing rules, The connections between the elements are along the two- and three-fold symmetry axes of the icosahedral structure. In addition, glue atoms are necessary. The canonical cell approach has been used to study binary quasicrystals like $Ca_{5.7}Yb$ and $Cd_{5.7}Ca$ Mihalkovič and Widom (2006) (see Fig. 3.18).

Whereas large unit cell approximants with the four canonical cells have been obtained Mihalkovič and Mrafko (1993), it is still an open question whether an icosahedral quasiperiodic tiling can be obtained. Recently N. Fujita succeeded in generating a quasiperiodic tiling with canonical cells, although with a symmetry that slightly departs from the icosahedral one and is in fact cubic (Fujita, 2017).

3.8 Summary

Basic properties of aperiodic crystals may be studied on simplified mathematical models, where there is more chance to prove properties rigorously. This has especially been done for aperiodic crystals for which the set of difference vectors is discrete. These are in the first place models for quasicrystals. In this context modulated structures are more complex, because their difference vectors do not form a discrete set.

Frequently studied types of models are tiling models. Tilings are interesting mathematical objects in themselves, but here they have mainly been considered as models for physical systems. This is not the place to give a comprehensive overview of the theory of tilings. Only very special properties have been addressed here.

Tilings as models for quasicrystals may be seen as sets consisting of a more general type of unit cell: for them one needs two or more different cells, which for a physical interpretation are decorated with atoms.

Summary

Tilings may be constructed in several ways in one, two, or three dimensions. Aperiodic, or better quasiperiodic tilings can also be constructed from a higher-dimensional space, or they may be embedded in such a higher-dimensional space, just as all quasiperiodic systems. For the construction starting in superspace, the symmetry plays an important role.

Quasicrystals may also be described by a covering of the space with one type of tile, if one allows overlaps of these tiles. This unit cluster approach may be formulated in physical space, and in superspace as well.

The stability of quasicrystals may be related to a minimization of the potential energy, or to the minimization of the free energy. In the first case, ideal quasicrystals are considered as quasiperiodic ground states of a Hamiltonian. In the other approach the configurational entropy stems from thermodynamic properties of random sets of tiles. Then the ground state in the thermodynamic sense is a quasiperiodic average over states. The theory of these random tilings has an extensive literature, but is here mentioned only briefly. These considerations about stability apply to incommensurate modulated phases as well. For those systems, phasons occur as well, and there is competition between the internal energy and the entropy. This will be discussed in Chapter 6.

. . . .

EXERCISES

3.1. Consider a Fibonacci structure with points at $x_n = n_1 + n_2 \Phi$ according to a series LSLLS... and the resulting two-dimensional tiling with identical particles at the vertices (x_n, x_m); $\Phi = (\sqrt{5} - 1)/2$.

   (a) Calculate the Fourier module.

   (b) Determine the shape of the atomic surfaces.

3.2. Consider a two-dimensional tiling with squares in the x-y-plane. In the centre of each square a dumbbell molecule of length ℓ is situated in either of two orientations: parallel or perpendicular to the x-axis depending on the sign of the function $\sin(2\pi\alpha x)$ in the centre of the square, where α is an irrational number.

   (a) What is the Fourier module of this structure?

   (b) Calculate the structure factor $F(\mathbf{H})$.

3.3. The Fourier module for the Penrose tiling is generated by the vector $(1, 0)$ and the other four obtained by rotations of a multiple of $72°$.

   (a) Choose a basis for the Fourier module and construct the corresponding basis for the (four-dimensional) reciprocal basis of Σ^*.

   (b) Construct as well the 4D unit cell in superspace.

   (c) What is its metric tensor? Determine the holohedry.

4
Structure

4.1 Diffraction

4.1.1 Diffraction from periodic and aperiodic crystals

One of the most powerful methods of structure determination is the diffraction technique. A monochromatic beam of X-rays, electrons, or neutrons is directed on a sample, and the direction and intensity of the scattered beams are measured. If the scattering power distribution is $\rho(\mathbf{r})$, the incoming beam has wave vector $\mathbf{k}_{in}$ and the outgoing beam $\mathbf{k}_{out}$, then the intensity for $\mathbf{k} = \mathbf{k}_{out} - \mathbf{k}_{in}$ is the square of the Fourier transform of $\rho(\mathbf{r})$, under the condition that multiple scattering and anomalous scattering may be neglected. This is called the *kinematic approximation*.

$$I(\mathbf{k}) = |F(\mathbf{k})|^2, \quad F(\mathbf{k}) = \int \rho(\mathbf{r}) \exp(2\pi i \mathbf{k}.\mathbf{r}) \, d\mathbf{r}. \quad (4.1)$$

For a (lattice periodic) crystal with atoms at positions $\mathbf{r}_j$ in the unit cell the integral may be restricted to the unit cell (with volume V_0). Because the scattering density is periodic with lattice periodicity Λ, its Fourier transform consists of delta peaks (*Bragg peaks*) at the positions $\mathbf{H}$ of the reciprocal lattice Λ^*.

$$F(\mathbf{H}) = \sum_{\mathbf{K} \in \Lambda^*} \delta(\mathbf{H} - \mathbf{K}) \int_{V_0} f(\mathbf{r}, \mathbf{H}) \exp(2\pi i \mathbf{H}.\mathbf{r}) \, d\mathbf{r}, \quad (4.2)$$

where $f(\mathbf{r}, \mathbf{H})$ is the atomic scattering factor for the atom at $\mathbf{r}$. The origin of the sharp peaks lies in the constructive interference of the radiation from the net planes. If we denote the angle between the incoming and outgoing beam by 2θ, it is limited to discrete values because of the condition that constructive interference from two parallel net planes with a distance d occurs if the difference in path satisfies

$$\Delta = 2d \sin \theta = n\lambda,$$

where λ is the wavelength of the radiation. This is *Bragg's law*.

Aperiodic Crystals: From Modulated Phases to Quasicrystals: Structure and Properties.
Second Edition. Ted Janssen, Gervais Chapuis, and Marc de Boissieu.
© Ted Janssen, Gervais Chapuis, and Marc de Boissieu 2018. Published in 2018 by Oxford University Press.
DOI: 10.1093/oso/9780198824442.001.0001

The *static structure factor* of a system of N particles is given by

$$F(\mathbf{H}) = \frac{1}{N} \sum_{n=1}^{N} f_n \exp(2\pi i \, \mathbf{H}.\mathbf{r}_n), \qquad (4.3)$$

where for a lattice periodic crystal the summation is over the N particles in the unit cell. For an aperiodic crystal, the formula still holds, but the summation runs over an infinite number of particles ($N \to \infty$). However, using the superspace formalism leads to formulae which are similar to those for lattice periodic crystals.

Constructive interference only occurs if there is long-range ordering. That is essentially what is measured by diffraction techniques, because the intensity is an integral over the space. Local ordering is, in principle, more evident with techniques like *scanning tunnelling microscopy* (STM) and *scanning electron microscopy*.

For a real crystal the peaks are not exactly delta peaks, because of size and disorder effects. The size effect may be described with a shape or characteristic function s($\mathbf{r}$), which has the value 1 inside the sample, and a value 0 outside. For convenience we put the scattering function equal to unity. Then the normalized structure factor (which has the value 1 for the origin) is

$$F(\mathbf{H}) = \frac{1}{V} \int \rho(\mathbf{r}) \, s(\mathbf{r}) \, d\mathbf{r},$$

and the intensity of the diffraction is then a convolution of the Fourier transform $\hat{s}(\mathbf{H})$ and the expression for the infinite crystal. This yields an array of peaked functions

$$I(\mathbf{H}) = \int \hat{s}(\mathbf{q} - \mathbf{H}) I(\mathbf{q}) \, d\mathbf{q}.$$

at the positions of the reciprocal lattice. In addition, disorder has an effect on the scattering intensity. It lowers the delta peaks by the *Debye–Waller factor*, and it creates a diffuse background. Therefore, the diffraction from a lattice periodic structure also has an intensity which depends smoothly on the scattering wave vector. However, there is an important difference from disordered systems, where no Bragg peaks are present. When the crystal becomes more and more ideal, and its size grows to infinity, mathematically the limit gives a series of delta peaks.

The intensity $I(\mathbf{H})$ is a real function with delta peaks on the reciprocal lattice. The inverse transform then gives again a lattice periodic function. This function is called the *Patterson function*.

$$\begin{aligned} P(\mathbf{r}) &= \sum_{\mathbf{H} \in \Lambda^*} I(\mathbf{H}) \exp(2\pi i \, \mathbf{H}.\mathbf{r}) = \sum_{\mathbf{H} ij} \exp(2\pi i \, \mathbf{H}.(\mathbf{r}_i - \mathbf{r}_j - \mathbf{r})) \\ &= \sum_{ij} \delta(\mathbf{r} - \mathbf{r}_i + \mathbf{r}_j). \end{aligned}$$

This is also called the *autocorrelation function*. It gives the distribution of the interatomic vectors. This may give important information for the structure analysis.

As seen in Chapter 2, a quasiperiodic function is the restriction of a lattice periodic function in superspace to the physical space. This means that there is a function in n dimensions ($\rho_s(\mathbf{r}_s)$), invariant under the n-dimensional lattice Σ such that the restriction to the physical space gives the quasiperiodic function:

$$\rho(\mathbf{r}) = \rho_s(\mathbf{r}, 0).$$

As a consequence, the Fourier transform of the quasiperiodic function is the projection of the Fourier transform of the periodic function in superspace, in the sense that

$$\hat{\rho}(\mathbf{H}) = \int_{V_I} \hat{\rho}_s(\mathbf{H}, \mathbf{H}_I) d\mathbf{H}_I.$$

This leads to the conclusion that the diffraction pattern of an ideal quasiperiodic crystal consists of Bragg peaks. A real crystal will also have a diffuse component, besides the Bragg peaks. In the diffraction there is no essential difference between periodic and aperiodic crystals. Quasiperiodic systems are sometimes described as 'between periodic and disordered' systems. Although the arguments for such a statement have to be formulated very carefully, it is an interesting question which ideal systems produce delta peaks in their diffraction. It is the well-known question 'Which distribution of matter diffracts?', meaning 'which ideal systems produce Bragg peaks in their diffraction?' There is not yet a general answer to this question.

For an aperiodic crystal, the structure factor $F(\mathbf{H})$ is the Fourier transform of a quasiperiodic function. Therefore, the normalized structure factor may be expressed as an integral over the unit cell in superspace:

$$F(\mathbf{H}) = F(\pi \mathbf{H}_s) = \frac{1}{V_0} \int \rho(\mathbf{r}_s) \exp(2\pi i \mathbf{H}_s . \mathbf{r}_s) d\mathbf{r}_s. \qquad (4.4)$$

For an aperiodic crystal consisting of point atoms, the embedding consists of atomic surfaces. Because of incommensurability, the projection of the lattice points of Σ on the internal space forms a uniform and dense set. The summation over all points in the unit cell may then be replaced by an integral over internal space. Suppose that the dimension of the internal space is 1, and that there is one atomic surface in the unit cell. Its projection on V_I has length L, and the points of the atomic surface have internal coordinates t and physical coordinates $\mathbf{r}(t)$. Then the structure factor becomes

$$F(\mathbf{H}) = \frac{1}{L} \int_0^L f(\mathbf{H}) \exp(2\pi i (\mathbf{H}.\mathbf{r}(t) + H_I t)) dt. \qquad (4.5)$$

In general, the dimension of internal space is higher (d), and there are several atomic surfaces per unit cell, numbered by μ. The volume of the projection of the atomic surface μ being Ω_μ, the *structure factor* becomes

$$F(\mathbf{H}) = \frac{1}{\Omega}\sum_\mu \int_{\Omega_\mu} f_\mu(\mathbf{H},\mathbf{r}_I) \exp(2\pi i(\mathbf{H}.\mathbf{r}_\mu(\mathbf{r}_I)+\mathbf{H}_I.\mathbf{r}_I)d\mathbf{r}_I, \qquad (4.6)$$

with $\Omega = \sum_\mu \Omega_\mu$. The atomic scattering factor depends, in principle, on the atomic species at the position $\mathbf{r}_I$ of the atomic surface μ, and on the scattering vector $\mathbf{H}$. The structure factor Eq. 4.6 is normalized in the sense that it gives unity for $\mathbf{H} = 0$ and $f_\mu = 1$. Multiplying with N gives an extensive expression. When this is translated into the expression in superspace, the factor $1/\Omega$ in Eq. 4.6 is replaced by the density ρ of points in the unit cell, corresponding to atoms in physical space.

The *Patterson function* may also be generalized for quasiperiodic crystals.

$$P(\mathbf{r}_s) = \frac{1}{V}\sum_{\mathbf{H}_s} |F(\mathbf{H}_s)|^2 \exp(2\pi\,\mathbf{H}_s.\mathbf{r}_s). \qquad (4.7)$$

It is a periodic function in nD superspace. For the Fibonacci chain the function is different from zero on finite line intervals parallel to V_I. The length of these intervals is $2+2\Phi$: for example, in the embedding there is an atomic surface at $r_E = 0$ from $r_I = -\phi$ until 1, and at $r_E = \Phi$ one from $r_I = 1-\Phi$ until 2. The maximal difference in r_I is $2+\Phi$, and the minimal one is $-\Phi$. The value of the function is zero at the end points and has a flat maximum of length $1+\Phi$ in the middle (Fig. 4.1).

Eq. 4.6 is a generalization of the usual expression in three-dimensional periodic crystals: instead of a summation over all particles in the unit cell we now have a summation over integrals on atomic surfaces. The integrals replacing the summations over all points. The equation is also a generalization in another sense. In the three-dimensional case the integral $\rho(\mathbf{r})$ over the unit cell is replaced by a summation of integrals over atomic distributions. These integrals are the form factors. In the same way the integral over the n-dimensional unit cell is replaced by a summation over the integrals over atomic surfaces, the latter corresponding again to form factors.

Consider a modulated phase. Then the unit cell contains one or more atomic surfaces, numbered by μ, which are on the average parallel to the space V_I. We consider the μ-th atomic surface and pick a point $\mathbf{r}_{\mu s}$ in V_s on it. The points $\mathbf{r}_{\mu s}$ in $V_s + \mathbf{a}_{si}$ then also belong to it for any $i = D+1,\ldots, n$. The atomic surface may then be parametrized by $x_{D+1},\ldots,x_n$ with $0 \le x_i < 1$. Eq. 4.6 becomes in this case

$$F(\mathbf{H}) = F(\pi_E \mathbf{H}_s) = \frac{1}{\Omega}\sum_\mu \exp(2\pi i\,\mathbf{H}_s.\mathbf{r}_{\mu s})G_\mu \qquad (4.8)$$

with

$$G_\mu = \int_0^1 dx_{D+1}\ldots \int_0^1 dx_n f(\mathbf{H},\mathbf{r})\exp[2\pi i(\mathbf{H}.\mathbf{u}_\mu(\mathbf{r})+\mathbf{H}_I.\mathbf{r})]d\mathbf{r}$$

with $\mathbf{r} = \sum_{i=D+1}^{n} x_i \mathbf{a}_i$.

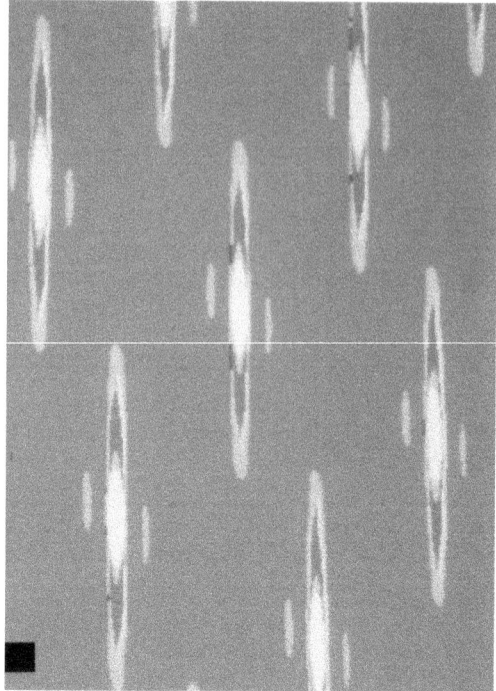

Figure 4.1 *Patterson function in 2D superspace of the Fibonacci chain. The width and the small streaks near the main objects are due to truncation effects.*

The second special case is the one encountered in quasicrystals. Suppose the atomic surfaces are flat and parallel to V_I. Take a point $\mathbf{r}_{\mu s}$ in superspace on the μ-th surface. Then Eq. 4.6 becomes

$$F(\mathbf{H}) = \frac{1}{\Omega} \sum_{\mu} \exp(2\pi i \, \mathbf{H}_s . \mathbf{r}_{\mu s}) \int_{\Omega_\mu} f(\mathbf{H}, \mathbf{r}) \exp(2\pi i \, \mathbf{H}_I . \mathbf{r}) \, d\mathbf{r}, \qquad (4.9)$$

where the vector $\mathbf{r}$ is the difference from a point on the atomic surface and $\mathbf{r}_{\mu s}$. This means that the integral is the Fourier transform of the characteristic function of Ω_μ, the function that has the value 1 on the surface and 0 outside. Again the integral may be seen as a form factor.

For an occupation modulation a position on an atomic surface is taken by a certain species with some probability, which depends on $\mathbf{r}_I$. The scattering factor $f_\mu(\mathbf{H}, \mathbf{r}_I)$ is the sum of the products of the probabilities $p_A(\mathbf{r}_I)$ and the atomic scattering factor for the species A. For an incommensurate phase with both occupation and displacive modulation, the *structure factor* is

$$F(\mathbf{H}) = \sum_j f_j(\mathbf{H}) \frac{1}{L} \int_0^L p_j(t_I) f_j(t_I) \exp[2\pi i(\mathbf{H}.\mathbf{r}(t_I) + H_I t_I)] dt_I \quad (4.10)$$

when we suppose that the internal space is one-dimensional, and that there is an occupation with probability p_j for the occupancy by species j ($\sum_j p_j = 1$); L is the projection of the atomic surface on V_I.

The *atomic density* of a particular type of atoms can be calculated from the d-dimensional volume of the projection of the specific atomic surface on V_I and the volume of the n-dimensional unit cell. The total density is the sum of the volumes of the projection of the atomic surfaces on V_I, divided by the volume of the unit cell. The length scale in $_I$ is arbitrary, but a dilatation with factor c changes both the area and the volume by the same factor. So, it is invariant under changes of c.

Summarizing, one starts with the determination of the Fourier module. Then this module is lifted to an n-dimensional superspace. The intensities of the peaks in V_3^* are the basis for the determination of the filling of the unit cell in V_s. Finally, the structure in the physical space is given by the restriction of the superspace structure to the physical space. See Fig. 4.2.

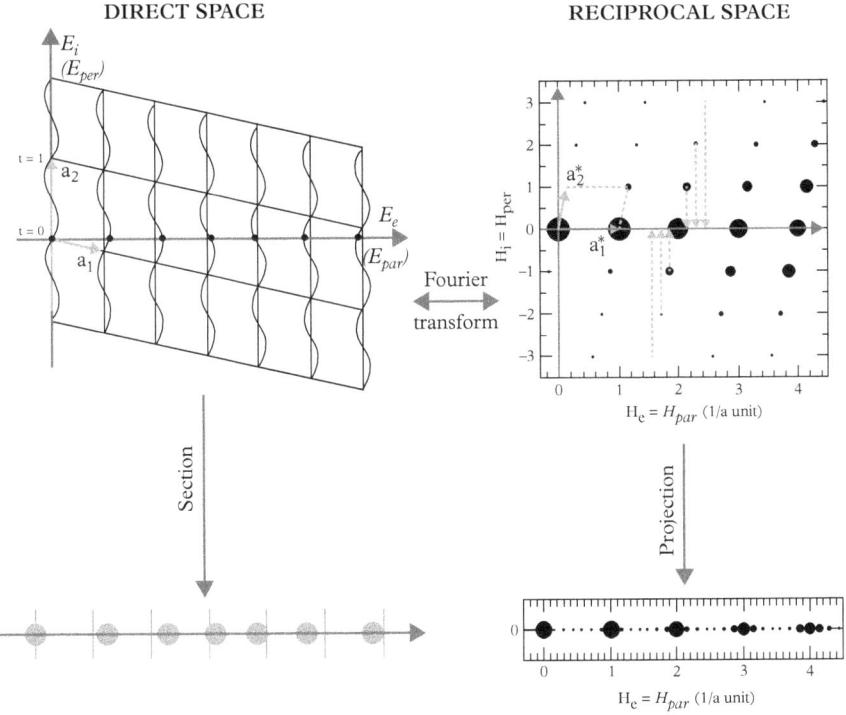

Figure 4.2 *Summary of the superspace procedure for an incommensurately modulated phase.*

Let us now consider a couple of specific examples.

1. A modulation with $n=4$, $d=1$. Consider a modulated phase with orthorhombic basic structure, modulation wave vector $\mathbf{q} = \sigma_1 \mathbf{a}_1^* + \sigma_2 \mathbf{a}_2^* + \sigma_3 \mathbf{a}_3^*$, and modulation function $\mathbf{u}_\mu$. If the average position of atom μ in the unit cell is $\mathbf{r}_\mu = \sum_i x_i \mathbf{a}_i$, then the atom positions are

$$\mathbf{r}_{\mathbf{n}\mu} = n_1 \mathbf{a} + n_2 \mathbf{b} + n_3 \mathbf{c} + \mathbf{r}_\mu + \mathbf{u}_\mu(\mathbf{q}.(\mathbf{n} + \mathbf{r}_\mu)).$$

Notice that in the definition of $\mathbf{u}_\mu$ we have made a choice for the phase different from earlier formulations by including $\mathbf{q}.\mathbf{r}_\mu$ in the argument. However, this is non-essential and leads to the generally used formula for the structure factor. The embedding of this structure is

$$\left(n_1 \mathbf{a} + n_2 \mathbf{b} + n_3 \mathbf{c} + \mathbf{r}_\mu + \mathbf{u}_\mu(\mathbf{q}.(\mathbf{n} + \mathbf{r}_\mu) + t), t\right).$$

The positions of particle μ in the four-dimensional unit cell are then $(\mathbf{r}_\mu + \mathbf{u}_\mu(x_4), x_4 - \mathbf{q}.\mathbf{r}_j)$ ($0 \le x_4 < 1$, cf. Fig. 4.3), and the modulation function has components

$$\mathbf{u}_\mu(x_4) = \sum_{i=1}^{3} u_\mu^i(x_4) \mathbf{a}_i.$$

The atomic surface μ has a projection of length 1 on V_I, and the structure factor for the vector $\mathbf{H} = h_1 \mathbf{a}_1^* + h_2 \mathbf{a}_2^* + h_3 \mathbf{a}_3^* + h_4 \mathbf{q}$ becomes

$$F(\mathbf{H}) = \sum_\mu f_\mu \exp(2\pi i \mathbf{H}.\mathbf{r}_\mu) \int_0^1 dx_4 \exp[2\pi i (\mathbf{H}.\mathbf{u}_j(x_4)) + h_4 x_4)]. \quad (4.11)$$

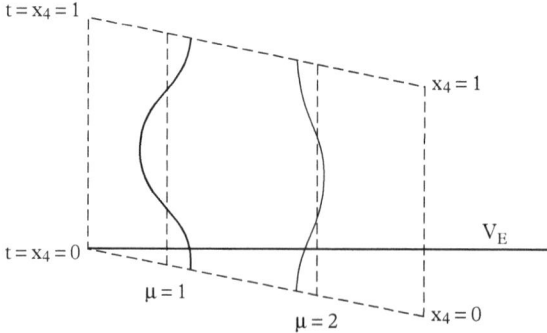

Figure 4.3 *The unit cell in superspace for a displacive modulation. The structure factor is obtained by integration over the atomic surfaces, using as integration variable either $t = r_I$ or x_4.*

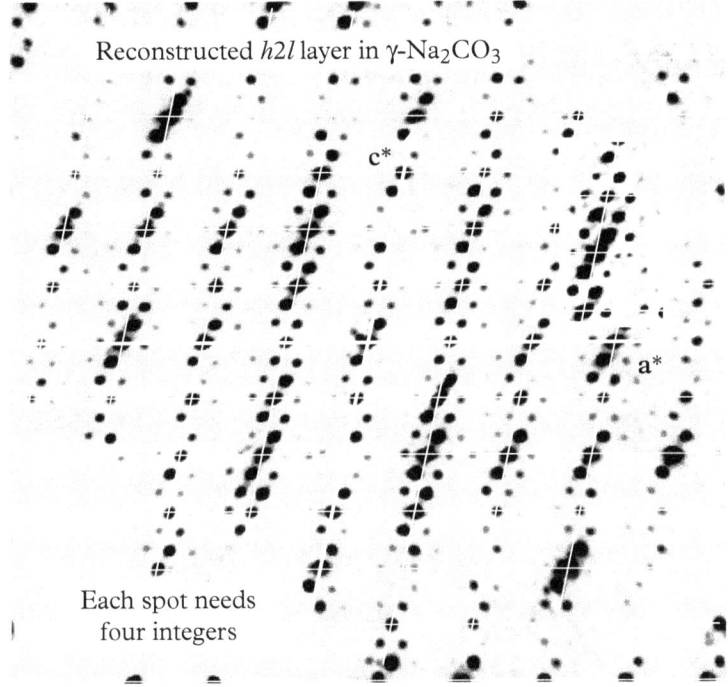

Figure 4.4 *The diffraction pattern in the a-c-plane of the displacive modulated structure of γ-Na_2CO_3.*

This can be rewritten as

$$F(\mathbf{H}) = \sum_\mu \exp\left(2\pi i \sum_{i=1}^{3} h_i x_i^\mu\right)$$
$$\times \int_0^1 f_\mu(t) \exp\left(2\pi i \left[\sum_{i=1}^{3}(h_i + m\alpha_i)u_\mu^i(t) + h_4 t\right]\right) dt.$$

This formula was derived in de Wolff (1974) (cf. Fig. 4.4). For the sinusoidal modulation along the **a**-axis in an orthorhombic basic structure,

$$F(\mathbf{H} = h\mathbf{a}^* + m\mathbf{q}) = \int_0^1 dt\, f \exp(2\pi i\,[\mathbf{H}.\mathbf{u}\sin(2\pi t) + mt]),$$

which again give the Bessel functions, as seen in Chapter 1.

For a displacive modulation, each atomic surface belongs to one type of atoms. For $n = 4$, the atomic surface is a line with a projection on V_I of length one. The projection of the 4D unit cell is the unit cell of the basic structure. Therefore, the total density is the number of atomic surfaces inside the nD unit cell, divided by the volume of the unit cell of the basic structure. So, it is equal to the density of the unmodulated structure, and this is obvious.

For an occupation modulation, the expression is

$$F(\mathbf{H}) = \sum_j \int_0^1 f_j p_j(t) \exp(2\pi i (\mathbf{H}.\mathbf{r}_j + h_4 t)) dt. \tag{4.12}$$

2. An incommensurate composite without modulation. For a system with atoms of species 1 at positions na and atoms of species 2 at mb, the embedded structure in two-dimensional space is given by the periodic array of (straight) atomic surfaces, which are lines in this case:

$$(na - \alpha t, t), \quad (mb - \beta t, t),$$

where α and β are arbitrary real numbers. The lattice of this structure is spanned by the two vectors

$$\mathbf{a}_1 = \frac{a}{\alpha - \beta}(-\beta, 1), \quad \mathbf{a}_2 = \frac{b}{\beta - \alpha}(-\alpha, 1).$$

A convenient choice is $\alpha = 0$ and $\beta = 1$. The reciprocal lattice is spanned by

$$\mathbf{a}_1^* = \frac{1}{a}(1, \alpha), \quad \mathbf{a}_2^* = \frac{1}{b}(1, \beta).$$

Then the Fourier module consists of vectors $H = h_1/a + h_2/b$. The atomic surfaces are two edges of the unit cell. Consequently, the structure factor is the sum of two integrals over the two atomic surfaces.

$$F(H) = f_1 \int_0^{\frac{b}{\beta-\alpha}} \exp(2\pi i(-\alpha H t + H_I t)) dt$$

$$+ f_2 \int_0^{\frac{a}{\alpha-\beta}} \exp(2\pi i(-\beta H t + H_I t)) dt.$$

The internal component of $\mathbf{H}_s$ is equal to $h_1\alpha/a + h_2\beta/b$. This sum of the integrals is equal to zero, unless $h_1 = 0$ or $h_2 = 0$. If $h_2 = 0$, then only the first term survives and is equal to $f_1 b/(\beta - \alpha)$ and, when $h_2 = 0$, then the structure factor is $f_2 a/(\alpha - \beta)$. If $h_1 = h_2 = 0$, then the structure factor is the sum of these. Therefore, there are Bragg peaks at 0, h_1/a, and h_2/b, but no satellites in this case. This is a consequence of the absence of modulations. For the general case, the atomic surfaces are no longer straight, and there are sum

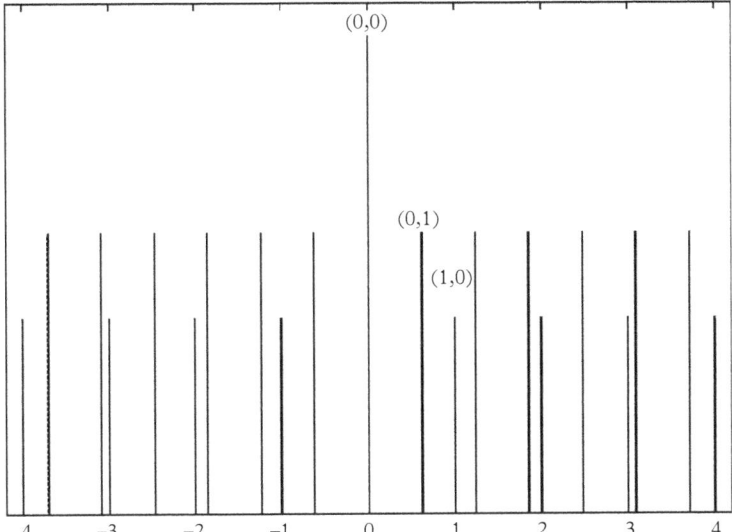

Figure 4.5 *The diffraction pattern of an incommensurate composite, with very weak interaction between the subsystems. Only the main reflections of the two subsystems are visible.*

peaks with both h_1 and h_2 different from zero. The intensity follows from the same expression with two integrals (see Fig. 4.5).

3. The Fibonacci chain. The two-dimensional embedding of the Fibonacci chain may be chosen with lattice generated by $(1, -\Phi)$ and $(\Phi, 1)$. Its reciprocal basis consists of the vectors $(1, -\Phi)/(2-\Phi)$ and $(\Phi, 1)/(2-\Phi)$. Therefore, the Fourier module consists of vectors $(h_1 + h_2\Phi)/(2 - \Phi)$. There is one atomic surface per unit cell, which is a straight line of length $1 + \Phi$. Consequently, the expression for the normalized structure factor becomes

$$F(H) = \frac{1}{\tau}\int_{-\tau/2}^{\tau/2} \exp(2\pi i H_I t)\, dt = \frac{1}{\pi H \tau} \sin(\pi H \tau).$$

The intensity becomes

$$I(H) = |F(H)|^2 = |\sin(\pi H_I \tau)|^2 / (\pi H_I \tau)^2,$$

for $H = (h_1 + h_2\Phi)/(2-\Phi)$ and $H_I = (-h_1\Phi + h_2)/(2-\Phi)$. Because the absolute value of the sine function remains smaller than 1, the intensity of the peaks decreases if the internal component H_I of the vector $\mathbf{H}_s$ increases (see Fig. 4.6). This is a very common feature. Generally, the intensity of

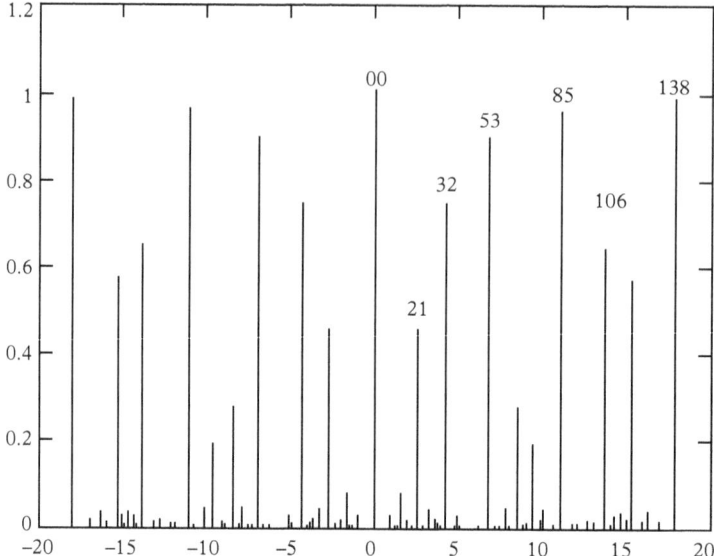

Figure 4.6 *The intensities of the diffraction peaks for the Fibonacci chain, with the indices of some peaks.*

peaks decreases for increasing internal (or perpendicular) components of the wave vector.

The density of vertices is simply the length of the atomic surface $(a(1+\Phi))$ divided by the volume of the 2D unit cell, which is $a^2(2-\Phi)$, that is, $(4+3\Phi)/(5a) = 1.1707\ldots/a$ (from which follows the average distance between vertices, which is $3\Phi - 1 = 0.854\ldots a$).

♯ The second, equivalent embedding has a lattice generated by the vectors $(3\Phi - 1, -\Phi)$ and $(0, 1)$. The reciprocal basis consists of the vectors $(1/(3\Phi - 1), 0)$ and $(\Phi/(3\Phi - 1), 1)$. The Fourier module has vectors $(H_1 + H_2\Phi)/(3\Phi - 1)$ corresponding to the two-dimensional reciprocal lattice vector $((H_1 + H_2\Phi)/(3\Phi - 1), H_2)$. The embedded structure is

$$(n(3\Phi - 1) + (\Phi - 1)\,\mathrm{Frac}(n\Phi + t), t).$$

Hence, the normalized structure factor equals

$$F(H_E) = \int_{-\frac{1}{2}}^{\frac{1}{2}} \exp(2\Phi(H_E(\Phi - 1)t + H_I t))\,dt.$$

This integral may easily be evaluated. It yields

$$|F(H_E)| = \sin(\pi((\Phi - 1)H_E + H_I)) / \pi((\Phi - 1)H_E + H_I),$$

which is the same as for the former embedding, of course. This can be seen using the relation between the two index schemes $H_1 = h_2$, and $H_2 = h_1 - h_2$, which follows from

$$H_E = \frac{H_1 + H_2 \Phi}{3\Phi - 1} = \frac{h_1 + h_2 \Phi}{2 - \Phi}, \quad H_I = H_2 = \frac{-h_1 \Phi + h_2}{2 - \Phi}.$$

A related chain with the same incommensurability is the chain with two intervals of length a and b according to a Fibonacci sequence. Then

$$x_n = nc + A \times \text{Frac}(n\Phi),$$

with average length $c = b + (a-b)\Phi$ and modulation amplitude $A = b - a$. For $A = 0$ this gives a periodic array with lattice constant $c = b = a$. For $A = \Phi - 1$ and $c = 3\Phi - 1$ the original Fibonacci is recovered. The structure factor for this generalization of the Fibonacci chain is

$$|F(H_E)| = |\sin(\pi(H_E A + H_I))/(\pi(H_E A + H_I))|.$$

The structure is a tiling with the same Fourier module ($1/c$ and Φ/c) as the standard Fibonacci chain, but with an arbitrary ratio between the lengths of the tiles. By varying A one gets a continuous transformation from a periodic chain to the Fibonacci chain.

4. An octagonal tiling. The vertices of an octagonal tiling of the plane may be obtained by intersection of the plane with a four-dimensional periodic structure. The four basis vectors of the lattice are

$$\mathbf{a}_1 = (1, 0, 1, 0), \quad \mathbf{a}_2 = (c, c, -c, c), \quad \mathbf{a}_3 = (0, 1, 0, -1), \quad \mathbf{a}_4 = (-c, c, c, c),$$

with $c = \sqrt{1/2}$, and the reciprocal basis is the same, up to a factor 2. Therefore, the vectors of the Fourier module are

$$\mathbf{H}_E = (h_1 + h_2/\sqrt{2} - h_4/\sqrt{2}, h_2/\sqrt{2} + h_3 + h_4/\sqrt{2})/2$$

and the corresponding

$$\mathbf{H}_I = (h_1 - h_2/\sqrt{2} + h_4/\sqrt{2}, h_2/\sqrt{2} - h_3 + h_4/\sqrt{2})/2.$$

The atomic surfaces are the projection of the unit cell of the lattice on V_I, which is an octagon with edge length 1 and diameter $1 + \sqrt{2}$. The normalized structure factor can then be calculated as an integral over the single atomic surface:

$$F(\mathbf{H}) = \frac{1}{2 + 2\sqrt{2}} \int_\Omega \exp(2\pi i \mathbf{H}_I . \mathbf{r}_I) \, d\mathbf{r}_I,$$

Figure 4.7 *The diffraction pattern of the Penrose tiling. The radius of the spots is proportional to the intensity. The pattern has ten-fold symmetry.*

where the integral is over the octagon. Compare this to the diffraction pattern of the Penrose tiling (Fig. 4.7). The expression for the *structure factor* is a special case of a system with a finite number of disjoint atomic surfaces parallel to V_I. Generally, models for quasicrystals are of this type. The structure factor then becomes

$$F(\mathbf{H}) = \sum_j \exp(2\pi i \mathbf{H}.\mathbf{r}_j) \int_{\Omega_j} \exp(2\pi i \mathbf{H}_I.\mathbf{r}_I) \, d\mathbf{r}_I, \quad (4.13)$$

where the sum runs over the atomic surfaces Ω_j in the unit cell, and $\mathbf{r}_j$ is the projection of Ω_j on V_E. The integral is just the Fourier transform of the characteristic function G of the surface, a function equal to 1 inside and 0 outside the surface. So,

$$F(\mathbf{H}) = \sum_j \exp(2\pi i \mathbf{H}.\mathbf{r}_j) \hat{G}(\mathbf{H}_I).$$

The structure factor for the octagonal tiling is given by

$$F(\mathbf{H}) = (S_1 - S_2)/(2 + 2\sqrt{2}),$$

with

$$S_1 = \frac{4 \sin((1+\sqrt{2})c/2) \sin(d/2)}{cd},$$

and

$$S_2 = \frac{4}{c(c-d)(c+d)}$$
$$\times \Big(c\cos((1+\sqrt{2})c/2)\cos(d/2) - c\cos(c/2)\cos((1+\sqrt{2})d/2)$$
$$+ d\sin((1+\sqrt{2})c/2)\sin(d/2) + d\sin(c/2)\sin((1+\sqrt{2})d/2)\Big),$$

where

$$c = 2\pi H_{Ix} \text{ and } d = 2\pi H_{Iy}.$$

This function $F(\mathbf{H})$ becomes unity for $\mathbf{H} = 0$, and approaches unity when the two internal components of the four-dimensional vector $\mathbf{H}_s$ are small. The pattern has eight-fold symmetry. Starting from a peak, and multiplying the length of its position vector by a factor $1+\sqrt{2}$, one arrives at a peak with a smaller internal component, and therefore with a higher intensity, if one neglects the atomic scattering function. The fall-off of the intensity with larger values of the internal component of the wave vector is given in Fig. 4.8. For small enough internal component, the intensity would go to the same value as that of the peak at the origin, but this effect is compensated for by the decrease of the atomic scattering factor for larger values of the scattering vector.

The expression for the structure factor, given in the examples above, neglects the effect of the thermal motion. The positions of the atoms oscillate around the equilibrium positions for which the structure factor has been derived. The scattering distribution then becomes

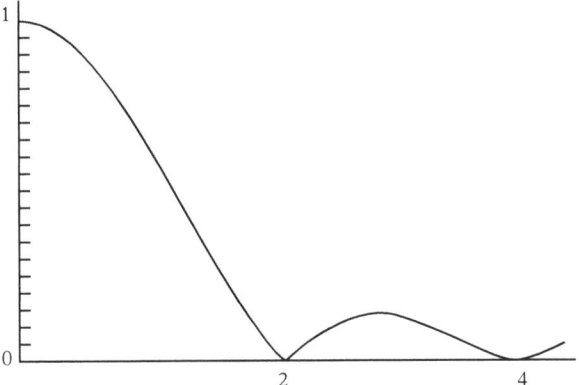

Figure 4.8 *The intensity of a diffraction peak as function of the length of the internal component for the octagonal Ammann–Beenker tiling, in one direction in reciprocal space V_I^*.*

$$\rho(\mathbf{r}) = \sum_{n=1}^{N} f_n \delta(\mathbf{r} - \mathbf{r}_n - \mathbf{v}) \exp[-\mathbf{v}^T B_n \mathbf{v}] d\mathbf{v}, \qquad (4.14)$$

where B_n is the *thermal factor*, a symmetric second rank tensor, which depends on the atom position and the temperature. The tensor is also quasiperiodic if the atoms labelled by n form a quasiperiodic crystal and, therefore, B may be embedded in superspace. For a displacive modulated structure the structure factor becomes

$$F(\mathbf{H}) = \sum_{j} f_j \exp[2\pi i \, \mathbf{H} \cdot \mathbf{r}_j] \qquad (4.15)$$

$$\times \int d\mathbf{r}_I \exp[2\pi i (\mathbf{H} \cdot \mathbf{u}_j(\mathbf{r}_I) + \mathbf{H}_I \mathbf{r}_I - \mathbf{H}^T B_j(\mathbf{r}_I) \mathbf{H}/4)].$$

The thermal vibrations lead to a decrease of the intensity of the Bragg peaks. The decrease is given by the so-called *Debye–Waller factor* (see Section 5.4.7).

4.1.2 Indexing the diffraction pattern

The diffraction pattern of an incommensurate modulated crystal has main reflections and satellites. The main reflections form the reciprocal lattice of the basic structure. The indices of the Fourier module are $h_1, \ldots, h_n$ or $h, k, \ell, m_1, \ldots, m_d$. As usual, the three first *basis vectors* may be chosen as basis for a primitive lattice or, in case the lattice is centred, one for a conventional lattice. In the latter case some of the intensities of main reflections may vanish as a consequence of centring extinctions.

An analogous situation may occur for the satellites. Choosing a basis for the main reflections, the basis vectors $\mathbf{a}_j^*$ for $j = D+1, \ldots, n$ may have rational and irrational components with respect to the basic structure. At least one component of each basis vector should be irrational, if expressed in the basic structure. If there are rational components, one may choose a new basis such that these components vanish. As an example, consider an orthorhombic basic structure, with basis $\mathbf{a}^*, \mathbf{b}^*$, and $\mathbf{c}^*$, with a modulation with wave vector $\mathbf{q} = \frac{1}{2}\mathbf{a}^* + \gamma \mathbf{c}^*$. The Fourier module has rank 4, and a possible basis choice is $\mathbf{a}^*, \mathbf{b}^*, \mathbf{c}^*$, and $\mathbf{q}$. A conventional basis would be given by the choice $\frac{1}{2}\mathbf{a}^*, \mathbf{b}^*, \mathbf{c}^*$, and $\gamma \mathbf{c}^*$. This corresponds to the choice of a centred lattice in superspace.

A *setting* for the basis structure may be given in the usual way. Space groups Pcmn and Pnma are equivalent by a basis transformation exchanging the a- and c-axes. For satellites the assignment of a satellite to a main reflection is not unique. Adding a reciprocal lattice vector to the modulation wave vector does not change the modulation. For example, a satellite may be given as $\gamma \mathbf{c}^*$ or as $(1-\gamma)\mathbf{c}^*$.

For incommensurate composites, the indexing is straightforward when the reciprocal lattices of the subsystems may clearly be identified, where a peak may belong to more than one reciprocal lattice, if the subsystems have common translation vectors. The basis of a reciprocal lattice of a subsystem should have a length scale that gives reasonable unit cell sizes. Generally speaking, the satellites, or summation peaks which are at a position which is a linear combination of basis vectors for different subsystems, should be weaker than the peaks belonging to one of the reciprocal lattices of subsystems.

For quasicrystals there is no natural way to distinguish main reflections and satellites. If there is a non-crystallographic symmetry in three dimensions, any choice of basis for a three-dimensional sub-module would break this symmetry. In this case a natural choice for the basis of the Fourier module would be one obtained by the symmetry operations on a chosen starting vector. As examples we give some choices for the basis of important *tilings or quasicrystals* in Table 4.1.

A second problem in the choice has its origin in the presence of scaling. A scaling transformation corresponds to an n-dimensional basis transformation or a basis transformation of the Fourier module. Any scaled basis is again a basis for the Fourier module. This gives a problem for the choice of length scale besides that of the orientation. However, there is a more or less natural choice.

Consider, for example, the Fibonacci chain with peaks at positions $H = (h_1 + h_2\Phi)/a$ The intensity of this peak is

$$I(H) = \left|\frac{\sin(\pi H_I)}{\pi H_I}\right|^2,$$

Table 4.1 *Bases for some Fourier modules for one-, two-, and three-dimensional tilings. (The basis vectors are given up to a common factor for each case.)*

Dimension	Rank	Case	Basis vectors
1	2	Fibonacci	$1, \Phi$
2	4	Octagonal	$(1,0), (\alpha,\alpha), (0,1), (-\alpha,\alpha)$
2	4	Penrose	$(\cos(2\pi j/5), \sin(2\pi j/5))$ $(j=1,\ldots,4)$
		alternative:	$(\cos(2\pi j/5), \sin(2\pi j/5))$ $(j=0,\ldots,4)$
3	5	Decagonal	$(\cos(2\pi j/5), \sin(2\pi j/5, 0), (0,0,1))$ $(j=1,\ldots,4)$
3	5	Octagonal	$(1,0,0), (\alpha,\alpha,0), (0,1,0), (-\alpha,\alpha,0), (0,0,1)$
3	6	Icosahedral	$(1,\tau,0), (-1,\tau,0), (0,1,\tau)$
			$(\tau,0,1), (\tau,0,-1), (0,1,-\tau)$
		alternative:	$(0,0,1),$
			$\left(\sin\theta\cos\left(\frac{2\pi j}{5}\right), \sin\theta\sin\left(\frac{2\pi j}{5}\right), \cos\theta\right)$ $(j=1,\ldots,5)$

$\Phi = (\sqrt{5}-1)/2, \quad \tau = 1/\Phi = (\sqrt{5}+1)/2, \quad \alpha = \frac{1}{2}\sqrt{2}$

where the internal component of the reciprocal lattice vector with physical component H is equal to $(h_1 - h_2\Phi)/a$. The intensities are normalized to $I(0) = 1$. If H_I is very small, then the intensity of the peak is almost equal to 1. This happens for $h_1/h_2 = F_n/F_{n-1}$, where F_n are the Fibonacci numbers. Now start with a high-intensity peak and scale the wave vector $H = (F_n + F_{n-1}\Phi)/a$ by a factor Φ. In the limit this vector will approach zero. But, before $\Phi^n H$ becomes very small, its intensity will decrease. This can be seen from the fact that the value of H_I increases with a factor τ. Plotting the intensity of the peaks $\Phi^n H$ against H or against H_I will give a sharp decrease at a value of H which is the inverse of the average distance $(3\Phi - 1)a$ (Fig. 4.9). In an experiment one may start with a strong peak and then scale it down with the scaling factor until the intensity drops sharply. This is a good choice for a basis vector of the Fourier module. The other basis vectors may then be obtained by the action of the symmetry operators.

A peculiarity of icosahedral quasicrystals is that the Cartesian components of the basis vectors of the Fourier module are, up to a common length scale, either 1 or τ. Also, scalar products of two basis vectors are of the form $n_1 + n_2\tau$. That gives the possibility to index the peaks in the diffraction pattern in a very compact way: if $\mathbf{H} = \sum_i h_i \mathbf{a}_i^*$, then the square $\mathbf{H}^2$ is of the form $N + M\tau$. Therefore, one often speaks of an N/M peak in the diffraction pattern. This indexing scheme was introduced in Cahn et al. (1986). A number of examples are given in Table 4.2.

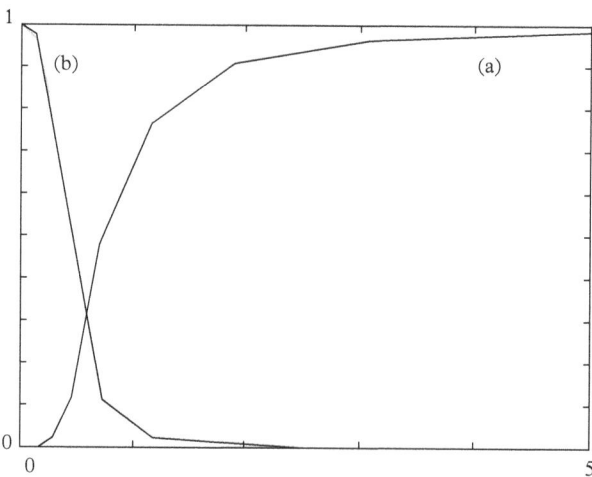

Figure 4.9 *The intensity of scaled diffraction peaks as a function of H_E and H_I. Starting at an intense peak and then scaling with a factor Φ in physical space, the intensity decreases (curve (a) moving to the left). The internal component H_I increases with a factor τ (curve (b) moving to the left).*

Table 4.2 *Comparison of primitive indices and Cahn–Gratias indices, as appearing in i-AlPdMn.*

N	M	h_1	h_2	h_3	h_4	h_5	h_6
4	4	1	1	0	0	0	0
8	12	1	1	1	1	0	0
20	32	2	2	1	0	0	1
52	84	3	3	2	0	0	2
2	1	1	0	0	0	0	0
3	4	$\frac{1}{2}$	$\frac{1}{2}$	$\frac{1}{2}$	$\frac{1}{2}$	$\frac{1}{2}$	$\frac{1}{2}$

4.1.3 ♯ Mathematical questions

Mathematically the intensity distribution function of the diffraction from a system is the Fourier transform of the autocorrelation function. Such a function may be decomposed uniquely into three parts. The first is the *pure point* part and consists of delta peaks (Bragg peaks). The second part is continuous in a special sense: it is the *absolute continuous* part. And the third part is the so-called *singular continuous* part. For a matter distribution of infinite extent and with lattice periodicity, there is only the first part. The diffraction consists entirely of Bragg peaks. For a real crystal, disorder creates a continuous background. This gives an absolute continuous contribution. A singular continuous component is 'all what remains' and is found in special models that are not quasiperiodic. The existence of Bragg peaks is a sign of long-range order. The International Union of Crystallography has defined a crystal as a matter distribution with diffraction which consists 'essentially' of Bragg peaks. This is supposed to mean that the idealized structure has only Bragg peaks. The Bragg peaks should appear in all three directions (not just on a line or in a plane). With that definition quasiperiodic structures are crystals, and the lattice periodic structures are a special type of crystals. A natural question then is 'What matter distribution diffracts with delta peaks?' This question has not yet been fully answered, but there are many partial answers.

A matter distribution given by a quasiperiodic function $\rho(\mathbf{r})$ is the restriction of a lattice periodic distribution function $\rho_s(\mathbf{r}, \mathbf{r}_I)$ to the n-dimensional physical space. The Fourier transform of ρ_s consists of delta peaks, and the Fourier transform of $\rho(\mathbf{r})$ is the projection of these peaks. A simple illustration is the following. Consider a function in two variables $\rho(x, y)$. Its Fourier transform is

$$\hat{\rho}(k_x, k_y) = \int \rho(x, y) \exp(i(k_x x + k_y y)) \, dx \, dy.$$

If we 'project' on the k_x-line by taking the integral

$$\int \hat{\rho}(k_x, k_y)\, dk_y = \int \rho(x, y) \exp(i(k_x x + k_y y))\, dx\, dy\, dk_y,$$

then this is equal to the Fourier transform of the function ρ on the line $y = 0$:

$$\int \hat{\rho}(k_x, k_y)\, dk_y = \int \rho(x, 0) \exp(ik_x x)\, dx = \hat{\rho}_x(k_x).$$

In other words, the *projection of the Fourier transform* on $k_y = 0$ is the Fourier transform of the 'intersection' (actually the restriction of ρ to the line $y = 0$).

This is fine if the functions are smooth enough, but in reality there are point atoms, atomic surfaces which may be bounded, and the Fourier transforms are distributions, in this case Dirac delta functions. Then the statement that the Fourier transform of one is the projection of the other is mathematically less obvious, and questions about convergence do not have a simple answer. Many investigations have been made of the problem, starting with atoms in physical space. Most of these are concerned with point sets for which the set of difference vectors is discrete, or which have inflation properties. For example, in a crystal with lattice periodicity the set of differences is the set of lattice translations plus the finite set of differences between atoms in a unit cell. For a Penrose tiling, the vectors connecting vertices form also a discrete set. This is called *finite local complexity*. But, for a modulated crystal phase (say, with positions $na + \Delta \cos(qna)$), the set of differences consists of continuous intervals (of length 2Δ in our example). Then many of the proven statements do not apply. We give here a short discussion. (For a deeper understanding see (Bombieri and Taylor, 1986; Hof, 1997a; Lagarias, 2000; Baake et al., 2003, 2015).)

Suppose that a system consists of point atoms. The scattering power is then a sum of delta peaks inside a finite volume B_L. If this volume has radius L, the autocorrelation function is the normalized convolution of the density $\rho(\mathbf{r})$:

$$C_L = \frac{\rho_1 * \rho_2(\mathbf{r})}{|B_L|}, \quad \rho_1(\mathbf{r}) = \rho(\mathbf{r}), \quad \rho_2(\mathbf{r}) = \rho(-\mathbf{r}). \tag{4.16}$$

Then

$$C = \lim_{L \to \infty} C_L. \tag{4.17}$$

If this limit exists, the diffraction intensity is given by the Fourier transform, and this can be decomposed into three parts:

$$I = \hat{C} = I_{pp} + I_{ac} + I_{sc}, \tag{4.18}$$

where the subindices mean pure point, absolute continuous, and singular continuous. A *pure point spectrum* consists of delta peaks. An *absolute continuous function* is a function on an interval such that for every (small) ϵ there is a length δ such that if one chooses a number of subintervals (a_i, b_i) with total length smaller than δ, then the sum of all $|f(b_i - a_i)|$ is smaller than ϵ. This is a stronger requirement than just continuity. A *singular continuous function* is a continuous function which has 'almost everywhere' (i.e. up to a set of points with measure zero) a derivative equal to zero. Or, in a more sloppy way, it has everything else. An example is the Cantor function. In an infinite lattice periodic system the diffraction is pure point: it has only Bragg peaks. As mentioned before, in real crystals there is always an absolute continuous contribution. One has to go to more exotic structures to find a singular continuous contribution. This happens, for example, for certain tilings. An example is the *Thue–Morse chain*. Consider the letters A and B, and a substitution rule $\sigma(A) = AB, \sigma(B) = BA$. Starting from A the substitution multiplies the number of letters by a factor 2. In the limit there is an infinite chain of letters. Replacing each A by an interval of certain length, and B by an interval of different length, the result is an infinite, aperiodic chain. It is perfectly ordered, by construction. Nevertheless, the diffraction pattern would not have any Bragg peaks (delta peaks). Its spectrum is singular continuous. This shows how difficult the formulation of the borderline of the notion of crystal is. Perfect order is not equivalent with a point spectrum. The Thue–Morse chain does not have Bragg peaks. Quasiperiodic structures always have a Fourier function consisting of delta peaks.

Hof showed that for the Fibonacci chain with equal atoms on the vertices the expression for the structure factor, as we have discussed in Eq. 4.5, is correct. (See (Hof, 1997b).) More recently, this has been generalized for other chains. Notice that in general there may be a difference in character between a crystal with identical atoms on the positions of a quasiperiodic structure, and an aperiodic crystal with atom positions on a lattice periodic structure, such as, for example, a displacively modulated chain and an occupation modulated chain.

Baake and Moody have shown that for a certain class of structures, the diffraction is pure point if the structure is 'strongly almost periodic'. The class contains quasiperiodic tilings obtained by the cut-and-project method. In this case the set of differences between points of the structure has a minimum length. For more general structures, like modulated structures, the question has not yet been answered.

4.2 Diffraction techniques

The field of aperiodic crystals has evolved in parallel with the development of new detectors for X-ray and neutron sources. The increasing availability of new synchrotron sources, the recent development of X-ray free-electron laser sources (XFEL) and the scarcity of neutron facilities are probably the main driving force behind the development of more efficient detectors. Currently, the use of *area detectors* for diffraction studies is not only common in large dedicated facilities but also in nearly all in-house research laboratories involved in diffraction. It is without doubt that the research on aperiodic crystals has greatly benefited from the availability of new area detectors and large facilities, for the simple reason that the number of intensity data involved is much larger than for conventional crystals and may sometimes approach numbers which are more familiar with macromolecular systems. Moreover, the identification and indexing of satellite or quasicrystalline reflections are much facilitated by the use of area detector measurements.

As for many diffraction experiments, the most crucial element is the detector. By detector we mean here the complex chain of systems that starts from a diffracted intensity impinging on a device and ends up in a form of digital information stored in a memory device. For the study of aperiodic systems, most of the detectors currently in use are area detectors and therefore in what follows we shall neglect other types.

4.2.1 X-ray area detectors

The development of X-ray area detectors is driven by the necessity to combine a large spectrum of required characteristics such as high detective quantum efficiency (DQE), wide dynamic range, linearity of response, high spatial resolution, large active area, uniformity of response, and high count-rate capability. We shall first describe an important type of area detector which is still in use in a large number of laboratories, namely the *charge-coupled devices* (CCD). More recently, new types of X-ray imaging detectors are spreading not only in large facilities but also in home laboratories. We shall present the most recent development of pixel detectors for synchrotron and XFEL sources.

The CCD type of area detectors are very often used for the study of aperiodic materials (Tate *et al.*, 2001). In general, they consist of several components: as energy converter, mostly phosphor; an optical relay with fibre optics taper, and the imaging CCD. The input signal is transformed several times while crossing the sequence of stages. Therefore, in order to maintain a high DQE, the number of quanta per X-ray must be well over unity at each stage of the detecting chain. The phosphor must have a good stopping power and therefore must consist of elements with high atomic number. $Gd_2O_2S{:}Eu$, for example, offers the required high light output by emitting more than 200 photons per 8 keV X-ray. The fibre optics stage consists of a bundle of optical fibres which are heated and then pulled. In this process each individual fibre becomes tapered thereby reducing the image scale from front to back. With a 3:1 reduction, they can typically transmit up to about 15 per cent of the light photons. Visible photons are converted to charge carriers in the silicon of the CCD with a quantum efficiency varying from 40 (in the red part of the spectrum) to 5 per cent (in the blue part). When coupled to an efficient phosphor screen, the CCD records 10–30 electrons per 10 keV X-ray. CCD are usually cooled well below the room temperature to minimize the thermally generated dark current which originates from surface defects. Since the dark current is temperature dependent, the temperature must be regulated within ±0.1 K. One should mention here that CCDs must be carefully calibrated especially due to the geometrical distortion of the taper. In addition, the dark current has to be subtracted and the radioactive decay events must be accounted for before the diffracted intensity can be evaluated.

In order to take full advantage of the high-brilliance and high-frequency X-ray sources currently in use or in development, a parallel race is taking place with the development of new and more efficient *pixel detectors*.

The detection process consists of two main steps. In a first step, the X-ray energy is absorbed by a photodiode, creating thus electron–hole pairs which drift apart under the effect of an electric field. In a second step, the readout electronic measures and registers the drifting charges. The interconnection between the photodiode and the signal processing electronic is at the origin of a number of difficulties for which different solutions are proposed. Two different tendencies can be observed in order to improve the efficiency, namely the hybrid and monolithic types of detectors (Hatsui and Graafsma, 2015).

The advantage of the hybrid detector type is that both the absorption and the detection components can be optimized separately. Each photon is counted individually, permitting thus the discrimination of higher-energy photons (harmonics) from lower-energy photons such as those generating fluorescence. There are, however, some disadvantages, namely the pixel size limit of the order of 170 μm for, for example, the Pilatus detector (Eikenberry *et al.*, 2003). This limit is, however, improving and currently reaches 75 μm with the latest developments in electronics. In the presence of large fluxes as encountered in storage rings, the counting electronic might not be able to reach the necessary frame rate. The binding between the two components, the so-called bump binding, is a very time-consuming process and also the cause of high costs.

Monolithic detectors combine the photodiode and the microelectronic for signal processing on a single chip which is practically limited to silicon. Thus, the detection efficiency is limited for hard X-rays. This is, however, greatly compensated by the small pixel size of the order of 20 μm and the low-noise characteristic of the detector. The current trend is to develop *monolithic active-pixel sensors* (MAPS). The transistors of the active pixels implement amplification and processing functions. Some readout rates of the order of 1 MHz have already been attained, which is an enormous step forward when compared to CCD.

4.2.2 Neutron area detectors

The tendency to collect the largest possible portion of scattered beam in order to minimize the exposure time in collecting a complete data set has also been the driving force behind the development of imaging technique for neutrons. The technique of the *image plate* (Amemiya, 1995), which was frequently used for X-rays, can also be applied for neutron diffraction. The LADI detector (Cipriani *et al.*, 1995), for example, is based on this technique and is able to collect a very large portion of the scattered beam. It consists of a cylindrical image plate with the cylinder axis oriented perpendicular to the neutron beam. The sample is located on this axis and can be rotated about this same axis in order to explore the reciprocal space. This device is able to cover more than 60 per cent of the full solid angle. The image plate consists essentially of the same photosensitive material as for X-rays, *BaFBr* doped with Eu^{2+}. In addition, a neutron-to-X-ray converter makes the image plate sensitive to neutrons. Gadolinium is a good converter exhibiting a large resonance to thermal neutrons. This converter gives a DQE value close to 20 per cent and a point spread function of approximately 100 μm. As for the X-ray image plate, the dynamic range reaches 10^5. This device is ideal for collecting Laue type diagrams, thus taking advantage of the full white-beam spectrum with a wavelength spread of the order $0.8 \leq \lambda \leq 4.5$ Å. Figure 4.26 represents a neutron *Laue diffractogram* of the composite structure $La_2Co_{1.7}$ (Dušek *et al.*, 2000). The Laue technique has the great advantage that the data collection time is accelerated by over an order of magnitude in comparison to the more conventional techniques and a full set of data can be collected within a matter of hours.

4.2.3 Measurement techniques

The ever-increasing demand for beam time on large facilities has largely contributed to the development of efficient techniques to minimize the recording time for the measurement of the diffraction pattern of a specific compound. The method which is mostly used is to register the full diffraction pattern by recording a sequence of two-dimensional frames from successive small rotations of an arbitrarily oriented sample about a diffractometer axis. The illustration in Fig. 4.10 based on the model presented in Estermann and Steurer (1998) indicates the reciprocal space mapping onto a sequence of two-dimensional frames. This technique is able to map the complete reciprocal space from which the crystal specific parameters can later be deduced along with the diffracted intensities. This technique also has the advantage to register any intensity not belonging to Bragg peaks, including diffuse scattering which can be extracted for later treatment of the data. Each pixel on the area detector collects scattered intensity from a small area on the Ewald sphere surface. After completing a sufficiently small rotation, the pixel contains the intensity of the diffracted reciprocal space volume. The total solid angle corresponding to a pixel is $\Delta\Omega = (1/F^2 \cos^3 \Psi \Delta x \Delta y)$ if $\cos^3 \Psi$ is sufficiently small over a pixel; Ψ is the angle between the normal of the detector plane and the scattered beam vector pointing to the detector, F is the crystal-to-detector distance, and Δx and Δy are, respectively, the vertical and horizontal sizes of the detector's pixel. Owing to the $\cos^3 \Psi$ term, the area on the Ewald sphere varies with the pixel position. This purely geometrical term is a consequence of the planarity of the area detector and could be eliminated with a spherical detector.

The diffractogram presented in Fig. 4.11 represents a reciprocal layer of a decagonal Al-Co-Ni crystal reconstructed from a large number of area detector frames measured on a synchrotron. As is often the case in diffraction patterns extracted from aperiodic crystals, Bragg peaks appear seldom alone. Specialists

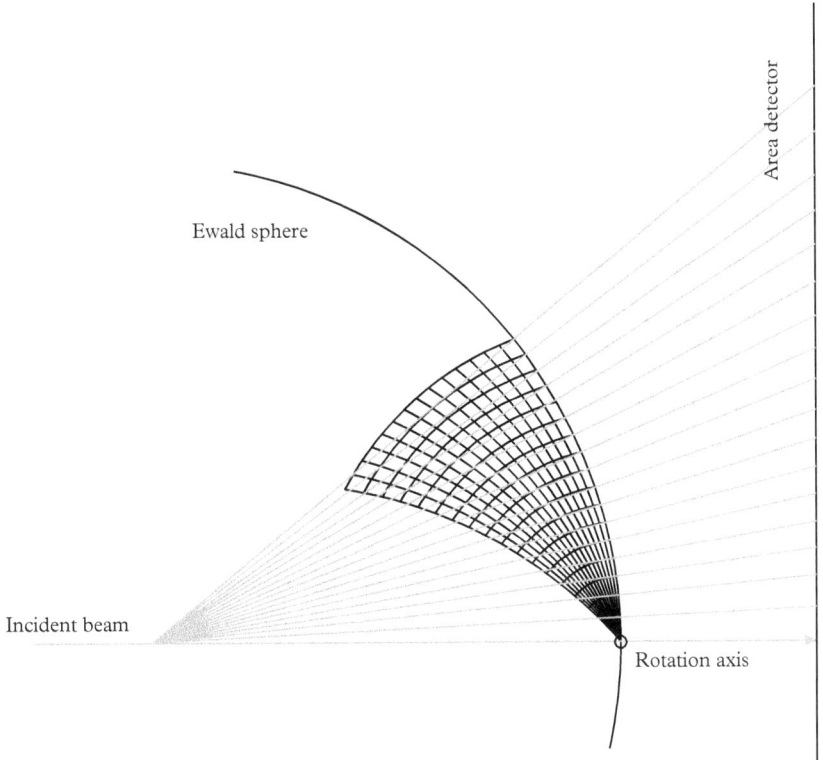

Figure 4.10 *Mapping of the reciprocal space onto an area detector. During the rotation of the sample, each pixel on the detector's plane collects scattering from a small area on the Ewald sphere.*

have developed software techniques to extract quantitatively not only the intensities of the Bragg peaks but also the non-Bragg diffuse intensities. The software necessary to index the diffraction pattern of conventional crystal has been adapted in order to index diffraction pattern of aperiodic crystals with rank of up to 6. Some of the commercially available software packages are now also able to index diffraction patterns of aperiodic crystals.

4.3 Determination of modulated phases and composites

4.3.1 Introduction

The methods of structure determination of modulated and composite crystals follow a pattern that is very similar to the resolution of conventional crystal structures.

172 *Structure*

Figure 4.11 *Decagonal Al-Co-Ni quasicrystal. This layer kindly provided by W. Steurer was reconstructed from a large number of area detector frames collected on a synchrotron.*

The essential steps consist of measuring the largest possible number of intensities and deducing the possible space group symmetries from the characteristics of the pattern and the systematic absences. After further reduction of the data, the solution and the refinement of the structure can be obtained by applying various automatic procedures depending on the complexity of the problem. A good number of freely and commercially available software tools are systematically used to solve most problems encountered in the structure determination process.

For modulated and composite crystals, the same pattern of analysis can be followed. However, as seen in Chapter 1, the requirement for additional dimensions in reciprocal space and consequently in direct space imposes some new conditions which we have to deal with. In addition, if the symmetry of conventional crystals is described by space group symmetry, we shall instead apply superspace group symmetry as described in Chapter 2. The structure factor expression which is the basis for modelling the crystalline structure also requires some adaptation in order to take into account that an atom in *superspace* is described by a periodic function of the internal space coordinate(s) which has to be determined. The steps towards the solution and analysis of modulated or composite crystals are more elaborate and the structure specialist does not benefit from the large number of methods and tools which exist for conventional crystals. The domain of aperiodic crystals

4.3.2 The structure factor of incommensurate structures

The resolution of a crystal structure is essentially a matter of comparing the experimental intensities with those deduced from a model of the structure. Once a satisfactory model is found, the structural parameters are optimized in order to best fit the observations. The square of the moduli of the calculated structure factors are directly related to the intensities. For incommensurate crystals, the structure factor is given in Section 4.1.1, Eq. 4.14. We can complete the structure factor with a probability p_j which is defined in Eq. 1.20. If the rank of the Fourier module is limited to 4, we obtain the following expression:

$$F(\mathbf{H}) = \sum_j f_j(\mathbf{H}) \exp(2\pi i \mathbf{H} \cdot \mathbf{x}_j) \qquad (4.19)$$

$$\times \int_0^1 dt\, p_j(t) \exp[2\pi i(\mathbf{H} \cdot \mathbf{u}_j(t) + h_4 t - \mathbf{H}^T B_j(t)\mathbf{H}/4)].$$

The sum is over all the j atoms with scattering factor f_j which are contained in the unit cell. The reciprocal vector $\mathbf{H}$ is defined in Eq. 2.29. The first part of the structure factor expression given in Eq. 4.19 preceding the integral sign is the same as for periodic structures in three dimensions. The integration over the internal space coordinate t depends on the *atomic modulation* expression $\mathbf{u}_j(t)$, which is not known a priori and has to be determined in the structure refinement process. The probability factor p_j usually describes *substitution modulation* or defects. The quadratic term describes the *anisotropic displacement parameters* (ADPs).

In general, the periodic displacements and probabilities are expressed in terms of a series of Fourier components. The number of terms which are used in the summation is, in general, limited by the number of experimental observations. In practice, this number is small. The conditions for real-valued modulation functions are also indicated below.

$$\mathbf{u}_j(t) = \sum_n \mathbf{u}_{jn} \exp(2\pi i n t); \qquad \mathbf{u}_{j,-n} = \mathbf{u}_{jn}^*$$

$$p_j(t) = \sum_n p_{jn} \exp(2\pi i n t); \qquad p_{j,-n} = p_{jn}^*. \qquad (4.20)$$

For structure refinements, the integral variable t applied in Eq. 4.19 is often preferred to the x_4 for practical calculations. The reason can obviously be found in Fig. 4.3, where real structures are observed on any line parallel to V_E, that is, with constant values of t. This is particularly convenient if chemical constraints are applied, for example, on bond or angle distances. Both formulations are, however, equivalent and are related by the expression (cf. Fig. 4.3):

$$x_{j4} = \mathbf{q} \cdot \mathbf{x}_j + t. \tag{4.21}$$

4.3.3 Possible expressions of the modulation functions

Let us first postulate that the modulation function in Eq. 4.20 consists of one single term. We have seen in Chapter 1 (Eq. 1.16) that by using the Jacobi–Anger relation, the structure factor for a modulated structure can be expressed in terms of *Bessel functions* which can be approximated numerically. For simple cases where the atomic modulations do not deviate too strongly from this approximation, this might be sufficient. Unfortunately, in most practical cases, this simple approximation is not valid and some additional terms should be included in the expression of the displacement and occupation modulations. Let us first define the truncated trigonometric series with complex coefficients:

$$\chi_N(\varphi) = \sum_{k=1}^{N} (Z_k \sin k\varphi + z_k \cos k\varphi). \tag{4.22}$$

This expression will appear in the structure factor for general modulations expressed by Eq. 4.20. The common approach is to use the following infinite sum of products involving ordinary Bessel functions:

$$\exp(i\chi_N) = \prod_{k=1}^{N} P_k(Z_k, z_k), \tag{4.23}$$

$$P_k(Z_k, z_k) = \sum_{n=-\infty}^{\infty} \mathcal{J}_n(\varpi_k) \Theta_k^n \exp(ikn\varphi), \tag{4.24}$$

$$\Theta_k = (Z_k + iz_k)/\varpi_k, \tag{4.25}$$

$$\varpi = (Z_k^2 + z_k^2)^{1/2}. \tag{4.26}$$

This approach to the analysis of anharmonic incommensurate structures has been developed in Petříček et al. (1985). However, we can take advantage of some recent theoretical works on a new class of functions called *generalized Bessel functions*, introduced in Dattoli et al. (1990). These functions share the same basic properties as ordinary Bessel functions, such as recurrences, addition and multiplication theorems, and expressions for derivatives. By adopting a particular type of generalized Bessel function, Eq. 4.23 can be reformulated as

$$\exp(i\chi_N) = \sum_{n=-\infty}^{\infty} \mathcal{J}_n(\{Z,z\}_N) \exp(in\varphi), \tag{4.27}$$

where the following notation is introduced:

$$\{Z, z\}_N = (Z_1, z_1, \ldots, Z_N, z_N). \tag{4.28}$$

By introducing this new function only the summation remains in Eq. 4.23 and the product disappears.

In the harmonic case (N = 1), the result is the familiar *Jacobi–Anger expansion* with the coefficients

$$\mathcal{J}_n(Z_1, z_1) = \mathcal{J}_n(\varpi_1)\Theta_1^n. \tag{4.29}$$

Because both even and odd terms are included in the trigonometric series, this coefficient is multiplied by a complex number to the power equal to its order. In our case, it is also more convenient to consider $\mathcal{J}_n$ as a function of Z_1 and z_2 rather than a function of ϖ and Θ_k.

The integral representation of the Bessel function

$$\mathcal{J}_n(z) = \frac{1}{2\pi i^n} \int_0^{2\pi} d\varphi \exp(iz \cos \varphi + in\varphi) \tag{4.30}$$

can be generalized and we obtain the following expression (Paciorek and Chapuis, 1994):

$$\mathcal{J}_n(\{Z, z\}_N) = \frac{1}{2\pi} \int_{-\pi}^{\pi} d\varphi \exp(i\chi_N - in\varphi). \tag{4.31}$$

This type of integral appears in the structure factor expression for any one-dimensional incommensurate structure. It can be interpreted as an integral representation of a new special function which is called an *N*-variable *N*-parameter *generalized Bessel function*. It has to be evaluated numerically for an anharmonic structure. However, numerical recipes for estimating this integral, which will eliminate the need for time-consuming numerical integration, are improving.

With the help of the generalized Bessel function, we can reformulate the *structure factor* of an incommensurate structure provided that the number of harmonic terms in the modulation function is finite:

$$F(\mathbf{k}) = \sum_{\mu, g} M^\mu f^\mu \Omega^g \overline{F}^{\mu, g} W^{\mu, g}. \tag{4.32}$$

The summation is over all the symmetry-independent atoms μ contained in the unit cell and the g symmetry operations of the space group. The terms M^μ and f^μ are the multiplicity and the complex scattering factor of atom μ and all quantities are evaluated for the reflections with four indices. The term Ω^g is a complex factor due to a non-primitive translation whereas $\overline{F}^{\mu, g}$ is the contribution of the structure factor to the basic structure, that is, the part on which the modulation is applied. The term $W^{\mu, g}$ contains all the terms due to the modulation expressed with the help of the generalized Bessel function defined in Eq. 4.31. The parameter P_n^μ

describes any atomic parameter affected by a modulation subject to the condition that it must be a complex conjugate of P^μ_{-n}:

$$W^{\mu,g} = \sum_{n=-M}^{M} P^\mu_n \mathcal{J}_{-m(g)-n}(\{Z^{\mu,g}, z^{\mu,g}\}_N). \quad (4.33)$$

The above expression of the structure factor constitutes an analytical solution of the structure factor and all its partial derivatives with respect to both the average structure and all modulation parameters for all one-dimensional incommensurate structures.

4.3.4 Additional expressions of modulation functions

The large number of modulated structures analysed to date by diffraction methods has induced a series of studies which has greatly enlarged the spectrum of possible modulation functions which can be applied for the modelling of atomic structures in the superspace description. Let us first introduce the *crenel function*, which is defined according to (Petříček et al., 1995) and represented in Fig. 4.12. This type of function has already been presented earlier in Fig. 2.15 and is useful, for example, to describe domains of structure where the same atomic position can be occupied by different types of atoms or when an atom can occupy one of two (or more) possible positions. It can also represent cases of missing atoms (defects) as illustrated in Fig. 4.12a.

The crenel function is periodic but discontinuous and is defined according to the following expressions:

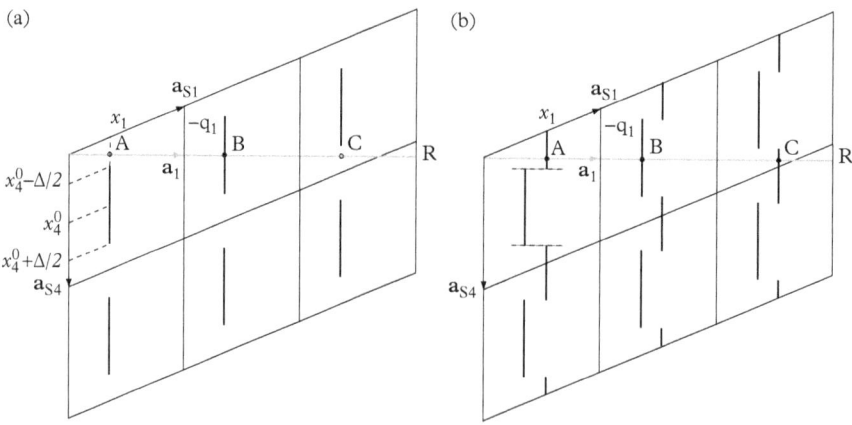

Figure 4.12 *Embedding of different cases of crenel functions in superspace. (a) Occupation modulation. For t = 0, the atom is present at position B but absent at A and C. (b) Displacement modulation. In this case the corresponding atom can occupy either one of two possible positions. A third possible case of substitution modulation is illustrated in Fig. 4.40.*

$$p(x_4) = 1 \quad x_4 \in \left\langle x_4^0 - \Delta/2, x_4^0 + \Delta/2 \right\rangle, \qquad (4.34)$$

$$p(x_4) = 0 \quad x_4 \notin \left\langle x_4^0 - \Delta/2, x_4^0 + \Delta/2 \right\rangle, \qquad (4.35)$$

where Δ and x_4^0 are the width and the centre of the crenel function. Formally, the crenel function can be defined as the difference between two Heaviside functions of amplitude 1. The Fourier transform of this function has the following form

$$P_m(\Delta, x_4^0) = \exp(2\pi i m x_4^0) \sin \pi m \Delta / \pi m. \qquad (4.36)$$

This type of modulation generates strong satellites up to very high order m. In principle, the number of satellites observed should be higher than for harmonic modulations. This relation indicates also that the intensities of satellites are decreasing only as $1/m^2$.

The crenel function has also been further generalized by combining both density and (linear) displacement modulations (Petříček et al., 1990). This is called the *sawtooth function* and is represented in Fig. 4.13. The vector **u** characterizing the displacement is given by

$$\mathbf{u} = 2\mathbf{u}_0(x_4 - x_4^0)/\Delta \quad \text{for} \quad x_4 \in \left\langle x_4^0 - \Delta/2, x_4^0 + \Delta/2 \right\rangle. \qquad (4.37)$$

The linear modulation function centred at x_4^0 is only defined within the interval Δ. The contribution to the structure factor is given by

$$F = f \exp(2\pi i \mathbf{k} \cdot \mathbf{r}) \exp(2\pi i m x_4^0) \Delta \frac{\sin \varphi}{\varphi}, \qquad (4.38)$$

where

$$\varphi = \pi m \Delta + 2\pi \mathbf{k} \cdot \mathbf{u}_0 \qquad (4.39)$$

Here again, the interval Δ of the modulation function and its slope can be adjusted. Figure 4.13 shows that within a limited domain, the atoms are periodically distributed along R but with a periodicity different from the basis vectors.

By combining the three types of modulation functions, harmonics, crenel, and sawtooth functions, the structure specialist can accommodate most of the cases which have been encountered in practice up to this point.

4.3.5 Practical aspects of structure determination and refinement

Many calculations related to incommensurate structures require elaborate numerical techniques to evaluate the integral terms contained in the structure factor expression 4.19. The first approach that was adopted to evaluate the integral expression was based on the *Gaussian integration method*. This method was used

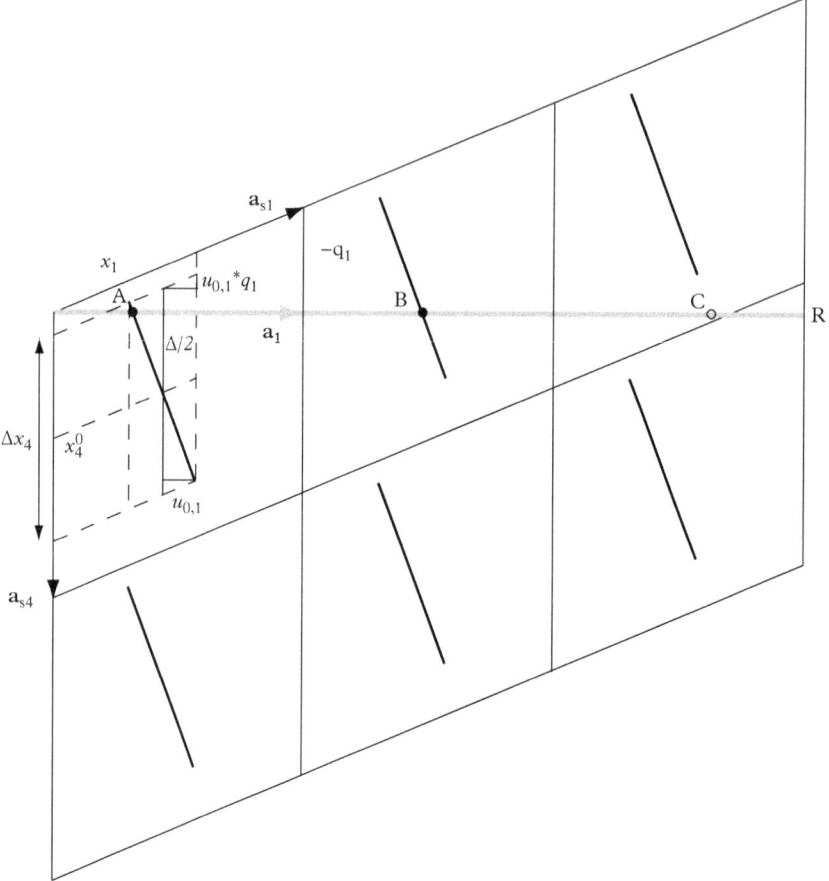

Figure 4.13 *Embedding of a sawtooth function in superspace. In real space and for $t = 0$, the atoms are present at positions A and B but absent at C.*

in the refinement software package *REMOS* (Yamamoto, 1982b). An alternative approach is to replace the integration over the internal coordinate by an infinite series of Bessel functions, obviously truncated for actual calculations. This method has been implemented in the software package *JANA* (Petříček and Dušek, 2000). The search for more efficient algorithms was induced by the ever-increasing complexity of the modulated structures which are studied. A new algorithm was proposed in Paciorek and Chapuis (1992) by applying the fast Fourier transform (FFT) as a substitute for numerical integration of the structure factor and all partial derivatives with respect to the modulation function parameters. Further improvement of this algorithm could be achieved by the inclusion of attenuation and end-point corrections (Paciorek and Chapuis, 1992). The FFT algorithm has two main advantages. First the accuracy is improved relative to other integration methods and second, the computing time is far less sensitive to the number of harmonic terms used to describe the modulation function. This algorithm has been included in the refinement program *MSR* (Paciorek and Kucharczyk, 1985) in complement to the integration method based on the *generalized Bessel function* and described in Eq. 4.32.

A large number of modulated structures known to date have been solved by single crystal diffraction. The powder method has also been used successfully in some cases (see e.g. (Morozov *et al.*, 2006)) but in combination with other methods, for example electron diffraction. We shall therefore focus here on the diffraction by single crystals. Nowadays, the intensities are obtained from area detectors, either on laboratory equipment or in synchrotron facilities.

The first difficulty in dealing with incommensurately modulated structures is to recognize that the collection of intensities cannot be indexed with the standard procedures developed for periodic crystals. This can be concluded from the large portion of intensities for which the indices deviate significantly from integer values. The second difficulty is to decompose the full spectrum of intensities between main and satellite reflections. The separation between the two categories is not always easy, especially if the rank of the diffraction spectrum is not known a priori or in the presence of complex twinning. Some practical experience is required to solve this task. Currently, most of the commercially available single crystal diffractometers have introduced the possibility of indexing intensities with up to six indices. Once the complete data set is available, the lattice constants and the modulation vector(s) with estimated errors can be obtained by an optimization procedure (e.g. *NADA* (Schönleber *et al.*, 2001)).

Another difficulty linked with the resolution of the measurements might arise when satellite reflections are very close to each other or when satellite reflections are very close to main reflections. In this case, the measuring software is in general not able to adequately evaluate the contributions from each individual reflection. However, the problem can be solved by integrating the contributions of non-resolved peaks and modifying accordingly the expression of the structure factors of the cumulated intensities. This method has been successfully applied, for example, for the five-dimensional refinement of melilite, a natural mineral (Bindi *et al.*, 2001).

4.3.5.1 Superspace group determination and structure solution

The tools available for the derivation of superspace groups are not nearly as complete as for conventional crystal structures. The first source of information can be found in Volume C of the *International Tables for Crystallography* (Janssen *et al.*, 1992) listing all possible superspace groups in (3 + 1) dimensions including systematic absences. For higher-dimensional superspace groups up to (3 + 3) dimensions, the reader can refer to the tables given on dedicated websites (Yamamoto, 1999; Stokes *et al.*, 2011). An additional source of information on the systematic absences in superspace group can also be found by using the

JANA software package (Petříček *et al.*, 2014). As with standard space groups, the systematic absences depend directly on the selection and orientation of the unit cell. Moreover, the choice of the modulation vector is also arbitrary, thus making a direct matching with an established table very unlikely. With regard to the very large number of possible cases, it is thus not practically possible to establish tables which would cover more than some predefined standard settings. Therefore, in dealing with the derivation of superspace groups and the corresponding systematic absences, the user is strongly urged to become familiar with the method for the derivation of systematic absences as listed in Appendix A.9. More detailed information along with some specific examples is given in Volume C of the *International Tables for Crystallography* (Janssen *et al.*, 1992).

The ambiguity regarding the presence or absence of the centre of inversion still remains and in general cannot be resolved from the systematic absences only. This ambiguity is mostly resolved from considerations of the convergence of the refinement or from geometrical or chemical criteria applied to the resulting structure model. It is surprising to note that the statistical model based on normalized structure factors E has rarely been implemented in the resolution of aperiodic crystal structures.

The solution of the structure occurs in two steps. First, the average structure is solved and in a second phase, the modulated structure is solved. Here, the *average structure* is the hypothetical structure which is obtained by neglecting the satellite reflections. Only the main reflections are used to solve this structure. In general, it is already known or can be solved by the usual techniques developed for conventional crystal structures, for example Patterson, direct methods, or charge flipping. The main features of this refined structure will be correct, but one can expect some unusually large ADPs pointing to the specific atoms subject to the displacive modulation. In many instances, this approximate structure is sufficient to initiate the refinement in superspace. A minimal number of harmonic terms are introduced first with zero amplitudes and phases for the atoms most affected by the modulation as indicated by the corresponding ADPs in the average structure refinement. The number of *harmonic terms* can be increased stepwise in order to improve the fit between observed and calculated structure factors.

4.3.5.2 Solution from a commensurate approximation

The method described above may fail in some cases where the modulation strongly deviates from the sinusoidal case or if the average structure is ill defined. For example, this can be the case if the modulation is due to an occupation distributed over two atomic positions. Other solution methods have to be adopted in order to solve such a modulated structure. One possibility is to solve a commensurate approximation or, in other words, a *superstructure*. This can be obtained by selecting a modulation vector **q** with the closest rational components

$$\mathbf{q} = q_1 \mathbf{a}_1^* + q_2 \mathbf{a}_2^* + q_3 \mathbf{a}_3^* \approx \frac{n_1}{m_1} \mathbf{a}_1^* + \frac{n_2}{m_2} \mathbf{a}_2^* + \frac{n_3}{m_3} \mathbf{a}_3^*, \qquad (4.40)$$

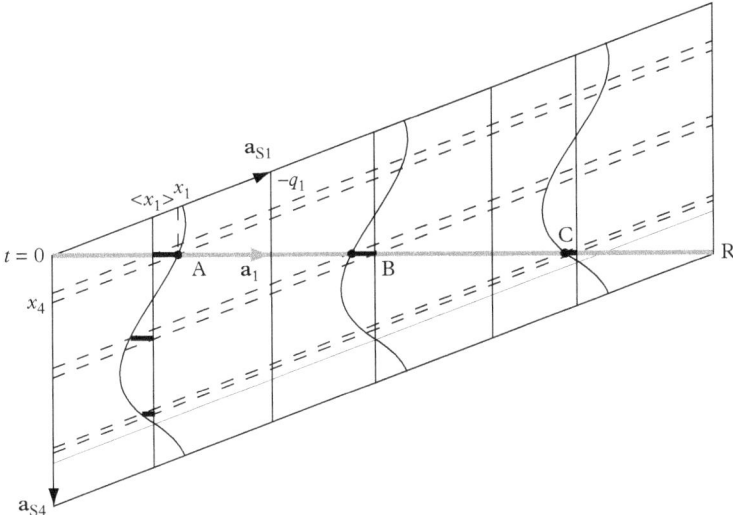

Figure 4.14 *Generation of the modulation curve in order to obtain an initial approximation for the refinement of an incommensurate structure. We suppose here that $q \approx q_{approx.} = \frac{1}{3}a_1^*$ is a commensurate but sufficiently close approximation for the mapping of the atomic positions.*

where n_i and m_i are integers. Figure 4.14 illustrates the case for $q_1 = \frac{1}{3}$. The line R crosses a node of the superspace lattice with a periodicity of $3\mathbf{a}_1$. The sequence of points A, B, and C located on R thus repeats periodically. This structure is therefore a much better approximation than the average structure.

The approximate shape of the modulation function can be obtained by mapping all the deviations of the individual atoms A, B, and C from their average position $\bar{x}_1$ as in Fig. 4.14. The amplitude and phase of the modulation curve can be introduced in the structure model in order to initiate the refinement of the structure in superspace. This method has been successfully applied for the modulated structure of *quininium (R)-mandelate* (Schönleber and Chapuis, 2004a). In this particular case, the modulation vector **q** was approximated by $\frac{1}{3}\mathbf{a}_1^* + \frac{1}{5}\mathbf{a}_3^*$. In other words, the supercell is 15 times larger than the original cell. The challenge remained, however, to locate 15 formula units which are structurally very close. Here again, the conventional methods for solving superstructures can be applied. In particular, the software *DIRDIF* (Beurskens et al., 1996) was best suited for solving this type of problem.

The heavy atom method, which is frequently used to solve conventional structures, can under some circumstances also be applied to find the amplitude of displacement of the heavy atom. This method has been successfully applied (Peterkova et al., 1998) and is illustrated in Fig. 4.15. In this particular case, the two atoms are related by a glide plane with a glide component along **a**. Moreover,

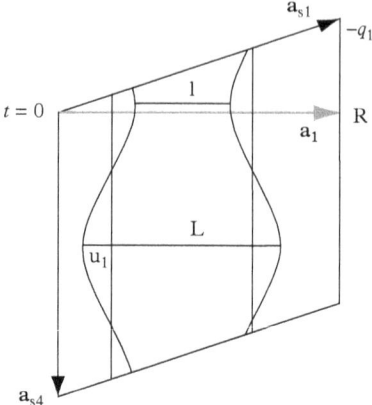

Figure 4.15 *Symmetry related modulations shifted by π. In this particular case, the amplitude u_1 of the modulation can be recovered from the Patterson function and is $\frac{1}{4}(L-l)$.*

the special position of the heavy atom induces a phase shift of the displacement modulation by π. The amplitude of the modulation function can be directly extracted from the four-dimensional *Patterson map* by measuring the smallest and largest extension of the peak along x_4. The displacement amplitude component of the corresponding heavy atom is a quarter of the difference between these extrema. This method is obviously not general but can be very helpful to obtain the initial displacement amplitude.

4.3.5.3 Structure refinement and validation

Once a suitable approximation is obtained for the modulated structure, the refinement of the structure can be initiated. If the shape of the modulation curve deviates from the sinusoidal shape, additional *harmonic terms* must be added in order to approximate the modulation curve which best fits the experimental data. However, the limitation is given by the number of observations, that is, the measured intensities. The ratio of the number of parameters to be adjusted relative to the number of observations should ideally be of the order of 10, but in any circumstance should not be below 5. As in conventional crystal structure determination, the quality of the refinement can be checked by a *difference Fourier synthesis*

$$\Delta\varrho(\mathbf{r}_S) = \frac{1}{V} \sum_{\mathbf{H}} \Delta F(\mathbf{H}) \exp(-2\pi i \mathbf{r}_S \mathbf{H}), \tag{4.41}$$

where $\Delta F(\mathbf{H})$ is the difference between the observed and calculated structure factors $F_{obs}(\mathbf{H}) - F_{calc}(\mathbf{H})$, and $\mathbf{r}_S$ indicates the position of an atom in superspace.

The two-dimensional section of the difference Fourier can either include an internal and an external dimension or two external dimensions. The first type of difference Fourier can be used to select the best type of modulation function as described in Section 4.3.4 which should be applied. At this point, one has to decide if a *crenel* or *sawtooth* function is better appropriate than a series of harmonic terms to describe the modulation function. Figure 4.14 illustrates an example where the shape of the modulation function representing atom N13 of *quininium (R)-mandelate* (Schönleber, 2002) is better represented by a sequence of three sawtooth segments.

In the course of refinement of modulated structures, it is often mandatory to restrain some parameters in order to compensate for the lack of resolution of the experimental data or to reduce the number of parameters to be refined. In organic materials, for example, it is often convenient to refine molecular parts or group of atoms. In conventional structure analysis, constrained refinements belong to the standard tools at the disposal of the structure specialist. For modulated structures, there is an additional subtlety which is linked to the nature of the superspace formalism. The geometrical constraints should be applied to the physical object which is given by the three-dimensional cut of the hyperspace. This is illustrated in Fig. 4.14, which shows that the constraints can only be applied for constant values of t. The algebraic background to deal with the constrained refinement has been treated in Paciorek *et al.* (1996).

The method of *maximum entropy* has also been applied in superspace (van Smaalen *et al.*, 2003). The method allows a reconstruction of the electron density

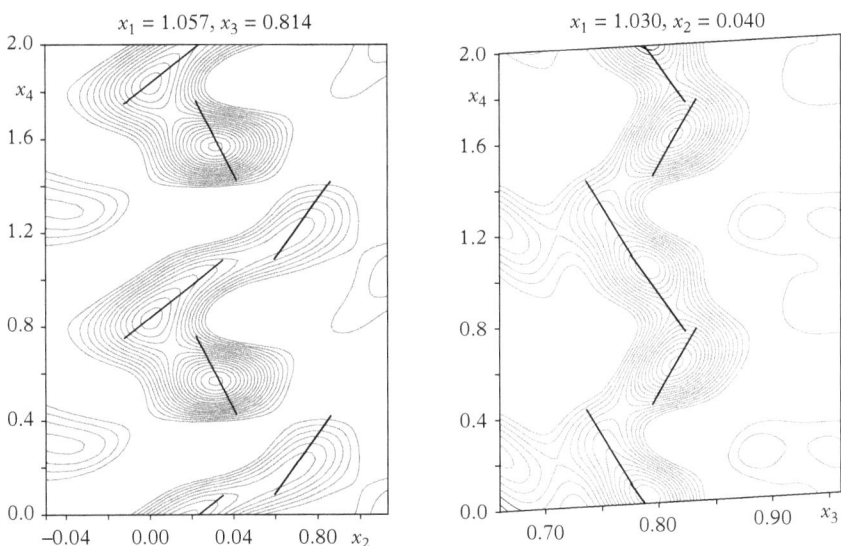

Figure 4.16 *Fourier maps of atom N(13) in quininium (R)-mandelate. Sections $x_2 - x_4$ and $x_3 - x_4$ are represented along with the corresponding segments of the sawtooth functions.*

from phased structure factors and provides a model-independent estimate of the modulation functions. This can be particularly useful as an a posteriori check of a refinement when the quality of the data is not of highest standard.

4.3.6 Ab initio methods

In the previous sections, we have seen that the *methods of resolution* proceed essentially in two steps: first, an average structure or a commensurate approximation is found and, in a second step, the modulations are determined. It is also interesting to note that *direct methods* which are so effective in conventional structure solutions have seldom been applied for the solution of aperiodic structures. (See, however, (Fan *et al.*, 1993; Fan, 1999).) Recently, new ab initio methods have been proposed which seem to be particularly suitable for the solution of aperiodic structures. In particular, the *charge-flipping* method proposed in Oszlányi and Sütő (2004, 2005) represents such a large step forward that it is worth describing in some details.

The power of the method lies in the very small number of prerequisites. Only a good and complete data set at atomic resolution and at a single wavelength is necessary. No further data like atomic contents of the average unit cell or symmetry considerations are required. The method is independent of the rank of the structure, it is iterative, and makes use of the discrete Fourier transform alternating thus between real and reciprocal space.

The method consists first of attributing arbitrary phases between 0 and 2π to the modules $|F_{\mathbf{H}}^o|$ obtained experimentally and without further consideration of scaling and origin determination. By inverse Fourier transform, the density $\rho(\mathbf{x})$ can be obtained. This density obviously contains positive and negative values. A new model $g(\mathbf{x})$ can be derived from this density by switching the sign (charge flipping) of any value below a specified threshold. All the values above this threshold remain unchanged. Later we shall give some more details on the selection of the threshold value. It is straightforward to calculate the Fourier transform of this new density $g(\mathbf{x})$ according to

$$G_{\mathbf{H}} = \int_V g(\mathbf{x}) \exp(2\pi i\, \mathbf{H}\mathbf{x})\, dV. \qquad (4.42)$$

The new structure factors $F_{\mathbf{H}}$ are then calculated from the expression

$$F_{\mathbf{H}} = |F_{\mathbf{H}}^o| \frac{G_{\mathbf{H}}}{|G_{\mathbf{H}}|}. \qquad (4.43)$$

In other words, the phases $\phi_{\mathbf{h}}$ derived from the calculation of G are attributed to the modulus of the observed structure factors. All unknown structure factors are reset to 0 except for F_0, which is set to $F_0 = G_0$. This procedure is recycled as many times as necessary until a convergence is obtained, at which step the density $\rho(\mathbf{x})$ should represent the solution of the structure. The iterative process can be schematically represented below, where *FFT* and FFT^{-1} represent the fast Fourier transform and its inverse, respectively:

$$\begin{array}{ccc} \rho & \xrightarrow{Inversion} & g \\ FFT^{-1} \uparrow & & \downarrow FFT \\ F & \longleftarrow & G \end{array} \qquad (4.44)$$

One may wonder why this method based on such a simple algorithm is able to solve (periodic or aperiodic) structures without any a priori information. The first experiences indicate that unit cells containing up to a few hundred atoms can be solved within a reasonable number of iterations, which can vary between ten or so to thousands. Apparently, this upper limit is still diminishing with further improvement of the algorithm.

The charge-flipping method considers that all structures are triclinic with space group $P1$ or its generalization for ranks larger than 3. This simplifies enormously the structure solution procedure. The identification of the proper symmetry of the structure is left to a later stage once all the atomic positions in the unit cell are known.

The computation of the electronic density $\rho(\mathbf{x})$ requires some particular considerations. In practice, only densities for values of $\mathbf{x}$ located on grid points can be calculated. As the number of observations is limited by $\|\mathbf{H}\|_{max} = 1/d_{min}$, the grid spacing necessary for the computation of $G_\mathbf{H}$ must be of the order of $d_{min}/2$. This means that the grid size can be limited to 0.4 Å with atomic resolution data. The efficiency of the current *FFT algorithm* along with the ever-increasing power of the CPU units makes the charge-flipping method very attractive.

The evolution of the charge-flipping algorithm presents, in general, four characteristic phases which are represented in Fig. 4.17, depending on the number of iterations. Each step is characterized by a reliability factor R which is defined by the following relation:

$$R^{(n)} = \frac{\sum_\mathbf{H} ||G_\mathbf{H}^{(n)}| - |F_\mathbf{H}^o||}{\sum_\mathbf{H} |F_\mathbf{H}^o|}. \qquad (4.45)$$

The first phase is, in general, very short and followed by an approximately constant phase, the length of which can vary enormously depending on each structure. The third step is a transition phase indicating that convergence is setting in. Finally, in the last phase, the R-value remains low and approximately constant. This is an indication that the iteration has converged to a stable solution.

In order to complete the description of the charge-flipping algorithm, some further considerations regarding the *density function* $\rho(\mathbf{x})$ should be mentioned. For crystalline structures, this density is practically zero everywhere except on the atomic positions. The computation of this density is limited by the number

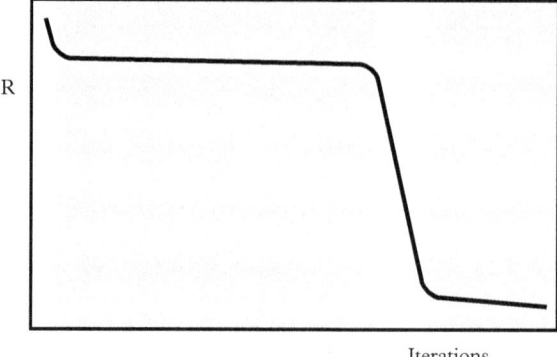

Figure 4.17 *Typical schematic evolution of the charge-flipping method with the cycle number. The four different regimes are indicated by alternating grey shadings.*

of terms $\|\mathbf{H}\| \leq \|\mathbf{H}\|_{max}$ accessible to the experimental measurements. The consequence for the Fourier series computation is the existence of series termination effects which induce some density fluctuations in the interval $[+\delta, -\delta]$. Therefore, the efficiency of the charge-flipping is greatly improved if all the values below the threshold $\rho(\mathbf{x}) < +\delta$ are inverted. The parameter δ is the only variable in the algorithm and its value can be expressed as $\delta = k \cdot \sigma$ where σ is the standard deviation of the electron density map. It appears that k is close to 1 and therefore δ can easily be approximated a priori. It turns out that approximately 80 per cent of the densities $\rho(\mathbf{x})$ are inverted in the iterative process. The charge-flipping algorithm can be further improved by introducing a special treatment of the weak intensities. The technique consists of applying an additional phase shift $\Delta\phi = 90°$ to the structure factor, thus yielding an improved expression $F_{\mathbf{H}} = G_{\mathbf{H}} \exp(i\Delta\phi)$.

The algorithm described above has been implemented as published in Palatinus (2004) and Palatinus and Chapuis (2007), and tested on experimental data of a number of incommensurate structures. The selection of structures covered both inorganic and organic materials with average unit cell varying from a few hundred to a few thousand cubic angstroms including also a composite structure. All the tested structures could be solved with great precision with the charge-flipping algorithm. For example, heavy atoms could be located within 0.05 Å in comparison to the refined structure. The values of δ vary typically between 0.1 and 1.0 for absolutely scaled intensities. The exact value of δ is not critical and only marginally influences the speed of convergence. The positions of all the non-hydrogen atoms for all tested cases could be located without exception on the Fourier maps thus reflecting the power of the solution method. Moreover, in $[(C_6H_5)_4P]_2[TeBr_6(Se_2Br_2)_2]$, for example, 4086 phases out of 4247 reflections had identical values compared to the refined model. The charge-flipping algorithm is able to solve very complex incommensurate structures like quininium (R)-mandelate (Schönleber and Chapuis, 2004*a*) consisting of 18 independent but nearly identical molecular units in the smallest reasonable supercell, each consisting of 35 atoms.

Thus, it appears that the structure derived from the charge-flipping algorithm is very close to the final model. Then it is an easy task to identify the symmetry elements of the resulting structure in order to initiate the refinement of the incommensurate structure directly in superspace.

We shall see later that the charge-flipping method can equally well be applied to the solution of both decagonal and icosahedral quasicrystal structures.

4.3.7 Relation between harmonics and satellite orders

The problem regarding the order of the observed satellite and the *number of harmonics* orders which are used to describe the shape of the modulation function in a structure refinement has been studied by various authors (Böhm, 1975; Korekawa, 1967) even before superspace groups and modulation functions were introduced. These studies revealed that the principle claiming that the satellites of a given order are caused by harmonics of the same order is only valid in the very special case of a purely occupational modulation. One might attempt to justify this rule for displacive modulations too by assuming that the amplitudes are small. By expanding the Bessel functions into series and retaining only the first term, one can obtain an expression for the structure factor formula similar to the occupational modulation (Paciorek and Kucharczyk, 1985). However, this approximation is so crude that it cannot be used in practice.

In all other cases, especially for displacive modulations with arbitrary amplitudes, this rule is not valid. On one hand, displacive modulations with one harmonic lead to diffraction patterns with satellites of all orders. This is illustrated in Fig. 4.18, where satellites of different orders are clearly present. On the other hand, an anharmonic displacive modulation (e.g. rectangular) can even give pseudo-extinction rules on satellites (Böhm, 1975).

It is usually impossible to obtain reasonable approximations if only the first few terms are used. It is also generally untrue that adding more harmonics will perturb the shape of the modulation functions and in many cases it is just the opposite. It is even desirable to add more harmonics, because they have a smoothing effect on unwanted ripples and other oscillations. This effect is equivalent to the *Fourier series truncation*. The result of this extension may be better visible on some other quantities (e.g. interatomic distances) rather than directly on individual functions.

In the refinement of incommensurate structures the only limitation for the harmonic range is a reasonable ratio between the number of parameters and the number of observations. The significance of the refined parameters can be studied using standard tests (e.g. Hamilton's test), as in a conventional structure refinement.

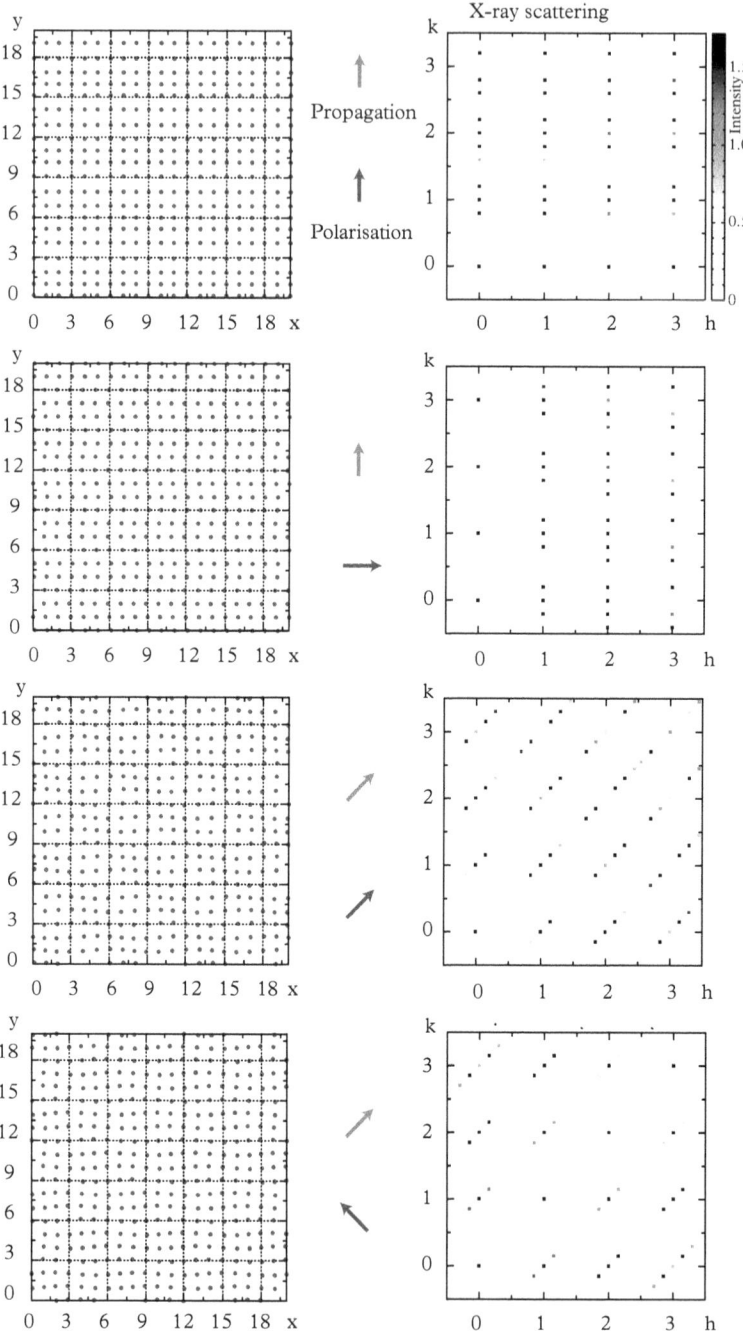

Figure 4.18 *Simulated diffraction patterns of various displacive modulations with a single harmonic wave combining different propagation and polarization directions.*

4.3.8 Composite structures

A general description of composite structures has been introduced in Sections 1.2.5 and 2.2.2. Numerous examples of composite structures have been described in the literature (see e.g. (Yamamoto, 1993)) and they represent an important category of incommensurate structures. In general, two different categories of composites can be distinguished. The first consists of a layer structure with a pair of alternating layers periodically stacked along the normal to the layers. Each layer is periodic, but two consecutive layers are mutually incommensurate. *Cannizzarite* $[(Pb, Bi)S]_x[(Pb,Bi)_2S_3]$ with $x \simeq 1.7$ or $[PbS]_x[VS_2]$ with $x \simeq 1.12$ belongs to this category. (The convention for the chemical formula is to enclose each subsystem in square brackets.) The other category of composites consists of host guest types of structures forming channels which are parallel in one direction. $La_2Co_{1.7}$ is an example of a pseudo-hexagonal structure where each atom type has its own independent periodicity along the c-direction.

A characteristic feature of the diffraction pattern of composite crystal structures is the prominent appearance of two or more sets of three-dimensional sublattices each caused by the corresponding substructures. These main reflections distinguish the composite crystals from the modulated structures. In addition, weak satellite reflections appear indicating the effect of interactions between the substructures. In general, these satellites are more evident from electron diffraction than from X-ray diffraction. As noticed already in Chapter 2, the embedding of the three-dimensional structure into the $(3 + d)$-dimensional space is not unique and there exist an infinite amount of equivalent solutions. We shall illustrate this by looking at the diffraction pattern of a composite structure and its embedding in superspace. Figure 4.19a represents a case with two independent periodicities along the reciprocal space dimension indicated by V_E. The embedding shows clearly the main reflections along the two main axes and additional satellite reflections. By deforming this figure with a vertical shear strain in the positive and negative direction, we can obtain the two diagrams indicated in b and c. All the representations give strictly the same pattern along V_E. Fortunately, it is always possible to find an equivalent embedding which corresponds to the standards of the modulated structures. Figure 4.19a illustrates another important difference between incommensurate and composite crystals: in the former, all the main reflections are located along V_E, which is not the case for composites.

The possible superspace group of the composite structure and consequently for the substructures can be obtained from the analysis of the average substructures and by considering the extinction rules including the satellite reflections. As the main reflections of one substructure are the main or satellite reflections of the other, the number of possible superspace groups are limited even in the absence of true satellites, namely those reflections which are not main reflections of any substructure.

For composites, the structure factor is the weighted sum of the individual substructures and is given by

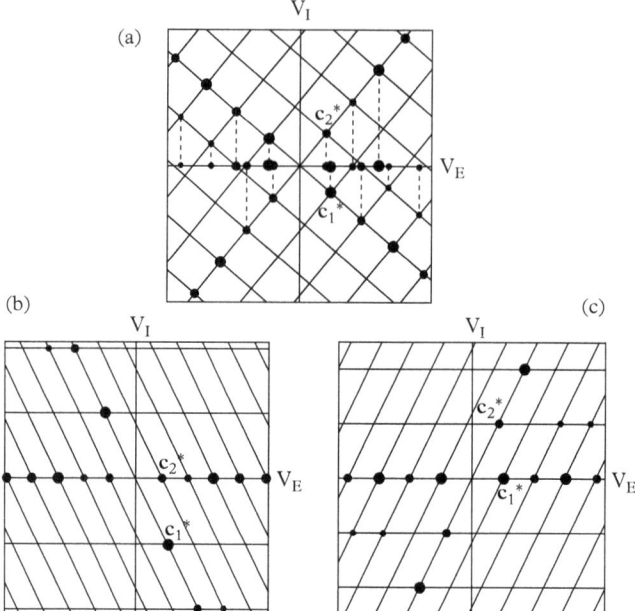

Figure 4.19 *Superspace extension of a two-component composite structure. (a) General embedding of the reciprocal space. The vertical dotted lines indicate the positions of the main (along c_1^* and c_2^*) and satellite (all the others) reflections on V_E. (b) and (c) Two possible vertical shear strain deformations bringing one of the reciprocal vectors c_1^* and c_2^* along the external space direction. In all cases, the positions of the reflections in V_E are invariant.*

$$F(\mathbf{H}) = \sum_{\nu=1}^{N} \frac{1}{v_\nu} F_\nu(\mathbf{H}). \tag{4.46}$$

Here N is the number of substructures each with its corresponding volume v_ν. The individual structure factor $F_\nu(\mathbf{H})$ corresponds to the structure factor given in Section 4.1 for a modulated structure (Yamamoto, 1992). The diffraction pattern of a single substructure is, however, not independent and each one can in principle contribute to the intensities of the full pattern, that is, the Fourier module M^*. Therefore, in order to take this characteristic into account, we can express the interdependence (van Smaalen, 1991) of the structure factors of composites by using Eqs 2.39 and 2.40. The transformation $\{R^\nu | \mathbf{t}^\nu\}$ of the individual substructures must satisfy the following conditions given the integer matrices Z^ν and transformation of the overall superspace space group symmetry $\{R|\mathbf{t}\}$ defined in Eq. 2.60:

$$R^\nu = Z^\nu R\,(Z^\nu)^{-1}, \qquad \mathbf{t}^\nu = Z^\nu \mathbf{t}. \tag{4.47}$$

4.3.9 Commensurately modulated structures

Following the introduction of the superspace formalism to describe incommensurately modulated structures, structure specialists soon realized that the distinction between incommensurately and commensurately modulated structures was very tiny. To a good approximation, it is always possible to represent any incommensurate component of a modulation vector $\mathbf{q}$ by a rational number. This can be illustrated from Fig. 4.14: a small vertical shear strain applied to the superspace lattice can shift any lattice node in order to cut the horizontal line R. This means that the component q_1 of the modulation vector $\mathbf{q}$ is commensurate. In this case, the structure is classically called a *superstructure* and many examples were already known long before the superspace concept was introduced. In general, the resolution of superstructures is not trivial and very often causes difficulties to the practitioner. This is generally signalled by the appearance of ill-conditioned matrices in the refinement procedure. This problem can best be solved (Prince, 1994) by changing the variables to be refined. The best solution is obtained by redefining the individual parameters of the structure and replacing them by an average term along with the individual deviations from the average. Interestingly enough, this is exactly the procedure that is adopted in incommensurately modulated structure refinements. Therefore, it is tempting to use the superspace formalism in solving superstructures, that is, commensurately modulated structures. All the current software dealing with the refinement of modulated structures in superspace permits this type of refinement. This has been applied in many instances, for example, it was used for the refinement of the lock-in phase of δ-Na_2CO_3 (Dušek et al., 2003) which is schematically illustrated in Fig. 4.20.

In a one-dimensionally modulated structure, the atomic positions are specified by the sum of the average position and a corresponding shift

$$\begin{aligned} x_{ij} &= \bar{x}_{ij} + u_{ij}(\bar{x}_4), \\ \bar{x}_4 &= t + \mathbf{q} \cdot \mathbf{x}, \end{aligned} \tag{4.48}$$

where $\bar{x}_{ij}$ represents the three coordinates of atom i in the unit cell of the average structure, and $u_{ij}(\bar{x}_4)$ its displacement resulting from the modulation. This function is periodic, that is, $u_{ij}(\bar{x}_4) = u_{ij}(\bar{x}_4 + 1)$. In order to set a reference point for the fourth coordinate, we can set $t = t_0$.

For incommensurate structures, the variable $\bar{x}_4$ can take any value between 0 and 1. However, in a *commensurate structure*, this variable is only meaningful at the N relevant positions of the atom in the N-fold superstructure illustrated in Fig. 4.20 for $N = 6$. The horizontal dotted lines represent the real space section t_0 and all its equivalents $t_0 + \frac{n}{6}$ with $1 \leq n \leq 6$. The intersection of the horizontal lines

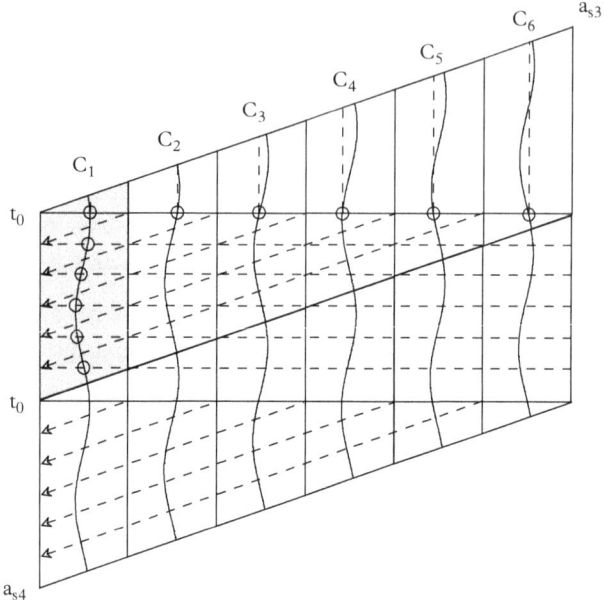

Figure 4.20 *Illustration of a six-fold commensurate structure embedded in superspace. The shaded area represents the $(\mathbf{a}_3, \mathbf{a}_4)$ projection of the four-dimensional supercell of δ-Na_2CO_3. The open circles are points for which the modulation function is defined. The reference point t_0 is not arbitrary and must be carefully estimated.*

with the modulation function defines six non-equivalent intersections representing six different atomic positions $C_1, C_2, \ldots, C_6$ located in the supercell, namely the six-fold cell of the superstructure. In the previous paragraph, the value of t_0 was arbitrarily selected along $\mathbf{a}_4$. It is clear from Fig. 4.20 that the positions of the atoms C_i are directly dependent of this choice. Therefore, it is the task of the refinement procedure to select the best value of t_0 in order to optimize the fit between the observed and calculated values of the structure factors. In addition the space group symmetry of the superstructure can be derived from the superspace group of the modulated structure and from the specific value of t_0. Section 2.4.5 illustrates the derivation of the superspace group for specific commensurate values of the modulation vector $\mathbf{q}$. The main advantage of this approach is that the space group of the superstructure can be directly obtained from the specific value of t_0. In other words, it is much easier to explore the possible space group symmetries of the superstructures in superspace than by testing individually each possible space group in three dimensions. Another advantage of this method is that the number of parameters can be better controlled by the appropriate number of modulation waves. In particular, correlations between atomic displacement parameters of the superstructure can be much better accounted for contrary to the conventional superstructure refinement. The superspace description of commensurately and incommensurately modulated crystalline phase is thus a powerful tool for a unified approach of, for example, sequences of temperature- or pressure-dependent phases. We have seen in Chapter 2 that one superspace group can be used to characterize a whole family of incommensurate and commensurate phases A_2BX_4, and we shall see later that the symmetries of both γ and δ phases of sodium carbonate can be deduced from a single superspace group. We shall also see later that the concept of superspace can be used for a unified description of complete families of compounds based on the building blocks concept.

4.4 Typical examples of modulated phases and composites

4.4.1 Introduction

Since the first descriptions of composite and incommensurate structures some decades ago, the number of examples has increased considerably in the specialized literature dedicated to structural studies. Incommensurate and composite structures have been observed in all possible classes of material, from organic to inorganic, from minerals to metals and alloys. Recently, many structures of metals under high pressures have been added to the list of non-periodic structures. The aim of the present section is not an attempt to give an exhaustive list of possible cases but rather an attempt to select a very limited number of representative cases illustrating the numerous facets of this recent field. We would also like to show that the new approach based on the superspace formalism has the capacity of providing new insight for a better understanding of the chemical interactions occurring in crystalline structures. We would also like to illustrate the unexpected application of the superspace approach in order to describe *modular structures* and its potential to predict new structures with some specific properties.

4.4.2 The modulated phases of Na_2CO_3

It is not surprising that the first example of an incommensurate structure presented in this context is γ-Na_2CO_3. The difficulties encountered in the indexation of a powder diffractogram followed by the observation of satellite reflections by de Wolff and co-workers (Brouns *et al.*, 1964) was at the origin of the new concept to describe incommensurate structures in a space of higher dimensions, namely the superspace. More than a decade later, after establishing the basic principles of the theory, a plausible model of the structure of γ-Na_2CO_3 was published (van Aalst *et al.*, 1976) based on modulation expressions with a single harmonic term. Later, the structures of the high-temperature phases α and β were published (Swainson *et al.*, 1995) followed by a new determination of phases γ and δ (Dušek *et al.*, 2003). By taking advantage of modern diffraction equipments in order to improve the resolution of the structure, the new determination of the γ-phase allowed to get some deeper insight into the origin and nature of the incommensurability of γ-Na_2CO_3.

Let us first look at the single crystal diffraction pattern of γ-Na_2CO_3 at room temperature. Figure 4.21 represents two reconstructed reciprocal layers, $h0\ell$ and $h1\ell$, which are generated from a large number of two-dimensional CCD-recorded frames from a rotating crystal placed in a monochromatic X-ray beam.

The analysis of the layers indicates that the diffraction pattern can be analysed in terms of main and satellite reflections. The main reflections are indicated by black (observed) and open (unobserved) spots forming a two-dimensional grid. A reflection is tagged unobserved if $I < 3\sigma(I)$. The additional spots are observed (dark grey) and unobserved (light grey) satellite reflections. The absences of reflections in some parts of the quadrants are due to the measuring strategy and the geometry of the diffraction system. By selecting the modulation vector $\mathbf{q}$ as indicated on the inset of Fig. 4.21, all the satellites can be indexed. The rank of this Fourier module is 4 and thus the structure of γ-Na_2CO_3 can be described in $(3 + 1)$-dimensional superspace. From the analysis of the diffracted intensities, we observe the *systematic absence* of groups of reflections including both main and satellite reflections. For γ-Na_2CO_3 the following conditions are observed:

$$(hk\ell m) : h + k = 2n,$$
$$(h0\ell m) : m = 2n.$$

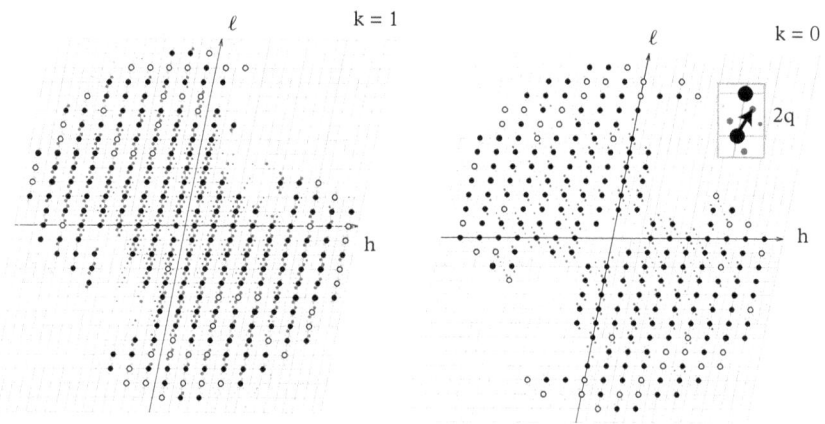

Figure 4.21 *Reconstructed $h0\ell$ and $h1\ell$ layers of γ-Na_2CO_3 from CCD measured diffraction data. Black and open dots are observed and unobserved main reflections, respectively. Observed and unobserved satellites are dark and light grey, respectively. The magnitude and orientation of the modulation vector $\mathbf{q}$ is given in the inset.*

The information derived from these two rules is not sufficient to determine the superspace group of the structure uniquely. However, as for conventional crystal structure determinations, the selection of possible superspace groups is quite restricted. In the present case, the highest symmetrical possibility is $C2/m(\alpha 0\gamma)0s$. This superspace indicates the presence of an inversion centre in three dimensions. Omitting the centre of inversion leads to $Cm(\alpha 0\gamma)s$, which exhibits the same systematic absences. In this case, the refinement of the average structure, that is, the refinement of the structure by omitting all satellite reflections, already shows that the centrosymmetric space group better describes the structure. Of course, the final choice of the superspace group has to be confirmed by a successful refinement of the modulated structure.

As for three-dimensional space groups, the symmetry transformations in superspace can be deduced from the superspace group symbol, as described in Section A.8. For $C2/m(\alpha 0\gamma)0s$, the four-dimensional transformations are

$$(0000), \left(\tfrac{1}{2}\tfrac{1}{2}00\right)+$$

x_1	x_2	x_3	x_4
$-x_1$	$-x_2$	$-x_3$	$-x_4$
x_1	$-x_2$	x_3	$\tfrac{1}{2}+x_4$
$-x_1$	x_2	$-x_3$	$\tfrac{1}{2}-x_4$

By applying the two translation operators given in parentheses, eight transformations are generated. In the present case, the refinement of the modulated structure is straightforward. First, the average structure is refined followed by a step-by-step inclusion of harmonic terms in the expression of the modulation function. The specification of arbitrarily small displacement parameters for every atom is sufficient to initiate the refinement in superspace. The final structure (Dušek et al., 2003) was refined to a high degree of accuracy from main and satellite reflections up to sixth order and harmonic terms up to fourth order.

Before going more into the details of the incommensurate phase γ, let us first summarize the sequence of phase transitions observed at various temperatures. Table 4.3 indicates the different phases with their (super)space group symmetries, lattice constants, and transition temperatures.

Figure 4.22 illustrates the structural aspect of Na_2CO_3 by omitting the oxygen atoms. The high-temperature hexagonal phase transforms to monoclinic below 754 K. The structure is essentially composed of hexagonal layers similar to BN type layers illustrated in Fig. 4.23, which are stacked in the third dimension. Above 754 K, the hexagonal c-axis is normal to the layers. Below this temperature, a monoclinic distortion occurs which results in a displacement of the layers relative to each other. For decreasing temperature, the translation vector along the stacking direction increasingly deviates from the normal to the layers, reaching 99° at the β-to-γ transition. In the layer, the O atoms are covalently bonded to the C atoms. The C–Na distances are close to 3 Å and the shape of the hexagons is close

Table 4.3 *The sequence of commensurate and incommensurate phases of Na_2CO_3 with transition temperatures (K), (super)space group symmetries, temperatures of measurement, lattice constants (Å), angles (°), and modulation vector q.*

α	β	γ	δ
754	605		170
$P6_3/mmc$	$2/m$	$C2/m(\alpha 0\gamma)0s$	$(C2/m\left(\frac{1}{6}0\frac{1}{3}\right)0s)$
			$P2_1/n$
770 K	703 K	295 K	110 K
$a = 5.207(1)$	$a = 9.010(2)$	$a = 8.920(7)$	$a = 19.91(2)$
$c = 6.471(1)$	$b = 5.231(2)$	$b = 5.245(5)$	$b = 5.237(6)$
	$c = 6.345(2)$	$c = 6.055(2)$	$c = 17.99(2)$
	$\beta = 96.06(2)$	$\beta = 101.35(8)$	$\beta = 119.01(5)$
		$q = (0.182(1), 0, 0.322(1))$	

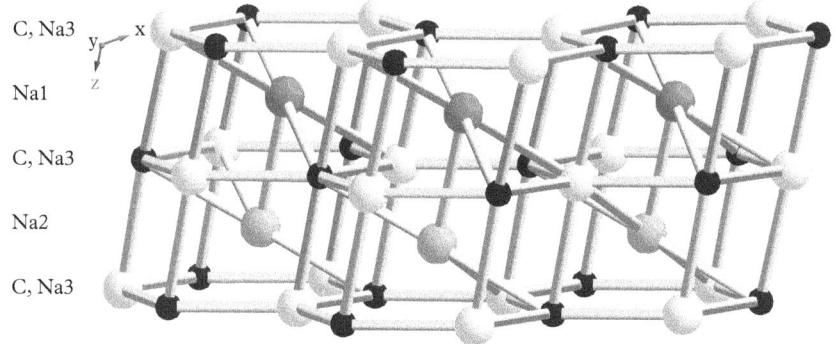

Figure 4.22 *Framework structure of Na_2CO_3 below 754 K without O atoms. C atoms are represented by small black spheres.*

to ideal. Along the third dimension the C atoms alternate with Na3. The centre of the hexagonal channels formed by two consecutive hexagons are alternatively occupied by Na1 and Na2 atoms. At high temperature, the C–Na1, 2 distances are larger than 3.4 Å and tend to reach the ideal value of 3 Å at low temperature.

The hexagonal to monoclinic transformation can be achieved along three possible directions corresponding to the three normals to the mirror planes sharing the hexagonal axis. At the transition only one of them remains (m_M) whereas the other two (m_V) are lost. The left-hand side of Fig. 4.24 represents the content of the

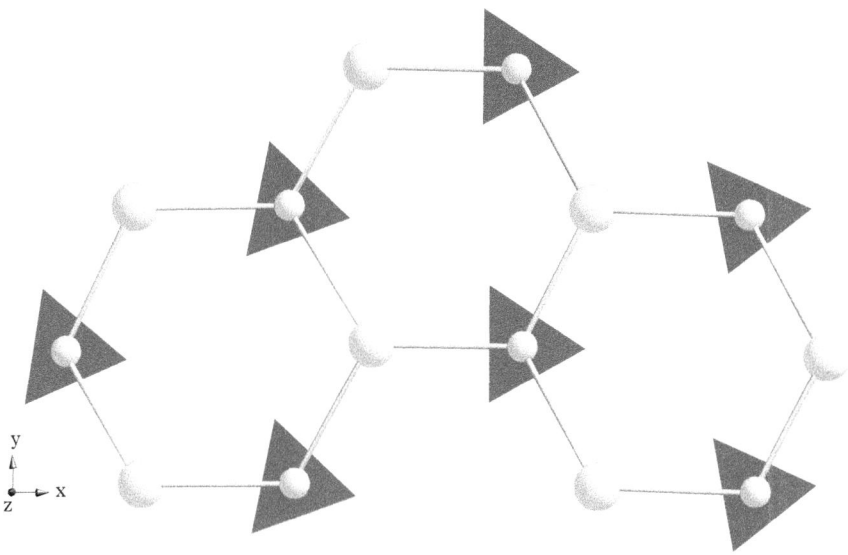

Figure 4.23 *Single layer of Na_2CO_3 consisting of Na3 atoms and CO_3 entities (triangles).*

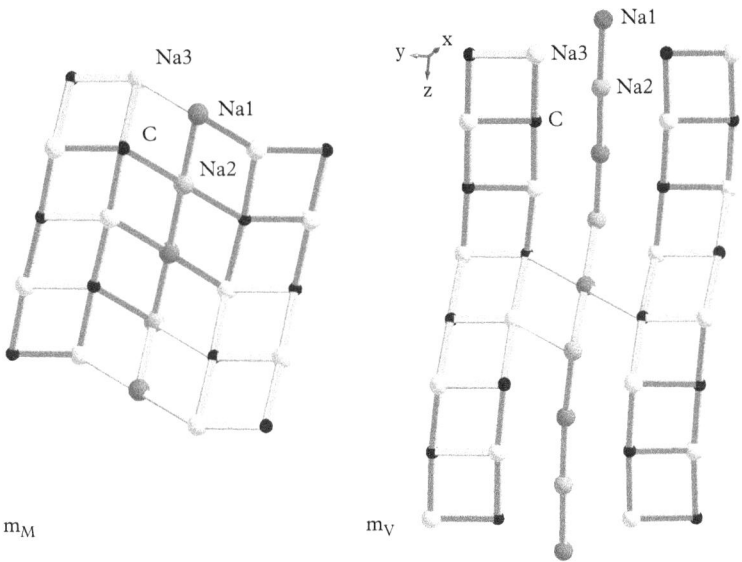

Figure 4.24 *Comparison between the content of the m_M plane of the monoclinic phase β and the m_V plane in the incommensurate phase γ. Oxygen atoms are omitted. Distances shorter than 3.1 Å are indicated.*

remaining mirror m_M which is reached at the β to γ phase transition with all distances smaller than 3.1 Å. This atomic arrangement has reached its own stability and remains invariant down to the lowest phase δ with negligible distance changes.

What are the origin and nature of the incommensurate phase? We have seen that the monoclinic distortion has favoured the formation of the stable unit represented by the content of the m_M plane. Below the β-to-γ phase transition, the natural tendency of the structure of Na_2CO_3 would be to reach a further degree of stability by forming other entities with the m_V planes similar to the m_M plane. Unfortunately, the important monoclinic distortion does not allow such a formation. However, if this cannot be reached throughout the structure, it can at least be realized locally within the crystal space. Thus, the Na_2CO_3 structure duly exploits the possibility to create an aperiodic structure, that is, to form an incommensurately modulated structure in order to recreate at least locally stable configurations similar to the m_M content. The right-hand side of Fig. 4.24, which represents a portion of the incommensurate phase γ, illustrates precisely the attempt to reproduce the stable configuration indicated on the left-hand side.

The *lock-in phase* δ stable below 170 K can be considered as a commensurately modulated structure with rational values for the coefficients of the modulation vector $\mathbf{q}$. At the γ-to-δ transition, the components of the modulation vector change slightly from (0.182, 0, 0.322), reaching ($\frac{1}{6}$, 0, $\frac{1}{3}$). By increasing the unit cell volume by a factor of 6, the three-dimensional periodicity lost at the β-to-γ transition can thus be recovered at the γ-to-δ transition. This phase is characterized by a slightly larger amount of closer Na–C and Na–Na contact distances.

In our discussion on the sequence of phase transitions of Na_2CO_3, we have intentionally set the emphasis on the Na–Na and C–Na interactions. Only these interactions are able to give a consistent model valid over the complete range of phases from high to low temperature. The essential role of the O atoms is to ensure electroneutrality. Moreover, one should note that the CO_3 units are not completely rigid. In particular, in phase β, we observe not only an increase of the C–O distances for decreasing temperature but also a different behaviour depending on their location on the m_M and m_V planes.

Finally, one should mention an interesting property of the superspace formalism in describing modulated phases. For the two phases γ and δ, a unique superspace group can describe both phases. Each phase is characterized by the components of the modulation vectors, rational or not. The small differences between the two modulation components express the close relationship between both phases. This is another illustration of the usefulness of the superspace concept not only to describe the symmetry of modulated phases but also to express the close relationship between different structures (cf. Section 2.4.5).

4.4.3 The composite structure of $La_2Co_{1.7}$

The structure of *lanthanum cobalt* is a very interesting case of a *composite structure* which has puzzled many structure specialists. The general features of the light

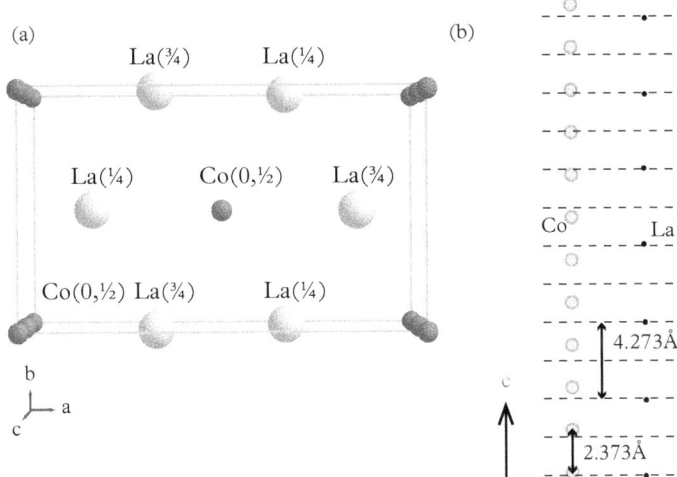

Figure 4.25 *(a) Average structure of $La_2Co_{1.7}$ along the pseudo-hexagonal axis. The z-coordinates of the atom are given in parentheses. (b) Schematic drawing of the composite structure showing the relative displacement of Co and La atoms along c.*

rare earth cobalt compounds were already known in the 1970s. It was recognized that the periodicity along the hexagonal axis of the closed packed structure of the rare earth element could not accommodate the Co atoms in the channels without introducing another periodicity. The crystallographic tools were not ready and the definitive solution to this problem had to wait for a few decades (Dušek et al., 2000).

The composite structure $La_2Co_{1.7}$ was solved from the combined methods of single crystal X-ray and neutron diffraction. The diffractogram obtained from neutron Laue white-beam technique (Fig. 4.26) reveals the presence of satellite reflections in the vicinity of strong main reflections.

The first approach used to solve this structure was to interpret the satellite reflections as originating from a six-fold twin specimen. The lattice parameters reported in Table 4.4 and Fig. 4.25a indicate clearly the pseudo-hexagonal character of the structure with a monoclinic angle of 90°. The second step was to devise a model which would be able to account for different periodicities between the Co and La atoms along the pseudo-hexagonal axis. This can be achieved by the introduction of a *sawtooth function*, as described earlier. The section obtained from the four-dimensional Fourier map in the vicinity of the Co atom represented in Fig. 4.27 might indicate a limiting case of a sawtooth function which is defined by its centre x_4^0, its width Δ, and its maximal displacement $\mathbf{u}_0$. When x_4 reaches point A, the atom is absent until x_4 reaches point B in the next cell. The width Δ is a direct measure of the occupation of the Co atom. When points A and B reach the edge of the unit cell, a gap in the presence of the atom may still exist

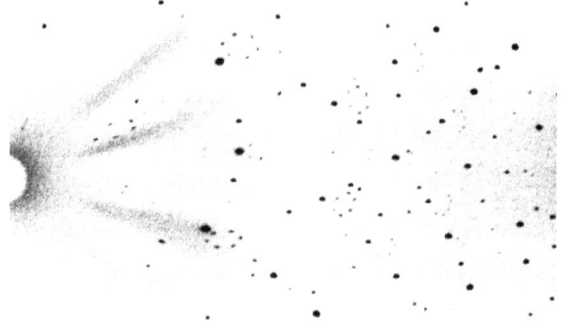

Figure 4.26 *Neutron Laue white-beam diffractogram of $La_2Co_{1.7}$. Satellite reflections forming hexagons can be observed in the vicinity of strong reflections.*

Table 4.4 *Crystallographic data for lanthanum cobalt. Lattice parameters are given in angstroms and degrees.*

Chemical formula:	$La_2Co_{1.7}$	Temperature (K):	378
Superspace group:	$C2/m(\alpha 0 \gamma)$		$C2/m(\alpha 0 \gamma)$
a	8.478(2)	a of Co composite	8.486
b	4.895(1)	b of Co composite	4.895
c	4.335(1)	c of Co composite	2.362
β	90.00(2)	β of Co composite	87.45
	$\mathbf{q} = (0.160(1), 0, 0.165(1))$		$\mathbf{q} = (0.087, 0, -0.455)$

unless A and B are equivalent. In this case the line is continuous in superspace and generates a sequence of equally distributed atoms in physical space. This limiting case corresponds to the *closeness condition* described in Chapter 2, Section 2.3.2.

The refined parameters of the sawtooth function for the description of the Co atom indicate that the extremities A and B (Fig. 4.27) reach the border of the unit cell. Moreover, points A and B represent equivalent positions of the Co atom. Therefore, the description of the structure as a composite seems more appropriate in the present case.

For this purpose, the structure requires a reinterpretation of the diffraction pattern in terms of two subsystems, Co and La. Figure 4.28 indicates the relation between the reciprocal subsystems of Co and La and the corresponding subsystems in direct space. The relation between the two subsystems is described by the matrix

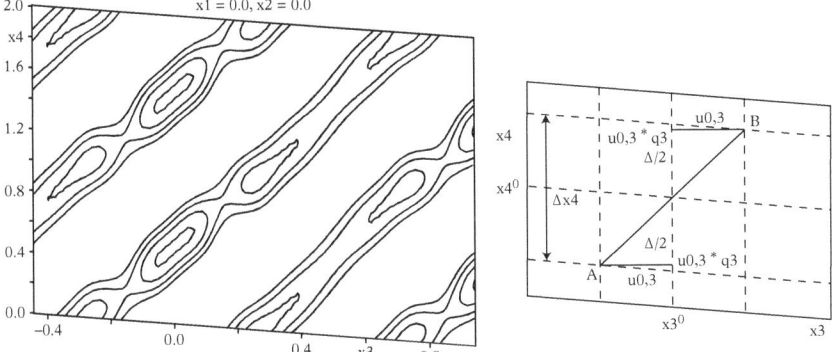

Figure 4.27 *Section of the four-dimensional Fourier map of $La_2Co_{1.7}$ in the vicinity of Co, suggesting a limiting case of a sawtooth modulation function. The definitions of the corresponding parameters are indicated on the right-hand side.*

$$Z = \begin{pmatrix} 1 & 0 & 0 & 0 \\ 0 & 1 & 0 & 0 \\ 0 & 0 & 2 & -1 \\ 0 & 0 & -1 & 1 \end{pmatrix}.$$

More precisely, the matrix of one of the subsystems, the La subsystem in the particular case, has been set to identity, in other words, as the reference system. The setting of the matrix Z can be established from the left-hand illustration in Fig. 4.28. In $La_2Co_{1.7}$ both subsystems have the same superspace group $C2/m(\alpha 0\gamma)$. The lattice constant c of the Co subsystem is 2.38 Å, a value close to the minimum allowed Co–Co distance, which justifies the choice of the reference subsystems. The final result based on the composite refinement of the structure is represented in Fig. 4.25b. No displacive modulation of the Co atom is observed, whereas, for La, a small modulation is observed, depending on the vicinity of the Co atoms.

The superspace description of $La_2Co_{1.7}$ is thus the only model which is able to accommodate both the chemical restrictions on the Co–Co distances and the hexagonal close-packed structure of the La atoms. Provided that the diffraction data are of sufficiently good quality, the superspace refinement also delivers the relative composition of the chemical elements included in the composite structure either in the form of the length of the sawtooth function Δ defined in Fig. 4.27 or indirectly by the volume ratio of the subsystems (Fig. 4.28). It should be mentioned here that the chemical composition of this alloy can be directly determined from the refinement of the composite structure. In the present refinement, the chemical composition is closer to $La_2Co_{1.8}$ although some fluctuations are observed between different samples. The sawtooth and the composite model refinements

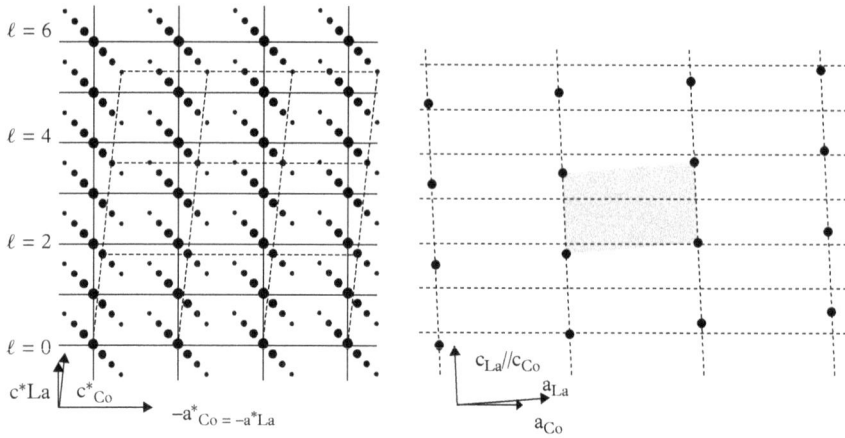

Figure 4.28 *Composite description of $La_2Co_{1.7}$ in reciprocal (left) and direct (right) space. The reciprocal La and Co subsystems show that main reflections of one subsystem are also satellites of the other, and vice versa. Experimentally, only first- and second-order satellites have been observed.*

are equivalent provided that the sawtooth model leaves no gap in the interval of definition of the atom in superspace. In the present refinement, the lack of a sufficient number of observed satellite reflections underlining the composite periodicity favoured the initial approach with the sawtooth function.

4.4.4 Alkane–urea compounds

An interesting and rich family of incommensurate composites is that of the alkane–urea compounds. These consist of a subsystem formed by urea molecules, and, in the channels of these structures, chains of alkane molecules are situated. A long series of such alkane chains has been investigated, with alkanes of different size. Most of them are incommensurate, and there are several phase transitions in the phase diagram (Fig. 4.29).

The systems (C_nH_{2n+2})–$(CO(NH_2)_2)$ vary from $n = 7$ to $n = 24$. The rank of the Fourier module varies between 3 and 5 in the n-p-T-phase diagram (Toudic et al., 2011b). The two basic structures together form an aperiodic crystal of rank 4, unless the lattice constants in the channel direction become commensurate or if the alkane system is disordered, as is the case for higher temperatures, in which case it is 3. Next to that, additional diffraction peaks appear at phase transitions, increasing the rank to 5.

For high temperatures, the alkane chains are disordered. Then the rank is 3, with Bragg peaks from the urea, and diffuse scattering from the alkanes. At lower temperatures the hexagonal framework is forced into an orthorhombic structure. At still lower temperatures, the chains order, initially in the channels, while remaining uncorrelated between the channels. Then correlation sets in, with

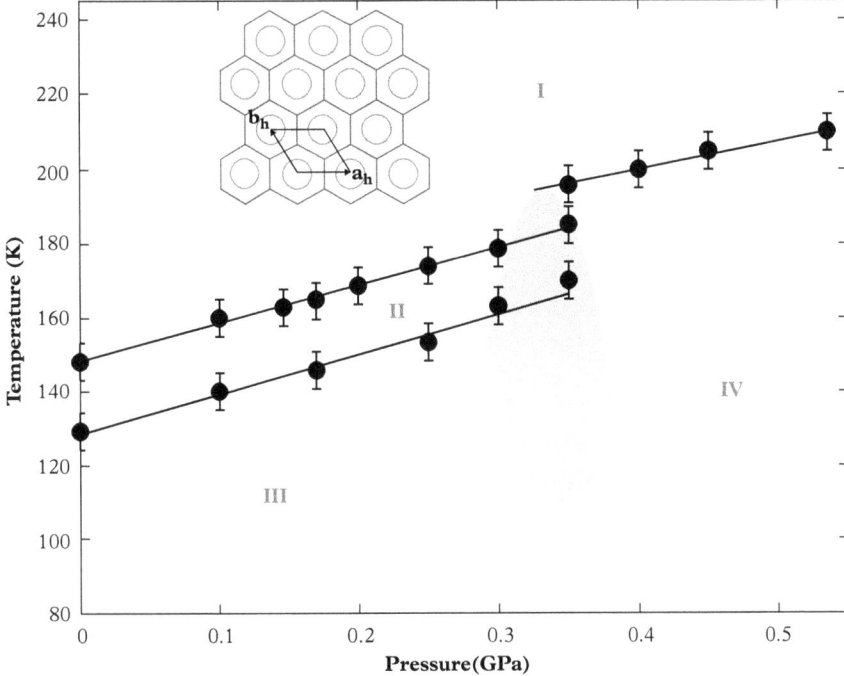

Figure 4.29 *Temperature-pressure phase diagram of the fully deuterated nonadecane–urea. (Courtesy B. Toudic, Rennes)*

the sharp spots belonging to a Fourier module of rank 4. Finally, for some of these systems, satellites appear when the temperature and/or pressure is changed, increasing the rank to 5.

An example is nonadecane–urea (Fig. 4.29). Its phase diagram shows four different structures. Their superspace groups are, respectively, I $P6_122(00\gamma)$, II $C222_1(00\gamma, 10\delta)$, III $P2_12_12_1(00\gamma, 00\delta)$, and IV $C222_1(00\gamma, 10\delta)$, or subgroups of these. (There is not yet a precise structure determination available.) The ranks are, respectively, 4, 5, 5, and 5. In the phase transition from I to II, the hexagonal channels are forced in to an orthorhombic arrangement: $\mathbf{a}_h, \mathbf{b}_h \rightarrow \mathbf{a}_h, \mathbf{a}_h + 2\mathbf{b}_h$.

4.4.5 Aperiodicity in the structures of elements

A large proportion of elements at normal pressure and temperature adopt the simplest possible structural models, consisting of closed-packed spheres with cubic and hexagonal variants or the body-centred cubic cell. These structures are characterized by high coordination numbers of 12 or 8 plus 6, respectively, typical for metallic bonds. It turns out that, as a result of increased pressure or temperature or under special growth conditions, some elements adopt a structure

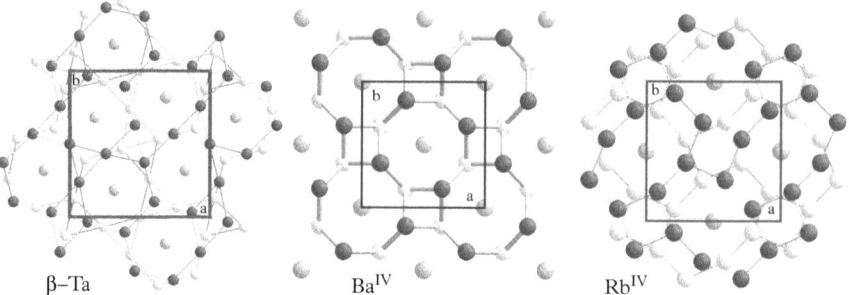

Figure 4.30 *Three variants of self-hosting structures. Black and white atoms form layers of host atoms whereas grey atoms are located in the channels perpendicular to the projection plane.*

of the *host–guest* type, which is mostly found in binary compounds. The host structure consists of a three-dimensional framework of atoms forming an infinite array of parallel channels in which the guest atoms are located. In typical binary host–guest structures, the channels are occupied by one type of atom whereas the framework is occupied by the second atom type. The very interesting case where both channels and framework are occupied by a single type of atom is called a *self-hosting structure*. A few typical examples of host–guest structures of elements are represented in Fig. 4.30. In all the represented structures, the bonds are not necessarily selected on the basis of the shortest distances but only for a better distinction between the host and the guest atoms.

If the nature of the bonds differs between the host and the guest atoms, it is conceivable that the guest atoms might have a periodicity independent of that of the host structure. Such an aperiodic model would correspond to the composite type of structure which we described earlier. This is precisely what was recently discovered by applying pressures of the order of a few tens of gigapascals on some metallic elements (McMahon and Nelmes, 2004). One of the first metallic elements studied by diffraction at high pressure was barium. Under normal conditions, Ba exhibits a body-centred cubic structure which transforms to Ba^{II}, a hexagonal closed-packed structure above 5.5 GPa. Between 12.6 and 45 GPa, another phase is formed, Ba^{IV}, which exhibits a self-hosting incommensurate structure consisting of host (H) and guest (G) components, as illustrated in Fig. 4.30. More specifically, it appears that different phases of Ba^{IV} exist depending on the pressure. They are called a, b, and c according to pressure increase. Ba^{IVa} was initially described as a composite model with two interpenetrating but independent 3D models (Nelmes *et al.*, 1999). Later, a model of Ba^{IVc} was proposed, with 99 independent Ba atoms (Loa *et al.*, 2012). Based on new experimental data, a recent publication (Arakcheeva *et al.*, 2017) presented a model of the unknown structure Ba^{IVb} by fully exploiting the symmetrical properties of the diffraction results. In the pressure range between 16.5 and 21.8 GPa, two different incommensurately modulated phases were observed

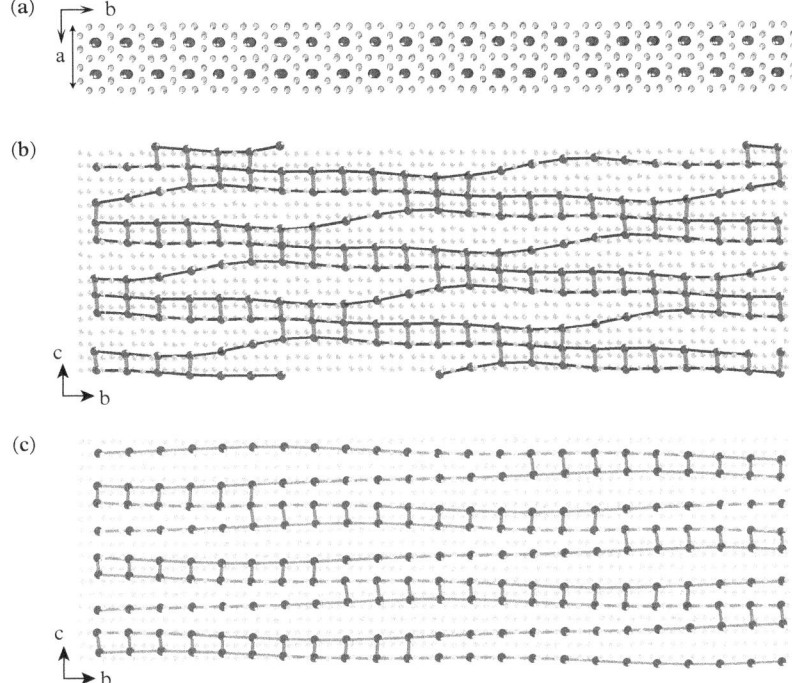

Figure 4.31 *Various projections of Ba^{IVb} at two different pressures. (a) View of the host–guest structure along the channels, indicating weak modulations in the ab plane. Projections of the modulated structures along a at (b) 19.6 GPa and (c) 16.5 GPa, respectively. The channel atoms are strongly modulated along c, forming dumbbells and triplets. At 16.5 GPa, only dumbbells are observed whereas, at 19.6 GPa, additional triplets are observed. Links between channel atoms characterize distances between 2.9 and 3.8 Å.*

(Fig. 4.31). Both of them were described with the same monoclinic superspace group $P2_1/b(0\beta\gamma)00$. The full diffraction pattern could be indexed with the reciprocal lattice vector $\mathbf{H} = h\mathbf{a}^\star + k\mathbf{b}^\star + \ell\mathbf{c}^\star + m\mathbf{q}$, where $\mathbf{q} = \beta\mathbf{b}^\star + \gamma\mathbf{c}^\star \approx 0.1\ \mathbf{b}^\star + 1.36\ \mathbf{c}^\star$ and lattice parameters $a \approx b \approx 11.5$, and $c \approx 4.6$ Å.

The recent refinements of the incommensurately modulated structure of Ba^{IVb} shed some new light on its high-pressure phases. The full exploitation of the diffraction data, including the identification of its superspace symmetry, reveals the existence of pressure-dependent density waves in the channels, and the formation of clusters with alternating dense and sparse regions, as illustrated in Fig. 4.32. The modulations affect essentially the channel atoms, and very minor displacements can be observed for the other atoms. The dense regions are due to the formation of clusters along the channels, with Ba–Ba distances ≤ 3.8 Å. They are essentially formed by pairs (dumbbells) of atoms and, to a lesser extent, by triplets, which appear only above 18 GPa (Fig. 4.33).

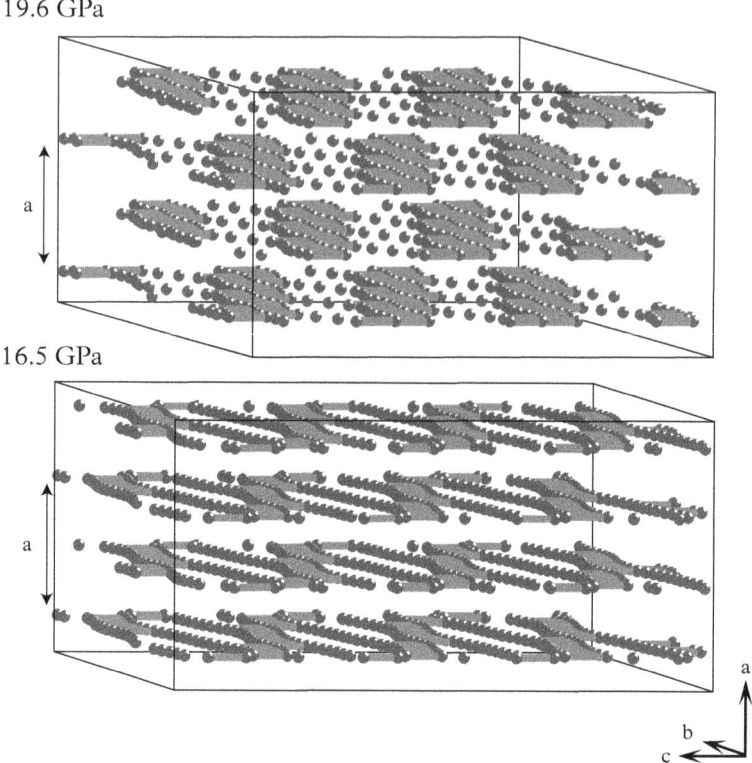

Figure 4.32 *Other projections of Ba^{IVb} at 19.6 and 16.5 GPa, illustrating the pressure-dependent formation of clusters of dense and sparse regions. For clarity, only channel atoms are indicated. Dense regions are characterized by short Ba–Ba contact distances below 3.8 Å.*

The phase transition observed above 18 GPa for Ba^{IV} (Arakcheeva et al., 2017) is worth mentioning. It might well be that this transition corresponds to the one observed before (Nelmes et al., 1999) between Ba^{IVa} and Ba^{IVb}. The pressure medium used in the two different experiments (Ne in the first reference, and mineral oil in the second reference) could possibly explain the difference in the observations. In this hypothesis, a single superspace group could explain both structures with different modulations. Of course, this hypothesis should be checked experimentally before confirmation. It is also worth noting that the published diffraction pattern of Ba^{IVc} seems to be closely related to that of Ba^{IVb}. It is thus not excluded that the 99 independent atoms could be reduced to a very small number by using the same superspace group symmetry obtained for Ba^{IVb} or one of its subgroups.

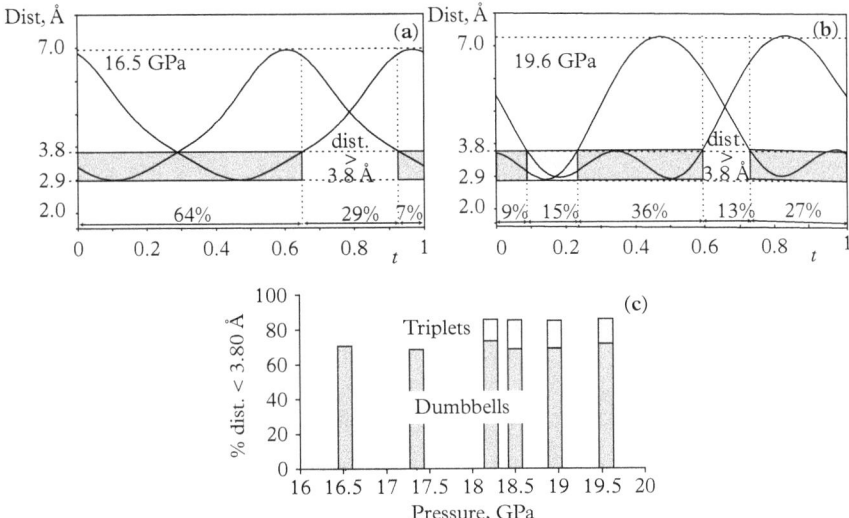

Figure 4.33 *(a, b) Closest interatomic distances in Ba^{IV} in function of the internal variable t at two different pressures. (c) The formation of triplets with links ≤ 3.8Å only occurs above 18 GPa.*

The high-pressure structures of Ba^{IV} are by no means exceptions. Sr^{V} above 46 GPa, and Bi^{III} between 3 and 8 GPa, have similar structures. In Bi^{III}, however, we observe some pairing of the guest atoms in the channels. This seems also the case for antimony (Sb^{II}) from 9 to 28 GPa. Rb^{IV}, which is stable between 17 and 20 GPa, on the other hand, exhibits a slightly different structure as seen from the projection along the channels (Fig. 4.30). In this structure, the guest atoms form a body-centered tetragonal cell with a and b lattice parameters which are identical with those of the host structure.

It is relevant to mention here that host–guest composite structures can also be commensurate. Both β-U (high-temperature modification) and β-Ta (non-equilibrium modification) exhibit this type of structure (Arakcheeva and Chapuis, 2005), which is closely related to the high-temperature phase of CrFe, the so-called Frank–Kasper σ-phase. A projection of the structure of β-Ta is represented in Fig. 4.30. In both structures, β-Ta and β-U, the ratio $c_G/c_H = \frac{1}{2}$ is rational. However, both structures exhibit a temperature variation of the symmetry of one substructure whereas the other substructure is stable. In β-Ta, the space group of H changes by lowering the temperature from $P\bar{4}2_1m$ (293 K) to $P\bar{4}$ (120K), reaching $P4_2/mnm$ (15 K), whereas the space group of G, P4/mbm, remains unchanged. In β-U, the space group of G varies from $P4_2/mnm$ at 955 K to P4/mbm at 1030 K, whereas the space group of H remains $P4_2/mnm$. Thus, only two out of the three attributes of composite structures can be observed in β-Ta and β-U, namely different space groups of the components and an independent behaviour of symmetry with temperature. Similar to the composite structure Ba^{IV}, in β-Ta and β-U, one substructure is stable and the other is variable. It is interesting to note that G is the stable substructure in β-Ta, and H is the stable structure in β-U.

The very close nature of the self-hosting structures of elements under high pressure and the host–guest structures of binary compound seems to suggest that the electronic configuration of the host and guest atoms are different even if the atoms are the same. The channel atoms would preferably have a d-like character whereas the host atoms would have an s-like character. This model has, however, been challenged recently (Haussermann et al., 2002). In this work, the authors propose instead a subtle interplay between the band energy and the electrostatic energy of the compound to explain the nature of the high-pressure structures of the elements. Obviously, the self-hosting structures of elements require some additional studies before a plausible explanation can be proposed in order to explain their stability. It might be that the new refinement in terms of the incommensurate phase, as applied for Ba^{IVb}, instead of the composite model, will permit the discovery of new structural properties for a better understanding of their origin.

In addition to composite structures, the high-pressure studies of elements exhibits also some interesting cases of incommensurately modulated structures as is the case for α-U below 43 K (van Smaalen and George, 1987). As an example, the structure of Te^{III}, stable between 7 and 11 GPa, is modulated with the superspace group $I'2/m(0q_20)s0$, with I' referring to the $\left(\frac{1}{2}\frac{1}{2}\frac{1}{2}\frac{1}{2}\right)$ centring. At 8.5 GPa, the lattice constants are $a = 3.9181(1)$, $b = 4.7333(1)$, $c = 3.0612$ Å, $\beta = 113.542(2)°$, and $q_2 = 0.2880(2)$. The refinement of the incommensurate structure yielded two components of the modulations: $B_{1x} = 0.0215(9)$ and $B_{1z} = 0.0925(7)$. This structure is illustrated in Fig. 4.34, where the effect of the largest component along **c** is seen on the modulation wave extending along **b**.

4.4.6 p-Chlorobenzamide

Organic crystals offer a large spectrum of modulated structures, and a large number of examples have been published in the specialized literature. In the present example, we would like to focus more precisely on the existence of superstructures, that is, commensurately modulated structures. In organic compounds, it is often the case that structures with large Z', in other words, with a large

Figure 4.34 *The incommensurately modulated structure of Te^{III} is stable between 7 and 11 GPa. The basic unit cell consists of an I-centred cell with two atoms.*

number of independent formula units in the asymmetric unit, lead to the formation of superstructures. Cases with Z' larger than 10 have been described in the literature (Schönleber and Chapuis, 2004a, b). The structure of p-chlorobenzamide (C_7H_6ClNO), described in Table 4.5, falls in this category. It exhibits two phases that are stable at room temperature and above (Schönleber et al., 2003).

The room-temperature phase α contains three independent molecules of p-chlorobenzamide in the asymmetric unit whereas the high-temperature phase γ contains only one independent molecule in the asymmetric unit. (The rather confusing notation of the phases corresponds to the published data.) Figure 4.35 shows two corresponding layers from each phase. The intensities of the α form which are absent in the γ form are clearly weaker than the others. It is thus natural to describe the α form as a commensurately modulated structure. Figure 4.36 illustrates the relationships between the two interpretations of the diffractograms, either as a commensurately modulated structure or as a superstructure description.

The superstructure of p-chlorobenzamide (Taniguchi et al., 1978) can be reformulated in terms of a modulated structure with a commensurate modulation vector $\mathbf{q} = \left(\frac{1}{3}, 0, 0\right)$. The refinements of the modulated structure either with first-order harmonic terms or with *sawtooth functions* (Fig. 4.37) are equivalent. Both types of refinement yield results that are identical to the superstructure refinement (Schönleber et al., 2003). This is not surprising, as only the three

Table 4.5 *Crystallographic data for p-chlorobenzamide with phase transition temperature* (K), *(super)space group symmetries, temperatures of measurement, number of independent molecules* Z', *lattice constants* (Å), *angles* (°), *and modulation vector* **q**.

γ		α	
	373		
$P\bar{1}$		$P\bar{1}$	$P\bar{1}(\alpha\beta\gamma)$
323 K		290 K	
Z' = 1		Z' = 3	Z' = 1
a = 4.992(8)		a = 14.99()	a = 4.997(4)
b = 5.47(1)		b = 5.467(6)	b = 5.467(4)
c = 14.71(3)		c = 14.43(1)	c = 14.43(1)
$\alpha = 98.4(1)$		$\alpha = 97.76(7)$	$\alpha = 97.76(7)$
$\beta = 112.9(1)$		$\beta = 111.91(6)$	$\beta = 111.91(6)$
$\gamma = 94.3(1)$		$\gamma = 95.16(5)$	$\gamma = 95.15(5)$
			$\mathbf{q} = (1/3, 0, 0)$

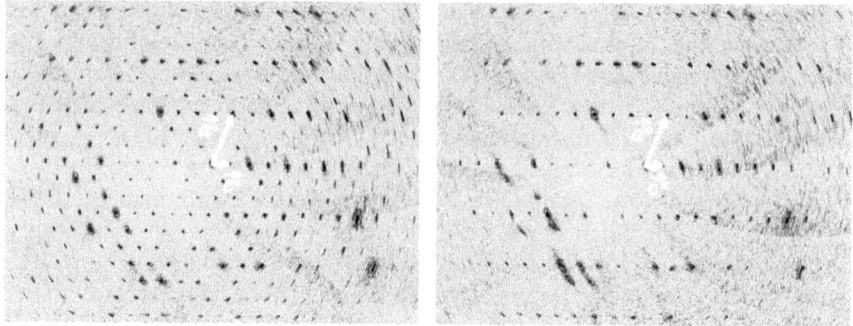

Figure 4.35 *Reconstructed diffractograms of p-chlorobenzamide. The h2ℓm layer of the γ form (left), and the h2ℓ layer of the α form (right).*

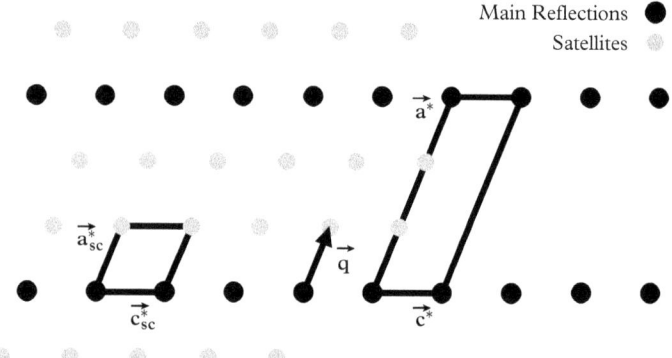

Figure 4.36 *Interpretation of the reciprocal space of p-chlorobenzamide in terms of a supercell (subscript sc on the left) and a commensurately modulated structure with satellite reflections defining the modulation vector **q** (right).*

points on the lines along the internal space dimension t which are separated by $\Delta t = \frac{1}{3}$ are determinant in the refinement. We see, that in both models, the points have identical coordinates. The essential difference in the three p-chlorobenzamide molecules concerns the torsion angles of the molecules relating the phenyl and the amide groups, which can be seen in Fig. 4.38. The three angles are 19.9(1)°, 33.7(1)°, and 29.7(1)°, with an average value of 27.7°. It should be mentioned that this average value is similar to the value found in the reference structure, that is, the structure on which the modulation is applied for the refinement in superspace. Moreover, the same value is found for the unique molecule of the high-temperature phase γ.

Although the superspace and superstructure refinements give similar crystallographic results, the superspace approach reveals better the relations between the two phases, first, by adopting the same unit cell in both forms (this is not the case

Typical examples of modulated phases and composites 211

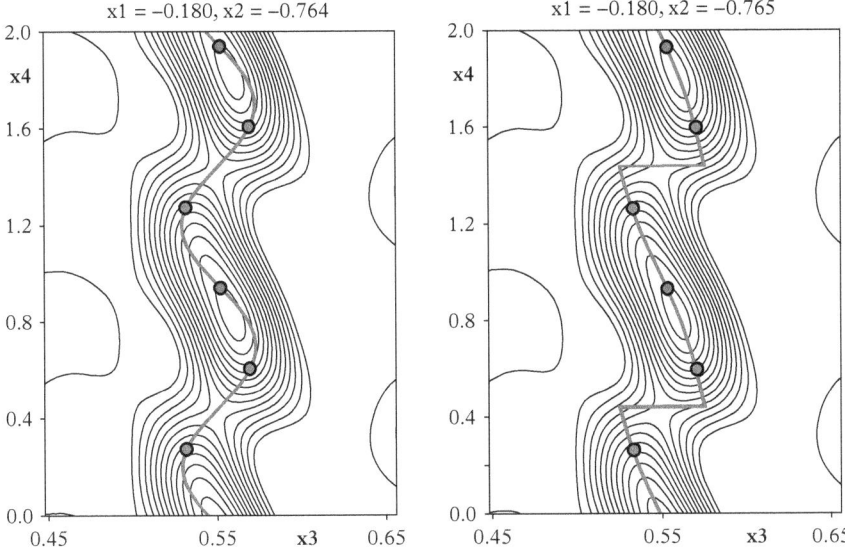

Figure 4.37 $x_3 - x_4$ *Fourier maps of the nitrogen atom N09 of p-chlorobenzamide, with harmonic terms (left) and sawtooth functions (right).*

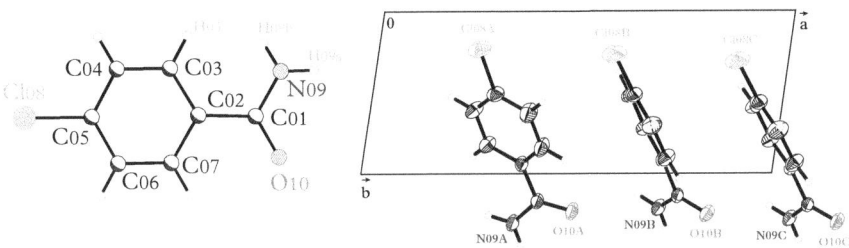

Figure 4.38 *The three independent molecules of p-chlorobenzamide in the superstructure (α form). In the superspace refinement, each molecule corresponds to different values of t separated by $\Delta t = \frac{1}{3}$.*

in the superstructure refinement and, second, by showing the close relation which exists between the room-temperature structure (through the reference structure) and the high-temperature structure.

4.4.7 Modular structures

In the preceding sections, we illustrated the use of the superspace formalism not only for the description of incommensurate but also for commensurate structures. For example, the lock-in phase of δ-Na_2CO_3 is characterized by the same superspace group as γ-Na_2CO_3. The only difference lies in the specific value of the modulation vector. Another extension of the use of the superspace theory was proposed recently (Elcoro et al., 2000) for the description of *modular structures*. Modular structures consist essentially of a limited number of different building blocks forming layers which are stacked in the third dimension. Many examples of modular structures are known and often present very challenging cases of structural resolution. The hexagonal perovskite-related series of phases $A_{1+x}A'_xB_{1-x}O_3$ (Elcoro et al., 2003) or the *Aurivillius phases* $Bi_{2m}A_{n-m}B_nO_{3(n+m)}$ can all be decomposed in terms of blocks stacked in the third dimension. Variations in the chemical compositions change the stacking sequences of the blocks, yielding a very large spectrum of possible structures each with their own space group symmetries. We shall describe how the superspace formalism was recently applied successfully to the description of a family of modular structures with the general composition $LaTi_{1-x}O_3$ (Elcoro et al., 2000) with $0 < x \leq 0.25$. The Ti valence of the end members $LaTiO_3$ and $La_4Ti_3O_{12}$ are +3 and +4, respectively. Any intermediate composition is thus necessarily characterized by mixed valence Ti ions. A single model in superspace is sufficient to describe the stacking sequence of the components for any particular chemical composition, provided that x is limited to the condition given above. The general aspect of these structures is illustrated in Fig. 4.39. We observe that any structure consists of an alternating sequence of LaO_3 and Ti layers. Given the origin of the unit cell, the LaO_3 layer can occupy three relative positions given by the usual letters A, B, or C describing the stacking sequences. The same possibilities are also valid for the Ti layer and are given by the small letters a, b, and c but subject to the condition that any Ti layer can only be surrounded by two different capital letters, both of them different from the small letter. The stacking sequence of $LaTiO_3$ is thus AcBaCbAc....

The use of superspace to describe the family of $LaTi_{1-x}O_3$ compounds is also suggested by electron diffraction observations. All members of the series exhibit a common subset of strong reflections. An additional set of composition-dependent

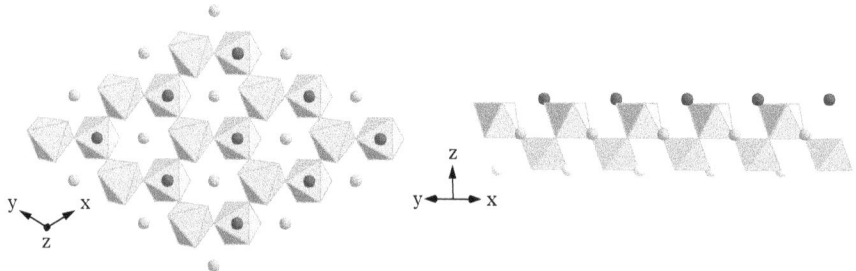

Figure 4.39 *Parallel and perpendicular view of a portion of the $LaTiO_3$ layer structure. Spheres represent La atoms whereas the octahedra consist of oxygen centred by Ti atoms. The structure can be interpreted as alternating sequences of LaO_3 and Ti layers.*

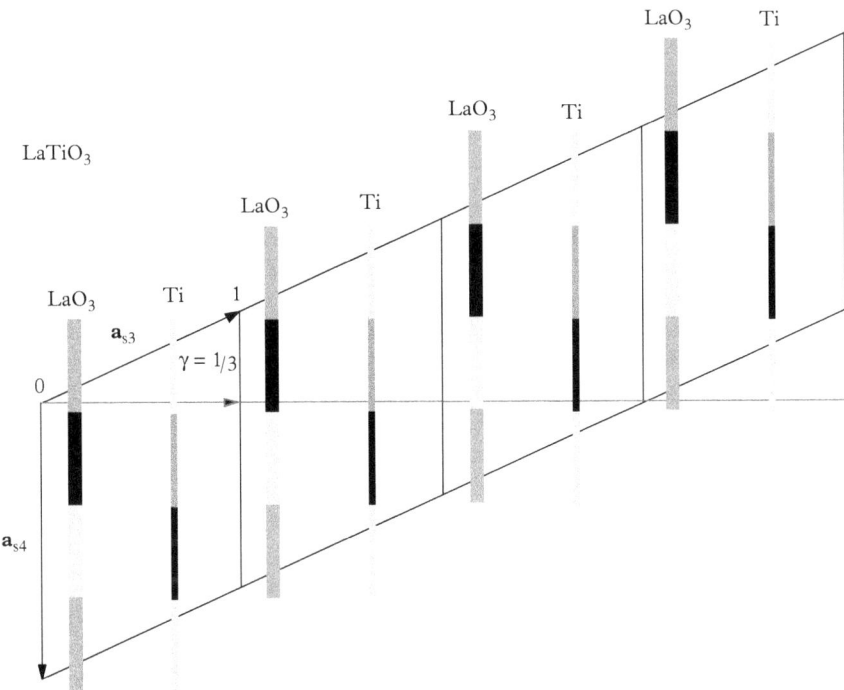

Figure 4.40 *Representation of the LaTiO$_3$ layer structure in the superspace formalism. Layers of LaO$_3$ alternate with layers of Ti atoms. Black, grey, and light-grey colours represent A, B, and C sequences and a, b, and c sequences, respectively.*

reflections is observed, which varies continuously with the composition, and can be interpreted as satellite or superstructure reflections.

Figure 4.40 illustrates the embedding of the LaTi$_{1-x}$O$_3$ structures in superspace for the particular case $x=0$. The broad vertical lines represent the three possible sequences of LaTiO$_3$ layers with the intercalated layers of Ti atoms, distinguished by different grey shadings. The same code is also applied for the narrow vertical lines. The sequence of LaTiO$_3$ mentioned above can be directly obtained on the horizontal line which represents the structure in real space. The modulation component, which is related to the angle between a_{s1} and a_{s4}, is directly related to the chemical composition x by the expression $\gamma = (1+x)/3$. We recall here that a vertical shift of the horizontal line only shifts the reference point in real space and does not represent a new structure.

The change of the chemical composition by increasing x induces a decrease of the Ti atoms. The structural analysis of the series of compounds indicates that this is translated into the corresponding structure by the complete disappearance of some of the Ti layers, with x expressing the ratio of the vacant interstitial

Ti layer. The purely cubic sequence of layer packing breaks down, and sets of hexagonal sequences appear accordingly. The missing Ti layers are always located between pairs of hexagonal layers. Crystal chemists have established the rules (van Tendeloo *et al.*, 1994) for the precise description of the sequences depending on the Ti ratio.

In order to describe the family of modular structures in superspace, we can introduce a gap in the atomic surfaces, that is, the vertical lines characterizing the Ti positions. The width of the titanium atomic surfaces is given by the composition and is equal to $(1 - x)/3$, thus leaving an interval of $x/3$ between the atomic surfaces. For example, Fig. 4.41a shows the correct sequence for $La_4Ti_3O_{12}$. It is clear that the missing sequence of Ti atoms in the structure induces some changes in the layer distances, especially for the nearest LaO_3 layers. This can be accommodated for in the superspace model by introducing slanted atomic surfaces, as illustrated in Fig. 4.41b. With this modification, two types of LaO_3 layer distances can be observed: one smaller, with the Ti layer (D_0) absent, and one larger, with the presence of a Ti layer (D_1).

The model presented here is able to reproduce the complete sequence of phases which are presently known, seven members with $x = \frac{1}{4}, \frac{1}{5}, \frac{1}{6}, \frac{2}{9}, \frac{2}{11}, \frac{3}{13}, \frac{11}{46}$. The detailed shape of the atomic surfaces can be further fitted to the experimental data by the introduction of additional modulation terms similar to the refinement of modulated structures. The interesting point here is that a single superspace group is able to characterize fully the complete series of modular structures. Moreover, this approach could detect some errors in one of the structures solved by diffraction. Another interesting aspect of the superspace formalism concerns the distribution of the mixed valence Ti ions. It appears that the Ti^{4+} ions are located at the extrema of the atomic surfaces whereas the Ti^{3+} ions are located in the centre of the atomic surfaces.

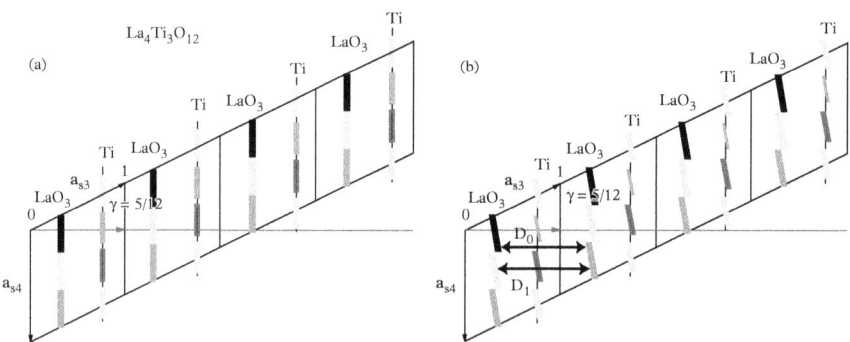

Figure 4.41 *Representation of the $La_4Ti_3O_{12}$ layer structure in superspace. (a) the absence of Ti atoms in some layers is indicated by gaps in the atomic shape. (b) The missing Ti layer influences the relative position of the neighbouring LaO_3 layers. This is indicated by the tilt of the corresponding atomic surfaces.*

4.4.8 Superspace and crystal chemistry

The use of the superspace concept described in Section 4.4.7 suggests that this formalism is reaching much beyond the pure description of modulated structures. We may wonder how the exploitation of superspace can contribute to a better understanding of crystal chemical concepts. Indeed, we will show with the examples of scheelite-like structures how to obtain a deeper understanding of the members of an extended family of structures and the origins of their similarities and differences (Arakcheeva and Chapuis, 2008). Scheelite-like structures are extensively studied, in particular for their luminescence properties. Their chemical compositions are of the form $A[XO_4]$, where A = K, Na, Li, Rb, Ag, Cs, Ca, Sr, Ba, Pb, Bi, Zr, Y, or Ln, and X = Mo, W, or V, among others. This composition can be further generalized to allow mixed A and X elements, in addition to the possible deficiencies $(A', A'')_{n-\delta A}[(X'X'')O_4]_{n-\delta X}$, where $\delta A \geq 0$ and $\delta X \geq 0$ characterize the vacancies of the corresponding elements.

Scheelites exhibit both classical and incommensurately modulated structures. Before the discovery of incommensurate members, a large number of them had been described in the literature, and many powder diffraction patterns have been published. Looking closer, many of them could be reinterpreted in terms of incommensurately modulated structures (Arakcheeva and Chapuis, 2008).

The structure of the incommensurately modulated structure KSm(MoO$_4$)$_2$ belongs to the superspace group $I2/b(\alpha\beta)00$, a portion of which is illustrated in Fig. 4.42. An analysis of the deformation and displacements of the octahedra relative to K and Sm gives an indication of the affinities between [SmMoO$_4$] and [KMoO$_4$] entities. In this particular case, [SmMoO$_4$] is preferred, and the subtle interplay between K and Sm complexes is at the origin of the modulation. It is clear that only a precise analysis of the incommensurately modulated structure can reveal the origin and nature of such interactions.

The resolution of the incommensurate scheelite-like structure can, however, be further extended. We observe first that all incommensurate members of the scheelite-like compounds can be described by the same monoclinic superspace group $I2/b(\alpha\beta)00$. The only essential difference resides in the two components of the modulation vector **q**. A large number of commensurate scheelite-like structures can be derived from rational values of the modulation vector components (see Chapter 2, Section 2.4.5). One can thus explore the possibilities of selecting various rational components of vector **q** with possible help from electron or X-ray diffraction patterns and by applying some simple crystal chemical rules. This has been done for a good selection of compounds, which are listed in Table 4.6. From the unique superspace group it is possible to characterize structures with cation and vacancy orderings in the A and X positions. Despite the large variation of chemical compositions, the structure of each member can be derived from a single superspace group by using the appropriate modulation vector and the variable t. This same procedure can also be exploited to predict new structures based on specific chemical components, in order to generate or improve new physical or

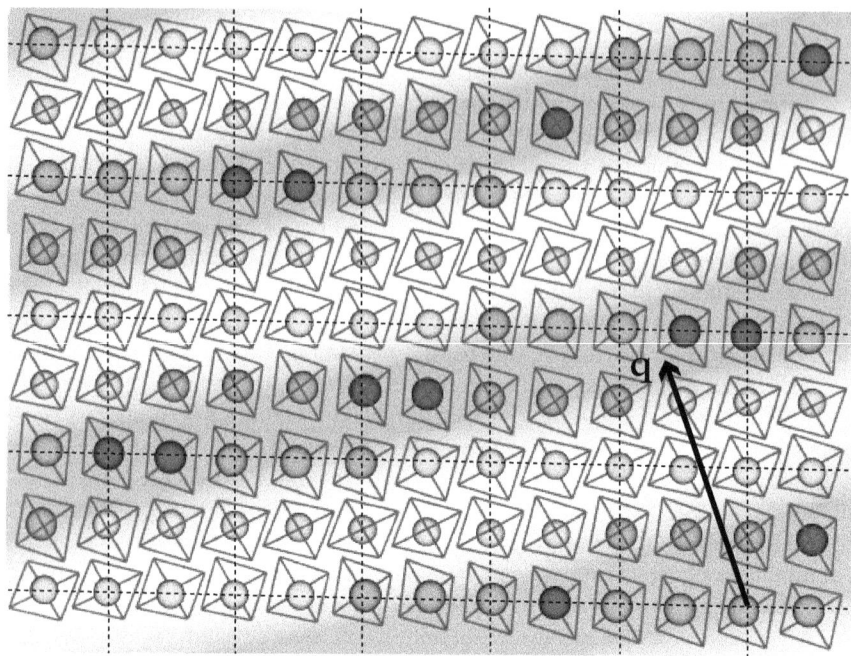

Figure 4.42 *Portion of an aperiodic layer of the scheelite-like structure $KSm(MoO_4)_2$. The MoO_4 tetrahedra correspond to K (shaded zone) and Sm atoms (clear zone). The orientation and the magnitude of the modulation defined by the two zones is specified by the modulation vector **q**.*

chemical properties. This is nicely illustrated for the luminescence properties of scheelites in Arakcheeva *et al.* (2012).

Crystal chemical properties of incommensurate structures like e.g. bonds, distances, or angles are usually described in terms of *t-plots*. The variable t is defined in Eq. 4.21 and illustrated in Fig. 4.33. A single *t*-plot of a given variable gives a concise overview of all the possible values which can occur in an incommensurate crystal. Therefore, it is easy to deduce the distribution of a specific variable and draw some conclusions on some crystal chemical aspects, such as coordination number or contact distances. One interesting study concerns the phases of $K_3In[PO_4]_2$ (Arakcheeva *et al.*, 2003). The origin of the incommensurate structure is due to various interactions between K, In, and P, each of which is attempting to reach an optimal geometric environment. The analysis based on *t*-plots makes it possible to distinguish between strong chemical interactions (stable distances along *t*) and weaker interactions and, in addition, to generate the frequency distribution of the variable, and the distance in the particular case. It is interesting to note here that, with the example of the incommensurate phase of $K_3In[PO_4]_2$, it is possible to reproduce very precisely the mean K–O, In–O,

Table 4.6 *Selection of commensurate members of the scheelite-like structures deduced from a single incommensurate phase. Vacancies are indicated by* □.

Compound	Space group deduced from $I2/b(\alpha\beta)00$	Modulation vector and t
RbBi [MoO$_4$]$_2$	$P2_1/a$	$\mathbf{q} = 0\mathbf{a}^* + \frac{1}{2}\mathbf{b}^*; t_0$
K$_2$Th[MoO$_4$]$_3$	$A2/a$	$\mathbf{q} = 0\mathbf{a}^* + \frac{2}{3}\mathbf{b}^*; t_0$
Eu$_2$□ [WO$_4$]$_3$	$A2/a$	$\mathbf{q} = \frac{2}{3}\mathbf{a}^* + \frac{2}{3}\mathbf{b}^*; t_0$
Bi$_2$□ [MoO$_4$]$_3$	$P2_1/a$	$\mathbf{q} = \frac{2}{3}\mathbf{a}^* + \frac{1}{3}\mathbf{b}^*; t_0$
La$_2$□ [MoO$_4$]$_3$	$A2/a$	$\mathbf{q} = \frac{2}{3}\mathbf{a}^* + \frac{8}{9}\mathbf{b}^*; t_0$
Bi$_3$[(FeO$_4$)(MoO$_4$)$_2$]	$A2/a$	$\mathbf{q} = 0\mathbf{a}^* + \frac{2}{3}\mathbf{b}^*; t_0$
Na$_4$Zr [□(MoO$_4$)$_4$]	$I2/b$	$\mathbf{q} = \frac{2}{5}\mathbf{a}^* + \frac{4}{5}\mathbf{b}^*; t_0$
Na$_4$Y [Na'(MoO$_4$)$_4$]	$I2/b$	$\mathbf{q} = \frac{2}{5}\mathbf{a}^* + \frac{4}{5}\mathbf{b}^*; t_0$

and P–O distances listed in the literature (Bergerhoff and Brandenburg, 1999). In other words, a close analysis of a single incommensurate structure might well act as a substitute for the missing data we are looking for!

Other interesting examples of crystal chemical applications can be cited. The large series of *hexagonal ferrites* can be described in terms of composition-dependent stacking of metal–oxygen layers. The various stacking sequences can be interpreted in terms of structural modulations of a common underlying average structure. The embedding of the hexagonal ferrites in superspace reveals the important property that all the members of the same family can be derived from a single superspace group model (Orlov and Chapuis, 2005a, b). The complete family of commensurate and incommensurate structures of *palmierite-like* structures can also be described from a single model in superspace (Arakcheeva and Chapuis, 2006).

4.4.9 Conclusion

The examples of applications based on the superspace formalism presented above are only a selection of possibilities. This formalism has many advantages in comparison to the classical formalism. First, a single superspace description is able to describe more than one phase of a single compound, depending on the temperature or pressure. Second, the new formalism improves the description of complex superstructures and represents a serious step forward for their refinements. It is also better adapted for the numerical methods of structure refinements. Third, the superspace is not only an elegant tool for the description of modular structures. It appears to be a very powerful tool for the prediction of new structures including their symmetry. It is also very helpful in pointing to the origin of chemical interactions between chemical units. Fourth, superspace groups can also extend the concept of structure types. The advantage is that new relations between structures can be discovered which would not be possible when only conventional 3D structures are used.

All these considerations are relatively new, but already show an enormous potential for a better understanding of crystal chemical properties and interactions. The extension capacity of the superspace formalism is such that we can expect, in the near future, to see many more useful applications in the field of structural sciences.

4.4.10 Structure determination of quasicrystals

Unlike incommensurately modulated structures, there is no periodic 'basic' structure in quasicrystals. It is thus not possible to start from the structure determination of the basic structure and add up a perturbative modulation. Structure determination of quasicrystals is thus, in general, more demanding. However, we will show in the following that similarities exist, especially when comparing with the structure determination of aperiodic composites structures.

Right after the quasicrystal discovery of Shechtman (Shechtman *et al.*, 1984), its interpretation in terms of a six-dimensional periodic crystal was proposed (Bak, 1985; Duneau and Katz, 1985; Elser, 1985; Kalugin *et al.*, 1985). It was also rapidly recognized that atomic clusters as found in periodic approximants could be used in modelling quasicrystals (Elser and Henley 1985; Guyot and Audier 1985)

In order to illustrate the principles of the structure determination of quasicrystals, we will first use a one-dimensional toy model. Although the one-dimensional case is peculiar and does not contain symmetry elements which are found in three-dimensional structures, it permits us to grasp most of the essential features of the structure determination of quasicrystals.

First we shall see how the two-dimensional periodic representation of the structure is intimately related to the resulting one-dimensional quasiperiodic structure. Next, we will illustrate how we can go from the X-ray (or neutron) diffraction pattern to a model structure. Finally, we will illustrate the various routes used in modelling the structure. We will then generalize this approach to icosahedral and decagonal symmetry.

4.4.11 A simple one-dimensional quasiperiodic model

The prototype of the one-dimensional quasiperiodic structure is the already introduced Fibonacci chain. The first modelling idea is to use the two tiles as basic units and try to find a decoration. It is the same procedure as used in crystallography of periodic crystals, where the unit cell is decorated with a motif. We shall see in the following that many models are based on such a decorated scheme. Whether real quasiperiodic structures can indeed be described by a decorated tiling is, however, still an open question.

In this tile decoration approach, the first step is to choose the proper underlying tiling, that is, not only the kind of tiling but also the tile length. Indeed, because of inflation rules, the diffraction pattern can be equally indexed with a Fibonacci chain having tile lengths $L, \tau L, \tau^2 L \ldots$ which gives some flexibility. For a given atomic density in real space, the larger the unit cell, the larger is the number of decorating atoms. Of course, a large unit cell allows a lot of flexibility, but the decoration becomes difficult to determine. On the other hand, too small a unit cell is, in general, not complex enough to fully explain the structure.

In the following we choose an embedding of the one-dimensional quasicrystal, where the periodic lattice is a square lattice with basis vectors (e_1, e_2) (Fig. 4.43). This defines the length scale in both the parallel space and the perpendicular space. (In agreement with the most common convention in the quasicrystal community, the spaces V_E and V_I are referred to as parallel space (E_{par}) and perpendicular space (E_{perp})). Notice that we use here a convention that is slightly different from that in Chapter 2. The figure is rotated so that e_1 is now horizontal, whereas the physical space is no longer horizontal. We also use a different basis and projection scheme. A vector of the 2D lattice **R** with coordinates n_1 and n_2, decomposes into the E_{par} and E_{perp} spaces, using the following relationship

$$\begin{pmatrix} R_{par} \\ R_{perp} \end{pmatrix} = \frac{a}{\sqrt{2+\tau}} \begin{pmatrix} \tau & 1 \\ 1 & \tau \end{pmatrix} \begin{pmatrix} n_1 \\ n_2 \end{pmatrix}. \tag{4.49}$$

In this relation, a is the lattice parameter of the square lattice, and τ the inverse golden mean equal to 1.618 ...; τ has the property that $\tau^2 = 1 + \tau$. As an example, let us consider that the two-dimensional lattice parameter is 0.6 nm. From the above expression, the L and S tiles will have lengths equal to 0.5104 and 0.3154 nm, respectively. If we consider that interatomic distances are of the order 0.27 nm, this is the smallest length that can be considered. To model this, we can choose a τ inflated tiling with two tiles having thus lengths equal to 0.8258 and 0.5104 nm. If we put one atom on each vertex of the Fibonacci chain, we see that there is now room for one more atom on the small tile and for two atoms on the larger tile: with this decoration, interatomic distances are around 0.26 nm. Of course, it is not necessary that all atoms be equidistant, and relaxation from the ideal position can be easily achieved. The corresponding two-dimensional periodic image of this structure is shown in Fig. 2.24b. This approach using large tiles has been used for both icosahedral and decagonal quasicrystals.

However, a more general approach is achieved by working directly in the two-dimensional space. First, we will see that it is the simplest way to analyse experimental data; second, this approach can generate more general structures. In the following we give the important ingredients and parameters necessary to fully describe the atomic structure, using this approach.

We first start with the Fibonacci chain generated in Fig. 4.43, top panel: the 2D periodic lattice is decorated by segment lines called 'atomic surfaces', which are parallel to E_{perp} and generate the L and S Fibonacci chains. There

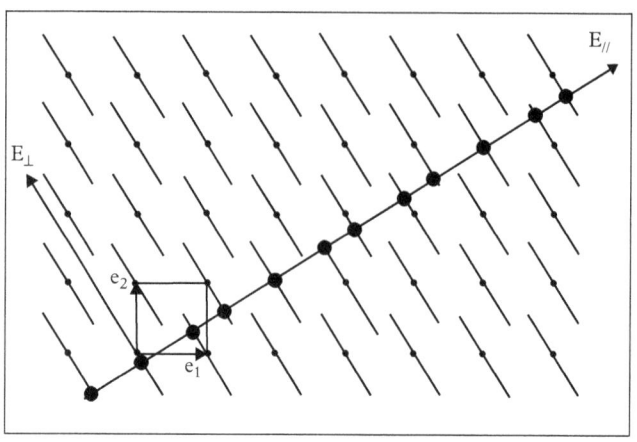
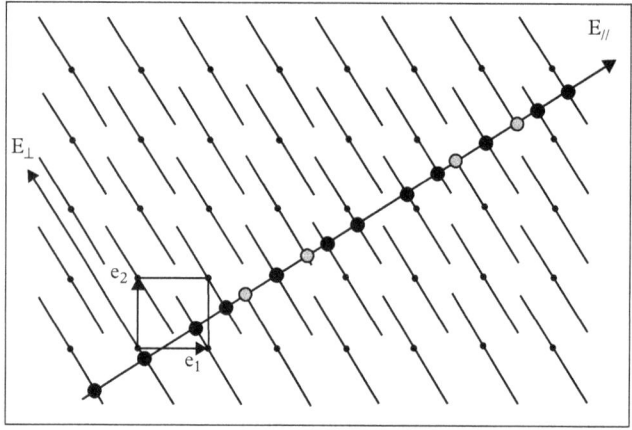
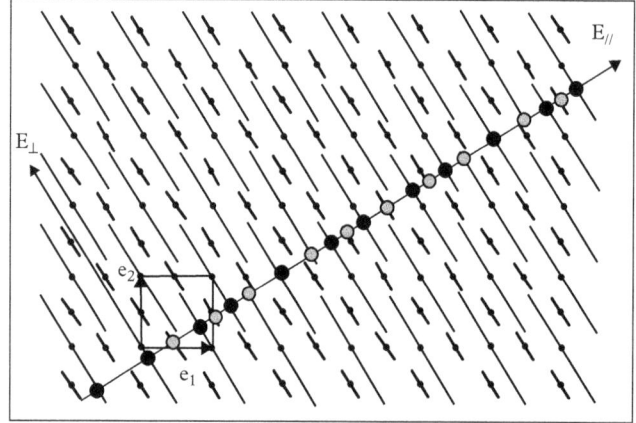

Figure 4.43 *Illustration of the different modifications which can occur in the periodic image of a one-dimensional quasicrystal. Top panel: Periodic image of the Fibonacci chain. Middle panel: The segment line has been increased; new atomic positions are generated in the one-dimensional quasicrystal, shown as light grey circles. Bottom panel: Two small atomic surfaces are located at mid-edge positions. This also generates a new local environment in the one-dimensional quasiperiodic structure, shown by the dark grey circles.*

are two main modifications that can be achieved on this structure and which will determine the resulting 1D quasiperiodic structure. We can change the shape of the atomic surfaces: here, the length of the segment lines. We can also decorate the periodic square lattice with new atomic surfaces. In short, we can say that the 1D quasicrystal structure will be determined by the position and shape of the atomic surfaces in the 2D periodic lattice. These modifications will have a direct impact on the local structure, which is, of course, most important. In fact, we will see later on that the tailoring of the atomic surfaces can be achieved in agreement with crystal chemistry.

The first important modification is exemplified in Fig. 4.43, middle panel, in which the length of the atomic surfaces has been increased. As can be easily understood, new atomic positions, shown as empty circles, are generated on some of the L tiles. It should be noted, however, that this modification does not produce a systematic decoration of the L tiles. The positions of the new atoms are based on the ratio $1/\tau$.

The second important modification can be achieved by a decoration of the two-dimensional square unit cell with new atomic surfaces. Instead of attaching atomic surfaces only to the nodes of the square lattice, we can also add atomic surfaces at new positions. Figure 4.43, bottom panel, illustrates this point, where two small atomic surfaces are located at the mid-edge positions (0, 0.5) and (0.5, 0). They will generate new atomic positions, shown as light grey circles. Again, the resulting structure is not a systematic decoration of the Fibonacci chain. Using Eq. 4.49, we can easily calculate what will be the resulting first neighbour interatomic distances in the one-dimensional quasicrystal. Let us assume that the 2D a parameter is equal to 0.97 nm. In that case, the first interatomic distances are equal to 0.255, 0.315, 0.413, and 0.510 nm. With this lattice parameter, the S and L tile lengths of the Fibonacci chain are equal to 0.51 and 0.83 nm, respectively.

So far, we have considered atomic surfaces that are flat in the complementary space. This is by no means a requisite, and Fig. 4.44 presents a situation in which a small part of the segment line has been given a component in physical space. This will not destroy the long-range quasiperiodic order, but local environments are in some places slightly changed. In a real structure, this is a way of describing the relaxation of some local environment. It turns out to be a crucial parameter in the structure refinement.

Finally, we have to consider the *chemistry* of the quasicrystal. Real quasicrystals are multi-component (at least binary, and most of the time ternary) alloys, and one should also describe their chemical order. This is achieved by assigning a chemical species to atomic surfaces or to part of them. In the previous example, the mid-edge positions could, for instance, be a B species, and the node an A one. However, we can also have more complex situations in which different parts of a single atomic surface correspond to different chemical components, with, for instance, the central part corresponding to a B atom, and the outer part to an A one. In fact, most of the structures determined so far correspond to complex cases, and it is one of the major difficulties of quasicrystal structure determination to assign properly each part of the atomic surface to the proper chemical species.

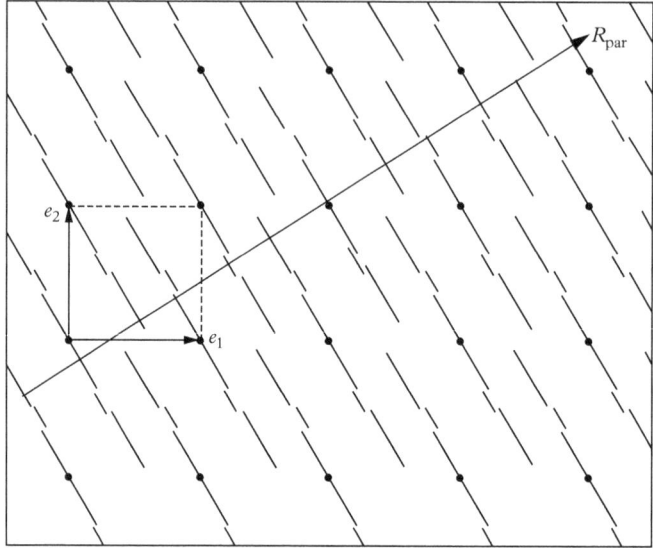

Figure 4.44 *Illustration of a general shape of an atomic surface. Part of the atomic surface has been given a parallel component. This will slightly modify the local environment.*

As usual in modelling, there are also some constraints given by the chemical composition, density, and unphysical short distances, which should be avoided. Chemical composition and density are easily calculated from the high-dimensional picture. The point density is given by $\rho = \sum \Delta_i/a^2$, and the chemical composition is obtained from the total size of the atomic surface associated with each chemical species. Thus, if both the density and the chemical composition are known, this will put some constraint on the total volume and repartition of the atomic surfaces. The short distances constraint imposes some tailoring of atomic surfaces, which is also one of the difficulties in structure determination.

We now have some feeling for what is going to be the work of structure determination. In the periodic description we have to specify the location and the shape (length) of the atomic surfaces. The simple example with one-dimensional atomic surfaces we have given can be generalized to the decagonal and the icosahedral cases. In those cases, atomic surfaces are two- or three-dimensional objects. We shall come back to this point later.

4.4.12 Structure determination of a one-dimensional quasicrystal

Among the various techniques used to probe the structure of materials, diffraction is the most powerful one. However, owing to the complexity of the structure of quasicrystals, using a combination of various techniques is often advisable.

As introduced in Section 4.1.1, the measurable quantity in the framework of the kinematical theory is the integrated intensity related to the Fourier transform of the structure. For a general atomic structure, the atomic density can be expressed as the convolution product between a lattice, here a 2D periodic lattice, and the cell decoration. The lattice is defined by the set of vectors $\mathbf{R}_{n1,n2}$. The decoration is defined by the set of atomic surfaces i, located at the positions $\mathbf{R}_i$ in the 2D unit cell and with the shape function $G_i(\mathbf{R}_{perp})$.

Its Fourier transform is easily calculated since the Fourier transform of a convolution is the product of the Fourier transform of each term. The reciprocal lattice of a square lattice is also a square lattice with a basis (e_1^*, e_2^*) and a lattice parameter $1/a$. Like the direct space, the two-dimensional reciprocal space decomposes into two one-dimensional subspaces, E_{par}^* and E_{perp}^*, where E_{par}^* is the physical reciprocal space in which the measurement is carried out. The position of a vector $\mathbf{H}_s$ in the two-dimensional reciprocal lattice with coordinates (h_1, h_2) might also be expressed by the basis (e_{par}^*, e_{perp}^*), using the following relationship:

$$\begin{pmatrix} H_{par} \\ H_{perp} \end{pmatrix} = \frac{1}{a\sqrt{2+\tau}} \begin{pmatrix} \tau & 1 \\ 1 & \tau \end{pmatrix} \begin{pmatrix} h_1 \\ h_2 \end{pmatrix}. \tag{4.50}$$

We can thus write the generalized structure factor as

$$F(\mathbf{H}_s) = \frac{1}{V} \sum_i G_i(\mathbf{H}_{perp}) \exp(2\pi i \mathbf{H}_s . \mathbf{R}_i), \tag{4.51}$$

where $G_i(\mathbf{H}_{perp})$ is the Fourier transform of the characteristic function of the atomic surface, and $\mathbf{R}_i$ its position in the high-dimensional unit cell. The 2D reciprocal space and intensity distribution is shown in Fig. 4.45. The strongest Bragg peaks have a small perpendicular component, and are labelled with Fibonacci numbers. The one-dimensional diffraction pattern is then simply obtained as a projection of the two-dimensional one onto E_{par}^*, with the integrated intensity being proportional to the square of the structure factor, i.e. $I(H_{par}) = F(H).F(H)^*$, as shown in Fig. 4.45, bottom panel, which displays the intensity distribution on a logarithmic scale. The positions of the Bragg peaks are indexed with two indices. The τ scaling is also displayed on the figure.

We have already used this equation in the simple case of the Fibonacci chain: in that case, there is only one atomic surface, and the $G(\mathbf{H}_{perp})$ function is the simple $\sin(x)/x$ function, which presents a smooth decrease with H_{perp}. When two or more atomic surfaces enter into play, Eq. 4.51 is analogous to the case of periodic three-dimensional structures: the function $G_i(\mathbf{H}_{perp})$ is equivalent to the atomic form factor, as we have seen in Section 4.1.1, and the phase factor $\exp(2\pi i \mathbf{H}.\mathbf{R}_i)$ leads to interferences between the different Fourier transforms $G_i(\mathbf{H}_{perp})$. This induces strong variation of the scattered intensity. As an illustration, the diffraction pattern of the structure of Fig. 4.43, bottom panel, is shown in Fig. 4.47. As can be seen, the intensity distribution is very different from the one for the Fibonacci chain. The

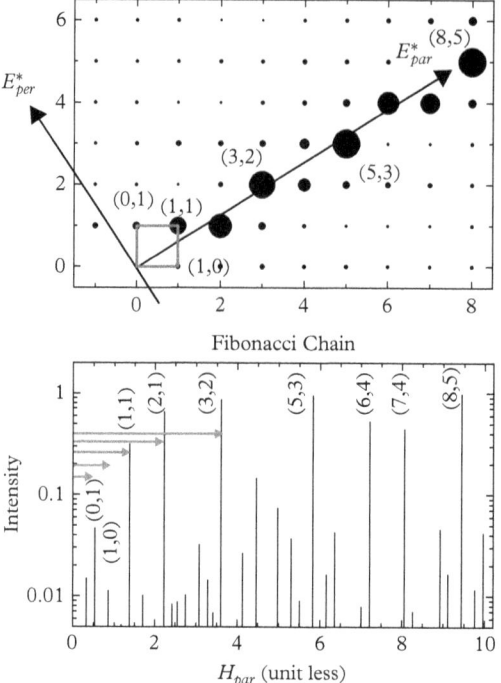

Figure 4.45 *Diffraction pattern of the Fibonacci chain as seen in the periodic 2D space (top) and projected once on the reciprocal physical space (bottom). The scale is such that $I(0) = 1$. There is a one-to-one correspondence between the two representations, which illustrates how the indexing is carried out.*

problem of the structure determination is to go from this intensity distribution to a model structure, which is not simple, since the phase of the structure factor cannot be measured.

4.4.12.1 Indexing the diffraction pattern

In a real experiment, the first step involves indexing the diffraction pattern and determining the space group. Since the diffraction spectrum is dense, this is not an easy task. Indeed, the position of a given Bragg reflection is known with some given accuracy, for the experimental resolution is finite. It is thus, in principle, possible to label this reflection with different reciprocal lattice vectors. However, this ambiguity can be lifted by considering the H_{perp} dependence of the 'form factor' $G_i(\mathbf{H}_{perp})$: it is maximal, and equal to the atomic surface volume, for $Q_{perp} = 0$ and then decreases rapidly. This implies that strong reflections must be indexed, with reciprocal vectors having a small perpendicular component. In practice, a list of calculated Bragg peak positions, with indices (h_1, h_2) and their corresponding value of the H_{par} and H_{perp} (Eq. 4.50) components, is compared to the experimental one. Once a unit cell is found, all peaks are indexed. As an important consistency check, all strong reflections must have a small H_{perp} component. Note that the reverse is not true: for complex structures, some small H_{perp} reflections might have a weak intensity, because of destructive interferences. Finally, because of inflation properties, the periodic unit cell parameter is only determined within the scaling factor τ and its power values. Once the indexing is properly carried out, we have a one-to-one correspondence between the one-dimensional quasiperiodic reciprocal lattice and its periodic embedding.

4.4.12.2 The phase problem and direct methods

At this point we are faced with the so-called phase problem: we have measured the intensity proportional to the square of the structure factor, but we do not know the phase of this structure factor. A common procedure in three-dimensional crystallography makes use of direct methods. Under the assumption of atomicity and positivity of the electron density, relationships between the phases of different Bragg reflections can be derived. This allows us to determine, at least approximately, a set of starting phases. A direct Fourier transform of the structure factor with their reconstructed phases leads then to the density distribution $\rho(R)$. From this map a starting model might be designed which is then compared and refined against diffraction data. During this procedure a few reasonable parameters (atomic coordinates, Debye–Waller temperature factors, etc.) are refined. The high-dimensional crystallography follows a similar procedure, although the problem of the proper parametrization of the model is complex. However, extending direct methods to the case of quasicrystals is not straightforward, and raises several theoretical problems.

The first simple phase reconstruction procedure was proposed in Jarič and Qiu (1993). It is not general, since it requires the existence of a periodic crystalline approximant to the quasicrystal. Although there are a few systems in which such a situation exists (AlLiCu, CdYb, etc.), it is not always the case. The method is based on the observation that the diffraction pattern of a quasicrystal and its corresponding approximant present strong similarities. In fact, the crystal approximant can be obtained by applying a linear phason (shear) strain, to the two-dimensional periodic structure. There is thus a relationship between the diffraction patterns of the quasicrystal and its approximant, and the rational indices of the crystal can be mapped out onto that of the quasicrystal. Jaric and Qiu proposed to assign the phases of the structure factor of the approximant to the corresponding one in the quasicrystal. The atomic density is then obtained by Fourier transform.

It is only recently that a phase reconstruction algorithm which applies for quasicrystals has been proposed, by several research groups. The first one was proposed in Elser (1999) and Brown et al. (2000). It is based on the positiveness of the electron density and on a minimization of the charge density: this leads to solutions having sharp maxima, as it should be since atomic surfaces are only confined to specific areas of the unit cell (this is the equivalent of the atomicity hypothesis for three-dimensional periodic crystals). From this algorithm it is possible to get a Fourier map of the structure, where about 80% of the phases are properly determined.

An alternative algorithm, using a low-density elimination procedure, leads to similar results (Takakura et al., 2001a).

Finally, the charge-flipping method, introduced earlier (Section 4.3.6) for the solution of incommensurate structures, can be used just as successfully for the solution of quasicrystalline structures. Here again, the method does not require any other prerequisite than the parameter δ defining the threshold for the charge-flipping limit. The method has been applied for the resolution of various decagonal and icosahedral quasicrystal structures. It can also deal with neutron diffraction data, where the scattering length is negative (this is not the case for the low-density elimination procedure, which relies on the positiveness of the electron density).

The three methods are thus now routinely used in the first step of the structure determination of quasicrystals. However, this approach does not avoid having to use a further modelling procedure. Indeed, only the phases of the strongest Bragg peaks are generally determined properly. The electron density is thus reconstructed with a truncated series, which means that the Fourier map does not contain details on the structure and that modelling is necessary. As an illustration of this phase reconstruction, let us consider a one-dimensional model quasicrystal whose two-dimensional embedding contains three atomic surfaces located at the origin and mid-edges of the square lattice (Fig. 4.46). The corresponding density map calculated with the 21 strong Bragg peaks with their phases is shown in Fig. 4.48. The position and rough size of the three atomic surfaces is easily recognized. In particular, the density profiles taken along the three atomic surfaces in the perpendicular direction show that the mid-edge atomic surfaces are much smaller than the one located at the origin. However, whereas the starting model is built up with atomic surfaces having sharp boundaries, the resulting density map shows smooth ones: this is because we used only a restricted set of Bragg reflections leading to Fourier series truncation effects. This will lead to so-called split positions, as shown by the black ellipse.

4.4.12.3 The Patterson analysis

Although now the structural solution is greatly simplified by the use of the above methods, allowing us to determine the phases of the structure factor, some of the structural analysis to date did not make use of this phase reconstruction. In this case, the first step in the structural analysis is carried out through the Patterson analysis. This function is obtained by Fourier transform of the measured intensity $I(Q)$. Since the Fourier transform of the product of two functions is the convolution of the Fourier transform of these two functions (here $\rho(R)$), the Patterson function $P(R)$, already introduced in Section 4.1.1, is equal to the autocorrelation of the density, that is, $P(R) = (\rho * \rho)(R)$. This function thus contains some information on the underlying atomic structure. For a one-dimensional

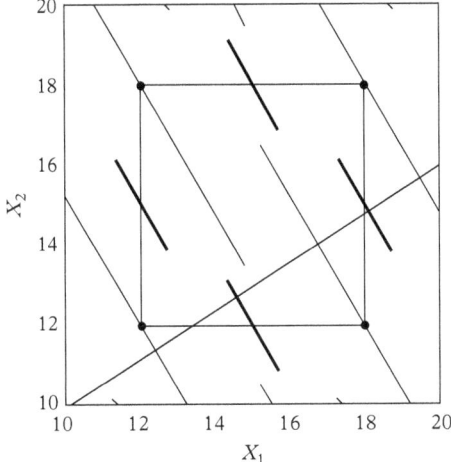

Figure 4.46 *Unit cell of a square lattice associated with a one-dimensional quasicrystal, with three atomic surfaces located at the origin and the mid-edges. This is an enlarged view of Fig. 4.43, bottom panel.*

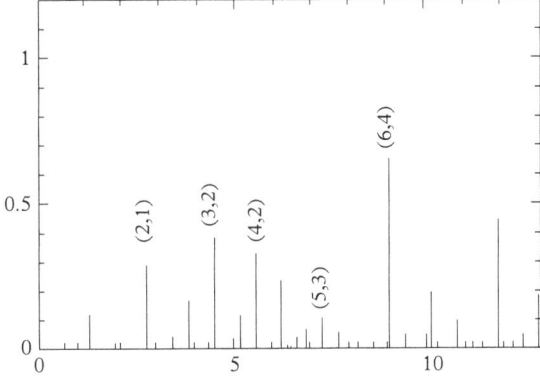

Figure 4.47 *Diffraction pattern of the 1D quasicrystal resulting from the decorated 2D lattice of Fig. 4.43, bottom panel, or Fig. 4.46. The intensity distribution is very different from the one of the Fibonacci chain.*

quasicrystal this function can be computed either in the one-dimensional E_{par} space or in the two-dimensional superspace. Of course, it is much easier to interpret it in the latter, where the structure is periodic. Readers not familiar with the Patterson function are referred to standard crystallographic books.

As an illustration of what might be encountered in a real experiment, we have computed the Patterson function of the structure shown in Fig. 4.46, by Fourier transformation of the 21 strongest reflections. This gives a good estimate of what is encountered in real quasicrystal structure determination. The Patterson map and corresponding profiles are shown in Fig. 4.48. The Patterson map is now more intricate to interpret. There are four contributions at the origin, mid-edge, and body-centre sites, and unravelling this Patterson map is not straightforward.

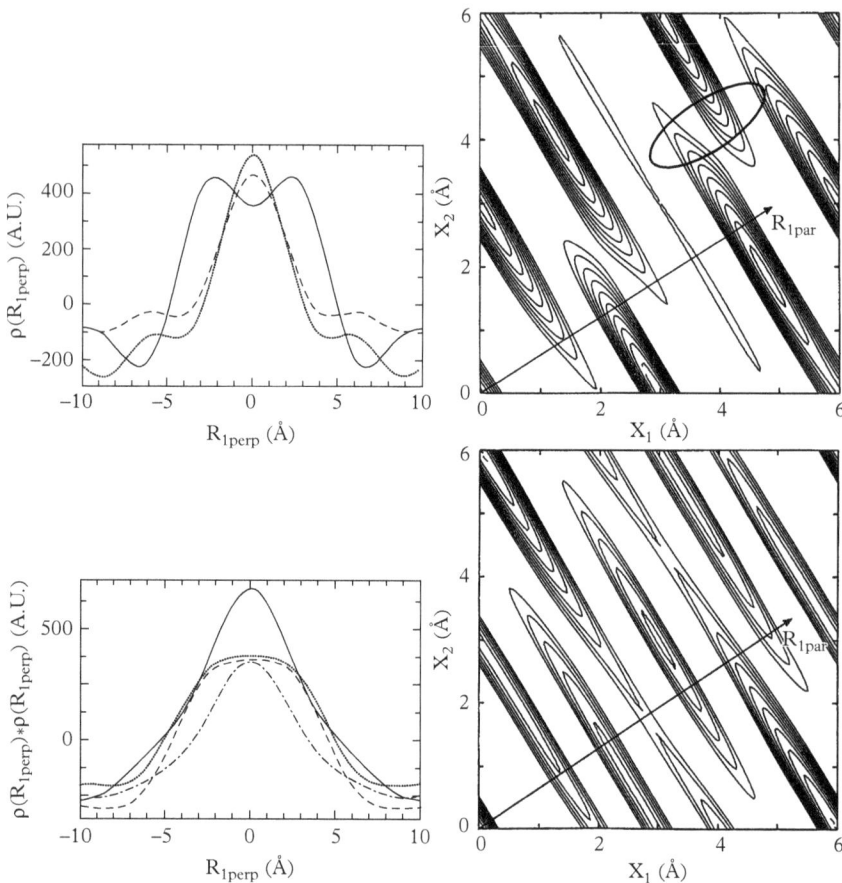

Figure 4.48 *Top: Density profile (left) and map (right) of the structure of Fig. 4.46, calculated by Fourier transform of the 21 strongest amplitudes with the correct phases. The positions of the three different atomic surfaces are clearly seen on the map. Their size can be inferred from the density profile (solid and dashed lines are for the origin and the mid-edges, respectively). Bottom: Density profile and map for the Patterson function of the same structure, obtained by Fourier transform of the 21 strongest intensities.*

At first sight, two structural solutions are compatible with this Patterson, but a deeper analysis shows that only the original structure is a possible solution. An inspection of the perpendicular profile of the different contributions might give some information on the size of the different atomic surfaces. In this example, the atomic surface at the origin has a length equal to 1.056 nm, and the mid-edge one has a length equal to 0.26 nm, the lattice parameter being equal to 0.6 nm. From the Patterson profile, it is possible to deduce that the mid-edge atomic surface is smaller than the one at the origin, but an exact determination of its length is difficult.

4.4.12.4 Contrast variation

The Patterson and structure analysis becomes much simplified if the different atomic contributions are separated. If we consider a binary alloy AB, the total structure factor is the sum of two contributions, $F(Q) = b_A F_A(Q) + b_B F_B(Q)$, where F_A and F_B are the partial structure factors, and b_A and b_B the corresponding scattering lengths. The partial structure factors are complex numbers, with a module and a phase if the structure is not centro-symmetric. It is a real number for centrosymmetric ones. It is possible to gain information on one of the sublattices by changing the scattering power of one element. It can be achieved using isotopic or isomorphic substitution when working with neutron diffraction, or by anomalous X-ray scattering. When the neutron and X-ray relative scattering power of the elements are different, combining two data sets will also lead to valuable information.

Extracting one partial structure factor is a more demanding task. In the case of a binary alloy, the measurement has to be carried out with three different settings allowing a change of the scattering power of one element: it can be achieved either by synthesizing three different samples having different B isotope concentration or by varying the wavelength of the incoming X-ray close to an absorption edge. Once this is done, it is possible to extract the modulus of the partial structure factors F_A and F_B and their relative phase difference. When such an experiment can be carried out, of course, the Patterson analysis gets much simplified and a unique structural solution is more likely to be found.

4.4.12.5 Modelling and refining the structure

Once a rough model is deduced, either from a density map after phase reconstruction or from the Patterson analysis, the final step in the structure determination is to construct a model having a few parameters which are then refined by comparison of the calculated and the measured data. This is the more demanding step, and many approaches have been used. Indeed, unlike the case for incommensurately modulated or composite structures, there is no 'standard' software which can be used routinely for quasicrystals. The most elaborated approaches have been developed in the high-dimensional description. However, although the description of the quasicrystalline structure becomes much more simplified, the number of parameters necessary to describe the atomic surfaces is, a priori, infinite: in some sense, a quasicrystal is a crystal with an infinite cell parameter, requiring an infinite number of position parameters. The problem is thus to find constraints (short distances, clusters, chemical order, etc.) that will allow us to reduce the number of parameters to a value compatible with the available data.

When they exist, periodic approximants are extremely useful, as they allow insight into the crystal chemistry of a complex structure. The notion of an approximant was introduced in Chapter 3, Section 3.5. If a strain is applied along E_{perp}, periodic approximants can be generated for special values of the strain. It is somehow (though not exactly) equivalent to changing the slope of the E_{par} space, as illustrated in Fig. 4.49. In this figure the slope defined by τ is replaced by the approximate Fibonacci ratios 1/1, 2/1, 3/2, etc. The approximants are thus labelled with respect to the Fibonacci ratio and so are called 1/1, 2/1, and 3/2, approximants. They have a periodic unit cell that increases by a factor τ, and a local environment that grows closer and closer to that of aperiodic tiling as the Fibonacci ratio approaches the golden mean τ. The procedure is illustrated in Fig. 4.49, which shows the construction of the 1/1 and 2/1 approximants for the case of the Fibonacci chain. The 1/1 approximant has a unit cell described by LS, whereas the 2/1 approximant is described by LSL. Figure 4.49 also displays the corresponding diffraction pattern as compared to the one of the quasicrystal: the strongest peaks almost match in both phases, as a signature of a similar local environment.

As an illustration of the procedure, let us return to the structure of Fig. 4.46. From the Patterson analysis we know that there are three atomic surfaces located at the origin and mid-edge positions, and that the mid-edge one has a smaller length. We have already seen in Chapter 2, Section 2.3.4, that density and chemical composition are directly related to the total length of the segment lines. We might also want to avoid unphysical short distances. This would be the case if some atomic surfaces were made too long. For instance, on the density map of Fig. 4.48, we can see that the origin and the mid-edge atomic surfaces have a small overlap shown by the ellipse: this would result in distances equal to 0.1 nm, which is much too small. However, the atomic density observed on the density map at this point is smaller than one: this could be interpreted as a disordered structure, with a statistical occupancy of the two sites, which is not the correct solution, since we know that we started with a structure with sharp boundaries.

A general approach is to model the structure by decorating the square lattice with three atomic surfaces (segment lines) which are then partitioned into smaller parts: for each small part we attribute a few parameters, such as the chemical composition (A, B, or a mix), a component along the parallel direction, or a Debye–Waller factor, both in the parallel direction and in the perpendicular (phason) direction. The way the partitioning of the segment line is carried out is not trivial and not unique. First the total number of subdomains must be consistent with the available data set: the total number of parameters must not be greater than 1/4 of the total number of independent observed reflections, so that a minimizing procedure is meaningful. Then the partitioning of segment lines can be carried out

Typical examples of modulated phases and composites 231

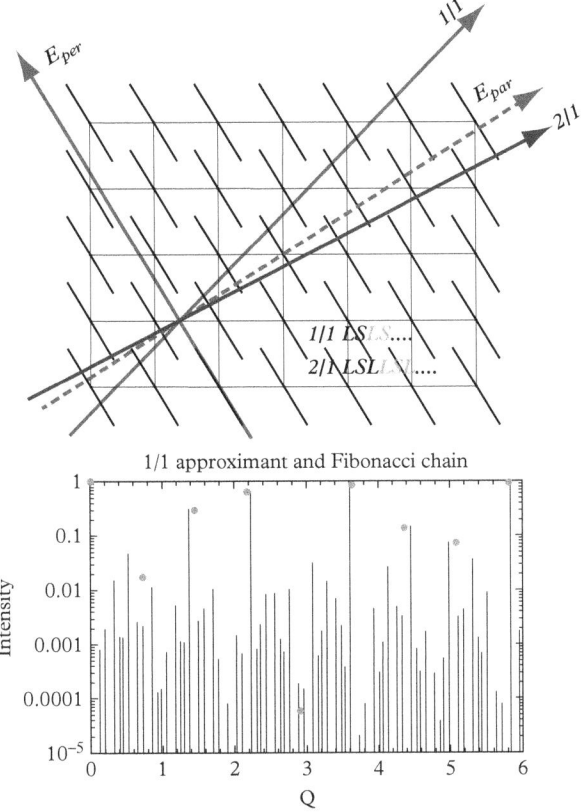

Figure 4.49 *Approximants of the Fibonacci chain. Top panel: 2D representation of the 1/1 and 2/1 periodic approximants, together with the Fibonacci chain. Bottom panel: Diffraction pattern of the 1/1 approximant and the Fibonacci chain. Strong peaks are only slightly displaced with respect to the Fibonacci diffraction pattern.*

by dividing each atomic surface evenly or by considering the possible local order, and, for instance, favouring one typical local configuration. In real icosahedral or decagonal structures, this procedure is one of the key points in the modelling and has been illustrated in more detail in Section 4.1.1.

Once this is achieved, parameters are refined by comparing the calculated intensities with the observed ones. In this procedure, weak reflections, with a large Q_{perp} component are particularly important. Indeed, they are very sensitive to the details of the structure. Experimentally, only intensities above a given threshold can be measured: since the intensity is rapidly decaying with Q_{perp}, this leads to a data set with Bragg peaks having a Q_{perp} component smaller than a maximum value. In some sense, the structure can only be determined with a resolution in the perpendicular direction linked to this maximum value: the higher the Q_{perp} value, the better will be the perpendicular resolution.

Of course, as is usual in any structure determination, the resulting model must be checked from the chemical point of view. If some inconsistencies are found, the model has to be modified.

Now that we are getting more familiar with the structure determination of a one-dimensional quasicrystal, we give in the following the general procedure for the two classes of quasicrystals for which most of the structure determinations have been achieved, namely the icosahedral and decagonal cases.

4.4.13 Structure determination of icosahedral and decagonal phases

As already shown, the periodic lattice associated with icosahedral symmetry is a six-dimensional hypercubic lattice. The six-dimensional space decomposes into two three-dimensional sub-spaces: the physical space and the perpendicular space. Now, atomic surfaces are three-dimensional objects, and the goal of the structure determination is to locate their position in the six-dimensional cubic unit cell and to determine their shape.

Of course, it is not possible to visualize the six-dimensional periodic space in one shot. To gain some insight into the six-dimensional structure, rational sections of the six-dimensional space, containing a high-symmetry direction (two-, three- or five-fold) both in the physical direction and in the perpendicular direction, are displayed. This is exemplified in Fig. 4.50, which shows a section containing a five-fold axis in both subspaces. The trace of the six-dimensional cubic unit cell is outlined as a rectangle and corresponds to the two directions [100000] and [01111$\bar{1}$]. The segment lines correspond to the trace of the three-dimensional atomic surface. This plane is particularly interesting because it contains two high-symmetry points, beside the origin: the 'mid-edge' one with coordinates (0.5 0 0 0 0 0), and the 'body-centre' one with coordinates (0.5 0.5 0.5 0.5 0.5 $\overline{0.5}$) visible on the diagonal of the rectangle. All icosahedral phases studied so far have a structure with atomic surfaces located only on these special points.

These rational sections are mainly used to locate the position of atomic surfaces, although a rough estimate of the size of the atomic surface is given by the length of the segment line. Atomic surfaces are best visualized in the perpendicular space only, either by a three-dimensional representation or by a two-dimensional section.

The route from diffraction data to a structural model follows the same path as the one explained for the one-dimensional quasicrystal. The first step involves indexing and space group determination. There are six possible symmorphic space groups: three with a centre of symmetry, and three acentric ones. All icosahedral phases studied belong to either the P$\bar{5}$3m or the F$\bar{5}$3m centrosymmetric space group. They may be distinguished in reciprocal space, looking at the scaling factor along a five-fold axis: whereas it is τ^3 for the P$\bar{5}$3m space group, it is τ for F$\bar{5}$3m.

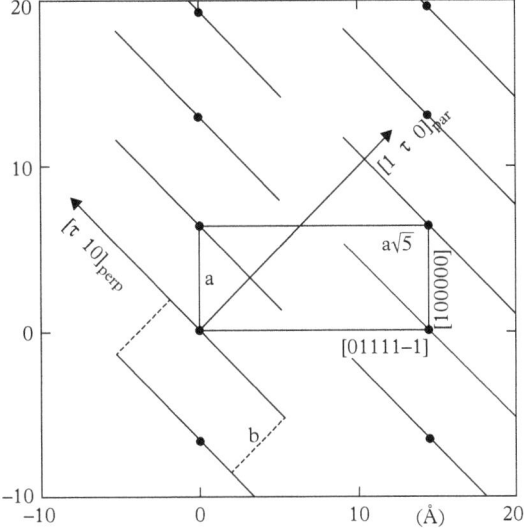

Figure 4.50 *Rational two-dimensional section of the six-dimensional cube, containing a five-fold axis both in the physical direction and in the perpendicular direction.*

The next step is either a Patterson analysis (Cahn *et al.*, 1988) or a phase-reconstruction approach. In both cases, density maps are calculated and displayed along one of the rational sections to locate the position of atomic surfaces. From this analysis a model is designed. We are now faced with the problem of parametrizing a three-dimensional object: this requires, in principle, an infinite number of parameters. As a simplification, three-dimensional atomic surfaces are modelled with polyhedra. Each polyhedron is then subdivided into smaller parts for the purpose of the refinement. As already said, there is no unique way to carry out this decomposition, and such a problem is still the object of current investigations.

The procedure is essentially the same for the decagonal case, except that the periodic space is five-dimensional (not cubic), because the decagonal quasicrystal is a periodic stacking of quasiperiodic planes. As a result, the perpendicular space is only two-dimensional, which makes the structure analysis somewhat easier.

Some symmetry elements and the projection matrices are given in Appendix A.

4.5 Examples of quasicrystal structures

4.5.1 Introduction

Since the discovery of quasicrystals, the problem of the atomic structure determination ('where are the atoms?') has remained a fascinating problem. More precisely, a detailed understanding of the crystal chemistry, i.e. of the chemistry where each atom is located, is one of the key questions if any physical property is to be determined. A lot of progress has been achieved in the field of quasicrystals, especially since the discovery of stable quasicrystals for which large single grains can be grown.

In the following, we present results obtained for three different phases, illustrating the various strategies employed to tackle structure analysis of quasicrystals. We start with the icosahedral i-AlPdMn phase, for which neutron and X-ray scattering have been used. This illustrates the use of contrast variation techniques, together with strategies for modelling. We will then continue with the structure of the binary icosahedral CdYb phase. It is the first binary quasicrystal, which of course simplifies the structural analysis problem. Moreover, there exists a crystal approximant to this phase, allowing a detailed study of clusters and their interconnection, which then serves as a template for the quasicrystal modelling. We will show that the understanding achieved for this icosahedral quasicrystal is very similar to that for complex intermetallic compounds. We will continue with the presentation of the structure of the decagonal d-AlNiCo phase, where the combined use of X-ray diffraction and high-resolution images has been widely used. We will finish with the newly discovered oxide quasicrystal.

4.5.2 Structure of the i-AlPdMn phase

The icosahedral i-AlPdMn phase was one of the first 'stable' quasicrystals for which a lot of structural analysis was carried out. A detailed study of the phase diagram shows that there is a small composition range for which the quasicrystal can be grown directly from the melt. The solidification is, however, non-congruent, which means that the composition of the solid phase is different from that of the liquid phase (Tsai *et al.*, 1990; Yokoyama *et al.*, 1991). Starting with a liquid alloy composition of $Al_{72.1}Pd_{20.7}Mn_{7.2}$, a single grain of the icosahedral phase with the composition $Al_{68.2}Pd_{22.8}Mn_{9.0}$ is obtained. Using the Czochralski technique, it is possible to grow centimetre-sized single grains (de Boissieu *et al.* 1992; Boudard *et al.* 1995) of very high quality, as shown in Fig. 4.51.

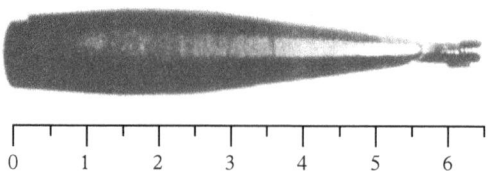

Figure 4.51 *Single grain of an i-AlPdMn quasicrystal grown by Czochralski. The seed is visible on the right part of the figure. The scale is in centimetres.*

4.5.2.1 Space group determination

Once a quasicrystal is grown, the space group and structural quality of the sample have to be checked before proceeding to the data collection of integrated intensities.

As for any aperiodic crystal, an easy check of the Laue symmetry of the i-AlPdMn diffraction pattern can be performed by X-ray Laue diffraction and rotating the sample around different symmetry axes. For instance, if one rotates the sample around a two-fold axis, one finds a three-fold and a five-fold symmetry diffraction pattern for a rotation angle equal to 20.9° and 58.3°, respectively. Figure 4.52 shows a typical five-fold Laue diffraction pattern. From this we thus deduce that the Laue group is $\bar{5}3m$.

A more accurate verification of the icosahedral symmetry is then generally achieved on a four-circle diffractometer, using monochromatized radiation and a point or two-dimensional detector.

Once the Laue group symmetry of the diffraction pattern is determined, the space group of the studied sample has to be found. This is best achieved by studying carefully the diffraction pattern on a four-circle X-ray or neutron diffractometer (using either a point detector or a two-dimensional one), or by electron diffraction. The choice of electron, X-ray, or neutron is essentially determined by the sample size. For instance, Shechtman (Shechtman *et al.*, 1984) obtained the first icosahedral phase by rapid quench from the melt. With such sample preparation single grains have sizes smaller than 1 μm, which makes

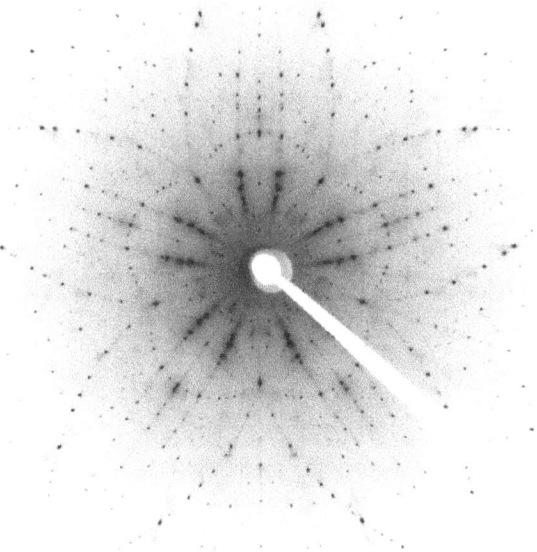

Figure 4.52 *Five-fold Laue pattern of the i-AlPdMn phase. (Courtesy W. Steurer)*

electron diffraction the only technique that can be used to characterize the symmetry properly. In the case of the i-AlPdMn phase, as already said, centimetre-sized single grains can be obtained, so that both X-ray- and neutron-diffraction techniques can be used. Using a four-circle type instrument, Q-scans along a high-symmetry axis can be carried out. Figure 4.53 shows two Q-scans measured along a two-fold and a five-fold axis, using synchrotron X-ray radiation: the high intensity of the incoming beam and the high resolution of the set-up allows one

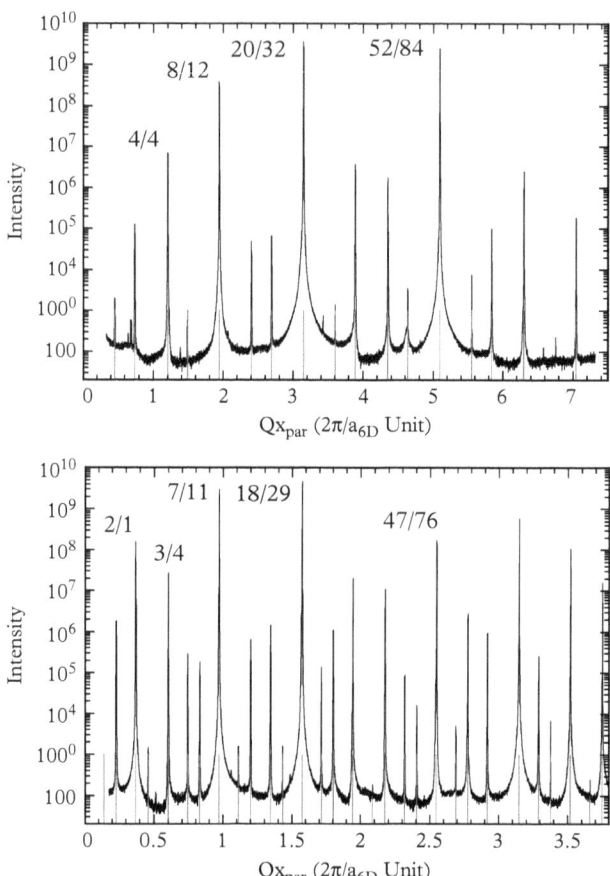

Figure 4.53 *Two-fold (top) and five-fold (bottom) Q-scans measured on a single grain of the i-AlPdMn phase using synchrotron radiation X-rays. The intensity scale is logarithmic in order to display both strong and weak reflections. The axis coordinates are in $2\pi/a_{6d}$ units or relative light units (see text). The vertical bars are the theoretical position of Bragg peaks having a Q_{perp} component smaller than 2.5. A few N/M indices are indicated.*

to measure weak and strong reflections and determine their positions with high accuracy. As can been seen in Fig. 4.53, where the scale is logarithmic, Bragg peak intensities span a wide range of values, with more than eight orders of magnitude between the strong and weak peaks. In fact, increasing the incoming flux will allow us to distinguish even weaker reflections: this is, of course, a consequence of the quasiperiodic long-range order. Nevertheless, Bragg peaks are easily identified and do not overlap too much. Moreover, we also clearly see on this diffraction pattern that there are only a few reflections which are strong and above a given threshold.

The space-group determination requires two steps: first, determine whether extinctions are present in the diffraction pattern, and then index the diffraction pattern to give the proper unit cell. These two steps, which are routinely (and generally automatically) carried out for periodic crystals, require some care in the case of quasicrystals. The Laue group $\bar{5}3m$ corresponds to two point groups with and without a centre of symmetry: $\bar{5}3m$ and 532, respectively. The comparison with periodic approximant crystals, anomalous scattering experiments (de Boissieu et al., 1994a), and arguments about the structure lead to the choice of the centro-symmetric $\bar{5}3m$ point group. For this point group there are three arithmetic classes corresponding to the three Bravais classes P, I, and F, that is, primitive, body centred, and face centred, respectively (see Appendix A). There are only three centrosymmetric symmorphic space groups corresponding to these Bravais classes, which are $P\bar{5}3m$, $F\bar{5}3m$, and $I\bar{5}3m$. As for the case of three-dimensional periodic crystals, there are extinction conditions in the diffraction pattern associated with these space groups. For instance, the diffraction pattern of the $F\bar{5}3m$ space group has only reflections such that all indices are of the same parity (that is, all odd or all even). The mixed parity indices are extinct in the diffraction pattern (see Appendix A). The distribution of reflections in reciprocal space is thus different for the three space groups. Moreover, the scaling properties are different, in particular along the five-fold axis. Whereas there is a τ^3 scaling for the P lattice, it is τ for the F or I lattice. This means that if a reflection is observed at the position H_{par}, a reflection must be observed at $\tau^3.H_{par}$ for the P lattice, whereas a reflection must be observed at $\tau.H_{par}$ for the I and F lattices. This scaling property (only valid in reciprocal space) already allows a simple identification (note that this is not true for the two-fold axis, where in this case the scaling is τ for the three space groups). Looking at the five-fold Q-scan in Fig. 4.53, we can verify that the scaling is τ: for instance, there is a Bragg peak at $Q_x = 0.372$, and another one at $Q_x = 0.602 = 1.618 \times 0.372$, then at $Q_x = 0.973$, $Q_x = 1.575\ldots$ (These reflections are labelled with their N/M indices 2/1, 3/4, 7/11, and 18/29, which is a shorthand notation for the six *Cahn–Gratias* indices (Cahn et al., 1986, see p. 164). N and M indices are calculated from the six indices and allow us to compute the modulus of H_{par} and H_{perp}). The Bravais lattice of the i-AlPdMn phase is thus either F or I.

Going further requires a proper indexing of the diffraction pattern and thus a determination of the six-dimensional unit cell parameter. Because of the above scaling properties, the unit cell parameter can only be given within a factor τ

and its powers. Moreover, we have seen that there are many Bragg peaks, so that the choice of a basis seems quite difficult at first sight. Indexing is done in noting that the intensity of a Bragg reflection decays roughly as $1/H_{\text{perp}}^6$, that means, very rapidly. This factor comes from the Fourier transform of the atomic surfaces. There are, however, also interference effects, so that this rule is only very approximate. Nevertheless, strong reflections must have indices such that their reciprocal lattice vector has a small perpendicular component. This property is used for indexing: a list of theoretical Bragg reflections is generated and one adjusts the calculated positions to those of the strongest observed reflections.

It is found that the i-AlPdMn diffraction pattern has extinctions such that mixed parity indices are absent. The corresponding space group of the structure is thus $F\bar{5}3m$. The lattice parameter a_{F6D} is equal to 2×6.45 Å. We will see in the following that the real six-dimensional space structure is best described as a superstructure on a primitive lattice with six-dimensional parameter $a_{P6D} = 6.45$ Å, so that the indexing can be carried out with a reciprocal lattice parameter equal to $1/a_{P6D}$ and six-dimensional indices that are either all integer or all half-integer. From now on, we choose this second convention. The above value of the cell parameter corresponds to the size of the unit cell necessary to generate a three-dimensional Penrose tiling (or icosahedral Ammann tiling: see p. 139) with an edge length equal to 4.56 Å. Thus, this is a reasonable length scale for describing the atomic structure. Indices can be expressed in a compact form by a pair of integers (N, M), following the *Cahn–Gratias* convention (see p. 164), from which the parallel and perpendicular components of the reciprocal lattice vector can be calculated. The vertical bars in Fig. 4.53 indicate the theoretical position of reciprocal lattice vectors whose perpendicular component is smaller than 2.5 (in relative light units): all reflections are perfectly indexed. A few reflections are labelled with their (N, M) indices, and a few indices are given in Table 4.2. Notice that indices with N odd correspond to the six-dimensional indices being all half-integer. As already mentioned, those odd N indices do not show up on the two-fold axis, but only on the five-fold one.

4.5.2.2 Structural quality

We now ask the question of how perfect the icosahedral symmetry is. This became the subject of a lot of controversy soon after the discovery of quasicrystals, raising questions which have since been quite clearly answered: in a few samples the icosahedral symmetry and the structural quality is only compatible with a quasicrystalline model. There are two questions to be addressed along these lines. (i) Is the symmetry of the diffraction pattern really icosahedral? Would it not be possible to describe the structure by a large crystal cell? (ii) How perfect or how close to the ideal quasiperiodic long-range order is the structure (that is, in terms of correlation length or Bragg peak width)?

For the first question, on the icosahedral symmetry, it is useful to remember that there are periodic approximants to a quasicrystal. They may be obtained by applying a shear strain (uniform phason strain) to the ideal icosahedral quasicrystal. Such a strain leads to a distortion of the diffraction pattern that is no longer of perfect icosahedral symmetry. The distortion of the diffraction pattern can be expressed as a shift from the ideal quasicrystal diffraction pattern. In this case, the magnitude of the shift is shown to be proportional to the perpendicular component of the Bragg peak considered. Precise measurement of the position of the Bragg peaks is thus a very powerful tool for such a study.

Peak shifts have been observed in several cases, for instance in some samples of the i-ZnMgY phase (Létoublon et al., 2000a). An example of a Q-scan measured along a five-fold axis of the i-ZnMgY phase is shown in Fig. 4.54. One clearly observes that Bragg peaks are shifted with respect to the ideal positions shown as vertical bars. Moreover, the sign of the peak shift, as determined with respect to the five-fold axis direction, can be positive (Bragg peak 23/4) or negative (Bragg peak 18/9). This is directly related to the sign of the corresponding H_{perp} wave vector, which is either positive or negative (for a five-fold axis the H_{perp} component also lies along a five-fold axis in perpendicular space; the sign of H_{perp} is defined with respect to this perpendicular five-fold axis). A systematic study of the peak shift as a function of H_{perp} leads to a linear dependence, illustrating that the structure is departing from a perfect icosahedral symmetry (right panel, Fig. 4.54). Note that the peak shifts are rather small, and can only be detected using a high-resolution set-up (resolution better than 10^{-3} Å^{-1}) available at synchrotron sources. This indicates a weak departure from the perfect icosahedral symmetry, and that most likely in this case a large periodic approximant would be a good description. Similar observations have been obtained in the i-AlCuFe phase, where there is a transition which takes place between a high-temperature icosahedral phase and a low-temperature phase and which is a multi-domain structure.

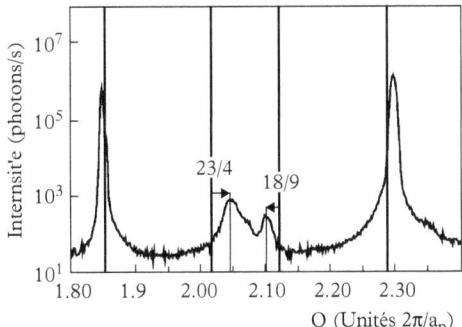
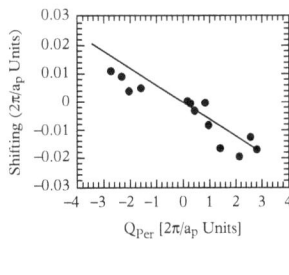

Figure 4.54 *Left panel: Enlarged part of a five-fold Q-scan measured on a single grain of the i-ZnMgY phase, using synchrotron radiation. The theoretical positions of the quasicrystal are shown by vertical bars up to the top of the figure. There are significant peak shifts, whose sign depends on the sign of H_{perp} (see text). Right panel: Evolution of the peak shift as a function of H_{perp}. There is a clear linear dependence, indicating the presence of a linear phason strain in the sample.*

This is at variance with the i-AlPdMn diffraction pattern, which does not show any departure from the perfect icosahedral symmetry, as shown in Fig. 4.53: the positions of the observed Bragg reflections are in perfect agreement with the theoretical ones.

The question of the quality of the quasicrystal is a bit more tricky since it essentially depends on the criterion used for its definition! The mathematical model that describes the quasicrystal leads in the kinematical theory of diffraction to Bragg peaks that are delta functions. In general the 'quality' is given by reference to this ideal delta function. An experimentally measured Bragg peak profile is the convolution of the 'true' Bragg peak profile (a delta function in the ideal case) with the instrumental resolution. As one can readily understand, the accuracy on the 'true' Bragg peak profile strongly depends on the Q resolution of the instrument used to record the diffraction pattern. Moreover, as for periodic crystals, the kinematical theory no longer holds once the quasicrystal is perfect enough, in other words, contains a small amount of defects. In this case the dynamical theory of X-ray diffraction must be used. Within this theory the shape of the Bragg peak is no longer a delta function, but a very narrow curve, whose width is given by the so-called Darwin width. The Darwin width depends on the value of the structure factor and can be evaluated if an atomic model is available. Thus, it was a breakthrough, when Kycia et al. (1993) reported the first evidence for dynamical X-ray diffraction in the i-AlPdMn phase. They observed the so-called Borman effect (or anomalous transmission), for which a beam is forward transmitted, although the thickness of the sample is such that the absorption would prevent any beam to go through. This only occurs at the Bragg peak position and is a sign that a standing wave field has been created in the sample. Other experiments have also been used to probe the dynamical theory of diffraction by quasicrystals: direct probe of the standing wave field by measuring the fluorescence of Al or Mn atoms (Jach et al., 1999) and topography images (Gastaldi et al., 1995, 2003). This last technique is particularly useful since it allows us to visualize defects such as dislocations. All these results indicate that the i-AlPdMn phase is 'perfect', although the Bragg peak width has always been found slightly larger than the theoretical Darwin width, because of the remaining defects. One has to notice that, among the millions of periodic crystals, only a few have rocking-curve widths corresponding to the Darwin width: these are, of course, semiconductors such as Si, Ge, quartz, and diamond. Metallic alloys are, most of the time, of worse quality than the i-AlPdMn phase and, in that respect, this alloy is close to being perfect.

Another check that has frequently been used is the H_{perp} dependence of the Bragg peak width. Imagine that the sample is built up of small domains, with each one distorted by a small uniform phason strain: each domain produces a diffraction pattern whose Bragg peaks are displaced from their ideal position. If there is a large number of such domains, and a random distribution of the strains, the resulting diffraction pattern is the superposition of all the displaced peaks, so that the resulting Bragg peaks are broadened and have a width increasing linearly with H_{perp} (Lubensky et al., 1986). Such a distribution of phason strain is equivalent to the distribution of strain in crystals, when dislocations are present in the sample, for instance. Indeed, the theoretical study of dislocations in quasicrystals predicts both a so-called phonon distortion field and a phason distortion field around the core of a dislocation, so that one might expect such a linear increase of Bragg peak widths as a function of H_{perp}. In fact, all quasicrystals studied so far display to a more or less important extent a Bragg peak broadening. An illustration is given in Fig. 4.55 for the case of the i-AlPdMn phase. The figure presents a high-resolution measurement using a two-dimensional CCD camera (Létoublon et al., 2001). Since the X-ray beam is partly coherent in this set-up, the broadened Bragg peaks appear with spikes. The width of the Bragg peak can, nevertheless, be extracted. On the left panel, which presents the measurement for the as-grown (by Czochralski) sample, a clear H_{perp}-dependent broadening is visible. Once the sample has been annealed (right panel), the low-H_{perp} reflection 20/32 is almost resolution limited and the 16/16 reflection present a slight broadening. When plotted as a function of H_{perp}, the peak width displays a linear increase, with a slope equal to 0.0004 in the case of the annealed sample. We stress again that this broadening is extremely small and can only be measured with synchrotron radiation and a high-resolution set-up.

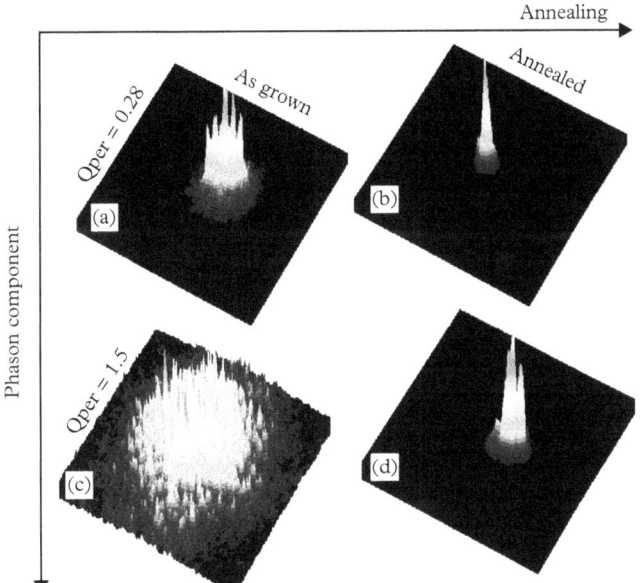

Figure 4.55 *Two-dimensional images of two Bragg peaks with different H_{perp} component measured in the i-AlPdMn phase. Images have been recorded with a high-resolution set-up. Spikes are due to the use of a coherent beam. The left panel is for the as-grown sample, whereas the right panel corresponds to the same measurement once the sample has been annealed and slowly cooled down.*

242 *Structure*

A last parameter that characterizes the sample quality is the measurement of diffuse scattering in the diffraction pattern. Indeed, the diffuse scattering intensity is sensitive to the two-body correlation function and thus to any disorder with respect to the average structure which is determined using integrated intensities of the Bragg reflections. Diffuse scattering may come from many different sources of disorder: it can be chemical disorder (including short-range order), lattice distortion, thermal lattice vibrations (so-called thermal diffuse scattering), phason fluctuations, etc. We shall present a more detailed presentation of this point in the section on phasons. In the case of the i-AlPdMn icosahedral phase, it turns out that the main source of disorder comes from phason fluctuations.

To conclude this part on structural quality, we can say that the quality of the i-AlPdMn phase has been studied very carefully and from different points of view. When all the studies are merged together (dynamical diffraction, Bragg peak width, diffuse scattering), it can be concluded that the i-AlPdMn phase has a high quality. Its diffraction pattern perfectly obeys icosahedral symmetry, the Bragg peak positions are in excellent agreement with those calculated, dynamical diffraction is observed, and Bragg peak broadening is weak. The correlation length which can be deduced from such measurements is of the order of 10 μm or larger. This means that we can confidently model the i-AlPdMn atomic structure using the quasicrystalline model.

4.5.2.3 *From diffraction data to a first-order model*

The i-phase of the ternary alloy AlPdMn is particularly well suited for a study using both X-ray and neutron diffraction. Indeed, comparing the Z number of the elements and the corresponding neutron scattering lengths, it is readily seen that X-ray scattering is sensitive to Al/Mn versus Pd (Z_{Al} = 13, Z_{Mn} = 25, Z_{Pd} = 46) whereas neutron scattering is particularly sensitive to Mn atoms because of their negative neutron scattering length (b_{Al} = 0.34, b_{Mn} = −0.37, b_{Pd} = 0.59). Using both X-ray and neutron scattering thus makes it possible to get detailed information on the atomic distribution of the three elements of the quasicrystalline phase (Boudard *et al.*, 1992). Moreover, it has also been possible to carry out anomalous scattering at the Pd edge, allowing the extraction of the partial structure factor of the Pd atoms alone, thus simplifying the structural analysis (de Boissieu *et al.*, 1994*b*, *c*).

We follow a procedure similar to the one explained in the simple one-dimensional example. Once integrated intensities have been collected and corrected for (absorption, Lorentz polarization), they are used to compute the Patterson function. From the analysis of the Patterson map, a simple model is extracted in which the three-dimensional atomic surfaces are modelled with spheres. The model is then refined from diffraction data. Of course, such a model is only a crude and first-order one and needs further refinement with more complex atomic surfaces.

Fig. 4.56 displays the intensity distribution of Bragg reflections in a two-fold plane for X-rays (bottom) and neutrons (top). In this case data collection has been

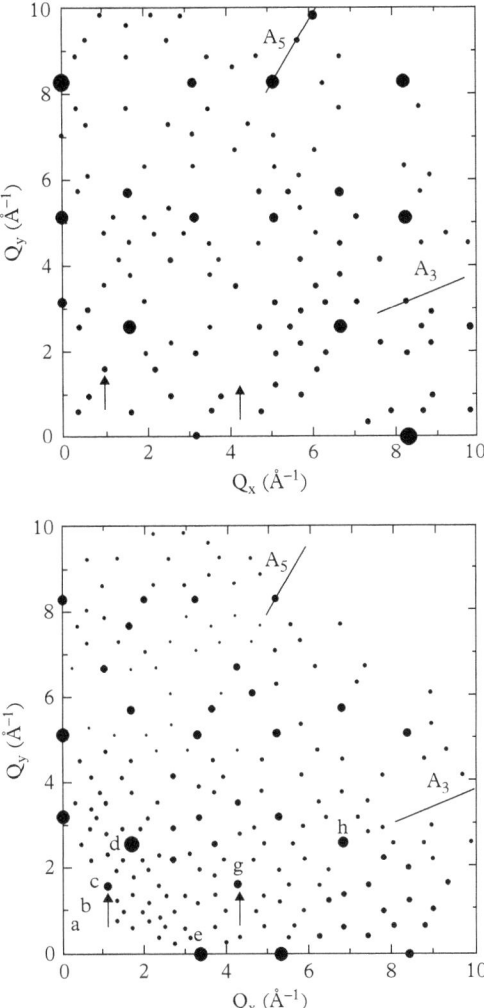

Figure 4.56 *Intensity distribution of Bragg peaks in a two-fold diffraction pattern of the i-AlPdMn phase. Top: neutron data; Bottom: X-ray data. The area of the dots is proportional to the intensity. Arrows point to regions where there is a large difference between the two measurements.*

carried out using an imaging plate for X-rays and a two-dimensional detector for neutrons. This allows us to collect several equivalent reflections, whose integrated intensities are merged together. About 400 independent reflections have been measured, with an internal agreement, which measures the quality of the data collection, equal to 0.03, which indicates the good match of integrated intensities with respect to icosahedral symmetry. The arrows in Fig. 4.56 point to regions where there is a large difference between the X-ray and the neutron data. In particular, the reflections with half-integer indices only are very weak in the neutron data set, whereas they are of medium intensity in the X-ray data set, indicating the role of Pd atoms.

If one analyses the intensity distribution of the reflections, it can be shown that they are grouped into two families: the all-integer indices reflections are the strongest ones, whereas reflections with only half-integer indices are generally weak. This is illustrated in Fig. 4.57, which displays the intensity distribution of the measured X-ray amplitudes F (square root of the intensity) as a function of Q_{perp}. This means that the real space structure is almost a P icosahedral structure with a parameter a_{P6D}, and that the F icosahedral lattice is generated by a small perturbation on this P lattice, leading to a doubling of the unit cell. This is referred to as a superstructure, by analogy to a similar situation in metallic alloys.

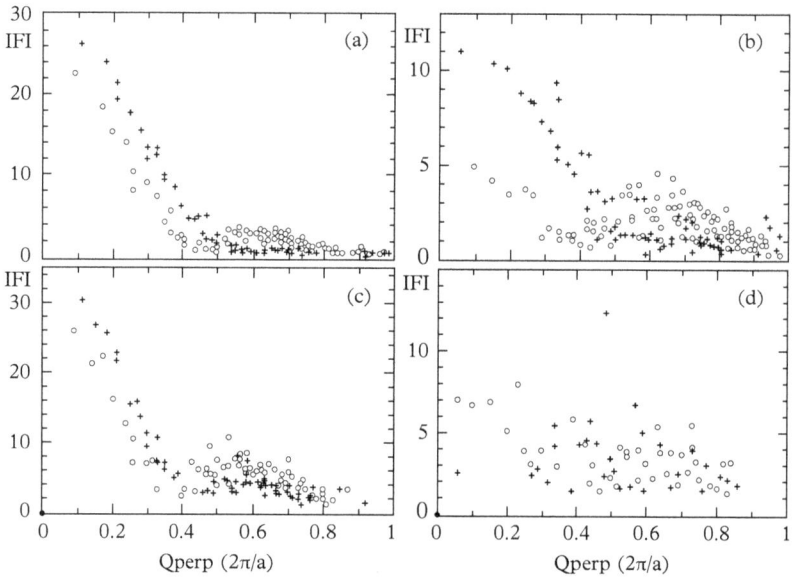

Figure 4.57 *Evolution of the measured structure factor F as a function of H_{perp} in the i-AlPdMn phase (top, X-ray data; bottom, neutron data). The left and right panels correspond to the main (all integer indices) and superstructure (all half-integer indices) reflection, respectively.*

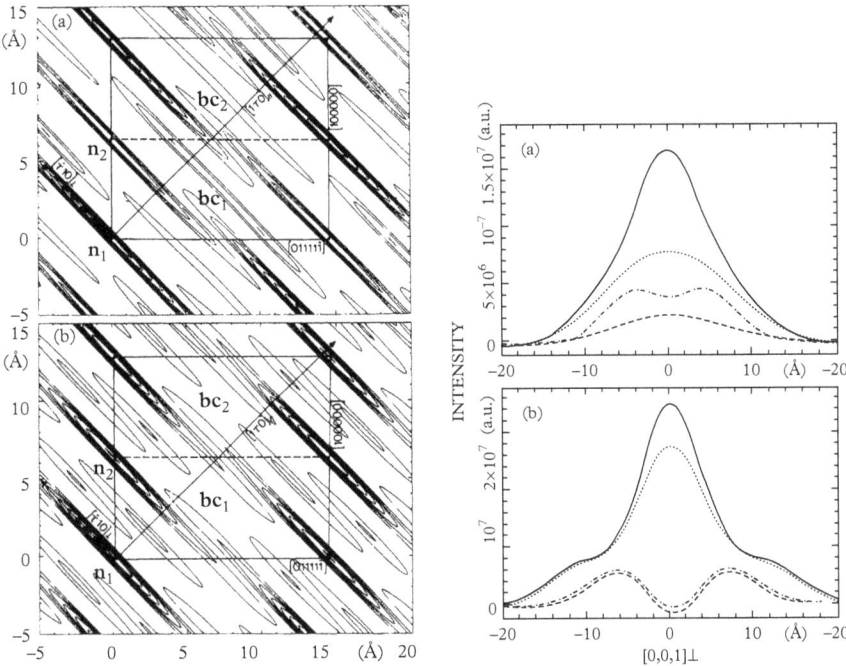

Figure 4.58 *Density contours of the five-fold section of the six-dimensional Patterson, calculated with X-ray (top) and neutron data (bottom). The right panel displays the perpendicular profile of the node and body-centre sites observed on the Patterson map. Note that the neutron data display negative values, due to the negative Mn scattering length.*

This is illustrated in the X-ray and neutron Patterson maps shown in Fig. 4.58 along a five-fold section. The map shows iso-intensity contours of the Patterson map which is computed by Fourier transform of the measured intensity. The density–density distribution is confined to a special area of the map, showing elongated cigarlike features. These are the traces of the atomic surfaces along the five-fold axis, in agreement with the hypothesis that the atomic surfaces are essentially parallel to the perpendicular space. The P- and F-unit cells are outlined. The F-unit cell is obtained as the superposition of several P-unit cells. A study of the Patterson map leads to the conclusion that atomic surfaces are located on only four inequivalent sites. Using the primitive P-unit cell as the basis (lattice parameter a_{P6D}) for the site labelling, the four sites are n_1, with six-dimensional coordinates (000000); n_2, with six-dimensional coordinates (100000); bc_1, with six-dimensional coordinates $\left(\frac{1}{2}\frac{1}{2}\frac{1}{2}\frac{1}{2}\frac{1}{2}\frac{\bar{1}}{2}\right)$; and bc_2, with six-dimensional coordinates $\left(\frac{3}{2}\frac{1}{2}\frac{1}{2}\frac{1}{2}\frac{1}{2}\frac{\bar{1}}{2}\right)$. These sites are indicated on the Patterson map. The reader can check that the other sites appearing in the F-unit cell are obtained by a symmetry operation on the F lattice. For instance, the site at $(11111\bar{1})$ is identical to the n_0 one. This is a consequence of the F centring of the lattice.

From the Patterson map it is straightforward to determine the positions of the atomic surfaces in the real space structure. There are, a priori, also four sites in the real structure located at n_1, n_2, bc_1, and bc_2. The difficulty here is to gain some information about the size and the chemical content of the different atomic surfaces. Some information is gained by looking at the perpendicular space profiles of the different Patterson contributions. When comparing neutron and X-ray profiles, there are marked differences. For instance, the body-centre contribution is negative for neutron data which arises from Mn–Al or Mn–Pd contributions, since Mn has a negative neutron scattering length. Moreover, the profile in the body centre is much weaker than that of the origin, indicating that the atomic surface at the body-centre site is certainly small. Finally, there is a larger difference between n_1 and n_2 sites in the X-ray data, indicating that Pd atoms contribute much to one of these atomic surfaces.

Since there are four atomic surfaces, it is also interesting to look at the H_{perp} dependence of the structure factor F. It is the sum of four contributions, each one carrying two terms: the Fourier transform of the atomic surface and a phase factor. The structure factor is

$$F(\mathbf{H}_s) = \frac{1}{V} \sum_{i=1,4} G_i(\mathbf{H}_{perp}) \exp(2\pi i \mathbf{H}_s . \mathbf{R}_i), \qquad (4.52)$$

where H_s is a vector of the reciprocal lattice in the six-dimensional superspace, V is the six-dimensional unit cell volume, $\mathbf{R}_i$ the position of the atomic surface i in the six-dimensional unit cell, and G_i the Fourier transform of its characteristic function.

Since the atomic surfaces are located on specific sites, the phase factors are either 1 or −1. Introducing the four positions of the atomic surfaces into Eq. 4.52 leads to

$$F(\mathbf{H}_s) = \frac{1}{V}(G_{n_1} + G_{n_2} \pm (G_{bc_1} + G_{bc_2})), \qquad (4.53)$$

$$F(\mathbf{H}_s) = \frac{1}{V}(G_{n_1} - G_{n_2} \pm (G_{bc_1} + G_{bc_2})). \qquad (4.54)$$

Equation 4.53 holds when the six-dimensional indices of the reflections are all integer whereas Eq. 4.54 is valid when six-dimensional indices are all half-integer. From the profiles of the Patterson function, it can be deduced that the atomic surfaces at the nodes are much larger than the ones at the body centre and thus contribute most to the structure factor. Since in Eq. 4.53 the two Fourier transforms add up, whereas in Eq. 4.54 they subtract, this explains why reflections with indices which are all integer have a larger intensity than those with half-integer indices. From these two expressions one thus expects four different curves for the evolution of the structure factor as a function of H_{perp}: the conditions on the

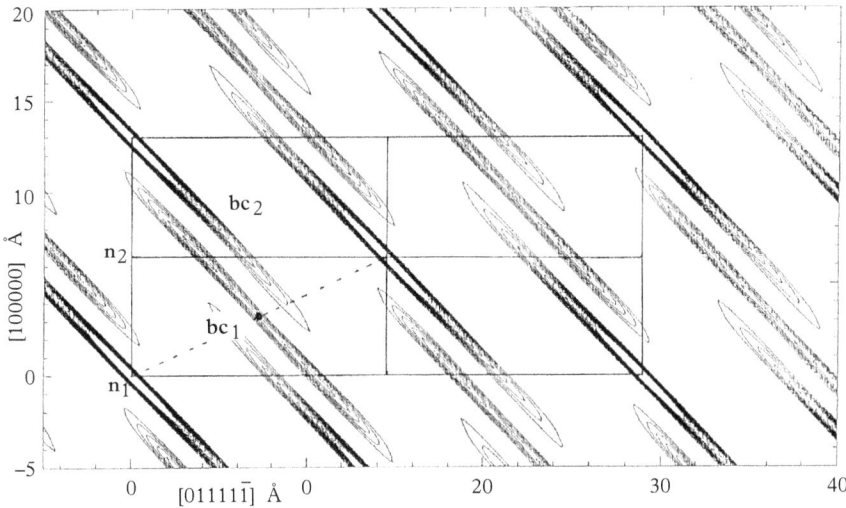

Figure 4.59 *Five-fold section of the six-dimensional Patterson calculated with the partial Pd intensities. The section contains a five-fold axis in both the parallel space and the perpendicular space. The traces of the primitive and superstructure cells are shown.*

indices are easily expressed as conditions on the (N, M) indices. Equation 4.53 applies when N is even and M even or odd, and Eq. 4.54 when N is odd and M even or odd. Plotting the variation of the structure factor for these four series of indices, and looking at the $H_{perp} = 0$ extrapolation, it is then possible to deduce the average volume of each of the atomic surfaces. (Note that the structure factors have to be corrected for the Debye–Waller factor and for the K dependence of the atomic form factor for X-rays.)

However, better information is obtained by performing anomalous scattering experiments at the Pd edge. When changing the energy of the incoming X-ray beam close to the Pd edge (below), a significant variation of the Pd scattering power can be achieved. The atomic form factor reads as

$$f_{Pd}(\mathbf{K}) = f_0(\mathbf{K}) + f' + if'', \qquad (4.55)$$

where f_0 is the $\mathbf{K}$-dependent part of the atomic form factor (with $f_0(0) = Z$), and f', f'' the real and imaginary parts of the anomalous correction. A few eV below the Pd edge, f' has a value around −6, whereas far below the edge it is equal to −2. Measuring the integrated intensities at three different energies, it is thus possible to change the contrast of the Pd atoms. If we neglect the imaginary contribution (which is small below the edge) the measured structure factor F can be written as

$$F(\mathbf{H}) = F_{tot}(\mathbf{H}) + f'_{Pd} F_{Pd}(\mathbf{H}) \exp(i\phi), \qquad (4.56)$$

where F_{tot} is the total structure factor and F_{Pd}, the partial structure factor, where only the Pd atoms contribute. The partial structure factor is the Fourier transform of the Pd atoms in the structure only. The phase factor $\exp(i\phi)$ is either $+1$ or -1 ($\phi = 0$ or π), since the structure is centrosymmetric. With a measurement at three different energies, it is thus possible to extract the modulus of the partial structure factor $|F_{Pd}|$.

The Fourier transform of I_{Pd} yields the Patterson map of the Pd atoms. Its five-fold section is displayed in Fig. 4.59: as can be seen and by comparison with the previous Patterson maps, its structure is much simplified. Only two sites are visible on the map: one at n_1 and another one at bc_1. Note in particular that there are no contributions on the n_2 node site. The interpretation of this Patterson map is straightforward: there are only two atomic surfaces located at n_1 and bc_1. Moreover since the bc_1 atomic surface is rather small, the profile of the Patterson feature at the bc_1 directly yields the density profile of the n_1 atomic surface: it is a spherical shell with internal and external radii equal to R_1 and R_2, respectively. Note that, in this case, the partial Pd Patterson map is really an F structure: it is not possible to interpret it as a superstructure on a P lattice. This is visible in the figure when looking at the n_1 contributions alone: they decorate the vertices of a rectangle together with its central face, in a way very similar to what would be obtained for a two-dimensional periodic face-centred lattice (although some care must be taken when making such a comparison, it is a useful one for getting some insight into high-dimensional lattices).

Combining these results with the ones obtained with the total Patterson map calculated with X-ray and neutron data, a simple six-dimensional model in which atomic surfaces are modelled by spheres and spherical shells is designed. A scheme of the resulting model is displayed in Fig. 4.60. Atomic surfaces are located at n_1, n_2, and bc_1 sites. The n_1 atomic surface is built up with three successive shells: a Mn core enclosed in a Pd shell and an external Al shell. The n_2 atomic surface contains a Mn core and an Al external shell. Finally, the bc_1 site is a sphere with Pd only. A schematic representation of this model is shown in Fig. 4.54. The top part of the figure displays the five-fold section of the six-dimensional periodic structure, and the trace of the three atomic surfaces. The bottom part displays the section of the three spherical shells. The size and distribution of the atomic surfaces is such that the chemical composition and the density of the model match the experimental ones. Indeed, the atomic density is simply given by $\rho = \sum_i V_i/2V_{6D}$, where V_i is the volume of the atomic surface i and V_{6D} the volume of the six-dimensional unit cell. The factor 2 comes from using the P-unit cell. Similarly, the concentration of atom A is given by the ratio of the volume of the atomic surfaces containing A, by the total volume. This is because, as shown in the one-dimensional example, the parallel space will span uniformly each atomic surface.

Note that an alternative route for obtaining the spherical model from neutron diffraction is by using the *charge-flipping method*, which allows one to get the proper phases of most of the measured structure factors (at least the strongest one). Although Mn atoms have a negative scattering length, a density map of impressive

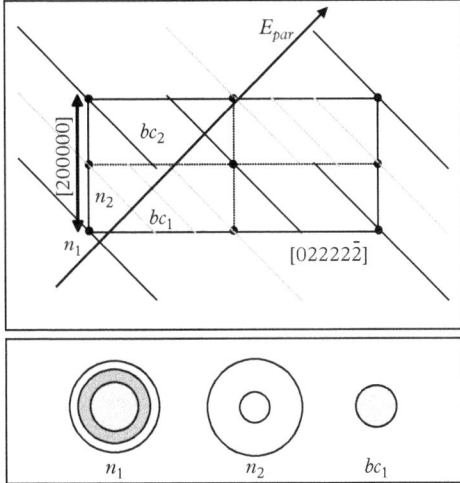

Figure 4.60 *Illustration of the spherical model. Top: Five-fold section of the structure: there are three atomic surfaces located at n_1, n_2, and bc_1. Bottom: Section of the three spherical atomic surfaces. They are built up with spherical shells having different atomic content: the white, light-grey and dark-grey colours stand for Al, Mn, and Pd atoms, respectively.*

quality can be obtained. In particular, the negative contribution of Mn atoms to the Fourier map is perfectly visible, which is particularly interesting. As already said above, this is now becoming a powerful and routine procedure for solving quasicrystalline structures.

The derived spherical model compares well with both X-ray and neutron diffraction data if one considers only the strongest reflections. Of course, we do not expect that such a crude model correctly reproduces the weak reflections, which are very sensitive to the details of the structure. Nevertheless, the present model is sufficient to give some insight into the resulting three-dimensional atomic structure.

The advantage of the six-dimensional modelling is that it allows an analysis of the resulting three-dimensional local environment. One can, as explained in Chapter 3, Section 3.4.2, for octagonal tiling (Fig. 3.11), determine which are the first interatomic distances, the first coordination shells, and the resulting atomic clusters. As an example, let us consider the five-fold section of the model. For the sake of simplicity we do not consider yet the atomic decoration of the different spheres (we consider the model with only one chemical species). Starting from the n_2 site we see in Fig. 4.61 that the E_{par} line intersects two n_1 atomic surfaces located at (000000) and (200000). This means that atom A is surrounded by two

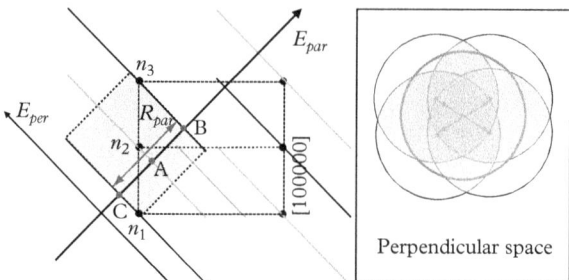

Figure 4.61 *Illustration of the determination of a local environment. Atom A is surrounded by two atoms B and C along a five-fold axis, if the E_{par} line is within the grey rectangle. The right-hand panel illustrates the intersection procedure in the perpendicular space for four atomic surfaces displaced along a five-fold axis.*

other atoms, B and C, located along the five-fold axis. The distance AB (or AC) is easily calculated and is equal to the parallel projection of the vector (100000), R_{par}, which is equal to $a_{P6D}/\sqrt{2} = 4.5$ Å in the present case. One may ask the following question: for which situation do we get the three atoms A, B, C together? By inspection of the figure, it is easy to show that for this to happen, the E_{par} cut space must lie within the grey rectangle. Indeed, if the E_{par} space is outside this grey rectangle, we may have A and B atoms or A and C atoms, but not the three together. The above geometrical condition can be expressed more rigorously looking at the intersection of the three atomic surfaces with six-dimensional coordinates (000000), (100000), and (200000) or n_1, n_2, and n_3. The three atomic surfaces are now in the perpendicular space, the n_3 atomic surface being displaced by the vector $\boldsymbol{R}_{perp}$, the perpendicular component of the six-dimensional vector (100000), and the n_1 one by a vector $-\boldsymbol{R}_{perp}$. The intersection of these three atomic surfaces defines the existence domain of the triplet ABC. This procedure can be generalized to a cluster of atoms. Since we have atoms along a five-fold axis, we may have an icosahedron (12 atoms along the five-fold axes) of atoms around the atom A. This will occur if there is an intersection of the atomic surface n_2 and the 12 perpendicularly translated n_1 atomic surfaces. The 12 perpendicular translations are calculated from the 12 equivalent six-dimensional vectors obtained by applying the icosahedral symmetry. In perpendicular space, the five-fold axis has coordinates proportional to $(\tau 1 0)$, on which we can apply circular permutation and sign changes, leading to the 12 vectors. Again, the intersection between the n_2 atomic surface and the 12 translated n_1 ones defines a volume called the existence domain of the icosahedron. Each time the E_{par} line goes through the existence domain, this ensures that an icosahedron is generated in the three-dimensional structure. This procedure is illustrated in Fig. 4.61, in the case of only four atomic surfaces.

Using the above procedure, it is thus possible to deduce which kind of atomic cluster is present in the three-dimensional structure. It can be shown that a *pseudo-Mackay cluster* is formed. The Mackay cluster is a 54-atom cluster (Mackay, 1962) formed with three successive icosahedral shells (Fig. 4.62): The first shell is a small icosahedron, the second shell is a large icosahedron (about two times larger), and the third shell is an icosidodecahedron (30 atoms on two-fold axes). The diameter of this atomic cluster is about 9.6 Å. In the spherical model, the second and third shell are generated, whereas the first shell is a partially occupied dodecahedron (seven atoms). Because the atomic content of the atomic surfaces is only approximated with spherical shells, it is not possible to give a detailed chemical decoration of this pseudo-Mackay cluster. However, the centre of this cluster originates from either an n_1 or an n_2 atomic surface, which implies that there should be at least two different pseudo-Mackay clusters, one of which only contains Pd atoms, whereas Al and Mn atoms are found in both clusters.

Another cluster is also generated from the body-centred site: the first shell is an icosahedron and the second one is a dodecahedron. It has a smaller size and is linked with the Mackay cluster. A more detailed study of the cluster distribution and their linkages will be presented in the next paragraph.

An alternative and useful description of the quasicrystalline atomic structure is in terms of *dense atomic planes* ((de Boissieu *et al.*, 1991; Boudard *et al.*, 1992)). At first sight, it might seem surprising that dense planes exist in the structure since there is no periodicity. The best way to show dense planes is to look at the projection of the structure onto high-symmetry planes. The two-fold plane is particularly well suited for this purpose, since it contains all high-symmetry axes, i.e. two-, three-, and five-fold axes. Figure 4.63 shows the projection of the electron density on a two-fold axis plane. It is calculated as the Fourier transform of the two-fold plane in reciprocal space since the Fourier transform of a section is a projection. The first important point is that the electron density is not spread uniformly in the plane, but shows a discrete distribution: this corresponds to

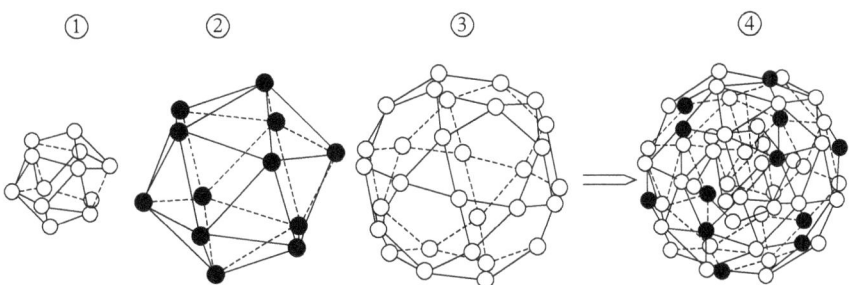

Figure 4.62 *The three successive shells constituting the Mackay cluster: an icosahedron, a double sized icosahedron and an icosidodecahedron (cf. Fig. 1.15).*

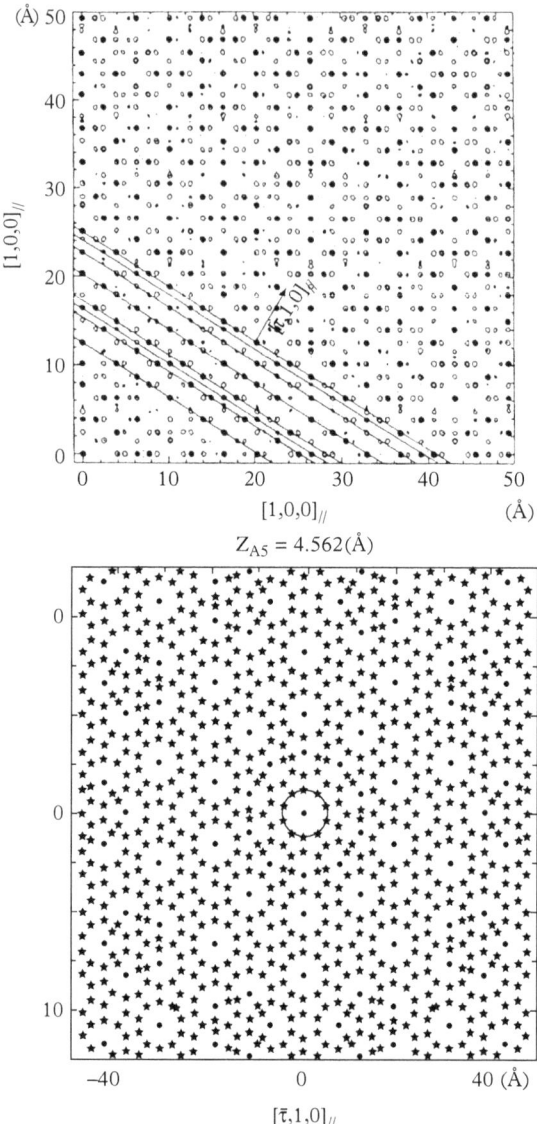

Figure 4.63 *Top: Iso-intensity contour plot of the projection of the electron density on a two-fold plane. There are dense planes as shown by the lines. They are perpendicular to a five-fold axis. Bottom: example of a dense plane. The circle is a trace of a pseudo-Mackay cluster.*

atomic columns. The second point to notice is that there are well-defined planes, some of which are highlighted by solid lines. They are best observed by looking at the figure at glancing angles. The planes are sometimes made of two or three layers 0.5 Å apart. The densest planes can be recognized by looking at the spacing between the lines: the larger the spacing, the denser the plane. The densest planes are perpendicular to a five-fold axis as shown in the figure. Other planes along two- or three-fold axis also exist, but as can be seen from the figure, the spacing in-between them is smaller so that they are less dense. The bottom part of Fig. 4.63 displays an example of a dense atomic plane perpendicular to a five-fold axis. Many ten-fold 'wheel' motifs, similar to the highlighted central one can be found in this plane. They even form a kind of hierarchical distribution. Most of these motifs correspond to a trace of a pseudo-Mackay cluster.

The description in terms of dense planes allows us to understand several quasicrystalline properties. For instance, single grains of the i-AlPdMn phase display facets, the largest of which are perpendicular to a five-fold axis. This can be understood with a simple model of 'cut bonds', in which the densest planes are the more stable, and are favoured in terminating the crystal. The same is true for the study of quasicrystalline surfaces: it has been observed that the surfaces are almost bulk terminated, and that five-fold surfaces are more stable, in agreement with the greatest density of five-fold planes. Finally, this also has consequences for the study of mechanical properties and dislocation motions. Indeed if the motion of a dislocation is accomplished via a glide, the glide plane might be a five-fold one. See also Chapter 7, Sections 7.1 and 7.2.

We have thus seen that already with a crude model, valuable information on the three-dimensional quasiperiodic structure can be gained. However, the spherical model suffers many drawbacks. The first is the existence of a number of unphysical short distances. The second is that the details of the chemical decoration of the structure cannot be obtained. This, of course, shows up in the comparison of the experimental data with the calculated ones, especially for the weak intensity reflections and for neutron data as shown in Fig. 4.64: there is a large discrepancy for weak reflections and N odd reflections.

It is thus necessary to improve the modelling and in particular to refine the shape of the atomic surfaces. Several strategies have been used to constraint the modelling, which ultimately must end up with a number of free parameters compatible with the experimental data set. For instance in the i-AlPdMn case, about 800 independent reflections have been measured (400 for X-rays, and 400 for neutrons) so that a 'reasonable' number of parameters should be smaller than 160.

In the following we present the two main routes which have been used in modelling the i-AlPdMn quasicrystal structure. The first is a six-dimensional modelling with some hypothesis about the shape of the atomic surfaces (Katz and Gratias model), whereas the second is based on an analysis of clusters in a decorated Penrose tiling scheme (Yamamoto model). Both approaches use six-dimensional modelling for calculating the Fourier transform.

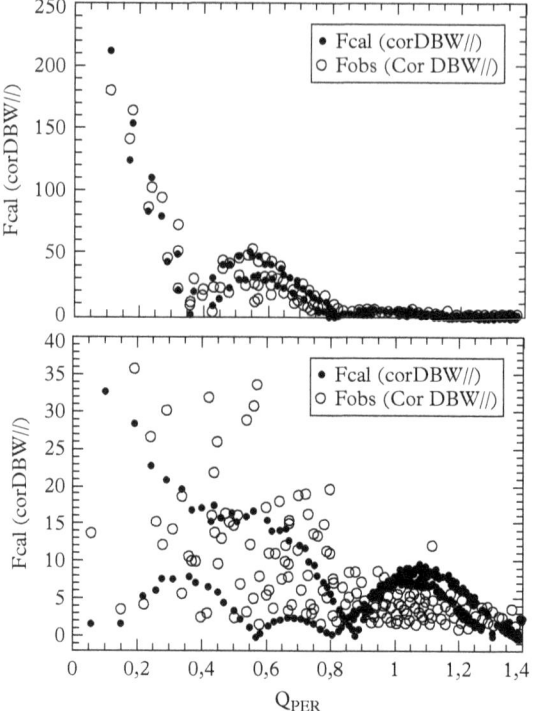

Figure 4.64 *Comparison between measured neutron structure factors (open circles) and calculated ones using the spherical model (full circles). The top panel is for reflections with all integer indices, whereas the bottom one is for all half-integer ones.*

4.5.2.4 Six-dimensional modelling: Katz and Gratias model

As we already pointed out, one of the main drawbacks of the spherical model is that it generates unphysical short distances of the order 1.5 Å. Of course such distances have to be suppressed in any realistic model, in particular for the calculations of physical properties. This is a first constraint on the shape of the atomic surfaces. A second comes from the density and chemical composition: the model must produce a structure which has a point density and a chemical composition close to the experimentally observed ones. Already these two constraints put strong limitations on the possible solutions, but are not sufficient.

In Katz and Gratias (1994) more constraints were added based on the two following hypotheses:

1. The shape of atomic surfaces should ensure the existence of local rules (or 'matching rules').
2. The shape and six-dimensional spatial distribution of atomic surfaces should fulfil the 'closeness' condition.

The existence of local rules or matching rules is related to the question of quasicrystal growth: 'How can quasiperiodic long-range order propagate?' 'Is it possible to grow a perfect quasicrystal with finite-range interatomic forces?' This is still a fascinating and open problem for which several theoretical solutions have been proposed. Local rules (called matching rules in the case of tilings) define a topological property of some quasicrystalline structure. Let us consider a generic quasicristalline structure and enumerate all the different local configurations having a radius R (called the R atlas) which exist in the structure. If the R atlas is finite, there are local rules if the following statement is obeyed: any structure whose R atlas is identical to the one previously defined is quasicrystalline. In other words, it is a way of 'recognizing' for sure a quasicrystalline structure by enumerating its local environments (the local rules). We stress that these local rules *are not* growing rules, in other words, it is not possible to grow a perfect quasicrystal from these rules. However, if local rules exist it might be possible to grow a defect quasicrystalline structure with finite-range atomic forces. If the number of defects is small enough they could be annealed after the growth process. In the case of icosahedral structures, Katz and Gratias showed that atomic surfaces whose shape is a polyhedron bonded by mirror planes always admit local rules.

The *closeness condition* is related to the phason degree of freedom of the structure. As will be presented later on, quasicrystals have a supplementary degree of freedom related to the position of the parallel or cut space. If one translates the parallel space in the perpendicular space direction, a different structure is generated, but its free energy is identical to the initial one. This property is the basis for the existence of phason modes. However, when translating the cut space, atomic rearrangements occur, called 'phason jumps' in the quasicrystal jargon. For instance, in the case of the Fibonacci chain some *LS* configurations are transformed into *SL* ones (this is equivalent to an atom jumping from one position to another: see Fig. 4.65). If the shapes of the atomic surfaces are made arbitrary (e.g. if the segment line describing the Fibonacci chain is made smaller), situations might occur for which the translation of the cut space will make an atom 'disappear' and 'reappear' away from its initial position. Such a situation is unphysical and/or might be too unfavourable from the energy point of view. Katz and Gratias proposed that the shape of the atomic surfaces should be such, that when displayed in the high-dimensional space, their surfaces should be 'connected' by planes parallel to the parallel space: this will ensure that the physical space may slide, that is, that any translation of this space creates energetically easy jump sites. The *closeness condition* is fulfilled for tilings such as the three-dimensional Penrose one for instance.

Combining the requirement of local rules and the closeness conditions, Katz and Gratias ended up with a set of eight different allowed polyhedra. This

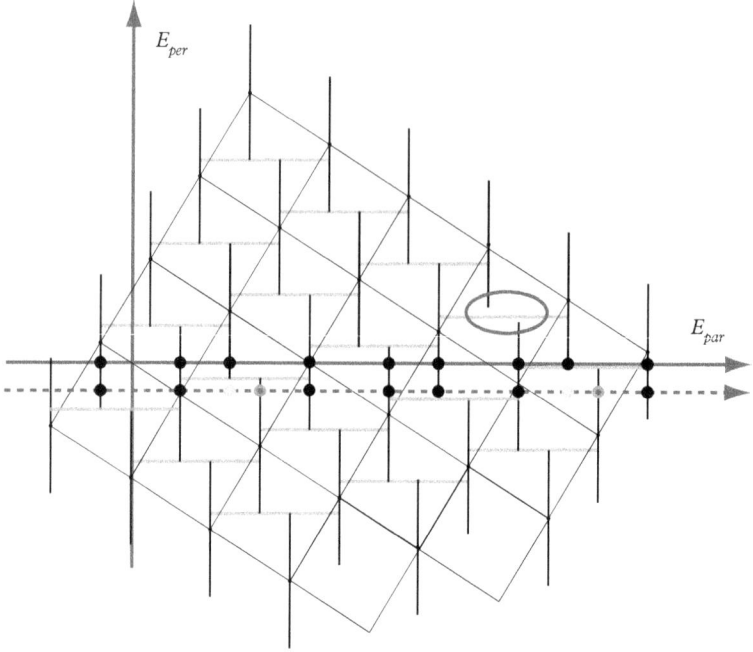

Figure 4.65 *Illustration of the closeness condition in the case of the Fibonacci chain. When the cut space E_{par} is moved to a new position (dashed line) some atomic rearrangements occur: two atoms have 'jumped' from one position to a new one (highlighted in dark grey). This situation occurs because the atomic surfaces are connected by lines (light grey) parallel to the cut space. The dark ellipse, on the right, encloses a situation where the closeness condition is not satisfied. Segment lines have been shortened: when the cut space goes through this position there will be atom 'annihilation' or 'creation'.*

already restricts the shape of possible atomic surfaces. However, since intersections between polyhedron are allowed, there is still a large range of possibilities.

Based on these considerations, and from the Patterson analysis, Katz and Gratias proposed that the three atomic surfaces located on n_1, n_2, and bc_1 be three polyhedra, which are shown in Fig. 4.66: a triacontahedron, τ times larger than the standard triacontahedron used to generate a Penrose tiling with edge length 4.5 Å; a truncated triacontahedron; and a small triacontahedron (τ^2 smaller) on the bc_1 site (Cornier-Quiqandon et al., 1991; Gratias et al., 2001). The relative positions of the three atomic surfaces in the six-dimensional unit cell are shown on the bottom part of the figure. With respect to the convention used for the spherical model there is a shift of the origin by (100000). As a consequence the n site corresponds to the n_2 site in the spherical model, and n' to n_1. With this in mind, there is an overall agreement between both models, as far as the size of the atomic surfaces is concerned.

The triacontahedron located on the n' (or n_1) site is truncated along its five-fold axis to avoid short distances which would occur with the n (or n_2) node and

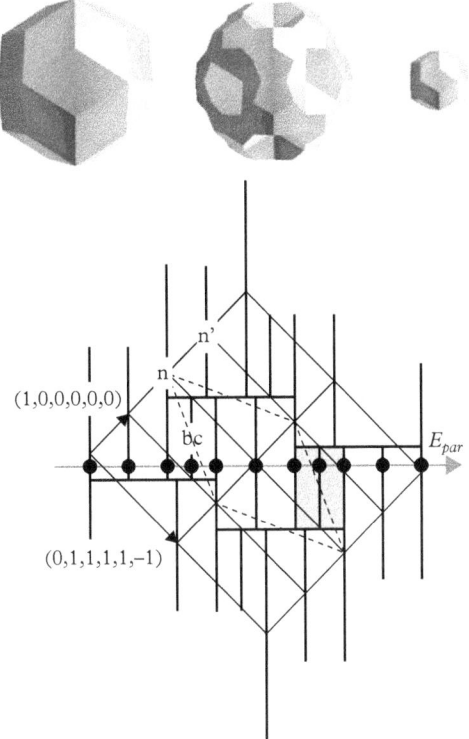

Figure 4.66 *The three atomic surfaces used in the Katz–Gratias model. The bottom panel shows the five-fold section. Atomic surfaces are 'connected' by horizontal lines along E_{par}. (From (Gratias et al., 2001))*

the bc site (called bc here). When looking at the intersection of the n_1 and n_2 sites corresponding to this short distance, the large triacontahedron just fits into the truncated one. The same is true for the short distance which would be generated by n' and bc. Actually, this sterical constraint imposes the relative position of the n' and bc atomic surfaces in the six-dimensional cube. This connection between the two atomic surfaces also ensures that the closeness condition is fulfilled. This is visible on the five-fold slice of the six-dimensional structure (Fig. 4.66, bottom) where the n, n', and bc atomic surfaces extremities can be joined by a line parallel to E_{par}. Note that along this section the n' atomic surface is significantly smaller than the n one. Connecting the atomic surfaces by segment lines generates a decomposition into 'cells', 'blocks', or 'Klötze', similar to what was done in the one-dimensional Fibonacci example (see Chapter 2). Cells, highlighted in grey in Fig. 4.66, are also useful in determining local environments, using a procedure similar to the one presented in the spherical mode. Indeed, the shaded area also shows the existence domain of the nearest neighbour interatomic distances occurring along a five-fold axis. There are two cells highlighted in light and dark grey which correspond to the two five-fold first nearest neighbour distances on this section. From a central atom generated by the bc atomic surface, two atoms are also generated along the five-fold axis by the n atomic surface. From a central atom generated by an n' atomic surface, two atoms are generated at a distance twice the previous one by the atomic surface n. This is illustrated by the black dots which give the atomic positions along a five-fold axis in real space. Of course, this figure is not the whole story because it concerns only a single direction in the physical space, but it gives a feeling of what the procedure is to determine local environments.

Gratias and co-workers also proposed a decomposition of the three polyhedra which fit the AlCuFe and the AlPdMn quasicrystal data (Gratias *et al.*, 2001; Quiquandon and Gratias, 2006). Although there are some slight differences in the chemical composition, the i-AlCuFe phase is almost iso-structural to the i-AlPdMn one, with Cu being equivalent to Pd, and Fe equivalent to Mn. By trying to minimize the different chemical decorations of the clusters that result from the six-dimensional model, Gratias *et al.* ended up with a decomposition of the atomic surfaces which fits both the i-AlCuFe and the i-AlPdMn chemical compositions and is in good agreement with the available X-ray or neutron data. The different shells are shown in Fig. 4.67. Again, there is an overall agreement with the spherical model, with, for instance, a shell of Cu atoms on the atomic surface n', whereas the n atomic surface contains mainly Al atoms. This decomposition was carried out trying to keep to a minimum the number of local environments, and taking into account the known information extracted from the Patterson analysis.

In order to compare the model with experimental data, it is necessary to calculate its Fourier transform. Although the final shape seems quite complicated, this can be achieved easily using the icosahedral symmetry and a decomposition into small tetrahedra. The model compares quite well with X-ray data, although the contrast between Cu and Fe is weak, so that it is difficult to get a proper determination of these two atom locations. A similar model that was proposed for the i-AlPdMn phase gives a reasonable agreement with neutron and X-ray data.

Once the polyhedra describing the structure have been determined, the three-dimensional local environment and clusters in the real space three-dimensional structure can be calculated from the six-dimensional model. The procedure is analogous to the one briefly outlined in the case of the spherical model. Let us consider, for instance, the first cluster shell around an atom generated by the body-centred atomic surface. From Fig. 4.66 we already saw that this atom is surrounded by two atoms along a five-fold axis, at a distance of about 2.82 Å, and generated by the node atomic surface. In fact, this atom originating from the bc atomic surface is surrounded by 12 atoms, forming an icosahedral shell. The existence domain of this cluster is obtained by the intersection of the 12-node atomic surface, once translated in perpendicular space by the perpendicular projection of the vector $\boldsymbol{R}_{6D} = 0.5(\bar{1}1111\bar{1})$ and its 11 equivalent positions under the icosahedral

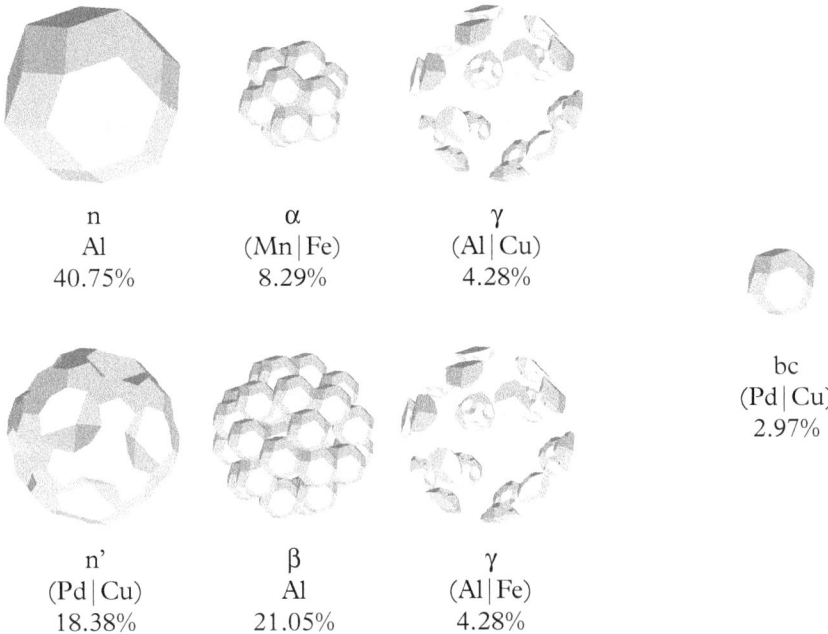

Figure 4.67 *Decomposition of the three atomic surfaces in the i-AlCuFe and i-AlPdMn phase, proposed by Gratias et al. (from (Quiquandon and Gratias, 2006))*

point group, with the *bc* atomic surface sitting at the origin (Fig. 4.68). Using the projection matrix it is easy to show that R_{6D} has a parallel component equal to $R_{par} = c(1/\tau, 1, 0)$, and a perpendicular one equal to $R_{par} = c\tau(\tau, \bar{1}, 0)$, where c is a normalizing constant equal to $a_{6D}/\sqrt{2(2+\tau)}$. Once the six-dimensional lattice parameter (6.45 Å) is entered in this expression, this leads to vectors with a modulus equal to 2.82 Å in the parallel space, and 7.38 Å in the perpendicular space. The intersection is, in fact, just equal to the *bc* atomic surface: this is illustrated in Fig. 4.68, which shows an equivalent image, where the *bc* atomic surface has been translated by the 12 R_{perp} vectors inside the *n* atomic surface. It is clear from this figure that all translations fall inside the *n* atomic surface. This means that each time the cut space goes through the *bc* atomic surface, an icosahedral shell with 12 atoms and a radius equal to 2.82 Å is systematically generated. Since the *n* atomic surface has an external shell made of Al atoms, this icosahedron is an Al icosahedron with a Pd central atom, since the *bc* atomic surface is Pd.

Using this procedure, Gratias *et al.* determined the different clusters, their proportions, and their linkages using their six-dimensional polyhedron model. This is not a trivial task, but because of the polyhedron shapes in the model, the calculation can be carried out exactly. The authors showed that three clusters are

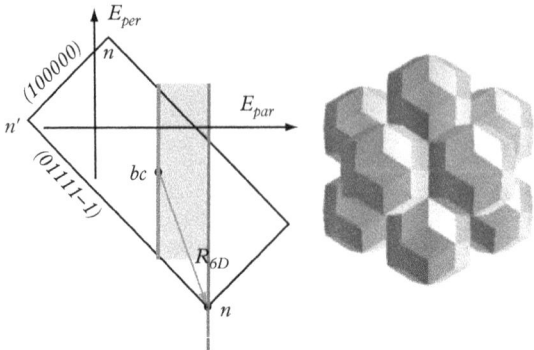

Figure 4.68 *Example of a decomposition to determine the various atomic clusters present in the structure. The left part shows the geometry of the* bc *and* n *atomic surfaces in the six-dimensional space. The right part displays the 12 translations of the bc atomic surfaces inside the n atomic one. They generate an Al icosahedron around a central Pd atom.*

generated with a central atom originating from the bc, n, or n' atomic surface. Around the bc atom, two shells are formed: an icosahedron, followed by a larger dodecahedron; this shell is called a B cluster. The central part of the n and n' atomic surfaces generate an atom surrounded by pseudo-Mackay clusters as already discussed; these are called M and M'. There are two different chemical decorations of these pseudo-Mackay clusters. The authors also calculated the ratio of atoms involved in each of these clusters. It is 0.79 for the B cluster, and 0.35 for the M and the M' ones. A more tricky problem is the one of the interconnection of the clusters. In fact, those three clusters are not at all isolated but share some atoms or interpenetrate each other. For instance, most B dodecahedra are connected through two atoms along a dodecahedron edge. Similarly, the M and M' clusters interpenetrate the B clusters. In fact, the cluster description is not unambiguous. Depending on the point of view, emphasis can be put either on the B clusters or on the M and M' ones. The structure is either an interconnection of M and M' clusters with B clusters playing the role of connection, or a B network, with incomplete M and M' clusters as linkages. What is most likely important is the coexistence of these three clusters to describe the structure. Figure 4.69 illustrates how the clusters are distributed in the structure. The left part shows an aggregate of clusters, whereas the right part shows a section perpendicular to a five-fold axis. The interconnection between B, M, and M' clusters is visible.

4.5.2.5 Six-dimensional modelling: The Yamamoto model

The model proposed by Yamamoto is the most elaborated one up to now. It is a generalization of procedures used in standard crystallography, with a few parameters to be refined from experimental data (Yamamoto *et al.*, 2003).

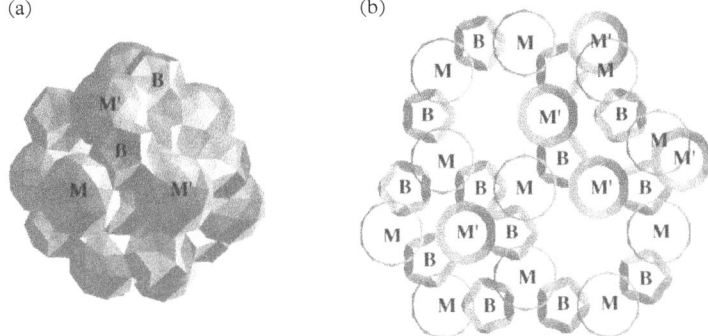

Figure 4.69 *(a) A portion of the full network of B, M, and M' clusters. (b) A typical slab of these clusters perpendicular to a fivefold axis showing how M and M' clusters intersect B clusters (from (Gratias et al., 2001)).*

The model is built up along three main ideas. (i) The six-dimensional model should be compatible with the spherical model, that is, with atomic surfaces located on nodes and body-centre sites. (ii) It should contain atomic clusters similar to the ones found in the 2/1 periodic approximant. (iii) The structure is based on a three-dimensional decorated inflated Penrose tiling.

When some Si is added to the AlPdMn alloy, it is possible to synthesize periodic approximants. In particular a 2/1 approximant could be grown as a single grain and its structure determined. It contains a large atomic cluster about 20 Å in diameter. Yamamoto used this cluster as a guide for the decoration of the i-AlPdMn structure.

The design of the quasicrystalline structure is then carried out on the basis of a τ^3 inflated three-dimensional Penrose tiling (see p. 138). The 3D Penrose tiling is made of two tiles: a prolate one and an oblate one. If we consider the six-dimensional cube with lattice parameter $a_{6D} = 6.45$ Å and decorated with a 'standard' triacontahedron atomic surface (i.e. obtained as the perpendicular projection of the six-dimensional cube), this generates a three-dimensional Penrose tiling with an edge length $A_{Pr} = a_{6D}/\sqrt{2} = 4.56$ Å. This is too small to accommodate the large atomic cluster, and thus Yamamoto *et al.* have chosen a τ^3 inflated tiling, that is, with an edge length equal to 19.3 Å. Indeed, we know that there are inflation rules, as already discussed for the indexing of the diffraction pattern. The atomic surface which is needed to generate this inflated Penrose tiling is a triacontahedron τ^3 times smaller than the standard one. It is thus a rather small atomic surface as compared to the overall size of the n_1 atomic surface. The two tiles of the inflated Penrose tiling are then decorated systematically by three clusters shown in Fig. 4.70: a dodecahedral star, a triacontahedron, and a rhombic icosahedron, each one made of 'standard' Penrose tiles. The packing of these 'clusters' is uniquely determined in the three-dimensional inflated Penrose tiling, as shown in Fig. 4.70.

262 *Structure*

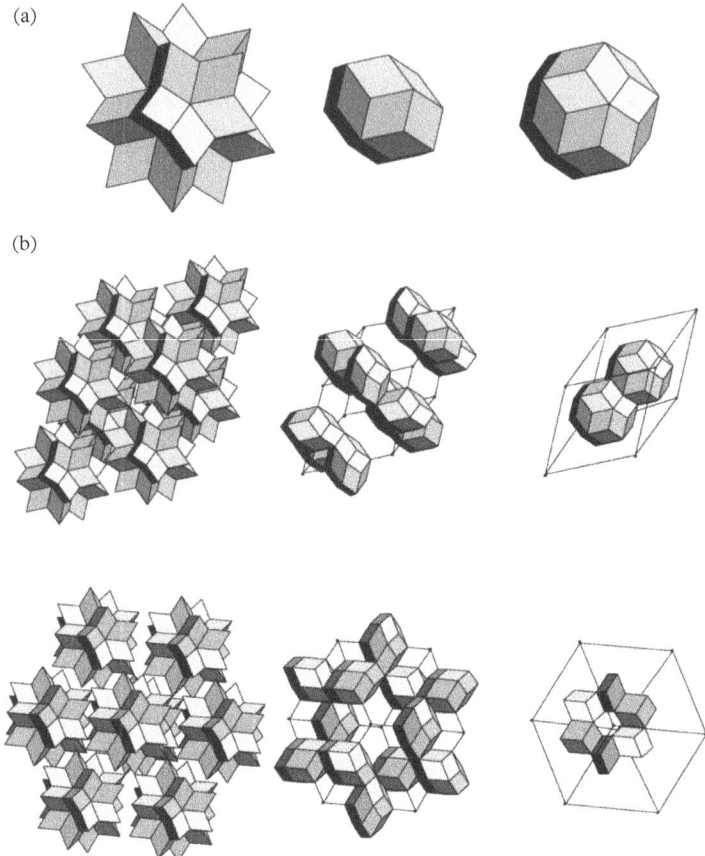

Figure 4.70 *(a) The three clusters which decorate the inflated three-dimensional Penrose tiling. (b) Illustration of the decoration of the three-dimensional inflated Penrose tiling.*

The three basic clusters are then decorated with atoms. The important point here is that each decoration is uniquely defined in the six-dimensional model by a small occupation domain, which is also one of the three clusters, but with an edge length τ^3 smaller. Figure 4.71 displays the occupation domain of each atomic decoration, shown in the asymmetric unit in the perpendicular space. Because of the icosahedral face-centred lattice, each cluster appears with two different decorations. The entire occupation domain decomposition is shown in Fig. 4.71. The grey levels in this figure give the atomic species constituting each domain, i.e. Al, Pd, or Mn atoms. The complete atomic surfaces are obtained by applying the icosahedral symmetry to the asymmetric unit. The external shape is shown in Fig. 4.72. As can be seen it is a rather complex shape. It should be noted that a

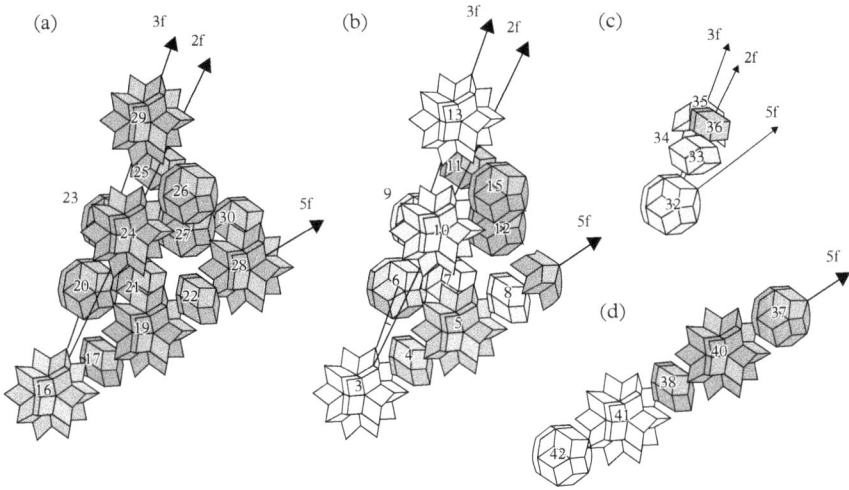

Figure 4.71 *Decomposition into small occupation domains of the four atomic surfaces in the six-dimensional model of Yamamoto. Only the asymmetric unit is shown. The number on each domain refers to the decoration. The level of grey indicates an atomic species (Al, Mn, or Pd).*

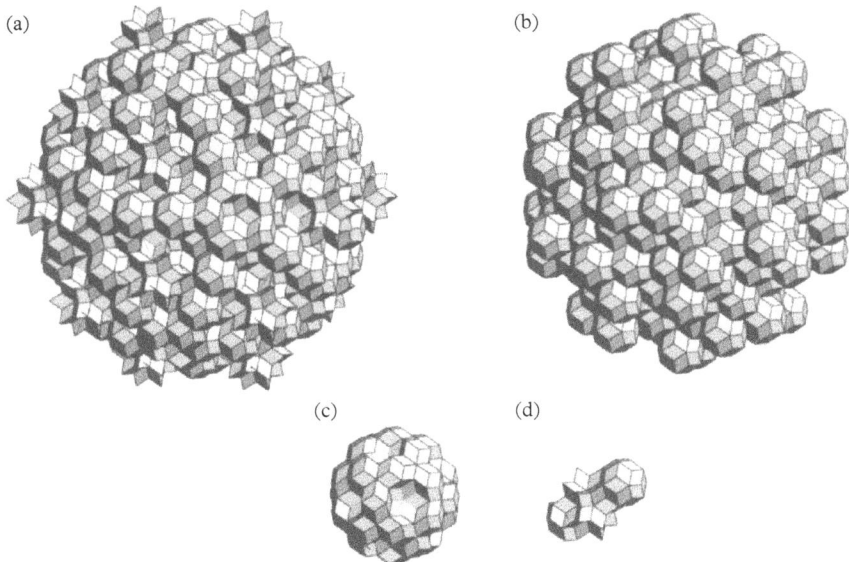

Figure 4.72 *Overall shape of the four atomic surfaces. This model is obtained after applying icosahedral symmetry to Fig. 4.71.*

supplementary fourth small atomic surface has been added, and is located on a mid-edge of the six-dimensional cube.

Once this decomposition is achieved, it is possible to refine the model from experimental data. Each subdomain is given three sets of parameters: the atomic probability occupation, a parallel space component, and an isotropic thermal Debye–Waller factor. The first two parameters have been shown to be of crucial importance in achieving a good agreement between the model and the experimental data. The occupation probability of each atomic domain is fixed to one atom only in an ideal model (i.e. Al, Mn, or Pd). However, in real systems, as is the case also for large unit cell crystals, some sites are disordered and either partially occupied or occupied by a mixture of two atoms (e.g. Pd/Al). Such a partial occupancy is thus introduced to the modelling. The parallel space component is also particularly important. It allows us to relax the atomic position from its ideal one, as has been illustrated in the one-dimensional example. This parameter is in fact directly related to the local environments of the atoms, so that one expects a slight shift for different local configurations. The resulting displacement in parallel space is, in general, small (of the order of 0.1 Å), but it influences significantly the quality of the fit. This parameter is completely equivalent to the atomic position parameter used in standard crystallography. Notice that the parallel component vector direction has some constraints given by the symmetry of the corresponding occupation domain. Finally, an isotropic thermal Debye–Waller is attributed to each domain. It is, in principle, possible to define an anisotropic thermal ellipsoid, but this parameter can only be refined once a quite good agreement is already achieved.

Altogether, there are 216 parameters, which have been refined from both X-ray and neutron data, consisting of 400 and 630 independent reflections, respectively. The joint use of X-ray and neutron data is important to locate both the Mn and the Pd atoms, due to the good X-ray versus neutron contrast. The obtained weighted R factor is 0.05 and 0.07 for X-ray and neutron data, respectively. The agreement is, of course, much better than with the spherical model, in particular for neutron data, where there is a quite significant improvement, as shown in Fig. 4.73. In particular, the structure factors of the superstructure reflections is now well reproduced. This is also visible in Fig. 4.74, which compares observed and calculated structure factors on a log-log scale for both X-ray and neutron data. This indicates that the model is now rather accurate, in particular for the Pd and Mn atom distribution, and that it can be used for further physical properties calculations.

As was done for the spherical model and the Katz and Gratias model, the Yamamoto model can be used to examine the three-dimensional quasiperiodic structure and, in particular, the local environment and the atomic clusters. First, it is important to realize that the spherical model is indeed a simplified version of the Yamamoto model. The overall shape of the atomic surfaces is similar (large atomic surfaces on the nodes and a smaller one on the body centre), but the chemical decoration is also similar with one node atomic surface having a core

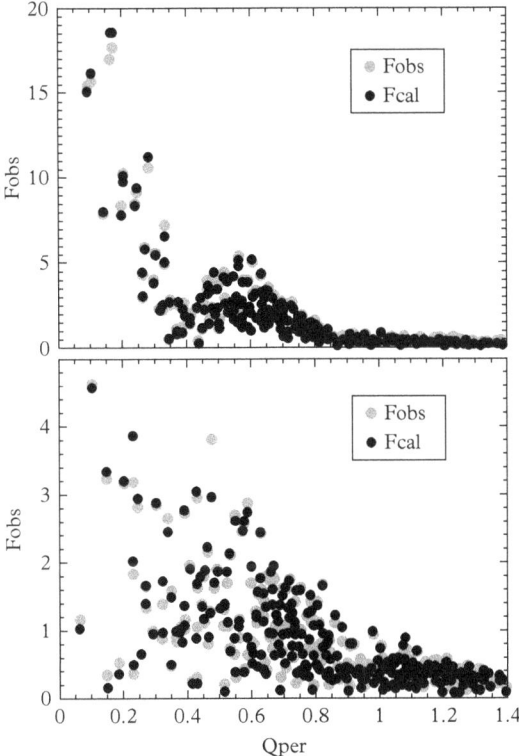

Figure 4.73 *Comparison of calculated and observed structure factors for neutron data in the i-AlPdMn phase. The top and bottom panel are for main and superstructure reflections, respectively. The agreement is much better than with the spherical model (cf. Fig. 4.64).*

of Mn, followed by a shell of Pd and, finally, a shell of Al, whereas the other node atomic surface is only built up with a core of Mn and an external shell of Al. Finally, the body-centre site is a Pd one. As far as the rough shape is concerned, one should notice that there is a supplementary mid-edge atomic surface in the Yamamoto model. Of course, now the details of the structure can be studied with much more confidence, since the R factor is quite low.

As for the case of the previous models, it is interesting to look at the structure in terms of both atomic cluster packing and dense planes packing.

There are two ways of studying local environments. First, it can be done through the use of decorated inflated three-dimensional Penrose tiling, since the model is based on this tiling. As already stated, each subdomain corresponds to a specific decoration of one of the three 'clusters' found in the inflated 3D

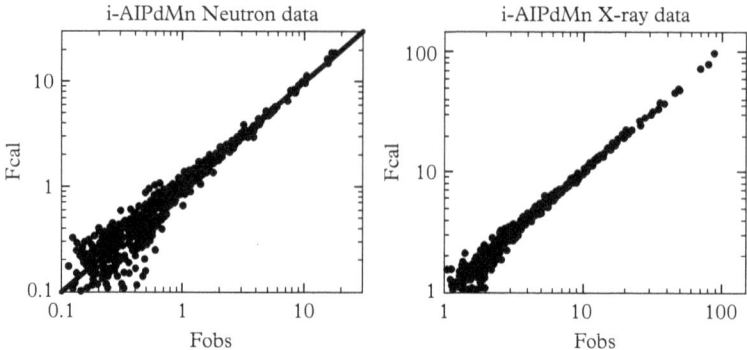

Figure 4.74 *Log-log comparison between observed and calculated structure factors for neutron (left) and X-ray (right) data. Notice the overall good agreement.*

Penrose tiling. In the model, the vertices of the inflated Penrose tiling are generated by the small triacontahedron labelled 32 and located on the bc_1 site and an equivalent empty one on the bc_2 site. This leads to the decoration already shown for one dodecahedral star, which is generated by translations of the rhombic triacontahedron in the perpendicular direction (this is visible in Fig. 4.71, where all subscripts are referring to one triacontahedron). Now the chemistry and parallel shifts, as determined experimentally, can be introduced to describe the structure properly.

The other route is to look at the atomic clusters in the structure. From the decorated inflated Penrose tiling, there are two large clusters sitting on the vertices of the three-dimensional Penrose tiling, one of which is shown in Fig. 4.75. The first three shells comprise an icosahedron followed by a dodecahedron and an icosahedron, and correspond to the Bergman cluster (B cluster) defined in the Katz and Gratias model. There are then other shells, some of which are only partially occupied. They define a cluster which is 20 Å in diameter and which is quite similar to the one observed in the 2/1 approximant. These clusters are located at the vertices of the three-dimensional inflated Penrose tiling. They occur with two different chemical decorations. In-between these large clusters there are Mackay type clusters, again with two kinds of chemical decoration, in particular with a strong chemical ordering of the Pd atoms. This is exemplified in Fig. 4.76, where the external shell of the Mackay cluster is made either of Al and Mn atoms or of Al and Pd atoms.

In this description, the density of Mackay clusters is small, their central position being given by the occupation domains 16 and 3, as shown in Fig. 4.71. But one has to realize that it is a matter of point of view, and that the existence domain of Mackay clusters could be extended, at the expense of the larger 20 Å one.

Finally, since some chemical disorder has been introduced, some sites of the clusters have mixed occupation.

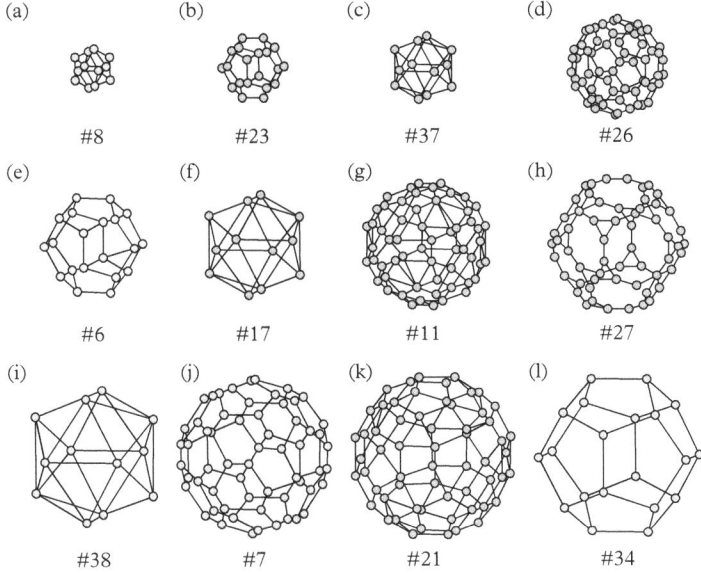

Figure 4.75 *Illustration of the successive shells of the large atomic cluster found in the model.*

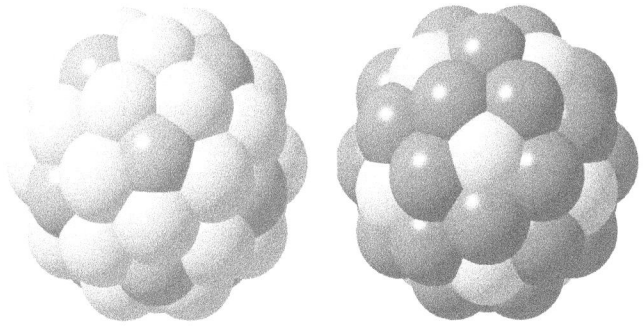

Figure 4.76 *External shells of the two Mackay clusters. The white, light-grey, and dark-grey colours stand for Al, Mn, and Pd atoms, respectively. The diameter is approximately 9.6 Å.*

As for the spherical model or the Katz and Gratias model, dense planes are observed in the resulting three-dimensional structure. The densest planes are found perpendicular to a five-fold axis (see Fig. 4.77). In general, planes are not flat but corrugated, and it is along the five-fold axis that the largest gaps between planes are observed. At a larger scale, one also observes a stacking of thick and thin slabs arranged in a Fibonacci sequence. This alternative presentation of the

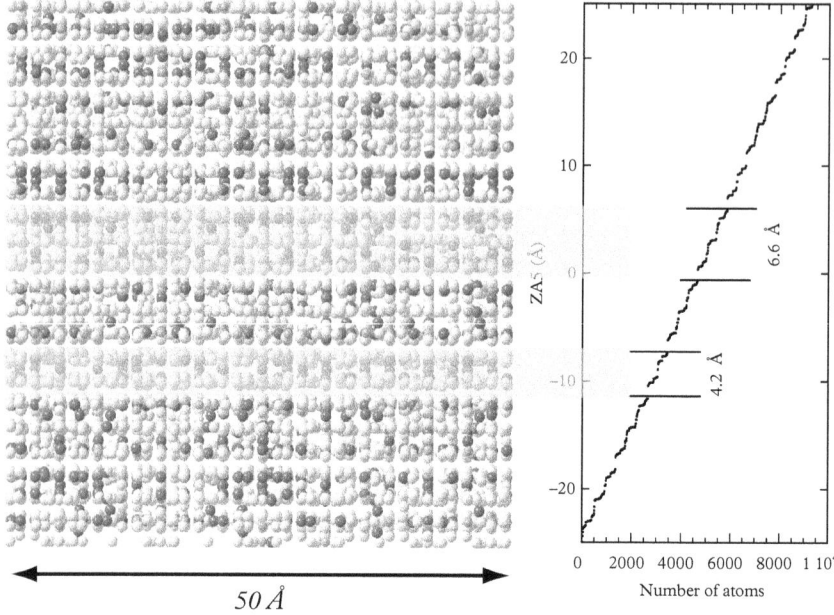

Figure 4.77 *Distribution of atomic planes along a vertical five-fold axis. There are thin and thick slabs of atoms disposed in a Fibonacci chain. The right panel illustrates the distribution of atoms along the five-fold axis. Relatively large gaps (1 Å) are observed in-between the slabs.*

structure is important for the study of quasicrystal surfaces, for instance, or the study of plastic deformation. One might expect, in line with the simple cut-bond rule, that planes above the largest gaps are the densest and the more stable. This has been confirmed by a careful study of the five-fold surface of the i-AlPdMn phase, which displays terraces in agreement with the succession of thin and thick slabs predicted by the model.

Both the description in terms of planes and that in terms of clusters may be used, depending on the problem to be studied.

It is worth mentioning some of the limitations of the present structure determination. One of the major problems in six-dimensional modelling is the determination of the exact shape of atomic surfaces. A step towards solving this problem is the partitioning of the large atomic surfaces into smaller occupation domains, as proposed by Yamamoto, each one playing a role equivalent to an atom position in a three-dimensional periodic structure determination. Of course, the smaller the subdomains are, the better is the accuracy of the structural fitting. However, the fitting procedure is limited by the number of available measured reflections, which in turn limits the number and thus the size of subdomains which can be used. From the available data one can, in principle, determine the average size of the subdomains which can be used. In some sense, this determines an experimental resolution in the perpendicular space. However, the problem of partitioning the large atomic surfaces is not simple, and the solution is not unique. One decomposition was used by Yamamoto, but others are possible. In particular, one would like to have a partitioning with subdomain sizes which are not too different from one another. From this point of view, the model proposed by Yamamoto is better than the Katz–Gratias one, which ends up with atomic subdomains having very different sizes. However, there are still some variations in the size of the different subdomains (e.g. the rhombic icosahedron is much smaller than the dodecahedral star), especially once icosahedral symmetry is applied. The size difference becomes important when one introduces chemical disorder or statistical occupancy. Indeed, if chemical disorder is found on a rather large subdomain, one may wonder if this is a real disorder or if it is due to the superposition of smaller occupation domains. In summary, there is a strong coupling between the size of the atomic surfaces and the resulting chemical order or disorder. One touches here upon one of the most difficult points in quasicrystal structure determination. It illustrates the importance of having some knowledge of the chemistry involved in the alloy. This can be tackled either by looking at the precise structure determination of approximant crystals or by doing ab initio and energy calculations as proposed in Mihalkovic et al. (2002).

In the next section, we present results recently obtained for the binary CdYb phase, where the crystal chemistry is very well understood.

4.5.3 Atomic structure of the CdYb icosahedral phase

The discovery of the first stable *binary icosahedral quasicrystal* by Tsai and co-workers was a real breakthrough (Tsai et al., 2000), allowing unprecedented detailed structural analysis for a quasicrystal. Indeed, this was the first example of a binary quasicrystal presenting a high structural quality. Moreover, whereas the quasicrystal is obtained for a chemical composition of $Cd_{5.7}Yb$, it is possible to synthesize a cubic 1/1 approximant with composition Cd_6Yb and a lattice parameter equal to 1.57 nm and a cubic 2/1 one with composition $Cd_{5.8}Yb$ and a lattice parameter equal to 2.53 nm from which detailed structural analysis has been carried out. These compounds are also characterized by a packing of a large (Yb) and a small (Cd) atom, so that very little chemical disorder is present in this system. Finally, there is a very good X-ray contrast between Cd and Yb atoms and high-quality synchrotron data could be collected. This allowed by far one of the most detailed understandings of a quasicrystal structure up to now.

Single grains of both the crystal approximant and the quasicrystal can be grown with a good quality, that is, with a narrow mosaic spread. A careful indexing of the diffraction pattern shows that the icosahedral lattice is primitive. This is particularly visible along the five-fold axis, where the scaling factor between reflections is τ^3, as opposed to τ in the case of the i-AlPdMn phase (cf. Fig. 4.53). All Bragg peaks are perfectly positioned according to the theory, with a six-dimensional lattice parameter a_{6D} equal to 7.15 Å. Bragg peaks are also narrow with a small phason strain broadening (de Boissieu *et al.*, 2002).

Using a high-quality single grain, data collection was carried out using synchrotron radiation. Because of the high quality of the CdYb phase, weak reflections with H_{perp} values up to 4 ($2\pi/a_{6D}$ unit) are visible, and span a large dynamical range. For this reason, a point detector was used, allowing collection of more than 5000 independent reflections whose intensity span more than 8 orders of magnitude. This is the first time that such a large data set could be collected.

Figure 4.78 displays the comparison between the two-fold diffraction pattern of the 1/1 approximant with that of the icosahedral quasicrystal. It is quite obvious that a much larger number of reflections is visible in the i-CdYb quasicrystal diffraction pattern. However, the position and intensity of the strongest Bragg peaks are rather similar: this is a signature of the fact that the two structures have a similar local environment and are built up with similar clusters.

The structural analysis is carried out in three steps. First, icosahedral clusters and their linkages are identified in the 1/1 and 2/1 approximants (Gomez and Lidin, 2001, 2003). Second, electron density maps are computed using the experimental data and the low-density elimination method for phase reconstruction (Takakura *et al.*, 2004b). Third, from this map a six-dimensional model is constructed and refined from experimental data.

The 1/1 approximant can be described as a body-centred cubic stacking of clusters with almost icosahedral symmetry. The successive shells and the cluster

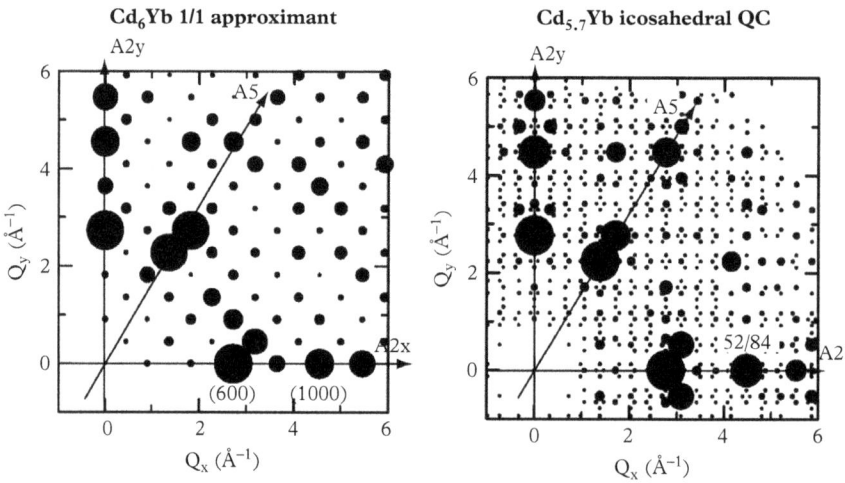

Figure 4.78 *Comparison of the two-fold diffraction pattern for the 1/1 Cd_6Yb cubic approximant (left) and the $Cd_{5.7}Yb$ icosahedral quasicrystal (right). Units are expressed for $Q = 2\pi k$. The position and intensity distribution of the strongest Bragg peaks are quite similar in both phases, indicating similar local environments.*

Examples of quasicrystal structures 271

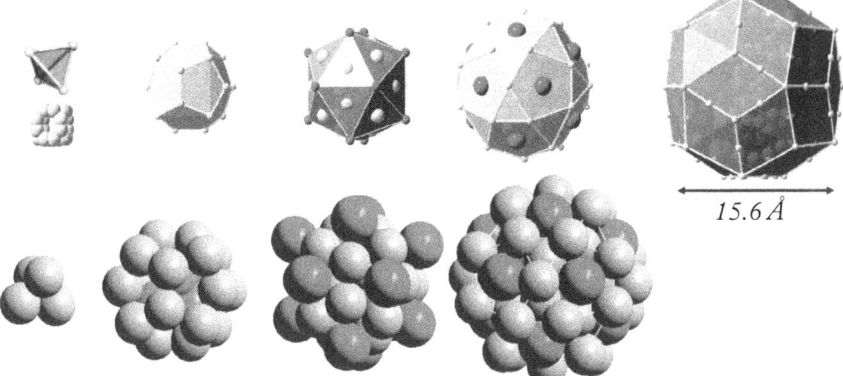

Figure 4.79 *Left: Successive shells of the cluster found in the Cd_6Yb 1/1 approximant and the quasicrystal. Cd and Yb atoms are displayed in light and dark grey, respectively. There is first a Cd tetrahedron (in different orientations), followed by a Cd dodecahedron, an Yb icosahedron, a Cd icosidodecahedron, and a large rhombic triacontahedron. The bottom panel displays the first four shells in space-filling mode to illustrate the close packing of the small (Cd) and the large (Yb) atoms.*

packing are shown in Fig. 4.79. The first inner shell is a Cd tetrahedron, which occurs in various orientations and is thus disordered. Only one tetrahedron position is shown in Fig. 4.79, but in reality there are six equivalent positions which are statistically distributed. It is believed that this disorder is dynamic in nature and leads to a superstructure ordering when cooling down. The next shell is a Cd dodecahedron (20 atoms). It has an almost perfect icosahedral symmetry with atoms 4.41 or 4.04 Å from the centre. The next shell is an Yb icosahedron (12 atoms) with radius 5.54 Å with an almost perfect icosahedral symmetry. The fourth shell is a Cd icosidodecahedron (30 atoms), obtained by placing atoms on the edges of the Yb icosahedron. This shell departs significantly from the perfect icosahedral symmetry, although all atoms are at a distance of about 6.4 Å from the centre. Thus, this forms a 66-atom cluster, which is different from the previously known Mackay and Bergman clusters. This cluster is packed on a body-centred cubic lattice, as shown in Fig. 4.79. The clusters are linked through a three-fold and a two-fold axis, the three-fold direction being the closest distance: it is this three-fold connection which produces the slight distortion of the icosidodecahedron. There remain 36 Cd atoms which occupy the vertices and mid-edges of a large triacontahedron. This large atomic triacontahedral cluster describes entirely the 1/1 approximant. It is connected along a two-fold axis (15.7 Å) by sharing a face and along the three-fold axis (13.6 Å) where they interpenetrate. The three-fold intersection of two clusters defines an oblate polyhedron.

The structure of the 1/1 approximant thus presents a well-defined chemical order, with Yb atoms forming an icosahedron. This is at variance with the approximants which have been obtained in other systems and which all display some chemical disorder. This strong chemical order is related to two points. First, there is a size difference between the Cd and Yb atoms, with a tetrahedrally closed packed structure referred to as Frank–Kasper type. This creates strong constraints for the dense packing of the atoms. The second point is that ab initio calculations show that there is a strong hybridization between the Yb and Cd electrons (Ishii and Fujiwara, 2001), with some kind of covalence of the linkages. Thus, one has most likely a very stable building unit, which might also be found in the quasicrystal.

What is also interesting is that a 2/1 approximant, with a unit cell τ times larger could also be synthesized. It contains the same atomic clusters, linked along three-fold and two-fold directions. However, the packing is different from the one found in the 1/1 approximant, and a new interstitial (or building block) has to be defined: it is formed by a double Friauf polyhedron which defines a decorated Penrose tile, as shown in Fig. 4.80.

Now that a clear understanding of the clusters and their connection has been achieved, the analysis of the icosahedral phase can be tackled. In a first step, six-dimensional electron density maps are obtained by Fourier transform of the structure factor. Using the low-density elimination method, it has been shown that the phases of the strongest reflections can be accurately determined. This procedure is a kind of generalization of the direct methods used for three-dimensional periodic crystals (Takakura *et al.*, 2001a).

Figure 4.81 displays a section of the six-dimensional density in a plane containing (a) a five-fold axis, (b) a three-fold axis, and (c) a two-fold axis. All high-symmetry Wyckoff positions of the six-dimensional cube are contained in the

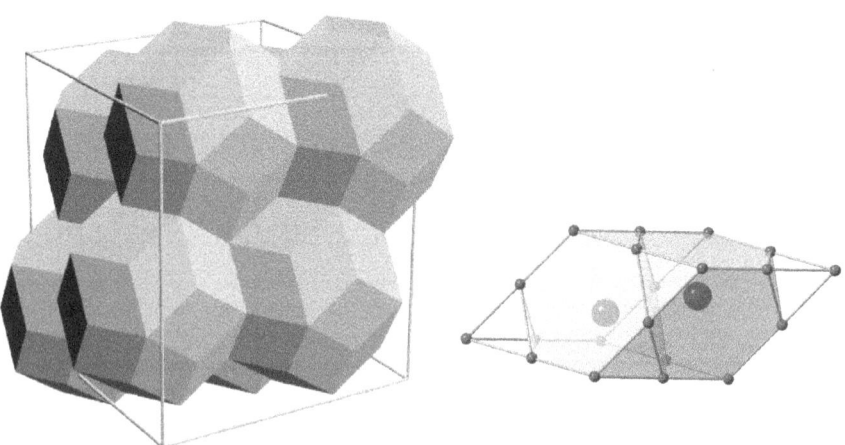

Figure 4.80 *Left: Packing of the triacontahedron clusters in the 2/1 approximant. The right-hand panel shows the double Friauf polyhedron which is found in-between triacontahedra.*

Examples of quasicrystal structures 273

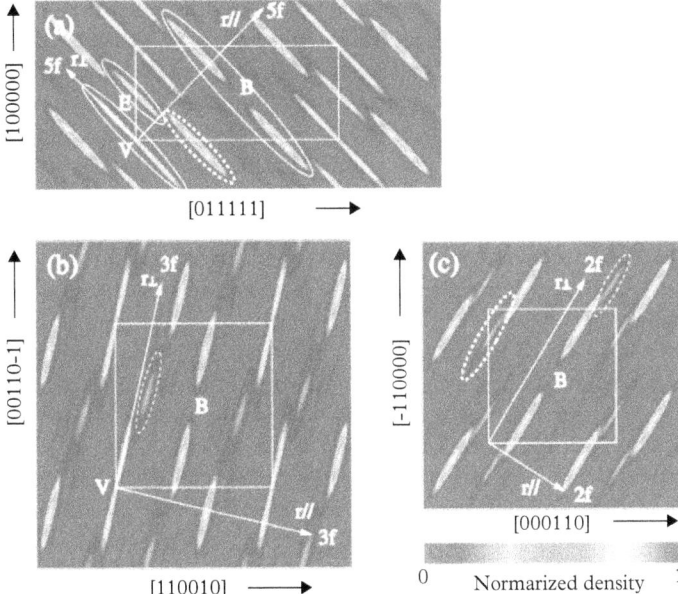

Figure 4.81 *Sections of the six-dimensional density map obtained by the low-density elimination method applied to the CdYb data. (a) is a five-fold section. There are three atomic surfaces located at the origin (V), the mid-edge (E), and the body-centre (B) site as shown by the white circles. (b) and (c) are three-fold and two-fold sections, respectively. (From (Takakura et al., 2004b))*

five-fold density map (Takakura *et al.*, 2004a, b). As can be seen in the figure, there are elongated density maxima which can be associated with three sites: the origin (V for vertex in the figure, coordinate (000000)), the mid-edge (E, coordinate $(\frac{1}{2}00000)$), and the body-centre (B, coordinate $\frac{1}{2}(111111)$) sites. From this figure it is also clear that the size of the three atomic surfaces is different, with a larger size for the body-centre atomic surface, a smaller one for the origin and a small mid-edge site. Finally, it is seen that the electron density of the body-centre site is larger than the other one, which means that this site contains a significant Yb contribution, which has a higher electron density than the Cd atom. One also sees on the two-fold section that the B site has a longer extension, with the extreme part having a shift along the parallel space (dashed ellipse), illustrating for the first time the importance of this component.

From this rough six-dimensional electron density map it is also possible to determine which atomic clusters are generated in the three-dimensional quasiperiodic structure. This is achieved by examining the three sections along five- three- and two-fold axes. Starting from the B site centre which is empty, the first shell is a dodecahedron (generated on the three-fold section by the dashed ellipse), followed by an icosidodecahedron (dashed ellipse on the two-fold section), and an Yb icosahedron. The atomic cluster generated by the calculated density map is thus similar to the one observed in the periodic approximant. Notice that the dodecahedron and the icosidodecahedron shells are generated by atomic surfaces having a parallel shift (i.e. which are not exactly in the prolongation of the main atomic surface). This shift is necessary to recover shell distances similar to the one of the approximant phase. This is very valuable information which can then be used for further modelling.

The modelling was achieved in Takakura *et al.* (2007) by considering the large triacontahedral cluster and its linkages along two-fold and three-fold axes, as observed in the periodic approximant. The basis of the model is a maximization of the number density of the clusters. This is achieved by placing clusters on the vertices of the so-called twelve-fold vertices of the 3D Penrose tiling. This is a subset of the Penrose tiling sites, where 12 edges meet. In the six-dimensional description, it is generated by a τ^{-2} deflated triacontahedron, slightly truncated along the five-fold directions and located at the vertices of a six-dimensional cubic lattice (Henley, 1986). The successive shells of the atomic cluster are generated by translating this atomic surface along the three-, five-, and two-fold six-dimensional vectors having parallel moduli equal to the radius of the different shells. This corresponds to the dashed ellipsoids shown in Fig. 4.81. As already pointed out, the three- and two-fold translated atomic surfaces have a significant component along the parallel direction. This constitutes the backbone of the model. However, there are some empty spaces, which have to be filled with atoms in order to reach the experimentally observed atomic density. It can be shown that the remaining space in-between the large triacontahedral clusters can be filled by putting a prolate rhombohedron and an oblate rhombohedron in them. The atomic decoration of these two supplementary building blocks has been chosen to be the same as the one observed in the 2/1 approximant: the already discussed double Friauf polyhedron constitutes the prolate one, whereas the oblate one has atoms on vertices and mid-edges, in a way similar to what is observed in the inter-penetrating triacontahedra. With these two supplementary building blocks, the model is completely defined and presents a density and chemical composition equal to the observed one. The resulting six-dimensional model is made of three atomic surfaces located at node, body-centre, and mid-edge positions in the six-dimensional cube. The shape of the different atomic surfaces is shown in Fig. 4.82.

These atomic surfaces have then been subdivided in smaller parts in order for the refinement to be successful. For each subdomain, a parallel component has been refined. The agreement with the experimental data is excellent, taking into account the very large number of measured reflections (R factor equal to 0.1).

Having refined the six-dimensional model, it is possible to describe the resulting three-dimensional quasiperiodic structure. As for the i-AlPdMn phase, one should choose the description which is the most appropriate for the physical properties under consideration. There are three complementary ways of looking at the structure: in terms of atomic clusters, in terms of atomic planes, and in terms of a hierarchical organization. The model was built in such a way that the quasicrystal

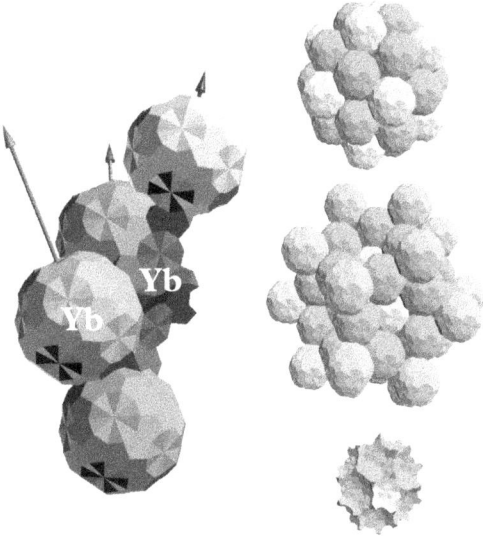

Figure 4.82 *Illustration of the six-dimensional model of the CdYb icosahedral phase. The left panel shows the decomposition of the body-centre atomic surface in the asymmetric unit. The truncated triacontahedron (with almost spherical shape) corresponds to the occupation domain that generates the atomic cluster. Atomic surfaces which generate Yb atoms are indicated. The other ones generate Cd atoms. The right-hand panel shows the external shape of the node, body-centre, and mid-edge atomic surfaces (from top to bottom). (From (Takakura et al., 2007))*

structure contains a large number of triacontahedral clusters with a perfectly defined chemical order. It can be shown that 94 per cent of the atoms belong to such a cluster. The atomic clusters are linked along the two- and three-fold directions (see Fig. 4.83), and present a distortion along the three-fold axis, similar to what has been observed in the periodic approximant, as will be explained hereafter. Of course, there are also new local environments, given by the remaining six per cent of atoms. They are found by associating several double Friauf polyhedra, as shown in Fig. 4.83. For instance, some of them form a 'Bergmann-type' cluster, similar to the one observed in the i-AlLiCu phase.

A second way to look at the structure is in terms of dense atomic planes. This is best observed by looking at the projected electron density, which can be computed easily from the Fourier transform of the structure factors, with their amplitude and phases. Figure 4.84 displays the electron density projected down a two-fold axis. As it was observed in the i-AlPdMn phase, there are dense atomic planes, with

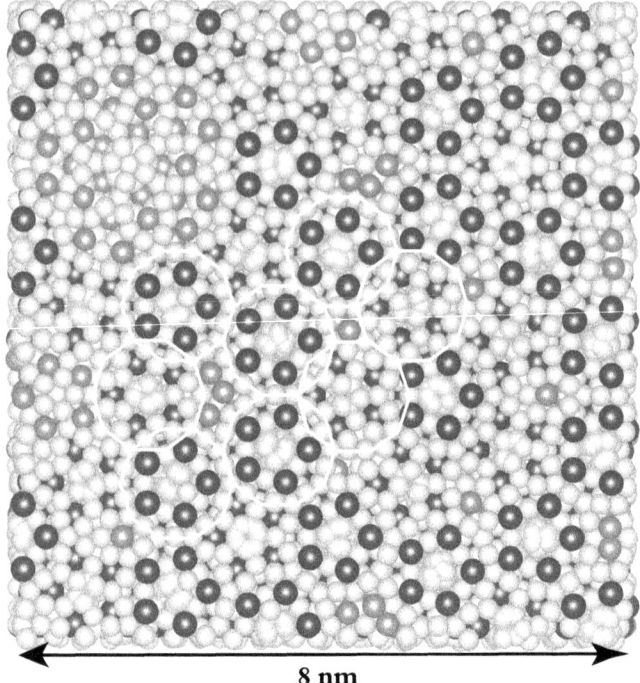

Figure 4.83 *Illustration of the resulting three-dimensional structure deduced from the six-dimensional refined model of the icosahedral CdYb phase, displaying an 8×8 nm slab viewed along a five-fold axis. The light-grey atoms stands for Cd atoms. The dark and dark-grey atoms stand for Yb atoms located inside the triacontahedron on the icosahedron, and inside the double Friauf polyhedra, respectively. The white decagons show the outlines of the large triacontahedra that are connected along the two-fold axis by sharing a face, and along the three-fold axis by overlapping. Note that, whereas the two-fold axes are in the viewed plane, the three-fold axes are inclined. Some of the spaces between the triacontahedra, filled by groups of Friauf polyhedra, are visible.*

significant gaps between them. The densest planes are observed along directions perpendicular to a two-fold or a five-fold axis, as highlighted in Fig. 4.84. This corresponds to the strongest Bragg reflections observed in the diffraction pattern, which also lie along two-fold or five-fold axis (see Fig. 4.78). A description in terms of dense planes is, for instance, well adapted for the understanding of quasicrystalline surface experiments.

The third way of looking at the structure is in terms of hierarchical properties. Indeed, as we have seen previously, one of the peculiar properties of the diffraction pattern is its scaling. In the case of the primitive icosahedral lattice, the scaling has

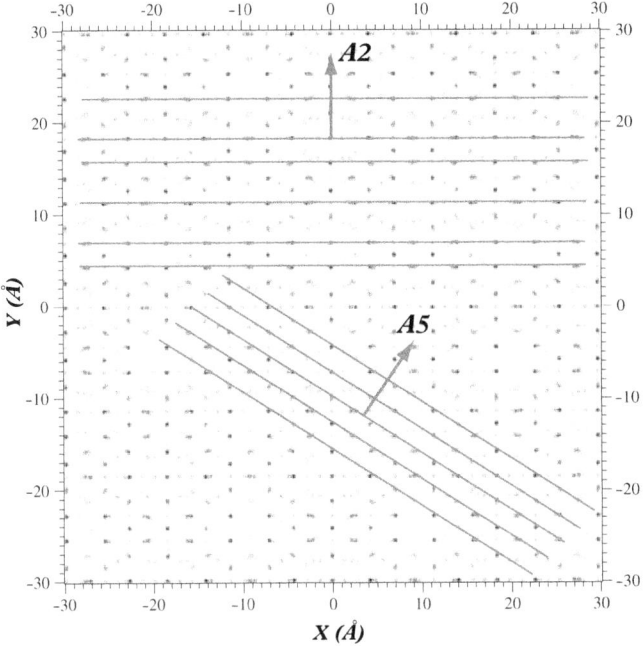

Figure 4.84 *Electron density of the i-CdYb phase projected down a two-fold axis. Dense atomic planes perpendicular to a two-fold or five-fold axis are outlined.*

a factor τ^3. This means that each time a reflection is observed at a position **Q** in reciprocal space, there is another one at a position τ^3**Q**. This scaling holds also in direct space for the three-dimensional quasiperiodic structure. This is exemplified in Fig. 4.85, which displays the location of the cluster. centres on a five-fold plane on a large scale (about 340 Å). Around each dot there is a triacontahedral cluster linked to others by sharing a face along a two-fold direction, as shown by the black line. The two-fold intercluster distance is equal to 15.7 Å. There are also clusters whose centre is just above and below the plane (2.5 Å) and which are connected along a three-fold axis, seen in the figure at empty spaces. Starting from the central cluster, it can be shown that a cluster of triacontahedra also is formed in the structure. Its external shell is an icosidodecahedron, with a radius of 34 Å, and whose five-fold section is shown in the insert (top left corner of the figure). Such a cluster of triacontahedra is highlighted by a shaded area in the figure. It can be seen that there is also a cluster of 'cluster of clusters', which is also an icosidodecahedron, illustrating the hierarchical distribution of atoms in the three-dimensional quasicrystal. The two length scales R_1 and R_2 are related by the scale inflation τ^3 and are equal, respectively, to 25 and 107 Å. Such a description in

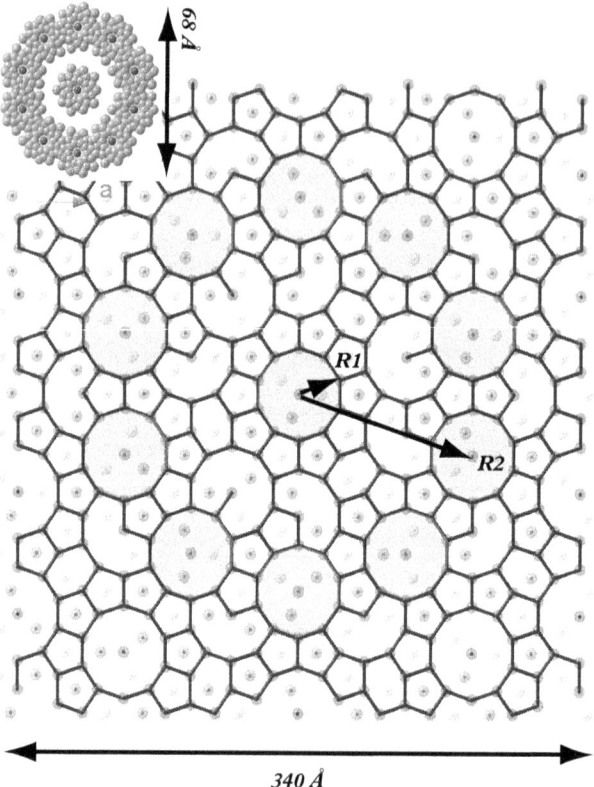

Figure 4.85 *Distribution of the cluster centres in a five-fold plane. The two-fold linkages are outlined with dark lines. Cluster centres above and below the plane (2.45 Å apart) are also shown as light grey: they are connected along three-fold axes with clusters in the plane. Each grey-shaded area highlights a 'cluster of clusters'; these, in turn, form a larger cluster of 'cluster of clusters'. The two arrows R_1 and R_2 indicate the two length scales of the inflated network of clusters. The insert in the upper left corner shows a section of the icosidodecahedron of triacontahedra.*

terms of hierarchical packing might be an alternative route for understanding the physical properties of quasicrystals (Janot and de Boissieu, 1994).

The crystal chemistry is also quite well understood in this system. There is a very good Cd/Yb chemical order: indeed it has not been necessary to introduce any Cd/Yb chemical disorder in the modelling. This is certainly related to the different Cd and Yb atomic size and to the Cd–Yb hybridization found in the electronic structure obtained by ab initio calculation (Ishii and Fujiwara, 2001). The 'glue atoms' also are perfectly chemically ordered. The way they interact with the clusters is certainly of importance for the understanding the stability of the CdYb icosahedral phase.

Another important point is the role of the tetrahedron (see Fig. 4.79) located at the centre of the cluster, which does not has any icosahedral symmetry. A temperature study of the cubic 1/1 approximant showed that a phase transition occurs in the temperature range 150–190 K and is related to an ordering of the central tetrahedron. In general, the low-temperature phase is characterized by a doubling of the unit cell along the [110] direction and by a slight monoclinic distortion leading to the low-temperature space group $C2/c$, whereas the high temperature phase is cubic with the space group $Im\bar{3}$ (Tamura *et al.*, 2002, 2003, 2004). A detailed study of the low-temperature phase in the iso-structural Zn_6Sc system showed that this phase is characterized by an anti-parallel ordering of the central tetrahedron, whereas in the high-temperature phase the tetrahedron is disordered (Ishimasa *et al.*, 2007; Yamada *et al.*, 2013). The atomic structure of the low-temperature phase could be solved accurately, showing that the central tetrahedron induces a very large distortion of the successive shells which no longer have icosahedral symmetry. This distortion is clearly related to the atoms close packing around the central tetrahedron, and propagates shell after shell up to the large triacontahedron. This distortion allows the interactions between the tetrahedra, which are far apart, to be mediated.

A very similar distortion has been observed in the refined i-CdYb structure. In the quasicrystal, the central tetrahedron was modeled with a disordered site and partial occupancy (12 atoms on the vertices of a dodecahedron). However, after the refinement procedure, the 12 partially occupied Cd atoms are grouped together and almost form a tetrahedron. At the same time, the successive shells depart significantly from the icosahedral symmetry in a way very similar to what has been determined experimentally for the 1/1 approximant. The very detailed crystal chemistry is thus similar in the icosahedral quasicrystal and in the 1/1 approximant. One of the most stunning results is that all shells have now lost their perfect icosahedral symmetry, even in the quasicrystal, as a result of the close-packing tendency around the central tetrahedron. This is illustrated in Fig 4.86. This thus contradicts the common belief that the periodic arrangement would somehow 'distort' the icosahedral shells, whereas this would not be the case in the icosahedral quasicrystal. The shell distortion is related to the three-fold connectivity of the clusters. Of course, in the quasicrystal, the overall icosahedral symmetry is preserved because each cluster will appear in the 3D structure with all the orientations generated by the icosahedral point group. We will see in the section on phonons (Chapter 5, Section 5.8.3) that this central tetrahedron behaves as a single molecule and, above a critical temperature, reorients dynamically on a time scale of the order of 1 ps (Euchner *et al.*, 2012, 2013), leading to a very surprising dynamical flexibility.

The local order around the tetrahedron is similar in the icosahedral quasicrystal and the 1/1 approximant; the main difference between the two lies in the cluster–cluster connectivity, with the distribution around a cluster having a two-fold or a

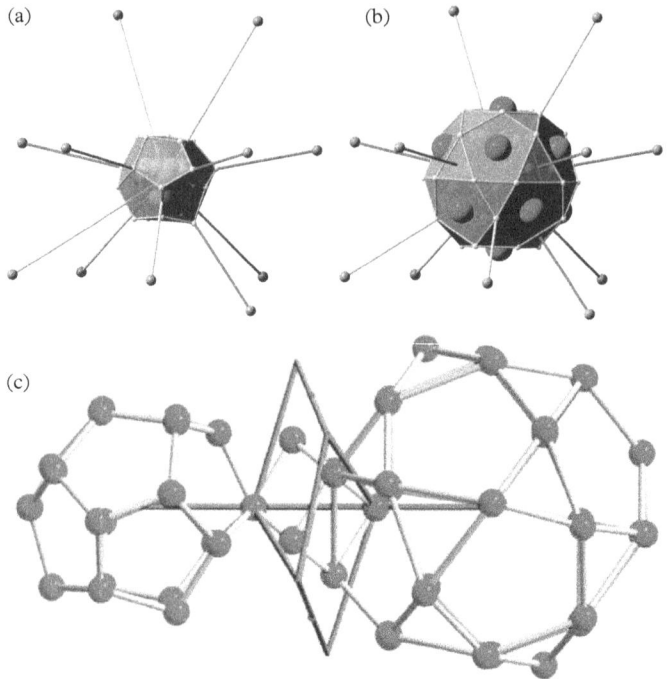

Figure 4.86 *Illustration of the cluster distortion in the refined model of the i-CdYb quasicrystal. (a) and (b) display the first two shells in a (12, 6, 6) environment, with two- and three-fold connections in light and dark grey, respectively. The dodecahedron distortion is clearly visible in (a). In (b) notice the larger triangles appearing on the icosidodecahedron when a three-fold axis is going through it. Panel (c) displays the connection framework along a three-fold axis, showing the dodecahedron and the icosidodecahedron.*

three-fold bound, respectively. The rhombic triacontahedron cluster coordination is best described by the (Z, N_b, N_c) notation proposed by Henley (1986), where Z is the coordination number, and N_b and N_c are the numbers of two- and three-fold linkages in Z, respectively. Whereas in the 1/1 approximant there is a single (14, 6, 8) environment, there are 18 different cluster–cluster local environments in the quasicrystal, the most frequent ones being (12, 7, 5) and (12, 6, 6), with a frequency larger than 50 per cent. In the icosahedral quasicrystal, there are a number of relatively low coordination numbers, where the double Friauf polyhedra are grouped together to fill the empty space, as illustrated in Fig. 4.83. As explained above, the cluster distortion is driven by the distribution of three-fold linkages. It is worth at this point to mention that an alternative description of the two- and three-fold icosahedral quasicrystalline networks can be made using the canonical cell tiles, as proposed by C. Henley (Henley, 1991a). Although there is no proof for the existence of an icosahedral quasicrystal using canonical cells, this is a very useful model that can be set up up to high-order approximants to carry out atomic scale simulations, as will be illustrated in the section on phonons (Chapter 5, Section 5.8.3).

As a further illustration of the very good understanding of the quasicrystal structure in this system, we present results obtained in a series of binary icosahedral phases found in Zn–Sc (Canfield et al., 2010) and Cd–RE (RE stands for Ho, Tb, and Sm) (Goldman et al., 2013): in this case, the stoichiometry is slightly different from that of the ideal i-$Cd_{5.7}$Yb icosahedral model and writes i-$Zn_{7.3}$Sc (or i-$Cd_{7.3}$RE), with, thus, a significant excess of Zn atoms with respect to the ideal composition. The data was obtained using 2D CCD detectors and high energy at the Cristal beam line of the Soleil synchrotron, resulting in more than 4000 independent unique reflections, with a large dynamical range. The model refinement leads to a structure where the majority of the large atom sites (Sc or RE) in the double Friauf polyhedra are replaced by a small atom (Zn or Cd). A residual disorder is also found on the Sc (RE) icosahedron within the large triacontahedral cluster (Yamada et al. 2016a, Yamada et al. 2016b).

In all these models, an overall phason Debye–Waller term is introduced. It represents 'fluctuations' of the atomic surfaces in the perpendicular direction, and is related to phason fluctuations (see p. 314). Phason modes are characteristic of the aperiodic crystal order and are diffusive modes with a polarization in the perpendicular direction. Similar to phonons, they give rise to both a Debye–Waller term and diffuse scattering. The phason Debye–Waller term is related to the mean-squared displacement in perpendicular space $\langle u^2_{perp} \rangle$. For all icosahedral phases, a clear diffuse scattering has been observed experimentally and interpreted as resulting from long-wavelength phason fluctuations. Absolute scale measurements did illustrate that the 'amount' of diffuse scattering is different depending on the systems, but the scattering is present. In the case of the i-$Zn_{7.3}$Sc quasicrystal, there is a rather large amount of diffuse scattering, and the phason Debye–Waller term has been found to lead to root-mean-square perpendicular fluctuations on the order of 0.1 times the radius of the larger atomic surfaces describing the 6D model.

The interpretation of the phason fluctuation at the atomic scale is still an open question. It might be related to the central tetrahedron in that case, although this is not clearly proven. One model could also be a random tiling model, as proposed in Henley (1991a, b).

In conclusion, on the structure determination of CdYb-type quasicrystals, one can say that, nowadays, the structural quality achieved in those systems is very much comparable to what is achieved in complex intermetallic phases. The central role played by the tetrahedron and the icosahedral symmetry breaking of the clusters seems not to be unique to this case and might be also encountered in other phases such as i-AlPdMn.

4.5.4 Structure of the AlNiCo decagonal phase

Decagonal quasicrystals are periodic along one direction and quasiperiodic in a plane orthogonal to it. They can thus be described as a periodic stacking of quasiperiodic layers. The period is of the order of 4 Å, or a multiple of this value. Determining their structure is thus particularly appealing, because the problem is somehow easier than for the icosahedral phase. Moreover, the periodicity along a single direction makes the use of high-resolution images obtained by transmission electron microscopy (TEM) extremely useful, resulting in almost a direct image of the structure. Finally, the periodicity also makes it easier to calculate stability and physical properties using ab initio calculations.

However, despite these advantages, and the very large number of studies, the question 'where are the atoms?' is still not yet fully answered. One of the main reasons is that, although it is possible to obtain single grains of the decagonal phase in a few systems (mainly AlCuCo and AlNiCo), they almost all show a large amount of diffuse scattering in their diffraction pattern, which has to be taken into account if one wants to understand the structure of these phases. Moreover, the phase diagram is extremely complex in the vicinity of the decagonal phase field: the decagonal phase is generally only stable at high temperature, the low-temperature phase being either a microcrystalline phase or a modulated phase. The AlNiCo phase diagram is particularly intricate in that respect: different variants around a main decagonal structure are obtained by changing the Ni/Co ratio, keeping the Al content close to 0.7. All these results strongly suggest that disorder is a key ingredient in the decagonal structure.

The most detailed structural information has been obtained for the $Al_{70.6}$-$Co_{6.7}Ni_{22.7}$ decagonal phase. This phase is only stable at high temperature and must be quenched-in for room temperature studies. The amount of diffuse scattering is very small in this phase and it presents a periodicity of 4 Å, with a weak superstructure of 8 Å. Its structure is referred to as the 'Ni-rich, basic' decagonal structure in the quasicrystal community. Studies have been carried out by X-ray diffraction and five-dimensional analysis, high-resolution electron microscopy (HREM) images, and Penrose tiling decoration, and also by total energy simulations.

4.5.4.1 X-ray analysis

The X-ray structure determination has been carried out by the groups of Steurer and Yamamoto. As already mentioned, diffuse scattering is sometimes quite strong, and using experimental techniques which allow an overview of reciprocal space is quite important. Using imaging plates, Steurer's group has developed software for accurately reconstructing reciprocal space layers including both diffuse scattering and Bragg peaks (cf. Section 4.2.3). Figure 4.11 shows the reconstructed zeroth layer of the AlNiCo decagonal phase. The ten-fold symmetry is clearly visible together with a distribution of weak diffuse scattering intensity.

As shown in the section on diffuse scattering (Section 4.6.1), the analysis of the structure using Bragg reflections only leads to an 'average' structure. The high-dimensional procedure is similar to the one developed for the icosahedral case. The decagonal quasicrystal is embedded in a five-dimensional space, with four dimensions being necessary to describe quasiperiodic planes, and the fifth dimension corresponding to the periodic axis c. The atomic surfaces are thus two-dimensional objects.

The analysis of extinctions in the diffraction pattern leads to the space group $P10_5/mmc$, which contains in particular a 10_5 screw axis along the periodic direction. This leads to extinction conditions similar to those found for three-dimensional periodic crystals.

Similar to the case of the i-AlPdMn phase, the position and rough shape of the atomic surfaces have been determined using a Patterson analysis. In the case of the d-AlNiCo phase, there are two independent atomic surfaces (two others are related by a symmetry operation) located at positions $\left(\frac{1}{5}\frac{1}{5}\frac{1}{5}\frac{1}{5}\frac{1}{4}\right)$ and $\left(\frac{2}{5}\frac{2}{5}\frac{2}{5}\frac{2}{5}\frac{1}{4}\right)$ in the five-dimensional unit cell (the fifth index corresponds to the periodic c-axis).

Instead of a Patterson analysis, a rough model can be extracted from electron density calculations using the charge-flipping method (or an equivalent one). This is illustrated for the decagonal phase $Al_{70}Ni_{15}Co_{15}$ which was published earlier (Steurer et al., 1993). The structure was solved in five-dimensional space from the X-ray data set kindly provided to us by the authors with the software 'superflip' (Palatinus and Chapuis, 2006). The illustration given in Fig. 4.87 represents a portion of the 22.7×22.7 Å^2 structure, which can be directly compared with Fig. 7a of the original article by Steurer et al. The x_1–x_2 plane is perpendicular to the ten-fold screw axis. The similarity between the two figures is striking and it is difficult to find any meaningful differences. The quality of the solution delivered by the charge-flipping algorithm is surprisingly good, and the transition metals can be clearly distinguished from the aluminium atoms. With current desktop computers, solution convergence is reached within minutes. The treatment of the symmetry is left at a later stage during the refinement once the crystal structure is solved.

Cervellino et al. (Cervellino et al., 2002) and Takakura et al. (Takakura et al., 2001b) used a similar strategy for modelling the decagonal structure. After creating a rough model of the atomic surfaces, they subdivided each atomic domain almost evenly in smaller parts. Each subdomain has a set of refined parameters: occupation probability, shift in parallel space, and phonon Debye–Waller term. Steurer et al. did a data collection on a synchrotron source, leading to about 1500 independent reflections, almost four times larger than the data set of Takakura collected on a laboratory instrument. In particular, they measured very weak reflections, which were important for the analysis of the details of the structure.

Although both models agree in average there are significant differences, whose origin is not as yet fully understood. This illustrates the difficulty in elaborating a model for quasicrystals, even with quite sophisticated fitting procedures.

Figure 4.88 shows the decomposition of one of the atomic surfaces in smaller subdomains for the Cervellino model for the Al atoms and Ni/Co atoms (because

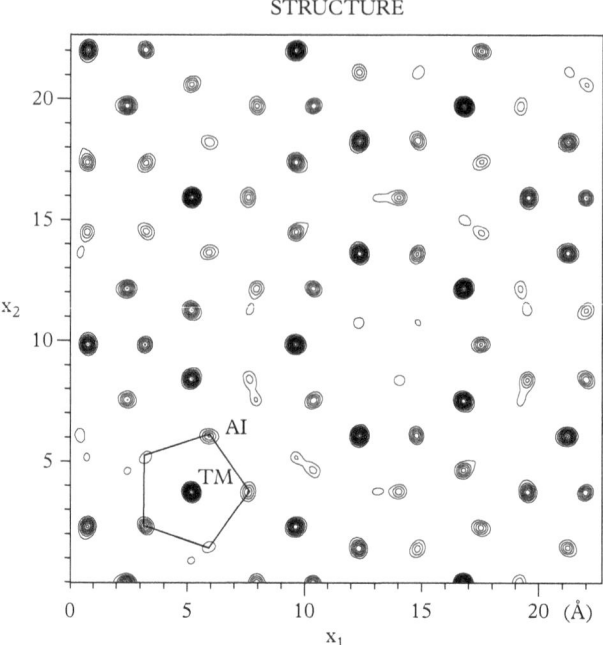

Figure 4.87 *Resolution of the structure of $Al_{70}Ni_{15}Co_{15}$ by the charge-flipping method. This 22.7×22.7 Å^2 section corresponds to Fig. 7a in Steurer et al. (1993). The similarity between the two figures is almost complete, including the split atoms.*

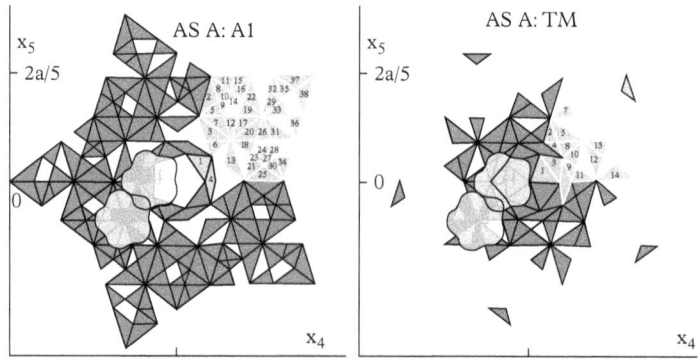

Figure 4.88 *Atomic surface A in the model structure of d-AlNiCo (from Cervellino et al.) The left-hand panel displays the Al atoms, and the right-hand panel the Ni/Co ones.*

X-rays have been used, it is not possible to distinguish Ni and Co atoms). The model is such that short distances are avoided, the *closeness condition* is fulfilled, and the density and stoichiometry is close to the experimental values. In the course of the refinement procedure, they could obtain a small reliability factor only by introducing a lot of disorder and partially occupied sites, mainly for Al atoms. The resulting structure is thus extremely disordered, which makes a description in terms of a decorated tiling almost impossible but for the Ni/Co subnetwork. Attempts to use a more ordered structure systematically gave large R factors. Cervellino *et al.* insist on the importance of weak reflections, which lead them to propose this disordered structure.

On the other hand, Takakura *et al.* proposed a model that presents a much larger degree of order, with far fewer partially or mixed occupied sites. Using this model and their structural data, Cervellino *et al.* could not obtain a satisfactory R factor. This might be due to the measurement of weak reflections, as already indicated. Another source of discrepancy is the slight difference in chemical composition of the two samples. Knowing that the diffraction pattern changes dramatically with minor changes in the composition, an effect on the intensity distribution would not be surprising. Finally, the high disorder introduced by Cervellino *et al.* is somehow equivalent to a description of a more ordered structure with a larger-phason Debye–Waller factor. Indeed, Takakura *et al.* found a phason Debye–Waller factor which is three times the one obtained by Cervellino *et al.*

It is interesting to try to describe the structure in terms of a decorated Penrose tiling or a variation on it. Indeed, this modelling has been widely used to interpret HREM images. Neither the Takakura nor the Cervellino models can be described by a systematic decoration of the Penrose tiling, since the decoration changes slightly from tile to tile. However, if we consider a low-resolution image of the structure, then a decorated tiling is a good approximation. Besides the standard Penrose tiling, two modifications are interesting to point out: the first is the so-called HBS tiling (for hexagon, boat, and star tiling), the second is the cluster covering proposed by Gummelt, which describes the Penrose tiling as a set of interpenetrating decagonal clusters (see Chapter 3, Section 3.6). Note that this description is in some sense a rephrasing of the matching rules. The HBS tiling and the Gummelt covering are the models which are best adapted for a description of the resulting structure. This is illustrated in Fig. 4.89.

The left-hand panel shows a low-resolution electron density projection, as is obtained by HREM images. When the resolution is low enough, a decoration by interpenetrating decagons (Gummelt covering) or by the HBS tiling is possible. Five interpenetrating Gummelt decagons, with a diameter of about 20 Å, have been outlined in Fig. 4.89. The decoration of these decagons is almost identical, although small variations can be observed. A more accurate view of the structure is shown as a density map of one of the layers (since the layer is slightly puckered along *c*, the electron density is projected). The Ni/Co atoms are displayed in black, whereas Al atoms are shown as dark grey, and contours indicate the occupation of the corresponding sites. The thick light-grey line indicates the HBS tiling or the

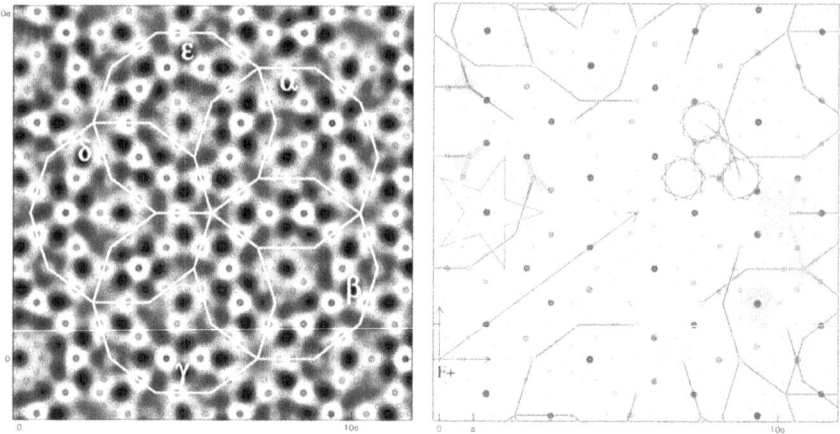

Figure 4.89 *Left: Low-resolution projection of the structure deduced from the five-dimensional analysis of Cervellino et al. Five Gummelt decagons have been superimposed, with a diameter of about 20 Å. Note that the decoration of this cluster changes slightly from one to another. Right: Electron density of one layer. The darker contours are for the Ni/Co atoms. The thick light-grey lines show the HBS tiling.*

interpenetrating Gummelt decagons. Again there is a kind of systematic decoration of these building blocks. For instance, hexagon tiles have only one Ni/Co atom in their inside, along the long diagonal. There are about six Ni/Co atoms on the edges of the hexagon. However, a careful examination of the figure indicates that there are some deviations from this decorating scheme.

4.5.4.2 HREM images

HREM images obtained by TEM are interesting in the case of a decagonal quasicrystal when looked down along the periodic axis, since some image interpretation is possible. One should not forget, however, that one is looking in this case to the projection of the structure, which is itself three-dimensional. Moreover, it is well known that conventional HREM pictures are not straightforward to interpret: the electron–matter interaction is strong, so that the kinematic approximation is no longer valid. Simulations thus have to be carried out in the framework of the dynamical theory of electron–atom interactions. There is some software available, using for instance the so-called multi-slice technique, which allows a simulation on a portion of decagonal phases. It is only when a careful set of various images taken with different focus and the corresponding images have been simulated that a one-to-one relation between the observed contrast (white or black) can be associated to an atomic column or a group of atoms.

Recently, a new technique called annular dark-field scanning transmission electron microscopy (ADF-STEM) has been introduced (Yan *et al.*, 1998; Abe *et al.*, 2004). Combining the scanning facility, to a measurement at high angle, this technique provides an image that is much easier to interpret and is directly related to the electron density projection of the structure. The resolution of this technique is now close to 1.5 Å, in other words, almost at the atomic level. Moreover, the differences in the observed contrast allow one to identify the chemical nature of the atoms, that is, Al versus Co/Ni in the case of the d-AlNiCo phase (this is why it is also referred to as the Z-contrast technique). Although it is not easy to carry out, this experimental technique thus provides unique information on the structure at the atomic level.

Figure 4.90a, b shows the diffraction pattern of the d-AlNiCo phase taken in parallel (a) and convergent geometry (b). The convergent beam diffraction pattern is particularly important for determining the space group, since this diffraction pattern allows one to distinguish between a centrosymmetric and a non-

Figure 4.90 *TEM study of the decagonal AlNiCo phase. (a) and (b) display the electron diffraction pattern and the convergent beam electron diffraction pattern, respectively. (c) is the HREM image obtained in phase contrast, and (d) is the HREM image obtained in Z-contrast. (Courtesy of E. Abe, Tokyo University)*

centrosymmetric space group. This is obtained by carefully examining the contrast of the diffraction images that are the result of dynamical diffraction (we remind the reader that, in the kinematical diffraction approximation, the diffraction pattern is necessarily centrosymmetrical due to the Friedel law). In the present case, the symmetry is ten-fold, in agreement with the centrosymmetrical space group $P10_5/mmc$.

The images in Fig. 4.90c, d correspond to phase and Z-contrast image, respectively (Abe et al., 2004). As already explained, the Z-contrast image is much easier to interpret. Typical decagonal 2 nm diameter clusters are easily distinguished on the images, with three of them being highlighted by white lines. An enlarged part of the Z-contrast image around one cluster is shown in Fig. 4.91. Remembering that the contrast is directly proportional to the electron density of each atom, the brightest dots correspond to the transition Ni/Co atoms. From this image analysis, Abe et al. proposed a decoration of the cluster which is shown on the right panel of the figure. As can be seen, there is a one-to-one correspondence with the Ni/Co atom positions in the model (shown in dark or light grey) and the observed image which is superimposed. In particular the central part of the cluster has a non-icosahedral symmetry, with an almost triangular shape highlighted by a thin line. Note, in particular, the double atom position visible on two edges of the triangle. Three subsets of the decoration are also highlighted by thick lines and light-grey shading. They correspond to the Gummelt type of overlap rules. In this scheme the entire quasicrystal structure can be described by a single covering cluster (called a quasi-unit cell in Steinhardt et al. (1998)), with overlap rules defined by the three shaded areas. Overlapping of two clusters is allowed only when patches are matching. Such a covering is shown on the bottom panel of the figure, superimposed on the original Z-contrast image. Such a description is appealing for several reasons. First, it gives a very compact description of the structure, with a kind of unit cell, corresponding to the decagonal cluster. Second, it also gives some arguments about the stabilizing mechanism of quasicrystals. Indeed, it has been shown that maximizing the number of decagonal clusters leads to a quasiperiodic structure, with a few defects. If one postulates that the cluster is a stable entity, then this would give a rather simple rule for a quasicrystal, which could be a ground state at 0 K.

Thus, we have in the decagonal AlNiCo case (for one specific chemical composition), two quite complete approaches to the atomic structure: one obtained from X-ray diffraction, and the other one obtained from HREM images. Both resulting models agree, as long as one does not look too closely into the details. An average description in terms of cluster coverings seems to apply. However, we have seen that X-ray scattering results lead to a fairly disordered Al subnetwork. Such a detailed analysis is far beyond the capacity of the Z-contrast images. One can notice, however, that a careful examination of the images shows that the quasi-unit cell is not strictly identical from one place to the other. The validity of a description of the long-range quasiperiodic order in terms of quasi-unit cell is thus highly questionable, since disorder also seems to be a key ingredient of this

Examples of quasicrystal structures 289

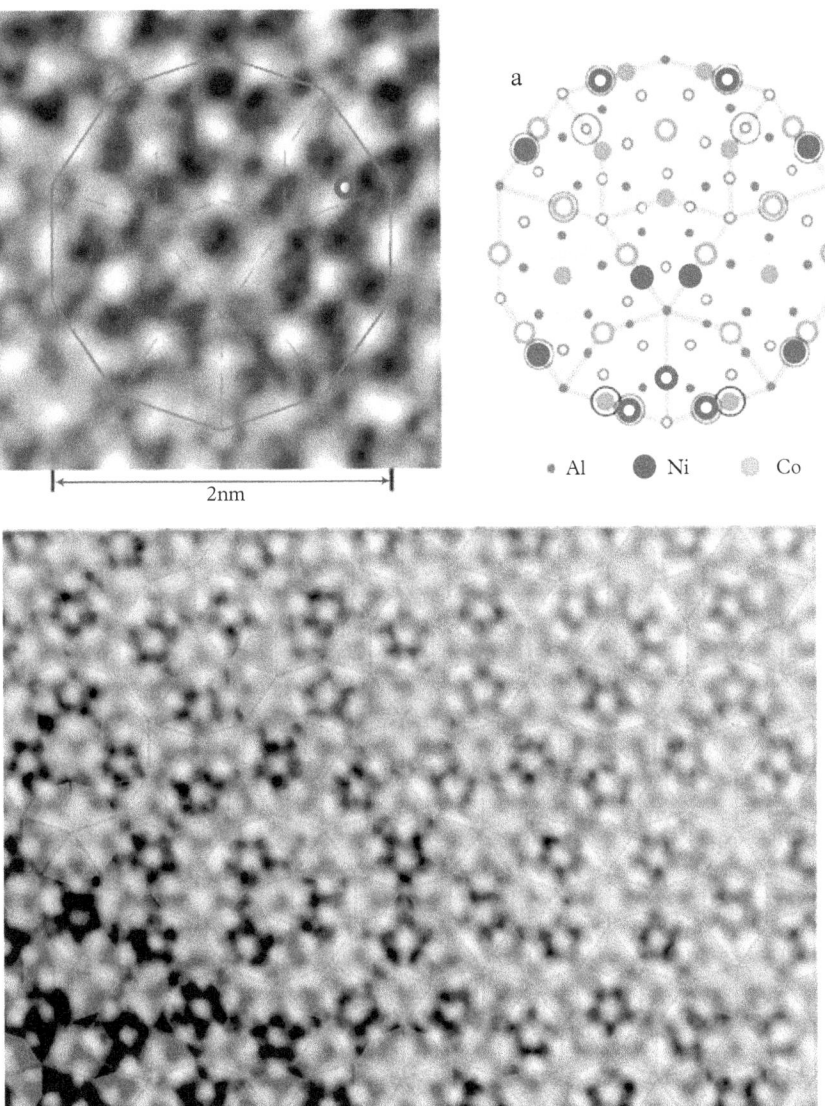

Figure 4.91 *Top: Enlarged part of the Z-contrast image showing the projection of one decagonal cluster. The right-hand panel corresponds to the atomic model proposed by E. Abe et al. There is a good correspondence with the image. The thick line highlights the 'Gummelt'-type decoration of the cluster. It is superimposed on a larger part of the image on the bottom panel as an illustration of the covering scheme. (Courtesy of E. Abe, Tokyo University)*

structure (note that attempts to fit X-ray data with the quasi-unit cell model failed to give a reasonable R factor). The quasi-unit cell model is thus more appropriate for a local description of the structure.

Finally, it is worth mentioning that the quasi-unit cell model is a rephrasing of matching rules. As such it is not a growing rule model, in other words, the overlapping rules do not allow one to grow a perfect quasicrystal at infinity. Starting from a seed, there is necessarily, some point where the overlapping rule cannot be fulfilled after a while. One can continue the growth of the quasicrystal only by using a position with a defect or matching rule violation. Moreover, the local character of the growth rule is questionable, since it generally requires the inspection of a large portion of the surface of the already grown quasicrystal. Nevertheless, using these matching rules it is possible to grow model quasicrystals with a limited set of defects, which might then be annealed. However, such a model should be considered more as a toy model rather than a realistic one, where electrons should play an important role.

To conclude this section on decagonal quasicrystals, we briefly mention their in situ temperature study (Steurer, 2005). These are crucial experiments to understand the stabilizing mechanisms of quasicrystals. As already mentioned, the phase diagram of the AlNiCo system is extremely complex, with superstructure phases showing up for different Ni/Co concentration and temperature ranges. Figure 4.92 (top panel) displays the diffraction pattern of the d-$Al_{70}Co_{12}Ni_{18}$ phase, with a vertical periodic axis, measured at room temperature and at 1120 K. There are diffuse layer located mid-way between Bragg layers, and visible at $h_5 = 1, 3, 5$. As can be seen in the figure, the diffuse scattering diminishes as the temperature is raised. The bottom panel shows the zeroth quasiperiodic Bragg layer of the diffraction pattern measured at room temperature and 1120 K. Various superstructure reflections labelled S1 or S2 are indicated in the figure. There are significant changes as the temperature is increased, both in the diffuse scattering and in the Bragg intensity distribution.

Thus, there is a quite complicated short-range ordering and superstructure ordering taking place in the sample. It is partly understood, and is the focus of current investigations. As a general rule, it is observed that the diffuse scattering diminishes as the temperature is raised, an indication of the importance of the entropic term in stabilizing the quasicrystal.

4.5.5 Dodecagonal quasicrystals

Dodecagonal quasicrystals are periodic in one direction and quasiperiodic with a twelve-fold symmetry in the plane orthogonal to it, and are thus layered quasicrystals. Recently, a new dodecagonal quasicrystal has been found in the Mn–Cr–Ni–Si system, with a five-dimensional space group $P12_6/mmc$ (Iwami and Ishimasa, 2015; Ishimasa *et al.*, 2015). Although its structural quality is not comparable with that for icosahedral quasicrystals, it constitutes a very interesting system for which a model can be inferred from a periodic approximant, and can be

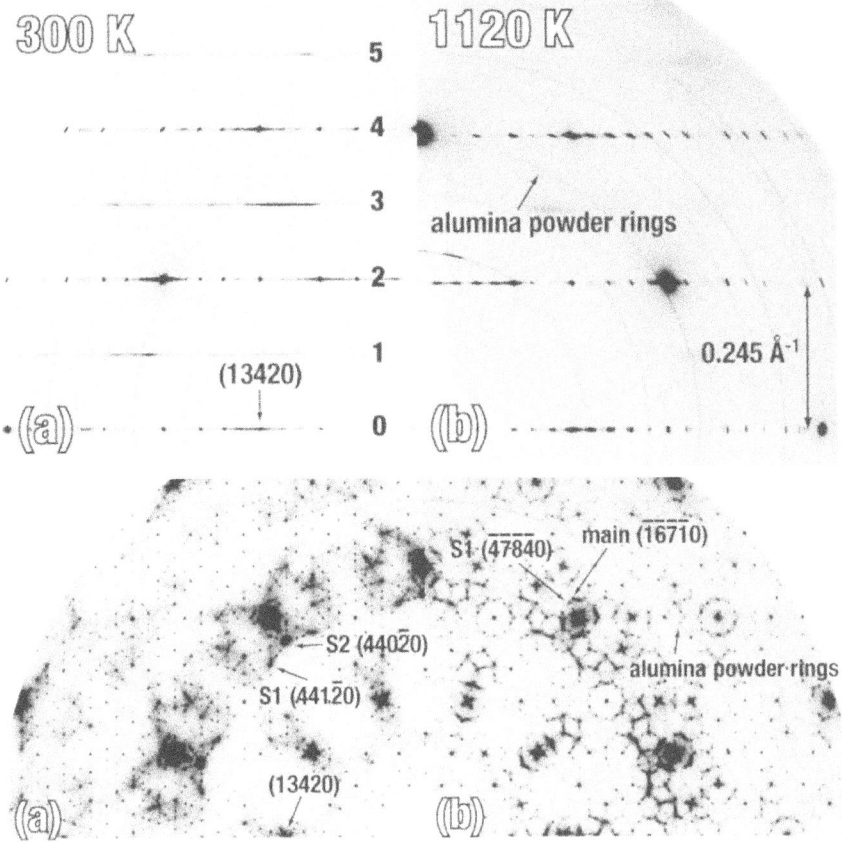

Figure 4.92 *In situ temperature study of the diffraction pattern of the d-AlNiCo phase. Top: Bragg layer containing the c^*-axis (vertical). Bottom: $h_5 = 0$ section of reciprocal space (quasiperiodic Bragg layer) taken at 300 K (left), and 1120 K (right). Note the large changes in the distribution of diffuse scattering. (Courtesy of W. Steurer, ETH Zürich)*

described with a decorated square and triangle tiling. Since this crystal is a layered quasicrystal, high-resolution TEM images can be interpreted and compared to the model. Figure 4.93 shows such a high-resolution TEM image. Dodecagonal clusters are clearly visible and form a quasiperiodic tiling, a model of which is shown in the right panel of Fig. 4.93. A detailed study of the tiling extracted from the high-resolution TEM images shows that there are small regions in the sample for which the quasicrystalline order is well preserved, whereas at a larger scale, phason strain and randomness occur.

A dodecagonal quasicrystal has also been found recently in an oxide quasicrystal. It is obtained as a $BaTiO_3$-derived layer deposited on a Pt(111) substrate. For

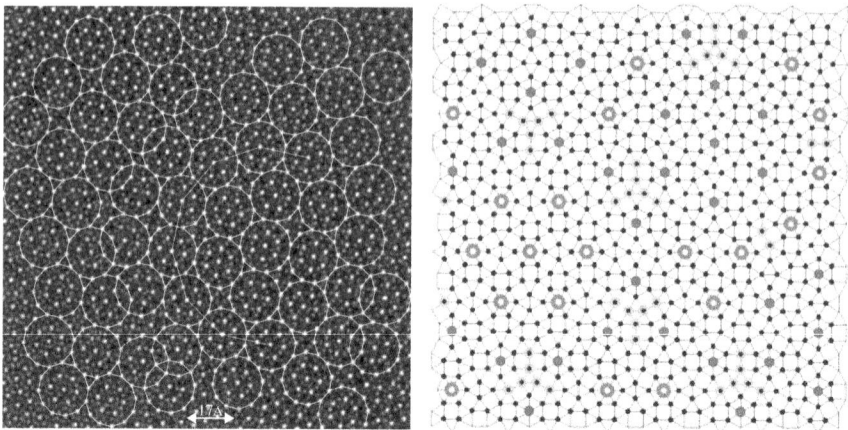

Figure 4.93 *Left panel: HREM TEM image of the Mn–Cr–Ni–Si dodecagonal phase. Large dodecagonal clusters are clearly visible. The right panel displays a possible dodecagonal tiling for the structure description. Its decoration can be inferred from the study of periodic approximant. (Courtesy of T. Ishimasa)*

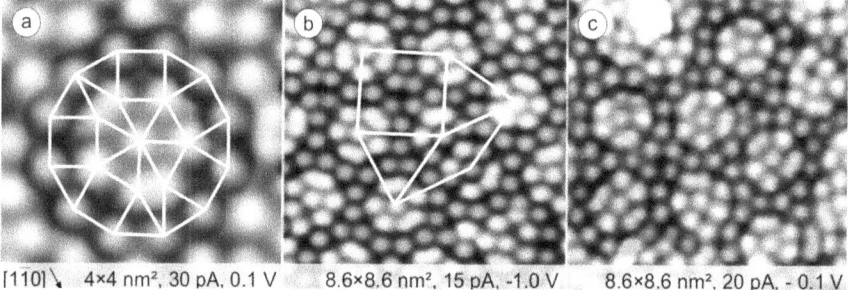

Figure 4.94 *STM images of the 2D oxide quasicrystal with atomically resolved Nüzeki–Gähler tiling. Fundamental lengths are 0.685 and 0.672 nm for the $BaTiO_3$- and the $SrTiO_3$-derived structures on Pt(111) in (a, b) and (c), respectively. Note the purely 2D character of the quasicrystal with a thickness of approximately 0.4 nm. (Courtesy of S. Foerster and W. Widdra)*

specific deposition conditions the first layer, with a thickness of approximately 0.4 nm, is a dodecagonal quasicrystal (Foerster *et al.*, 2013, 2016), whereas after several deposited layers the structure returns to the bulk perovskite $BaTiO_3$ one. The dodecagonal structure has been investigated using scanning tunnelling microscopy (STM). As illustrated in the Fig. 4.94, the structure can be described with square, triangle, and rhombus tiling with an edge length equal to approximately 0.7 nm. The detailed decoration and, in particular, the atom type distribution in this tiling are still under investigation, but inflation properties are clearly visible. The tiles group themselves to form large dodecagon clusters, which themselves form a tiling, as illustrated in Fig. 4.94. A similar structure has been found for a deposition of $SrTiO_3$. This is the first covalently bonded quasicrystal and opens thus new possibilities for physical property studies, since all known hard condensed matter quasicrystalline systems have been found in intermetallic compounds.

Dodecagonal quasicrystals are also found in soft condensed matter, where they form structures with length scales that are much larger than the one for intermetallic compounds and are presented in a special section (Section 7.6).

4.5.6 Reversible phase transitions

The occurrence of reversible phase transitions in aperiodic crystals is very similar to that in periodic crystals. Phase transitions can be first or second order, displacive or diffusive, related to phonon softening of a specific mode, etc. However, in aperiodic crystals there are new phason variables, so that transitions related to these phason modes are now possible. In the case of quasicrystals, this has been studied within the Landau theory, which gives some constraints on the possible transitions following the group-to-subgroup classification and possible phason-driven transitions, in particular, from a quasicrystal towards a periodic approximant (Ishii, 1989, 1990). See also Chapter 6, Sections 6.2–6.5.

Indeed, the reversible transition from a quasicrystal to a periodic approximant is particularly interesting, since this is a signature of the entropic stabilization mechanisms put forward within the random tiling scenario. The transition might be driven by soft-phason modes (Ishii, 1990, 1992; Widom, 1991), and is somewhat similar to a soft-phonon-mode-driven transition, although here phason modes are diffusive modes and thus associated with diffusive phase transitions.

Reversible phase transitions have been observed only in a few icosahedral and decagonal phases.

For icosahedral phases, the AlCuFe system has been studied thoroughly. This was the first example of a high-quality quasicrystal that could be obtained using standard growth techniques from the melt (Tsai *et al.*, 1987). The phase diagram of this phase is such that the i-AlCuFe single quasicrystal, which displays the shape of a dodecahedron, shows a transformation towards a rhombohedral phase with a lattice parameter equal to 3.2 nm and $\alpha = 36$, if the cooling from the high temperature (820 °C) is slow enough (Audier and Guyot, 1989; Audier *et al.*, 1990). In fact, the low-temperature rhombohedral phase appears with all the variance imposed by the transformation from the icosahedral phase to the rhombohedral one, with a rather small domain size of the order of 20 nm, so that even TEM diffraction patterns show a peak splitting resulting from the domain distribution. The low-temperature phase has been referred to as a microcrystalline state. A careful temperature, TEM, and X-ray study has shown that this phase transition is mediated by phason modes, leading to a transient modulated phase and a pentagonal phase (Menguy *et al.*, 1993*a,b*). In particular, a high-resolution X-ray study showed that the modulated phase is characterized by the appearance of 12 satellite reflections around each main Bragg peak, with a position **q** parallel to the five-fold axes. It can simply be interpreted as resulting from a sinusoidal phason phase with a propagation along a five-fold axis in the physical space and a polarization in the perpendicular (phason) space also along a five-fold axis (Menguy et al., 1993b). The phason wave is characterized by a modulation wavelength equal to approximately 20 nm, leading to satellite reflections very close to the main Bragg peak. In a first-order approximation, the first-order satellite intensity around a main Bragg peak can be computed in a way similar to what is done for a displacive modulated phase and writes $I_{sat} = 2\pi^2 (\mathbf{U}_{perp}.\mathbf{H}_{perp})^2 I_B$, where $\mathbf{U}_{perp}$ is the phason wave polarization vector oriented along a five-fold axis in the perpendicular space, $\mathbf{H}_{perp}$ the perpendicular component of the main Bragg peak, and I_B the main Bragg peak intensity. For orientations where $\mathbf{U}_{perp}$ and $\mathbf{H}_{perp}$ are almost orthogonal, the satellite intensity vanishes. Indeed, this is what is observed experimentally, as is shown in Fig. 4.95, for instance for the two-fold reflection. The intensity variation of the satellites also follows the above expression, and the phason wave amplitude has been extracted using more than 10 Bragg peaks and their 12 associated satellite reflections, from which the amplitude U_{perp} is found to be equal to about 0.04, the radius of the large atomic surfaces located on the nodes of the 6D lattice (the structure is similar to the one of the AlPdMn icosahedral quasicrystal). The modulated phase is not a stable phase, but rather a transient state, which is followed by a pentagonal phase, which is then followed by the low-temperature rhombohedral phase.

The second icosahedral quasicrystal system for which much temperature studies have been carried out is the i-AlPdMn one. As explained previously, large single-grain quasicrystals can be grown from the melt. However, the temperature evolution of the quasicrystalline phase strongly depends on the exact composition. The quasicrystalline phase is stable down to room temperature when the composition is $Al_{68.8}Pd_{22.0}Mn_{9.2}$ whereas a reversible transition takes place for the composition $Al_{69.3}Pd_{22.0}Mn_{8.7}$, i.e. a slight Al–Mn exchange. Above T_C equal to 710 °C, the structure is an icosahedral quasicrystal. A reversible transition takes place at 710 °C, where the icosahedral phase transforms in the so-called F2M phase (Ishimasa, 1995; de Boissieu et al., 1998a). The diffraction pattern of the F2M phase is extremely complex and rich, with two sets of supplementary satellite reflections called S1 and SF2. The two wave vectors $\mathbf{q}_{S1}$ and $\mathbf{q}_{SF2}$ have directions parallel to a three-fold icosahedral axis. There are thus 20 satellites associated with a main Bragg peak and with each satellite order; $\mathbf{q}_{S1}$ and $\mathbf{q}_{SF2}$ can, in fact, be expressed as the parallel components of a rational value of 6D vectors, so that the F2M structure is a superstructure of the high-temperature icosahedral phase (Ishimasa, 1995; de Boissieu et al., 1998a). Figure 4.96 shows the diffraction pattern around a strong two-fold reflection (with N/M indices equal to 20/32). Whereas only diffuse scattering is observed above T_C, a large number of satellite reflections are visible in the F2M phase. First- and second-order satellites of the S1 and SF2 types are observed, together with linear combinations, as shown in the figure (de Boissieu et al., 1998a). It has not been possible yet to derive a model for this complicated F2M phase. However, the S1 satellites are also related to a phason

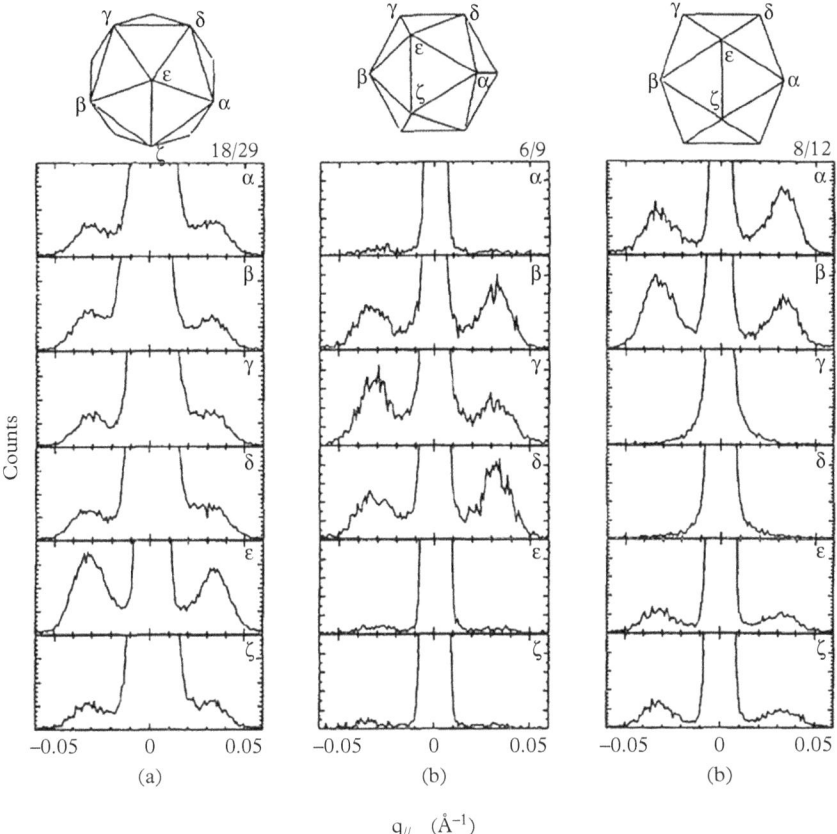

Figure 4.95 *Diffraction pattern of the modulated AlCuFe phase around different Bragg reflections. The six pairs of satellite intensity distribution are shown around a five-fold (18/29), three-fold (6/9), and two-fold (20/32) main Bragg reflections. The relative orientations of the main Bragg reflection and the satellite are shown on the upper icosahedron. There is strong intensity variation related to the relative orientations of U_{perp} and H_{perp}. In particular, for the two-fold reflections, some satellite intensity vanishes.*

wave, in a way very similar to the case of the i-AlCuFe modulated phase. A detailed TEM study showed that the structure is in fact a multi-domain one, with only one of the 20 three-fold axis satellites visible in each domain (Audier *et al.*, 1999). The structure remains quasiperiodic but with a cubic symmetry. Possible symmetry groups have been derived from these TEM studies. This F2M phase remains one of the most complicated phases, and structural determination for it remains a very difficult task. Finally, if the cooling from the high-temperature phase is rapid, only the SF2 satellites appear, as the S1 satellites are replaced by a large amount of diffuse scattering (Ishimasa, 1995; Létoublon *et al.*, 2000*b*).

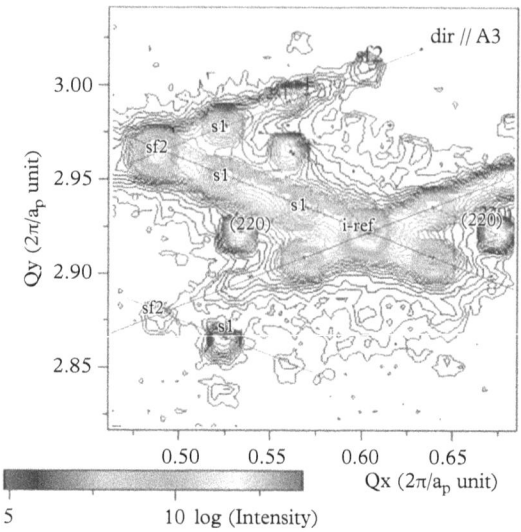

Figure 4.96 *Diffraction pattern of the F2M phase measured around the 20/32 two-fold reflection, labelled i-ref. It is the only Bragg peak that survives in the high-temperature icosahedral phase. The supplementary satellite reflections, characteristic of the F2M phase, are labelled as s1 and sf2.*

We have seen, in this section, that the high-dimensional approach is a very efficient tool which allows one to refine a structure in a way very similar to what is done for three-dimensional crystallography of periodic crystals. For a few phases (i-AlPdMn, i-CdYb, d-AlNiCo), accurate models are available, which, in turn, should allow a detailed analysis of their physical properties. The situation is, however, more complex than for incommensurately modulated phases or composites, for there is no standard procedure to solve a structure. We have seen in the three examples that there is no unique solution for designing a six- or five-dimensional model. It is thus important to incorporate as much as possible the crystallochemistry of these compounds. This can be achieved, for instance, by studying approximant crystals. Another possibility is to carry out total energy calculations (Mihalkovic *et al.*, 2002), which might provide a guide for a structural solution. Finally, we have not discussed random tiling model in this section. It will be dealt with in the section on phasons (Section 5.9).

4.6 Diffraction by an imperfect crystal

The previous sections dealt with the *ideal* perfect crystal, periodic or aperiodic. In both cases there is a perfect long-range order with a unit cell (in three or more dimensions) which repeats identically at infinity. In real structures there are most of the time defects (point defects, dislocations, short-range order, local lattice distortions, thermal vibrations, etc.) which perturb this long-range order. When the defect density is not large, the description in terms of the ideal model is still a good approximation. However, there are cases where the deviation from strict long-range order is such that the contribution from disorder has to be taken into account (Welberry, 2004; Keen and Goodwin, 2015).

Generally, defects are grouped in two different categories. This distinction is related to their effect on the diffraction pattern. The observed diffracted intensity is given by the Fourier transform of the Patterson function, which is the autocorrelation function of the real space structure (Section 4.1.1). If the Patterson function is described by a periodic function, we recover the result of the ideal crystal. If the distorted structure can be described by an average periodic function plus a deviation $\Delta\rho(R)$, where the deviation is such that its autocorrelation function remains bounded as R goes to infinity, then its diffraction pattern contains two parts: a Bragg part and a diffuse part. If the deviation is such that its autocorrelation diverges as R goes to infinity, then the long-range order is lost and the diffraction pattern no longer shows Bragg reflections but rather broadened reflections. Figure 4.97 (from Guinier) exemplifies these two kinds of defects in a simple one-dimensional example. The top part shows atoms displaced randomly around their equilibrium positions: in this way an average lattice can be defined. The bottom part shows a structure where atoms are added one after the other, with an interatomic distance which fluctuates around a. As can be seen, the deviation from the average lattice increases as R goes to infinity: it is not possible to define an average lattice.

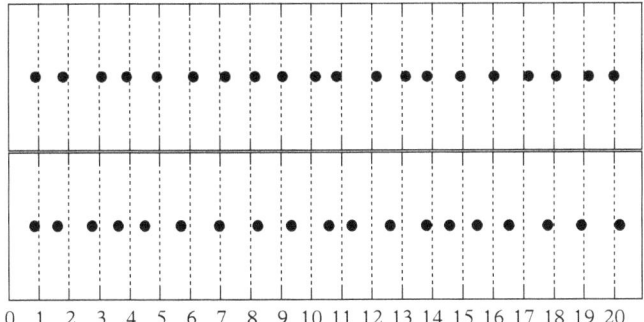

Figure 4.97 *Illustration of the two kinds of disorder in a one-dimensional system. The top panel shows fluctuations around an average structure. The bottom panel exemplifies a situation where the fluctuations diverge.*

4.6.1 Diffuse scattering when an average lattice exists

In the case where an average lattice exists, the diffraction pattern contains both a Bragg part and a diffuse part. If the fluctuations are such that their mean is zero, the intensity reads:

$$I(\mathbf{K}) = \int exp(2\pi i P_{av}(\mathbf{R})\mathbf{K})\, d\mathbf{R} + \left\langle \int (exp(2\pi i (\Delta\rho(\mathbf{R}) * \Delta\rho(\mathbf{R}))\mathbf{K})\, d\mathbf{R} \right\rangle_{r,t}, \quad (4.57)$$

where $P_{av}(R)$ is the Patterson function of the average structure, and $\Delta\rho(R)$ is the difference density. The first term in Eq. 4.57 is the Bragg component of the diffraction pattern, whereas the second term is the diffuse component. The diffuse component contains a statistical average over space and time. The average over time has to be carried out for fluctuations which are much faster than the measuring time. This is the case, for instance, for thermal vibrations which are much faster than the X-ray measurements. The spatial average comes from the X-ray or neutron beam coherency being much smaller than the sample size interacting with the radiation. (Here, the coherency refers to the longitudinal and transverse coherency of the beam defined in the same way as for visible light and laser radiation, for instance.) It means that the measured intensity is the *incoherent* summation of the contribution of smaller domains whose size is equal to the coherence length of the incoming radiation. In X-ray laboratory equipment, for instance, the coherence length is smaller than 1 μm, and sample sizes are of the order 1 mm, so that the above average is really a statistical average. This is no longer the case with third-generation X-ray sources where the coherence length can reach 0.1 mm in some cases. If the beam is fully coherent (i.e. acts as a laser), then the diffuse intensity is no longer a smooth curve, but presents strong intensity fluctuations called speckles.

Since the fluctuations are bounded, and do not show many correlations, the diffuse scattering part is, in general, a slowly varying function in reciprocal space. It is important to note that it is this function which carries the information on the short-range correlation occurring in the disordered structure. More precisely, the diffuse scattering intensity is related to the two-body correlations, whereas the Bragg component only contains the information on the one-body correlation. This is why measuring the diffuse scattering is so important.

A complete treatment of the diffuse scattering in periodic and aperiodic systems is beyond the scope of this book. We only give here some general principles and their application to the case of aperiodic crystals.

The function $\Delta\rho(R)$ contains two components: the first describes the chemical disorder (an atom A may be replaced by an atom B) and may be expressed as a density fluctuation, and the second describes the displacement field. The diffuse scattering is related to the spatial evolution of these two terms and to their autocorrelation function.

A classical example of chemical disorder is the one occurring in metallic alloys such as β-brass CuZn. The Cu and Zn atoms occupy at random the sites of a body-centred cubic lattice at high temperature, whereas they are ordered at low temperature, with Zn atoms at the corner and Cu atoms on the centre of the cube. When approaching the transition temperature from above, a short-range order appears, which is similar to the low-temperature ordered structure, but only on a limited length scale. This shows up in the diffraction pattern as a broad diffuse scattering intensity, whose shape and intensity is related to the Fourier transform of the so-called Warren–Cowley short-range order parameters. This diffuse scattering intensity is proportional to $\langle f^2 \rangle - \langle f \rangle^2$ and sharpens as the short-range order increases. In the case of second-order phase transitions, such as the one observed in the CuZn alloy, it is possible to describe the transition in a mean-field approach. In this case, the shape of the diffuse scattering is a Lorentzian. The peak maximum and the correlation width (inverse of the Lorentzian width) diverge as T tends to T_C, with a power law characteristic of the universality class of the Ising model. This is a typical example, where there is a direct connection between the observed diffuse scattering and the physics of the system.

A displacement field gives rise to a large variety of diffuse scattering phenomena. Generally, if displacements are small, a Taylor expansion of the exponential term in Eq. 4.57 is performed and leads to first-order and second-order terms.

The exact calculation of the diffuse scattering requires knowledge of the displacement field: it can be the displacement field related to equilibrium phonons in the system, the displacement field around a defect (point defect, precipitate, dislocation), a displacement field related to chemical disorder (so-called size effect), a displacement field related to the orientation of some rigid unit (octahedron in perovskites), etc. Two approaches are used to calculate the diffuse scattering: a calculation in direct space, where large disordered models are constructed, and the diffuse scattering directly computed by a Fourier transform of the model or a calculation in Fourier space, where the displacement field is decomposed into its Fourier components $u(q)$ (this approach is sometimes named the modulation approach).

In this latter approach, the diffuse scattering intensity is calculated as $S(\mathbf{H}+\mathbf{q})$, where $\mathbf{H}$ is a reciprocal vector of the average lattice and $\mathbf{q}$ a wave vector of the modulation describing the displacement field. For instance, in the case of thermal vibrations, $\mathbf{q}$ is the wave vector of a phonon mode. The advantage of this approach is that it allows us to connect the observed diffuse scattering intensity distribution and the corresponding physical parameters, especially in the framework of the hydrodynamic theory of solid state. It also allows one to include important constraints imposed by symmetry. This is very similar to the case of incommensurately modulated structure, where the displacement field of the modulation is restricted by the symmetry of the average structure.

If one combines both a density fluctuation and a displacement field, the diffuse scattering intensity can be split into three parts corresponding to the zero-, first-, and second-order terms in the Taylor expansion as

$$I_{diff}(\mathbf{H}+\mathbf{q}) = I_{chem} + I_{size} + I_{TDS}. \tag{4.58}$$

The first term is related to chemical disorder, the second term is the so-called size effect, which is a coupling between chemical disorder and the distortion field, and the third term corresponds to thermal diffuse scattering (TDS), or Huang scattering. It is useful to give some general trends of the intensity distribution and the wave vector dependency of the three terms.

The first term, I_{chem}, is proportional to $\langle f^2 \rangle - \langle f \rangle^2$, the Laue diffusion. Chemical short-range order results in a modulation of the intensity around the Laue term. It has the same distribution in all Brillouin zones, and presents a decay due to the fall-off of the atomic form factor in the case of X-rays (for neutrons the scattering length is K independent) and to the Debye–Waller term.

The second term I_{size} is proportional to $f \Delta f K$, where $\Delta f = f_A - f_B$. The Fourier transform of the displacement field gives rise to $\sin(\mathbf{H}+\mathbf{q})$ terms, which result in an asymmetry of the diffuse scattering at position $\mathbf{H}+\mathbf{q}$ and $\mathbf{H}-\mathbf{q}$ around the Bragg peaks of the average structures, some intensity being 'transferred' from one side to the other.

The third term is proportional to $f^2 K^2$. In the case of TDS, the diffuse scattering intensity decays as $1/\omega^2$ or $1/q^2$ in the acoustic regime, where ω is the frequency of the phonon mode with wave vector q. It thus peaks around the Bragg reflections of the average lattice. Moreover, the intensity is proportional to $(\mathbf{u}.\mathbf{K})^2$, where $\mathbf{u}$ is the polarization of the phonon mode with wave vector $\mathbf{q}$: this gives rise to extinction of the diffuse scattering signal when $\mathbf{u}$ and $\mathbf{K}$ are orthogonal.

One sees that the K dependence of the diffuse scattering is quite different for the three contributions and thus gives a means of distinguishing them. For instance, the TDS is growing as K^2, and is thus best measured at high values of K, whereas the chemical diffuse scattering is independent of K. The TDS is also stronger close to the Bragg reflection since it goes as $1/q^2$. The diffuse scattering is nevertheless weak and requires a special care to be measured properly. A schematic presentation of the distribution of diffuse scattering is shown in Fig. 4.98.

The above formulation is also valid for aperiodic crystals. Real aperiodic crystals also contain defects which give rise to diffuse scattering. There are, moreover, fluctuations that are characteristic of the aperiodic long-range order. These are known as phason modes (see Section 5.9). These long-wavelength fluctuations give rise to a diffuse scattering similar to the TDS. Its shape and intensity distribution can be calculated in the framework of the hydrodynamic theory, and is also related to the perpendicular component H_{perp} of the Bragg peak position of the average aperiodic structure (where average here means average with respect to the defects).

Diffuse scattering measurements are also important when studying phase transitions and their precursors. Most incommensurate phases are only stable in a limited temperature range, between a periodic crystal with high symmetry and another one with lower symmetry at low temperature. For instance, incommensurate displacive modulated phases are frequently the result of the condensation

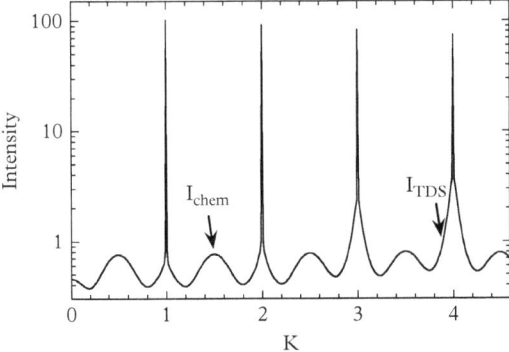

Figure 4.98 *Schematic illustration of the various contributions to the diffuse scattering; I_{chem} indicates the contribution from chemical short-range order, whereas I_{TDS} indicates the contribution from TDS. Note that the TDS contribution increases as K increases.*

of a soft-phonon mode. Of course, the best technique to study a phonon mode is by means of inelastic neutron or X-ray scattering. However, a soft-phonon mode is also clearly visible in the diffuse X-ray or neutron scattering pattern, since the intensity of the diffuse signal goes as $1/\omega^2$. These measurements are usually easier to carry out and can be realized on much smaller single-grain sizes than the ones required for inelastic neutron scattering.

Understanding the diffuse scattering in the high-temperature phase is also important in understanding the mechanisms stabilizing the incommensurate phases. Figure 4.99 shows such an example in the case of the high-temperature phase of cristobalite. The diffuse scattering is clearly seen in the transmission electron diffraction pattern. It shows up as streaks, which are in fact sections of diffuse planes. This indicates that there is correlation in only one direction. The model of the disordered structure is shown in the right panel. There is a cooperative rotation of the SiO_4 tetrahedra which are connected by sharing corners. This explains why the diffuse scattering is especially strong perpendicularly to the [110] direction.

4.6.2 Diffuse scattering when there is no average lattice

When a mean (periodic or aperiodic) lattice cannot be defined, the diffraction pattern no longer presents Bragg reflections. This is the case, for instance, for liquid or amorphous materials, where the diffraction pattern presents only broad maxima. There are cases, however, where the diffraction pattern is closer to that of an ideal crystal. This is the case of the so-called paracrystal depicted for the one-dimensional case in Fig. 4.57. It is constructed by considering atoms located with a mean interatomic distance a which randomly fluctuates, and that there

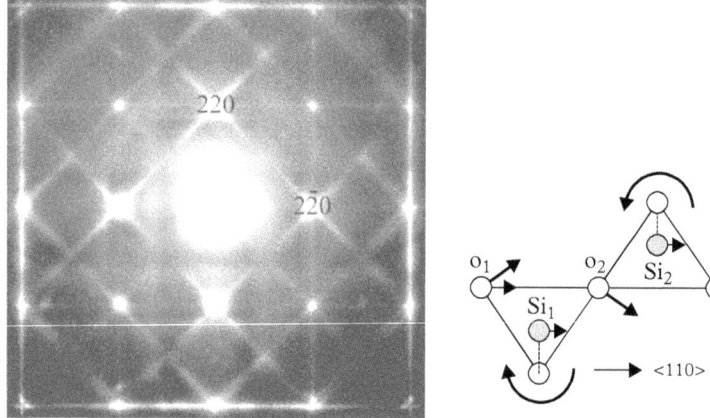

Figure 4.99 *Transmission electron diffraction pattern of the high-temperature phase of cristobalite. The streaks of diffuse scattering result from a cooperative rotation of the rigid sharing corner SiO$_4$ tetrahedra, as shown on the right panel. (From R. Whithers)*

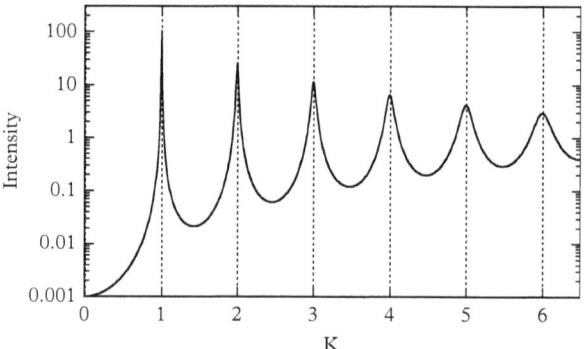

Figure 4.100 *Diffraction pattern of a one-dimensional paracrystal; K is labelled in units of 1/a where a is the average interatomic distance.*

is no correlation with the next nearest neighbour. Its diffraction pattern can be calculated analytically and is shown in Fig. 4.100.

As can be seen, it shows maxima for values of K equal to n/a where a is the average interatomic distance. However, the peaks are not delta functions but are broad peaks whose half-width at half-maximum increases as the square of K.

This example is also interesting since it corresponds to the van Hove theorem which states that long-range order cannot exist in a one-dimensional crystal with finite-range interactions. Phonon fluctuations destroy the long-range order as soon as T is non-zero. The diffraction pattern of this one-dimensional system is similar to the case of a paracrystal.

Such one-dimensional diffraction patterns have been observed, for instance, in Hg chains or in the quasi-one-dimensional conductor TTF-TCNQ. Most of the time, however, there are interactions between the chains, so that the system is not really a pure one-dimensional one. This leads to incommensurate structures at low temperature.

A similar example occurs for alkane–urea compounds. At high temperature the alkane subsystem is disordered and gives rise to diffusive layers in the diffraction with a spacing that is incommensurate with the Fourier module for the ordered subsystem. The intensity distribution across the diffuse plane is very similar to the one of Fig. 4.100 (Mariette *et al.*, 2015).

In the case of two-dimensional structures, phonons also destroy the long-range order, although in this case the Bragg peaks are extremely peaked, with a logarithmic divergence.

Another typical example where long-range order is partially destroyed is the case where there is a distribution of strain in a crystal. This can be the result of a plastic deformation of the sample and/or a distribution of dislocations. Calculating the complete diffraction pattern in this case is not straightforward, but a strain distribution always leads to a Bragg peak broadening, with an HWHM (expressed in reciprocal space units) which is most of the time linear in K. The defect distribution also leads to a characteristic decay of the intensity close to the Bragg peak maxima.

The same also applies to aperiodic crystals. Here, however, one has to take into account the effect of phason modes on the perpendicular component of the broadened Bragg reflections. For two-dimensional random tiling models, for which phason modes destroy the perfect long-range order, with diffraction peaks no longer being delta functions. For three-dimensional random tilings, however, phason fluctuations do not destroy long-range order.

There are also defects (vacancies, discommensurations, dislocations, etc.) in aperiodic crystals. The case of dislocations is again more complex since both a phonon and a phason distortion field appear around the core of the dislocation. The phason strain field is related to a change of the slope of the physical space. The phason component leads to a broadening which is proportional to H_{perp}, the perpendicular component of the reciprocal lattice vector. This will also be the case for a distribution of phason strain. Experimentally, most of the quasicrystals, even the best ones, have Bragg peak widths which depend linearly on H_{perp}.

Finally, we also mention the icosahedral glass model, which is a packing of icosahedral clusters, with a linkage along a particular axis (for instance, a three-fold axis). The clusters are packed randomly. The resulting structure has a diffraction pattern with icosahedral symmetry, but with broadened Bragg peaks for which the width increases as H_{perp}^2 (in some sense similar to the case of the paracrystal).

In conclusion, the diffuse scattering component of the diffraction pattern gives key information for the understanding of the detailed structure of aperiodic crystals. Its physical origin can be the same as that for periodic crystals, but a supplementary contribution comes from the phason fluctuations in the structure, either long-wavelength phason modes or some phason strain distribution.

5
Physical properties

5.1 Introduction

The study of physical properties of conventional periodic crystals is strongly based on the three-dimensional lattice periodicity. Physical properties of aperiodic crystals are more difficult to tackle theoretically because of the lack of this three-dimensional periodicity. For the description of the structure there is a periodic representation in a higher-dimensional space, but, generally, this scheme is difficult to use for physical properties because a mapping between interatomic forces and the high-dimensional representation is not straightforward. The calculation of physical properties has thus to be carried out in the three-dimensional physical space, where the Bloch theorem does not apply any more: results on electronic properties or lattice dynamics are thus difficult to obtain and are analytically tractable mainly in one-dimensional cases. In three-dimensional aperiodic systems, the main approach to studying physical properties is by simulations on large periodic approximants to the aperiodic structure. Looking at the asymptotic evolution of these simulations as a function of the system size, one may then extrapolate to the aperiodic case.

An obvious question one may ask is whether the aperiodic long-range order leads to specific physical properties. We shall show in the following that electronic states and vibrational spectrum do indeed show a characteristic behaviour, although its experimental verification is sometimes difficult to carry out. There are two phenomena which are strictly connected to the aperiodicity. The first is restricted to conductive materials. It is the pseudo-gap in the electronic density of states. The other plays a role in all aperiodic crystals, and is related to the fact that these systems may be embedded in a superspace. The existence of the superspace does not increase the number of degrees of freedom. This remains three times the number of particles. But in the dynamics certain motions are more conveniently described as motions in superspace. Such excitations have been named phasons. A simple example is a shift of the phase of the modulation in an incommensurate modulated phase. This affects the positions of all the particles in the crystal that can be described as a displacement in physical space but more conveniently as a displacement of the atomic surface in internal space. An eigenmode of the system of this type is called a phason, because it is related to a shift of the phase. When the atomic surfaces are smooth and unbounded the displacements go to zero if the amplitude goes to zero. Such an excitation can be studied in the framework of the long-wavelength or hydrodynamic elasticity theory of aperiodic crystals. When the atomic surfaces have discontinuities, as in composites with a discontinuous modulation function, or in quasicrystals, an infinitesimal shift of the atomic surface may lead to finite jumps, as already mentioned in Chapter 3. These jumps are also called *phasons*, or better *phason jumps*, because they are connected to shifts in internal space as well (Coddens et al., 1999; Edagawa et al., 2000). It is an interesting question whether these jumps may occur collectively, as a mode, correlated or only independently. In this chapter we shall discuss, besides other physical properties, the dynamics of these types of phasons. Notice that the term 'phason' is also used for static properties. We have seen that an approximant of an aperiodic crystal may be obtained by a strain in the higher-dimensional structure. This strain is called '*phason strain*', although no dynamics is involved, contrary to what is expected for a word like 'phason' (cf. phonon, photon, exciton, etc.). In the field of aperiodic crystals the word has obtained the meaning of a degree of freedom related to internal space. Sometimes this could lead to confusion.

5.2 Tensorial properties

We start with a reminder of old but perhaps not too well-known facts. Physical properties are often described by tensors. A *tensor* is a function of vectors and reciprocal vectors that is linear in each of its arguments. The arguments are n_1 vectors and n_2 reciprocal vectors. For example, the scalar product of two vectors $\mathbf{r}_1$ and $\mathbf{r}_2$ is $\mathbf{r}_1.\mathbf{r}_2 = g(\mathbf{r}_1, \mathbf{r}_2)$, which is linear in each argument. If one writes the vectors as linear combinations of the basis vectors, $\mathbf{r}_1 = \sum_i \xi_i \mathbf{a}_i$ and $\mathbf{r}_2 = \sum_j \eta_j \mathbf{a}_j$, one obtains

$$g(\mathbf{r}_1, \mathbf{r}_2) = \mathbf{r}_1.\mathbf{r}_2 = \sum_{ij} \xi_i \eta_j \mathbf{a}_i.\mathbf{a}_j = \sum_{ij} \xi_i \eta_j g(\mathbf{a}_i, \mathbf{a}_j) = \sum_{ij} \xi_i \eta_j g_{ij}.$$

The tensor g is a bilinear function in its arguments, and the tensor elements are $g_{ij} = g(\mathbf{a}_i.\mathbf{a}_j)$. For an orthonormal basis one has $g_{ij} = \delta_{ij}$. Because $\mathbf{r}_1.\mathbf{r}_2 = \mathbf{r}_2.\mathbf{r}_1$, one has the relation $g_{ij} = g_{ji}$: the tensor is symmetric. This *metric tensor* is a symmetric tensor of rank 2, the *rank* being the number of arguments of g. A tensor of rank 0 is a scalar, a vector is a *covariant tensor* of rank 1, and a reciprocal vector is a *contravariant tensor* of rank 1. If both n_1 and n_2 are non-zero, the tensor is mixed.

Another example is the *dielectric tensor*. $D = \epsilon E$ or $D_i = \sum_j \epsilon_{ij} E_j$. The scalar product $D.E = \sum_i E_i \epsilon_{ij} E_j = \epsilon(E, E)$ is the energy, and ϵ is a symmetric tensor of rank 2. The scalar product $\mathbf{k}.\mathbf{r}$ of a vector $\mathbf{r}$ and a reciprocal vector $\mathbf{k}$ is a mixed tensor of rank 2.

A basis transformation S such that $\mathbf{a}'_i = \sum_j S_{ji}\mathbf{a}_j$ leads to a transformation of the tensor T. If T is a covariant tensor of rank n, then

$$T'_{i_1\ldots i_n} = T(\mathbf{a}'_{i_1},\ldots,\mathbf{a}'_{i_n}) = \sum_{j_1\ldots j_n} S_{j_1 i_1}\ldots S_{j_n i_n} T_{j_1\ldots j_n}.$$

Vectors transform under a point group K according to a representation of K. For $S\mathbf{a}_i = \sum_j D(S)_{ji}\mathbf{a}_j$ (S in the point group K) the matrices $D(S)$ satisfy $D(S_1)D(S_2) = D(S_1 S_2)$. This representation is called the vector representation of K. In general, the vector space may be decomposed into subspaces such that each vector in such a subspace transforms with an irreducible representation of K. For example, if K is the three-dimensional point group 4mm generated by a 90° rotation along the z-axis and a reflection from a mirror through the z-axis, the axis is left invariant, but also the plane perpendicular to the axis. An arbitrary vector is the sum of components transforming with irreducible representations. (In our example each vector is the sum of a vector along the rotation axis and a vector in the invariant plane perpendicular to the axis.) Then a tensor transforms according to a (symmetrized, anti-symmetrized, or unsymmetrized) power of the vector representation. An invariant tensor transforms with the trivial representation, where each element of the group is represented by '1'. This gives the possibility of determining the general form of an invariant tensor.

In the example of the group 4mm the vector representation has two irreducible components: the one-dimensional trivial representation and a two-dimensional representation. Its character is given by $\chi_v(R) = \text{Tr}(R)$ for any element R of K. A general tensor of rank 2 transforms with a representation with character $\chi(R) = \chi_v(R)^2$, the symmetric rank 2 tensor has character $\chi_v^{2+}(R) = \frac{1}{2}(\chi_v(R)^2 + \chi_v(R^2))$, and the anti-symmetric tensor has character $\chi^{2-}(R) = \frac{1}{2}(\chi_v(R)^2 - \chi_v(R^2))$. These characters may then be decomposed into irreducible characters. The number of times the trivial representation occurs is given by the expression

$$n_1 = \frac{1}{N}\sum_{R\in K} \chi(R). \tag{5.1}$$

Here N is the order of the group K. The decomposition of the symmetrized and anti-symmetrized tensors from the example with point group 4mm are

$$\chi^{2+} = 2\chi_1 + \chi_3 + \chi_4 + \chi_5, \quad \chi^{2-} = \chi_2 + \chi_5,$$

where χ_j ($j=1,\ldots,5$) are the five irreducible characters (see Table 5.1).

Here χ_v is the character of the *vector representation*, $\chi_v^{2\pm}$ that of the (anti-)symmetric rank 2 tensor, and $(\chi_v^{2+})^{2+} = \chi_{elast}$ that of the elasticity tensor of rank 4. This implies that an arbitrary symmetric rank 2 tensor for symmetry 4 has two independent elements, and the anti-symmetric rank 2 tensor none, with the multiplicities of χ_1 in the decomposition.

Physical properties

Table 5.1 *Irreducible character representations of group 4mm*

	E	A, A³	A²	B, A²B	AB, A³B
χ_1	1	1	1	1	1
χ_2	1	1	1	−1	−1
χ_3	1	−1	1	1	−1
χ_4	1	−1	1	−1	1
χ_5	2	0	−2	0	0
χ_v	3	1	−1	1	1
χ_v^{2+}	6	0	2	2	2
χ_v^{2-}	3	−1	3	3	3
$(\chi_v^{2+})^{2+}$	21	1	5	5	5

The elasticity tensor is a rank 4 tensor which satisfies the symmetry relations

$$T_{ijkl} = T_{jikl} = T_{ijlk} = T_{klij}.$$

For the point group 4mm the corresponding character may be written in terms of irreducible characters as

$$\chi_{el} = 6\chi_1 + \chi_2 + 3\chi_3 + 3\chi_4 + 4\chi_5.$$

Therefore, there are six independent elastic constants in this case.

This is all well known for periodic crystals. What changes are needed in the formulation for quasiperiodic crystals? In the first place, the point groups are no longer limited to the 32 three-dimensional crystallographic point groups. For quasiperiodic crystals the point group may be any three-dimensional point group. The way to determine the invariant tensors remains the same. Let us take the examples of the three-dimensional cyclic point group 5 and the three-dimensional icosahedral group 532. The characters of the (physical) irreducible representations are shown in Table 5.2.

For the point group 5, the last three representations (below the horizontal line) contain the trivial representation one, two, and seven times, as follows from Eq. 5.1. For the point group 532 these numbers are 0, 1, and 5. That means that there is one invariant vector under the point group 5, the invariant symmetric tensor of rank 2 has two independent parameters, and the invariant elasticity tensor seven. For the icosahedral point group there is no invariant vector, the symmetric rank 2 tensor has one parameter, and there are five elastic constants.

For a quasiperiodic surface with N-fold symmetry the symmetry group is the rotation group N. In this case, what is the number of independent parameters of a tensor of rank n? As discussed before, it is the number of times the trivial

Table 5.2 *Characters of the point groups 5 and 532*

	Point group 5						Point group 532				
	E	A	A^2	A^3	A^4		E	A	A^2	B	AB
χ_1	1	1	1	1	1	χ_1	1	1	1	1	1
χ_2	2	Φ	$-\tau$	$-\tau$	Φ	χ_2	3	τ	$-\Phi$	0	-1
χ_3	2	$-\tau$	Φ	Φ	$-\tau$	χ_3	3	$-\Phi$	τ	0	-1
χ_v	3	τ	$-\Phi$	$-\Phi$	τ	χ_4	4	-1	-1	1	0
χ_v^{2+}	6	1	1	1	1	χ_5	5	0	0	-1	1
χ_{elast}	21	3	3	3	5	χ_v	3	τ	$-\Phi$	0	-1
						χ^{2+}	6	1	1	0	2
						χ_{elast}	21	1	1	0	5

representation is present in the n-th power of the vector representation, which has character $\chi_v(A^m) = 2\cos(2\pi m/N)$, or $2\cos(\phi)$. For the isotropic case ($N \to \infty$) the number of parameters in the tensor of rank n is given by

$$p_\infty^n = \frac{1}{2\pi}\int_0^{2\pi} d\phi\,(2\cos(\phi))^n,$$

which is 0 when n is odd and $n!/[(n/2)!]^2$ when n is even. For finite N it is

$$p_N^n = \frac{1}{N}\sum_{m=1}^{N}(2\cos(2\pi m/N))^n.$$

It may be shown that $p_N^n = p_\infty^n$ for all values $n < N$. Therefore, a surface with N-fold symmetry is isotropic for all tensor properties of rank smaller than N. This is relevant for measuring tensorial properties of quasicrystal surfaces. The theorem is due to C. Hermann (1934).

A lattice in superspace is determined by its *metric tensor*, with elements equal to the scalar products of pairs of basis vectors. The metric tensor in superspace is a symmetric rank 2 tensor, which is invariant under the higher-dimensional point group. For example, if the point group is the cyclic group of order 5 generated by the five-dimensional matrix

$$A = \begin{pmatrix} 0 & 1 & 0 & 0 & 0 \\ 0 & 0 & 1 & 0 & 0 \\ 0 & 0 & 0 & 1 & 0 \\ 0 & 0 & 0 & 0 & 1 \\ 1 & 0 & 0 & 0 & 0 \end{pmatrix},$$

the vector representation has three irreducible components: $\chi = \chi_1 + \chi_2 + \chi_3$. The symmetric square has character $\chi(A^n) = 0$ (if $n \neq 0$) or 15 (if $n = 0$). It contains the trivial representation three times. Therefore, the metric tensor has three independent components. Its tensor elements are given by

$$\begin{pmatrix} a & b & c & c & b \\ b & a & b & c & c \\ c & b & a & b & c \\ c & c & b & a & b \\ b & c & c & b & a \end{pmatrix},$$

where a, b, c are arbitrary real numbers. This is the general form of the metric tensor for a five-dimensional lattice that is invariant under point group 5.

A second feature occurring for quasiperiodic crystals is the fact that one may distinguish external (physical) and internal (perpendicular) components, sometimes called phonon and phason components. Displacements of particles are displacements in physical space, and forces act in this space as well. Nevertheless, some displacements may more conveniently be described in internal space. For example, a local change in the phase of the modulation function for a modulated phase induces displacements in physical space, but it may be more convenient to describe the displacement in internal space. This is, in particular, the case when the energy changes continuously with the internal displacement. This means that there are no jumps when the physical space shifts along the internal (perpendicular) space. This argument also plays a role when describing the dynamics of a quasiperiodic systems by hydrodynamic modes (see Section 5.2).

The displacement of a particle at $\mathbf{r}$ in physical space then has a physical and an internal component: $\mathbf{u}(\mathbf{r}) = (\mathbf{u}_E(\mathbf{r}), \mathbf{u}_I(\mathbf{r})) = (\mathbf{v},\mathbf{w})$. The displacements lead to a strain described by the *strain tensor*, and this may also be decomposed into a physical and an internal part:

$$\mathbf{e}_{ij} = (\partial_i v_j + \partial_j v_i)/2 \;\; (i,j = 1,2,3); \;\; \mathbf{f}_{ik} = \partial_i w_k \;\; (i = 1,2,3; k = 4,\ldots,n). \quad (5.2)$$

As usual, the anti-symmetric part of the strain tensor corresponds to a rotation and is not relevant. The tensor $\mathbf{e}$ is the phonon or physical strain tensor, and the tensor $\mathbf{f}$ is the phason or internal strain tensor (cf. (Bak, 1985)).

There are two problems with this approach. We have seen already that the phason degrees of freedom are not new degrees of freedom. This has to be kept in mind when one has to sum over all degrees of freedom. A second problem is the occurrence of discontinuities in the atomic surface. A shift in internal space may then lead to local jumps, which do not become small when the shift in internal space becomes small, with the consequence that one may question the validity of a harmonic approximation. This is certainly not valid for a single, local jump. However, when one considers a function $\mathbf{w}(\mathbf{r})$, which is sufficiently smooth, the amplitude determines the density of jumps. Each jump changes the energy of the system, and in the long-wavelength approximation one may consider energy changes in a power series of this density or rather its derivative, the phason strain.

Strain may be caused by an external field. Consider the piezoelectric effect in a quasiperiodic crystal, the coupling between strain and an electric field. The free energy is given by

$$F = \sum_{ijk} p^e_{ijk} e_{ij} E_k + \sum_{ijk} p^f_{ijk} f_{ij} E_k. \quad (5.3)$$

The *piezoelectric tensors* p^e and p^f should be invariant under the point group of the system. The tensor e transforms as the symmetrized square of the vector representation in physical space, and the tensor f according to the product of the vector representations in physical and internal spaces. Then the piezoelectric tensors transform as the product of these two representations and the physical vector representation. It is the three-dimensional vector representation, because electric fields exist in physical space.

As an example we consider the icosahedral case (see Table 6.2). The physical and internal vector characters are χ_2 and χ_3, respectively. The symmetrized square of the former has character $\chi_1 + \chi_5$, and the product of the two has $\chi_4 + \chi_5$. That means that p^e transforms as the product of χ_2 and $\chi_1 + \chi_5$, which does not contain the trivial character. The tensor p^f transforms as the product of χ_2 and $\chi_4 + \chi_5$. This does not contain the trivial representation, either. Therefore, an icosahedral crystal has neither a phonon piezoelectric effect nor a phason piezoelectric effect.

As our next example we take the *elasticity tensor*. In terms of the strain tensors e and f the free energy is

$$F = \frac{1}{2} \sum_{ijkl} c^E_{ijkl} e_{ij} e_{kl} + \frac{1}{2} \sum_{ijkl} c^J_{ijkl} f_{ij} f_{kl} + \sum_{ijkl} c^{EI}_{ijkl} e_{ij} f_{kl}.$$

The elastic tensor consists of three components: c^E, c^J, and c^{EI}. The first transforms as the symmetrized square of the representation of e, the second as the symmetrized square of the f representation, and the third as the product of the representations of e and f. For the icosahedral group 532, considered already above, the characters of the representations to which the various tensors belong are given in Table 5.3, together with the number of times the trivial representation occurs, i.e. the number of independent parameters in the tensor considered.

In Table 5.2 the first two characters are those of the vector representations in physical and internal space, and the last three characters are those of the phonon, phason, and coupled phonon–phason part of the elasticity tensor. Using again Eq. 5.1, one derives that the trivial representation occurs twice in χ_E, twice in χ_I, and once in χ_{EI}. Therefore, there are five independent elastic constants, two phonon constants (the Lamé constants λ and μ), two phason constants, denoted by K_1 and K_2, and one phonon–phason constant, K_3 (Lubensky, 1988).

Table 5.3 *Tensor characters for the point group 532*

	E	A	A^2	B	AB	Number of invariants
χ_{phys}	3	τ	$-\Phi$	0	-1	0
χ_{int}	3	$-\Phi$	τ	0	-1	0
χ_E	21	1	1	0	5	2
χ_I	45	0	0	0	5	2
χ_{EI}	54	-1	-1	0	2	1

For given elastic strain one may calculate the elastic energy, or the elastic energy density if the strain is not uniform. Using the elastic constants one may calculate the dynamics as well. The equations of motion involve the stress and the time evolution of the strain. This is valid as long as the microscopic structure does not come in. It is a long-wavelength approximation, and the elastic excitations are those of hydrodynamic theory.

5.3 Hydrodynamics of aperiodic crystals

5.3.1 Hydrodynamic theory of fluids and periodic crystals

The number and nature of long-wavelength excitations in a given system can be derived from *hydrodynamic theory*. This theory deals with the slow, long-wavelength degrees of freedom and the modes for which the frequency ω vanishes as some power of the wave vector q of the mode in the limit $q \to 0$, that is, in the long-wavelength limit. Two key notions in this theory are those of broken symmetry and order parameter. The general framework goes as follows (Martin et al., 1972): first identify the broken symmetry that distinguishes two systems, for instance a fluid and a crystal. Then define the order parameter, that allows to define the new state. Finally, predict all elementary excitations in a system using symmetry and 'mode counting' arguments . An introduction to this approach can be found in Chaikin and Lubensky (1995).

Let us first consider what distinguishes a fluid from a three-dimensional crystal. The fluid has full rotation and translation symmetry. Suppose we are observing water molecules inside a large container. The time averaged atomic density will look identical if someone rotates or translates the container, we thus say that the fluid has a full rotation and translation symmetry. The situation is quite different for a crystal. It is possible to distinguish a crystal located in A from the same one translated to B (we are talking here of real translations, which are not related to the lattice periodicity). We say that the crystalline state has broken a continuous translational symmetry.

We can go a step further by considering the free energy of the system. The free energy of a crystal is invariant under any translation. Indeed, whether the crystal is in A or B, it has the same free energy. We thus have a continuous symmetry or a continuous degeneracy of the ground state with respect to a translation. However, since we can distinguish a crystal in A from the same one in B, the ground states break that continuous symmetry. It is the interplay between the broken symmetry and the continuous degeneracy of the free energy with respect to translation that leads to the phonon *Goldstone modes*.

Next, we must find a parameter, called the *order parameter*, which allows us to distinguish between the fluid and the crystal. This is in general achieved in Fourier space by considering the Fourier components $F(Q)$ of the electron or atomic density, with wave vectors Q from the reciprocal lattice. This also allows one to take into account the crystal symmetry in the further calculations.

The final step is to identify the hydrodynamic variables. In a mono-atomic fluid there are five variables related to a conservation law: the particle number, the three components of the momentum, and the energy. Hydrodynamic theory tells us that the number of hydrodynamic modes is equal to the number of hydrodynamic variables. Using this *mode counting* argument and the associated hydrodynamic variables, five hydrodynamic modes are expected in a simple fluid: (a) a longitudinal sound mode which is propagating and therefore counts for two modes (one propagating to the right and one to the left, the two being related by time reversal); (b) two non-propagating diffusive shear modes: this means that any fluctuation associated with these modes will decay exponentially with time; in this case, time reversal does not apply; and (c) one diffusive entropy mode. We get dispersion relations in the form $\omega = vq$ for propagative modes, and $-i\omega = Dq^2$ for diffusive ones.

Considering now the crystal case: in addition to the five hydrodynamic variables of the fluid, we must add three variables related to the continuous translation broken symmetry discussed previously. Indeed, since the free energy of the system is invariant under a uniform translation, a local displacement field $T(R)$ cannot cost much energy as long as it is a continuous and slowly varying function of space coordinates. Furthermore, that energy should vanish in the long wavelength limit. In the crystal we thus have eight hydrodynamic variables instead of the five in the fluid. This leads to eight hydrodynamic modes.

1. One longitudinal and two transverse propagating acoustic modes (each one counting for two modes).
2. One mass diffusion mode.
3. One heat or entropy diffusion mode.

As compared to the fluid case, we see that the two shear diffusive modes become two transverse propagating modes, and a new vacancy diffusive mode has appeared. The frequency of the propagative acoustic modes goes to zero when the wavelength goes to infinity (or when q goes to zero) which is expressed by the dispersion relation $\omega = vq$. Propagating phonons are associated with the continuous symmetry breaking and are related to the Goldstone theorem so that they are sometimes called *Goldstone modes*.

5.3.2 Hydrodynamic theory of aperiodic crystals and phason modes

As for the case of periodic crystals, aperiodic crystals have their free energy invariant under translation in physical space. Hence, one expects to get the same eight hydrodynamic modes as in the crystal case, including three pairs of propagating acoustic phonons.

In addition, the free energy of aperiodic crystals is invariant under translations along the perpendicular or internal space, sometimes referred to as 'new degrees of freedom'. However, changing the internal coordinates leads to an equivalent, but different realization of the aperiodic crystal. The translations along the perpendicular direction thus break the continuous symmetry in a way similar to that for translations along the physical space. Shifts in the internal space may also be considered as gauge transformations (see Chapter 2, Sections 2.2.5 and 2.4.6). Aperiodic crystals have lost their continuous gauge invariance. From this, hydrodynamic theory predicts new modes which are called phasons (Lubensky, 1988). These are the Goldstone modes for the loss of continuous gauge invariance, just as there appears a new sound mode at the superfluid phase transition. There should be as many phason modes as there are dimensions in the internal space: for instance, for a one-dimensional incommensurate modulation, the theory predicts one phason mode whereas, for the icosahedral quasicrystal, three phason modes should appear. However, mode-counting arguments lead to the prediction that, unlike phonon modes, phason modes are diffusive in the long-wavelength limit.

For all aperiodic crystals, hydrodynamic theory predicts that in the long-wavelength limit the phason modes are not propagative, but are diffusive and their 'dispersion relation' reads

$$-i\omega = D_{phason}q^2, \quad (5.4)$$

where D_{phason} is the phason diffusion constant. Once this relation is inserted in the equation of motion, one finds that the phason mode decays as $\exp(-t/t_0)$, with a characteristic time t_0 given by

$$1/t_0 = D_{phason}q^2. \quad (5.5)$$

The diffusive character is clearly expressed in Eq. 5.4, with a purely imaginary component, and in the exponential decay of the mode. The hydrodynamic character is somewhat incorporated into Eq. 5.5, which says that the characteristic time t_0 scales with the square of the phason-mode wavelength. This is the signature of a process that is similar to a diffusive process.

Hydrodynamic theory is a continuum theory which does not give a direct microscopical interpretation of phason modes. Nevertheless, a simple picture of a phason mode can be given in the higher-dimensional space. It is a sine wave, with a wave vector $\mathbf{q}_{par}$ along the physical space and a polarization $\mathbf{e}_{per}$ in the perpendicular (or internal) direction, as exemplified in Fig. 5.1. As already stated, the phason mode is not propagative in the long-wavelength limit, and this perturbation decays exponentially with time. Consequences at the atomic level of a phason mode can be derived from this simple picture. One readily understands that the situation is quite different for the various classes of aperiodic crystals, and strongly depends on the 'shape' of the atomic surfaces. In the case of a displacive incommensurate modulated structure, atomic surfaces may be continuous. Then, a phason mode leads to small shifts of atoms away from their equilibrium position. In these systems there exists a regime for which phason modes are propagative, although with a finite lifetime. The situation is quite different for modulated phases with a discontinuous modulation or for quasicrystals, in which phason modes result in a series of atomic rearrangements such as the $LS \to SL$ one in the Fibonacci chain. Such atomic rearrangements have been referred to as 'phason jumps' or 'phason flips' in the literature. Obviously, the physics which governs such modes is quite different from the one in modulated phases with continuous modulation functions, and in this case phason modes are diffusive for all wave vectors.

It is worth pointing out here some of the hypotheses underlying this theory, which are not obvious a priori. The first one relates to the ability of the system to move slowly and without energy cost along the perpendicular direction. Moving the cut space along the perpendicular direction has, at the microscopic scale, a quite different effect if one considers an incommensurately displacive modulation (with smooth atomic surfaces), in which case the perpendicular translation results

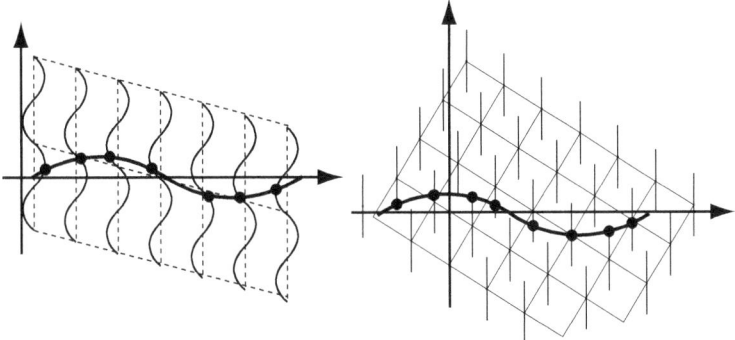

Figure 5.1 *Illustration of a phason mode in a one-dimensional aperiodic crystal. The wave vector q_{par} is along the physical space, whereas the polarization is along the perpendicular (or internal) space. (Left) An incommensurate modulated phase. The mode is propagating, but overdamped. (Right) A quasicrystal. In hydrodynamic theory a phason mode is not propagative and decays exponentially with time.*

in small shifts of the atomic position. This is clearly not the case when atomic surfaces are discontinuous, as is the case for the Fibonacci chain, for instance. In that case, a motion of the cut space along the perpendicular direction results in a series of atomic 'jumps' (see Fig. 5.1). Hydrodynamic theory does not deal, in fact, with those differences. The second hypothesis necessary for hydrodynamic theory and for the generalized elasticity leading to Eq. 5.5, is that the system energy has a square gradient response with respect to a phason strain. This is not obvious a priori, but has been verified, for instance, in the case of quasicrystals for the random tiling model, in which case the restoring force is entropic in nature.

In the following sections atomistic models are presented and developed. In the case of modulated phases, there is a frequency regime for which phason modes appear as propagative modes, which might seem to contradict hydrodynamic theory. This has been also observed experimentally, with a phason linear dispersion relation that goes to zero with the phason wave vector, as with an acoustic phonon. There is, however, an important difference with acoustic phonon modes, for which the same property holds. In acoustic modes with very-long-wavelengths the particles in a unit cell barely change their mutual distances. In a phason mode the distances change with the derivative of the modulation function, and these are not necessarily small. This leads to non-linearity and a finite damping at $q = 0$, which means that phasons with small wave vector are overdamped or diffusive. In the long-wavelength limit, phason modes are, indeed, diffusive (this has been shown also experimentally), raising an apparent contradiction (Zeyher and Finger, 1982).

On the other hand, when atomic surfaces are discontinuous, as in quasicrystals, one does not expect any regime for which phasons are propagative, for phason modes imply atomic rearrangments which are diffusive in nature. Moreover, it is expected that it is only above a critical temperature that hydrodynamic theory will hold.

5.4 Phonons and phasons: Theory

5.4.1 Introduction

Lattice vibrations are oscillations of the atomic positions around their equilibrium values. The displacements $\mathbf{u}_n(t)$ from the equilibrium positions are solutions to the equations of motion. In the harmonic approximation these are the equations for a set of harmonic oscillators. The solutions then are linear combinations of eigenmodes, each of which for a periodic crystal may be characterized by a wave vector in the Brillouin zone, and a branch index. The number of eigenmodes is the number of degrees of freedom, this is, three times the number of particles. In the quantum mechanical problem the phonons are the quanta of the eigenmodes. Eigenmodes correspond to an eigenvector of the dynamical matrix. If the potential energy in the harmonic approximation is written as

$$E = \frac{1}{2}\sum_{n_1\alpha_1 n_2\alpha_2} D\begin{pmatrix} n_1 & n_2 \\ \alpha_1 & \alpha_2 \end{pmatrix} u_{n_1\alpha_1} u_{n_2\alpha_2},$$

where n_j numbers the particles, and α_j the Cartesian coordinates of the displacement $\mathbf{u}_j$, then the equations of motion are

$$m_i \ddot{u}_{i\alpha} = -\sum_{j\beta} D\begin{pmatrix} i & j \\ \alpha & \beta \end{pmatrix} u_{j\beta}. \qquad (5.6)$$

For a crystal with lattice periodicity the labels n are replaced by a pair $(\mathbf{n}, j)$, where $\mathbf{n}$ denotes the unit cell, and j the particle in the unit cell. Then eigenmodes may be written as

$$u_\alpha^\mathbf{q}\begin{pmatrix} \mathbf{n} \\ j \end{pmatrix} = \frac{1}{\sqrt{m_j}} u_\alpha^\mathbf{q}\begin{pmatrix} 0 \\ j \end{pmatrix} \exp(i(2\pi \mathbf{q}\cdot\mathbf{n} - \omega(\mathbf{q})t)), \qquad (5.7)$$

with $\mathbf{q}$ in the Brillouin zone. The dynamical matrix is defined as

$$D_{\alpha_1\alpha_2}(j_1 j_2 | \mathbf{q}) = \sum_\mathbf{n} D_{\alpha_1\alpha_2}\begin{pmatrix} \mathbf{n} \\ j_1 & j_2 \end{pmatrix} (m_{j_1} m_{j_2})^{-\frac{1}{2}} \exp(2\pi i \mathbf{q}\cdot\mathbf{n}). \qquad (5.8)$$

When the number of particles in the unit cell is equal to s, then the eigenmodes are the $3s$ solutions to the equations

$$\omega(\mathbf{q})^2 u_{\alpha_1}^\mathbf{q}\begin{pmatrix} 0 \\ j_1 \end{pmatrix} = \sum_{\alpha_2 j_2} D_{\alpha_1\alpha_2}(j_1 j_2 | \mathbf{q}) u_{\alpha_2}^\mathbf{q}\begin{pmatrix} 0 \\ j_2 \end{pmatrix}. \qquad (5.9)$$

An arbitrary lattice vibration then is given by

$$u_\alpha\begin{pmatrix} \mathbf{n} \\ j \end{pmatrix}(t) = \frac{1}{\sqrt{m_j}} \sum_{\mathbf{q}\nu} Q_{\mathbf{q}\nu} \mathbf{e}(\mathbf{q}\nu|\alpha j) \exp(i(2\pi \mathbf{q}\cdot\mathbf{n} - \omega_\nu(\mathbf{q})t)) + c.\,c, \qquad (5.10)$$

where $\mathbf{e}(\mathbf{q}\nu)$ is the ν-th eigenvector of the dynamical matrix $D(\mathbf{q})$.

The reduction of the number of coupled equations from infinity to three times the number of particles in the unit cell is possible due to the lattice periodicity, which means the existence of a Brillouin zone. For quasiperiodic crystals there is no three-dimensional Brillouin zone of non-vanishing volume, and the dynamical problem remains infinite-dimensional. The fact that there is lattice periodicity in superspace does not help here, because the dynamics plays in the physical space. Formulated in superspace, one may say that there are an infinite number of points on the atomic surfaces, and therefore the number of degrees of freedom per unit cell, which is equal to the dimension of the dynamical matrix, is infinite.

Although there is no Brillouin zone for an aperiodic crystal, one may introduce a *pseudo-Brillouin zone* (PBZ). To see this, consider the long-wavelength limit and make a continuum approximation. The equations of motion in this case become

$$\frac{\partial^2}{\partial t^2} u(x,t) = T(x) \frac{\partial^2}{\partial x^2} u(x,t), \qquad (5.11)$$

where T is the local force constant divided by the local mass density. Consequently, T is a quasiperiodic function, which can be embedded in n dimensions. The n-dimensional function is $T(x, x_I)$. The equations of motion lead for $u(x,t) = U(x, x_I) \exp(i\omega t)$ to

$$-\omega^2 U(x, x_I) = T(x, x_I) \frac{\partial^2}{\partial x^2} U(x, x_I). \qquad (5.12)$$

The functions U have the Bloch form in n-dimensional space, $U(x, x_I) = V(x, x_I) \times \exp(2\pi i q x)$, and can be expanded in a Fourier series:

$$U(x, x_I) = \sum_{K \in \Sigma^*} c^q(K) \exp(2\pi i (q x + K_E x + K_I x_I)),$$

whereas the function T is equal to

$$T(x, x_I) = \sum_{K \in \Sigma^*} A(K) \exp(2\pi i (K_E x + K_I x_I)).$$

Substitution into the equations of motion gives the eigenvalue problem

$$\omega^2 c^q(K) = \sum_{K' \in \Sigma^*} (q + K_E + K_I')^2 A(K') c^q(K - K'). \qquad (5.13)$$

The squares of the frequencies are the eigenvalues of an infinite-dimensional matrix. If the function T is a constant, the solutions are plane waves. These are coupled by the interaction terms A. The most important Fourier components for A are those with a wave vector K for which the structure has a strong Fourier component, because T is determined by the mass and force constants. Coupling between two modes is strong when the modes have the same frequency, and their wave vectors differ by a reciprocal lattice vector for which the structure factor is strong. This happens for modes with a wave vector with equal distances to two strong Bragg peaks. Around the origin one considers the planes perpendicular to the connection line between the origin and a strong Bragg peak and bisecting that line. These planes enclose a PBZ (Niizeki, 1989; Niizeki and Akamatsu, 1990).

What happens in aperiodic crystals may be seen by comparison with approximants for which the unit cell goes to infinity. Consider a simple harmonic chain

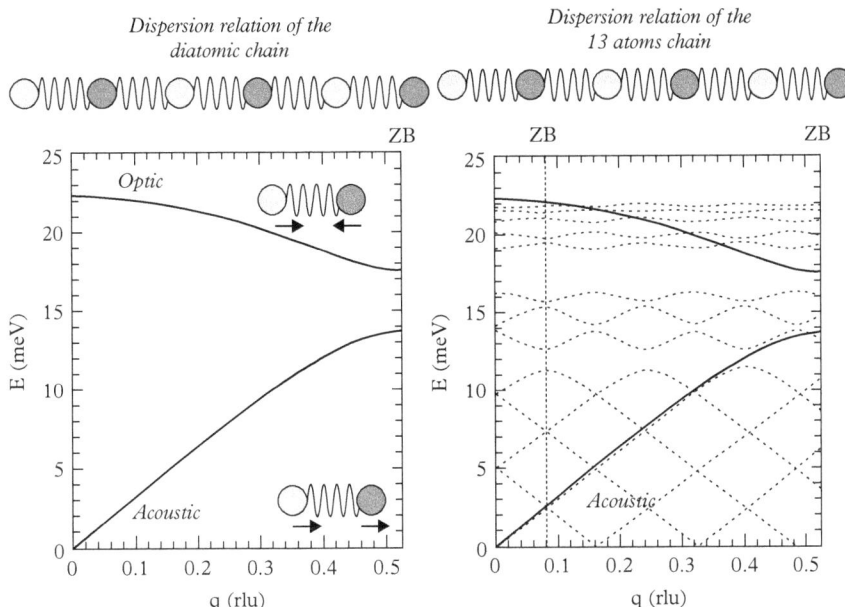

Figure 5.2 The harmonic chain with period $N = 2$ has two branches (the acoustic and the optic branch); with period $N = 13$, there are 13 branches which are, to a first approximation, the folding back of the dispersion curves for $N = 1$. The lower curves correspond to the acoustic modes for $N = 2$; the higher ones, which are more flat, correspond to the optic modes. The Brillouin zone is $1/N$ of the original Brillouin zone.

with equal masses m and equal spring constants α. The eigenmodes are plane waves $\exp(2\pi i k n a)$, where a is the lattice constant, and k belongs to the Brillouin zone $(-1/2a < k \leq 1/2a)$, and the frequency is $\omega(k) = 2\sqrt{\alpha/m}\sin(\pi k a)$. When the masses are not the same, but have a period Na, the Brillouin zone shrinks by a factor N, and there are N phonon branches $\omega(k, \nu)$ ($\nu = 1,\ldots,N$). If the mass differences are very small, the branches are just the folding back of the original dispersion curves (see Fig. 5.2). For larger differences, gaps open up at the centre and the border of the Brillouin zone. For an aperiodic chain ($N \to \infty$) the Brillouin zone shrinks to zero. Nevertheless, the modes will be very similar to those in the original chain, as long as the modulation is small. The diffraction from such an aperiodic system is determined by the *dynamic structure factor* (defined in Eq. 5.21) $S(Q, \omega)$, which is different from zero only if $\omega = \omega(k, \nu)$ for some branch ν. For small mass differences, however, the main contribution will be along the line $\omega = \omega(k)$ for the unmodulated chain.

For periodic crystals phonons in the harmonic approximation are propagating plane waves. Including higher-order terms leads to a coupling between the phonons. That means that exciting a phonon leads to excitation of other phonons via this coupling, and to a decrease of the energy stored in the original phonon. This leads to a finite lifetime. A damping term $\exp(-\gamma t)$ appears in the solution, and, generally this damping depends on the wave vector. In the long-wavelength approximation one may show that the damping of acoustic modes depends as q^2 on the wave vector. Besides non-linearity, defects may also contribute to the damping. This damping may become so big that the modes become overdamped. This occurs if the frequency of a mode tends to zero, except when the mode is acoustic. In that case the damping tends to zero also when the wave vector, and therefore the frequency, tends to zero.

Because, for periodic crystals, the displacements of atoms in an eigenmode get simply a phase factor in going from one unit cell to another (cf. Eq. 5.7), the amplitude of the oscillation is the same in each unit cell. The mode has an extended character. If defects are present, other modes may appear for which the amplitude decays exponentially with the distance to the impurity. Such a mode is called 'localized'. In aperiodic crystals both *extended* and *localized modes* may appear. Moreover, there is a third possibility: a mode for which the amplitude decays with a power law ($\sim r^{-\alpha}$) or for which the atoms participating in the mode form a fractal. Such modes are called critical. If one considers a sphere of radius R around an atom with the maximal displacement in a mode, then the sum of the squares of the displacements over the atoms inside the sphere depends on R. For an extended mode one has

$$\sum_{n \text{ inside the sphere}} |u_n|^2 \sim R^d,$$

if d is the dimension of the space and R tends to infinity. For a localized state the expression is

$$\sum_{n \text{ inside the sphere}} |u_n|^2 \sim R^0,$$

and for a critical state

$$\sum_{n \text{ inside the sphere}} |u_n|^2 \sim R^{d-2\delta},$$

where $0 \leq \delta \leq d/2$. The three types of modes may occur in aperiodic crystals. Only the first behaviour is present in ideal periodic crystals without defects.

5.4.2 Simple models

5.4.2.1 Introduction

We shall discuss the special properties of phonons in aperiodic crystals in some simple models. These do not describe the specific properties of a material, but they exemplify a number of differences from periodic crystals, and in some cases the differences between different types of aperiodic crystals. We shall consider the spectrum of phonons, their density of states, and the character of their eigenvectors. In particular, we shall be concerned with phonons related to the quasiperiodic structure, and the possibility of embedding these in a higher-dimensional space.

After that we shall consider more realistic calculations for specific materials. That will be a brief discussion, because it goes beyond the scope of this book. For a more specialized treatment we refer to review papers (Currat and Janssen, 1988) for incommensurate phases, and (Quilichini and Janssen, 1997), (Hafner and Krajčí, 1999), (Suck, 2002), and (Elhor et al., 2003) for quasicrystals.

5.4.2.2 Modulated chain

A very simple model for studying excitations in a quasiperiodic system is the modulated chain model. Consider a linear one-dimensional chain with harmonic interactions. The potential energy may then be written in terms of the displacements u_n from the equilibrium positions as

$$V = \sum_n \frac{1}{2}\alpha_n(u_n - u_{n-1})^2. \tag{5.14}$$

The chain is quasiperiodic if the function α_n is quasiperiodic. An example is the dependence $\alpha_n = \alpha[1 + \Delta \cos(2\pi n\gamma + \phi)]$ for irrational values of γ ($0 \leq \Delta < 1$). The equations of motion for $u_n(t) = u_n \exp(i\omega t)$ lead to

$$m\omega^2 u_n = \alpha_n(u_n - u_{n-1}) + \alpha_{n+1}(u_n - u_{n+1}). \tag{5.15}$$

For $\Delta = 0$ the solutions are

$$u_n(t) = u_0 \exp(i(2\pi k n - \omega_k t)), \quad \omega_k^2 = 4\alpha \sin(\pi k)^2/m,$$

but for $\Delta \neq 0$ the mode with wave vector k couples to that with wave vector $k + \gamma$, and one has to solve an infinite set of coupled equations for irrational values of γ. No direct way is known, and one approach is to take a series of rational approximants to γ. The value γ has an infinite continued fraction expansion

$$\gamma = \cfrac{n_1}{m_1 + \cfrac{n_2}{m_2 + \cfrac{n_3}{m_3 + \cdots}}}$$

with integers n_i and m_i. Truncation at level n gives a rational approximant to γ: $\gamma_n = L_n/N_n$ such that $\lim_{n \to \infty} \gamma_n = \gamma$. For each rational value L_n/N_n the number of coupled equations is N_n and this can be solved in the usual way. The limit of the spectra when n tends to infinity will give the spectrum of the aperiodic chain, if this limit exists. For $\gamma = \Phi$, the golden mean, this would give a series of approximants with rational wave vectors

$$\gamma_n = \frac{1}{2}, \frac{2}{3}, \frac{3}{5}, \frac{5}{8}, \frac{8}{13}, \frac{13}{21}, \ldots.$$

Generally, for wave vector $\frac{L}{N}$ the Brillouin zone has a width $1/N$ and there are N branches of dispersion curves. As long as γ is rational, the spectrum depends on the phase ϕ in α_n, but for incommensurate values this is no longer the case. There are, generally, $N-1$ gaps in the spectrum. The distribution of gaps is interesting. This is most clearly seen for $\Delta = 1$. This is an unphysical value, because the force constant may become arbitrarily small, but for the proper choice of ϕ it never strictly becomes zero, and the gap distribution is clearly seen (Fig. 5.3). In the figure the spectra are plotted for rational values $k = L/N$, where L/N are *Farey numbers* between 0 and 1 (de Lange and Janssen, 1981). Farey numbers between 0 and 1 are fractions p/q, obtained in the following way. Write 0 as 0/1, and 1 as 1/1. This is the first generation. If $p_1/q_1, \ldots, p_n/q_n$ are the fractions in a certain generation, then the next generation consists of all these fractions to which are added all fractions $(p_i + p_{i+1})/(q_i + q_{i+1})$. So, in the second generation one obtains 0/1, 1/2, and 1/1, and in the third 0/1, 1/3, 1/2, 2/3, and 1/1. If one continues, all

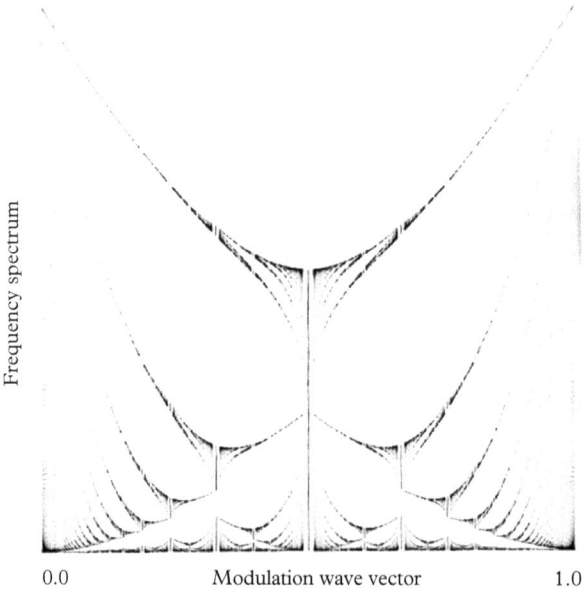

Figure 5.3 *The spectra of the modulated chain model for approximants L/N with $N < 50$; $\Delta = 1$. (de Lange and Janssen, 1981)*

rational numbers between 0 and 1 are obtained in this way. In the figure Farey numbers of the first generations are easily distinguished. The figure shows the same motif on every scale, albeit deformed. It resembles the analogous figure for spectra of electrons in a crystal in an external magnetic field, as first studied by Hofstadter (the *Hofstadter butterfly*: (Hofstadter, 1976)). The figure shows *self-similarity*.

The spectrum for a given irrational value is the limit of the spectra for its series of approximants, and has an infinite number of gaps and its measure is zero. It is a fractal structure with self-similarity properties. This is, of course, a caricature of the real situation, for two reasons. The modulation strength of $\Delta = 1$ is unphysical and, for a three-dimensional structure with a one-dimensional modulation, the fractal spectrum has to be combined with the spectra for vibrations with wave vectors which are not parallel to the modulation wave vector, and few of the gaps will survive. For values $0 < \Delta < 1$ there are still gaps, but the measure of the spectrum is no longer zero. Because the potential energy is invariant under a displacement which is the same for all particles, there is a zero frequency mode for $k = 0$. This is just the $k = 0$ acoustic mode.

The gaps for small values of Δ may be obtained by perturbation calculations. Because the uncoupled modes for $\Delta = 0$ become coupled to modes differing by (a multiple of) the modulation wave vector γ, the most important gaps appear at $\gamma/2$. The first-order gaps are easily recognizable in the ω-wave vector plot. The higher-order gaps are smaller in magnitude. For $\Delta = 0.31$, the spectra are given as function of the wave vector γ in Fig. 5.4.

For given approximant L/N to an irrational value the eigenfrequencies may be calculated by determining the eigenvalues of an $N \times N$ dynamical matrix. Putting these in order of increasing frequency, one gets the dispersion curves, which show gaps. In the case that $\Delta = 1$, there are an infinite number of gaps. The *vibrational density of states* (VDOS) then also has an infinite number of gaps. In one dimension the VDOS is always singular. In this case the *integrated density of states* (IDOS) forms a 'devil's staircase', a function that is non-decreasing, has derivative zero almost everywhere, and increases from 0 to 1 (Fig. 6.5). A periodic chain with N particles per unit cell then shows simply at most $N - 1$ gaps in the VDOS and $N - 1$ plateaus in the IDOS.

5.4.2.3 Frank and Van der Merwe *model*

Another model to study the dynamics of a quasiperiodic system is the model studied by *Frank and Van der Merwe*, often referred to as the *Frenkel–Kontorova model*, although, in their paper, incommensurability is not mentioned; that will be discussed in more detail in Section 6.3. The model consists of a linear chain in a sinusoidal background potential, while the lattice constant of the chain is incommensurate with the periodicity of the background. The equations of motion become

$$-m\omega^2 u_n = -\alpha(2u_n - u_{n+1} - u_{n-1}) - \Delta\alpha \cos(2\pi \bar{x}_n/a)u_n. \quad (5.16)$$

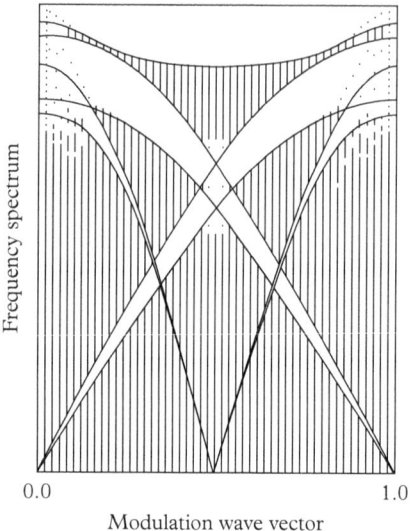

Figure 5.4 *The spectra of the modulated chain model as function of the wave vector;* $\Delta = 0.31$.

The ground state is a modulated chain. In the formula, $\bar{x}_n$ is the equilibrium position of the nth particle without the background potential, u_n is the displacement of the nth particle from its equilibrium position, and $\Delta\alpha = 2\pi\lambda/b$. The second term gives the coupling between the eigenmodes of the unmodulated chain. The simple dispersion relation then becomes a curve with discontinuities, resulting in a VDOS which has gaps. Because of the lack of translation invariance, a constant displacement does not correspond to an eigenmode with zero frequency. We can study the spectrum in the same way as done for the modulated chain, by considering a series of approximants, or all approximants with a nominator smaller than a given value. The latter gives a picture that is similar to that of the modulated chain. Also, here the hierarchy of gaps is recognizable for small $\Delta\alpha$ in the plot ω versus b/a.

Although there is no zero frequency acoustic mode, there may be a zero frequency mode with another character when the interaction between chain and substrate is small enough. An analysis of the displacements in this mode shows that it corresponds to a rigid shift of the modulation function with respect to the chain. As we have seen for weak interaction there is a continuous modulation function, but for stronger interaction it becomes discontinuous. Aubry has rigorously shown that there is a critical value λ_c of the coupling parameter λ such that for $\lambda < \lambda_c$ the modulation function is continuous, and for $\lambda > \lambda_c$ it is discontinuous (Aubry and Le Daeron, 1983). He called this the *breaking of analyticity*. This transition from

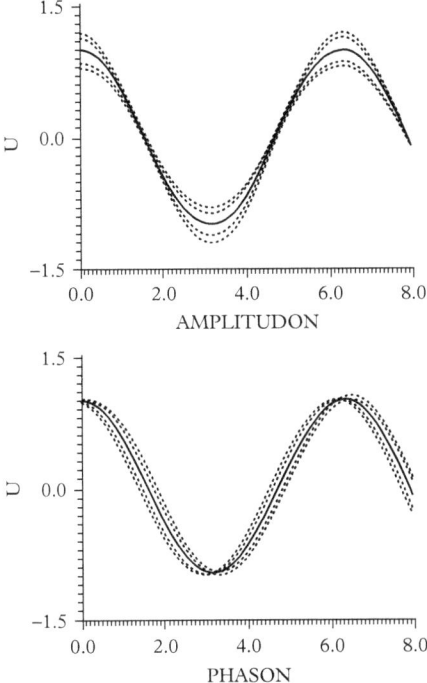

Figure 5.5 *When the soft mode becomes unstable at an incommensurate wave vector* **q**, *then the two modes at* ±**q** *combine in the incommensurate phase to two modes: the phason and the amplitudon.*

continuous to discontinuous coincides with the opening of a gap in the spectrum. For $\lambda < \lambda_c$ there is an excitation with the character of a phason and with zero frequency. For $\lambda > \lambda_c$ the minimum frequency in the density of states is positive, and there is a phason gap.

The reason for the existence of a zero frequency mode is the following. The positions of the particles in the chain are given by

$$x_n = x_0 + na + f(x_0 + na), \quad f(x) = f(x+b).$$

If the particles are all translated over a distance Na and renumbered, one obtains a chain with almost the same distances between the particles:

$$x_0 + (n+N)a + f(x_0 + (N+n)a - Na + Lb)) = x_0 + n'a + f(x_0 + n'a + \delta),$$

where δ may be made arbitrarily small by choosing N and L, because a and b are incommensurate: $\delta = Lb - Na$. A shift δ in the modulation function variable therefore leads to a chain with the same potential energy. When the function f is continuous, this means that the phase of the modulation function may be varied arbitrarily without changing the potential energy. This means that an infinitesimal shift in the modulation function is a zero frequency oscillation, a phason. If the phase shift is not constant, but has a long-wavelength periodicity, the corresponding excitation has a non-zero, but nevertheless low frequency. It is an excitation of the phason branch. In the higher-dimensional superspace such an excitation may be described as a displacement in internal space (see Fig. 5.1). These modes are exactly the excitations discussed in the hydrodynamic context. Actually these phasons are phonons with a special character. Their displacements are displacements in V_E, but may also be viewed as displacements in V_I. If the internal space is one-dimensional, a uniform shift along V_I moves the atoms from positions $\mathbf{n} + \mathbf{f}(\mathbf{Q}.\mathbf{n})$ to $\mathbf{n} + \mathbf{f}(\mathbf{Q}.\mathbf{n} + \epsilon)$. Therefore, the displacement in V_E is given by $\epsilon \mathbf{f}'(\mathbf{Q}.\mathbf{n})$. The eigenvector of a phonon which may be considered as a phason is the derivative of the modulation function (cf. Fig. 5.5). That means that the oscillation becomes more localized if the modulation function squares up.

Phasons are not new modes in the strict sense. This can be seen easily from the situation of an incommensurate modulated phase. Above the phase transition modes at $\pm \mathbf{q}_c$ become soft (see Fig. 5.7). Below the phase transition the modes at $(1+\epsilon)\mathbf{q}_c$ and $(-1+\epsilon)\mathbf{q}_c$ are coupled. Their combinations give the modes with phason and amplitudon character. Therefore, two soft modes give one phason and one amplitudon mode. So the number of modes does not increase in the aperiodic structure.

When the modulation functions have discontinuities a displacement in internal space may give rise to phason jumps. But also the harmonic excitations may have a phason character. However, the frequency of the latter is no longer zero: there is a phason gap. For the Frank–Van der Merwe model the 'harmonic' phason mode is the phonon which describes a sliding of the chain over the substrate. Besides these are the phason jumps which may occur localized or as a collective excitation. However, these jumps are solutions to the non-linear problem, the 'harmonic' excitations are solutions to a linear problem.

When in the Frank–Van der Merwe model the interaction parameter λ is varied, three quantities have a characteristic behaviour. There is no phason gap for $\lambda < \lambda_{rmc}$ and a phason gap above the critical value. The participation ratio is large for small and for large values of λ but drops strongly around the critical value, and, a discontinuity in the modulation function sets in at λ_c. The participation ratio behaviour is anomalous. If the periodic substrate is replaced by a quasiperiodic substrate, there is also a critical interaction parameter, but the participation ratio remains small (localized excitations) for $\lambda > \lambda_c$ (van Erp et al., 1999). The three quantities are shown in Fig. 5.6. Phason jumps do no show up here, because the phonons have small displacements.

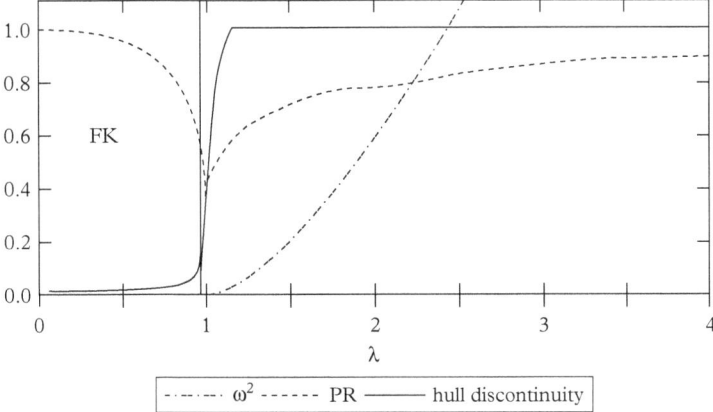

Figure 5.6 *The behaviour of the discontinuity, the phason gap, and the participation ratio for the Frank–Van de Merwe model, as a function of the interaction parameter. (van Erp et al., 1999)*

5.4.2.4 The DIFFFOUR model

The DIFFFOUR (for Discrete Frustrated Phi to the Fourth) model is a one-dimensional model of particles with first- and second-neighbour interactions. The ground state of the model is a modulated chain, as will be discussed in Chapter 6, Section 6.3.3. The phonons in this model may be calculated in the harmonic approximation as oscillations around the ground-state positions.

For the dynamics it is interesting to consider two versions of the standard DIFFFOUR model. One has potential energy for the one-dimensional system

$$V = \sum_n \left[\frac{A}{2}x_n^2 + \frac{1}{4}x_n^4 + Bx_n x_{n+1} + Dx_n x_{n+2} \right]. \quad (5.17)$$

The other version is a chain of particles with up to third-neighbour interactions, situated at positions $na + y_n$:

$$V = \sum_n \left[\frac{\alpha}{2}(y_n - y_{n-1})^2 + \frac{1}{4}(y_n - y_{n-1})^4 \right.$$
$$\left. + \frac{\beta}{2}(y_n - y_{n-2})^2 + \frac{\delta}{2}(y_n - y_{n-3})^2 \right]. \quad (5.18)$$

The main difference is that Eq. 5.18 has translational invariance, which means a zero frequency acoustic mode, and Eq. 5.17 has not. (Substitute a constant for x_n and y_n, respectively, in the equations to see the difference.) The two versions correspond to two different physical situations. The DIFFFOUR model considers only one degree of freedom per unit cell of the unmodulated system. To go from a physical system with more degrees of freedom to this model, one eliminates irrelevant degrees, which do not play a role in the phase transition. Very often the remaining degrees correspond to normal coordinates, the coefficients of eigenmodes. If these eigenmodes belong to the acoustic branch, then the long-wavelength excitations have frequency close to zero, as a consequence of translational invariance of the whole system. Then the differences of the positions are relevant, and Eq. 5.18 is the one to choose. If the relevant degrees belong to an optic mode, then Eq. 5.17 is the appropriate choice.

We first continue with the first expression. If the equilibrium positions are $\bar{x}_n$, and the displacements are given by u_n, then the equations of motion are

$$m\omega^2 u_n = (2A + 3\bar{x}_n^2)u_n + B(u_{n+1} + u_{n-1}) + D(u_{n+2} + u_{n-2}). \quad (5.19)$$

If the ground state is given by $\bar{x}_n = 0$ (all n), then the solutions are

$$u_n \sim \exp(i(2\pi kn - \omega t)), \quad \text{with} \quad \omega_k^2 = 2A + 2B\cos(2\pi k) + 2D\cos(4\pi k).$$

For varying values of the parameter A, the squared frequency ω^2 may become negative, which implies an instability. There are arguments to consider A as a function of temperature. Then the lowest frequency shifts with temperature: it is a soft mode. The minimum of the dispersion curve is at $k_0 = 0$, $k_0 = \frac{1}{2}$, or $k_0 = \cos^{-1}(B/4D)/2\pi$. In the third case the first mode becoming unstable when A is, decreased is, in general, incommensurate, with a wave vector at an arbitrary point in the Brillouin zone. The instability occurs for a value of A for which $2A + 2B\cos(2\pi k_0) + 2D\cos(4\pi k_0) = 0$. The mode at k_0 becomes soft and reaches the stability limit for the critical value of A. This behaviour has been found, for example, in ThBr$_4$ (Fig. 5.7).

For values of A below the critical value, the set of coupled equations of motion becomes infinite, and the eigenmodes can only be calculated numerically, using approximants. For an approximant there are a finite number of branches. Just as in the Frank and Van der Merwe model, there appears a branch which tends to a very low frequency if the wave vector goes to zero, and the minimum value tends to zero when the approximant goes to infinity. It is a combination of the two degenerate soft modes at $\pm k_0$. It may be described as a fluctuation in the phase of the modulation function. There is a second combination which corresponds to an oscillation in the amplitude of the modulation function. The two elementary excitations are called *phason* and *amplitudon*, respectively (see Fig. 5.5). The frequency of the phason remains zero when A is lowered below the critical value, but the frequency of the amplitudon rises sharply. As will be shown in Section 6.3.3, a commensurate phase may appear at lower temperature. Then there is no longer a zero frequency mode, because the phason has a gap, and there is no acoustic mode.

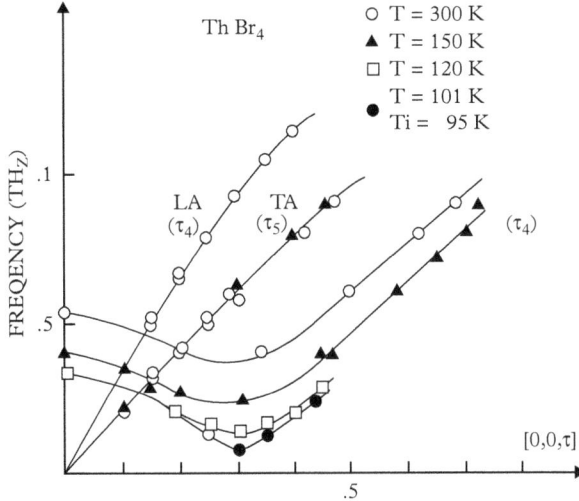

Figure 5.7 *The soft mode in ThBr$_4$ as measured with inelastic neutron scattering. The branch reaches the stability limit for an incommensurate wave vector. (Bernard et al., 1983)*

The superspace approach has been very successful for the description of quasiperiodic structures. This is much less the case for physical properties, although it may support the understanding and description of other phenomena as well. An important difference between a quasiperiodic structure and excitations of the latter is that an excitation may be quasiperiodic as well, but generally the rank is higher than that of the structure. Suppose the modulation is very weak. Then elementary excitations are mainly plane waves. A plane wave with wave vector **q** in a modulated chain with modulation wave vector **Q** may be embedded in a $(D + 2)$-dimensional space (D being the dimension of the physical space). There is, however, an embedding in $(D + 1)$-dimensional superspace. For a simple example, consider a chain with lattice constant a, wave vector q of the modulation $u_n = A\sin(2\pi qna)$, and modulation u_n at position na. Then the embedding in the two-dimensional superspace of the crystal structure consists of the lines $(na + A\sin(2\pi qna + 2\pi t_I), t_I)$, and the vibrating chain oscillates around these lines with frequency ω. The oscillations occur in this case horizontally. This is, however, not unique. It is useful if the oscillation occurs in the amplitude of the modulation. For oscillations that are more like oscillations in the phase of the modulation (phasons), the idea is better visualized with vertical oscillations (in V_I), but this is an arbitrary choice. The total number of degrees of freedom remains the product of the physical dimension (D) and the number of particles in the sample.

The description of excitations with a motion in V_I becomes different in the case that the atomic surfaces are discontinuous, as in quasicrystals and modulated phases with discontinuous modulation functions. In this case, arbitrary small displacements in V_I may create jumps of atoms over small but not arbitrary small distances. The description of such phason modes may well be given as fluctuations in V_I (Fig. 5.1).

For the other (translationally invariant) version (Eq. 5.18) there is, of course, always a zero frequency acoustic mode at wave vector 0. The oscillations around equilibrium positions $y_n = 0$ have the frequencies

$$m\omega_k^2 = 4\left(\alpha \sin^2(\pi k) + \beta \sin^2(2\pi k) + \delta \sin^2(3\pi k)\right). \qquad (5.20)$$

When this expression is positive, the solution $y_n = 0$ is stable, or at least a local minimum. We have seen that the mean-field equations are qualitatively the same. This means that for decreasing temperature the effective value of α decreases, and this leads to a soft mode. Below a certain value of α the structure becomes unstable, and the oscillations then are around the new equilibrium points $\bar{y}_n$. The eigenfrequencies are then given by the eigenvalues of the equation

$$m\omega^2 y_n = \left(2\alpha + 2\beta + 2\delta + 3(\bar{y}_n - \bar{y}_{n+1})^2 + 3(\bar{y}_n - \bar{y}_{n-1})^2\right) y_n$$
$$- \left(\alpha + 3(\bar{y}_n - \bar{y}_{n+1})^2\right) y_{n;1} - \left(\alpha + 3(\bar{y}_n - \bar{y}_{n-1})^2\right) y_{n-1}$$
$$- \beta(y_{n+2} + y_{n-2}) - \delta(y_{n+3} + y_{n-3}).$$

If the incommensurate phase is approximated by a rational approximant with modulation vector L/N there are N branches of dispersion curves. One of the branches is linear in the wave vector of the oscillation k. It is the acoustic mode. Near the transition from the unmodulated to the modulated phase, the model gives a second branch going to zero at $k = 0$. Its eigenvector describes a shift of the phase of the modulation function: if the equilibrium positions are $na + \bar{y}_n = na + f(2\pi Qn)$, then the eigenvector of the mode is given by $u_n = df(x)/dx$ in $x = 2\pi Qn$. The excitations of this branch are the phasons. In addition, another branch is that of the amplitudon, in which the eigenvector corresponds to a change in the amplitude of the modulation. An example of the branches for an approximant is given in Fig. 5.8. The lowest branch which is linear in k is the phason branch, the other linear branch is the acoustic branch.

For still lower values of α, corresponding to lower temperatures in the mean-field approximation, a gap opens up in the phason branch. The phason no longer has zero frequency for $k = 0$. An analysis of the modulation function shows that the gap opens up as soon as the modulation function becomes discontinuous. This has been found for the DIFFFOUR model in Janssen and Tjon (1983) and

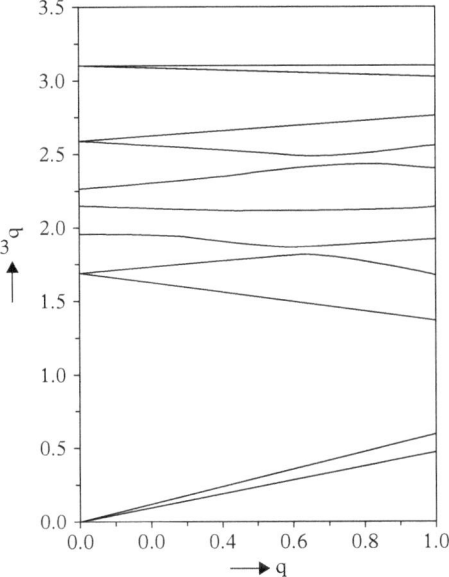

Figure 5.8 *Phonon branches for an $N = 11$ approximant in the regime with a smooth modulation function. Plotted are the curves of the frequencies ω_q of the 11 phonons as function of the wave vector q. The two lowest branches are the phason and the acoustic branch. (Janssen and Tjon, 1982)*

Janssen et al. (2002). In this discommensuration regime the eigenvectors change character. The modulation functions are no longer continuous and the phason eigenvector (the phonon eigenvector with phason character) now involves an oscillation of the discommensuration walls (see Fig. 5.9). It is concentrated on the atoms in the domain walls, and describes a movement of the wall to the right or to the left. The domain wall could move by a lattice constant. However, this is not a phason in the sense we have studied here. It is more like a phason jump.

Using the potential energy of Eq. 5.17 in a mean-field approximation gives a similar result: the typical behaviour for an incommensurate phase is the following. Above T_i there is a soft mode with a frequency tending to zero at T_i. Between T_c and T_i there are a phason branch and an amplitude branch. The frequency of the latter increases sharply from zero below T_i. The frequency of the phason branch tends to zero for a wave vector tending to zero for temperatures between T_d and T_i, but goes to a finite value at the zero wave vector for $T_c < T < T_d$. The transition at T_d is very smooth. There is only a discontinuity in the third derivative of the free energy (cf. Fig. 5.10).

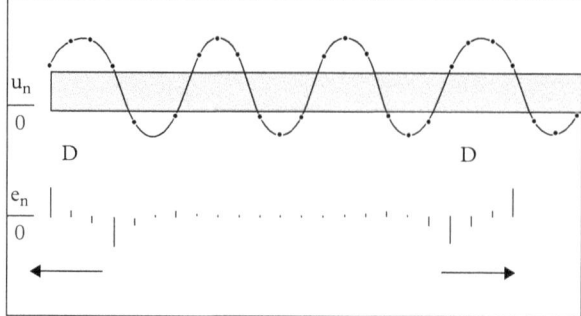

Figure 5.9 *The lowest phason vibration in the discommensuration regime, where the modulation function is discontinuous. Shown is a period 19 approximant. Above: The modulation function u_n. Two discommensurations are indicated by 'D'. The shaded area is avoided. Below: The eigenvector e_n of the phason. It is concentrated in the 'D' regions. There it describes a displacement of the wall to the left or to the right (arrows). (Janssen and Tjon, 1982)*

5.4.2.5 Measuring phonons

When there are s particles in a three-dimensional unit cell, there are $3s$ phonon branches, and a phonon is characterized by $\mathbf{q}$ and the branch index ν ($= 1, \ldots, 3s$). When the soft mode leads to an N-fold superstructure, the Brillouin zone shrinks by a factor N and the branches are numbered $1, \ldots, 3sN$. When the soft mode is at an incommensurate wave vector, the Brillouin zone shrinks to zero and the number of branches goes to infinity. However, when the amplitude of the modulation is small, the phonons are still similar to those of the original basic structure. When the wave vector of the instability is $\mathbf{q}_0$, the modes at $\mathbf{q}$ couple with those at $\mathbf{q} + m\mathbf{q}_0$, and the coupling is weak for small amplitude. This can be seen in the expression for the inelastic neutron scattering. The scattering of a neutron with initial wave vector $\mathbf{k}_i$ and final wave vector $\mathbf{k}_f$ such that the momentum transfer is $\mathbf{Q}$ and the energy difference is $\hbar\omega$ is determined by the function

$$S(\mathbf{Q}, \omega) = \sum_{\mathbf{q},\mathbf{K},\nu} \left| \sum_j \frac{b_j}{\sqrt{M_j}} e^{-W_j(\mathbf{Q})} \mathbf{Q} \cdot \mathbf{e}_{\mathbf{q}\nu|j} e^{2\pi i \mathbf{Q} \cdot \mathbf{r}_j} \right|^2 \delta(\mathbf{Q} - \mathbf{q} - \mathbf{K}) \delta(\omega - \omega_{\mathbf{q}\nu}) \frac{n(\mathbf{q}\nu)}{\omega}. \tag{5.21}$$

This is the *dynamic structure factor* or *scattering function*. Here the sum over $\mathbf{q}$ runs over the Brillouin zone and the sum over $\mathbf{K}$ over the reciprocal lattice. The sum over j runs over the particles in the unit cell, which have positions $\mathbf{r}_j$. Here W_j is

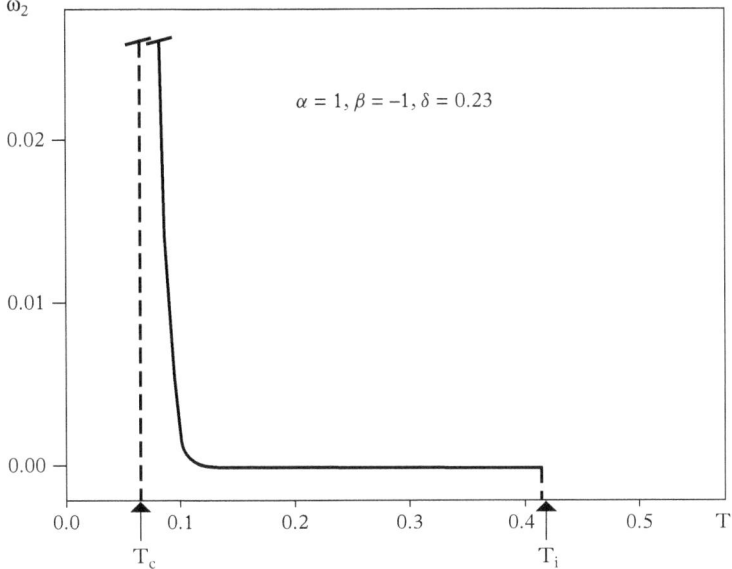

Figure 5.10 *The frequency of the lowest phason mode, in the DIFFFOUR model, using the mean-field approximation. As a function of temperature, the frequency remains zero between T_i and T_d. At T_d the phason gap opens. At T_c there is a phase transition to the superstructure. (Janssen and Tjon, 1983)*

the Debye–Waller factor, and $n(\mathbf{q}\nu)$ is the temperature-dependent occupation of the mode labelled by $\mathbf{q}\nu$. The scattering therefore shows the dispersion curves: the maximum intensity occurs along these curves. However, the intensity is strongly dependent on $\mathbf{q}$ and ω. In Fig. 5.11 the intensity is plotted for $N=2$ for a relatively low amplitude. The intensity is maximal along the lines of the original dispersion curve, which is a sine function. However, in principle there is intensity at N values of ω for each $\mathbf{q}$. For an incommensurate phase this means that there is intensity for an infinite number of frequencies at each $\mathbf{q}$, but nevertheless generally dispersion curves will still be discernible in the scattering, because of the different contribution of modes to $S(\mathbf{Q},\omega)$. This difference also plays a role for the distinction between phason and amplitudon. In Fig. 5.11 both branches are plotted for the DIFFFOUR model, for a certain choice of the parameters, for which one has an incommensurate phase. The intensity of the scattering is indicated by the length of the lines at a point $\mathbf{q}$, ω. The lowest frequency of the phason branch is zero, while that of the amplitudon branch is positive. The phason contributes more strongly between the modulation wave vector (the minimum) and zero, whereas the amplitudon branch scatters more strongly between the modulation wave vector and the Brillouin-zone border.

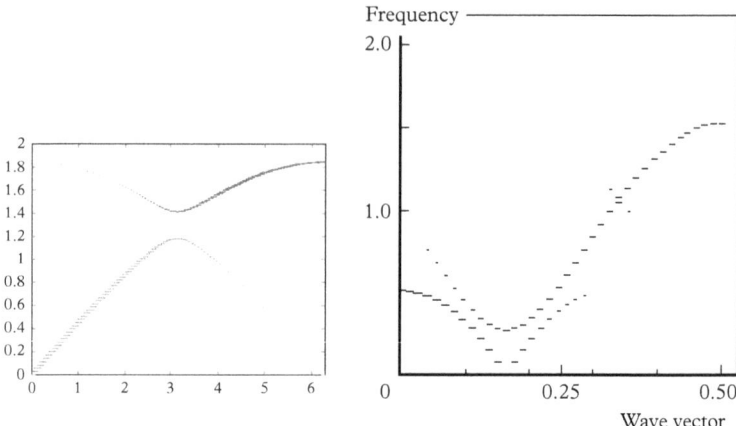

Figure 5.11 *Left: Intensity of the scattering in the* **q**, ω *plane for the case of a twofold superstructure. Right: The intensity in the incommensurate phase for the two modes (amplitudon and phason) into which the two soft modes combine. (Intensity is proportional to the length of the traits.)*

5.4.2.6 Double chain model

A simple model for composites is the *double chain model*. It consists of two subsystems, each consisting of one chain or array of chains. All the chains are parallel and the lattice constants of the chains in different subsystems are incommensurate. Each subsystem is modulated because of the interaction with the other subsystem. We take here the simple case of two parallel chains. The equilibrium positions of the atoms are $x_0 + na + f(x_0 + na)$ and $y_0 + nb + g(y_0 + nb)$, respectively. From the potential energy (cf. Chapter 6, Eq. 6.19)

$$V = \sum_n V_1(x_n - x_{n-1}) + \sum_m V_2(y_m - y_{m-1}) + \frac{\lambda}{2} \sum_{nm} V_3(x_n - y_m),$$

where V_1 and V_2 are the nearest-neighbour interactions between particles in the first, and second chain, respectively, and V_3 the interaction between particles in different chains, follow the equations for the phonons. When the displacements are $u(n)\exp(i\omega t)$ and $v(m)\exp(i\omega t)$, respectively, the eigenmodes are solutions of

$$m_1 \omega^2 u_n = [V_1''(x_n - x_{n-1}) + V_1''(x_{n+1} - x_n)]u_n$$
$$- V_1''(x_n - x_{n-1})u_{n-1} - V_1''(x_{n+1} - x_n)u_{n+1} + \frac{\lambda}{2}\sum_m V_3''(x_n - y_m)v_m,$$

(5.22)

$$m_2\omega^2 v_m = [V_2''(y_m - y_{m-1}) + V_2''(y_{m+1} - y_m)]v_m$$
$$- V_2''(y_m - y_{m-1})v_{m-1} - V_2''(y_{m+1} - y_m)v_{m+1} - \frac{\lambda}{2}\sum_n V_3''(x_n - m)u_n$$
(5.23)

or, when the interactions in the chains are harmonic and the spring constants α and β,

$$m_1\omega^2 u_n = \alpha(2u_n - u_{n-1} - u_{n+1}) + \lambda/2 \sum_m V_3''(x_n - y_m)v_m, \quad (5.24)$$

$$m_2\omega^2 v_m = \beta(2v_m - v_{m-1} - v_{m+1}) - \lambda/2 \sum_n V_3''(x_n - y_m)u_n, \quad (5.25)$$

where the second derivatives are taken at the equilibrium positions.

For vanishing value of λ the chains are decoupled, and the dispersion relations are just the superposition of those for the two chains (Fig. 5.12). In that case each chain has a zero frequency acoustic mode, corresponding to a uniform displacement of the chain. When $\lambda \neq 0$, the phonons in both chains are coupled (Fig. 5.13). However, there remains one zero frequency mode, with displacements $u_n = v_m =$ constant (all n, m). This is a linear combination of the two zero frequency modes in the separate chains. In the same way, the phonons in the double chain model are combinations of phonons in the two chains. Phonons may propagate in one of the chains but, in general, both chains are involved. This can be seen from the participation of each chain in a mode. This is defined as the quantity

$$P_1 = \sum_n |u_n|^2, \quad \text{and} \quad P_2 = \sum_m |v_m|^2 \quad (P_1 + P_2 = 1):$$

$P_1 + P_2 = 1$ means normalization of the mode. For wave vectors for which the dispersion curves of both chains intersect, the modes mix and the participation of each chain is of the order of a half (Fig. 5.12).

The two zero frequency modes for the uncoupled chains give the zero frequency mode of the coupled system as a symmetric combination. The asymmetric combination which leaves the centre of gravity of the whole system invariant is the motion of one chain with respect to the other. As in the Frank–Van der Merwe case, this motion may also have zero frequency. Then it is not simply a rigid shift of one chain with respect to the other. Because of the interaction there will be an incommensurate modulation. Again the mode corresponding to a rigid shift of the higher-dimensional periodic structure in internal space has zero frequency, provided the atomic surfaces are smooth, that is, the modulation functions continuous. This phonon mode therefore, is a phason, which is in the framework of composites usually called a *sliding mode*.

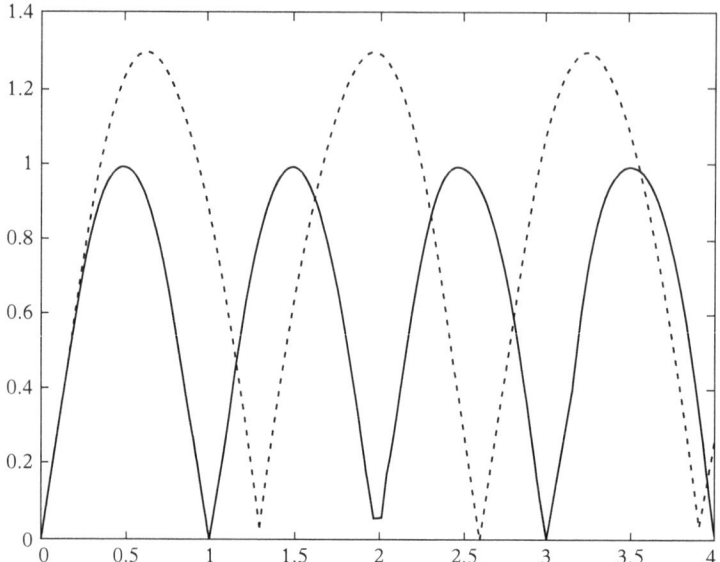

Figure 5.12 *The dispersion curves when there is no coupling between the two chains. When $\lambda \neq 0$, the modes corresponding to intersections of the curves are particularly coupled. For the modes originating from mixing at the intersection points, the participation ratio is of the order of 1/2.*

The dynamical properties of aperiodic composites were first investigated on the mercury-chain $Hg_{3-\delta}AsF_6$ compound (Pouget *et al.*, 1978; Heilmann *et al.*, 1979). Above the phase transition temperature of 120 K, the mercury chains are disordered and show liquid-like behaviour. Below the phase transition, parallel Hg chains order and form 3D incommensurate modulated structures in the x-y plane. (The chains run either in the x- or in the y-direction.) One observes a small gap in the Hg-chain dispersion. Longitudinal acoustic modes have been found around the reflections of the AsF_6 subsystem, the acoustic branches around the Hg subsystem reflections have been interpreted as the sliding mode.

Another material is the layered structure $Bi_2Sr_2CuO_{8+\delta}$ or $Bi_2Sr_2CuO_{6+\delta}$ (Etrillard *et al.*, 2001, 2004). (See, for the structure, Section 5.8.2.) Here, the Bragg peaks are at positions $h\mathbf{a}^* + k\mathbf{b}^* + \ell\mathbf{c}^* + m\beta\mathbf{b}^*$. In this case, the distinction between guest and host is not as clear as in the case of the Hg-chain compound. Here the results of the neutron scattering results could be explained by a model with two weakly interacting subsystems. For lower energies (frequency below 10 MHz), one observes a single acoustic mode.

More recent are the experiments on *n*-nonadecane–urea (Toudic *et al.*, 2011a). In this case in the plane perpendicular to the chains' longitudinal acoustic and transverse acoustic modes were observed, with the damping tending to zero for

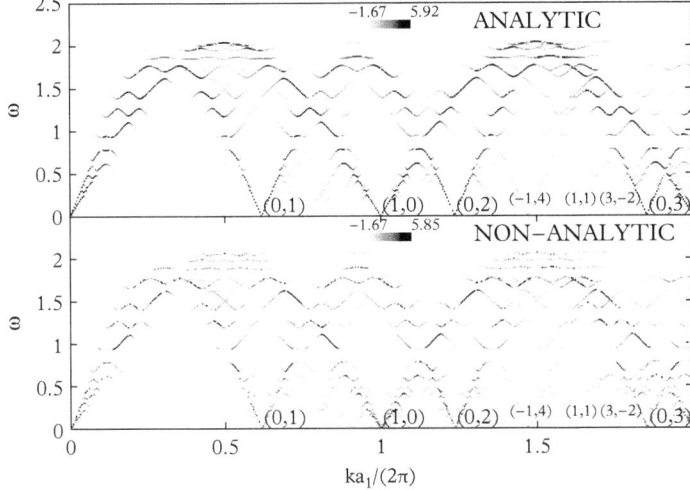

Figure 5.13 *The scattering function S(Q, ω) when the coupling between the two chains is small (continuous modulation function) or strong (discontinuous modulation function, respectively. (Radulescu et al., 2002)*

the wave vector going to zero. There are some indications that, in this case, the intermodulation, the modulation of the alkane system by the interaction with the urea system, is discontinuous. In the planes through the alkane chains, a transverse acoustic mode was observed, with a damping that follows the q^2 law predicted by hydrodynamic theory.

In the incommensurate spin-ladder system $Sr_{14}Cu_{24}O_{41}$, inelastic neutron scattering experiments have shown sliding modes with a small gap (Chen et al., 2016). The compound has 1D chains incommensurate with a 2D spin-ladder substructure. As well as phonons, magnons have been measured, with a steep dispersion curve. The $S(\mathbf{Q},\omega)$ shows a behaviour with several pseudo-acoustic branches, with polarization along the chain direction, similar to the curves in Fig. 5.13.

However, for the relation between theory and experiment for the dynamics of aperiodic composites there remain several unresolved questions.

5.4.2.7 Fibonacci chain

A special case of the modulated chain model is the *Fibonacci chain*. In this case, the energy is given by

$$E = \sum_n \left(p_n^2/2m_n + \alpha_n(u_n - u_{n-1})^2/2 \right), \quad (5.26)$$

where the masses m_n, the spring constants α_n, or both have two values in the order of a Fibonacci series of L's and S's. The formula is

$$m_n = \begin{array}{l} m \text{ if } Frac(n\Phi) < \Phi \\ M \text{ else} \end{array}; \quad \alpha_n = \begin{array}{l} \alpha \text{ if } Frac(n\Phi) < \Phi \\ \beta \text{ else} \end{array}. \quad (5.27)$$

This is a modulated chain with a discontinuous modulation function.

For the eigenvibrations with frequency ω, the equations of motion lead to

$$m_n \omega^2 e_n = (\alpha_n + \alpha_{n+1})e_n - \alpha_n e_{n-1} - \alpha_{n+1} e_{n+1}.$$

For an approximant $\Phi \approx L/N$ the problem becomes periodic with N particles in the unit cell. Then there are N branches of phonons for $-\frac{1}{2} \leq k \leq \frac{1}{2}$. The resulting VDOS has $N-1$ gaps, but their values differ significantly. When the masses and the spring constants become $m = M$ and $\alpha = \beta$, then there are no gaps. For small differences the main gap is located at $k = L/2N$ but, in fact, an infinite number of gaps open. This can be seen from the IDOS. In a gap the IDOS is constant. The resulting curve is a devil's staircase. The gaps turn up in the dispersion curves as well (cf. the corresponding dispersion curves for electrons in Fig. 5.24). Taking free boundary conditions for a finite chain with N particles instead of periodic boundary conditions, there are N modes. If we plot the frequency against the number of the mode (the modes are numbered in order of frequency), the result is a curve with gaps. For the periodic limit (equal masses and spring constants) this is just the familiar dispersion curve for a linear harmonic chain. For non-vanishing differences the result is a curve with gaps. The gaps correspond to wave vectors which are multiples of the modulation wave vectors (modulo the reciprocal lattice). In this way the gaps may be labelled by the order of the wave vector to which they belong.

The dynamical structure factor $S(\mathbf{Q}, \omega)$ for the Fibonacci chain (Fig. 5.14) shows the essential features of scattering from quasicrystals. One distinguishes phonon branches emanating from the strong Bragg peaks in the (static) structure factor $F(\mathbf{H})$, the broadening of these branches for higher values of ω, because of the critical character of the eigenvectors, the gaps, and relatively flat bands for higher frequencies (the optic modes). This picture is in agreement with the experimental findings, to be discussed later in this chapter.

In the periodic limit the eigenvectors of the dynamical matrix are simply harmonic waves. When the differences in the spring constants do not vanish, the low-frequency eigenvectors keep a similarity to those of the periodic case. There is always an acoustic zero frequency mode, but no zero frequency phason, because the system is modulated with a discontinuous modulation. For higher frequencies the vibrations are much more localized in the unit cell. The vibrations are of the form $\exp(2\pi i k x) U(x)$, where U has the periodicity of the lattice, that is, its Fourier transform contains the vectors of the reciprocal lattice. If the modes are strongly localized inside the unit cell, the number of Fourier components is large. Many neighbouring reciprocal lattice vectors may contribute in that case. In the aperiodic limit the spacing of the reciprocal lattice tends to zero, and the Fourier decomposition becomes a broad distribution. Instead of the delta

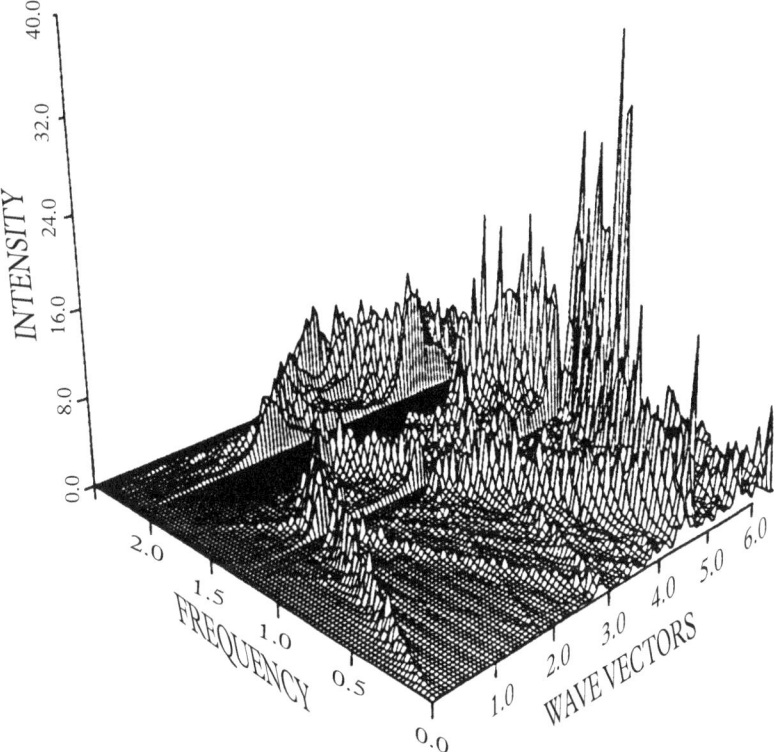

Figure 5.14 *The scattering function $S(\mathbf{Q}, \omega)$ for a Fibonacci chain with spring constant 1 and masses m and M.*

peaks at the original reciprocal lattice one finds a broad peak. This is one of the possible mechanisms for a broadening of the dispersion curves observed in inelastic neutron scattering.

In these calculations for the Fibonacci 'quasicrystal', there are no modes with phason character. The reason is, of course, that in the harmonic approximation only small displacements are considered, and phason jumps are certainly out of the range of the harmonic approximation. To study the dynamics of phason jumps in a simple model one has to incorporate non-linear terms. This will be done in a later section.

♯ Besides the direct diagonalization, other methods for finding the spectrum exist. We will describe here the *trace map* method, especially useful for studies of the mathematical properties in one dimension. Consider the case that the spring constants are all α, and that the masses m_n, assuming two values, follow a Fibonacci sequence. Then the equations of motion are

$$m_n\omega^2 u_n = \alpha(2u_n - u_{n-1} - u_{n+1}),$$

from which follows that

$$u_{n+1} = \left(2\alpha - m_n\omega^2\right)/\alpha - u_{n-1}.$$

Introduce next a two-dimensional vector $\mathbf{v}_n$ with components u_n and u_{n-1}. Then the equation of motion may be written as a linear transformation:

$$\mathbf{v}_{n+1} = T_n \mathbf{v}_n; \quad \text{with } T_n = \begin{pmatrix} 2 - m_n\omega^2/\alpha & -1 \\ 1 & 0 \end{pmatrix}.$$

Iteration of this map gives a sequence of vectors, which remain bounded only if the product of all the matrices T_n keeps the trace smaller than 2 in absolute value. The condition is that

$$|\text{Trace} \prod_j T_j| \leq 2. \tag{5.28}$$

The Fibonacci sequence may be obtained by concatenation: $w_1 = S$, $w_2 = L$, and w_n is the concatenation of w_{n-1} and w_{n-2}. The limit of $n \to \infty$ gives the Fibonacci chain. If we introduce a matrix M_n for each Fibonacci number F_n as

$$M_n = \prod_{j=1}^{F_n} T_j,$$

then we have the relation

$$M_{n+1} = M_n M_{n-1}.$$

Two-dimensional matrices A with determinant ± 1 satisfy

$$A^2 = \text{Tr}(A)A - \text{Det}(A)E,$$

where E is the unit matrix. Applied to the recurrence relation for the matrices M_n this gives

$$\text{Tr}(M_{n+1}) = \text{Tr}(M_{n-1}M_n) = \text{Tr}(M_{n-1}^2 M_{n-2})$$
$$= \text{Tr}(M_{n-1})\text{Tr}(M_{n-1}M_{n-2}) - \text{Tr}(M_{n-2}).$$

If $x_n = \text{Tr}(M_n)$, then the relation between the traces becomes

$$x_{n+1} = x_{n-1}x_n - x_{n-2}. \tag{5.29}$$

This can be formulated as a three-dimensional non-linear map, if one writes $x = x_{n-1}, y = x_n, z = x_{n+1}$:

$$(x, y, z) \rightarrow (y, z, yz - x). \tag{5.30}$$

Actually, the motion is two-dimensional because there is an invariant:

$$I = x^2 + y^2 + z^2 - xyz.$$

The map is called the trace map (Kohmoto et al., 1983). For given value of ω an orbit is created. If the absolute value of x_n remains inside the interval from -2 to $+2$, the value ω is in the spectrum. ♮

5.4.2.8 Two-dimensional tiling models

The results for the Fibonacci chain are peculiar because the system is one-dimensional. Higher-dimensional analogues, however, share some of these properties. The singularities, which appear due to the fact that the system is one-dimensional, disappear, of course.

The procedure for studying lattice vibrations in higher-dimensional quasicrystalline materials is similar to that in one dimension, where the Fibonacci chain was studied. In two dimensions an example is a tiling, where atoms are placed in the vertices or in the centres of the tiles. Let us consider an octagonal tiling with atoms on the vertices. The Hamiltonian in the harmonic approximation is

$$H = \sum_{\mathbf{n}} \left(\mathbf{p_n}^2/2m + \sum_{\mathbf{m}} V(\mathbf{u_n} - \mathbf{u_m}) \right), \tag{5.31}$$

where $\mathbf{m}$ runs over all vertices connected to $\mathbf{n}$ via an edge of the tiling. The number of interacting particles depends on the site, with the coordination number not being a constant. The potential V is the sum of a longitudinal and a transversal component, and becomes, in the harmonic approximation,

$$V(\mathbf{u_n} - \mathbf{u_m}) = \frac{1}{2}\alpha_{\parallel}(\mathbf{u_n} - \mathbf{u_m})_{\parallel}^2 + \frac{1}{2}\alpha_{\perp}(\mathbf{u_n} - \mathbf{u_m})_{\perp}^2,$$

with $\mathbf{u} = \mathbf{u}_{\parallel} + \mathbf{u}_{\perp}$, and $\mathbf{u}_{\perp}.(\mathbf{n} - \mathbf{m}) = 0$. As discussed earlier an approximant to a quasiperiodic tiling is obtained by a strain in superspace. An example of a unit cell for an approximant for the octagonal tiling is given in Chapter 3, Fig. 3.15. It has N particles in the unit cell, and the number of phonon branches is $2N$. In the literature one finds such simple vibrational models on octagonal, pentagonal, and dodecagonal tilings.

The most frequently used technique is that of rational approximants. For the octagonal, pentagonal, and dodecagonal cases, this comes down to using the continued fraction expansions of $\sqrt{2}$, Φ, and $\sqrt{3}$. This leads to the solution of the eigenvalue problem for a finite dynamical matrix. Because only nearest-neighbour interactions are considered in these models, the matrices are sparse. There are many zeros. This property can be used for making the diagonalization more

342 Physical properties

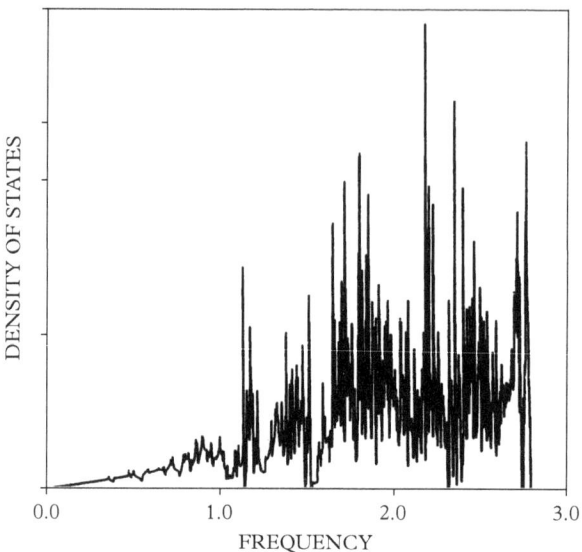

Figure 5.15 *The low-frequency VDOS for an approximant to the octagonal tiling.(Los et al., 1993a)*

efficient. The VDOS has much more structure than that for a periodic crystal (Fig. 5.15). There are several gaps or pseudo-gaps. It is interesting that the VDOS becomes linear for low frequencies. The slope can be related to the average spring constant. The influence of the quasiperiodicity here disappears and the long-wavelength excitations with low frequency are just the sound waves.

An even simpler model is the combination of two one-dimensional aperiodic chains. Suppose we have solved the problem for an aperiodic chain:

$$M\omega^2 u_n = \alpha_n(2u_n - u_{n-1} - u_{n+1}),$$

where the α_n are the spring constants in the aperiodic chain with first-neighbour interactions. We assume that there are two different spring constants: one for the long intervals and one for short intervals. If x_n are the positions in the chain, we construct a two-dimensional tiling with vertices on the positions $\mathbf{u}(n,m) = (x_n, x_m)$. Our first model assumes one degree of freedom for each site, $u(n,m)$. The Hamiltonian then is

$$H = \frac{1}{2}\sum_{n,m}\Big(p(n,m)^2/M + \alpha_n(u(n,m) - u(n-1,m))^2 \\ + \alpha_m(u(n,m) - u(n,m-1))^2\Big).$$

If u_n is a one-dimensional mode with frequency ω_1, and v_m another with frequency ω_2, then the eigenvibrations of the two-dimensional system are $u(n,m) = u_n v_m$ with frequency ω such that

$$\omega^2 = \omega_1^2 + \omega_2^2.$$

The density of states is the convolution of the one-dimensional VDOS:

$$DOS_{2D}(\omega^2) = \int DOS_{1D}(z) DOS_{1D}(\omega^2 - z)\,dz. \qquad (5.32)$$

When there are three degrees of freedom, and the particles have displacements from an orthorhombic lattice in three directions, the Hamiltonian is

$$H = \frac{1}{2} \sum_{\mathbf{n}} \sum_{i=1}^{3} \left(p(\mathbf{n})_i^2/M + \alpha(\mathbf{n})_i (u(\mathbf{n})_i - u(\mathbf{n} - \mathbf{a}_i)_i)^2 \right)/2. \qquad (5.33)$$

The Hamiltonian is separable. There are oscillations in the x-direction, with frequency ω_1; in the y-direction with frequencies ω_2; and in the z-direction. The VDOS is the sum of three copies of a one-dimensional VDOS. Numerically, these problems are easier then the two-dimensional problems on tilings. And they give a first idea of the properties of a two-dimensional system. An example is given in Fig. 5.16 for the modulated chain embedded in three dimensions. In more dimensions the singularities in the VDOS of one dimension disappear. In Ashraff et al. (1990) such two-dimensional arrays of Fibonacci chains with scalar phonons are studied. The IDOS shows plateaus, as in the one-dimensional case, at values Frac($m\Phi$) for integer values of m. The latter may be used to label the

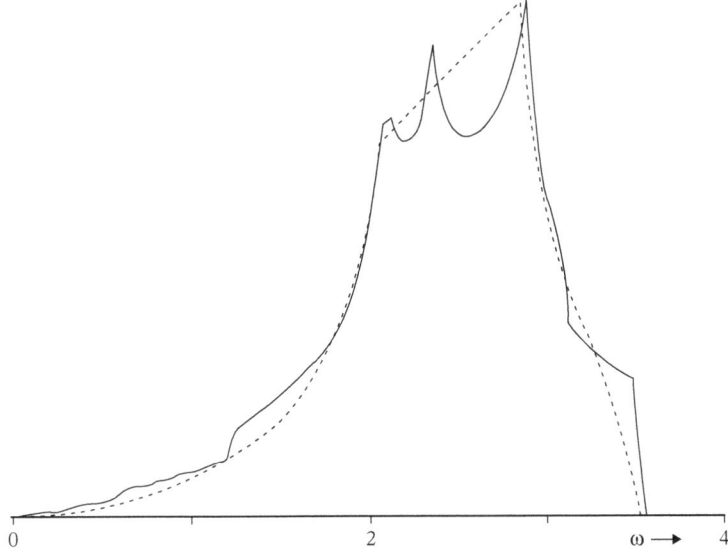

Figure 5.16 *The VDOS for a three-dimensional array of modulated chains, if the coupling between particles in parallel chains is a first-neighbour, harmonic interaction. The dashed line gives the VDOS for the unmodulated chains.*

gaps. For low frequencies the density of states is linear, as for a periodic system in two dimensions. This means that for long-wavelength phonons the underlying structure is less relevant.

5.4.2.9 The three-dimensional icosahedral tiling model

The standard model for studying vibrations in three-dimensional icosahedral quasicrystals is the icosahedral Ammann tiling, composed of tiles of two types. Here, as in the octagonal example, atoms are placed on the vertices or in the centres of the tiles. There are harmonic interactions between vertex atoms and their neighbours connected by tile edges, or between centre atoms and the atoms in adjacent tiles. The icosahedral situation is approximated by choosing an approximant with the golden number Φ in the internal coordinates replaced by the fraction L/N. A technical problem here is the rapid growth of the number of atoms per unit cell for higher approximants. The dimension of the dynamical matrix grows as N^3. This limits the numerical possibilities, but usually the spectra and densities of states converge rather rapidly to what is assumed to be the aperiodic limit. The results of these calculations in three dimensions confirm tendencies found in two dimensions, but in less pronounced fashion.

A particular property is the scaling property. If one considers a series of approximants, that is, using truncations of the continued fraction expansion, and one looks at the lowest 50 branches, the dispersion curves of two consecutive approximants are very similar, up to a change of frequency scale. This is a consequence of the self-similarity of the underlying structure (Fig. 5.17).

5.4.2.10 Models for phasons

Phasons may occur in any aperiodic crystal but, as we have seen, there is a fundamental difference between systems with continuous and those with discontinuous atomic surfaces. If the atomic surfaces are smooth and connected, besides the

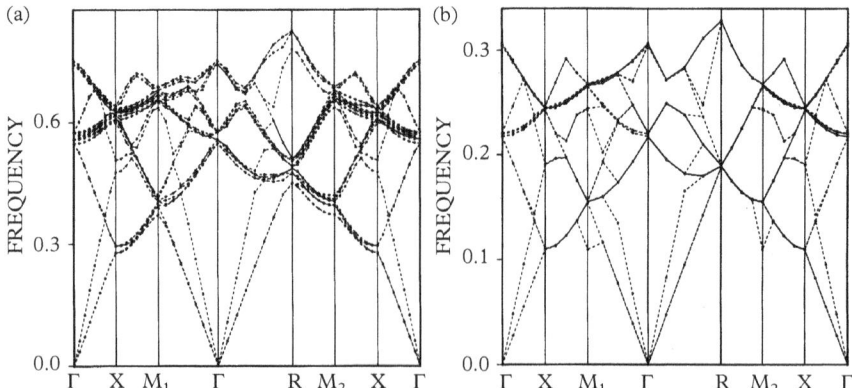

Figure 5.17 *Dispersion curves for two consecutive approximants of the three dimensional icosahedral tiling. The next approximant has practically the same curves at the lower part, up to a change in energy scale. (Los et al., 1993b)*

linear acoustic branch (if the system is translation invariant) there is a linear phason branch. To the other class belong composites with strong interaction between the subsystems, and quasicrystals. In these the phasons are either phason jumps or phonons with phason character. As already stated, it is not obvious, a priori, that hydrodynamic theory applies to quasicrystals. Even if it applies, one has to understand what the possible forces in play are. In particular, one may ask, what is the restoring force that brings back the system to the quasicrystalline state when a phason strain is applied? What are the interatomic forces that play a role in this process? Of course, answering these questions requires modelling at the atomic scale. We present in the following two frequently used 'toy' models: random tiling and matching rule tiling. In these models, phason jumps are accessible for study.

5.4.2.11 Random tiling models

A model for which extensive theoretical and numerical studies have been carried out is the random tiling model (Henley, 1991*b*). The main assumption of this model is that the quasicrystalline structure can be described by a decorated quasiperiodic tiling. The simplest one is the three-dimensional Penrose or the two-dimensional Amman–Beenker tiling constructed with two rhombi: a fat one and a thin one. This tiling is then decorated by atomic clusters, so that the edge length is at least of the order of the intercluster distance. A more complex tiling scheme has been proposed by Henley, called the *canonical cell tiling* (Henley, 1991*a*). Decorated tilings are sometimes useful for studying the dynamics, but they are not necessarily realistic models for real quasicrystals. Moreover, if icosahedral phase models can be mapped onto decorated tilings, it has not yet been proven that the decoration scheme is uniquely defined: it seems that a certain amount of chemical disorder has to be introduced which changes the decoration from one tile to the other. This being said, let us consider what has been obtained so far for random tiling models. The effect of a local distortion of the physical space on the resulting three-dimensional structure depends on the geometry of the atomic surfaces. We have seen that in the case of a Fibonacci sequence this results in permutations *LS* to *SL* in the sequence. For the three-dimensional Penrose tiling this results in a change inside a rhombic dodecahedron made of two fat and two thin rhombuses: the two possible sites are related by mirror symmetry along a two-fold direction (see Fig. 5.18), and are called dodecahedron sites hereafter. The simplest three-dimensional Penrose random tiling is obtained by a randomization of each dodecahedron site by a Monte Carlo move (it has been called 'maximally random tiling' by Henley, corresponding to the high-temperature limit). When performing these moves, new dodecahedron sites will appear and other disappear (cf. Fig. 3.17). In this model, the only energy coming into play is the configurational entropy.

It has been shown numerically that the free energy of this model random tiling is minimal in the absence of phason strain and increases quadratically with the amplitude of such strain, thus establishing the applicability of phason elasticity theory. In particular, the diffuse scattering intensity can be modelled by using a positive elastic constant ratio K_2/K_1 (Tang, 1990). An example of a simulated

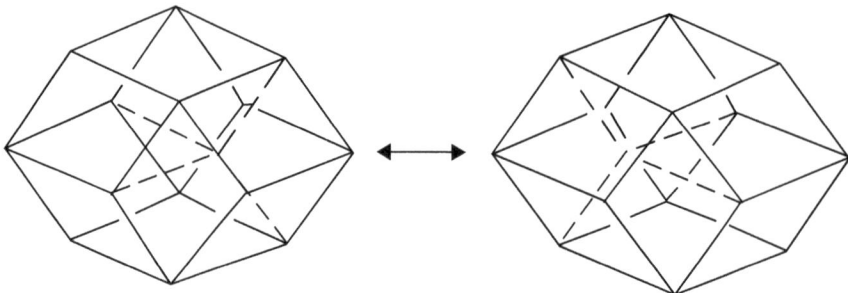

Figure 5.18 *Illustration of the two equivalent sites (or 'flip' sites) inside the rhombic dodecahedron of the icosahedral Ammann (3D Penrose) tiling. A three-dimensional random tiling is obtained by a Monte Carlo randomization of all these dodecahedral sites.*

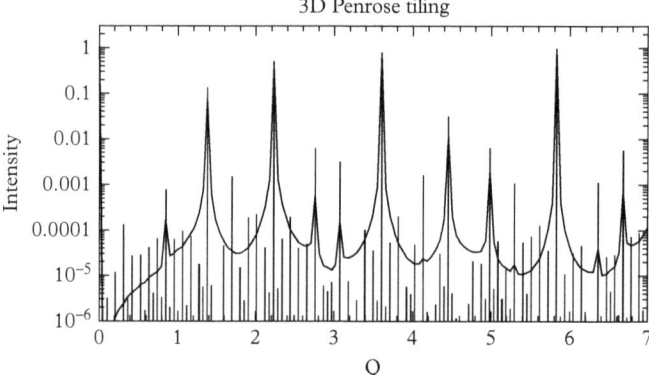

Figure 5.19 *Simulated diffraction pattern along a two-fold axis for a perfect three-dimensional Penrose tiling (vertical bars) and the corresponding three-dimensional random Penrose tiling (continuous line). Most of the weak peaks have disappeared, whereas a diffuse scattering contribution occurs between the Bragg reflections. (Courtesy M. Mihalkovič)*

diffraction pattern is shown in Fig. 5.19 whereas the ideal three-dimensional Penrose tiling only displays Bragg peaks, shown as vertical bars, the random tiling diffraction pattern has both a Bragg and a diffuse scattering component. It can also be observed that most of the weak Bragg reflections of the ideal tiling have disappeared in the diffraction pattern of the random tiling. This is because weak reflections have a large Q_{per} component and thus have their intensity strongly diminished by the perpendicular Debye–Waller factor. A measure of the 'degree' of disorder present in the random tiling model is given by the mean-square perpendicular 'fluctuations' of the atomic surfaces $\langle u_{per}^2 \rangle$ whose value is of the order of half the average radius of the atomic surface describing the three-dimensional Penrose tiling (a triacontahedron). In this random tiling model, the restoring force is the configurational entropy. It means that the quasicrystalline state has the largest configurational entropy and any homogeneous or long-wavelength phason distortion driving the system away from the quasicrystalline state (as, for instance, towards a periodic approximant) will have smaller entropy: roughly speaking, the density of dodecahedron sites is maximal in the quasicrystal, allowing more flexibility for configurational rearrangements.

It might seem surprising at first sight that elasticity theory applies to this model. Indeed the elementary process involved in this randomization is discontinuous, and the geometry of the high-dimensional description is rather analogous to the Frank–Van der Merwe model in the regime of strong coupling between the atomic chain and the substrate, in which the free energy is non-analytic. However, this comparison may be misleading, because there is no direct link between the high-dimensional random tiling picture and its entropy variation with strains, contrary to the case of the Frank–Van der Merwe model where the modulation function directly describes the coupling between the two subsystems. In the case of the random tiling quasicrystal the link with hydrodynamics is done via a coarse graining of the fluctuating physical space (see Henley's review (Henley, 1991*b*)).

5.4.2.12 Matching rules model

The applicability of hydrodynamic theory has also been studied numerically on energetically stabilized quasicrystals or on a 'matching rules' model (Ingersent, 1991). This model assumes that the quasicrystal is a ground state at 0 K. A way to implement this hypothesis on a tiling model is to consider the 'matching rules', defined as the set of local environments which uniquely characterize the quasicrystal. In this model any violation of the matching rule has an energy cost. By considering a simple Hamiltonian on a three-dimensional Penrose tiling model, Dotera and Steinhardt (1994) succeeded in favouring a quasicrystalline ground state at 0 K. As the temperature is increased, flipping inside the dodecahedron sites (Fig. 5.18) becomes possible and produces a matching rule mismatch. The temperature study of this simulation showed that hydrodynamic theory only applies above a critical temperature T_c. Although not proven, it has been conjectured that below T_c the free energy is non-analytic with respect to phason strain and would vary as $|u_{per}|$ instead of $\nabla^2 u_{per}$. This transition from the non-analytic to the analytic regime bears some similarities to the transition studied by Aubry in the Frank–Van der Merwe model. But again, contrary to the case of the Frank–Van der Merwe model, there is no direct connection between the shape of the atomic surfaces in the quasicrystal and the analyticity breaking. The analyticity breaking is most likely related to the spatial distribution of dodecahedron sites inside the tiling as the temperature is raised, which might be more continuous above a given temperature: the associated perpendicular coordinate and its variation as a function of site position represents a surface in high-dimensional space, which would be continuous above a given temperature, allowing all long-wavelength phason fluctuations to develop. In any case, for temperatures higher than T_c, the situation is similar to the one of the random tiling model and the restoring force with respect to phason strain comes from the tile configuration entropy. It should be kept in mind that the tile configuration entropy is not the only possible restoring force. Indeed the only requirement for hydrodynamic theory to apply concerns the free energy $F = U - TS$. Koschella et al. (Koschella et al., 2002) have recently carried out simulations on a two-dimensional binary tiling using pair potential interactions that stabilize a quasicrystalline ground state at 0 K. They investigated the variation of the energy (U) of the system and found that U increases quadratically with phason strain. This can be understood if we note that a phason strain will imply a change both in the density of the different local environments and in the local chemical composition of the system, which in turn will change the total energy of the system. In this simulation, the restoring force is strongly related to the chemistry of the system.

5.4.3 Eigenvectors and spectrum

From the general expression of the harmonic motion in a periodic crystal one sees that the displacements in an eigenmode are the same in every unit cell, apart from a phase shift. This means that the phonons are extended over the whole crystal. Moreover, the spectrum of vibrations of an infinite periodic chain consists of bands. The spectrum is (absolute) continuous. This no longer true for aperiodic crystals. The latter differ from periodic crystals in character of the eigenvectors and the spectrum.

One quantitative measure for the number of particles participating in a mode is given by the *participation ratio*. If the displacement of particle n in a given mode is given by u_n, and there are N particles in the crystal, then the participation ratio is

$$P = \frac{\left(\sum_n |u_n|^2\right)^2}{N \sum_n |u_n|^4}. \tag{5.34}$$

Then, in the limit of N going to infinity, the participation ratio is 0 for a strictly localized mode (with one $u_n = u$, and $u_m = 0$ for $m \neq n$), and 1 for an extended mode (all u_n equal). The long-wavelength acoustic modes are extended, and some modes of higher frequency have a smaller participation ratio. Localized modes which fall off exponentially with distance have a participation ratio equal to 0. A generalization is the 2p-norm of an eigenmode. It is defined as

$$P_{2p} = \frac{\left(\sum_n |u_n|^2\right)^p}{N^p \sum_n |u_n|^{2p}}.$$

The (generalized) participation ratio gives an idea of the character of the vibration modes. It correlates with the way the eigenvectors behave in a sphere with diameter tending to infinity, as discussed before. For the long-wavelength phonons the character is usually (close to) extended, the modes for higher frequencies have a critical character. However, these statements are not mathematical theorems. Only in one dimension are there rigorous results. Nevertheless, the impression exists that higher frequency modes in aperiodic crystals, especially those with discontinuous atomic surfaces, have a critical character, and this may lead to broadening in the scattering function, and to peculiarities in (thermal) conductivity measurements.

The frequencies of eigenmodes in periodic crystals form bands. Each band is given by a function $\omega_\nu(\mathbf{k})$, and the projection on the energy axis is an interval. Such a spectrum is called absolute continuous. The spectrum of any Hermitean operator is the sum of three parts, of which one or two may be absent as in the case of periodic crystals: an absolute continuous part, a singular continuous part, and a discrete part. Localized vibrations around defects have a discrete spectrum. The particular property of aperiodic crystals is that their vibrational (and as we shall see later their electronic) spectra may have three components. For certain one-dimensional substitutional chains it has been shown that the spectrum has a singular continuous component. For the general case, and in more than one dimension, there is numerical evidence that indeed the three components exist.

5.4.4 Damping

Lattice vibrations in the harmonic approximation have an infinite lifetime. In that approximation inelastic neutron scattering would have sharp lines. Non-linear interactions between phonons lead to finite lifetimes and broadening of the lines. Moreover, a contribution to the damping stems from coupling to other degrees of freedom, e.g. the electron system. For acoustic modes, where the frequency is proportional to the wave vector, one may argue that the damping is proportional to the square of the wave vector. For other low-frequency excitations, the phasons, other mechanisms come in as well, as we shall see in Section 5.5. The difference between acoustic phonons and phasons in the long-wavelength limit is that the distances between neighbouring atoms change very little for the phonons, but vary proportional to the derivative of the modulation function for phasons. This is not always small, which implies that the non-linearity of the forces becomes relevant. The reference (Zeyher and Finger, 1982) shows that for incommensurate phases the dynamical structure factor becomes

$$S(\mathbf{q}, \omega) = \frac{2T\gamma\Delta^2}{(\omega^2 - c^2 q^2) + \omega^2 \gamma^2}.$$

Here γ is the damping, Δ is the average value of the normal coordinate $Q_{\mathbf{q}_0 \nu}$ of the modulation function, and c the phason velocity. This implies that, for non-zero T and for sufficiently small values of q, the mode is diffusive, whereas for larger values of q it is a damped oscillator. Thus, one finds a result in agreement with hydrodynamic theory: phason modes are diffusive in the long-wavelength limit.

In Currat *et al.* (2002) the damping is studied for an incommensurate composite assuming a phenomenological damping constant. In addition to the damping because of the energy transfer from a mode, the scattering function may also be broadened because of the character of the eigenvectors. A critical state may lead to many closely grouped peaks in the scattering function, which appears as a broadening.

5.4.5 Calculation of phonons for real incommensurate phases

Calculations of the lattice vibrations in simple models, simple in structure and simple in interactions, give an insight into the special properties of the lattice dynamics of aperiodic crystals, but, of course, they are not enough to explain details in measured phonon frequencies and densities of states. There are many publications on calculations with structural models that agree with the real structure, and with realistic interactions. For incommensurate phases most of the work is simpler in the modelling (Axe *et al.*, 1980; Garcia *et al.*, 1989; Etxebarria *et al.*, 1992; Mischo *et al.*, 1997). The interactions used in these models are fits of the model parameters to experimental data or to ab initio calculations.

A rather recent example is the incommensurate ground state of Ni_2MnGa (Bungaro *et al.*, 2003). This is at higher temperatures a cubic Heusler compound (a ferromagnetic face-centred cubic structure). For this compound the lattice dynamics has been calculated with density functional perturbation theory. This shows a softening of the transverse acoustic mode at the commensurate wave vector $q_1 = 2\pi/a(1/3, 1/3, 0)$ for T = 260 K. At 220 K another instability shows up at $q_2 = 2\pi/a(0.43, 0.43, 0)$. This is a so-called martensitic transition. The main contribution comes from the electron–phonon coupling and the fact that the Fermi surface shows considerable flat pieces, coupled by the modulation vectors. Compare Fig. 6.22 in Chapter 6. (There are more vectors in the star of q_2.) This leads to an incommensurate tetragonal phase. The results of these calculations agree well with the experimental neutron scattering. (For details, see Shapiro (2010).)

5.4.6 Dynamics of quasicrystals

In particular, for quasicrystals detailed calculations have been carried out (Hafner and Krajčí, 1999; Elhor *et al.*, 2003; Mihalkovič and Widom, 2004).

For quasicrystals usually low-order approximants to the real structure are used. For icosahedral quasicrystals there are differences between the class of Mackay cluster compounds, like AlMnPd, and the class of Frank–Kasper type compounds, like AlZnMg. As models for the first class are taken those of Katz and Gratias, Cockayne and Widom, and Boudard and de Boissieu. Approximants are often limited to 1/1 approximants, which means that the golden mean τ which plays a role in the structure is approximated by 1/1. A few higher-order approximants have been used, for example, a 3/2 approximant. Besides, there are calculations of the vibrations in decagonal phases. The methods are usually a direct diagonalization

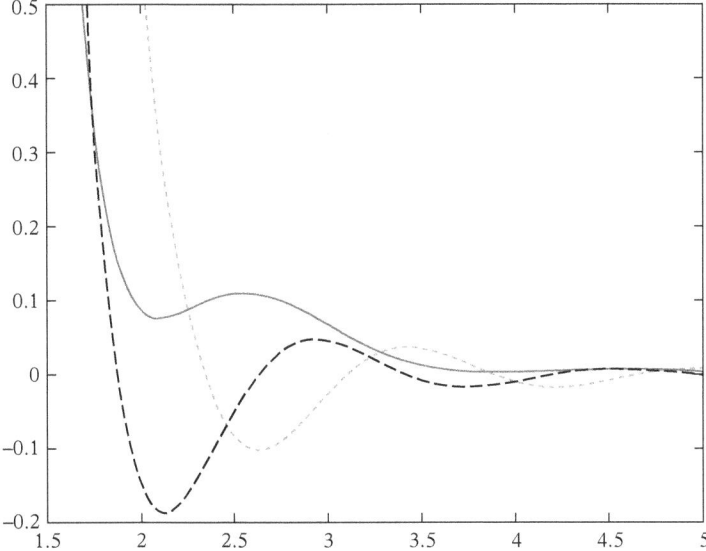

Figure 5.20 *Oscillating potentials between the atoms of B-Fe: between the B atoms, between the Fe atoms, and between B and Fe atoms. (Mihalkovič and Henley, 2012)*

of the dynamical matrix or application of a recursion method. For details we refer to the literature mentioned above.

More realistic potentials were determined by Mihalkovič and Henley (Mihalkovič and Henley, 2012). These are oscillating potentials as exemplified in Fig. 5.20. They were derived by ab initio calculations. Such double-well potentials had been used for a two-dimensional model for quasicrystals in Engel and Trebin (2007). Similar potentials were also used to calculate the dynamic structure factor for binary quasicrystals. (Mihalkovič *et al.*, 2008). The theoretical and experimental results agree quite well (see Fig. 5.29).

5.4.7 Diffuse scattering and Debye–Waller factors

The diffraction of X-rays and neutrons by aperiodic crystals has been discussed here mainly for perfectly ordered systems. Excitations and defects usually break the perfect order. Here, we briefly discuss these effects on scattering. These influence the intensity of the Bragg peaks and represent the origin of diffuse scattering.

The intensity of the Bragg peaks diminishes as a consequence of thermal motion. It is remarkable that, notwithstanding the randomness due to thermal disorder, there are still sharp Bragg peaks. The intensity decrease is a consequence of the motion and is given by the Debye–Waller factor (e.g. Kittel (1996)). The structure factor becomes

$$F(\mathbf{H}) = \sum_{\mathbf{n}j} f_j e^{-i\mathbf{H}\cdot\mathbf{r}_j} \left\langle e^{(-i\mathbf{H}\cdot\mathbf{u}_j(t))} \right\rangle \approx \sum_{\mathbf{n}j} f_j e^{-i\mathbf{H}\cdot\mathbf{r}_j} \left(1 - \frac{1}{2}\left\langle (\mathbf{H}\cdot\mathbf{u})^2 \right\rangle \right). \quad (5.35)$$

This leads to the intensity

$$I = I_0 e^{-\frac{1}{2}\langle u^2 \rangle H^2}. \quad (5.36)$$

Here, the displacements $\mathbf{u}_j$ are replaced by uniform values $\mathbf{u}$. The brackets $\langle$ and $\rangle$ mean a thermodynamic average; I_0 is the intensity when the atoms are stationary at their average positions.

For incommensurate phases the expressions have been discussed by Overhauser Overhauser (1971) and Axe Axe (1980). The main difference between commensurate and incommensurate phases for the Debye–Waller factor is due to phase fluctuations. We start here with a continuous modulation function, so that there are no phason jumps. In an incommensurate modulated phase with a continuous modulation function, there is a phason branch in the phonon spectrum. For a phason, not all particles have the same oscillation amplitude. Axe compared two simple models with a sinusoidal modulation and one particle per unit cell. One approach is to consider a Gaussian oscillation of the phase, and another is to consider a Gaussian distribution of the displacements. Both show that the Debye–Waller factor due to phasons is one for the main reflections and the first satellites. In Perez-Mato et al. (1991) the influence of phason fluctuations on the Debye–Waller factor is discussed. It is argued that in the sinusoidal region of the incommensurate phase these fluctuations give a second-harmonic modulation of the Debye–Waller factor. In Aslanyan et al. (1998) the Debye–Waller factor is calculated again. The result differs from that of Axe and Overhauser as a consequence of an anomalous increase of the phase fluctuations $\langle \phi^2(\mathbf{r}) \rangle$. This problem has also been considered in van Smaalen (2012): rather than introducing a phason Debye–Waller factor, anharmonic Debye–Waller terms are applied to the modulation function. This makes it possible to take into account the different local environments resulting from the modulation. However, it seems that, comparing theory and experiment, the situation is still not completely clear.

For incommensurate modulated phases with a discontinuous modulation function and for quasicrystals the atomic displacements, responsible for the decrease in intensity of the Bragg peaks, are no longer arbitrarily small. There are phason jumps and this changes the considerations for the Debye–Waller factor. This has been discussed in Chapter 4, Sections 4.1 and 4.6. The Debye–Waller factor cannot be calculated as is done for harmonic oscillators, for which one has (cf. Eq. 5.35)

$$\langle \exp(-i\mathbf{H}\cdot\mathbf{u}) \rangle \approx 1 - \frac{1}{2}\left\langle (\mathbf{H}\cdot\mathbf{u})^2 \right\rangle, \quad (5.37)$$

because phason jumps are not harmonic oscillators. For phason jumps $\mathbf{u}$ is constant, or one of a small number of values, and the number of jumps increases with temperature. The jump vectors **u** are the external space vectors between two atomic surfaces sharing an edge in their projection on V_I.

For quasicrystals the theory of the phason Debye–Waller factor has also been derived within the hydrodynamic and generalized elasticity theory. Although, in this case, atomic surfaces are discontinuous, a phason Debye–Waller factor can be defined. It is intimately related to the hydrodynamic behaviour of phason fluctuations or phason modes that give rise to the phason diffuse scattering and to a decrease of the Bragg peak intensity. In a simplified view, which has yet to be verified on icosahedral Amman tiling calculations, the phason Debye–Waller factor might be viewed as resulting from Gaussian fluctuations of the atomic surfaces, leading to a reduction of the Bragg peak intensity given by

$$I = I_0 e^{-B_{perp}^2 H_{perp}^2} \quad (5.38)$$

where $B_{perp}^2 = \langle R_{perp}^2 \rangle$, and R_{perp} correspond to the perpendicular space fluctuation of the atomic surface, which is considered to be Gaussian. As already mentioned, this has been applied for the structure determination of quasicrystals. The treatment here is very similar to that for phonons in periodic crystals. The phason modes lead to a decrease in Bragg peak intensity, which in turn leads to phason diffuse scattering near the Bragg peak, as shown not only within hydrodynamic theory but also experimentally. In decagonal AlNiCo, anomalous behaviour of the Debye–Waller factor has been found (Abe *et al.*, 2003*b*). This is due to an order–disorder transition when the number of phason flips increases at higher temperatures.

Often, there is some disorder in incommensurate modulated phases. An example is mullite, an aluminum silicate compound. The disorder creates diffusive lines, which have been modelled in Welberry (2004). See also Fig. 4.99 in Chapter 4 and Withers (2015). This is an example of structured or correlated disorder. This structured disorder does not occur for aperiodic crystals only. It has in general been discussed in Keen and Goodwin (2015).

5.5 Non-linear excitations

The lattice vibrations in the harmonic approximation are linear combinations of the eigenmodes. Taking higher-order terms into account leads to coupling between these modes. This is one of the reasons why phonons have a finite life time. This is also the case in aperiodic crystals. However, the non-linear terms also have other consequences. As we have seen, the soft-mode behaviour and the appearance of a phase transition are due to such non-linear terms. For example, in the DIFFFOUR model the fourth order in the potential energy is essential for having a stable ground state as soon as an instability sets in. Another consequence is that non-linear excitations may occur in aperiodic crystals in a way that is different from that for periodic crystals (Janssen, 2005).

The frequency of an acoustic mode with wave vector 0 vanishes, because of the translation invariance. This is independent of the amplitude of the excitation, because any translation keeps the distances the same. Another possible excitation with zero frequency, as we have seen, is a phason excitation. This may appear in a modulated phase if the modulation function is continuous. If the modulation is shifted arbitrarily with respect to the lattice, the energy does not change. This implies that there is a zero frequency linear excitation. Does it imply that an initial phase velocity is conserved? If $x_n = na + f(na)$ are the ground state positions, then the configuration with positions $y_n = na + f(na + vt)$ has the same potential energy, but the kinetic energy of a particle depends on v and the position. This is a non-linear problem.

For acoustic phonons, the damping of the mode goes to zero if the wave vector tends to zero. For phasons there is no gap if the atomic surfaces do not have discontinuities, but the damping remains finite if q goes to zero (Zeyher and Finger, 1982). (See also (Lyons et al., 1982).) If the atomic surfaces are discontinuous, there is a phason gap, and both frequency and damping remain finite when q goes to zero. This leaves the possibility that the mode does not become overdamped.

When the amplitude grows, the displacement is no longer infinitesimal, as in the harmonic approximation. One can start from an initial state with positions $na + f(na)$ and initial moments $p_n = mvf'(na)$, and integrate the equations of motion numerically. These calculations show that the positions at time t are given by $na + f(na + vt)$, provided the value of v is small enough. The modulation function glides through the crystal undeformed. In the DIFFFOUR model one can make an estimate of the value of this critical velocity (Janssen et al., 2002). If the potential energy of the DIFFFOUR model is given by

$$\sum_n \left(-\frac{1}{2}ax_n^2 + \frac{1}{4}x_n^4 + (x_n - x_{n-1})^2 + d(x_n - x_{n-2})^2 \right),$$

then the condition for having an undeformed moving modulation function is

$$mv^2 < -8d - 2.$$

That is the condition for having a solitary wave which is stable. When v exceeds the critical value, then the mode consisting of a moving modulation function loses energy, and the motion stops (Fig. 5.21) It should be noted that the DIFFFOUR model is rather special, because in the continuum limit it is integrable. For general interaction potentials the damping may become larger. This is also the case when instead of Hamilton equations a Langevin equation is taken, to introduce the effects of the temperature. As we have seen in the section on damping, this depends strongly on T.

In incommensurate composites a similar behaviour has been found. The phason mode here is the sliding mode. If the interaction between the subsystems is small enough, then the subsystems have smooth modulation functions. Starting with an equilibrium configuration and a relative velocity of the subsystems which

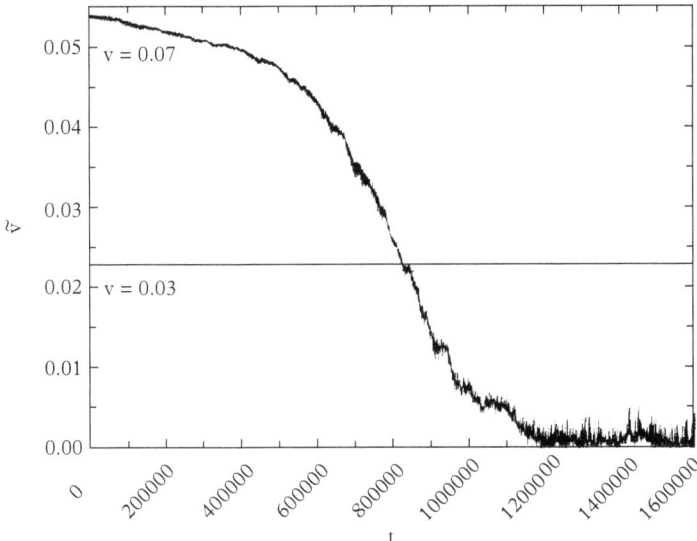

Figure 5.21 *The velocity of a smooth-moving modulation function in the DIFFFOUR model for two initial velocities: one above and one below the critical value. (Janssen et al., 2002)*

corresponds to a shift in internal space, there is no loss of centre-of-mass kinetic energy, provided the initial speed is below a critical value. When the initial velocity is too high, the relative motion of the subsystems diminishes in time (Janssen, 2005).

Phasons in modulated phases are related to another type of non-linear excitation, the discommensurations. When the temperature in an incommensurate phase decreases and approaches the lock-in transition, such discommensurations appear. Between the discommensurations the phase of the modulation function, with respect to the lock-in phase, is constant. With respect to the lock-in phase structure the discommensurations are non-linear excitations which lower the energy. Their density remains finite because they repel each other.

We can again discuss these excitations in the framework of the DIFFFOUR model. We make a continuum approximation for the equations of motion

$$\ddot{x}_n = -Ax_n - x_n^3 - B(x_{n+1} + x_{n-1}) - C(x_{n+2} + x_{n-2})$$

near the lock-in transition to an N-fold superstructure. There x_n can be written as a slowly varying coefficient times an oscillation with period N. For simplicity we take $N = 2$. Then we write $x_n = (-1)^n Q_n$ and replace Q_n with the continuous function $Q(z, t)$. The difference equations of motion are then transformed into the differential equation

$$-\ddot{Q} + 2(B - 2C)Q'' = (A - 2B + 2C)Q + Q^3.$$

Here Q'' is the second derivative with respect to z. Such a differential equation has exact solutions, and the latter are solitary wave solutions, a function that moves without deformation:

$$Q(\xi, t) = \sqrt{-(A - 2B + 2C)} \tanh\left((z - z_0 - vt)\sqrt{\frac{-A + 2B - 2C}{4B - 8C - v^2}}\right). \quad (5.39)$$

A series of equidistant kinks of this shape in the underlying structure with period N transforms the latter in an incommensurate phase. The general case of arbitrary N has been treated in Slot and Janssen (1988). The solitary waves are non-linear excitations which may also occur below the lock-in transition. In that case they have positive energy. If the temperature is above the lock-in transition, kinks have negative energy. In the period N structure kinks are then created until the positive interaction energy between the kinks stops the decrease in total energy. The equilibrium density of kinks then transforms the superstructure N to the incommensurate phase, which is the ground state at the given temperature.

In addition, these kinks (or domain walls) may move. In the continuum approximation they are really solitary waves, and the velocity remains equal to

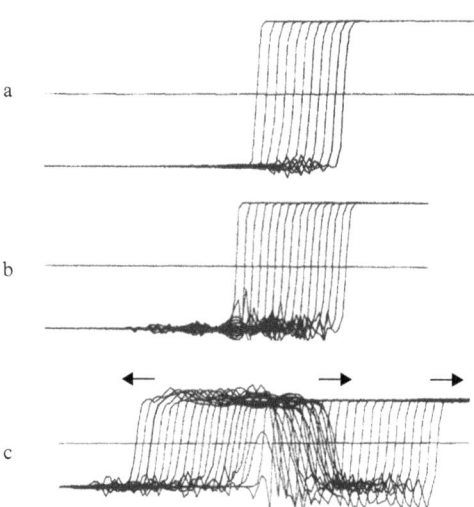

Figure 5.22 *The motion of a solitary wave for three different initial velocities. In the first, there is practically dissipation-free motion; in the second, some dissipation of the motion into phonons can be seen; in the third, the initial speed is so high that there is energy to create an additional kink-anti-kink pair. (Slot and Janssen, 1988)*

the initial velocity. This changes when the discrete structure is taken into account. At small initial velocities, the motion is almost dissipationless. At higher velocities, phonons are created and the motion is damped. For even higher velocities the kink may even break apart into two kinks and an anti-kink (Fig. 5.22).

5.6 Electrons

5.6.1 Introduction

Just as for phonons, the properties of electrons in aperiodic crystals are different from those in periodic ones. Also here, the absence of a Brillouin zone is the reason why the usual techniques cannot be used, the eigenstates are not necessarily extended, and spectrum and density of states are quite different. This is reflected in physical properties like electrical conductivity. We shall see that the formation of a pseudo-gap in the density of states may play a role in the stabilization of quasicrystals.

As for phonons, rigorous results are largely limited to one dimension. In two and three dimensions one has to rely on numerical calculations, either on simplified models to investigate the fundamental properties or on more sophisticated models with more or less realistic interactions. The numerical effort needed for the latter case means that one is limited to relatively small sizes. Much insight in the properties of electrons in aperiodic crystals has been obtained by the study of simple models. Here, characteristic properties like the character of the wave functions, the types of spectra, and the way to characterize them have been developed first.

As with phonons, electrons behave in a more varied way in quasiperiodic systems than in periodic ones. Wave functions may be extended, localized, or critical. A function Ψ is extended if asymptotically

$$\int_{|\mathbf{r}|<R} |\Psi(\mathbf{r})|^2 d\mathbf{r} \sim R^d,$$

where d is the dimension of space; it is localized if

$$\int_{|\mathbf{r}|<R} |\Psi(\mathbf{r})|^2 d\mathbf{r} \sim \text{constant}$$

and it is said to be critical if

$$\int_{|\mathbf{r}|<R} |\Psi(\mathbf{r})|^2 d\mathbf{r} \sim R^{d-2\delta}, \quad 0 \leq \delta \leq d/2.$$

Critical wave functions occur at the metal–insulator transition in disordered systems, and in quasiperiodic crystals. *Critical wave functions* may fall off with a power law, or have a support with a fractal dimension smaller than d.

5.6.2 Simple models

Most of the models used are *tight-binding models* leading to a discrete Schrödinger equation, and models with a one-electron Schrödinger equation with an aperiodic potential. The simplest version of a tight-binding model is a one-dimensional chain of particles with one electron state per site ($|n>$). The creation operator for an electron in the state $|n>$ is $b_n^\dagger$, and the corresponding annihilation operator is b_n. The energy of the state at particle n is ϵ_n, and the hopping frequency between sites n and m is t_{nm}. Then the Hamiltonian operator is

$$H = \sum_n \left(\epsilon_n b_n^\dagger b_n + \sum_m t_{nm} b_n^\dagger b_m \right). \tag{5.40}$$

If the hopping frequency is only non-zero if n and m are first neighbours, and if we write the eigenstate Ψ as $\sum_n c_n |n>$, the Schrödinger equation becomes

$$\epsilon_n c_n + t_{n+1} c_{n+1} + t_n c_{n-1} = E c_n, \tag{5.41}$$

where the hopping frequency $t_{n(n-1)}$ is written as t_n. Physical solutions are those for which the coefficients c_n are polynomially bounded. The values of E for which there is a solution belong to the spectrum. In general, there are two solutions for every eigenvalue in the spectrum. For a chain where ϵ_n and t_n are independent of n the spectrum runs from $\epsilon - 2t$ to $\epsilon + 2t$. For each value of E in this range there are two solutions: $\exp(\pm 2\pi i q n)$ and $E = E(q) = \epsilon + 2t \cos(2\pi q n)$. For a quasiperiodic sequence in ϵ_n and/or t_n the solution can, in general, not be written down explicitly. Just as for phonons an approach is via periodic approximants with unit cells which tend to infinity. Take for simplicity the case that $t = 1$ and $\epsilon_n = f(Qn)$, where f is a periodic function with unit period. Then a series of approximants L_m/N_m with limit equal to Q is considered. This gives a finite problem: N_n coupled equations which can be solved numerically by standard methods. For the approximants the solutions are characterized by a wave vector q in the Brillouin zone $|q| \le 1/2N_m$. A generalization to more dimensions is straightforward.

A second class of simple models is based on the Schrödinger equation

$$H\Psi = E\Psi, \quad H = -\frac{\hbar^2}{2m}\Delta + V(r), \tag{5.42}$$

where $V(r)$ is a quasiperiodic potential. A well-known example is the *Kronig–Penney model*, a one-dimensional system with

$$H = -\frac{\hbar^2}{2m}\frac{d^2}{dx^2} + V(x),$$

with $V(x) = \sum_n f_n \delta(x - na)$ or $V(x) = \sum_n f \delta(x - na - \epsilon \cos(2\pi q n a + \theta))$. For quasiperiodic f_n or irrational q and $\epsilon \ne 0$, this model has been used to study the properties of electrons in quasiperiodic materials. The first potential $(V(x) = \sum_n f_n \delta(x - na))$ may be used for the study of electrons in quasicrystals by choosing the values f_n according to the Fibonacci sequence ... LSL..., with $f_n = a$ if the n-th symbol is an L, and $f_n = b$ if it is S. The second potential is based on a modulated chain. It is called the *modulated Kronig–Penney model*.

One of the first studied models is *Harper's equation*. Harper's equation is the discrete Schrödinger equation for an electron in a crystal potential in an external magnetic field. It is a two-dimensional model with a square lattice and an energy band given by

$$E(\mathbf{k}) = 2E_0(\cos(2\pi k_x a) + \cos(2\pi k_y a)).$$

The magnetic field is given, in the Landau gauge, by a vector potential $\mathbf{A} = B(0, x, 0)$ and is perpendicular to the two-dimensional crystal. The wave function Ψ may be written in the lattice points as

$$\Psi(ma, na) = \exp(2\pi i n \nu) g(m),$$

where the function g satisfies the equation

$$g(m+1) + g(m-1) + 2\cos(2\pi m\alpha - \nu)g(m) = Eg(m), \quad \alpha = \frac{eBa^2}{\hbar c}.$$

This is a special case of Eq. 5.41. A generalization is the *almost-Mathieu equation*:

$$g(m+1) + g(m-1) + 2\lambda \cos(2\pi m\alpha - \nu)g(m) = Eg(m). \quad (5.43)$$

This is the discrete version of a Schrödinger equation with the Laplacian $\Delta g(m) \sim g(m+1) + g(m-1) - 2g(m)$. This equation has been studied in detail and it is one for which rigorous results are available. (Aubry and André, 1980) The model is special in that it is self-dual. Taking the Fourier transform of it, one gets the same equation with another coupling constant. It has been proven that for $\lambda < 2$ the solutions are extended states, for $\lambda > 2$ the states are all localized, and for $\lambda = 2$ the states are critical. The special property is that all states change their character at the transition (see Fig. 5.23). Usually, when such a transition occurs, there is a mobility edge. Depending on the energy, the states have one or the other character. The mobility edges are those values of the energy which separate extended from localized states.

Another example of a model of class 1, is the *Fibonacci tight-binding model*. Take a function on the interval from 0 to 1: $f(x) = a$ if $0 \le x < \Phi$, and $f(x) = b$ if $\Phi \le x < 1$. Then take $t_n = 1$ in Eq. 5.41, and $\epsilon_n = f(n\Phi)$. This gives a Fibonacci series of letters a and b. A generalization is to take a substitutional chain with p letters and to put ϵ_n equal to a value a_j if the n-th letter is A_j ($j = 1, \ldots, p$). This is an aperiodic tight-binding model.

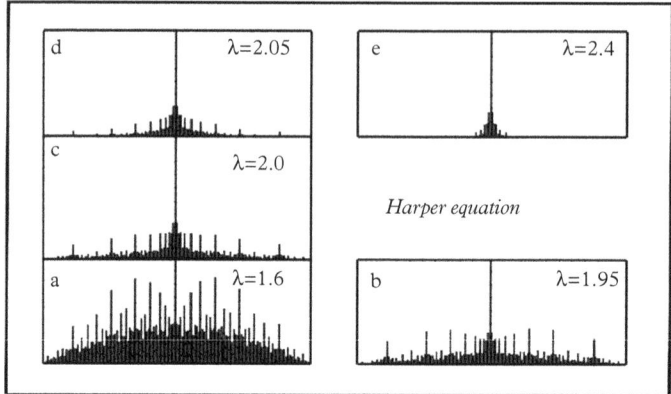

Figure 5.23 *The highest-energy electron state for the almost-Mathieu equation, for various values of the coupling parameter λ. The system is normalized such that the maximum values are equal. For $\lambda < 2$ (a, b), the state is extended; for $\lambda > 2$, it is localized (d, e); and, for $\lambda = 2$ (c), it is critical, and has multi-fractal character. (Cf. (Schellnhuber and Urbschat, 1987))*

For one dimension, the transfer technique can be used. Equation 5.41 can be rewritten as

$$\begin{pmatrix} c_{n+1} \\ c_n \end{pmatrix} = T_n \begin{pmatrix} c_n \\ c_{n-1} \end{pmatrix} = \begin{pmatrix} (E - \epsilon_n)/t_{n+1} & -t_n/t_{n+1} \\ 1 & 0 \end{pmatrix} \begin{pmatrix} c_n \\ c_{n-1} \end{pmatrix}. \quad (5.44)$$

For t_n independent of n, the matrices T_n have determinant equal to $+1$. Then consider a series of approximants L_m/N_m to the irrational number Q, for example, in the case of the Fibonacci chain, $L_m = F_m$ and $N_m = F_{n+1}$. Using the properties of the Fibonacci numbers, the trace x_p of the product of N_p matrices T_n satisfies a non-linear recurrence relation, the trace map, as we have seen for phonons in Section 5.4.

For one-dimensional chains it has been shown that for a substitutional chain the spectrum of the tight-binding Hamiltonian is singular continuous. For the Fibonacci chain the dispersion curves are given in Fig. 5.24. There are discontinuities for all wave vectors which are multiples of Φ. The density of states has infinities, because it is a one-dimensional model, and the IDOS is a Cantor function (in the aperiodic limit).

For generalizations to more dimensions, consider a two-dimensional or three-dimensional tiling (e.g. a two-dimensional octagonal tiling or the icosahedral Ammann tiling), put the site energies $\epsilon_\mathbf{n} = 1$, and put the hopping frequencies equal to t between vertices connected by an edge, and otherwise zero. In this case the density of states has much structure, but does not diverge. The IDOS is a smoother curve for higher dimensions. The two-dimensional curve still has plateaus, but the three-dimensional curve is smooth. The character of the wave

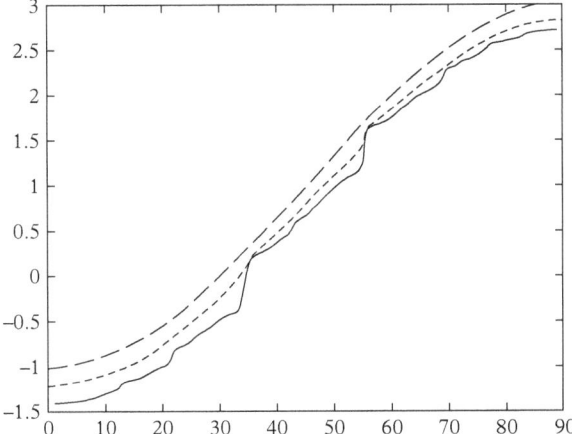

Figure 5.24 *The dispersion curve in the extended zone scheme for a periodic approximant to the Fibonacci tight-binding model; $t = 1; \epsilon_n = 1$ for L, and 1, 0.7, or 0.1 for S sites. When the ratio between the two values of ϵ increases, gaps open more and more. In the gaps, the IDOS is constant. This leads to its character as a devil's staircase.*

functions may be found by calculating the participation ratio, or by a multi-fractal analysis. Conclusions are less certain than for one dimension, but it seems that next to critical states there are also extended states. In some case there are also localized states concentrated on a special configuration of the tiling, the confined states (see Fujiwara *et al.*, 1996).

Two-dimensional model systems have been studied that are products of 1D aperiodic systems. One example is the product of two Fibonacci chains; another is based on a labyrinth model obtained from the product of two orthogonal octonacci chains. When the positions of the chains are x_n, the vertices of the labyrinth tiles are the points (x_n, x_m) in the plane, with $n + m =$ even.

Although the fact that the structure may be embedded in superspace does not make solving the Schrödinger equation easier, one may also describe the wave functions in superspace. From group-theoretical arguments it is easy to show that in superspace the Bloch theorem applies as well and the embedded wave functions may be written as

$$\Psi(r, r_I) = \exp(2\pi i \mathbf{k}.\mathbf{r}) U(r, r_I),$$

where the function U has lattice periodicity in superspace. Therefore, the restriction of U to the physical space is a quasiperiodic function, the absolute value of which is the electron density in physical space. The values of $U(\mathbf{r}, 0)$ can be folded back in the unit cell in superspace. This is a way to visualize the wave functions. The difference between the various types of states is clearly seen then. In Fig. 5.25 a number of examples are given for the modulated Kronig–Penney model.

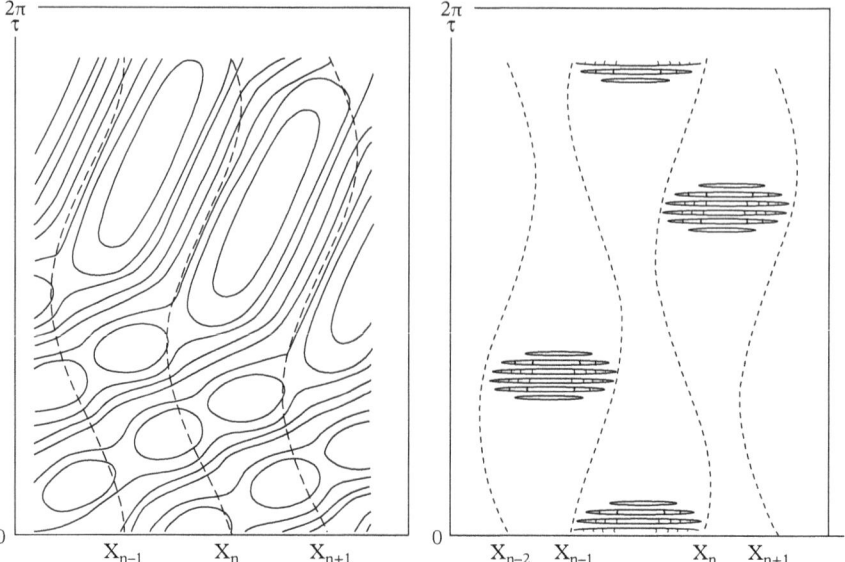

Figure 5.25 *Two examples of the (absolute value of the) function U in superspace, corresponding to two wave functions of the modulated Kronig–Penney model. Left: An extended state. Right: A critical state. (de Lange and Janssen, 1984)*

Although some very important questions cannot be answered using these simple models, such as the conductivity of charge-density wave systems, and the existence of a pseudo-gap in quasicrystals, for other fundamental properties like the structure of the spectrum and the character of the eigenstates, the simple models give quite a good insight.

What one can see from the calculations is the following.

1. In one-dimensional systems dispersion curves can still be defined. For modulated phases with continuous modulation function these have gaps at the wave vectors belonging to the Fourier module. These gaps open up for increasing modulation amplitude. When the amplitude goes to zero the dispersion curves of the unmodulated chain are recovered. For systems with discontinuous modulation functions the number of gaps goes to infinity.
2. The density of states shows much structure.
3. For modulated phases the participation ratio varies from 0 to 1, and the wave functions may be extended, localized, or critical. For some pure substitution chains which correspond to systems with a discontinuous modulation function, it has been proven that all states are critical.
4. For higher dimensions, the density of states also shows a lot of structure, but the function is smoother than for one dimension (Zijlstra and Janssen, 2000a, b; Grimm and Schreiber, 2003).

5.6.3 Electrical conductivity

Quasicrystals have quite interesting electric transport properties. Among these one can mention the following:

1. Although consisting of rather good conducting elements, the resistivity at low temperatures is high.
2. The conductivity as function of temperature is remarkable. It increases with increasing temperature:

$$\sigma(T) = \sigma_0 + \Delta\sigma(T),$$

where σ_0 does not depend on T. This is called the *inverse Mathiessen rule*.
3. The conductivity becomes lower when the structural quality increases. The best quasicrystals are the worst conductors.
4. For decagonal quasicrystals, which consist of periodically stacked aperiodic layers, the conductivity is very anisotropic. In the periodic direction it is metallic and has normal temperature dependence; in the aperiodic directions it is much lower, and behaves as in the inverse Mathiessen rule.

The explanation for this behaviour is not yet clear. It is likely to be due to the special electronic band structure. Usually, a pseudo-gap is found at the Fermi surface. This reminds one of a Hume–Rothery mechanism for the formation, and probably has important consequences for the conductivity, and for other physical properties. For example, the wetting is special. Water on a quasicrystalline surface does not spread; it is not wetting. Since the wetting depends on the interaction between fluid and substrate, for which the electrons are important, the electronic density of states determines the wetting to a considerable extent.

5.6.4 Realistic potentials

Just as for phonons, the simple models give insight into the peculiar properties of electrons in quasiperiodic structures, but they cannot explain the specific shape of the dispersion curves and the densities of states. For realistic structures and potentials, the numerical effort needed becomes very high. Therefore, one is limited to calculations on rather small approximants. An often used method is the linear muffin-tin orbital tight-binding method. For details we refer the reader to Fujiwara (1999), Hafner and Krajči (1999), Krajči et al. (2000), Krajči and Hafner (2002), and Solbrig and Landauro (2003).

The calculations show always a very spiky structure of the density of states. This might be due to the existence of an analogue of van Hove singularities. However, in experiments such a spiky structure was never found. This has been explained by the finite resolution one should use in calculations for comparison with experiments (see (Zijlstra and Janssen, 2000*b*)).

An interesting feature is the appearance of a pseudo-gap at the Fermi level. This is a considerable lowering of the density of states at that point, and this plays an important role for the stability of quasicrystals (see Peierls instability, to be discussed in Chapter 6, Section 6.6). This pseudo-gap has been found both in theory (Hafner and Krajči, 1992; Hafner and Krajči, 1999; Krajči and Hafner, 2002) and in experiment (Nayak *et al.*, 2012, 2015).

5.6.5 Quantum criticality in a magnetic quasicrystal

Although there is still little experimental evidence, it is expected that, in general, electrons in a quasicrystal will behave differently than in a lattice periodic crystal. A difference has been found in the icosahedral $Au_{51}Al_{34}Yb_{15}$, in the family of the Tsai-type binary quasicrystal $Cd_{5.7}Yb$. In the present case the AuAlYb quasicrystal could be compared to an approximant $Au_{51}Al_{35}Yb_{14}$. The difference found was in the behaviour in the resistivity near $T=0$. For the approximant the behaviour was as $\rho(T)-\rho(0) \approx T^2$, whereas this was $\approx T$ for the quasicrystal. The phase transition at $H=0$ and $T=0$ is an example of *quantum criticality* (Deguchi *et al.*, 2012). In this case, the phase transition is not driven by thermal fluctuations, which vanish at $T=0$, but by quantum fluctuations. The magnetic susceptibility at $H=0$ goes as T^{-p}, with $p \approx 0.51$.

The origin of this criticality has been discussed in Watanabe and Miyake (2016). A model calculation for an approximant of the quasicrystal AuAlYb shows that the transfer mode of the $4f$ electrons of the Yb and the $3p$ electrons of the Al has an almost flat wave-vector dependence. Then, mode-coupling theory shows that magnetic and valence susceptibility shows $T^{0.5}$ behaviour in the zero-field limit and a single scaling dependence of the temperature over magnetic field over several decades. It gives an argument for the quantum criticality of the quasicrystal.

Also, pressure may lead to quantum criticality. This has been shown by magnetic susceptibility measurements under pressure (Matsukawa *et al.*, 2016). Below the critical pressure p_c the susceptibility changes as $\chi(T) \approx T^{-0.5}$. Above p_c the approximant shows a magnetic phase transition at $T \approx 100$ mK, whereas the quasicrystal is robust in the quantum criticality.

5.7 Summary of the theoretical situation

1. The understanding of the physical properties of aperiodic crystals is much less than that of their structure.
2. Phonon eigenvectors and electron states may occur as extended, as found in all periodic crystals, but also as localized or as critical. A Bloch theorem holds in superspace, but not in physical space. Non-extended states have been found in 1D and 2D model calculations.

3. In model calculations, electron and phonon dispersion curves may have an infinite number of gaps, although these may be negligible for modulated phases and composites.
4. The experimental verification of the points above is extremely challenging and has not yet been achieved.
5. The possibility of embedding a quasiperiodic crystal into a higher-dimensional superspace is directly related to phason degrees of motion. In incommensurate modulated systems there are phonons which correspond to a shift in internal space, called phasons. In incommensurate composites there are sliding modes. In quasicrystals collective phason modes are possible and are diffusive. For the dynamics, the difference is not so much between quasicrystals and the other aperiodic crystals, but between aperiodic crystals with continuous atomic surfaces and those with discontinuous atomic surfaces.
6. A partial understanding of the damping of phasons may be obtained using hydrodynamic theory.

5.8 Phonons: Experimental findings

5.8.1 Scattering

We first briefly present the experimental methods used for the study of lattice dynamics in crystals and aperiodic crystals. We then give a few experimental results on lattice dynamics in quasicrystals.

There are several experimental methods which give access to the phonon spectrum of a crystal. In the first place, there are macroscopic techniques where thermodynamic properties such as the specific heat or the thermal conductivity are measured. In such a case, the experimentally measured quantity is mainly sensitive to the overall phonon spectrum through the VDOS or to acoustic modes. However, their temperature dependence gives valuable information.

The very-long-wavelength phonons, related to the elastic tensor of the system, are best measured by ultrasonic techniques, which are extremely accurate and make it possible to measure the weak anisotropy of the elastic constants.

However, the most detailed experimental information is obtained by inelastic neutron or X-ray scattering, which allows direct measurements of the phonon spectrum. A detailed presentation of this technique is beyond the scope of this book and we refer the interested reader to specialized books.

An inelastic neutron or X-ray experiment is characterized by both a momentum transfer $\mathbf{Q} = \mathbf{k}_f - \mathbf{k}_i$ and an energy transfer $\hbar\omega$ between the incident particle and the crystal. The energies of phonons are in the range 1–100 meV, and it is in this range that the experiment has to be designed.

For neutron scattering two cases have to be considered, namely incoherent and coherent scattering. In the case of incoherent scattering, a measurement of all scattered neutrons in a Q range and over a wide energy range allows one to extract the so-called generalized vibrational density of state (or GVDOS), in which the incoherent neutron scattering cross section of each element enters.

Coherent neutron or X-ray scattering allows the measurement of a single phonon in favourable cases. The measured quantity is the double differential cross section $\frac{d^2\sigma}{d\Omega d\omega}$ related to the one-phonon scattering function $S(\mathbf{Q}, \omega)$ (cf. Eq. 5.21).

$$S(\mathbf{Q}, \omega) = \sum_{\mathbf{q},\mathbf{K},\nu} \left| \sum_j \frac{b_j}{\sqrt{M_j}} e^{-W_j(\mathbf{Q})} \mathbf{Q}.\mathbf{e}_{\mathbf{q}\nu|j} e^{2\pi i \mathbf{Q}.\mathbf{r}_j} \right|^2 \delta(\mathbf{Q} \pm \mathbf{q} - \mathbf{K}) \delta(\omega - \omega_{\mathbf{q}\nu}) \frac{n(\mathbf{q}\nu)}{\omega}.$$

(5.45)

Here the summation runs over the s atoms in the unit cell of the crystal, M_j is the mass of the atom j, and n is the Bose occupation factor: $(\exp(\hbar\omega/k_B T) - 1)^{-1} + \frac{1}{2} \pm \frac{1}{2}$. In the above expression, $\mathbf{K}$ is a vector of the reciprocal lattice. The total scattering function is obtained by summing up all the individual contributions. Inspecting this expression, we can see that each phonon mode gives rise in the scattering function to a delta peak, whose intensity is weighted by the inelastic structure factor corresponding to the sum over j. It is worth noticing that it corresponds to a Fourier transform of the $\mathbf{Q}.\mathbf{e}$ scalar product, and is intimately related to the atom displacement field. This also indicates that, for a properly chosen position in reciprocal space where $\mathbf{Q}$ and $\mathbf{e}$ are orthogonal, the contribution of a mode might be zero. This is particularly useful for selecting a restricted number of modes in the measurement.

In the case of long-wavelength acoustic phonons, where all atoms in the unit cell are vibrating in phase, the above expression is written in the high-temperature limit:

$$S_\nu(\mathbf{K}+\mathbf{q}, \omega) \propto |F_{elast}|^2 \sum_{3 modes} (\mathbf{Q}.\mathbf{e}_{T,L})^2 \frac{kT}{\omega_{T,L}^2} \delta(\mathbf{Q} - \mathbf{K} \mp \mathbf{q}).\delta(\omega - \omega_{\mathbf{q}\nu}), \quad (5.46)$$

where F_{elast} is the elastic structure factor measured at the point $\mathbf{K}$ of the reciprocal lattice, and the subscripts T and L stand for transverse and longitudinal acoustic modes, respectively. The scalar product $\mathbf{Q}.\mathbf{e}_{T,L}$ allows one to select only one of the three acoustic modes during the measurement. It is thus possible to measure the dispersion relation experimentally. Moreover, we also see from the above expression that the intensity of the measured signal is proportional to Q^2 and to F^2/ω^2; in other words, acoustic modes are best measured close to strong Bragg reflections and at large distances in reciprocal space.

When dealing with aperiodic crystals, Eq. 5.45 is still valid, but now the summation runs over an infinite number of terms. It is thus difficult to use the expression as such. Nevertheless, Eq. 5.46 for the long-wavelength limit is also valid for aperiodic crystals. One expects thus to observe a strong inelastic signal close to strong Bragg reflections.

5.8.2 Modulated phases and composites

The structure and phase diagrams of many compounds with an incommensurate phase have been measured, and are qualitatively understood. The situation with the dynamics is less satisfying. What are the features predicted by theory which are experimentally verified? There are many observations of a soft mode leading to an incommensurate phase. However, often the frequency does not go to zero. Very often the soft mode becomes overdamped for lower frequencies, but it has also been found that there remains a gap. The predicted change in behaviour at the discommensuration transition has not yet been verified. There are, however, new optically active modes appearing in the incommensurate phase, as found already in theoretical models. Finally, the situation about phasons and amplitudons is puzzling. We shall come back to that point in Section 5.9.

Soft-mode behaviour occurs, for example, in the case of the ThBr$_4$ phase. Below $T_i = 95$ K, the structure is incommensurate and is shown in Fig. 5.26. The grey and white spheres stand for the Br and Th atoms. The modulation occurs along the c-axis, and affects mainly the Br atoms. The modulation is visible as a wavy oscillation of some Br chains. The soft-mode behaviour in the case of the ThBr$_4$ phase is shown in Fig. 5.7. Above 95 K an entire optic branch softens and condenses at $q = 0.31c^*$, where c^* is the reciprocal lattice parameter of the high-temperature phase.

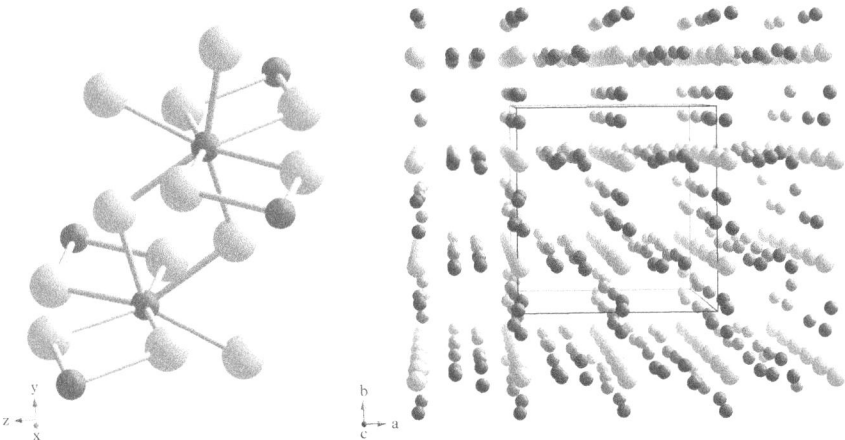

Figure 5.26 *Atomic structure of the ThBr$_4$ phase in the incommensurate modulated phase. Left: Basic structure (Th dark, Br light). Right: The c-axis is orthogonal to the plane of the figure. Th and Br atoms are shown as dark and light spheres, respectively. The modulation is visible for some Br chains, displaying a wavy line along c.*

The compound BCPS [ClC$_6$D$_4$)$_2$SO$_2$] shows a soft mode as well. In this case the damping is unusually low. In this case one can follow the frequency till the phase transition. It has been found that a gap remains, which is in agreement with the existence of a central peak (Ollivier et al., 1998).

The difference in the behaviour of phonons between incommensurate modulated phases and composites has been used to show, using inelastic neutron scattering, that the superconductor Bi$_2$Sr$_2$CaCu$_2$O$_{8+\delta}$ behaves more as a composite than as a modulated phase (Etrillard et al., 2001). Also the incommensurate crystal Bi$_2$Sr$_2$CuO$_{6+\delta}$ shows the character of a composite: one may distinguish the acoustic branches of each of the two subsystems (Etrillard et al., 2004).

New modes, compared to the high-temperature phase, have been found in the incommensurate phase of Rb$_2$ZnBr$_4$ by Raman scattering (Rasing et al., 1982). This shows that the selection rules in the incommensurate phase are different from those in the high-temperature phase, because of the change in (superspace group) symmetry.

5.8.3 Quasicrystals

Experimentally, the lattice dynamics of quasicrystals has been studied on large single grains using inelastic neutron or X-ray scattering in various icosahedral phases (i-AlLiCu (Goldman et al., 1991, 1992), i-AlCuFe (Quilichini et al., 1990, 1992), i-AlPdMn (de Boissieu et al., 1993; Boudard et al., 1995), i-ZnMgY (Shibata et al., 2002), i-CdYb and i-ZnMgSc (de Boissieu et al., 2007)), and in the d-AlNiCo phase (Dugain et al., 1999).

High-precision measurements of the elastic constants of icosahedral phases have shown that, as expected from the icosahedral symmetry, they are isotropic. This is different than the 1/1 cubic periodic approximant, for instance, for which a weak anisotropy has been observed.

Inelastic neutron scattering on single grains has shown that a similar trend apply for all the studied quasicrystals. Close to strong Bragg reflections there are well-defined acoustic excitations, whose width is limited by the instrumental resolution. This is illustrated in Fig. 5.27, which displays the experimental results for the i-AlPdMn phase when measured close to the strong 52/84 Bragg peak chosen as the zone centre (de Boissieu et al., 1993). The figure shows the measurement as observed in the transverse geometry for different values of the phonon wave vector **q**. For low values of the wave vector there is a well-defined peak in the energy spectrum, which corresponds to the acoustic mode which disperses as expected from sound velocity measurements. For wave vectors q larger than 0.3 Å^{-1} there is an abrupt broadening of the scattering function, although the intensity still obeys the acoustic mode relation as given by Eq. 5.46. When q becomes larger than 0.6 Å^{-1}, the intensity distribution displays a rapid increase, which indicates that the observed signal corresponds to a mixing of several excitations.

At higher energy, the signal is very smooth and can be decomposed into several 'bands' of excitations, whose width is of the order 4 meV.

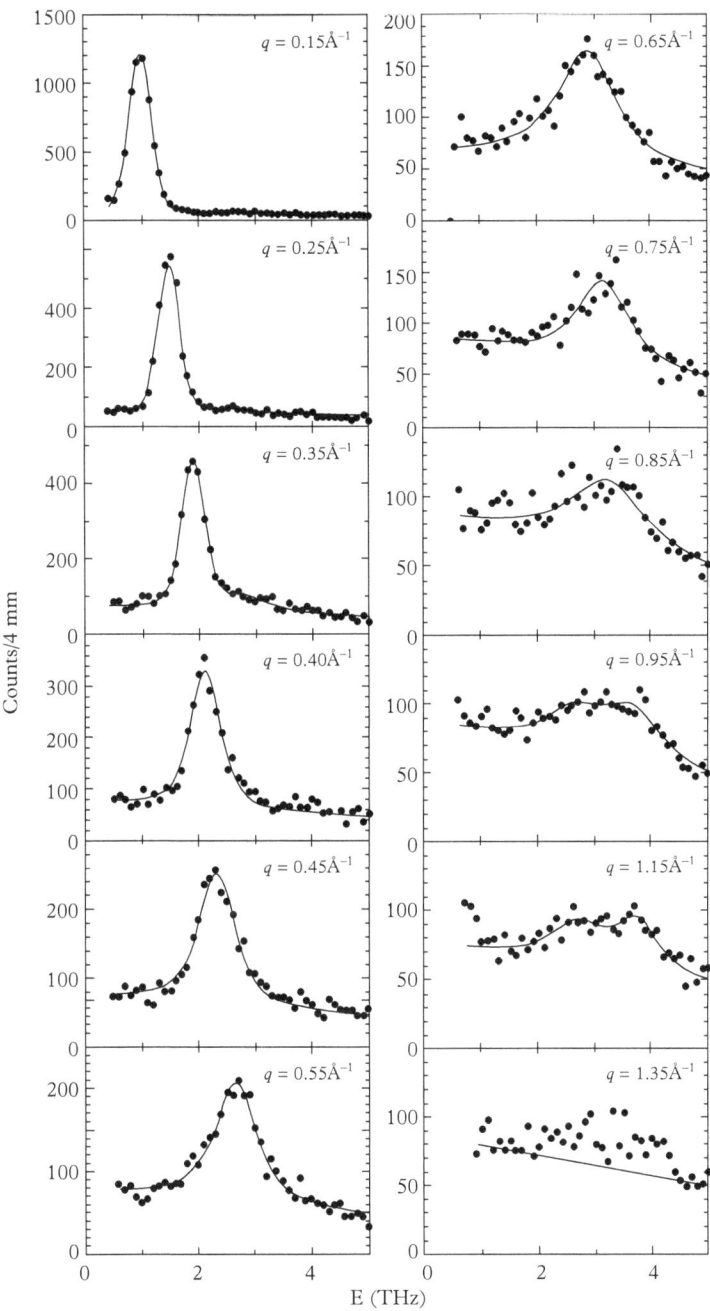

Figure 5.27 *Constant-Q energy scans measured by inelastic neutron scattering in the i-AlPdMn phase. Measurements are carried out in the transverse geometry starting from the strong two-fold Bragg reflection with N/M indices 52/84.*

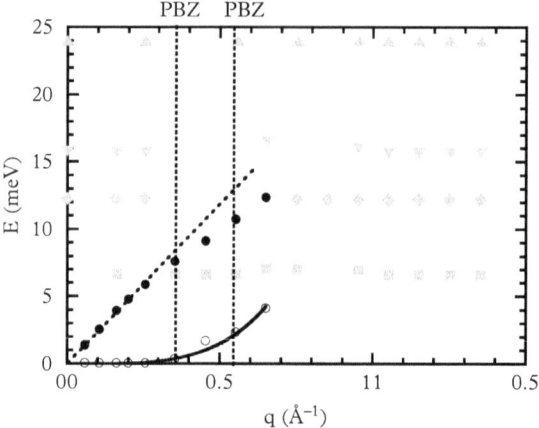

Figure 5.28 *Dispersion relation for a transverse mode propagating along a two-fold axis, as measured in the i-AlPdMn phase (Fig. 5.27). Full and open circles correspond to the acoustic mode position and width, respectively. Light-grey symbols stand for the various optic-like bands. The vertical dashed lines indicate the different PBZs.*

The resulting dispersion relation is shown in Fig. 5.28. The slope of the acoustic mode dispersion is equal to the one extracted by ultrasonic measurements. Above 0.35 Å^{-1} the acoustic dispersion relation departs from linearity, whereas the width of the signal (shown by open circles) displays a rapid broadening. The broadening rate is indeed fitted in this case by a q^4 law. The position of the 'optic' like bands are shown by grey symbols in the figure: they are centred at 7, 12, 16, and, 24 meV, respectively.

Detailed understanding of these experimental findings has not yet been achieved but the following observations can be made:

1. The broadening occurs for a characteristic wave vector and in a short interval which is similar for all icosahedral phases, whatever their structure. In the acoustic regime it is possible to extract the mean free path of a phonon mode. For a wave vector equal to 0.6 Å^{-1}, the mean-free path is equal to 12 Å in the i-AlPdMn phase, i.e. it is equal to the size of the atomic clusters that build up the structure. In the i-ZnMgY phase, this is equal to 24 Å, which is twice the size of the atomic clusters. This result points to the importance of this characteristic length, which is typical of quasicrystalline material.

2. The positions of the two or three lower-energy optical bands of excitations are related to the crossing of the acoustic branch and the first PBZ boundary, as indicated in Fig. 5.28. No gap opening has been observed in the i-AlPdMn phase dispersion relation close to the PBZ boundary (within the experimental resolution). However, at the position where the acoustic

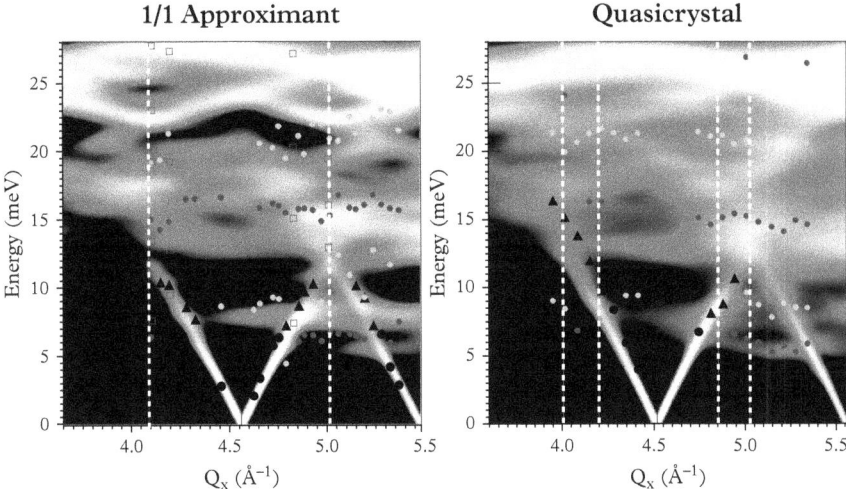

Figure 5.29 *Comparison of the dynamics of the i-ZnMgSc quasicrystal and its 1/1 ZnSc periodic approximant. The filled circles and triangles stand for the acoustic excitations, whereas the grey ones correspond to broad optical excitations. The lines are a guide for the eyes. The greyscale background corresponds to the intensity distribution of $S(\mathbf{Q}, \omega)$, as calculated with the simulation carried out using oscillating pair potentials. The vertical dashed lines stand for the Brillouin zone and the PBZ. The plain vertical line is a zone centre. (From (de Boissieu et al., 2007))*

branch crosses the PBZ boundary, there is always a dispersionless optic band which might be considered as an indirect trace of this gap opening.

The two main results are thus that there is a characteristic length scale related to the cluster size, and that the concept of PBZ is indeed a pertinent one for describing the lattice dynamic of quasicrystals.

The most complete set of results has been obtained in the ZnMgSc system (Francoual et al., 2004; de Boissieu et al., 2007). As already presented in the structure section (Section 5.8.2), these systems are particularly interesting since both the 1/1 cubic periodic crystal and the icosahedral quasicrystal can be obtained for small composition changes. Both phases are built up with the same clusters so that it allows one to study the effect of both the short-range order (clusters) and the long-range periodic or quasiperiodic order. Moreover, oscillating pair potentials have been developed, allowing a detailed simulation of the dispersion relation but also of the $S(\mathbf{Q}, \omega)$ one-phonon scattering function. The intensity distribution of $S(\mathbf{Q}, \omega)$ contains the information on the phonon eigenvectors and is a very demanding test when compared to the experimental data.

Figure 5.29 presents the dispersion relation as measured for the transverse excitations, whereas Fig. 5.30 shows individual constant-Q energy scans for

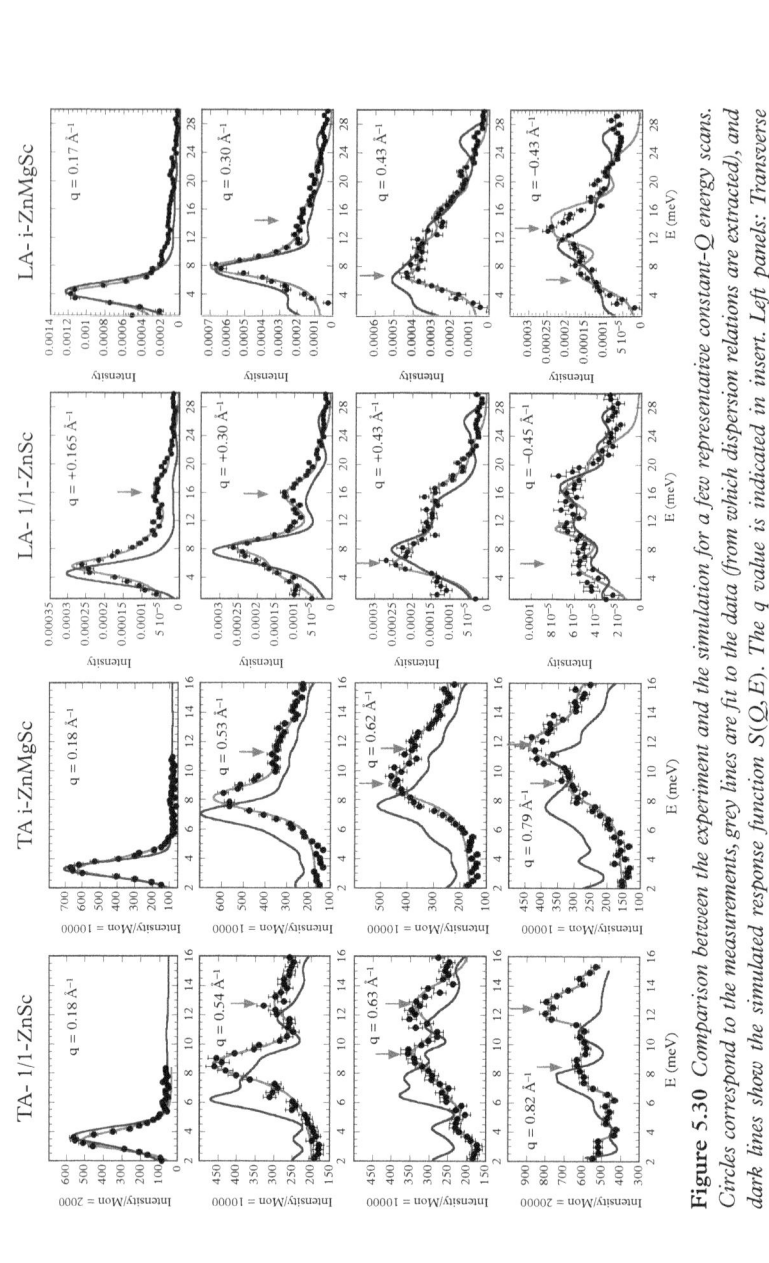

Figure 5.30 *Comparison between the experiment and the simulation for a few representative constant-Q energy scans. Circles correspond to the measurements, grey lines are fit to the data (from which dispersion relations are extracted), and dark lines show the simulated response function S(Q,E). The q value is indicated in insert. Left panels: Transverse acoustic (TA) modes measured in the 1/1 approximant and the quasicrystal by inelastic neutron scattering. Right panels: Longitudinal acoustic (LA) excitations measured in the 1/1 approximant and quasicrystal by inelastic X-ray scattering (for clarity, the elastic contribution has been subtracted from the data). The simulated data have been convoluted by a Gaussian with a full width at half maximum of 1 meV and 3 meV for neutron and X-ray data, respectively. The energy position of the two transverse calculated spectra at 0.18 Å^{-1} have been artificially translated at higher energy in order to illustrate the good reproduction of excitation, whose width is limited by the instrumental resolution. Notice the visible broadening as q increases in transverse geometry. The intensity distribution is well reproduced by the simulation. Arrows point to remarkable similarities and differences. (From (de Boissieu et al., 2007))*

transverse and longitudinal excitations. In both figures, results for the 1/1 cubic approximant and the icosahedral quasicrystal are compared.

The overall results are very similar in both the crystal and the quasicrystal. As for other quasicrystals, there is a limited q range for which there is a well-defined acoustic mode, followed by a rapid broadening, which also occurs for a characteristic length scale of the same order as the cluster size. However, contrary to what was observed in other quasicrystals, the optic bands are better defined. This is likely due to the very high degree of chemical order which exists in these phases. Moreover, the behaviour is very reminiscent of what is occurring in a simple crystal, with a pseudo-gap opening at the Brillouin zone boundary, where a minimum is observed in the scattering function as illustrated in Fig. 5.30 (TA panel $q = 0.54$ and 0.53 Å^{-1}) (Francoual, 2006; de Boissieu *et al.*, 2007). However, one of the main differences is the width of the phonon pseudo-gap opening, which is significantly larger in the crystal than in the quasicrystal, as is visible in Figs 5.29 and 5.30. This can be interpreted using the concept of PBZ boundaries in quasicrystals, as introduced earlier. Whereas in the 1/1 cubic crystal approximant there is a single Brillouin zone boundary, which is shown as a vertical dashed line in Fig. 5.29, there are two PBZ boundaries in the quasicrystal: those two PBZ boundaries can be defined using the 6D indexing of the quasicrystal diffraction pattern and their intensity distribution. The two consecutive PBZ boundaries are located at 0.33 and 0.53 Å^{-1}, corresponding to the 4/0 and 4/4 (N/M indices) two-fold axis reflections, whereas the first Brillouin zone boundary of the 1/1 approximant is located at 0.45 Å^{-1}. Moreover, the Fourier component associated with these two PBZ boundaries are smaller than the one associated with the 1/1 approximant, as can be inferred from the absolute value of the associated Bragg reflections. As a consequence, the double PBZ boundaries is, first, more extended in reciprocal space and, second, less effective, leading to a weaker phonon Bragg scattering in the quasicrystal as observed experimentally.

A very detailed atomic scale simulation has been also carried out in this system and compared to the experimental details. In fact, this constitutes the first realistic atomic scale simulation in a quasicrystal. Several difficulties had to be overcome to achieve a good comparison with the experimental data, both for the dispersion relation and for the intensity distribution. First, as illustrated in Section 5.8.2, whereas the structure is well described by a large triacontahedral cluster packing, the central tetrahedron induces a very strong distortion of the successive icosahedral shells describing this large cluster, both in the quasicrystal and in the 1/1 approximant. In particular, at room temperature, the tetrahedron appears in a disordered orientation in both phases. Second, one has to find a suitable periodic model for the quasicrystal. A sophisticated procedure has thus been used to design model systems as close as possible from the real ones. For both structures, oscillating pair potentials, fitted on a set of ab initio simulations, have been used. The pair potentials are shown in Fig. 5.31, and turn out to be a very efficient model Hamiltonian in many different intermetallic systems. Although a direct interpretation of the pair potentials shape is difficult, the following salient

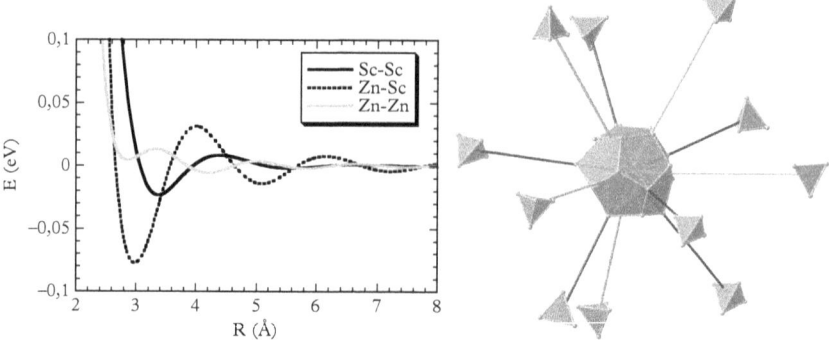

Figure 5.31 *Left: Oscillating pair potentials for the ZnSc system, as fitted from ab initio data. Right: Representative local cluster environment in the 3/2 approximant to the quasicrystal, as obtained after a molecular dynamics run, followed by a quench and 0 K relaxation, using the oscillating pair potential. The central tetrahedron induces a strong dodecahedron distortion. (From (de Boissieu et al., 2007))*

features can be extracted: (i) the large Sc and small Zn atoms are clearly indicated by the different minima for the Zn–Zn and Sc–Sc pairs; (ii) the deeper ZnSc minima is an indication of a ZnSc first-neighbours preference, most likely related to the ZnSc sp–d hybridization, as observed in the CdYb quasicrystal (Ishii and Fujiwara, 2001); (iii) the oscillations are related to the Friedel oscillations, although their period is different for each pair; and (iv) the decay of the potential goes roughly as R^{-3}, so that a potential cut at R about 0.8 nm is sufficient for modelling the systems properly.

For the 1/1 approximant, a supercell containing eight clusters was used. To simulate the tetrahedron disorder, molecular dynamic runs were carried out at 300 K and then quenched to 0 K, where the structure is fully relaxed under the pair potentials. Using this representative model, the dynamical matrix was then diagonalized, allowing the calculation of the dispersion relation, the eigenmodes, and the $S(\mathbf{Q}, \omega)$ scattering function to be compared with experimental data. For the icosahedral quasicrystal a cubic 3/2 approximant was chosen, containing 32 clusters in the unit cell, about 3000 atoms, and with a lattice parameter equal to 3.61 nm, to be compared to 1.38 nm for the 1/1 approximant. The 3/2 approximant was obtained via decorated canonical cell tiling, which is equivalent to a rational cut of the 6D model. The procedure was then similar to the one used for the 1/1 supercell: molecular dynamic run were carried out at 300 K and then quenched to 0 K. A few sites were chosen, corresponding to the energy minima of the system. The diffraction pattern of this 3/2 approximant already grasps the important features of the quasicrystal diffraction pattern, in particular, its intensity distribution. In both cases, the proper description of the tetrahedron disorder turns out to be a crucial point. One configuration present in the 3/2 approximant is shown in Fig. 5.31, with a very visible tetrahedron-induced cluster distortion.

A comparison of the simulations with the experimental dispersion relations is shown in Fig. 5.29, where the simulation is shown as a greyscale background. There is a very good agreement between the simulation and the experiment: both the 1/1 approximant and the quasicrystal dispersion relations are well reproduced: the acoustic regime, together with the almost dispersionless optical excitations, are well accounted for. Close to the Brillouin zone the pseudo-gap between the acoustic and optical excitations is also larger in the 1/1 approximant, as compared to the one simulated in the quasicrystal close to the PBZ. This is even more true when the intensity distribution of the scattering function is compared to the experiment. Indeed, this is a much more constraining test than a mere dispersion relation. In addition, there is an energy shift equal to 1.4 and 1.2 in the approximant and the quasicrystal, respectively. As can be seen in Fig. 5.30, the intensity distribution is extremely well reproduced by the simulation, both in the quasicrystal and in the approximant. The simulation has been convoluted with a Gaussian, taking into account the instrumental resolution. It is striking that the simulation reproduces quite well the observed experimental acoustic mode broadening, which is thus, in this case, the result of several modes of mixing: the large number of low-energy optical excitations are coupling to the acoustic mode for wave vectors larger than 0.3 Å^{-1}. However this acoustic–optic mode mixing, in the region 0.3 to 0.6 Å^{-1}, can likely be interpreted as a single damped acoustic mode.

The very good agreement between simulation and experiment thus validates the model used, both from the structural point of view and from the pair potential one. From this simulation it is thus in principle possible to carry out a detailed study of the eigenmodes of the different excitations. However, this would be a tremendous task, since there are, for instance, 9000 modes in the 3/2 approximant modelling the quasicrystal. A few general trends have been extracted, however. One may ask, for instance, if the geometrical cluster description plays a role in the lattice dynamics. For this purpose, a simulated VDOS has been decomposed on the different cluster shells. This demonstrates that the tetrahedron involves relatively low-energy modes, whereas the icosidodecahedron is involved in three main optical bands, with energies centred at 7, 13, and 17 meV.

That the tetrahedron plays a peculiar role in the lattice dynamics has been also evidenced by quasi-elastic neutron scattering experiments combined with an atomic scale simulation. As explained in Section 5.8.2, in the 1/1 cubic approximant the central tetrahedron appears disordered above T_C, whereas it orders in an anti-parallel direction along the (110) direction. That the high-temperature disorder is dynamic in nature has been shown using quasi-elastic neutron scattering techniques (Euchner *et al.*, 2012). This technique makes it possible to measure relatively fast atomic motion or rearrangements. Figure 5.32 shows the measured signal in the 1/1 Zn_6Sc 1/1 approximant from 150 K to 550 K. In this system the tetrahedron ordering takes place at T_C equal to 160 K. As shown in the first panel, whereas the signal is resolution limited at 150 K, there is a clear contribution at 175 K, demonstrating that the tetrahedron disorder is

376 *Physical properties*

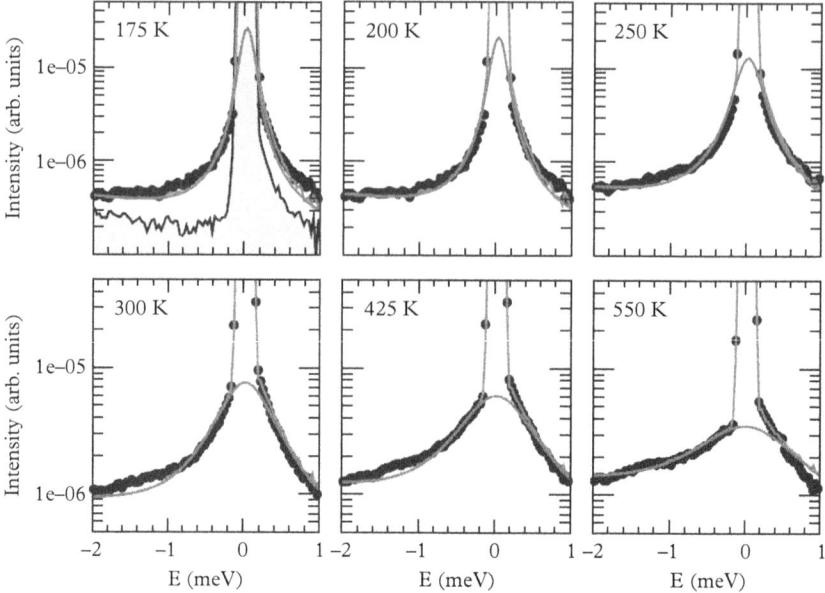

Figure 5.32 *Evolution of the quasi-elastic signal measured at T = 150, 175, 225, 300, 425, and 550 K in the 1/1 ZnSc approximant. Filled circles stand for the measured data. Continuous lines stand for the Lorentzian contribution and the total fit. In the first panel the grey curves correspond to the measurements carried out at 150 K. There is a clear supplementary contribution at 175 K. As the temperature rises, the width of the quasi-elastic signal increases but the maximum intensity decreases. (From (Euchner et al., 2012))*

indeed dynamic in nature. The exact dynamical behaviour could be extracted, combining the Q and T dependence of the measured signal together with atomic scale simulation and molecular dynamic simulation using oscillating pair potential. The observed signal has a width which is Q independent but changes rapidly with temperature, whereas its intensity shows a maximum as a function of Q. Comparing these results with atomic scale simulations lead to the rather surprising result that the central tetrahedron behaves as a single molecule and reorients on a timescale on the order of 1 ps. Figure 5.33 displays an example of such a tetrahedron reorientation. As the tetrahedron reorients, the dodecahedron and the successive shells are strongly distorted, with displacement as large as 0.5 Å for the dodecahedron and the icosidodecahedron, whereas displacement are limited to 0.1 Å for the Sc icosahedron. A similar behaviour has also been found for the quasicrystal, although here the central tetrahedron freezes progressively rather than abruptly, as can be expected from the larger number of cluster–cluster configurations (18 instead of only 1 in the 1/1 approximant) (Euchner *et al.*, 2013). In fact, the distribution of the local environment might lead to a kind of 'spin glass' transition, assimilating the two tetrahedra orientations to spins. Nevertheless, here

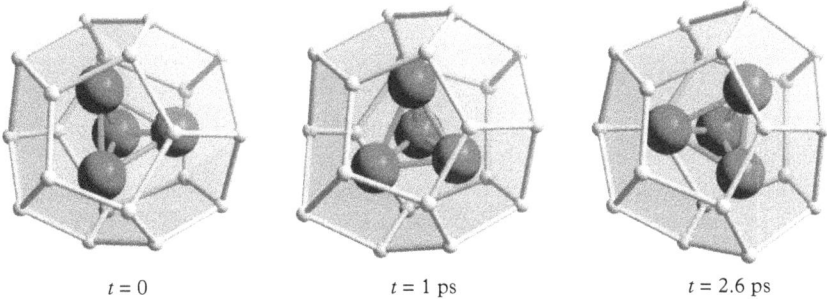

Figure 5.33 *Illustration of the atomic configuration of the tetrahedron and the dodecahedral shell taken at $t = 0, 1$, and 2.6 ps from the simulation. The displacement of the tetrahedron is almost a clockwise rotation around a three-fold axis normal to the projection plane and passing through the atom in the back. Notice the strong distortion of the dodecahedron as the rotation proceeds. (From (Euchner et al., 2012))*

also the tetrahedron reorient dynamically with a timescale of the order of 1 ps at room temperature.

We have thus a very peculiar property for an intermetallic compound which can be characterized by the tetrahedron behaving as a single molecule, reorienting dynamically above a critical temperature, and leading to an *exceptional dynamical flexibility*.

As already mentioned in Section 5.8.2, a central unit breaking the local icosahedral (or decagonal) symmetry seems to be quite frequent, so that such dynamics could also be encountered in other quasicrystals.

In conclusion, the lattice dynamics of aperiodic crystals presents particular features. Whereas the Bloch theorem does not apply, a PBZ can be defined, allowing an analysis of the experimental results. A characteristic length scale of the order of the diameter of the clusters plays a crucial role: for this wavelength there is a strong coupling between the acoustic mode and some dispersion less optical excitations. One fascinating theoretical property of phonons in aperiodic systems, is their critical character. It can be understood with hand-waving arguments as being the result of the quasiperiodic and hierarchic packing of the same structural units. The eigenvectors are never strictly localized on a single cluster, but they 'crosstalk' with neighbouring ones. Understanding this property in three-dimensional systems and trying to detect it experimentally remains a challenging task. The accurate and realistic atomic scale simulations now at hand will certainly play a role in understanding those properties, where their accurate analysis call for further theoretical development.

The complete experimental data now at hand in a few systems, together with sophisticated atomic scale simulations, has also been a very powerful input for the study of other structurally complex materials such as clathrates, for instance (Euchner *et al.*, 2012; Pailhes *et al.*, 2014).

5.9 Phasons: Experiment

5.9.1 Introduction

In theoretical studies of the dynamics of aperiodic crystals, modes have been found which may be described with a polarization direction in the internal space. They are not a sign of additional degrees of freedom, but they have a particular character. The theoretical predictions of their properties, however, have only been partially confirmed. There are questions left both for the systems with smooth modulations and for those with discontinuities. In the following sections we concentrate on the experimental findings (de Boissieu *et al.*, 2007).

5.9.2 Phason modes in modulated crystals

We first focus on the class of quasiperiodic structures for which the properties of phasons are best understood, namely the incommensurate modulated crystals. These systems can be viewed as resulting from a continuous phase transition at, say, some finite temperature T_i. As illustrated in Fig. 5.34, the modulated phase is stable, typically, between a high-temperature parent periodic phase, which plays the role of the high-symmetry phase, and a low-temperature commensurately modulated phase, which is again periodic (the corresponding sequence for quasicrystals would be liquid–quasicrystal–crystalline approximant). In the quasiperiodic state there is at least one new periodicity, the inverse of the modulation wave vector q_s, each Bragg reflection of the high-temperature periodic phase becomes a main reflection and is accompanied by a series of satellite reflections, $m = 1, 2, 3$, for first-, second-, and third-order satellite reflections.

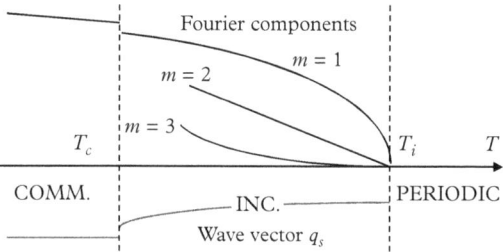

Figure 5.34 *A typical phase diagram for a modulated crystal. The temperature evolution of the modulation wave vector q_s and of the intensity of the $m = 1, 2, 3$ Fourier components is shown schematically. The high-temperature phase is a periodic high-symmetry parent phase; then, the incommensurately (INC.) modulated phase appears; finally, at low temperatures, the modulation has a lock-in on a commensurate (COMM.) wavenumber.*

The modulated variable may be a number of different physical quantities. One may have a displacive structural phase transition at T_i and modulated atomic positions below. One may have an order/disorder-type transition and modulated site occupation probabilities below T_i. One may have a modulated chemical ordering process as in minerals and alloys, or modulated magnetic moments, or mixtures of several types of modulations.

Hydrodynamic theory predicts new modes, phason modes. Using Landau theory, we may address the practical question of where to go in Fourier space in order to observe these long-wavelength phasons. The answer is that we should look near first-order satellite reflections, because phasons are part of the fluctuation spectrum of the order parameter and that is where these fluctuations are visible above T_i. That defines the scattering geometry for an inelastic neutron scattering measurement or for an X-ray diffuse scattering experiment.

In $ThBr_4$ the soft mode leads to an instability towards an incommensurate phase (Fig. 5.26). In the incommensurate phase the theory predicts that, near the satellite reflection, two excitations will be observable: a phason and an amplitudon (Bernard et al., 1983).

The amplitudon mode is more or less related to the previous soft excitation, with a frequency that goes to zero at T_i and then rises sharply. The phason mode, however, is a new excitation. It can be measured by inelastic neutron scattering and should display a dispersion which is similar to an acoustic mode. However, and this is one of the experimental difficulties, the slope should be different than the one of the acoustic mode: around the satellite reflection and at low energy one thus expects in principle two excitations: the acoustic mode and the phason mode. Differences in the intensity of the observed signal allow one to distinguish both excitations, at least in principle.

Figure 5.35 shows what the dispersion curves look like below T_i for $ThBr_4$. Instead of one optic branch, we now have two branches: the amplitudon branch, with a finite frequency at the satellite position, and the phason branch, with a linear dispersion, somewhat like an acoustic mode emanating from a first-order satellite position, but with a different slope. The two branches are well defined only in a neighbourhood of the satellite. As we move away from the satellite position, one of the two branches disappears gradually while the other picks up strength, and, eventually, one recovers the picture of a single optic branch as seen in the high-temperature phase (the broken curve in Fig. 5.35 shows the optic mode dispersion at some temperature $T > T_i$.)

What is the displacement pattern corresponding to a phason mode? The phason mode can be represented in the high-dimensional picture as a wave, with a polarization in the perpendicular direction. For q equal to zero, the displacement field is equivalent to a rigid translation and we thus have a shift of the phase. For q different from zero, the phason mode has a displacement field equivalent to a kind of frequency modulation of the modulation function.

One expects their damping coefficients Γ all to be very similar and slowly varying with wavelength and temperature. For phasons, this implies a damping

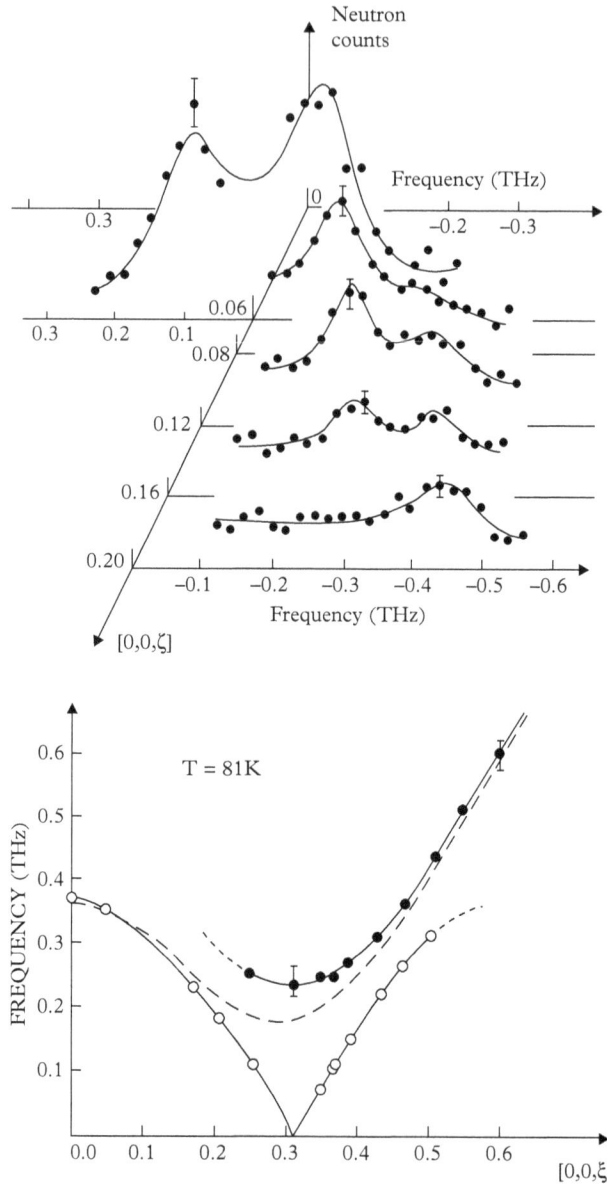

Figure 5.35 *Top: Constant-q energy scans measured by inelastic neutron scattering near the satellite reflection at 81 K. The phason mode is clearly visible. Bottom: Phason (open circles) and amplitudon (closed circles) mode dispersions in $ThBr_4$ at 81 K. The broken curve shows the soft-mode dispersion at 150 K. (After (Bernard et al., 1983))*

coefficient that is slowly varying in q and T and finite at $q = 0$. In the long-wavelength, low-q limit, the phason mode thus has a real part of the frequency which vanishes for continuous modulation function, while the imaginary part (damping) stays finite. This leads to a damped harmonic oscillator response which in the overdamped regime is equivalent to a diffusive mode.

There are only a few systems for which damped propagative phason modes have been observed experimentally, for example biphenyl, ThBr$_4$, and BCPS ((ClC$_6$D$_4$)$_2$SO$_2$). The presence of defects or impurities pin the phase of the modulation and drive the system out of the hydrodynamic regime. This shows up as a non-vanishing frequency of the phason mode as q goes to zero. This is a phason gap.

Another interesting system is the NaNO$_2$ crystal. In this case, the modulation is related to an order/disorder transition. The NO$_2$ molecules can have two alternate orientations, up or down, corresponding to opposite electric dipole moments. At low temperatures ($T < T_c$), all the electric dipoles point in the same direction, downwards in the figure, and we have a ferroelectric phase. At high temperatures ($T > T_i$) the up and down orientations are occupied randomly (with some short-range order, of course) and, in-between these two phases ($T_c < T < T_i$), there is a modulated phase with mixed displacive-order/disorder character.

The dynamics of such a system is dominated by the large-amplitude orientational jumps of the nitrite ions with a typical energy barrier of 0.4 eV at a temperature close to $T_i \approx 450$ K. Therefore, the dynamics is of the slow relaxational type, closer to what is found in quasicrystals than to the usual soft-phonon behaviour observed near displacive phase transitions.

Here the order–parameter fluctuations are entirely relaxational, not just in the long-wavelength limit, as in the displacive case, but at all wavelengths. Nevertheless, the results from the displacive case can be transposed to the order/disorder case by simply replacing $(\omega^2/\Gamma)_{phonon}$ by the relaxation rate or the inverse relaxation time τ^{-1}.

So, above T_i, one observes the critical slowing down of the order-parameter fluctuations, characterized by a T- and q-dependent relaxation rate with a parabolic minimum at the critical wave vector.

In Fig. 5.36 we show the results obtained by Durand et al. on sodium nitrite using the neutron spin echo technique (Durand et al., 1990). The upper right-hand frame shows a series of spectra taken at the critical wave vector q_s as a function of temperature for $T \to T_i^+$. The corresponding relaxation rates are shown in the bottom frame. One notices that the relaxation rate seems to extrapolate to zero at T_i, not quite linearly, as would be expected in a strictly mean-field picture—but there is no indication of a gap at T_i. By continuity this implies that one does not expect to find a phason gap in the modulated phase just below T_i. Direct measurements in the modulated phase are more difficult to interpret because the neutron spin echo technique is less accurate when several relaxation rates are present. But the data are certainly consistent with a diffusive, temperature-independent phason relaxation rate.

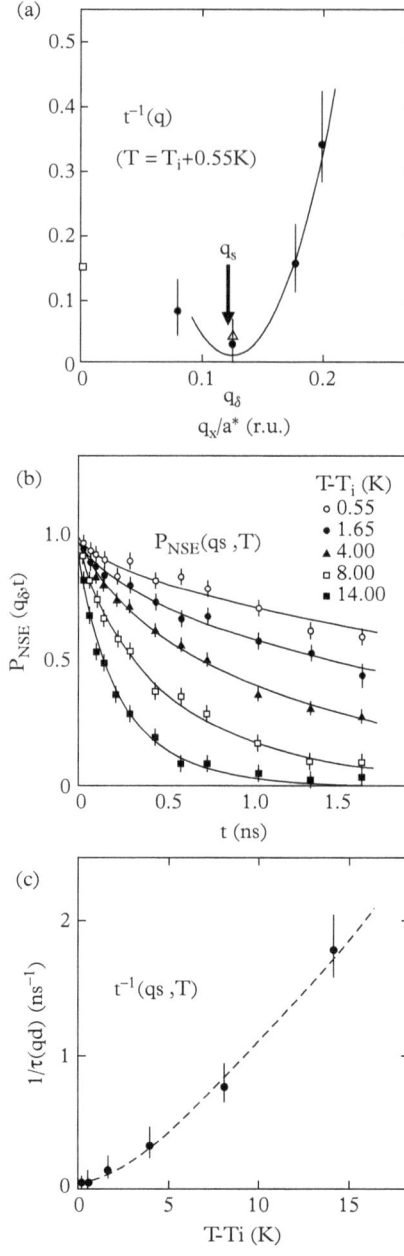

Figure 5.36 *Critical slowing down of NO_2-reorientations in $NaNO_2$. (a) Relaxation rate of $\tau^{-1}(q)$ for q along a∗ and $T = T_i + 0.3$ K (b) Neutron spin echo spectra at the critical wave vector q_s as a function of temperature. (c) Temperature dependence of $\tau^{-1}(q_s)$. (After Durand et al. (1990))*

We end this section with some results on composite structures. In this case two subsystems are interacting, as in the double chain model. The phason mode is often referred to as the sliding mode (cf. p. 335) This is because, in the long-wavelength limit, the phason mode in a composite leads to a relative motion of the subsystems, as if one were sliding on top of the other. This is, of course, only true if the coupling between the two subsystems is not so strong that pinning occurs. So far, attempts to detect the sliding mode by spectroscopic techniques has been rather unsuccessful because of the difficulty of obtaining high-quality samples. On a phenomenological basis one expects a very long-wavelength regime, for which the two systems are coupled and only one sound velocity is measured, and a shorter wavelength regime, for which the two subsystems decouple with the appearance of the sliding phason mode. This is still an open experimental problem.

5.9.3 Diffuse scattering and phason modes in icosahedral quasicrystals

Phason modes are collective diffusive excitations which might also be studied by spectroscopic techniques such as inelastic or quasi-elastic neutron scattering. However, some information can be gained by measuring the signal using X-ray or neutron diffuse scattering. The elasticity theory of quasicrystals allows one to predict the shape and the Q dependence of the measured signal. This technique has been so far mainly used to study icosahedral quasicrystals for which we give the main results. The results are analogous for other aperiodic crystals.

At a given temperature and at thermodynamic equilibrium there is a distribution of phonon and phason modes in the quasicrystal. These modes lead to changes of atomic positions with respect to the ideal (i.e. without any fluctuations) structure, and thus have consequences for the scattered intensity in a diffraction experiment. In general, the scattered intensity may be decomposed into two components:

$$S(\mathbf{Q}) = S_{Bragg}(\mathbf{Q}) + S_{Diff}(\mathbf{Q}). \tag{5.47}$$

The first term is the Bragg component, which remains a set of delta functions if fluctuations are not diverging; the second term corresponds to the diffuse scattering observed in-between Bragg reflections. The calculation of these two terms can be carried out following a procedure analogous to what is done for the calculation of the diffraction pattern of a periodic crystal with phonons at equilibrium: in this case it is well known that the Bragg peak intensity remains a delta function but with an intensity decreased by the Debye–Waller factor, while TDS appears in-between the Bragg reflections. In the case of a quasicrystal the same is true for phonons but an additional component arises from phason fluctuations.

The Bragg component is related to the Fourier transform of the time and ensemble average of the density $\langle \rho(R) \rangle$, whereas the diffuse scattering component is related to the Fourier transform of the two-body density correlation function $\langle \rho(\mathbf{r}_i)\rho(\mathbf{r}_j) \rangle - \langle \rho(\mathbf{r}) \rangle^2$.

The Bragg component is computed in considering that the average structure is obtained by the convolution of Gaussian fluctuations with the 'ideal' (i.e. without fluctuations) decorated lattice. In the case of quasicrystals, the Bragg component is easily calculated in the high-dimensional periodic lattice decorated by atomic surfaces. Phonon fluctuations will lead to a Gaussian broadening of the atomic surfaces along the parallel direction, with a mean-square amplitude $\langle u_{par}^2 \rangle$. Phason fluctuations lead to a Gaussian broadening in the perpendicular direction with mean-square amplitude $\langle u_{per}^2 \rangle$.

A phason fluctuation may be described equivalently either by an undulation of the parallel space or by a fluctuation $u_{per}(R)$ of atomic surfaces along the perpendicular direction. The Bragg peak intensity measured on a point of the Fourier module with a vector $\mathbf{H}_{par}$ is thus written:

$$S_{Bragg}(\mathbf{H}_{par}) = S_{Ideal}(\mathbf{H}_{par}) \exp\left(-\langle u_{par}^2 \rangle H_{par}^2\right) \cdot \exp\left(-\langle u_{per}^2 \rangle H_{per}^2\right), \quad (5.48)$$

where $S_{Ideal}(\mathbf{H}_{par})$ is the Bragg component of the ideal quasicrystal without any fluctuations and where we have neglected the phonon–phason coupling term. The values of $\langle u_{par}^2 \rangle$ and $\langle u_{per}^2 \rangle$ can be computed from the hydrodynamic matrix $C(q)$. When the phonon–phason coupling term is neglected the matrix $C(q)$ is block diagonal with two 3 × 3 matrices acting on the parallel and perpendicular component of the reciprocal vector $\mathbf{H}$ and denoted $C_{par,par}(q)$ and $C_{per,per}(q)$. The two 3 × 3 matrices contain matrix elements related to the Lamé coefficient (par, par matrix) and K_1 and K_2 (per, per matrix).

For three-dimensional quasiperiodic systems it can be shown that the fluctuations are bounded and thus Bragg peaks of icosahedral quasicrystals remain delta functions with their intensity reduced by both a 'phonon' and a 'phason' Debye–Waller term.

The diffuse scattering intensity measured at the point $\mathbf{H}_{par}+\mathbf{q}$ is also computed using the hydrodynamical matrix C. We have

$$S_{Diffus}(\mathbf{H}_{par} + \mathbf{q}) = S_{Bragg}(\mathbf{H}_{par}) \langle \mathbf{H}_{6D} | C^{-1}(q) | \mathbf{H}_{6D} \rangle, \quad (5.49)$$

where the Bragg peak position has coordinates $\mathbf{H}_{6D} = (\mathbf{H}_{par}, \mathbf{H}_{per})$ and the measurement is carried out at the position $\mathbf{H}_{par} + \mathbf{q}$. In the case where the K_3 phonon–phason coupling term can be neglected, this expression reads

$$S_{Diffus}(\mathbf{H}_{par} + \mathbf{q}) = S_{Bragg}(\mathbf{H}_{par}) \langle \mathbf{H}_{par} | C_{par,par}^{-1}(\mathbf{q}) | \mathbf{Q}_{par} \rangle \quad (5.50)$$
$$+ S_{Bragg}(\mathbf{H}_{par}) \langle \mathbf{H}_{per} | C_{per,per}^{-1}(\mathbf{q}) | \mathbf{H}_{per} \rangle.$$

The first term of this expression corresponds to the usual TDS, whereas the second term corresponds to phason diffuse scattering. The TDS component is isotropic and presents the same shape around all Bragg peaks: it is an ellipsoid elongated in the transverse direction. The well-known characteristics of the TDS component are summarized as follows. Along a given direction $\mathbf{q}$ from a Bragg reflection the TDS intensity decays as q^{-2}. The term $\langle \mathbf{H}_{par} | C^{-1}_{par,par}(\mathbf{q}) | \mathbf{H}_{par} \rangle$ induces a selection rule as a function of the relative orientation of $\mathbf{q}$ (or $\mathbf{e}$) and $\mathbf{H}_{par}$: for instance, if the two vectors are parallel, only longitudinal phonons will contribute to the signal, whereas only transverse phonons contribute when the vectors are orthogonal. In these two directions the TDS intensity is inversely proportional to the longitudinal and transverse squared sound velocities V_L^2 and V_T^2 and, since V_L is larger than V_T the TDS intensity is smaller along a direction collinear with $\mathbf{H}_{par}$ than for a direction perpendicular to it, giving rise to a characteristic ellipsoidal intensity distribution around each Bragg reflection. This is illustrated in Fig 5.37, left panel. Finally, the TDS intensity is proportional to the Bragg peak intensity, and to H_{par}^2. The TDS is better measured at high H_{par} and around strong Bragg peaks.

The same reasoning can be applied for the phason part of the diffuse scattering. When measured along a given direction $\mathbf{q}$ from a Bragg reflection, the diffuse scattering intensity decays as q^{-2}. The term $\langle \mathbf{H}_{per} | C^{-1}_{per,per}(\mathbf{q}) | \mathbf{H}_{per} \rangle$ contains the product of the matrix $C^{-1}_{per,per}(\mathbf{q})$ and the perpendicular component of the Bragg scattering vector H_{per}, leading to a selection rule as a function of the relative orientation of $\mathbf{e}_{per}$ (the polarization of the phason mode) and $\mathbf{H}_{per}$. Different

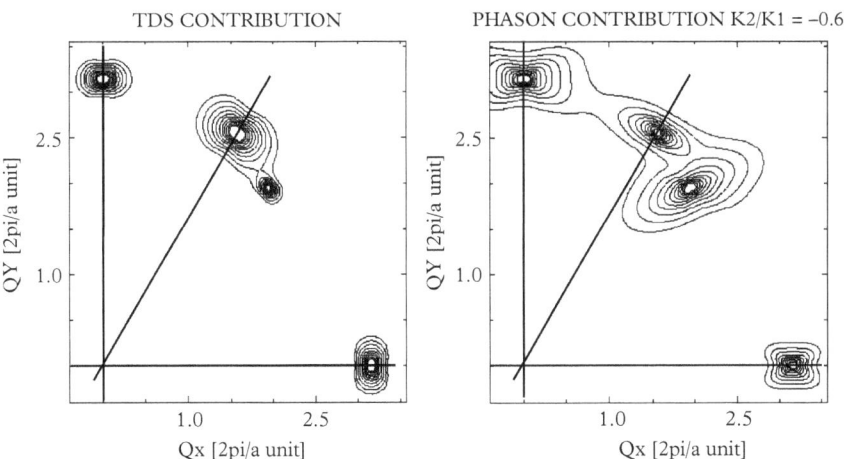

Figure 5.37 *Comparison between the phonon (TDS, left panel) and the phason (right panel) contributions to the diffuse scattering, as deduced from elasticity theory. The shape of the intensity distribution of the TDS contribution is similar around every Bragg peak, whereas strong anisotropies are observed for the phason contribution. The simulation of the phason contribution has been carried out with a ratio K_2/K_1 of -0.6.*

386 *Physical properties*

phason polarizations are thus enhanced in the measured diffuse scattering signal for different **q** and $\mathbf{H}_{per}$ orientations. The term which drives the anisotropy of the signal is the ratio K_2/K_1, whose value lies between −0.75 and +0.75. As a consequence, the phason diffuse scattering intensity can display strong anisotropies, even for reflections which have similar $\mathbf{H}_{par}$ directions, if their $\mathbf{H}_{per}$ components are different (Jarič and Nelson, 1988; Widom, 1991; Ishii, 1992).

This is illustrated in Fig. 5.37, right panel, where the phason part of the diffuse scattering has been simulated in a two-fold scattering plane, as it would appear around a few Bragg reflections of an icosahedral quasicrystal, for a K_2/K_1 ratio equal to −0.6. The two reflections on the five-fold axis and nearby show a diffuse scattering anisotropy having elongations almost orthogonal to each other, because their $\mathbf{H}_{per}$ components are almost orthogonal: this is a characteristic signature of phason diffuse scattering and is very different from that of TDS (Fig. 5.37, left panel). Finally, the phason diffuse scattering intensity scales as the Bragg peak intensity and as Q_{per}^2. This relation can easily be checked by measuring the diffuse scattering around several reflections lying along the same high-symmetry axis. As a conclusion, diffuse scattering measurements are a very efficient tool to show and characterize long-wavelength phason fluctuations in the case of icosahedral quasicrystals.

5.9.4 Phason modes in the i-AlPdMn icosahedral quasicrystal

A very complete study on phason modes have been carried out for the i-AlPdMn icosahedral quasicrystal for which we give the results in the following.

5.9.4.1 *Phason modes and room temperature diffuse scattering measurements*

As explained previously, an efficient way to detect long-wavelength phason modes is by means of diffuse scattering measurements. This experimental approach is completely analogous to the one used for the study of phonons in crystals reported in Laval (1938), Olmer (1948), Curien (1952), and Walker (1956).

Data analysis is much simplified if the TDS contribution can be suppressed. This can be achieved by using inelastic neutron scattering and energy analysis on a three-axis spectrometer. With the proper settings the measured elastic signal only contains the contribution from static distortions and with 'slow' time dependencies, the phonon contribution being eliminated. Measurements have thus been carried out around a few strong Bragg reflections in the neutron diffraction pattern shown in Fig. 5.38. In the case of elastic (phonon-like) distortions, the diffuse intensity is expected to display the same shape around all Bragg reflections, because the icosahedral phase is elastically isotropic. This will not be the case if phason fluctuations are present: depending on the ratio of K_2/K_1, differently shaped anisotropies are expected around the different Bragg reflections as already shown. Since the phason diffuse scattering scales as Q_{per}^2, reflections having

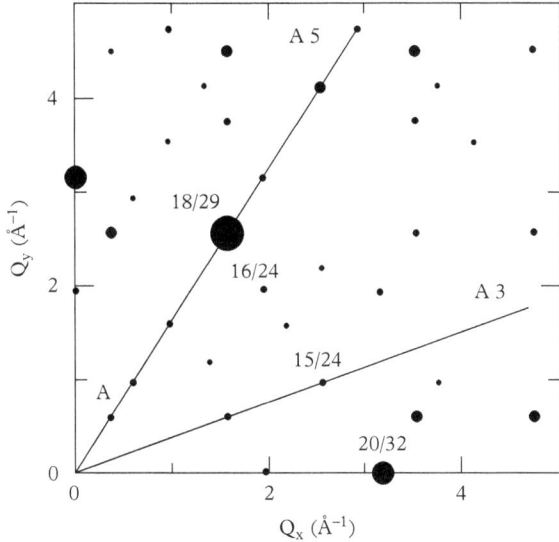

Figure 5.38 *Neutron scattering intensity distribution in a two-fold plane for the i-AlPdMn phase. The area of the dots is proportional to the intensity of the reflections. A few of them are indexed with their N/M indices*

different perpendicular components were studied. They are labelled by their N/M indices, which is a short-hand notation for the full 6D indexing. The 18/29, 15/24, and 20/32 reflections lie on five-, three-, and two-fold axes, respectively (Fig. 5.38).

The results of these measurements are shown in Fig. 5.39, where iso-intensity contours are shown. The same intensity range has been selected for all four reflections, so that the different panels can be directly compared. As can be seen, the intensity distribution displays strong anisotropies. Several observations point towards a phason contribution to this diffuse scattering. First, the 16/24 and 15/24 Bragg peaks have almost the same intensity, but different values of the H_{per} components of the scattering vector, the 16/24 one being about three times larger than the 15/24 (de Boissieu *et al.*, 1995). As can be observed in Fig. 5.39b, c, there is a much larger diffuse scattering intensity around the 16/24 reflection as would be expected from Eq. 5.50. Second, the 18/29 and 16/24 reflections are very close to each other in reciprocal space: any elastic (phonon-like) distortion would lead to a very similar shape of the diffuse scattering whereas the experiment displays almost orthogonal anisotropies. This can be understood if we look at the perpendicular component of these two reflections: they are almost orthogonal, and will thus lead to a very different scalar product $\mathbf{H}_{per}.\mathbf{e}_{per}$ 'selection rule' in Eq. 5.50. Finally, measuring two collinear reflections, it is possible to show that the intensity of the diffuse scattering decays as q^{-2} and scales with the product $I_{Bragg}.H_{per}^2$. These

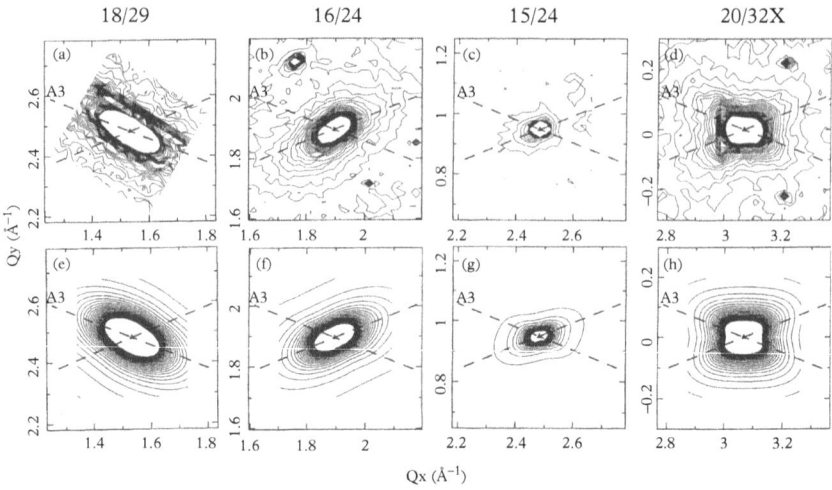

Figure 5.39 *Top panel (a–d): Iso-intensity contour plots of the diffuse scattering signal measured by elastic neutron scattering around different Bragg peaks in the i-AlPdMn phase at room temperature. The indices are given in the N/M notation. All contour lines are on the same scale. Bottom panel (e–h): Simulation of the diffuse scattering, using as a single parameter the K_2/K_1 ratio set to -0.52. (From (de Boissieu et al., 1995))*

results are confirmed by a computation of the diffuse scattering using Eq. 5.50. If we consider that the diffuse scattering is due mainly to phason fluctuations and if we neglect the coupling constant K_3, then the shape anisotropy depends only on the ratio K_2/K_1. Using the measured Bragg peak intensities it is then possible to simulate the diffuse scattering. This is what has been carried out in Fig. 5.39 e–h, using a ratio $K_2/K_1 = -0.52$. As can be seen in the figure, there is a good agreement between simulation and experiment. A quantitative agreement was also achieved by comparing the intensity decay observed along different directions and various Bragg peaks.

If the diffuse scattering is measured on an absolute scale it is also possible to get the individual values of the two elastic constants K_1 and K_2. This kind of measurement is completely analogous to the determination of elastic constants or phonon dispersions using X-ray diffuse scattering. From the experimental point of view, an absolute scale measurement is more demanding, so that it has been only carried out in a few systems. Using both X-ray and neutron scattering leads to similar results, that is, to values of K_1/k_BT and K_2/k_BT equal to 0.1 and -0.052 atom^{-1}, respectively (Létoublon et al., 2001). An example of the results obtained from absolute X-ray diffuse scattering measurements along a two-fold axis is shown in Fig. 5.40. The different contributions to the signal are indicated, namely the phason diffuse scattering component (only the contributions from the strong 20/32 and 8/12 reflections are shown), the Compton scattering component, and

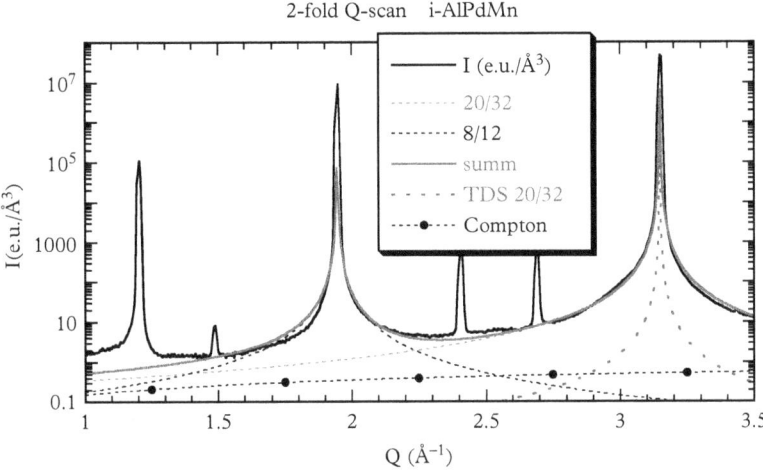

Figure 5.40 *X-ray diffraction pattern measured along a two-fold direction in the i-AlPdMn phase. The intensity is set on an absolute scale, and the various contributions to the signal are highlighted (Compton and TDS parts). The solid line is the result of the simulation for the sum of all contributions: the phason contribution dominates the diffuse scattering intensity. (From (Létoublon et al., 2000b))*

the TDS contribution originating from the 20/32 reflection. It is clear from this figure that the contribution from TDS in X-ray measurements is indeed much smaller than the phason one, which makes the X-ray data easy to analyse in terms of phason diffuse scattering. However, this cannot be inferred a priori, and neutron elastic measurements or absolute scale X-ray measurements are crucial for the characterization of the phason diffuse scattering. The $1/q^2$ decay of the diffuse scattering is also well reproduced: it extends over a rather wide wave vector range, roughly from 0.03 to 0.6 Å^{-1}, corresponding to a phason mode wavelength between 200 and 10 Å. In fact high-resolution measurements showed that the lower q bound for which the $1/q^2$ decay is observed is 0.01 Å^{-1}, corresponding to about 600 Å. Absolute scale measurements of the diffuse scattering are also important for any quantitative comparison of the amount of diffuse scattering present in different samples. In particular, it was shown that the amount of diffuse scattering is independent of the sample annealing treatment for i-AlPdMn single grains (Létoublon *et al.*, 2001). As-grown samples have a larger Q_{per} dependence of their Bragg peaks width when compared to annealed ones, indicating that the amount of 'phason strain' is larger in the as-grown samples. Yet, both samples have exactly the same amount of diffuse scattering, which demonstrates that the observed phason fluctuations are thermal equilibrium fluctuations and are not related to a distribution of structural defects (as dislocations, for instance) giving rise to the broadening. As we will show further with the temperature dependence

390 Physical properties

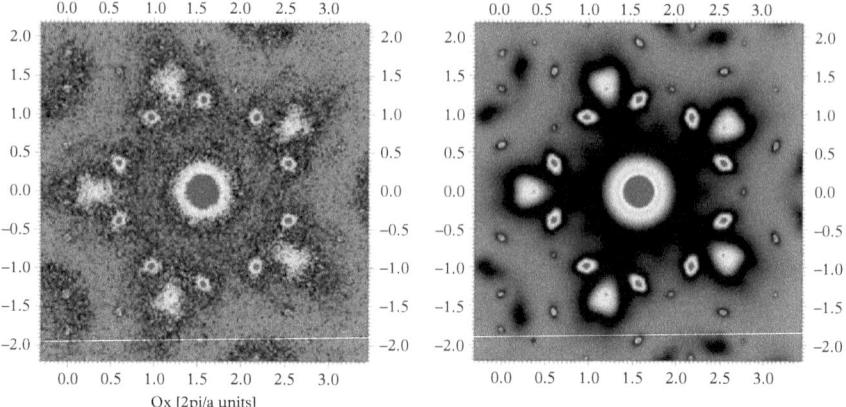

Figure 5.41 *Diffuse scattering in the i-AlPdMn phase. Left panel: Diffuse scattering measured around the strong five-fold 18/29 reflection, in a plane perpendicular to the five-fold axis. Measurements were performed on a rotating anode (Cu target) four-circle diffractometer. Right panel: Simulation of the diffuse scattering, taking into account the contribution from in- and out-of-plane Bragg reflections. Both the shape and the intensity distribution of the diffuse scattering are well reproduced, such as, for instance, the lobe-like distribution seen on the outer part of the diffraction pattern.*

of the diffuse scattering intensity, it is because phason fluctuations are indeed an intrinsic property of the i-AlPdMn quasicrystal.

Phason fluctuations also allow one to explain the observed diffuse scattering when a wide survey of reciprocal space is carried out. Figure 5.41 displays such a measurement in a diffracting plane orthogonal to a five-fold axis and passing through the strong 18/29 reflection. The diffraction pattern (left panel), displays characteristic distributions of the intensity with arcs of diffuse scattering, for instance, and a five-fold symmetry. Such an intensity distribution is fairly well reproduced (Fig. 5.41, right panel) if the diffuse scattering arising from different Bragg reflections and, in particular, out-of-plane reflections, is added up together (de Boissieu and Francoual, 2005). With such a calculation, phason fluctuations can explain 90 per cent of the observed diffuse scattering.

5.9.4.2 Temperature dependence of the diffuse scattering and phason elastic constants

Room temperature measurements do not give any hint of the nature of the microscopic mechanisms associated with phason fluctuations. Indeed, at room temperature, these fluctuations are most likely frozen-in, rather than equilibrium fluctuations. Moreover, the random tiling and matching rule models predict a different behaviour of the diffuse scattering as a function of temperature. These two models have been frequently referred to, respectively, as the entropically and the energetically stabilized quasicrystal models (Henley, 1991*b*). Before presenting the experimental results obtained in the i-AlPdMn phase, we first give a brief overview of theoretical results.

We turn now to the predicted temperature dependence of the phason elastic constants as a function of T, and its consequences for diffuse scattering. We compare two simple models which have been put forward to explain the stability of quasicrystals: the energy- and the entropy-stabilized models, which predict a quite different temperature dependence for the phason elastic constants. In the first model, the quasicrystal is considered as being a ground state at 0 K and is energetically stabilized. Whether this can be achieved using only finite range interatomic interactions or that infinite-range interactions are required is still an open question. Since phason fluctuations are increasing with temperature, we should thus observe an increase of the diffuse scattering when the temperature increases. The shape of the diffuse scattering provides a clue to the applicability of hydrodynamic theory. We have already discussed the three-dimensional 'matching rules' Penrose model, which presents a locked–unlocked phase transition. Although the diffuse scattering intensity has not been computed below T_c, it would most likely resemble a broad Lorentzian, whose width would decrease as the temperature is raised, as phason fluctuations build up. Above T_c the $1/q^2$ decay of the diffuse scattering is recovered, with an diffuse intensity increasing proportionally to $k_B T$ (Dotera and Steinhard, 1994).

The entropy-stabilized model was developed in the random tiling scenario. It postulates that the quasicrystal is stabilized by configurational entropy, which is essentially associated with short-wavelength phason fluctuations. In this scheme the quasicrystal is only stable at high temperatures and undergoes a phase transition towards a crystalline phase as the temperature is lowered (Henley, 1991b). This transition is accompanied by some pre-transitional phason softening and hence the first-order phase transition should be preceded by changes in the phason elastic constants and corresponding changes in the diffuse scattering. This has been studied theoretically within the Landau theory for an icosahedral quasicrystal, considering various phason-driven phase transitions (Ishii, 1989; Widom, 1991; Ishii, 1992). The soft-phason mode to be considered depends on the nature of the first-order transition and the corresponding phason strain involved: for instance, we expect a soft-phason branch along three-fold axes for an icosahedral to tetrahedral transition. This also gives some limits for the ratio K_2/K_1, which goes from -0.6 (three-fold instability) to 0.6 (five-fold instability). This temperature dependence has been verified on a simple canonical cell model: large variations of the elastic constants are observed in the simulation when going from the high- to the low-temperature region, the ratio K_2/K_1 going from positive to negative in the low-temperature region, and the 0 K ground state being a crystalline phase (Mihalkovič and Henley, 2004).

To summarize, the two models can be distinguished by the relative weight they give to the energy and entropy terms stabilizing the quasicrystal. Both models predict a high-temperature region, where phason fluctuations are hydrodynamic and the diffuse scattering can be computed. However, the two models predict a different temperature dependence of the diffuse scattering. In the case of the energetically stabilized scenario, the diffuse scattering should increase as the temperature is increased and phasons are hydrodynamic only above a critical temperature T_c. In the random tiling scenario, there should be a low-temperature crystalline phase, the quasicrystal being stable only in the high-temperature region. Starting from the high-temperature quasicrystalline phase and going down in temperature, soft-phason modes should appear, leading to an increase in diffuse scattering.

5.9.4.3 Experimental results

The temperature dependence of the diffuse scattering has been studied by neutron scattering between 25 °C and 770 °C, using a three-axis spectrometer to suppress the phonon contribution. This is particularly important, since the mean-square atomic displacement due to phonons is roughly proportional to temperature, so that the TDS contribution might not be negligible in the high-temperature region (Boudard et al., 1996).

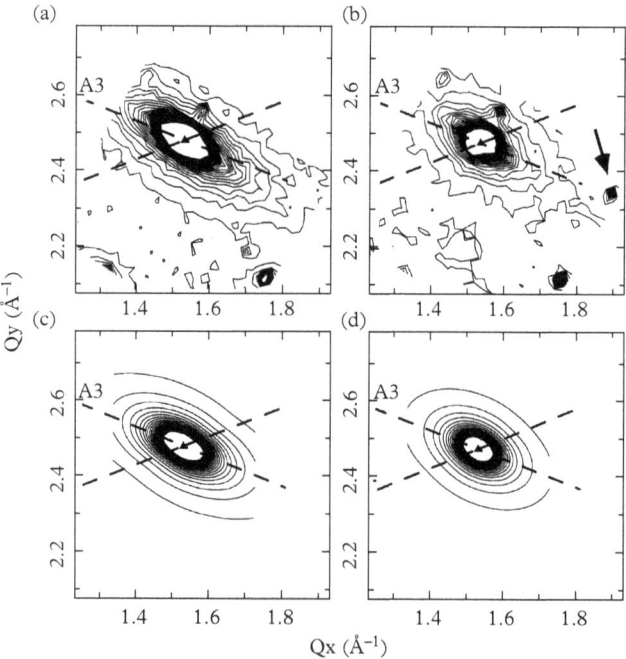

Figure 5.42 *Temperature dependence of the diffuse scattering, measured around the 18/29 five-fold reflection at (a) 200 °C and (b) 770 °C. The evolution is well reproduced by changing the phason elastic constant ratio K_2/K_1 from (c) −0.52 to (d) −0.4. (From (Boudard et al., 1996))*

Figure 5.42 displays iso-intensity contours of the diffuse scattering measured around the 18/29 reflection at 200 °C (Fig. 5.42a) and at 770 °C (Fig. 5.42b). Two observations can be made: first, there is a clear decrease of the diffuse scattering intensity at high temperature; second, the shape of the diffuse scattering is more 'circular' at high temperature. The decrease of the diffuse scattering intensity when temperature is increased has been checked on systematic scans taken around different reflections as a function of the temperature: almost no intensity change is observed up to 500 °C; from this temperature and up to 770 °C there is a continuous decrease of the diffuse scattering intensity, by a factor of almost 2 for some directions. In the meantime, the intensity of some weak high-Q_{per} reflections increases. All these results can be interpreted if we consider a change of the ratio of the phason elastic constants as temperature is increased. It goes from $K_2/K_1 = -0.52$ at room temperature to $K_2/K_1 = -0.4$ at 750 °C. Such an evolution reproduces quite well the observed diffuse scattering, in particular the more 'circular' part of the diffuse scattering in the high-temperature region, as shown in Fig. 5.42c,d. In fact, we can even give a more detailed evolution of the phason elastic constants, since all data are measured on the same scale. Using Eq. 5.50, we can deduce that $K_2/k_B T$ is temperature independent and only $K_1/k_B T$ decreases as temperature is decreased. We thus have a softening of the K_1 phason elastic constant as the temperature is decreased.

These results are clearly in agreement with the random tiling scenario. However, in this scheme, we should observe a transition towards a crystalline phase at low temperature, which is not the case experimentally. The evolution of the phason elastic constants points towards a three-fold type instability. From the Landau theory of phase transitions this should lead to a phase with $\bar{3}m$ symmetry. One possible explanation is that the transition is frozen-in because the kinetics is too slow. The ground state of the system would be a crystalline phase with tetrahedral symmetry, but because atomic diffusion becomes too slow below 500 °C, the system is blocked in a metastable icosahedral state with frozen-in phason fluctuations.

Some hint that this is indeed the correct interpretation is given by a temperature study of the phase transition occurring for icosahedral phases with different Pd/Mn contents. If the Mn concentration in the single grain goes from 0.09 to 0.08, the icosahedral phase is only stable at high temperature. At about 700 °C, there is a first-order transition towards a phase of tetrahedral symmetry of extreme complexity, called F2M. Although the symmetry is $\bar{3}m$, the F2M phase is quasiperiodic (de Boissieu et al., 1998b; Audier et al., 1999; Létoublon et al., 2000b). It can be described as a superstructure of the icosahedral phase, with satellite reflections along directions parallel to three-fold axes. An example of the diffraction pattern measured around a strong high-temperature two-fold reflection is shown in Fig. 4.96. In the high-temperature phase, only the reflection labelled 'i-ref' survives. The position of the reflections S1 and SF2 with respect to the main reflection can be expressed as rational values of reciprocal lattice vectors of the parent high-temperature phase, having a wave vector q_{par} of 0.0369 and 0.120 Å^{-1}, respectively. A detailed atomic description of this F2M phase has not yet been given, but the intensity distribution of the supplementary S1 superstructure reflections is relatively well reproduced by considering a cosine wave distortion of the high-temperature icosahedral phase with a wave vector pointing along a three-fold axis in the parallel (physical) space and a polarization along a three-fold axis in the perpendicular (phason) space. From the measured intensity it can be shown that the amplitude U_{per} of this phason wave is of the order of 0.55 Å, with the wavelength of the modulation being equal to 170 Å. This F2M phase may thus be considered as resulting from a phason-driven phase transition, and would be the stable low-temperature phase.

However, we should keep in mind the complexity of the chemistry, of the phase diagram, and of the stabilization mechanism of quasicrystals. The two models (energetic and entropic) are certainly only first-order approximations. From experimental results, it is clear that the entropy term has an important weight on the stability of the i-AlPdMn quasicrystal, since the diffuse scattering intensity decreases as the temperature is raised. However, dramatic changes in the phase diagram with small chemical composition changes indicate that both the configurational geometric entropy and the chemical potential (which contains an energy and an entropy term) are playing a role. The observed hydrodynamic behaviour of phason fluctuations would thus be the result of a squared gradient behaviour of the free energy, with an entropy part and an energetic part. The observed diffuse scattering, as we have shown, can be explained by long-wavelength pre-transitional phason fluctuations. Hydrodynamic theory does not give a microscopic interpretation of these long-wavelength phason fluctuations. Some hint on the resulting three-dimensional structure can be given by taking the 'undulating cut' picture and considering that the three-dimensional structure is obtained as the convolution of an ideal quasicrystal with perpendicular fluctuations. The effect on the resulting three-dimensional structure will be both some partial site occupancy (as the one illustrated in Fig. 6.1), and chemical disorder at specific location: this can be, for instance, Mn/Pd disorder or Al/Pd disorder, the term disorder being understood by comparison with an 'ideal' quasicrystal. Because of the geometry of the high-dimensional model, the disorder will not be randomly distributed on the whole structure, but will occur on specific sites.

5.9.4.4 Dynamics of phason modes

So far, we have only considered the static aspect of phason modes in quasicrystals. As explained previously, phason modes are collective diffusive modes. It means that a phason mode with a wave vector q, has exponential time decay, with a characteristic timescale t_{char} varying as q^{-2}. Expressed in real space, this is equivalent to a characteristic timescale varying as λ^2.

Long-wavelength phason fluctuations are expected to be too slow to be visible in the neutron timescale window. Their dynamical study has thus been carried out using coherent X-ray and photon correlation spectroscopy techniques. These techniques allow measuring slow dynamics (10^{-3} to 1000 s) at the atomic scale.

When a coherent X-ray beam is used to measure the diffuse scattering, the q dependence of the intensity no longer shows a smooth variation (resulting from an ensemble average) but rather displays strong intensity fluctuations called speckles (Sutton, 2001). These speckles are due to constructive and destructive interference between the scattered waves from the illuminated sample volume. This process is analogous to what is obtained when a disordered medium is illuminated with laser light.

Reasonable intensity coherent X-ray beams can be obtained at third-generation synchrotron sources like the European Synchrotron Radiation Facility, using high-brilliance undulator sources. For instance, using the focusing optics of the ID20 beam line (at the European Synchrotron Radiation Facility), a flux of several 10^9 photons/s is obtained through a 10 μm pinhole located just before the sample. The diffuse scattering spectrum is recorded using a directly illuminated CCD camera, located 1.85 m from the sample and acting as a two-dimensional photon detector using a droplet algorithm (Livet et al., 2000). With such a set-up, the partial coherence of the beam β was found to be equal to 0.05 and 0.03 for low- and high-angle reflections, respectively. A typical two-dimensional image of the diffuse scattering recorded close to a Bragg reflection, together with its cross section, is shown in Fig. 5.43. Spikes and large-intensity fluctuations are clearly visible. The shape of the diffuse scattering is well reproduced using only phason fluctuations, which ensures that the observed signal is the one which is relevant, in other words, that it is related to phason modes.

A dynamical fluctuation in the atomic distribution $\rho(R)$ due to phason fluctuations will lead to a different speckle distribution, so that any time evolution of phason fluctuations will result in a time dependence of the speckle pattern. This is best evidenced by computing the intensity correlation function, which is defined as

$$F_{cor}(\mathbf{q}, t) = \langle I(\mathbf{Q}_B + \mathbf{q}, t') I(\mathbf{Q}_B + \mathbf{q}, t' + t) \rangle_{t'} / \langle I(\mathbf{Q}_B + \mathbf{q}, t'), \rangle^2, \quad (5.51)$$

where $\mathbf{Q}_B$ is the Bragg peak position, and $\mathbf{q}$, the phason wave vector. We have the relation $F_{cor}(q, t) = [1 + \beta g(q, t)]$, where $g(q, t)$ is the function accounting for the time dependence of a phason mode with wave vector q, and β is the coherent fraction of the beam.

The i-AlPdMn sample was placed in a furnace under secondary vacuum and in situ measurements were carried out between room temperature and 650 °C. Measurements have been carried out around the five-fold 7/11 reflection, and for different position of the detector whose central spot probed phason modes with wave vectors along the $(\tau, -1, 0)$ direction. When performing the measurement below 500 °C, the correlation function does not show any time dependence up to timescales of the order of 30 minutes, confirming that phason modes are 'frozen-in' on this timescale. Above 500 °C the correlation function did show a slow time evolution. Results obtained at 650 °C and for two values of the wave vector q are displayed in Fig. 5.44. The time decay of the correlation function is clearly visible, and has been fitted to an exponential law $g(q, t) = \exp(-t/\tau)$, from which

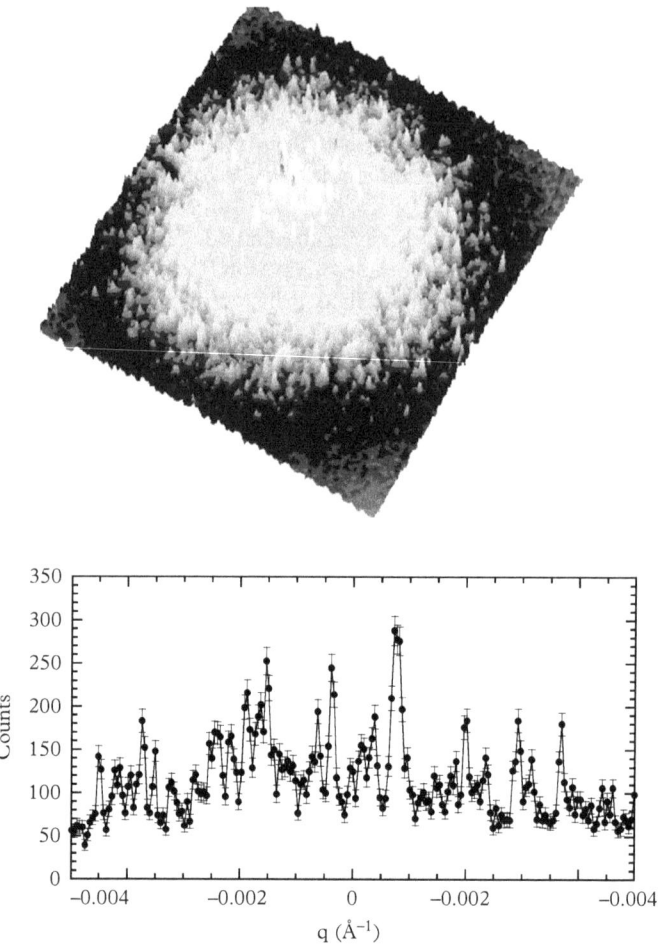

Figure 5.43 *Top: Two-dimensional CCD image of the diffuse scattering measured close to a Bragg reflection by using a coherent X-ray beam. Bottom: Section through the diffuse scattering. The 'spikes' are much larger than the statistical uncertainties, as shown by the error bars: they are due to the coherence of the X-ray beam. (From Létoublon et al. (2001))*

a characteristic time t_{char} is extracted. The top and bottom parts of Fig. 5.44 correspond to a wave vector q of 0.064 and 0.13 nm^{-1}, respectively. As can be seen in the figure, the characteristic time is much shorter for the larger wave vector, in agreement with the expected diffusive character of phason modes. The diffusive character of the phason mode is shown in a more quantitative way in Fig. 5.45, where the characteristic time measured along the direction $(\tau, -1, 0)$ for different wave vectors is plotted as a function of q^{-2}. As can be seen, there is a

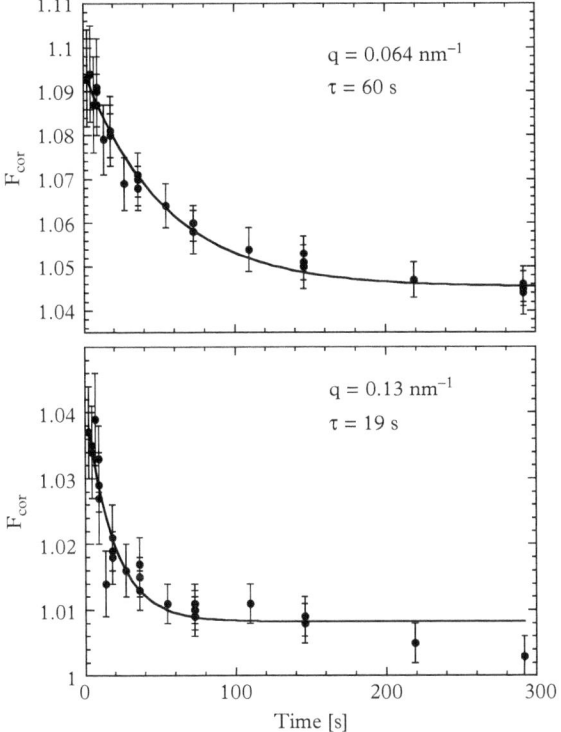

Figure 5.44 *Time-dependent intensity correlation function measured at 650 °C in the i-AlPdMn phase. The solid line is a fit to an exponential decay. The top and bottom figures correspond to two different wave vectors; the decay is much faster for the larger wave vector. (From Francoual et al. (2003))*

linear dependence, from which a phason diffusion constant, D_{phason}, is evaluated and found equal to 1.5×10^{-16} m^{2}s^{-1} (Francoual et al., 2003).

A few measurements were also carried out at 600 °C. At this temperature phason fluctuations have much larger time decays. An estimate of the phason diffusive constant is difficult, because characteristic times are now close to the maximum value which can be measured. Nevertheless, we find that characteristic times are between five to ten times larger than at 650 °C. Assuming an Arrhenius law for the variation of the diffusion constant as a function of the temperature, we estimate that the activation energy of phason modes is of the order of 3($\pm$1) eV.

Such results are compatible with what is known for atomic diffusion in these systems. Mn, which is the slow diffuser in the i-AlPdMn phase, has a diffusion constant equal to 10^{-14} m^{2}s^{-1} at 650 °C and an activation energy equal to 2 eV. We found a smaller phason diffusion constant: this might be understood as the

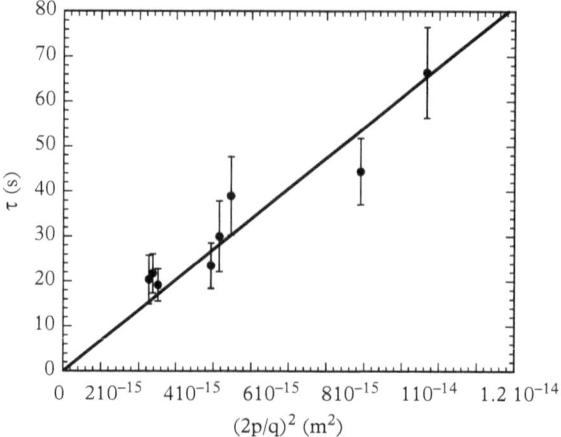

Figure 5.45 *Evolution of the characteristic time as a function of q^{-2} measured at 650 °C, along the $(\tau, -1, 0)$ direction. The solid line is a linear fit which yields the phason diffusion constant. (From Francoual et al. (2003))*

result of a relaxation process which involves 'exchange' of chemical species on neighbouring sites, and thus with a longer timescale.

Phason modes play a major role in plastic deformation and dislocation motion in quasicrystals. Indeed a dislocation, generalized to quasicrystals, generates both a phonon- and a phason-like distortion field in the system.

5.9.5 Phason modes in other quasicrystals

The study of phason modes in other quasicrystals by means of diffuse scattering measurement has been carried out mainly at room temperature. For icosahedral phases, diffuse scattering measurements were carried out in the i-AlPdRe (de Boissieu et al., 2002), i-AlCuFe, i-CdYb, i-ZnSc (Yamada et al., 2016b), and i-ZnMgSc (de Boissieu et al., 2005) systems. The first two systems are isostructural to i-AlPdMn. The i-CdYb and i-ZnMgSc phases are isostructural and present an atomic structure with a different atomic cluster as a building block.

All icosahedral quasicrystal phases display diffuse scattering around Bragg peaks in their diffraction pattern. The amount of diffuse scattering is related to the perpendicular component of the reciprocal lattice vector, thus indicating a relationship with phason modes. However, the intensity distribution of the diffuse scattering presents a differently shaped anisotropy, even for isostructural alloys. For instance, the i-AlCuFe phase displays diffuse scattering which does correspond to a positive K_2/K_1 ratio, in agreement with the five-fold intermediate modulated phase which appears during the icosahedral to rhombohedral phase

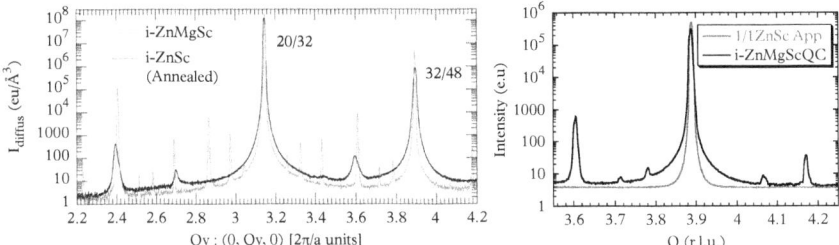

Figure 5.46 *(left) Comparison of the diffraction pattern taken along the two-fold axis of the i-$Zn_{80.5}Mg_{4.2}Sc_{15.3}$ and the i-$Zn_{88}Sc_{12}$ phases. Both diffraction patterns have been put on the same absolute scale units, so that the amount of diffuse scattering can directly be compared. It is quite visible that there is a larger amount of diffuse scattering in the i-$Zn_{88}Sc_{12}$ phase. As a result, the number of weak high-H_{perp} reflections is much smaller. (right) Comparison of the diffuse scattering distribution in the 1/1 ZnZc approximant and the i-ZnMgSc quasicrystal: there is a clear excess of phason diffuse scattering in the quasicrystal.*

transition which occurs upon cooling down the sample. The relative amount of diffuse scattering can only be evaluated when absolute scale measurements are performed. This has been carried out in several icosahedral quasicrystalline phases for which there is now a quite complete set of data. For instance, whereas the amount of diffuse scattering is similar in the i-AlPdMn and i-AlPdRe phases (although the shape anisotropy is different), the phason diffuse scattering is significantly smaller in the ZnMgSc icosahedral phase.

An interesting comparison has been carried out in the ZnSc and ZnMgSc systems, for which icosahedral quasicrystal can be obtained with different composition, as well as 1/1 approximant. It has been found that the phase with the smallest amount of phason diffuse scattering is the ZnMgSc icosahedral phase (de Boissieu et al., 2005). As a result, there is a tremendous number of weak Bragg peaks which otherwise are suppressed by the phason Debye–Waller term associated to the phason diffuse scattering. In the case of icosahedral phases, the number of weak reflections can be somewhat 'quantified' by looking at the largest H_{per} value necessary for the diffraction pattern indexing when measured in the same experimental conditions. Weak Bragg peaks are best measured at synchrotron sources using a reflection geometry and a point detector so as to minimize the background and maximize the signal-to-noise ratio. For instance, the maximum H_{per} value necessary for indexing the diffraction pattern is equal to 2.5 in the case of i-AlPdMn, whereas it is equal to 7.5 in the i-$Zn_{80.5}Mg_{4.2}Sc_{15.3}$ phase (de Boissieu et al., 2005), illustrating the high quality of this phase, as far as phason diffuse scattering is concerned. This is also illustrated when the diffraction pattern of this phase is compared to that of the binary i-$Zn_{88}Sc_{12}$, as shown in Fig. 5.46: there is a much larger amount of diffuse scattering in the binary ZnSc quasicrystal, together with a dramatic decrease of the number of Bragg reflections. The diffuse scattering of the ZnSc binary quasicrystal is better seen

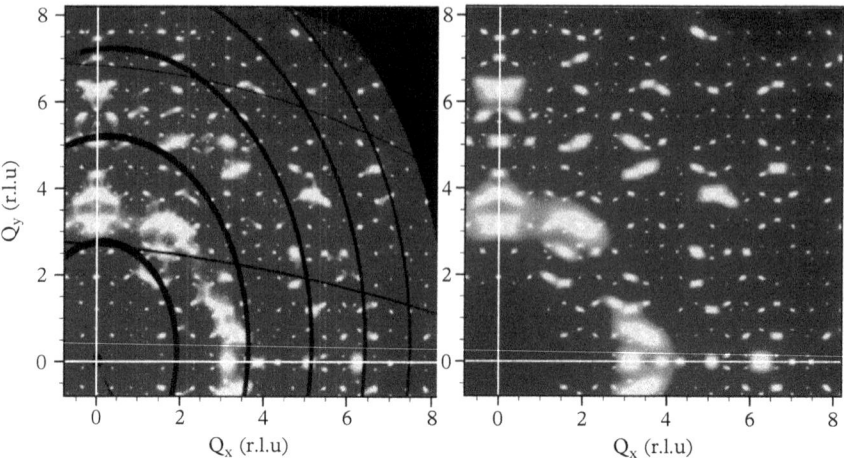

Figure 5.47 *Comparison of the two-fold plane diffraction pattern of the i-$Zn_{88}Sc_{12}$ phases (left) with the simulation taking into account phason diffuse scattering and the two phason elastic constants (right). The black lines in the experimental diffraction pattern are due to the 2D pixel detector structure. Notice the very good reproduction of the entire reciprocal space, demonstrating that all the diffuse scattering arises from phason fluctuations. (From Yamada et al. (2016a))*

in the 2D reciprocal space map shown in Fig. 5.47 (Yamada *et al.*, 2016a). The figure displays an extended portion of reciprocal space, where the distribution of diffuse scattering is clearly visible. Using both absolute scale measurements and accurate maps around a selection of Bragg peaks, the diffuse scattering is fully accounted for by considering the phason fluctuations and phason elastic constants K_1 and K_2 to be equal. In fact, the two-fold diffraction pattern of i-$Zn_{88}Sc_{12}$ can be fully reproduced by taking as input ZnMgSc one, applying a phason Debye–Waller, and computing the diffuse scattering, using the phason elastic constants $K_1/k_BT = 10^{-2}\text{Å}^3$, $K_2/K_1 = -0.53$, and $K_3 = 0$ (Yamada *et al.*, 2016a). Similarly to the case of the i-AlPdMn phase, the ratio of the two-phason elastic constant is close to the value for three-fold instability, leading to an elongated shape of the diffuse scattering parallel to three-fold axes. Using this value of the elastic constant and the measured Bragg peak intensity the simulation reproduces perfectly the observed distribution of diffuse scattering, as shown in Fig. 5.47: even the details are perfectly reproduced and, again, in this case at least 90 per cent of the observed diffuse scattering is due to phason diffuse scattering. Since the measurements are done at room temperature, those phason fluctuations are most likely quenched-in. The microscopic origin of those phason fluctuations has not yet been determined. However, one can see that the ZnSc binary quasicrystal presents a significant Zn/Sc substitutional disorder along the diagonal of the double Friauf polyhedra (see Section 5.8.2), whose direction is parallel to the three-fold axis. As already explained, the large amount of phason diffuse scattering is associated with a phason Debye–Waller which suppress weak reflections. The phason Debye–Waller has been determined in the process of the structure determination of the i-$Zn_{88}Sc_{12}$ quasicrystal and was found to correspond to Gaussian fluctuations of the atomic surfaces, with a root mean square equal to about 10 per cent of the radius of the large atomic surfaces describing the structure. Applying this Debye–Waller factor to the i-ZnMgSc diffraction pattern, together with the found phason elastic constant, makes it possible to accurately simulate the ZnSc diffraction pattern and, in particular, the small number of weak reflections. This can be easily understood by computing the phason Debye–Waller for a Bragg peak with H_{per} equal to 5, leading to a reduction by a factor of 10^{-6}.

The i-ZnMgSc quasicrystal diffuse scattering has also been compared to that of the Zn_6 1/1 cubic approximant. In principle, the phason diffuse scattering only occurs for aperiodic crystals. A significant difference is thus expected for the 1/1 periodic approximant for which the phason degrees of freedom does not exist: moving the parallel space along the perpendicular direction in the 6D space does not lead to an infinite degeneracy of the structure in the case of an approximant. As a result, phason modes do not exist in the 1/1 cubic approximant. This is, indeed, what is observed experimentally when comparing the diffuse scattering in the 1/1 approximant and the i-ZnMgSc quasicrystal. Comparing the tails of two Bragg peaks shown in the right panel of Fig. 5.46, there is a clear supplementary diffuse scattering in the case of the icosahedral quasicrystal. In fact, only the usual TDS is observed in the 1/1 approximant (de Boissieu et al., 2005). This nicely demonstrates that phason diffuse scattering are indeed a characteristic signature of the long-range aperiodic order.

Phason modes in decagonal phases have only recently been studied. Indeed, the decagonal diffraction pattern displays, in general, a quite complicated structure. In decagonal phases, the diffraction pattern consists of a periodic stacking of quasiperiodic 'Bragg layers' along the c^* axis. In the AlNiCo decagonal phase, which has been extensively studied as function of the Ni/Co content, diffuse scattering is observed both in the quasiperiodic Bragg layers and in-between them, at a position halfway between two quasiperiodic Bragg layers. The temperature dependence of the two diffuse scattering phenomena is different, and they have been given a different microscopical interpretation. The in-between Bragg layer diffuse scattering is due to a disordering of the columnar cluster. For the in-Bragg layer diffuse scattering, there are indications that it is related to phason modes. In this case the phonon–phason coupling term is essential, for it drives the anisotropic distribution of the diffuse scattering intensity.

The periodic stacking occurring in decagonal phases also allows an easier interpretation of HREM images obtained in transmission mode. Although it should not be forgotten that what is observed is a projection of a slab of thickness varying between 5 and 50 nm, a lot of experimental work has been done using this technique. Detecting long-wavelength phason modes on images is not an easy task. Indeed, simulation showed that the difference from a perfect quasicrystal

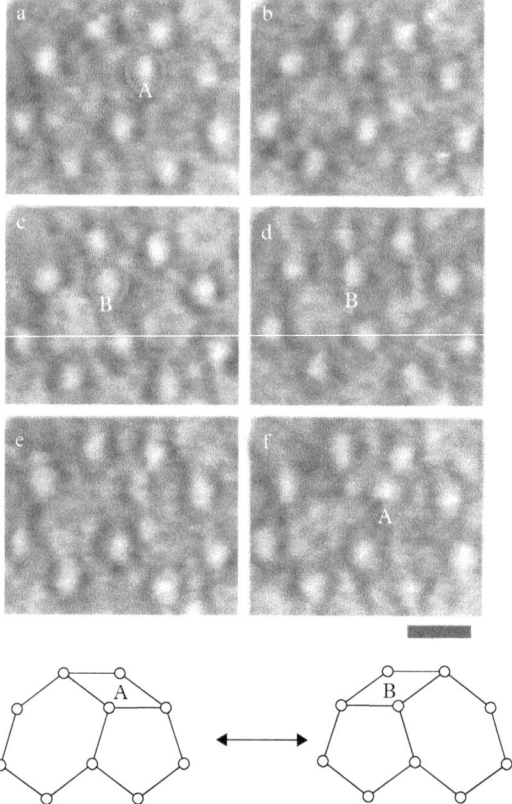

Figure 5.48 In situ *transmission HREM images measured at 1123 K by Edagawa et al. in the d-AlCuCo phase. (a) to (f) are time sequences of the same area taken at 0, 5, 8, 110, 113, and 115 s. The scale bar indicates 2 nm. Some column jumps are highlighted. The bottom panel indicates the tile 'flip' configuration, which may be overlayered on the top panels.*

image is quite subtle. In situ images by Edagawa *et al.* have, however, given a hint to the presence of collective diffusive phason modes (Edagawa *et al.*, 2000). When looking at thick slices of their d-AlNiCo specimen, they could map the contrast observed on the images to a Penrose tiling with an edge length of 2 nm (Fig. 5.48). In this case, the observed image contrast is not on an atomic scale, but most likely corresponds to 'cluster' columns with a diameter of the order of 1 nm. When a series of images is taken at 1123 K, there are variations in the distribution of Penrose tiles, corresponding to 'flips' from one configuration to another (bottom panel). These changes in the images correspond to atomic columns containing several tens of atoms because the images were taken on a thick specimen (more than 200 Å). Moreover, the 'jump' distance is of order 10 Å, and can by no means be related directly to a single atomic jump. A more detailed analysis of the time dependence of the images has been carried out. By analysing changes in the tiling pattern, the spatial and temporal evolutions of the phason field could be extracted. There is a decay which bears some resemblance with the predictions of the random tiling model; however, a large phason strain seems to remain in the sample.

Local atomic jumps have been revealed by temperature in situ studies of the $Al_{72}Ni_{20}Co_8$ decagonal phase. Using in situ atomic-resolution annular dark-field images, Abe *et al.* have observed an anomaly in atomic vibrations at 1100 K (Abe *et al.*, 2003*a*). This anomaly occurs at specific sites of the structure, and is interpreted as the result of atomic jumps between two sites separated by 1 Å. These sites correspond to the border of atomic surfaces in the high-dimensional description of the decagonal phase, which is precisely the theoretically expected location for easy jump sites. As already stated above, diffuse scattering phenomena and phase transitions are extremely complex in decagonal phases. Roughly speaking, it can be stated that most phase transitions can be explained by the application of a linear phason strain onto the high-temperature decagonal phase. In a recent work, Kobas *et al.* made an attempt to interpret the observed diffuse scattering in the framework of the phason elasticity of decagonal phases (Kobas *et al.*, 2005). In situ X-ray measurements carried out at 1123 K showed a large amount of diffuse scattering, displaying characteristic anisotropic shapes. Part of this anisotropy is well accounted for by a superposition of phonon and phason contributions. As pointed out in Ishii (2000), only a phonon–phason coupling term can introduce anisotropy in the diffuse scattering of decagonal phases. Simulation using a phonon–phason coupling, although not perfect, give a reasonable agreement with experimental data. The same authors also proposed an atomic model to interpret their results. For a first approximation, they considered a fivefold orientational disorder of the columnar cluster. This orientational disorder, obtained by successive 72° rotations of the initial cluster, produces two kinds of disorder: occupational disorder (i.e. Al-Ni/Co), and displacive 'split' positions as shown by arrows and labels in the figure. Although the simulated diffuse scattering obtained with this model is only approximately similar to the observed one, it is an interesting step towards a better understanding of phason fluctuations at the atomic scale.

To conclude this overview of results obtained in various quasicrystals, phason fluctuations have been detected in many different systems. However, their microscopic interpretation is far from being achieved and certainly requires further elaborated simulations at the atomic scale. All temperature studies carried out so far lead to the same results: there is a phase transformation at low temperature towards a crystalline phase, and/or the diffuse scattering increases as the temperature is lowered. This is a strong indication of an entropy term stabilizing the quasicrystalline structure.

In conclusion, in this section, we have seen that phason modes have been observed in various aperiodic systems. Their microscopic interpretation is certainly better understood in the case of incommensurately displacive modulated structures. For composites the experimental situation is not as well established. In quasicrystals, phason modes have been observed and interpreted in the framework of hydrodynamic theory, but its microscopic interpretation still remains an open problem.

5.10 Summary of the experimental findings

1. The characteristic signature of the aperiodic long-range order for electrons or phonons is still lacking experimentally, although some differences have been evidenced between, for instance, a 1/1 approximant and a quasicrystal.
2. For phonons, the acoustic regime is rather limited and shows an abrupt crossover to mode mixing or damped modes for wave vectors with a wavelength of the order of 1–2 nm.
3. The concept of PBZ boundaries for quasicrystals can be used to interpret those results qualitatively.
4. The spikiness of the electron density of states is difficult to show.
5. Electron state dispersion relations have been measured in a few systems.
6. Long-wavelength phason modes are characteristic of the aperiodic state. They can be viewed as a perturbation with a polarization in the internal or perpendicular space.
7. As predicted by hydrodynamic theory, diffusive phason modes have been observed in incommensurately modulated phases and quasicrystals. In the latter case this shows up as a characteristic distribution of diffuse scattering observed in all icosahedral quasicrystals for which phason elastic constants have been extracted. The diffusive character has been shown in at least one system.
8. For harmonic displacive incommensurate modulation, there is a frequency regime for which damped phonons with a linear dispersion have been observed experimentally. However, the damping remains finite as the wave vector goes to zero, leading to a diffusive mode in the long-wavelength limit, in agreement with hydrodynamic theory.
9. For incommensurate composites, where phason modes are called sliding modes, the experimental situation is still unclear.

6
Origin and stability

6.1 Introduction

The question of why the ground state of a system is sometimes quasiperiodic is difficult to answer. After all, it has never been proven that the ground state of matter would be crystalline either, in the sense of lattice periodic. One has sought reasons for a lattice periodic ground state, but it has become clear that incommensurate phases are ubiquitous, Also the existence of quasicrystals is at variance with the assumption of periodicity of the ground state, but here there is still discussion about the stability at low temperature. Moreover, quasicrystalline phases usually occur in relatively small corners of a phase diagram. Very often the term 'competitive forces' or 'frustration' is used in the explanation of aperiodic crystals, but this does not clarify much, although it may suggest a physical picture. Anyway, frustration may explain why a lattice periodic ground state may become unstable and transform to an aperiodic ground state.

Aperiodic crystals appear in very diverse families of compounds. One cannot really speak of 'the' mechanism for creating aperiodicity. There are several types of such mechanisms. The ground state is the minimum of the free energy, where entropy and internal energy come in, and is therefore temperature dependent. An aperiodic structure may be found in a corner of the phase diagram, where it may be stabilized by thermal effects, pressure, or composition.

Let us briefly mention a number of important mechanisms. If electrons play a role, the position of the Fermi surface is important. If the Fermi level is not at the Brillouin zone boundary, changing the atomic positions may create a lowering of occupied electron states. When this overcomes the increase in elastic energy, a so-called Peierls instability occurs. When the wave vector of the modulation is incommensurate, an aperiodic ground state may appear. A related mechanism is the Hume–Rothery mechanism in alloys. A cooling liquid solidifies into a structure where the Fermi surface is in a favourable position. This mechanism is supposed to play a role in quasicrystals.

Sterical hindrance is another interaction that might lead to an incommensurate phase. There are many compounds with a framework of connected polyhedra.

The spaces between these are occupied by atoms which may be too small or too big for the space. This kind of frustration may be resolved by a modulation.

For inclusion compounds, which are non-stoichiometric, with incommensurate lattice constants of the subsystems, the aperiodicity is created by the interaction between the subsystems. Depending on the strength of this interaction, the subsystems become modulated with continuous or discontinuous modulation functions. Other mechanisms are vacancy ordering in non-stoichiometric compounds or valence fluctuations.

There are simple, semi-microscopic models for these situations. A first model is the Frank–Van der Merwe model, also called the Frenkel–Kontorova model, consisting of a linear chain on a periodic substrate. A second one is the axial next-nearest-neighbour Ising (ANNNI) model developed for aperiodic spin systems, but which may be used for structural phase transitions as well. A related model with continuous degrees of freedom is the DIFFFOUR model. There are a range of variations on these semi-microscopic models. More specific than these semi-microscopic models are calculations on the basis of realistic potentials. And on the other side of the semi-microscopic models there are the models in the framework of the phenomenological Landau theory of phase transitions. We shall give a brief discussion of the various models which play a role for aperiodic crystals.

6.2 The Landau theory of phase transitions

Landau introduced a theory of continuous phase transitions independent of the underlying mechanism. It is a very general theory, where symmetry plays an important role. In the early years of incommensurate modulated phases the experimental results were analysed mainly by the use of this theory. An extensive literature on the subject exists, and we want to mention here only a few points relevant for aperiodic crystals. For a more comprehensive overview we refer readers to the literature (general Landau theory (Landau and Lifshitz, 1959; Kocinski, 1985; Toledano and Toledano, 1987); especially applied to incommensurate phases: (Blinc and Levanyuk, 1986a, b; Ishibashi and Dvorak, 1978; Ishibashi and Shuba, 1978)). Later, Landau theory was generalized to the case of first-order transitions. We shall limit ourselves here to the case of continuous phase transitions.

The Landau theory describes a second-order phase transition in terms of an order parameter. For magnetic systems this might be the magnetic polarization; for other systems, it is the density or the lattice constant. It is supposed to be zero in the most symmetric phase, which is usually above the phase transition temperature, and non-zero and uniform (not position dependent) in the other phase. Near the phase transition the free energy is expanded in a Taylor series in this order parameter:

$$F = F_0 + A\eta^2 + B\eta^3 + C\eta^4 + \cdots . \tag{6.1}$$

The linear term is zero because the minimum free energy is obtained for $\eta = 0$ above the phase transition. The coefficients $A, B, C, \ldots$ depend, generally, on temperature and pressure. The phase with $\eta = 0$ is stable for a temperature where $A > 0$. For $A < 0$ it becomes unstable. In the neighbourhood of the critical temperature T_c the dependence of A on T is taken to be linear: $A = \alpha\,(T - T_c)$. For a second-order phase transition there is no jump of the order parameter at the phase transition. This implies that $B = 0$. In the simplest case there is stability for $C > 0$.

The theory is based on the change of symmetry at the phase transition. For convenience we shall call the most symmetric phase the high-temperature phase and the other the low-temperature phase. If the symmetry group of the former is G, that of the latter (H) is lower: H is a subgroup of G. If the high-temperature phase has a density $\rho(\mathbf{r})_0$, this is invariant under G. The low-temperature phase has a density $\rho(\mathbf{r}) = \rho(\mathbf{r})_0 + \delta\rho(\mathbf{r})$. The second term is not invariant under G, but according to the theory of representations, it may be decomposed into a sum of terms each transforming with an irreducible representation of G. In fact, one may limit oneself to a single component close to the phase transition. Then $\delta\rho$ transforms under G according to one of its irreducible representations. Since this representation is, in general, more-dimensional one gets

$$\delta\rho(\mathbf{r}) = \sum_{i=1}^{d} c_{\nu i}\phi_{\nu i}(\mathbf{r}),$$

where ν denotes the irreducible representation, and d is the dimension of this representation. Then the free energy may be written in terms of the coefficients $c_{\mu i}$:

$$F = F_0 + A_\nu \sum_{i=1}^{d} |c_{\nu i}|^2 + \cdots .$$

If two or more irreducible representations are involved, a sum over these representations appears.

For structural phase transitions in crystals, the group G is a space group. As seen before, the irreducible representations of space groups are characterized by a wave vector $\mathbf{k}$ and a representation of the point group of the group $G_\mathbf{k}$, the group of $\mathbf{k}$. The representation of the point group can again be labelled with ν. For example, if the phase transition occurs via a phonon that becomes unstable, the order parameter is the thermal average of the coordinate $Q_{\mathbf{k}\nu i}$ of the phonon. We recall that a phonon is characterized by a wave vector, an irreducible representation, and, in case the representation is not one-dimensional, by a row index. The free energy now can be written as

$$F = F_0 + \sum_{\mathbf{k}\nu} A_{\mathbf{k}\nu} \sum_{i=1}^{d} |\eta_{\mathbf{k}\nu i}|^2 + \cdots .$$

For simplicity we suppose from here on that the representation is one-dimensional, that is, that $d=1$. Generalization to $d \neq 1$ then is rather straightforward.

The coefficients $A_{\mathbf{k}\nu}$ do not depend on the index i, because all the terms differing in i only belong to the same representation. $A_{\mathbf{k}\nu}$ is the square of the frequency of a phonon branch. The branch becomes unstable if the value of $A_{\mathbf{k}\nu}$ becomes zero. The dispersion curve then crosses the zero frequency line. The instability leads to the phase transition. For the second-order phase transition the structure below the transition is given by the displacement in the unstable mode, which is limited by the higher-order terms in the free energy. For positive value of $A_{\mathbf{k}\nu}$ the frequency approaches zero if the temperature tends to the phase transition temperature. This is called a *soft mode*.

The behaviour may be compared to that of a simple pendulum with varying mass. For an upright harmonic oscillator in a gravity field the potential energy is

$$V = \frac{1}{2}\alpha u^2 + mg \cos u.$$

Its equilibrium is given by the solution of $dV/du = \alpha u - mg \sin u = 0$. For $mg < \alpha$ there is only one solution ($u = 0$), in the other case there are two more, the non-zero solutions of $u = mg \sin(u)/\alpha$. The stability of the first solution follows from the squared frequency $\omega^2 = d^2V/du^2 = \alpha - mg \cos u = \alpha - mg$. When the mass m is increased the frequency goes to zero, and for $m > \alpha/g$ the first solution is unstable, and the system now oscillates around one of the two new equilibrium points. If one increases the mass the oscillations become slower and finally there is a 'phase transition' to a new equilibrium point (Fig. 6.1).

The phase where the order parameter is zero is called the paraphase, because in a ferromagnetic phase transition it corresponds to the paramagnetic phase, and for a ferroelectric structure it corresponds to the *paraelectric phase*. Very often the dispersion curve touches the zero frequency line at a symmetry point, for example the centre of the Brillouin zone. Then the new structure is again lattice periodic. This can be, for example, a ferroelectric phase. In general the lattice periodic phase is called a *ferroic phase* if the instability occurs at the zone centre, *anti-ferroic* if this happens at the zone boundary, and *superstructure* if the new phase

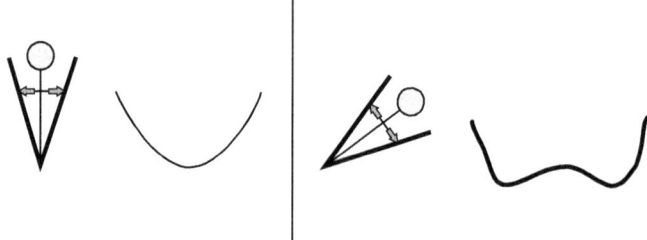

Figure 6.1 *An inverse pendulum changes its average position when the mass exceeds a critical value, and oscillates around that position. The potential changes from a single well to a double well.*

is lattice periodic, but not (anti)ferroic. An aperiodic new ground state occurs if the curve touches zero at a point inside the Brillouin zone that does not have rational coefficients with respect to the reciprocal lattice. The amplitude of the modulation is then determined by the minimum of the free energy, which is close to the phase transition

$$F = F_0 + \alpha(T - T_c)|\eta_{\mathbf{k}\nu}|^2 + C|\eta_{\mathbf{k}\nu}|^4 + \cdots.$$

Because the transition is mainly determined by the change of sign of the second-order term the coefficient C may be considered to remain constant. The minimum of the free energy for $T < T_c$ is obtained for

$$|\eta_{\mathbf{k}\nu}|^2 = \frac{\alpha(T_c - T)}{2C},$$

where the phase of η remains undetermined.

Limitation of the sum over $\mathbf{k}$ to the wave vector of the unstable mode is actually not correct: if a mode becomes unstable, the modes with a vicinal wave vector become unstable as well. Actually one has to keep the sum over $\mathbf{k}$ in the neighbourhood of $\mathbf{k}_0$, the position of the minimum of the curve. In the neighbourhood of the phase transition one may approximate the dependence of $A_{\mathbf{k}\nu}$ by a parabola. Considering a narrow range of wave vectors around the minimum is the same as considering an order parameter that varies slowly in space. The free energy polynomial becomes a free-energy functional. The free energy is an integral over the spatial variables of a polynomial in the space-dependent order parameters and its derivatives. The function under the integral sign is the free energy density. The first terms give

$$F = F_0 + \int \left[\frac{A}{2}|\eta(z)|^2 + \frac{iL}{2}\left(\eta\frac{\partial \eta^*}{\partial z} - \eta^*\frac{\partial \eta}{\partial z}\right) + \frac{\kappa}{2}\left(\frac{\partial \eta}{\partial z}\right)\left(\frac{\partial \eta^*}{\partial z}\right) \right.$$
$$\left. + \frac{C}{4}|\eta(z)|^4 + \frac{D}{4}\left(\eta^4 + \eta^{*4}\right) \right] dz. \qquad (6.2)$$

The first term under the integral sign is the term that leads to the instability, the second is the so-called *Lifshitz term*, which is important for a transition towards an aperiodic state, the third term is an elastic term tending to limit the space dependence of the order parameter, and the last two terms are stabilizing terms. Terms may be absent because of symmetry reasons. The order parameter is a slowly varying variation of an order parameter with fixed wave vector $\mathbf{k}_0$. If the order parameter is written as $\eta(z) = \int \eta_q \exp(2\pi i(k_0 + q)z)\, dz$, then the free energy becomes

$$F = F_0 + \int \left[\frac{A}{2} + Lq + \frac{\kappa}{2}\right]|\eta_{k_0+q}|^2 + \cdots. \qquad (6.3)$$

The coefficient between square brackets is the parabola approximation to the dispersion curve. If $\mathbf{k}_0$ is chosen at the minimum of the curve, the contribution of

L vanishes. Sometimes, the Lifshitz invariant vanishes because of the symmetry of the free energy. Then the instability occurs at $\mathbf{k}_0$.

The free energy (Eq. 6.2) is minimized by a solution of the Euler–Lagrange equations. If we introduce ρ and ϕ as the absolute value and the phase of η, respectively, then the free energy becomes

$$F = F_0 + \int \left[\frac{A}{2}\rho^2 + \frac{C}{4}\rho^4 + \frac{D}{2}\rho^4 \cos(4\phi) + L\rho^2 \frac{\partial \phi}{\partial z} \right. \quad (6.4)$$
$$\left. + \frac{\kappa}{2}\left(\frac{\partial \rho}{\partial z}\right)^2 + \frac{\kappa}{2}\rho^2\left(\frac{\partial \phi}{\partial z}\right)^2 \right] dz.$$

The solution of the Euler–Lagrange equation for small absolute values of A is

$$\rho^2 = -\frac{A - L^2/\kappa}{C}, \quad \phi(z) = \lambda z = -\frac{Lz}{\kappa}. \quad (6.5)$$

This is a sinusoidal wave with wave vector $\mathbf{k}_0 - L/\kappa$, which is the minimum of the dispersion curve A_k, and, in general, at an incommensurate position. The phase of the modulation function is undetermined in this case. There is an infinite degeneracy of the ground state. The amplitude grows with the square root of $T_c - T$.

For lower temperatures, higher harmonics come up in the modulation function. The displacements of the atoms up to second harmonics are given by

$$\mathbf{u_n} = \eta_1(z)\mathbf{a}\exp(2\pi i \mathbf{k}_0.\mathbf{r}) + \eta_2(z)\mathbf{b}\exp(4\pi i \mathbf{k}_0.\mathbf{r}) + \cdots .$$

The two functions $\eta_i(z)$ are again slowly varying functions of z ($\mathbf{k}_0$ is taken parallel to the third axis). Additional terms in the free energy density then are

$$\frac{B}{2}|\eta_2|^2 + g(\eta_1^2 \eta_2^* + \eta_2^2 \eta_1^*),$$

if these are permitted by symmetry. The coupling between η_1 and η_2 may cause a squaring-up of the modulation function. The relation between the two order parameters in the minimal energy solution is

$$\eta_2(z) = \frac{-2g}{B}\eta_1(z)^2.$$

Still further away from the transition, more and more harmonics couple. If the m-th harmonic of $\mathbf{k}_0$ becomes a vector at the Brillouin zone boundary, there are pinning effects. The phase of the commensurate wave vector $\mathbf{k}_0$ then becomes fixed, because the order parameter at the zone boundary is real. Suppose that ϕ, the phase of $\eta_{\mathbf{k}_0}$, becomes π/m. Using the *constant amplitude approximation*

(McMillan, 1976), which means that ρ is assumed to be independent of z, the Euler–Lagrange equations may be rewritten as an equation for $\psi = m\phi$. They have the well-known form

$$\frac{d^2}{dt^2}\psi = -\sin\psi, \qquad (6.6)$$

the pendulum equation. Low-energy solutions are just harmonic oscillations, corresponding to a sinusoidal modulation function. For higher energies, the pendulum rotates, and stays relatively long in the upright position. This means that, for the modulation function, there are domains with an almost constant phase alternating with domain walls where the phase changes rapidly. This may be made more explicit by solving the differential equation exactly in terms of elliptic functions. The phase is the phase of the mode with commensurate wave vector $\mathbf{k}_0$. In the domains with constant ψ the positions of the atoms are like those in a periodic crystal. These are alternating with domain walls between the domains, which have as effect that the full structure becomes incommensurate. The domain walls are called *discommensurations* or *solitons*. The equidistant domain walls are the reason that the structure is aperiodic. If their mutual distance is b, the modulation wave vector is $\mathbf{k} = \mathbf{k}_0 + \hat{\mathbf{z}}/b$, where $\hat{\mathbf{z}}$ is a unit vector in the z-direction. For decreasing temperature b grows, and finally there is a *lock-in transition* where the modulation wave vector becomes $\mathbf{k}_0$. There is a phase transition to a commensurate phase. The situation may be described in terms of the *soliton density*. It is the ratio between the width d of the soliton and the distance b between them: the soliton density is d/b and varies from 0 at the lock-in to 1 at the temperature T_i. The commensurate phase has a superstructure of index m. If $\mathbf{k}_0 = 0$, then the lock-in phase has the same lattice as the high-temperature parent phase. Generally, the symmetry breaking occurs via a lowering of the point group symmetry.

An incommensurate phase may occur if there is a Lifshitz invariant. The Lifshitz term in the free energy shifts the minimum of the dispersion curve from the commensurate $\mathbf{k}_0$ to an incommensurate value. An incommensurate phase of this type is called a *Type I incommensurate phase*. However, also when there is no Lifshitz term, an incommensurate phase may be found. (Levanyuk and Sannikov, 1975; McConnell, 1978). The appearance of an incommensurate phase in that situation is due to coupling to other degrees of freedom. This is called a *Type II incommensurate phase*. Consider an incommensurate phase transition near the centre of the Brillouin zone. Take into account two coupled order parameters, η and ζ, and suppose that both belong to a one-dimensional irreducible representation. The free energy is written as

$$F = F_0 + \int \left[\frac{A}{2}\eta^2 + \frac{B}{4}\eta^4 + \frac{C}{2}\zeta^2 + \frac{D}{4}\zeta^4 + \frac{G}{2}\eta^2\zeta^2 \right. \qquad (6.7)$$

$$\left. + \frac{\kappa_1}{2}\left(\frac{d\eta}{dz}\right)^2 + \frac{\kappa_2}{2}\left(\frac{d\zeta}{dz}\right)^2 + \gamma\left(\frac{d\eta}{dz}\zeta - \frac{d\zeta}{dz}\eta\right) + \cdots \right] dz.$$

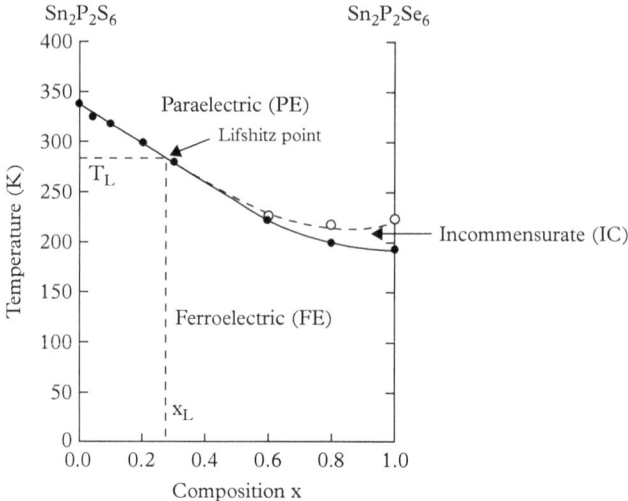

Figure 6.2 *The phase diagram of the solid solution $Sn_2P_2(S_{1-x}Se_x)_6$ shows a Lifshitz point. (Eijt et al., 1998)*

Levanyuk and Sannikov showed that for $T = T_0$ an instability occurs, and a phase transition occurs. The wave vector of the instability is zero if $4\gamma^2 < \kappa_1 C$, but unequal to zero for $4\gamma^2 > \kappa_1 C$. The two branches interact, and one pushes the other away. When the interaction (γ) is strong enough, this results in the minimum of the lower curve being pushed to zero at a point not in the centre. If $4\gamma^2 = \kappa_1 C$, the incommensurate phase meets the (anti-)ferroic phase, at the phase transition from the paraphase. Thus three phases come together. Such a point in the phase diagram is called a *Lifshitz point*. Examples of Type II phases are BCCD (Brill and Ehses, 1985), $NaNO_2$ (Yamamoto, 1985), thiourea (Moudden et al., 1978), and $Sn_2P_2S_6$. The latter is special because it is the end point of a series of solid solutions $Sn_2P_2(Se_{1-x}S_x)_2$ in which a Lifshitz point occurs for $x = 0.28$ (see Fig. 6.2) (Gommonai et al., 1981; Eijt et al., 1998).

The description of the phase transitions towards an aperiodic crystal phase has been discussed here mainly for modulated phases, and partially for composites. Also the formation of quasicrystals has been described using Landau theory. This always involves a first-order transition and we shall not go here into the details. Landau theory explains in a phenomenological way the main features of phase diagrams with an incommensurate modulated phase.

6.3 Semi-microscopic models

6.3.1 Substrate models

One of the possible reasons for having an incommensurate modulation is interaction with another system that has an incommensurate fundamental length. For

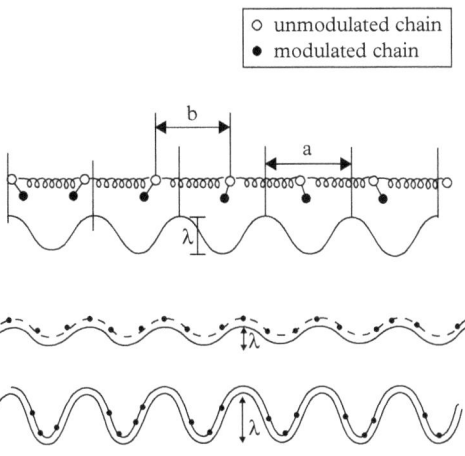

Figure 6.3 *The Frenkel–Kontorova or Frank–Van der Merwe model consists of a harmonic chain on a periodic substrate. First row: A chain of particles of natural periodicity* b *is adsorbed on a sinusoidally corrugated substrate of strength* λ *and periodicity* a, *incommensurate with* b. *Second and third rows: Chain configurations in the unlocked and locked regimes.*

example, a film on a periodic substrate where the natural lattice constants of film and substrate are incommensurate. A very simple model for this situation in which aperiodic ground states may be studied is a linear harmonic chain on a periodic substrate. The potential energy is given by

$$V = \sum_n \left[\frac{A}{2}(x_n - x_{n-1} - b)^2 + \frac{\lambda}{2\pi}[1 - \cos(2\pi x_n/a + \phi)] \right], \quad (6.8)$$

where x_n denotes the position of the n-th particle in the chain, A is the spring constant of the harmonic nearest-neighbour interaction, b is the lattice constant, and a is the periodicity of the substrate potential with strength λ. This model was first introduced by Dehlinger for the study of lattice deformations, with $a = b$. Frank and Van der Merwe used the same model and introduced dynamics. Incommensurability appeared in the work by Frank and Van der Merwe (see Fig. 6.3). They considered the case of irrational a/b and solved the dynamical equations in a continuum approximation. The model is usually called the *Frenkel–Kontorova model*, but in the paper by these two authors incommensurability did not appear. Therefore, we prefer to call the aperiodic model by the names of the authors who introduced aperiodicity into the model and so refer to it as the Frank–Van der Merwe model (Dehlinger, 1929; Frenkel and Kontorova, 1938; Frank and der Merwe, 1949).

The ground state of the model satisfies the equation

$$A(2x_n - x_{n-1} - x_{n+1}) + \frac{\lambda}{a}\sin(2\pi x_n/a + \phi) = 0. \tag{6.9}$$

The first term is a discrete version of a second derivative. In the continuum approximation this equation leads to the sine-Gordon or pendulum equation, which is exactly solvable. The discrete equation has to be solved numerically. In the space of parameters λ/A and b/a, both commensurate and incommensurate ground states occur.

Define an average distance between neighbours by

$$\ell = \lim_{N\to\infty} \frac{x_N - x_0}{N}.$$

If ℓ/a is irrational, then the ground state is incommensurate and the positions of the particles are given by

$$x_n = n\ell + x_0 + f(n\ell).$$

The modulation function f is smooth if λ is small. If ℓ/a satisfies certain conditions (♯ namely, for any fraction p/q the difference between ℓ/a and p/q should be larger than a factor times q^{-m} for some m: ℓ/a is 'sufficiently incommensurate' ♭), then there is a critical value λ_c such that f is continuous for $\lambda \leq \lambda_c$ and discontinuous for $\lambda > \lambda_c$. This was called the transition by breaking of analyticity by Aubry. In the case of a weak interaction parameter λ the deviations of the atom positions from the equidistant positions will be small and f smooth. When λ increases the particles are more pulled into the substrate wells, and the function f becomes discontinuous, because the maxima of the substrate potential will be avoided (see Fig. 6.4). That this happens suddenly, like a phase transition has been proven rigorously. Other rigorous statements about the ground state also exist (Aubry and Le Daeron, 1983). Whether the modulation function is continuous or not plays an important role in the dynamics, as we shall see in Chapter 6. For small λ the particles are mobile (unlocked), for $\lambda > \lambda_c$ they are locked.

♯ There is an interesting relation with non-linear dynamical systems. We introduce new variables ϕ_n and p_n by

$$x_n = na + \frac{\phi_n}{2\pi}a; \quad p_n = \phi_n - \phi_{n-1}.$$

The condition for the ground state then becomes

$$2\phi_n - \phi_{n-1} - \phi_{n+1} + \gamma\sin(\phi_n) = 0.$$

Semi-microscopic models 415

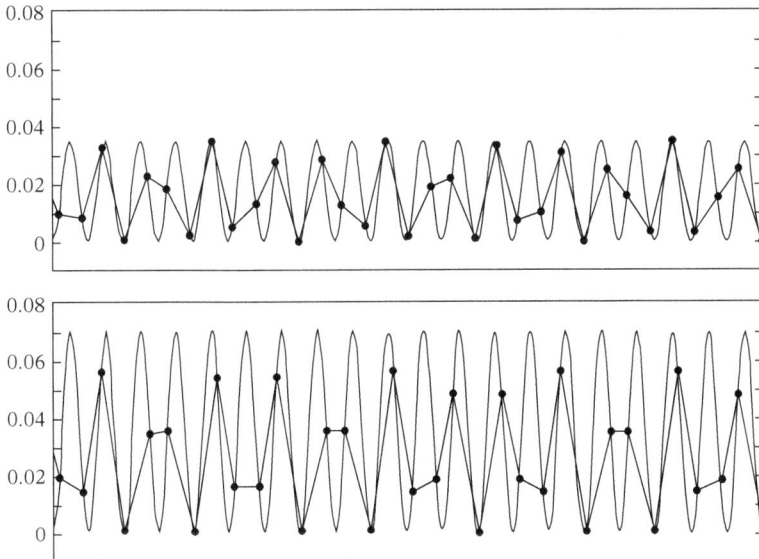

Figure 6.4 *For small values of the interaction parameter of the Frenkel–Kontorova model the amplitude of the modulation is small. For $\lambda > \lambda_c$ the summits are avoided and the modulation function becomes discontinuous.*

This is equivalent with the following pair of difference equations

$$\phi_{n+1} = \phi_n + \gamma \sin(\phi_n) + p_n$$
$$p_{n+1} = p_n + \gamma \sin(\phi_n).$$

This is the well-known standard mapping in the theory of non-linear systems, and used in the study of chaotic phenomena. It is a mapping of a square with edge 2π onto itself. For $\gamma = 0$ all the orbits fall on lines p_n = constant. For non-zero values chaotic orbits appear, and if γ is larger than a critical value, an orbit covers the whole space densely. Many results are known for this mapping.

When ℓ is not kept fixed, but follows from the value of $\mu = (a-b)/b$, then ℓ is a monotonic non-decreasing function of μ. For $\lambda = 0$, it is a straight line, but for non-zero values of λ there are plateaus: ℓ 'locks in' at rational values. For a critical value the function $\ell(\lambda)$ has the strange property, that it has almost everywhere derivative equal to zero, but nevertheless it is a continuous function increasing from 0 at 0 to 1 at 1. It is a 'devil's staircase' (Fig. 6.5). The modulation wave vector changes with the parameters and locks in at every commensurate value.

Clearly, the model is especially suited for modelling composite structures. It explains the appearance of the modulation, and the change in the character of the modulation function.

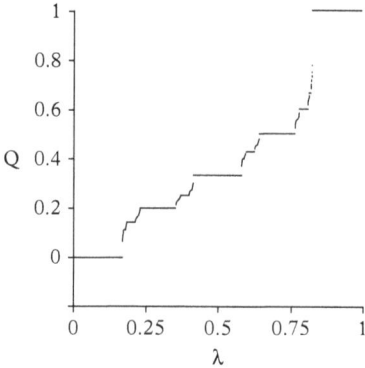

Figure 6.5 *The complete devil's staircase: A function (the wave vector) grows from 0 to 1 with almost everywhere a zero derivative.*

6.3.2 Spin models

In the Frank–Van der Merwe model, one of the subsystems is rigid, which creates an asymmetry between the subsystems. Incommensurate structures may also have their origin in the interactions inside a crystal with a lattice periodic basic structure. Spin models are then simple models which may describe incommensurate spin structures, but which may also model structural phase transitions. The first model was proposed in Elliott (1961) for a discussion of the spin ordering in heavy rare-earth metals. He considers a Hamiltonian for a chain of spins with interaction

$$\sum_n \sum_{m=0}^{2} \left[A_m M_n^z M_{n+m}^z + B_m (M_n^x M_{n+m}^x + M_n^y M_{n+m}^y) \right].$$

As shown in Yoshimori (1959), m should take at least three values in order to have an incommensurate spiral. A similar model was used for the study of the phase diagram of Eu-chalcogenides (Janssen, 1972), where long-period magnetic and structural superstructures were observed. A simpler Hamiltonian which allows for incommensurate spin configurations is a generalization of the well-known ANNNI model. Suppose the spins are positioned on the sites **n** of a tetragonal lattice with basis vectors **a**, **b**, and **c**. The Hamiltonian is

$$H = -\sum_n \left[\mathcal{J}_0 (S_\mathbf{n} \cdot S_{\mathbf{n}+\mathbf{a}} + S_\mathbf{n} \cdot S_{\mathbf{n}+\mathbf{b}}) + \mathcal{J}_1 S_\mathbf{n} \cdot S_{\mathbf{n}+\mathbf{c}} + \mathcal{J}_2 S_\mathbf{n} \cdot S_{\mathbf{n}+2\mathbf{c}} \right] \quad (6.10)$$

There is a ferromagnetic coupling between first neighbours in the xy-plane, and couplings between first and second neighbours in the z-direction. This model has

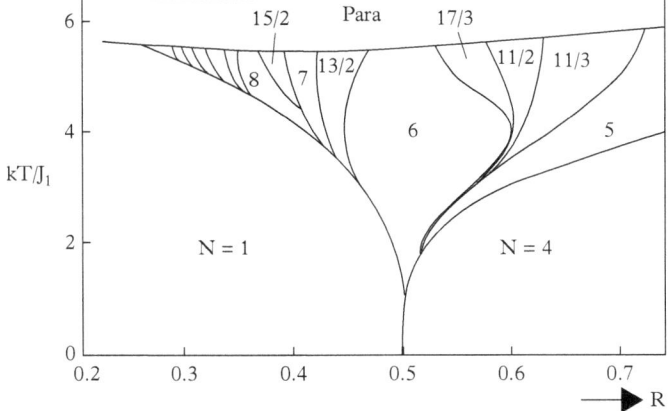

Figure 6.6 *The phase diagram of the ANNNI model.*

been studied extensively using mean-field approaches and Monte Carlo simulations (Bak and von Boehm, 1980; Fisher and Selke, 1980; Selke and Duxbury, 1984; Selke, 1988). For high temperatures, the ground state is a paramagnetic state; for lower temperatures, the ground state depends on the ratio $R = \mathcal{J}_1/\mathcal{J}_2$, and may be incommensurate (Fig. 6.6). This means that the period of the spin system is incommensurate with respect to the lattice constant.

The incommensurate and commensurate phases have a modulation wave vector along the *c*-axis. For commensurate phases, solutions of the ground-state equations are characterized by the number of consecutive spins of the same sign. For example, the period 4 solution $++--$ has two spins up and two spins down in the unit cell. Such a structure is denoted by $<22>$ or, more simply, by $<2>$. A period 5 solution would be $<32>$, and a period 8 solution with the wave vector 3/8 ($++-++-+-$) is $<212111>$ or, shorter, $<(21)^2 1^2>$. The sum of all digits gives the period, half the number of digits the number of waves, and their ratio the wave vector. This notation makes it possible to study structures in the phase diagram with building blocks similar to those in adjacent phases. For example, between the phases with wave vectors 1/4 ($<2>$) and 1/5 ($<32>$), one may find the phase $<32^3>$ with the wave vector 2/9 (9 spins in 4/2 waves). This wave vector may be obtained by adding the nominators and denominators of the neighbouring phases. In general,

$$\frac{p_1}{q_1} \oplus \frac{p_2}{q_2} = \frac{p_1 + p_2}{q_1 + q_2}.$$

This is called a Farey summation. Starting from 0/1 and 1/1 one may obtain any rational number in a *Farey tree*: each generation consists of the preceding generation plus the Farey sums of all neighbours of that generation.

0/1														1/1
0/1							1/2							1/1
0/1			1/3				1/2				2/3			1/1
0/1		1/4	1/3		2/5		1/2		3/5		2/3	3/4		1/1
0/1	1/5	1/4	2/7	1/3	3/8	2/5	3/7	1/2	4/7	3/5	5/8	2/3	5/7	3/4	4/5	1/1

...

At the end every rational number is obtained. Any irrational number may be approximated by rational numbers in consecutive generations of the Farey tree. The tongues (the regions ending in a sharp point) in the ANNNI model phase diagram (Fig. 6.6) correspond to the various terms in the Farey tree. A tongue for one rational value becomes narrower when the paraphase is approached. At the line where this paraphase becomes the ground state they have become points and in-between there are incommensurate phases.

6.3.3 Models with continuous degrees of freedom

In principle, there is no lattice dynamics in models like the ANNNI model. In incommensurate modulated phases, however, very often a soft mode is observed, leading to the phase transition. This shows that it is useful to introduce continuous degrees of freedom, as this allows the study of phonons in the model. This is already the case in the Landau theory, but there the discrete nature of the lattice is not considered. A model that combines continuous degrees of freedom and a discrete basis structure, inspired by the Elliott model, is the following. Consider a tetragonal lattice with a unique axis z and basis vectors $\mathbf{a}$, $\mathbf{b}$, and $\mathbf{c}$. At each vertex of the lattice an atom has one degree of freedom. This may be a displacement in a given direction, a rotation angle, or something else. Actually, any displacement field may be developed in normal modes. If one normal mode is dominant in a phase transition, one may choose its normal coordinate as model parameter. The only freedom is the choice of amplitude and phase. Taking an amplitude which depends on the unit cell, one gets one variable per unit cell, which can be chosen as the relevant variable in the model. The corresponding variable is denoted by $x_\mathbf{n}$ for the lattice site $\mathbf{n}$. Then the Hamiltonian is

$$H = \sum_\mathbf{n} \left[\frac{1}{2m} p_\mathbf{n}^2 + \frac{A}{2} x_\mathbf{n}^2 + \frac{1}{4} x_\mathbf{n}^4 + C \sum_\mathbf{m} x_\mathbf{n} x_\mathbf{m} + B x_\mathbf{n} x_{\mathbf{n}+\mathbf{c}} + D x_\mathbf{n} x_{\mathbf{n}+2\mathbf{c}} \right]. \quad (6.11)$$

The sum over $\mathbf{m}$ is a sum over the four neighbours in the xy-plane. This is the DIFFOUR model, which is a discrete version of the continuous model with a fourth-order polynomial term and first- and second-neighbour interactions. The latter may have conflicting tendencies, leading to 'frustration', hence the use of the term 'frustrated' in the name of this model (Janssen and Tjon, 1981, 1982). (See also Axel and Aubry (1981).)

The potential energy may be rewritten, by choosing new units for energy and length, to

$$V = \sum_{\mathbf{n}} \left[-\frac{a}{2} u_{\mathbf{n}}^2 + \frac{a}{4} u_{\mathbf{n}}^4 + \sum_{\mathbf{m}} \frac{1}{2}(u_{\mathbf{n}} - u_{\mathbf{n}+\mathbf{m}})^2 + d \sum_{\mathbf{m}'} \frac{1}{2}(u_{\mathbf{n}} - u_{\mathbf{n}-2})^2 \right]. \quad (6.12)$$

The summation with $\mathbf{m}$ runs over the nearest neighbours, that with $\mathbf{m}'$ runs over the second neighbours along the unique axis. The ground state at $T = 0$ minimizes the energy. If we limit the solutions to those which are constant in planes perpendicular to the unique axis, then the model is a chain with particles in double well potentials (if $a > 0$), connected by harmonic springs. For a tending to infinity, the depth of the double wells increases, with minima at ± 1. This limit then corresponds to the ANNNI model. If a tends to zero, the double wells are shallow. The parameter a characterizes the system: for large values of a, it is an *order–disorder* model; for a tending to zero, it is in the *displacive limit*. The interaction between first neighbours favours a structure with the same value for $u_{\mathbf{n}}$ and $u_{\mathbf{n}+\mathbf{m}}$, whereas the coupling between second neighbours depends on d. This may lead to frustration if the two interactions compete. So, d characterizes the frustration. The two parameters a and d determine the ground state.

The ground state for the one-dimensional chain satisfies the equation

$$-au_n + au_n^3 + 2u_n - u_{n-1} - u_{n+1} + d(2u_n - u_{n-2} - u_{n+2}) = 0. \quad (6.13)$$

An obvious solution is $u_n = 0$ for all n. To see whether this is the ground state we check its stability by a linear stability analysis. The frequency of an oscillation around the positions given by $u_n = 0$ with wave vector k is given by

$$m\omega_k^2 = -a + 2 + 2d - 2\cos(k) - 2d\cos(2k). \quad (6.14)$$

The minimum of this curve occurs at $k = 0$, $k = \pi$, or $k = \arccos(\frac{1}{4d})$. The latter is the position of the minimum if $\text{abs}(1/4d)$ is smaller than 1. Only for a sufficiently strong frustration parameter ($|d| > 0.25$) will it be possible for a phase transition to an incommensurate phase to occur. The value of ω^2 at the minimum depends on a. If a is large enough, then the equidistant array is unstable. Because the value of k is, in general, at an arbitrary point of the Brillouin zone, the instability has an incommensurate wave vector, and will lead to an incommensurate modulated phase. The structure of the ground state follows from the set of coupled non-linear equations (Eq. 6.13). The infinite set may be truncated by looking at periodic solutions with period N. The wave vector of the modulation is then the rational number L/N. After finding the solutions one

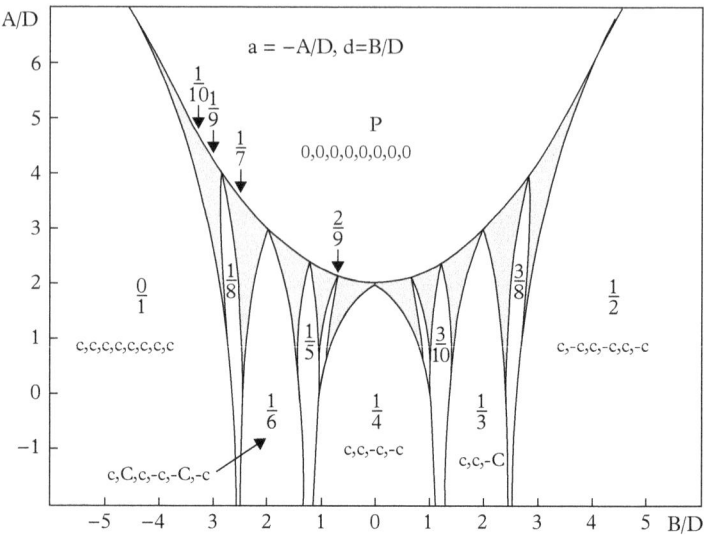

Figure 6.7 *The phase diagram of the DIFFFOUR model ($x_n = x_m$ if n and m are in the same plane in Eq. 6.11). It contains a ferroic phase (modulation wave vector 0/1), an anti-ferroic phase (1/2), a paraphase (P: $u_n = 0$), superstructures (L/N), and incommensurate phases (hatched regions). There are two Lifshitz points, where the paraphase, the incommensurate phase, and the $\frac{0}{1}$ or $\frac{1}{2}$ phase meet.*

has to compare the energies of various stable solutions. This results in a phase diagram in the ad-plane. In Fig. 6.7 this phase diagram is shown. It shows a *paraphase* ($u_n = 0$), a *ferroic phase* ($u_n = c \neq 0$ independent of n), an *anti-ferroic phase* ($u_n = c(-1)^n$), and commensurate and incommensurate phases. These are characterized by the wave vector k or, in the case of commensurate phases, by a string of digits giving the number of consecutive values of u_n with the same sign. If a digit occurs several times in a row it is given by an exponent. For example, the ferro-phase has $k = 0$ and the string <1>, the anti-ferro-phase has $k = 1/2$ and the string <1^2>, a period 4 superstructure has $k = 1/4$ and the string <2^2>, and a period 7 solution has $k = 1/7$ and the string <43> or $k = 2/7$ and the string <$2^3 1$>. There are two *Lifshitz points* where the 0/1 (or 1/2) phase, the paraphase, and the incommensurate phases meet. The incommensurate-ferroic and incommensurate-paraphase phase transition lines go smoothly over into the (anti-)ferroic-para phase transition.

Decreasing the value of A while keeping B and D constant in Fig. 6.7 leads to a typical series of phase transitions: from the paraphase, via an incommensurate phase, to a superstructure. The latter is called the *lock-in transition*. Such a series is often found in real systems with incommensurate phases. Near the line where the paraphase becomes unstable, the modulation function is sinusoidal, with a small amplitude. The wavelength is that of the unstable mode in the paraphase. For larger values of a (smaller for A) the modulation function gets more harmonics: it squares up.

If the incommensurate phase starts with a wave vector k, and the lock-in transition is at k_0, then the modulation function may be written as $f(z) = U(z)\sin(2\pi k_0 z + \phi(z))$. Near the incommensurate transition, where the phase varies as kz, the function $\phi(z)$ is $(k - k_0)z$, for the superstructure it is a constant. In Fig. 6.10 the development of the function $\phi(z)$ is given for two values of the parameter a. Near the incommensurate transition it is a straight line, in the incommensurate phase near the lock-in transition it shows plateaus: the phase remains approximately constant in domains, and changes rapidly in the transition regions (domain walls). Without these rapid changes the structure would be a superstructure. The domain walls (earlier also called solitons) are the reason why the structure becomes incommensurate. Therefore, they are called *discommensurations*. The phase $\phi(z)$ is constant in a domain. If the wave vector k_0 is that of an N-fold superstructure, there are N values for the plateaus. They correspond to the N realizations of the superstructure.

In the ground state at $T = 0$ the discommensurations are usually flat. At higher temperatures this is no longer true. Each domain corresponds to a certain value of the phase of the N-fold superstructure. The domain walls (discommensurations) may become rough, and they may intersect. Because of topological constraints the discommensurations meet in points where a fixed number of discommensurations come together (*stripples*). This may give the impression of a spaghetti-like type of structure. In the diffraction pattern, however, even such wild structures produce Bragg peaks (see Fig. 6.8.).

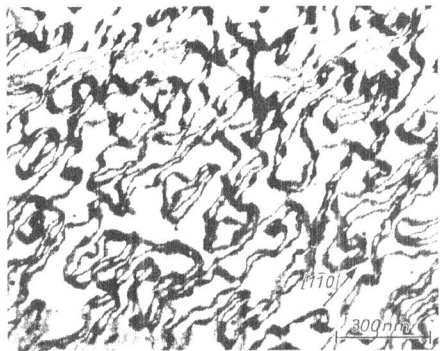

Figure 6.8 *Discommensurations separate domains with a fixed value of the phase of the N-fold superstructure. They may meet in points, and they may have a meandering structure.*

For a solution x_n of the ground state, and lattice constant ℓ, the positions of the atoms are $n\ell + x_n$. The modulation function follows from this relation as

$$f(nq\ell) = x_n.$$

The solutions near the incommensurate transition are in a good approximation sinusoidal, and the amplitude is small. For larger values of a, that is, deeper wells, the amplitude increases, and the shape gets more harmonics. As in the Frank–van der Merwe model, the modulation function becomes discontinuous if a exceeds a critical value a_c. This is the *discommensuration transition* or *transition by breaking of analyticity*, as it was called in Aubry (1978). The transition is sharp, and leads to effects in the dynamics which will be discussed later. Examples of the modulation function for various values of a are given in Fig. 6.9 (Janssen et al., 2002).

For large values of a, there are only commensurate phases (large negative values of A/D in Fig. 6.7), but on the transition line connecting the two Lifshitz points, the wave vector of the modulation varies continuously from 0 to $\frac{1}{2}$. For the commensurate values the tongues open up, the hatched regions become narrower, and finally disappear. As a function of d the wave vector changes, smoothly if one follows the incommensurate transition line, and with plateaus below that line. The wave vector behaves as an incomplete devil's staircase: it is an approximation to the complete devil's staircase, which would consist of an infinite number of regions with constant and commensurate values of the wave vector.

The phase diagram in Fig. 6.7 was constructed by minimizing the potential energy for the various choices of a and d (or A/D and B/D). It is a picture that is valid for zero temperature. To introduce temperature in the model, one could use a mean-field approximation. Another approach is to use Monte Carlo techniques (Janssen and Tjon, 1983; Rubtsov and Janssen, 2001; Savkin et al., 2003). The phase diagram for the DIFFFOUR model in the mean-field approximation is topologically the same as the one given in Fig. 6.7. The parameter A in that diagram may be identified with the temperature T. Decreasing A/D, increasing a, or decreasing T leads to the same sequence of phases. A simple argument for the relation $a - T$ is given in Janssen and Tjon (1982). For high-enough values of a, the ground state at $T = 0$ is commensurate; for high temperatures, the structure is lattice periodic; and, in-between, there is an incommensurate phase. The latter only exists typically for non-zero temperature. Therefore, one may say that the potential energy favours a commensurate phase, and that the incommensurate phase is partly due to entropy. For quasicrystals a similar argument plays a role. Then the quasicrystalline state could be stabilized by entropy terms.

When using Monte Carlo techniques to study thermal effects, the dimension of the system is important. This is in contrast to mean-field calculations. It is only in three dimensions that the phase diagram as found with Monte Carlo is similar to that presented above, and that the incommensurate phases have long-range order. In two dimensions the solutions of the models are of one of three

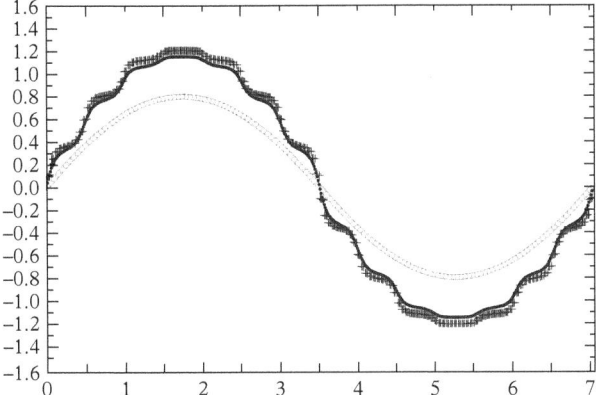

Figure 6.9 *Modulation function in the DIFFFOUR model for three values of the parameter* a: *0, 0.5, and 0.6. (Those for 0.5 and 0.6 almost coincide.) For* a $= 0.6$ *the function is discontinuous. (Janssen et al., 2002)*

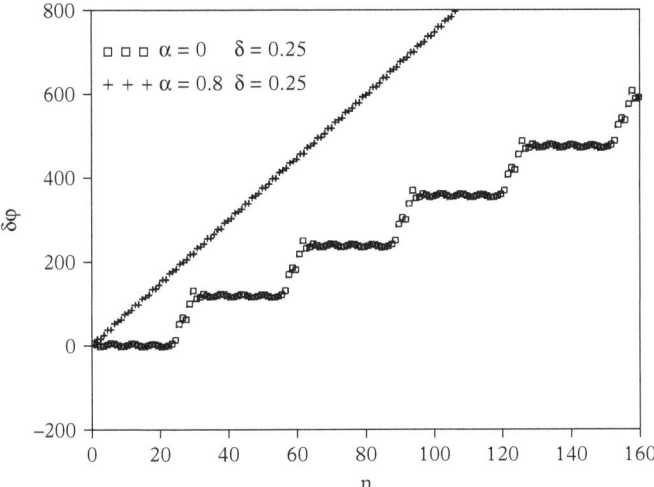

Figure 6.10 *The phase of the modulation function with respect to that of the lock-in phase, as a function of* z, *for a value close to the commensurate-incommensurate transition (upper line) and close to the lock-in transition (lower curve). The regions of constant phase are commensurate domains, and the transitions correspond to discommensurations. Plotted for solutions of the DIFFFOUR model. (Janssen and Tjon, 1982)*

types. These types are distinguished according to the behaviour of the correlation function

$$C(\mathbf{r}) = \langle u(\mathbf{r}')u(\mathbf{r}'+\mathbf{r})\rangle,$$

where there is an ensemble average over $\mathbf{r}'$. The correlation function does not decay for $r \to \infty$ in an incommensurate phase but $C(r) \sim \cos(qr)$; it decays exponentially like $C(r) \sim \exp(-r/\xi) \cos(qr)$ in a floating-fluid case, and with a power law $C(r) \sim r^{-\alpha} \cos(qr)$ in a *floating-incommensurate phase*. In two dimensions floating phases occur in the phase diagram. This had already been found for the ANNNI model.

The DIFFFOUR model (Janssen, 1992) can be applied to incommensurate phase transitions in A_2BX_4 ferroelectrics. It is closely related to the model for commensurate–incommensurate phase transitions presented in Chen and Walker (1991a), where the degrees of freedom in the model are derived from symmetry considerations.

The model in more than one dimension has ground states with a richer domain structure. Because an incommensurate ground state has infinite degeneracy, because the zero point of the phase is not determined, one may have an infinite number of domains. In the discommensuration region near the lock-in transition, the domains correspond to N different values, where Nk_0 is a reciprocal lattice vector of the basic structure. Because of topological constraints the discommensuration lines either extend to infinity or come together in multiples of N. From these points the discommensuration lines leave, and may eventually again come together, forming typical domain wall structures, called *stripples*. Examples are given in the section on numerical simulations. See also Fig. 6.8.

This is an example of an incommensurate phase in more than one direction. Similar models have been developed for the onset of incommensurate phases in two or three dimensions. A new feature is then the transition from 1q to 2q or 3q modulated phases (rank D + 2 or D + 3; see Section 6.3.4.2). An example of a compound with such a $1q - 2q$ *transition* is biphenyl (Cailleau et al., 1979).

6.3.4 Specific models for incommensurate phases

6.3.4.1 Introduction

The models discussed in the previous sections throw light on the principal, general features of incommensurate modulated phases and composites. Similar models exist for quasicrystals, but because these cannot be obtained by a simple phase transition from a periodic crystal phase, the models are more ad hoc, and show mainly that quasicrystalline phases, in the sense of decorated tilings, may be stable.

For the study of specific materials, or of families of specific materials the models discussed above may also be used. A major problem is an estimate of the relevant parameters compared with the material properties. In any case, the family of A_2BX_4 compounds (e.g. the phase diagram of tetramethylammonium-

Figure 6.11 *Simulated model of particles placed on the nodes of a hexagonal lattice. First- and second-nearest neighbours are indicated by the circles. Two additional neighbours above and below the central particle contribute also to the interactions.*

chlorometallates $(N(CH_3)_4)_2BX_4$, with $BX_4 = ZnCl_4$, $ZnBr_4$, SeO_4, $CoCl_4$, etc.) may be studied from a unifying point of view using the DIFFFOUR model (Janssen, 1986), and the rich phase diagram of BCCD has been modelled quite accurately with similar models (Chen and Walker, 1991b). Quartz has been studied in Dmitriev *et al.* (2003, 2004).

We have already seen Monte Carlo studies of the correlation functions in incommensurate phases in two and three dimensions (see Section 6.3.3). In the following, we present some further studies of specific modulated phases.

6.3.4.2 *Phase transitions in a hexagonal model*

In a first step, we present a numerical simulation of a simple three-dimensional model of particles that are located at the nodes of a hexagonal array (Parlinski and Chapuis, 1993, 1994), as illustrated in Fig. 6.11. The selection of this model is justified by the numerous examples of incommensurate phases which appear in hexagonal or related structures as in quartz (Bachheimer, 1980) or Al_2PO_4 (Snoeck *et al.*, 1986). In these systems, we observe essentially two classes of modulated phases. Either the modulation is one-dimensional ($1q$) and forms a series of parallel planes along a symmetry direction, or the modulation is two-dimensional and consists of columns aligned along the hexagonal axis ($3q$). The ability to form different types of incommensurate phases combined with various types of commensurate phases makes the hexagonal array of particles a very interesting study case for simulation.

In the present model, each particle $z_{j,l,n}$ can move independently but only along the z-direction parallel to the hexagonal axis. Here j, l, n are the indices of the lattice nodes along **a**, **b**, and **c**. Moreover, each particle interacts only with its nearest and next-nearest neighbours. In other words, each particle interacts with 12 neighbours in the hexagonal plane and two additional particles located directly above and below the plane. The equations of motion are from Newton's law, with potential energy given by

$$V = \frac{1}{2} \sum_{j,l,n} \Big\{ A z_{j,l,n}^2 + B z_{j,l,n} \Big(z_{j+1,l,n} + z_{j-1,l,n}$$
$$+ z_{j,l+1,n} + z_{j,l-1,n} + z_{j+1,l+1,n} + z_{j-1,l-1,n} \Big)$$
$$+ z_{j,l,n} \Big(z_{j+1,l-1,n} + z_{j-1,l+1,n} + z_{j+2,l+1,n}$$
$$+ z_{j-2,l-1,n} + z_{j+1,l+2,n} + z_{j-1,l-2,n} \Big)$$
$$+ C z_{j,l,n} \Big(z_{j,l,n+1} + z_{j,l,n-1} \Big) + H z_{j,l,n}^3 + z_{j,l,n}^4 \Big\}, \qquad (6.15)$$

with the four independent parameters A, B, C, and H. This potential is very close to the potential applied in Eq. 6.11. The coefficients of the harmonic terms between the second-nearest neighbours and the fourth-order anharmonic terms could be set to 1 by an appropriate choice of displacement and energy units (Parlinski et al., 1992).

In order to illustrate some characteristic phase transition mechanisms obtained by molecular dynamics, it is necessary to establish the phase diagram in terms of the four independent parameters. The phase diagrams presented in Fig. 6.12 have been established by performing a number of heating and cooling simulations by changing A and B for specific values of C and H. The intensity of the satellite reflections has been monitored in order to detect the boundaries of the phase diagram. Phase boundaries between characteristic $1q$ and $3q$ phases have been set up at points where the diffraction pattern with two satellites ($1q$) present along a symmetry direction changes to six satellites ($3q$) located along **a***, **b***, and **b*** − **a***.

In the absence of a third-order term $H = 0.0$, the commensurate phases $\frac{1}{4}$, $\frac{1}{3}$, and $\frac{1}{2}$ and the incommensurate phases are all of the $1q$ type and stable. The fractions refer here to the magnitude of the modulation vector k. For finite H, the phase diagram becomes more complicated since $1q$ and $3q$ phases appear along with commensurate and incommensurate ones. One can observe that the $3q$ incommensurate phase occurs at the phase boundary of the normal phase.

In the following, we shall show a selection of the results obtained by molecular dynamical simulation between phases identified on the phase diagrams (Fig. 6.12). In hexagonal structures, the $1q$ modulated phase with wave vector **k** normal to the hexagonal axis can be interpreted as a sequence of *discommensuration planes* normal to the modulation vector. The $3q$ modulated phase can be viewed as a

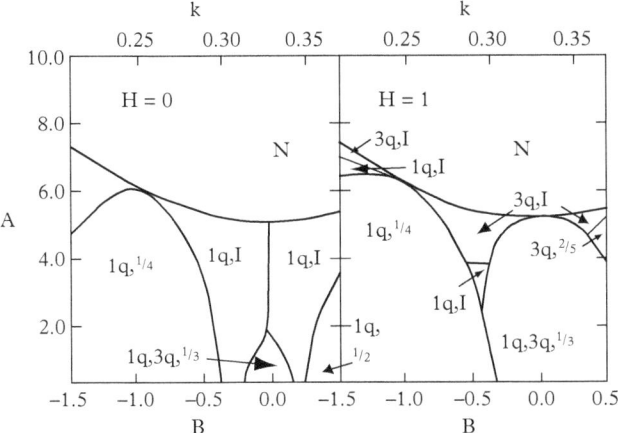

Figure 6.12 *Portion of the phase diagram of the hexagonal model for C = −1 and for (left) H = 0.0 and (right) H = 1.0 as a function of a limited range of the parameters A and B. The modulation vector k is related to B by the expression* $B = -1(1 + 2\cos 2\pi k)$.

set of *discommensuration columns* parallel to the hexagonal axis. They can also be considered as the result of three sets of discommensuration planes, each oriented along a reciprocal lattice direction normal to the hexagonal axis. The discommensuration planes and columns form *stripes* and columns of discommensuration lattices, respectively. The lattices may contain topological defects which are responsible for some of the phase transition mechanisms. In particular, we shall observe the formation of *stripples*, a type of topological defect which resembles a disc and drives the phase transition from a commensurate to an incommensurate phase (see Chapter 5, Section 5.3.3). A stripple consists of a definite number of discommensuration planes which meet at a closed *deperiodization loop*. The number of discommensuration planes depends on the number of domains formed by the commensurate reference phase. We can see here the analogy between a classical dislocation line which refers to the lattice periodicity of the crystal and a deperiodization line which refers to the discommensuration lattice.

The results of the simulations will be represented by *discommensuration* and displacement maps which are characteristic for each type of phase transition. For the discommensuration maps, a shading depending on the absolute value of displacements was used:

$$d_{j,l,n} = a \frac{|\langle z_{j,l,n}\rangle - z_{av}|}{|z|_{av}}, \tag{6.16}$$

where $\langle z_{j,l,n}\rangle$ represents the particle displacements averaged over a short period of time, $z_{av} = \frac{1}{N}\sum_{j,l,n}\langle z_{j,l,n}\rangle$ is the average position of particles, and

428 *Origin and stability*

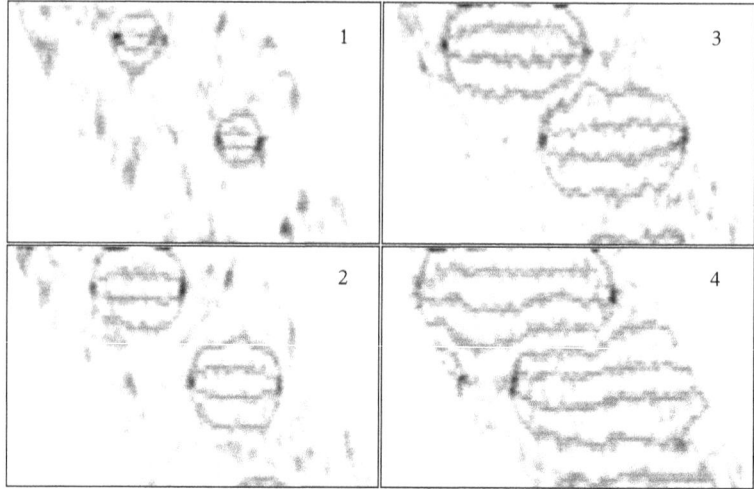

Figure 6.13 *Sequence of discommensuration maps normal to the hexagonal c-axis illustrating a transition from the commensurate $\left(1q, \frac{1}{4}\right)$ to the incommensurate $\left(1q, \frac{3}{11}\right)$ phase. The panels show the growth of two stripples, which ultimately form an incommensurate structure of type 1q. The size of the crystallite is $88 \times 88 \times 5$ unit cells.*

$|z|_{av} = \frac{1}{N} \sum_{j,l,n} |\langle z_{j,l,n}\rangle - z_{av}|$ is the absolute value of the fluctuations around z_{av}. We take the constants equal to $a = 2.0 - 3.0$ and $z_o = 0.4 - 0.5$ for best contrast.

The map representing the particle displacements indicates a distribution of the quantity

$$c_{j,l,n} = \frac{(\langle z_{j,l,n}\rangle - z_{av} + 2|z|_{av})}{4|z|_{av}}, \tag{6.17}$$

which guarantees that only relevant fluctuations are indicated.

The process of phase transitions is caused by system fluctuations which exist at finite temperatures. Topological defects can initiate the growth of a new phase if the system is forced into metastable conditions. In the following, we shall illustrate some of the possible phase transitions occurring in the systems under consideration.

6.3.4.3 Simulation of a phase transition from $\left(1q, \frac{1}{4}\right)$ to $(1q, I)$

The mechanism of the phase transformation from a commensurate $1q$ to an incommensurate $1q$ can be conveniently described in terms of discommensuration planes and stripples. The commensurate phase $\frac{1}{4}$ may exist in four antiphase domains D_1 to D_4. The incommensurate modulation consists of an ordered sequence of the four domains. A stripple capable of generating a layer of an incommensurate phase in domain D_1 must contain the three remaining domains D_2 to D_4. Figure 6.13 illustrates the formation of stripples with four discommensuration planes at the commensurate-to-incommensurate transition. The simulation was obtained by increasing B from the starting value $B = -0.52$. A, C, and H were fixed to 4.0, -1.0, and 0.0, respectively. Under metastable conditions, two stripples nucleated from thermal fluctuations (Fig. 6.13 (1), (2)) and increased their size isotropically. After reaching their maximal size, a further evolution took place laterally (Fig. 6.13 (3), (4)). The final incommensurate structure is modulated with the wave vector $\frac{3}{11}$.

It is clear from the preceding simulation that each stripple adds only one new modulation period to the system. In order to fill the volume of the crystal completely with a definite density of discommensurations, a large number of stripples must be created. From the nucleation point of view, it is of interest to study the mechanism related to their appearance. For this simulation, we shall use a crystallite elongated along one direction with dimensions 44 × 400 × 10 but subject to periodic boundary conditions. The modulation is oriented along the **b**$\star$ direction and contains over 100 modulation periods. With $A = 4.0$, $B = -0.52$, $C = -1.0$, and $H = 0$, the simulation was performed by changing B to -0.30 in order to induce a metastable state. From a small temperature increase it was possible to generate the series of stripples illustrated in Fig. 6.14. This figure illustrates in 12 sequences the successive appearance of numerous stripples illustrating the transition from the commensurate $\frac{1}{4}$ to the incommensurate phase. This serial mechanism is illustrated by the appearance of stripples in random layers normal to the modulation direction. The discommensurations generated by each stripple diffuse away thus creating space for the next stripples.

Three series of discommensurations indicated by a, b, and c in Fig. 6.14 could be observed in this simulation. The c series starts already in Fig. 6.14 (1) whereas series a and b appear in Fig. 6.14 (2). A new stripple appears in Fig. 6.14 (3) in the middle of four discommensuration planes and this process continues sequentially over the entire series until Fig. 6.14 (12) with a total of seven stripples. In the other series, four new stripples are created in the same way in each of them, with one exception. In series b, one observes in Fig. 6.14 (3) a new stripple growing outside the preceding one. At the end of the equilibration process, the system produces a single incommensurate structure of the type $1q$ with the modulation wave vector $k = 0.29$. This example illustrates nicely the leading role of the stripple formation in the phases transition from the commensurate $\left(1q, \frac{1}{4}\right)$ to the incommensurate phase.

6.3.4.4 Simulation of a transition from an incommensurate $(1q, I)$ to a commensurate $\left(1q, \frac{1}{4}\right)$ phase

One would expect the transition from an incommensurate to a commensurate phase to be driven by the inverse mechanism leading to the formation of stripples that is, the formation of *anti-stripples* by which four discommensurations would be removed from the system. Surprisingly, the simulation results exhibit a different mechanism which is nevertheless able to remove the discommensurations. The simulation started with an incommensurate $1q$ system with a wave vector $\frac{3}{11}$ and the parameters $A = 4.0$, $B = -0.71$, $C = -1.0$, and $H = 0$. The system was

430 *Origin and stability*

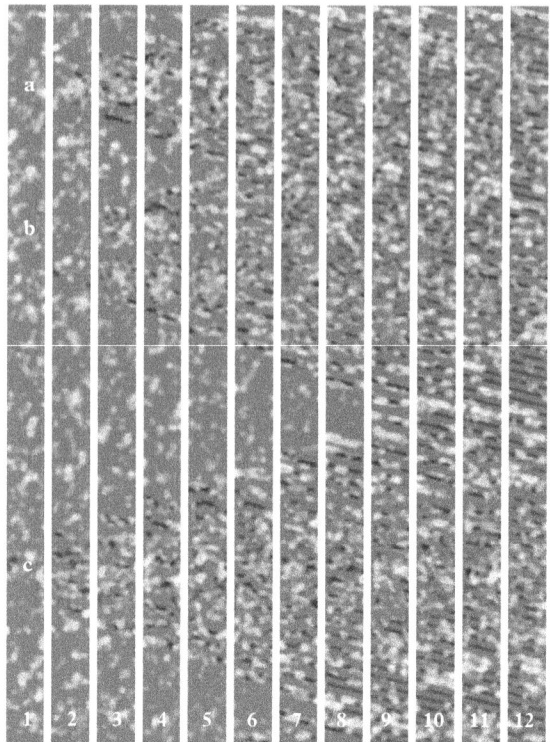

Figure 6.14 *Sequence of 12 discommensuration maps in a crystallite of 44 × 400 × 10 units, illustrating from left to right the transition from the commensurate (1q, $\frac{1}{4}$) to the incommensurate phase, with* k = *0.29 and a constant time interval.*

first moved to a metastable state by decreasing B to −1.16. In Fig. 6.15 (1), we observe that one of the four possible domains, D_1, becomes metastable and slowly rearranges to a domain D_3 (Fig. 6.15 (2),(3), and (4)). This produces a new sequence of domains D_1, D_2, D_3, D_4, D_3, D_2, D_3, and D_4 and we observe that two domains D_2 and D_4 are confined inside D_3. With this configuration, only two discommensurations should meet in order to form an anti-stripple. Indeed, the anti-stripple illustrated in Fig. 6.15 (5) removes domain D_4 and later also D_2. We also observe that a domain D_4 is slowly rearranging to a domain D_2 (Fig. 6.15 (6), (7), (8)).

The same phase transition has been simulated with a larger crystallite as represented in Fig. 6.16. The system was brought to a metastable state in order to initiate the phase transition by changing A from 4.0 to 5.1 and B from −0.6252 to −1.2252. From Fig. 6.16 (1) to (5), we observe first a reordering of the

Semi-microscopic models 431

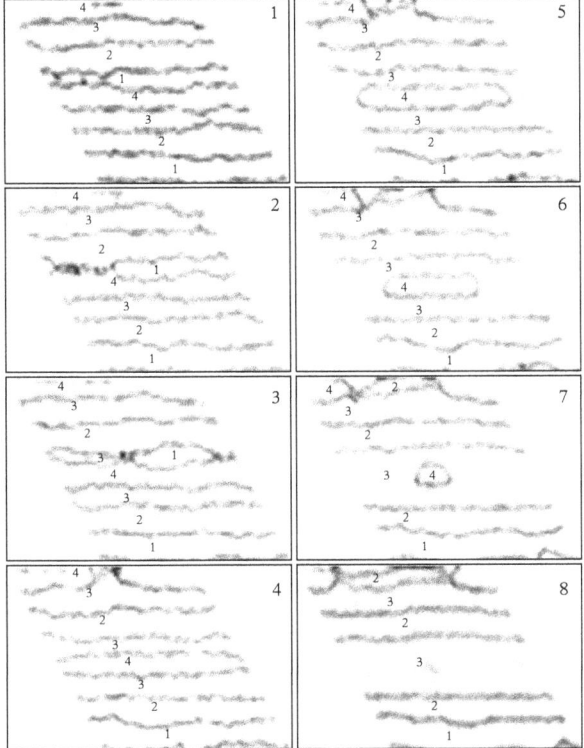

Figure 6.15 *Evolution of a series of discommensurations from an incommensurate to a commensurate 1q phase. The maps are normal to the hexagonal axis and contain 88 × 88 × 5 unit cells.*

displacements according to the rule $D_1 \rightleftharpoons D_3$ and $D_2 \rightleftharpoons D_4$. This process creates a number of cases where one specific domain is confined within two other identical domains. The transition mechanism occurs with the formation of simple antistripples (Fig. 6.16 (6) to (11)). At the end of the process, however, a fraction of the discommensurations still remain owing to the ever-increasing distances between the interacting domains.

6.3.4.5 Simulation of a transition from an incommensurate 3q to a 1q phase

The phase diagram illustrated in Fig. 6.12 indicates that the column-like phase $3q$ exists for higher values of H. The simulation of a phase transition from $3q$ to $1q$ can thus be obtained by decreasing the value of H. Starting from a crystallite size of $88 \times 88 \times 5$ and the parameters $A = 4.0, B = -0.555$, and $H = 0.925$, the phase transition mechanism could be induced by decreasing H to 0.5. The modulation

432 *Origin and stability*

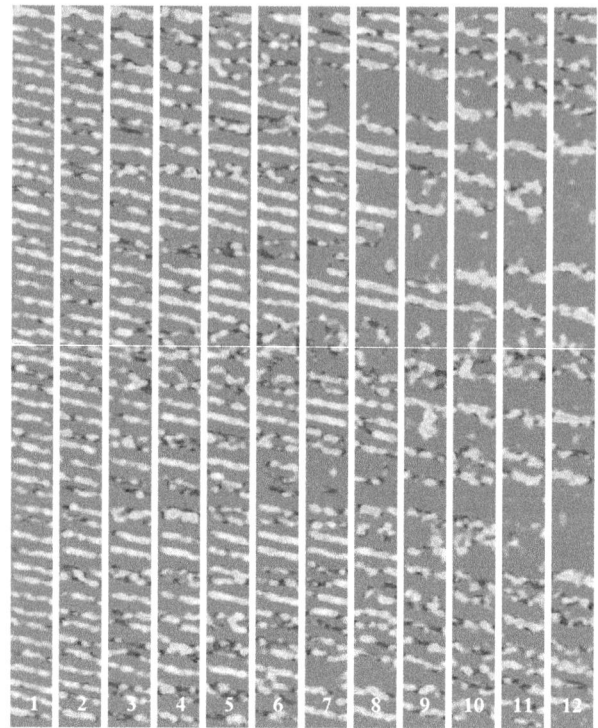

Figure 6.16 *Sequence of discommensuration maps from the incommensurate phase with* k = 0.28 *to the commensurate phase* $\frac{1}{4}$. *The maps normal to the hexagonal axis contain* $44 \times 400 \times 5$ *unit cells.*

Figure 6.17 *Maps of particle displacements normal to the hexagonal axis, illustrating the transition from an incommensurate 3q to an incommensurate 1q phase in three sequences. The wave vector* k *is constant and equal to* 0.2727.

vector $k = 0.2727$ did not change during the transition but only the dimensionality of the modulation. The evolution of the phase transition mechanism is illustrated in Fig. 6.17. Some of the neighbouring columns merge together, forming stripes of modulation. Other merging columns in the vicinity have the tendency to form parallel stripes, thus forming a single domain cluster of the 1q type. At the end of the simulation, the model consists of three families of parallel stripes, each one oriented along the three reciprocal directions **a★**, **b★**, and **b★ − a★**.

6.3.4.6 Conclusion

The simulations obtained from the hexagonal model close to the commensurate phase $\frac{1}{4}$ not only allow one to establish the phase diagram of the system but also reveal clearly the details of the phase transition mechanisms between commensurate and incommensurate phases and, in particular, the formation of *domain structures*. If the transition mechanism from the commensurate $(1q, \frac{1}{4})$ to the incommensurate $(1q, I)$ is driven by stripples consisting of four discommensuration planes, the reverse transition does not occur simply by the inverse mechanism. We observe, on the contrary, a reordering of the domains which results in different local sequences. This reordering favours the formation of antistripples consisting only of two discommensurations, which are removed during the phase transition mechanism. This process, however, leaves some residual discommensuration planes which do not disappear completely during the simulation. The transition from an incommensurate $3q$ to $1q$ could also be analysed by the simulation. The resulting structure is a combination of three types of domains, each consisting of parallel stripes oriented along the three directions normal to the hexagonal axis.

6.3.4.7 Simulation of organic incommensurate crystal structures

Many of the known examples of aperiodic structures belong to the class of organic crystals. This is often associated with the so-called large Z' structures, that is, structures containing a series of identical molecules or entities in the asymmetric units. Many such structures have been analysed by diffraction methods (see, for example (Schönleber and Chapuis, 2004a; Bussien-Gaillard *et al.*, 1996)) and it is thus tempting to use the very efficient molecular dynamical tools which have been developed specifically for the simulation of organic structures. The forces in Newton's equation of motion are calculated from the potentials which are here referred to as *force fields*. The commonly used force fields for the description of molecules include essentially two types of terms: bonding and non-bonding terms. Bonding terms include, among others, the bond distances and the bond and torsion angles, whereas non-bonding terms include van der Waals and electrostatic interactions between atoms. In general, the force field is deduced from a small set of molecules but can be applied to an extended set of related molecules and structures. In the following example, the consistent force field best adapted for the type of molecular interactions was applied for the simulations (Maple *et al.*, 1994). The applied force field is, however, not able to reproduce precisely the experimental lattice constants measured by diffraction within a few per cent. In order to increase the fit between the experimental and simulated values, a practical method based on the pressure tensor can be applied thus avoiding the need to tamper with the parameters of the consistent force field. The equation of state of the N-atom system can be expressed by the virial expression. A compensating pressure tensor can be introduced in the following expression

$$V \equiv V_{\exp} = \frac{\sum_{i=1}^{N} m_i \dot{\mathbf{r}}_i \dot{\mathbf{r}}_i}{P + P_{compen} - P_{bond} - P_{nonbond}} \quad (6.18)$$

in order to adjust the volume to the experimental observations. This pressure tensor can be estimated in the first steps of the simulation and is kept unmodified for the remaining part of the simulation (Pan et al., 2003).

As an illustration we present the simulations of the modulated phases of hexamine suberate. This organic compound has been studied by diffraction and the incommensurately modulated phase stable between 120 and 300 K has been fully characterized within the superspace formalism (Bussien-Gaillard et al., 1996). The structure consists of alternating layers of suberic acid and hexamine linked by hydrogen bonds between the O and N atoms. A commensurate approximation of this structure is represented in Fig. 6.18.

In this model, we observe that the zigzag chains of the suberate layers are disposed in pairs each with different orientations of the chain axis and zigzag planes.

In order to simulate this incommensurate structure and to investigate the mechanism of the commensurate to incommensurate phase transitions, a (commensurate) starting model with dimensions 180 × 50 × 180 Å and representing

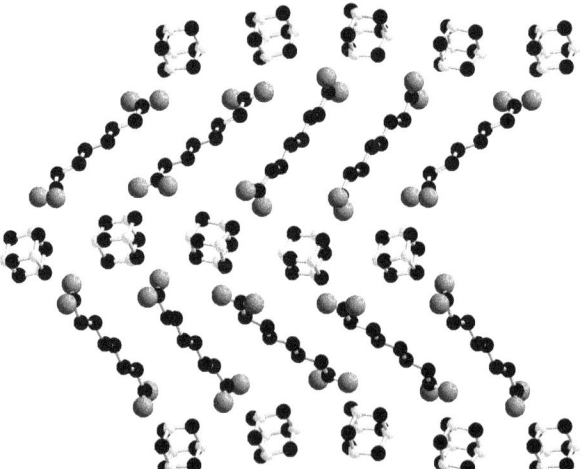

Figure 6.18 *Commensurate crystal approximation of hexamine suberate. Layers of suberic acid alternate with layers of hexamine along **b**. The layers are linked together with H-bonds between the oxygen atoms and the nitrogen of the hexamine.*

more than 196,000 atoms was first equilibrated under constant volume (NVT) conditions in order to determine the parameters of the pressure tensor, which was then applied throughout the simulations. The lattice constants and angles obtained by simulations in the temperature range from 15 to 580 K are very close to the experimental values. At 123 K, they differ by at most 0.6 per cent.

The phase transition temperatures can be detected from the changes in the physical properties of the simulated systems. In particular, we can observe the variations in the lattice constants and angles, the density, the mean-square displacement and the torsion angle energy over the whole temperature range. Two phase transitions could be detected: one at about 290 K, and another at 150 K. It is most convenient to represent each phase by simulating its single crystal diffraction pattern, as is illustrated in Fig. 6.19 for three different temperatures. The resulting diffraction patterns reveal the high-temperature commensurate phase, followed by the incommensurate phase and then reaching finally the low-temperature lock-in phase.

Figure 6.19 *Simulated diffraction patterns at three different temperatures.*

436 Origin and stability

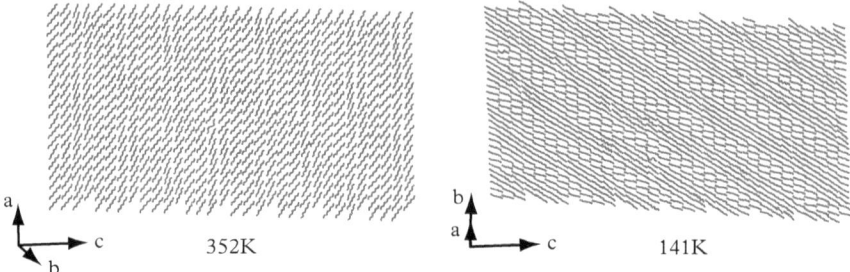

Figure 6.20 *Long-range ordering of the aliphatic chains in simulated high- and low-temperature phases.*

The characteristic diffraction pattern with a four-fold superstructure along c^* corresponds to the structure at 352 K, as illustrated in Fig. 6.20. The structure of the aliphatic layer indicates clearly the four-fold periodicity along **c**. This characteristic pattern is due to the existence of H-bonds mentioned above between three consecutive rows and leaving the fourth vacant. Below 290 K new satellites appear in the diffraction pattern. This corresponds to a new incommensurate periodicity with approximately eight chains, which is closely related to the model simulated at 141 K and represented in Fig. 6.20. This incommensurate modulation is superimposed with the commensurate modulation present at high temperature.

Both X-ray diffraction and simulation studies point to the origin of the modulation, which is located at the interface between the hexamine and the suberate layers. The packing of the hexamine and suberate layers requires specific lattice parameters which are not compatible with each other. The pattern of hydrogen bonds which glue the layers together adopts the best possible configuration under frustrated periodic conditions and evolves towards greater complexity with decreasing temperature. One should emphasize at this point that the simulation methods did reveal a new pattern of H-bond which was not identified in X-ray diffraction measurements. This is probably due to the low diffracting power of the H atoms in the diffraction experiment which might have been overseen in the resolution of the incommensurate structure. This particular H-bond scheme relates to two adjacent suberate molecules one of which being bonded to the other which is itself bonded to the hexamine layer.

It appears that the molecular dynamical methods based on a consistent force field are perfectly able to reproduce the structures of both commensurately and incommensurately modulated organic structures. The introduction of an adjusting pressure tensor which is constant over the whole temperature range is sufficient to compensate for the specificities of each compound. This is a remarkable result which can be exploited for the simulation of other non-periodic structures.

6.4 Composites

The incommensurability of composites is usually caused by non-stoichiometry. There are two or more subsystems interacting such that both subsystems become modulated. The type of the modulation depends on the strength of the interaction. The essential features can be seen on a simple model, consisting of two parallel chains, with a distance d, and with incommensurate lattice constants a and b. It is called the *double chain model*. The Hamiltonian then gets the form

$$H = \sum_n \frac{p_{1n}^2}{2m_1} + V_i(x_n - x_{n-1}) + \sum_m \frac{p_{2n}^2}{2m_2} + V_2(y_m - y_{m-1}) + \frac{\lambda}{2} \sum_{nm} W(x_n - y_m). \tag{6.19}$$

The intra-chain interactions V_1 and V_2 have minima at a and b, respectively. Then for $\lambda = 0$ the positions of the atoms in the chains are $x_n = x_0 + na$ and $y_m = y_0 + ma$. For $\lambda \neq 0$ the chains become modulated:

$$x_n = x_0 + na + f(na), \quad y_m = y_0 + mb + g(mb),$$

with $f(na + b) = f(na)$ and $g(mb + a) = g(mb)$, because a shift of one chain over a distance equal to the lattice constant of the other does not change the situation. If one of the chains is rigid, for example the second, it creates a rigid substrate potential with periodicity b. If this is mimicked by a cosine potential one comes back to the Frank–Van der Merwe model. Then $\sum_m W(x - mb) \sim \cos(2\pi(x - y_0)/b)$.

The quasiperiodic system of rank 2, with Fourier module spanned by $1/a$ and $1/b$, has smooth atomic surfaces in the superspace if λ is smaller than a critical value λ_c. If the interaction becomes stronger, the modulation functions become discontinuous. This transition does not so much depend on temperature as on other external conditions, in particular pressure and composition.

6.5 Quasicrystals

The formation and stability of quasicrystals are still not understood. A first type of model is a tiling or a decorated tiling, where the tiles have a certain distribution of atoms. Two types of mechanism for stability have been discussed. In one, it may be energy that plays the most important role. Another mechanism is based on random tilings. Atomic positions are given by the intersection of atomic surfaces with the physical space. Small oscillations of the physical space will give rise to phason jumps. In general, these will increase the internal energy, but they may increase the entropy and decrease the free energy. Researchers have studied these two effects by using simple models.

One model is a two-dimensional tiling model with one or two types of atoms and certain interaction potentials. One choice for the latter is the LJG (for Lennard–Jones–Gauss) potential proposed in Engel and Trebin (2007). It is given by

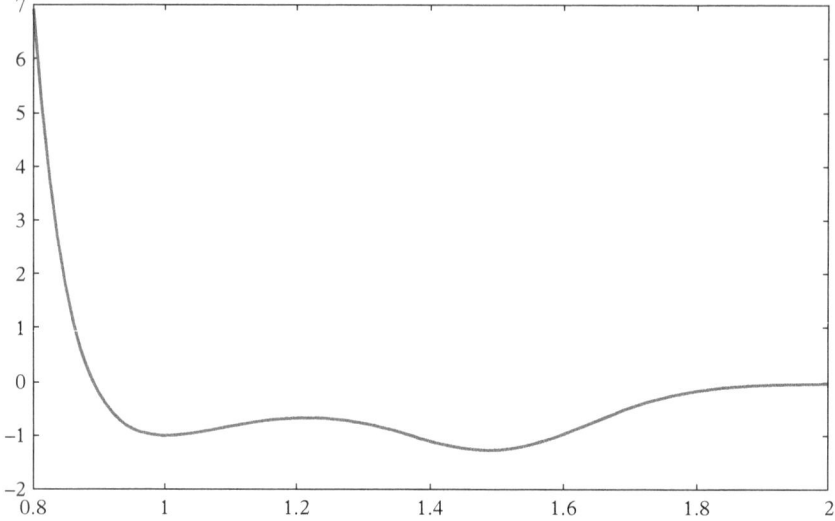

Figure 6.21 *Double well potential in Eq. 6.20, with $\epsilon = 1.1, \sigma^2 = 0.02, r_0 = 1.3$.*

$$V(r) = \frac{1}{r^{12}} + \frac{2}{r^6} - \epsilon \exp\left[-\frac{(r-r_0)^2}{2\sigma^2}\right]. \tag{6.20}$$

It is a combination of a Lennard–Jones potential and an additional Gaussian term. Depending on the choice of parameters, it may be a two-minima potential (Fig. 6.21), as one has used as well for calculating phonons (see Fig. 5.20 in Chapter 5).

The potential has been used in Engel and Trebin (2007) to find lowest-energy configurations. Depending on the values of the parameters, the solution with one type of particles is a tiling with lattice periodicity or an aperiodic tiling. The potential has been used to determine the phase diagram of a two-dimensional model, consisting of a single type of atoms with this potential for varying values of r_0. For certain parameters, one finds a decagonal tiling.

6.6 Electronic instabilities

6.6.1 Charge-density and spin-density systems

In the foregoing, the role of the electron system was not taken into account explicitly. There are also compounds where the coupling between electrons and the lattice is the principal reason for an incommensurate instability. Electrons in periodic systems form energy bands, and the states are characterized by a wave vector **k**. Consider as a simple example a one-dimensional tight-binding system with one state per site. A wave function ψ then is a linear superposition of states localized at a site n. The contribution to the wave function at the position n is c_n, which satisfies the Schrödinger equation

$$Hc_n = Ac_n + B(c_{n-1} + c_{n+1}) = Ec_n, \qquad (6.21)$$

where A is the site potential and B the hopping frequency between two nearest neighbours. The wave functions are waves $\exp(2\pi ikna)$ and the energy levels are characterized by the wave vector k and are

$$E_k = A + 2B\cos(2\pi a), \quad \left(-\frac{1}{2a} < k \leq \frac{1}{2a}\right).$$

The energy levels are occupied up to the Fermi level. Now suppose that there is one conduction electron per site. Then the up- and down-spin pairs fill the levels to the Fermi level which is up to the wave vector $1/(4a)$. The band is half-filled. It turns out that the energy of the electron may be lowered by a dimerization of the chain: the lattice constant becomes $2a$. Then the Brillouin zone boundary lies at $1/(4a)$, which is exactly the Fermi level. Because the dispersion is flat at the zone boundary, occupied levels go down in energy, and unoccupied levels go up (see Fig. 6.22). On the other hand the dimerization costs elastic energy in the chain. The total effect may be calculated easily and shows that a dimerization diminishes the total energy of lattice plus electrons. This is the *Peierls mechanism*. If the lattice has a filling L/N the Brillouin zone shrinks with a factor N and the lattice instability occurs at a point $k = L/(2Na)$. In general, for an incommensurate filling the lattice instability occurs at an incommensurate wave vector.

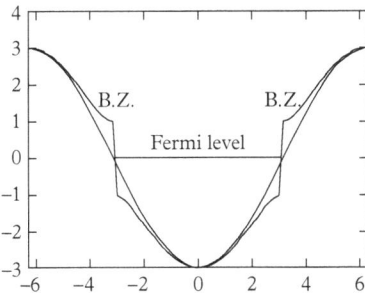

Figure 6.22 *An electron system with a half-filled band may lower its energy by the formation of a two-fold superstructure; smooth curve: without electron–phonon coupling; discontinuous curve: with coupling; B.Z., Brillouin zone boundary.*

In two and three dimensions the situation is more complicated. Now, the lattice instability does not always occur, but it may occur if the shape of the Fermi surface (two points in one dimension, a curve in two dimensions, and a surface in three dimensions) has special properties, called *nesting properties*. This occurs if pieces of the Fermi surface are connected to other pieces by the same wave vector, which then is the wave vector of the instability, and if pieces of some extent are flat and parallel to another flat piece of the Fermi surface. For example, in a three-dimensional system with parallel conducting chains and with low conductivity between parallel chains, the Fermi surface may consist of two flat surfaces connected by the wave vector $2\mathbf{k}_F$. This is perfect nesting. The resulting structure has a modulated lattice and a modulated electron density. It is called a *charge-density-wave* system. Because of the coupling between electron system and lattice, the latter becomes modulated, giving rise to an incommensurate phase. The role of the charges is in other systems taken over by the spins. Then the spin-density may have a periodicity that does not belong to the basic structure. Then a *spin-density wave* develops.

6.6.2 Hume–Rothery compounds

Quasicrystals are usually intermetallic alloys. For these materials a mechanism has been suggested that is related to the Peierls instability underlying charge-density-wave systems, and that is well known for periodic alloys. It is known as the *Hume–Rothery mechanism*.

The Hume–Rothery rules for intermetallic alloys have been known for a very long time. One of the rules says that a certain phase is stable for a constant value of the e/a ratio, the ratio between the number of valence electrons and atoms. A problem is what exactly the number Z of valence electrons is for an element. There are standard values for the various elements. For example, $Z = 0$ for Fe, Co, Ni, and Pd; 1 for Cu, Ag, and Au; 2 for Zn; and 3 for Al. A face-centred cubic structure has $e/a \approx 1.36$, a body-centred cubic structure $e/a = 1.5$, and the ϵ-phase has $e/a = 1.75$. The rule can be understood from the density of states in the rigid-band approximation, where one assumes that the density of states does not change if the Fermi level varies by a varying filling factor. For a smooth density of states, like that of free electrons, it does not cost much energy to add electrons. In the neighbourhood of the Brillouin zone boundary this cost is much higher. The system may decrease this by changing the shape of the Brillouin zone, which leads to a structural change. The Fermi level then falls in a pseudo-gap, a dip in the electron density of states, because then the total kinetic energy of the valence electrons will be lowered. In other words the Fermi wave vector $\mathbf{k}_F$ must satisfy $2\mathbf{k}_F = \mathbf{H}$, a reciprocal vector in the Fourier module. So, in the diffraction pattern there are strong peaks at a position of twice the Fermi wave vector.

Quasicrystals have been considered as Hume–Rothery alloys since the discovery by Tsai of the thermally stable families Al-Cu-TM (TM = transition metal) and Al-Pd-TM. The rather heuristic explanation given above has since been put on a firmer basis by ab initio calculations (Takeuchi and Mizutani, 1995). These families have e/a values of approximately 1.75. A continuing argument is about the relative role of internal energy and configurational entropy. The latter is increased by so-called phason jumps, described already in Chapters 3 and 5. The question is whether a quasicrystal is stable for zero temperature, or whether it is only stabilized by entropy. One should keep in mind the situation in modulated phases, where the ground state at low temperatures is often commensurate, and the incommensurate modulation appears only at higher temperatures, although there are structures as well where no lock-in transition has been found. There, both energy and entropy are at the basis of the incommensurate phase. Numerical calculations on approximants have revealed a pseudo-gap at the Fermi level (Fujiwara, 1999).

6.7 Growth of quasicrystals

Incommensurate modulated phases usually originate from a phase transition. The new phase grows from the unmodulated phase through displacements. For quasicrystals, this is not the case. The mechanism of their growth is still unknown. Because there are close relations between quasicrystals and aperiodic tilings, one could get some insight into quasicrystal growth from the construction of tilings. For tilings, one may decorate the edges of the tiles such that, if in the tiling two neighbouring tiles share edges with the same decoration, the tiling is quasiperiodic. These are called the *matching rules*. These, however, are not local growth rules. Starting from a kernel of tiles, there are sites where only one possibility for continuation exists (forced tiles), and others where one has to make a choice. This means that there are always some defects present. Ways to construct a perfect tiling have been discussed in Chapter 3, Section 3.4.1, and these are not local growth rules either. So, the growth of quasicrystals remains puzzling. An in situ study of the *growth* of a quasicrystal is presented in Nagao et al. (2015).

Theoretically, one has simulated the growth of quasicrystals. An early example was given in Minchau et al. (1987), where a simple model is presented for the growth of a decorated Penrose tiling. The model uses particles of two types interacting with hard core potentials. For a proper choice of radii, structures were found that look locally like decorated Penrose tilings.

In fact, it is possible to construct quasicrystalline structures with Penrose tiles. Each vertex in a Penrose tiling is of one of eight possible configurations. If one has reached a piece of tiling, there are vertices where only one type of tile can used (forced configuration), and others with unforced configurations. If one plans the continuation in unforced vertices carefully, it is possible to construct a tiling that is a Penrose tiling with a small number of defects.

Other approaches to the problem use physical arguments, interactions between particles of one type or of a small number of types. With Lennard–Jones potentials this was not very successful. Two-dimensional models with an inter-particle interaction with two minima (Fig. 6.21) showed regions in the parameter space

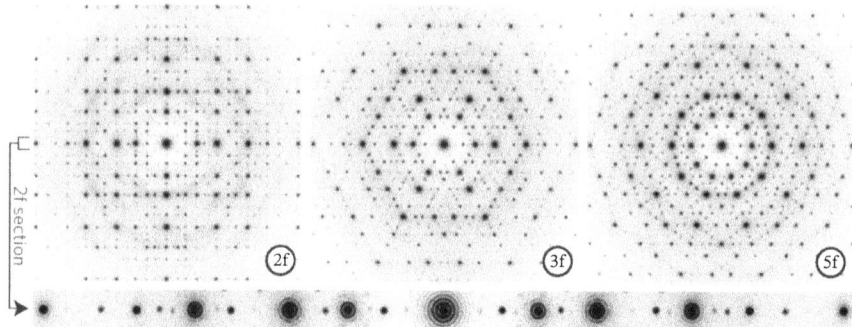

Figure 6.23 *Sharp diffraction along three axes (two-, three- and five-fold) for the 3D quasicrystal simulation, as in Engel et al. (2015).*

where periodic or aperiodic structures appear after annealing (Engel and Trebin, 2007).

In Keys and Glotzer (2007), a Monte Carlo simulation led to a decagonal phase. The potential used in that study was an adapted Lennard–Jones potential. Outside a certain radius, a repulsive potential was added. Starting with a seed, clusters formed to which atoms become attached, giving finally a dodecagonal quasicrystal.

A three-dimensional simulation of a system with identical particles and a interaction potential

$$V(r) = r^{-15} + r^{-3} \cos(k(r - 1.25) - \phi)$$

shows icosahedral structures (Engel *et al.* 2015; see Fig. 6.23). The growth is fast, and annealing improves the quasicrystal structure. It is interesting that this model uses only particles of one kind. Quasicrystals known till now are all at least binary.

For more details about the search for a solution to this problem, see the review article (Grimm and Joseph, 2002). Although quasicrystalline structures have been found in some simple model systems, the questions 'when will a quasicrystalline phase be the ground state' and 'how does this state grow' have not yet been answered. Actually, this is not just an open problem for quasicrystals. The same questions may be asked for lattice periodic systems with huge unit cells with tens of thousands of atoms, such as the Samson β-phase.

6.8 Summary

Incommensurability is a very common feature in materials, whether artificially made or occurring in nature. There are many different mechanisms. For nonconducting materials, structural effects like sterical hindrance, influence of nonstoichiometry, and competing interactions each favouring a different configuration may be responsible. Changing the circumstances may influence the balance between the structures, leading to a phase transition. Magnetic interactions may lead via a spin-lattice coupling to an instability if the interaction parameters change. When electrons are present, the position of the Fermi surface with respect to the important Bragg peaks becomes important. This interaction may lead to a Peierls instability, charge-density waves, and in the case of intermetallic alloys to Hume–Rothery phases, even with crystallographically 'forbidden' symmetries. For all quasiperiodic systems the ground state is a minimum of the free energy, where both internal energy and entropy contribute. For incommensurate phases this has been studied in the context of thermodynamics. For quasicrystals this aspect has mainly been studied in the framework of tilings. Phenomenologically the situation may very often be well described in terms of the Landau theory of phase transitions. The ground state, the modulation, the phase transitions and their kinetics in modulated phases may very well be described with simple models. To understand the details of specific materials more sophisticated calculations are needed, using more or less realistic potentials. This is also true for quasicrystals, where only the simplest general features are found in model systems. For a better understanding the study has to be based on more complex calculations, using, for example, ab initio methods.

7
Other topics

7.1 Morphology of aperiodic crystals

Up to this point, the study of aperiodic crystal structures has been essentially based on elastic and inelastic scattering. The characterization of diffracted intensities with a number of integer indices equal to its rank rather than its dimension led to the concept of superspace which is now systematically applied in the description of aperiodic crystals. The consequence of the rank being higher than the dimension of the system is that the crystal structure exhibits additional periodicities which are in general incommensurate with the others. Consequently, it should be possible to observe the additional periodicities on macroscopic crystal samples. It has been known since Haüy that the faces of a crystal are directly related to the periodicity of their structure. The three Miller indices used to characterize each face of a crystal are integer components of the reciprocal vector which is normal to that face. Therefore in aperiodic crystals, it should be possible to detect faces associated with reciprocal vectors requiring more than three indices, provided that they remain small. Indeed many examples of crystalline samples exhibiting faces associated with reciprocal lattice vectors with a number of components equal to the rank of the system have been discovered. We would like first to illustrate this property with the mineral calaverite. The story of the numerous attempts by different authors to index its faces extends over a long period; it took more than 80 years of research before a satisfactory and definitive solution could be proposed.

7.1.1 The puzzling habit of the mineral calaverite

The first morphological descriptions of calaverite $Au_{1-p}Ag_pTe_2$ ($p < 0.15$) were published in the early 1900s by Smith, Penfield, and Ford (Penfield and Ford, 1902; Smith, 1903). These authors were puzzled by the rich variety of forms and wondered why many of them could not be indexed by three small integers. They concluded that calaverite crystals contradict the basic law of crystallography. The problem of the indexation of the calaverite crystal forms was again considered in the early 1930s (Goldschmidt et al., 1931) with an impressive study based on more than 100 samples of the finest quality and from different origins. They could

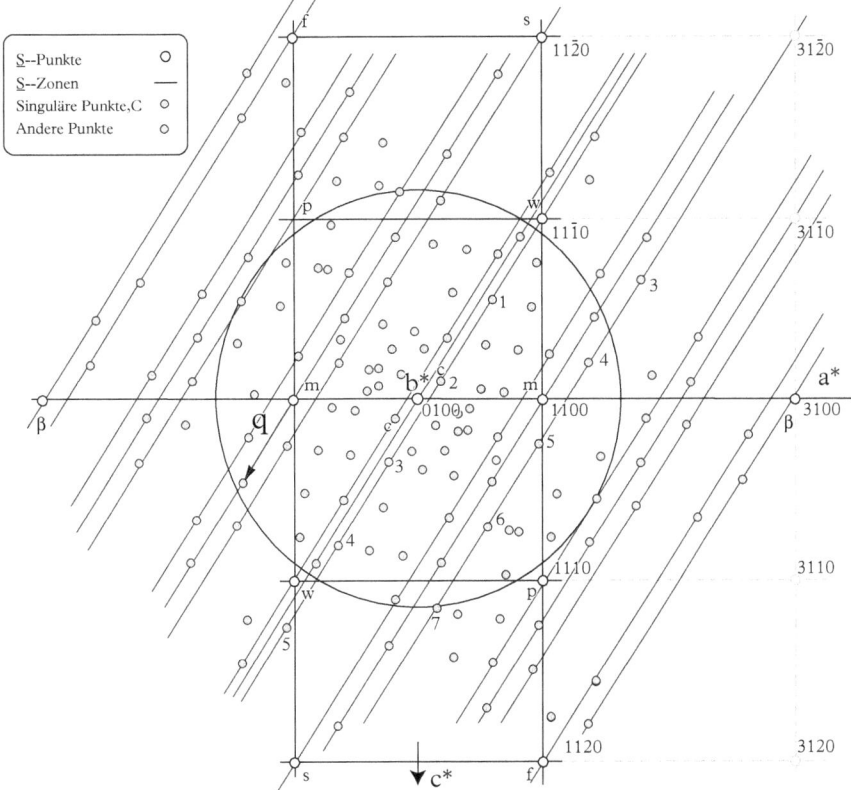

Figure 7.1 *Gnomonic projection of calaverite according to Table 1 in Goldschmidt et al. (1931). The modulation vector **q**, the oblique lines parallel to this vector and the four-integer indices of the reciprocal lattice points have been added.*

essentially confirm the work of the previous authors. After attempting to index most of the faces with four distinct and independent lattices, one monoclinic and three triclinic, they realized that the proposed solution was highly unsatisfactory and consequently declared that the law of rational indices is not generally valid. This statement was, of course, true, as we now know. However, at that time the scientific community was not ready to take these considerations seriously.

Based on Table 1 from (Goldschmidt *et al.*, 1931), Fig. 7.1 represents a combined gnomonic projection of the possible faces found by the authors for the whole series of samples under study. One of the interesting properties of the gnomonic projection is that it represents an undistorted description of the normals to the crystal faces, that is, of the reciprocal lattice points. Indeed, by looking at Fig. 7.1, we clearly observe the so-called S-Punkte (S-points) represented by small open circles forming a two-dimensional lattice. There are in addition a very large number of face normals called Singuläre and Andere Punkte (singular and other points) lying outside the lattice nodes and represented by solid grey circles. It appears that all the grey dots are located along a series of lines parallel to a single vector indicated by **q** on the figure. We observe also that each parallel line crosses some S-points and that all the grey dots are distributed equidistantly along the lines with distance $\|\mathbf{q}\|$. Consequently, each S-point can obviously be characterized by three (Miller) integer indices, and all the other dots lying outside can be uniquely identified by the corresponding triplet of the node localized on the parallel line and in addition by a non-zero integer specifying the position of the dots on that line in terms of the vector **q**. With this indexing procedure, all the dots represented in Fig. 7.1 can be uniquely characterized by four indices. As an example, we can observe the S-point indicated by $(31\bar{2}0)$ and the associated solid dots with indices varying between $(31\bar{2}3)$ and $31\bar{2}7)$. Points with indices $(31\bar{2}1)$ and $(31\bar{2}2)$ are, however, missing in the particular case. The possibility of using a fourth integer index was apparently overlooked by the authors of the 1931 paper. The need for a number of indices which is larger than three indicates that the number of fundamental periodicities is also larger than three. The scientists of that period were probably not ready to admit such a revolutionary concept and thus the puzzle of indexing the calaverite faces remained unexplained for more than half a century.

The first hints towards the existence of a modulated structure of calaverite were published by different authors (Sueno *et al.*, 1979; van Tendeloo *et al.*, 1983) from X-ray and electron diffraction studies of single crystals. These publications revealed the modulated nature of calaverite from the presence of satellite reflections around the main diffraction spots. Recognizing the implication of the modulated character of calaverite, A. Janner and B. Dam managed to reconsider the indexation of the complete set of faces which were recorded in Goldschmidt *et al.* (1931), provided that four indices were used but with a single lattice. The elegant solution of this long-standing puzzle is illustrated in Fig. 7.2. On the left, a typical twinned sample of calaverite is represented according to Table 10 in Goldschmidt *et al.* (1931). On the right, an illustration of the same crystal sample but reindexed according to Fig. 3 in Janner and Dam (1989) using four fundamental periodicities. The zone axis formed by the important prismatic faces is aligned along the monoclinic axis of the structure.

We know that from the stereographic or gnomonic projections of the crystal faces, the angles and the relative magnitudes of the corresponding lattice constants can be obtained. We have seen earlier in this book that the components of the modulation vectors **q** are defined relative to the reciprocal lattice constants. Therefore, it should be possible to derive the components of the modulation vectors from the gnomonic projection described in Goldschmidt *et al.* (1931). This is precisely what was done in Janner and Dam (1989) with the following results compared to the diffraction measurements (Schutte and de Boer, 1988) and reported in Table 7.1. This remarkable result confirms the high quality of the goniometric measurements obtained in Goldschmidt *et al.* (1931) although more than half a century separates the two studies!

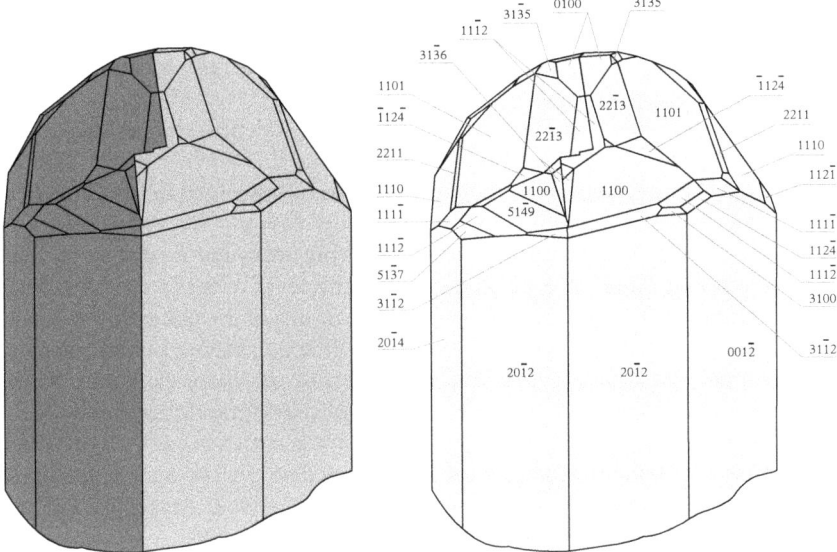

Figure 7.2 *Twinned sample of calaverite redrawn according to Table 10 in Goldschmidt et al. (1931). On the right, the same sample but reindexed with four indices according to Fig. 3 in Janner and Dam (1989).*

Table 7.1 *Components q_i of the modulation vector $\boldsymbol{q} = q_1\boldsymbol{a^*} + q_2\boldsymbol{b^*} + q_3\boldsymbol{c^*}$ of calaverite, as deduced from the gnomonic projection (Goldschmidt et al., 1931) and from X-ray diffraction data (Schutte and de Boer, 1988).*

From morphological data:	− 0.4095	0	0.4492
From X-ray data:	− 0.4076(16)	0	0.4479(6)

The complete identification of the full set of 92 crystal forms published in Goldschmidt *et al.* (1931) with a single set of four basic vectors $\boldsymbol{a^*}$, $\boldsymbol{b^*}$, $\boldsymbol{c^*}$, and $\boldsymbol{q}$ is a very elegant proof of the incommensurately modulated structure models. The same study (Janner and Dam, 1989) could also correct some obvious misprints (Goldschmidt *et al.*, 1931) as a sign change of an angle or misreadings of goniometer values.

The structure of calaverite determined from single crystal X-ray diffraction measurements (Schutte and de Boer, 1988) revealed clearly its incommensurate nature. The superspace group $C2/m(\alpha 0\gamma)0s$ obtained from diffraction is fully compatible with the morphological observations. The analysis of the structure refined in superspace indicates that the modulation is essentially caused by a valence fluctuation of the Au atom. The linear coordination of Au^I by Te alternates with a square-planar co-ordination of Au^{III} by Te with a periodicity independent of the three fundamental periodicities giving thus rise to the incommensurate nature of the structure. The valence fluctuation induces some strong Te displacements of the order of 0.5 Å, breaking up Au–Te bonds and creating isolated Te–Te pairs.

The indexation of calaverite on the basis of four independent periodicities is still fully compatible with the Friedel's morphological law which states that the morphological importance of the crystal faces can be expressed in terms of small wave vectors **h**. This law is still satisfied with relatively large values of m as long as $\|\mathbf{h}(hklm)\|$ is small. This method of indexation is obviously more satisfactory than using at least four different lattices and still leaving some of the faces unindexed!

In conclusion, the remarkable fit between the results of the electron diffraction observations, the structural analysis by X-ray diffraction methods, and the morphological observations is a further confirmation of the aperiodic nature of the crystal structure of calaverite. We shall see in the next paragraphs that this does not only apply to calaverite but to other aperiodic crystal structures as well. Actually, calaverite was not the first compound where satellite phases were recognized. Earlier, these faces had been found in Rb_2ZnBr_4 by de Wolff (unpublished) and in Rb_2ZnBr_4 and Rb_2ZnCl_4, as reported in Janner et al. (1980).

7.1.2 The morphology of the TMA Zn phases

TMA Zn (for tetramethylammonium tetrachlorozincate ($C_4H_{12}N_2\ ZnCl_4$)) is an organo-metallic material for which six different phases have been identified between 160 and 300 K (Dam and Janner, 1986). Table 7.2 lists four of them, which occur above 181 K, including one which is incommensurately modulated (phase II). This phase is stable above 279 K and extends over a range of 14°.

The temperature evolution of the crystal morphology has been carefully observed on perfectly spherical samples which were grown in supersaturated solutions between 274 and 298 K thus covering the existence of four different phases. This process yielded well-formed crystal faces which could be indexed from goniometric measurements, cell constants, and space group symmetry Pcmn of phase I.

The temperature variation of the modulation vector $\mathbf{q}(T) = \frac{1}{3} + \delta(T)$ established on the basis of morphological observations is represented in Fig. 7.3. This diagram, which compares well with the data derived from X-ray observations, indicates clearly the incommensurate nature of phase II with a smooth variation of $\mathbf{q}(T)$. We also observe the very short temperature range of existence of the commensurately modulated phase III with $\mathbf{q} = \frac{2}{5}$.

In this context we shall limit our considerations to crystal forms that are related to modulation vectors only and not to main reflections. This is justified by the fact that the main faces are not affected by the phase transitions. Table 7.2 lists the crystal forms associated with commensurate and incommensurate modulations. Obviously, their visual aspects using oblique illumination cannot be distinguished from the forms appearing in paraphase I. However, the reflected light observed

Table 7.2 *Crystal forms of* $C_4H_{12}N_2ZnCl_4$ *associated with satellites for different phases according to Tables 1 and 5 from (Dam and Janner, 1986). The left-hand column uniquely characterizes each form.*

T Phases	>181 K IV	> 276.5 K III	> 279 K II	> 293 K I
{hklm}	$q = \frac{1}{3}c^*$	$q = \frac{2}{5}c^*$	$q \approx 0.42c^*$	para
{10 1̄ 2̄}	{101}	{101}	{10 1̄ 2̄}	–
{1002}	{102}	–	–	–
{1101}	{111}	{112}	{1101}	–
{111 1̄}	{112}	{113}	{111 1̄}	–
{0101}	{011}	{012}	{0101}	–
{0102}	{012}	–	–	–

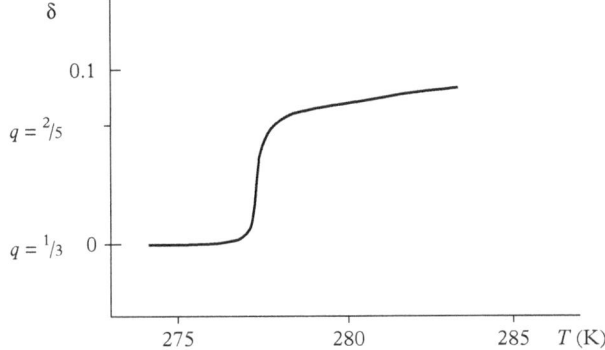

Figure 7.3 *Morphologically derived temperature variation of the modulation vector* $q(T) = \frac{1}{3} + \delta$, *according to Fig. 4 in Dam and Janner (1986).*

at the goniometer is rather weak. The forms associated with the incommensurate modulations are also observed in both of the commensurately modulated phases III and IV with a slightly differing orientation according to the magnitude of the modulation vector given in Fig. 7.3.

Morphological observations of TMA Zn are another elegant way of confirming the modulated nature of the different phases. The position and appearance of the satellite faces are directly related to the modulation wave which can be monitored from the relative position of the satellite faces with respect to the main faces. Moreover, these observations also confirm the systematic absences that are associated with the corresponding superspace group Pcmn(00γ)0s0. This is related to the morphological importance of crystal faces ($hklm$) which are zero if the corresponding reflection is absent.

7.1.3 The morphology of icosahedral and decagonal quasicrystals

The discovery of quasicrystals was based on diffraction observations where a non-crystallographic symmetry pattern could be observed. Here non-crystallographic symmetry must be understood in the classical sense of three-dimensional space where only two-, three-, four-, and six-fold rotation symmetries are compatible with translation symmetry. The most striking feature of quasicrystalline material is the perfect five- or ten-fold rotation symmetry exhibited by diffraction patterns along some specific directions. Shortly after the discovery of quasicrystals, various studies by scanning electron microscopy revealed some unusual crystal forms with the non-crystallographic symmetry $\bar{5}3m$. In particular, perfect single crystals forming rhombic triacontahedra with dimensions of about 100 μm could be observed in an AlCuLi quasicrystal (Dubost et al., 1986). The triacontahedron is a non-regular polyhedron with 30 rhombic-shaped faces, 32 vertices, and 60 edges with icosahedral symmetry. Figure 7.4 is a schematic representation of the triacontahedron superimposed with the pentagonal dodecahedron and the icosahedron in order to reveal the close relationship between these forms. Triacontahedral morphology is not observed in conventional crystals.

After the discovery of quasicrystalline phases and the subsequent search for structure models on the atomic level, researchers have realized the relationship with so-called approximants, that is, periodic structures with very similar chemical compositions. It is known that a cubic crystalline phase with a composition similar to the icosahedral quasicrystalline AlCuLi consists of icosahedral clusters of atoms forming a body-centred cubic lattice. It is thus not surprising to find the coexistence of cubic and quasicrystalline forms as schematically represented in Fig. 7.5. This characteristic model of epitaxial growth represented according to Fig. 10 in Jaszczak (1994) shows the perfect alignment of the quasicrystalline and crystalline forms and thus their close relationship on the atomic level.

Other forms related to icosahedral symmetry are also observed in quasicrystals. AlMnSi alloys exhibit the pentagon dodecahedral form (Robertson et al., 1986) which is closely related to the triacontahedron, as shown in Fig. 7.4. Most interesting is the shape of the so-called perfect quasicrystals observed in i-AlCuFe, i-AlPdMn, or i-ZnMgY. In these cases the dodecahedron is the most prominent polyhedron, although more complex shapes have been observed in i-AlCuFe and i-AlPdMn.

Relating these macroscopic shapes to the atomic structure is a fascinating problem. Soon after the first structural models were published it was shown that dense planes also exist in quasicrystals (de Boissieu et al., 1991). For instance, the atomic structure of the i-AlLiCu phase has dense planes that are perpendicular to the two-fold axes. It is thus logical to take these directions as the ones corresponding

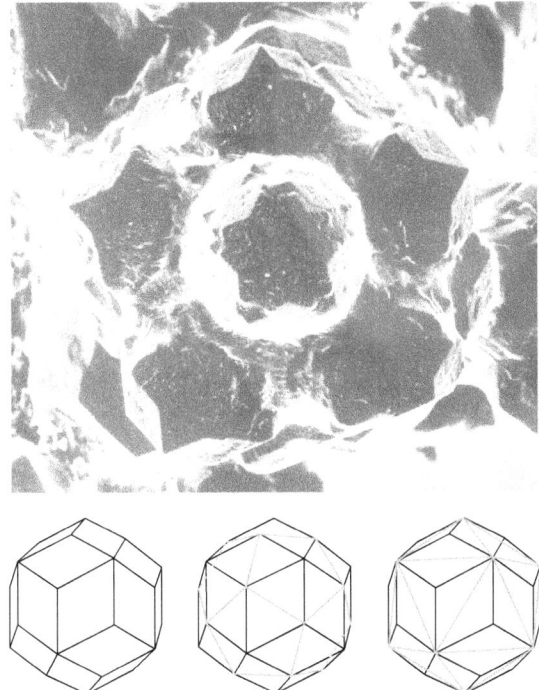

Figure 7.4 *Top: View of the interior of a geode-like cavity, where several single grains in the shape of a triacontahedron are formed in the i-AlCu system (Courtesy of M. Audier) (Dubost et al., 1986). The rhombic faces are clearly visible. Bottom: Relationship between the rhombic triacontahedron observed in quasicrystalline morphology with the pentagonal dodecahedron and the icosahedron. The triacontahedron has symmetry $\bar{5}3m$; it consists of 30 rhombic faces, 32 vertices, and 60 edges and is observed, for example, in AlCuLi quasicrystals (Dubost et al., 1986).*

to stable facets: one obtains a triacontahedron, which is a polyhedron bounded by two-fold facets.

We have already seen this in more detail in the case of the i-AlPdMn phase (Chapter 4, Section 4.5), where the densest planes and the larger spacing have been identified as perpendicular to a five-fold axis (Boudard *et al.*, 1992). Looking back at Figs 4.63 and 4.77 in Chapter 4, one can recognize that, indeed, five-fold planes are the densest ones, in agreement with the geometrical shape of a dodecahedron that is frequently observed and which is bounded by five-fold planes.

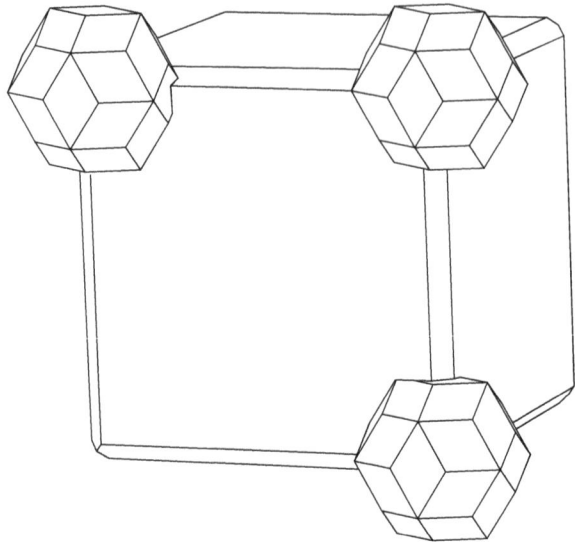

Figure 7.5 *Schematic representation of an epitaxial growth of rhombic triacontahedral forms of an AlCu quasicrystal on a cubic crystalline form represented according to Fig. 10 in Jaszczak (1994).*

The other family of quasicrystals, the decagonal quasicrystals also present some interesting aspects from the morphological point of view (Steurer and Cervellino, 2001). Decagonal quasicrystals exhibit crystal faces perpendicular to the periodic axis and, in addition, a series of faces that are orthogonal to the quasiperiodic planes. The growth morphology of decagonal quasicrystals also presents many facets inclined to the ten-fold axis as represented in Fig. 7.6. These facets correspond to planes relating to periodic and quasiperiodic directions suggesting the existence of dense atomic layers in directions inclined to the decagonal axis. In a study on the decagonal phases $Al_{71}Co_7Ni_{22}$ (Steurer and Cervellino, 2001) the authors use the concept of *periodic average structures of quasicrystals* to determine the correlation between quasiperiodic and periodic directions in decagonal quasicrystals (Steurer and Haibach, 1999). Notice that this concept of average structure is different from that introduced earlier. Here, it is the time and space average of a structure. It appears that all quasiperiodic structures fulfilling some minimal conditions have periodic average structures. Based on the superspace approach, average structures can be obtained by some oblique projections of the hyperstructure describing the quasicrystal. Under favourable conditions, each atom of the quasiperiodic structure can be assigned to one projected atomic surface of the average structure. Based on this concept, the net planes (rather than lattice planes!) related to the densest atomic layers can be easily found by considering the lattice planes of the average structure with the

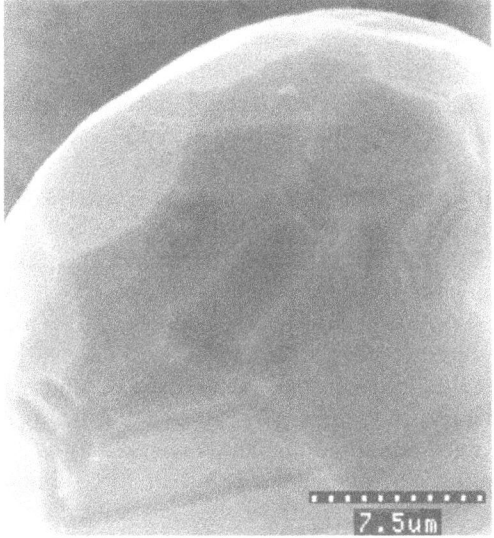

Figure 7.6 *Growth morphology of a decaprismatic needle of decagonal AlCoNi, reproduced with permission from Fig. 1 in Steurer and Cervellino (2001). The small facets correspond to planes relating periodic and quasiperiodic directions.*

lowest indices. Obviously, the atomic layers associated with these planes are not completely flat and exhibit some degree of corrugation. The authors estimate the largest deviations from the plane to be less than 0.15 Å. This result can well be compared to the degree of corrugation found in conventional crystal structures!

Before generalizing the morphological laws observed for periodic crystals, it is interesting to look at what is known of facets at the atomic scale, i.e. the study of the surfaces of aperiodic crystals. Indeed, understanding the stability and morphology of aperiodic surfaces is intimately related to the problem of predicting which morphology a single grain will adopt.

7.2 Surfaces

7.2.1 Introduction

As discussed in Section 7.1, the facets of an aperiodic crystal may be characterized by vectors of the Fourier module, with n indices. Such a vector corresponds to a unique reciprocal lattice vector in n dimensions. A net plane is an $(n-1)$-dimensional hyperplane in the n-dimensional structure perpendicular to a reciprocal lattice vector. This net plane will intersect the physical space along a $(D-1)$-dimensional plane. For simplicity we take in this section $D=3$, unless specified otherwise. Then the intersection is a two-dimensional plane which may be the surface of the crystal.

The surface of a crystal with an aperiodic structure will, generally, reflect the incommensurability of the bulk. But an incommensurate structure may also appear on the surface of a crystal with lattice periodicity. This happens in reconstruction of the surface, or when a thin layer is adsorbed on the surface of a periodic crystal. The latter cases are examples of incommensurate composites, and have, in principle, been discussed in earlier chapters. There are many examples. Let us just mention alkali metal on a Si(111) surface, Cs on a Cu(111) surface, and octadecyl thiol on Ag(111), as well as, incommensurate magnetism in a thin Cr film on Fe(100), epitaxial $Bi_2(Sr,Ca)Cu_2O_x$ on $SrTiO_3(001)$, and many more.

The net planes of the n-dimensional embedding of a quasiperiodic structure may be characterized by an n-dimensional reciprocal lattice vector $\mathbf{H}_s = (\mathbf{H}, \mathbf{H}_I)$. Such a net plane intersects the D-dimensional space along a two-dimensional plane. The plane in V_E is perpendicular to the three-dimensional vector $\mathbf{H}$. Therefore, one may use $\mathbf{H}_s$ to characterize the faces of the quasiperiodic structure. There are, however, two technical problems. In general, there is an infinite number of vectors $\mathbf{H}_s$ with a parallel component perpendicular to the plane, and they may form dense sets. Also in the usual three-dimensional crystallography there is an infinite number of reciprocal lattice vectors perpendicular to a net plane, but they are all multiples of some vector and one chooses the shortest one for the characterization. This is impossible when the set is dense, and, in general, the infinite number of vectors $\mathbf{H}_s$ are, generally, not collinear. The n-dimensional reciprocal lattice vectors characterizing the facets are the generalization of the Miller indices.

Consider the case of a tetragonal lattice with a modulation vector along the $\mathbf{c}^*$-axis: $\mathbf{q} = \gamma \mathbf{c}^*$. In four dimensions all vectors $[00h_3h_4]$ are perpendicular to a net plane that intersects V_E along a plane parallel to the $(a-b)$-plane. The three-dimensional vectors $(h_3 + h_4\gamma)\mathbf{c}^*$ form a dense set on the c-axis. A face perpendicular to these vectors may be characterized by any of the $[00h_3h_4]$ vectors with h_3 and h_4 co-prime (without common divisors, because otherwise one may choose a shorter one in the same direction). However, there is a shortest among the four-dimensional vectors $[00h_3h_4]$, provided a length scale is chosen in V_I. For faces perpendicular to a vector with h_1 or h_2 different from zero, there is always a shortest one which may be used. For example, $[1001]$ and $[h_1h_2h_3h_4]$ are only parallel if $h_2 = h_3 = 0$ and $h_1 = h_4$. The shortest one is $[1001]$.

For isotropic quasicrystals the whole reciprocal space is filled densely by vectors from the Fourier module. Consider an icosahedral quasicrystal, and use the basis from Eq. 2.54. A face perpendicular to the five-fold axis is perpendicular to the Fourier module vector $[100000] = (1, \tau, 0)$. Because of the scaling property the projection of the scaled six-dimensional vector $S[100000] = [211111]$ on V_E has the same orientation as the three-dimensional vector $[100000]$. It has coordinates which are τ^3 times larger than those of $[100000]$. Because τ is irrational, the linear span of the two vectors fill a line densely without a smallest vector. The whole two-dimensional lattice in six-dimensional reciprocal space spanned by [100000] and [211111] consists of vectors with a projection parallel to the three-dimensional [100000]. It is thus quite difficult to choose one Fourier module vector to characterize a five-fold facet as observed in the case of the i-AlPdMn phase, for instance.

Usually, in surface science one uses coordinates where the normal is along the z-axis, which may be achieved by a simple rotation of the axes.

7.2.2 Structure of surfaces of aperiodic crystals

For three-dimensional crystals the most stable facets are those with small Miller indices. This statement has to be modified, because extinctions for Bragg peaks correspond to the absence of the corresponding faces. The most stable faces correspond to the largest distances between the net planes and the highest density in these planes. This is explained by the Hartman–Perdok theory and is related to the energy it costs to break bonds in the so-called bond chains (Hartman and Perdok, 1955). This theory has been generalized to aperiodic crystals in Bennema *et al.* (1991) and van Smaalen (1993, 1999). Then one may understand why satellite surfaces (facets with a satellite and not a main reflection characterizing the wave vector) in modulated phases may become prominent, as in Rb_2ZnBr_4 and calaverite, and why quasicrystals may show icosahedral symmetry.

The stability of facets and their morphology has been much studied in the case of icosahedral quasicrystal. Indeed, the possibility of growing large single grains in the i-AlPdMn phase opened the route of a large number of experiments.

Most of the experiments have been carried out on surfaces prepared under ultra-high vacuum by ion bombardment and subsequent annealing. This is the usual surface preparation employed for crystals. In the case of the i-AlPdMn phase it was shown that the most stable surfaces are the five-fold surfaces, whereas in general two-fold surfaces decompose into five-fold ones. Indeed in the case of the five-fold surface preparation one obtains flat terraces at the atomic scale level. Five-fold low-energy electron diffraction (LEED) patterns clearly demonstrated that the obtained surfaces have a quasicrystalline nature. One has thus here again a sign that at the atomic level, dense five-fold planes are favoured.

If one cuts an icosahedral quasicrystal by a plane perpendicular to an arbitrary vector of the Fourier module, the result will always be aperiodic. For a decagonal phase the same holds for a plane perpendicular to the unique axis. The question remains whether in a realistic situation the structure of a surface will be that of a plane cut through the quasicrystal or that reconstruction and diffusion will lead to a different structure at the surface, known as surface reconstruction.

Several experiments (quantitative LEED analysis, surface X-ray diffraction, etc.) have shown that the surface is 'bulk' terminated and Al rich. This is what is expected in a 'simple' picture of a surface which should be a dense plane, and Al rich because of the differences in vapour pressure of the different elements.

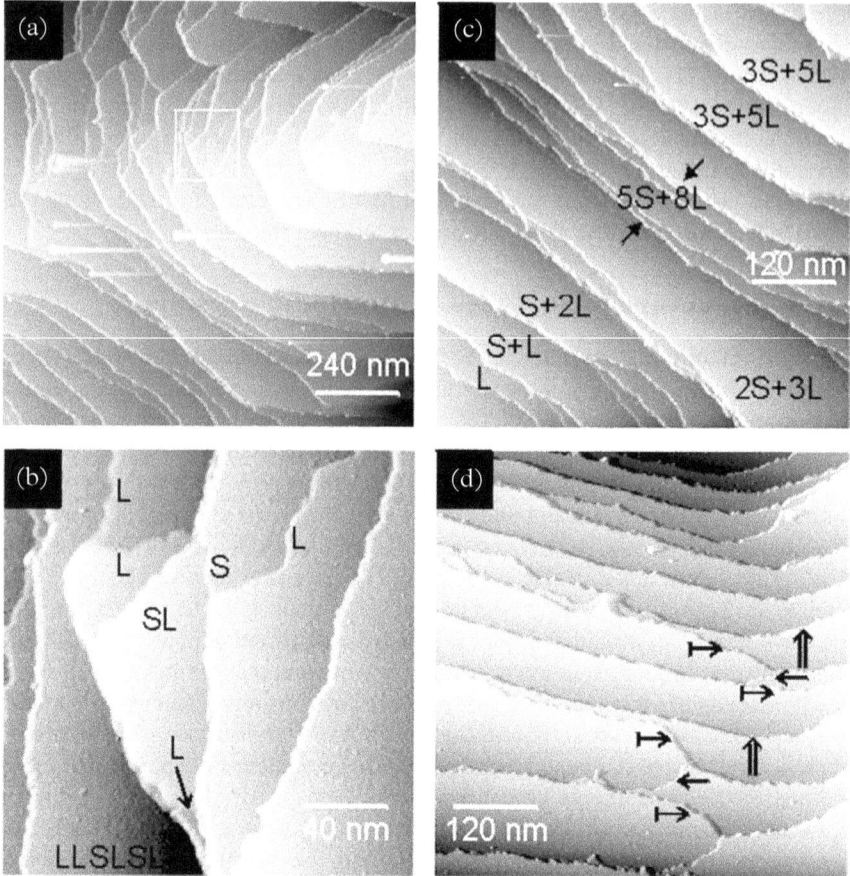

Figure 7.7 *Terraces on the surface of i-AlCuRu. The letters S and L indicate step heights: L is τ times longer than S. (Courtesy Shimoda, Sharma, and Tsai (Shimoda et al., 2005))*

Quantitative comparison with bulk models derived from X-ray diffraction (see Chapter 4, Section 4.5) are in good agreement with these hypotheses. See (Thiel, 2004) and references therein.

Using these well-prepared flat terraces it is even possible to image their atomic structure using STM. In high resolution one may distinguish motifs on the faces of an icosahedral quasicrystal. These consist of filled or empty rings. These have been observed by several groups. See, for example, Fig. 7.8. The images collected agree very well with atomistic models, if one assumes that the surface is just bulk terminated. (Papadopolos et al., 2002; Barbier et al., 2002; Yamamoto, 2004; Shimoda et al., 2005)

The behaviour of steps on a vicinal surface is very interesting. When the surface deviates slightly from a major crystallographic direction, steps appear

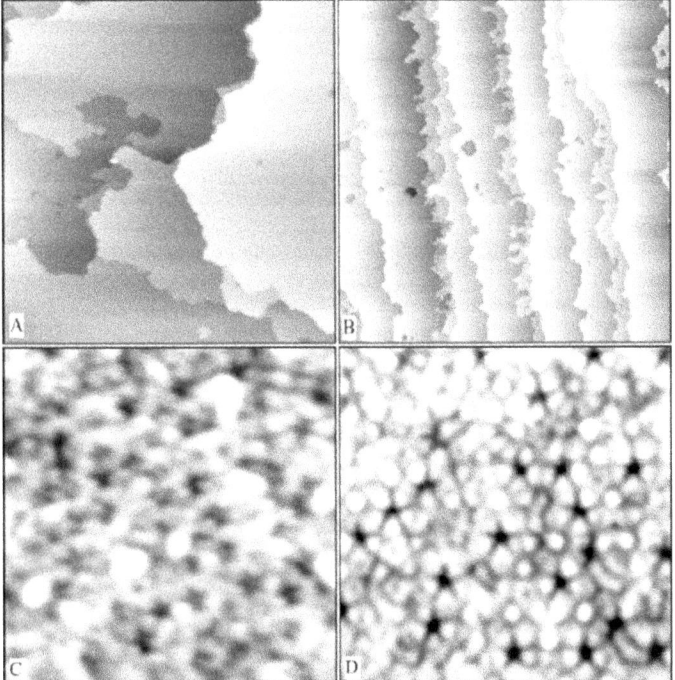

Figure 7.8 *Structure of a surface of i-AlMnPd observed with STM. (From (Papadopolos et al., 2002)) (a) STM image shows flat terraces (150 × 150 nm); (b) 1750 × 1750 nm atomically flat surface; (c) 10 × 10 nm image after a first preparation; (d) the same after a second preparation.*

on the surface. In a periodic crystal the step height is that of the interplanar distance and for simple structures corresponds to an interatomic distance. One may thus wonder what the situation would be in a quasicrystal. For an icosahedral quasicrystalline five-fold surface, these steps have heights which are mutually related by a factor which is a power of τ (see Figs 7.7, 7.9, and 4.77). The smallest step height is equal to 0.21 nm and occurs very rarely. The most frequent heights are 0.37, 0.58, and 0.98 nm.

The observed step heights and their frequencies can be explained by looking at the bulk structural model. We have already seen that there are dense planes and 'gaps' in-between thick layers in the bulk model (Chapter 4, Fig. 4.77). Making the assumption that the most favoured termination is one which is close to a large gap, having a high density and being rich in Al content, allows one to reproduce the distribution and frequencies of the step heights. The bulk model is also in agreement with the images observed by STM.

If one cleaves a quasicrystal in ultra-high vacuum, the surface is generally rough (Ebert *et al.*, 2003), instead of being flat as when obtained by ion bombardment.

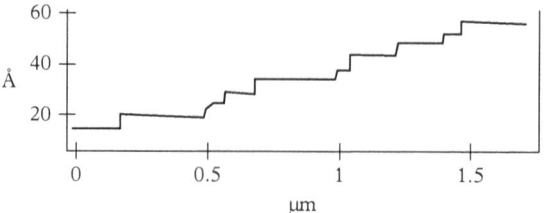

Figure 7.9 *The steps in the height curve have ratios τ^n.*

On the rough surfaces one may distinguish geometrical features which have been interpreted as clusters and sub-clusters. The smaller ones have a size of approximately 1 nm. These would correspond to the (Mackay and Bergman) clusters in the structure. This has been observed on faces of i-AlPdMn, but the interpretation in terms of clusters is still being debated.

On the faces of the decagonal d-AlCoNi with ten-fold symmetry the observed features correspond to the columns formed by the atoms in this phase.

For incommensurate modulated phases, smooth faces have been found in, for example, Rb_2ZnCl_4 and calaverite (see Section 7.1), with optical methods. There are no STM data on these surfaces, as far as we know.

As soon as a quasicrystal surface is not treated in ultra-high vacuum, it is covered rapidly by an oxide layer that changes, of course, the properties.

Satellite surfaces (faces perpendicular to a satellite reflection vector) are not necessarily smooth. Similar to facets on periodic crystals, there may be a roughening transition: above the roughening temperature the facet becomes rough. Such a roughening transition has been observed, for example, in TMA-ZC, that is, $((CH_3)_4N)_2 ZnCl_4$, on a satellite face (Dam, 1985).

7.2.3 Generalization of the morphological laws

We have seen that, both at the macroscopical scale (shape of single grains) and at the atomic level (stable facets and their structure), there are facets, which are stable and correspond to dense planes in the bulk atomic structure, at least for quasicrystals.

For three-dimensional crystals the most stable facets are those with small Miller indices. This is known as the Bravais–Friedel law which assigns a greater morphological importance to a face with larger interplanar distance. If

$$d_{hkl} = 1/\|\mathbf{H}\| \qquad (7.1)$$

then

$$d_{hkl} > d_{h'k'l'} \quad \text{implies} \quad P_{hkl} > P_{h'k'l'}, \qquad (7.2)$$

where P_{hkl} is a measure for the morphological importance of faces $\{hkl\}$. This relation expresses the fact, that in general, most of the crystal forms are characterized by low indices.

Following the discovery of aperiodic crystals and the consequent extension of crystal structures, it should also be possible to generalize the morphological law accordingly. The key concept here is that the number of fundamental periodicities in this case is higher than the classical 3. The Fourier wave vectors of a structure are integral linear combinations of the $3 + d$ fundamental ones. The general form of a Fourier wave vector is expressed by

$$\mathbf{H} = h_1 \mathbf{a}_1^* + h_2 \mathbf{a}_2^* + \cdots + h_{3+d} \mathbf{a}_{3+d}^*. \tag{7.3}$$

Due to the lack of lattice periodicity in aperiodic crystals, the concept of lattice planes has to be reformulated in terms of Fourier wave planes. The face associated with the corresponding Fourier wave has the indices $(h_1, h_2, \ldots, h_{3+d})$. With this formulation, the classical case is included and corresponds to $d = 0$.

Such an approach has been used to generalize the Bravais–Friedel law to aperiodic crystals in Dam and Janner (1986), Janssen and Janner (1987), Lei and Henley (1991), Bennema et al. (1991), van Smaalen (1993), Kremers et al. (1994), and van Smaalen (1999). The most important facets can be found using a generalization of the Hartman–Perdok theory and is based on the energy of broken bonds when a net plane is cut in the crystal (Bennema et al., 1991; van Smaalen, 1993, 1999). The equilibrium shape of a crystal is determined by the free energy of its facets. The free energy of a net plane is the energy needed to split the crystal along the plane into two semi-infinite parts and depends on the energy of the cut bonds. This may be generalized to aperiodic crystals in a superspace description. A net plane in superspace cuts a number of bonds. These bonds are the same as those cut by the array of parallel net planes in the physical space (see Fig. 7.10). Having a model for the bond energies, the equilibrium shape may be calculated. The results are in good agreement with the observed shapes of modulated phases and quasicrystals.

Another approach has been suggested in Lei and Henley (1991). One has to take into account not only the length of the reciprocal vector $\mathbf{H}_s$ but also the amplitude of its Fourier component F_{H_s}. One can restate the Bravais–Friedel law for aperiodic crystals as 'The most important facets are those which have both a small H_s component and a large Fourier amplitude'.

Indeed, saying that F_H is large, is related to the fact that dense planes and large spacings are formed perpendicularly to H, so that one recovers the simple cut bound model. Determining the most stable facets can thus be readily achieved by looking at the intensity distribution in the X-ray diffraction pattern. (Note that this formulation in terms of Fourier components is an equivalent way of determining the 'average lattices' as proposed by Steurer).

Let us come back to two icosahedral examples, namely the i-AlLiCu phase and the i-AlPdMn one.

In the i-AlLiCu phase there are two strong Bragg peaks with a d spacing of about 0.2 nm: one along the two-fold direction and the other along the five-fold direction. Taking a smaller H Bragg peak results in significantly smaller intensity, so that these two reflections are optimal. Comparing now the relative intensity of these two reflections one finds that the two-fold one is more than

460   *Other topics*

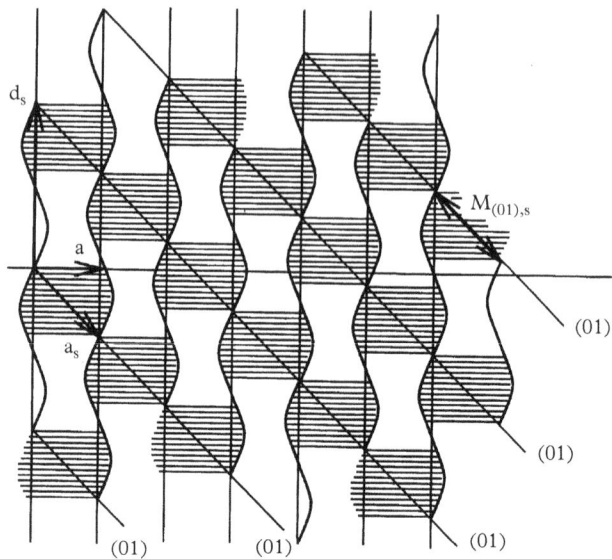

Figure 7.10 *A 1 + 1 dimensional crystal in superspace. The (01) net planes cut a selection of bonds (dense sets of horizontal lines). Generally, not every bond is cut by a given net plane. (Courtesy Marcel Kremers)*

two times stronger than the five-fold one. One can thus predict that the most stable facets are the two-fold ones leading to a triacontahedral shape as observed experimentally.

For the i-AlPdMn phase, there are also two strong reflections in the diffraction pattern at a distance d of about 0.2 nm. However, in this case the situation is reversed as compared to the i-AlLiCu case, the stronger reflection being the five-fold one (about twice as strong), in good agreement with the observation of a dodecahedral shape.

Of course this is only a simple rule, and a deeper understanding requires a careful analysis of the models at the atomic level either qualitatively or rigorously looking into the details for the consequences of the shape of the atomic surfaces (Papadopolos *et al.*, 2004; Shimoda *et al.*, 2005).

7.2.4 Physical properties of quasicrystalline surfaces

We mention here briefly some physical properties of quasicrystalline surfaces. For a more extensive overview of the state of the art see the special volume of 'Progress in Surface Science' on quasicrystals (Vol. 75, August 2004), edited by P.A. Thiel. The electronic properties of quasicrystalline surfaces have been measured on the ten-fold symmetry faces of decagonal phases. These show strongly dispersive

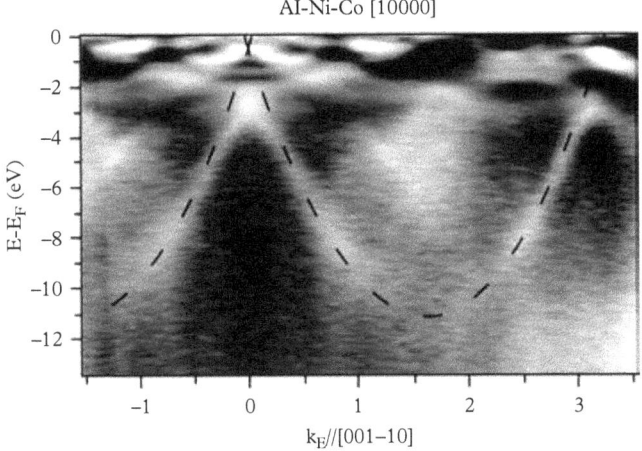

Figure 7.11 *Band map at the [10000] surface of d-AlNiCo. (Theis et al., 2003a)*

valence bands (Theiss et al., 2003a, b). Below the d-bands there are sp-bands with a dispersion resembling free electrons (Fig. 7.11).

Phonons localized at the surface have been found in theoretical models (Zijlstra et al., 1999) and in He scattering (Theis et al., 2003b). The measured surface phonons agree with dispersion curves measured in the bulk.

The physical properties, especially of quasicrystalline surfaces, include hardness, non-wetting, and low friction. Non-wetting means that the adhesion of the water to the surface is weak. A model for this phenomenon is given in Dubois (2004). Low friction has been seen on an atomic level (with atomic force microscopy) and on a macroscopic level, with pin-on-disc measurements. These properties have not yet been explained in detail but are believed to have their origin partly in the electronic density of states, partly in the incommensurability. Low friction as a consequence of incommensurate substructures has been discussed in the section on non-linear excitations and is related to the existence of low-frequency phason modes. A simple model is the Frenkel–Kontorova (Frank–Van der Merwe; see p. 323) model, where a periodic system lies on top of a periodic substrate with an incommensurate lattice constant. For aperiodic crystals the situation is slightly more complicated, because the substrate is already aperiodic. This increases in principle the rank of the Fourier module. The consequences of this are discussed in van Erp et al. (1999) and Vanossi et al. (2000). The role played by the incommensurability is shown by friction experiments on decagonal quasicrystals. Friction, measured with an atomic force microscopy tip, on a face through the ten-fold axis shows a large anisotropy. In the periodic direction (the axis direction) the friction is much higher than in the perpendicular quasiperiodic direction (Park et al., 2004, 2005). Bulk quasicrystals are brittle, but thin coatings combining hardness, low corrosion, non-wetting, and low friction may find their way in applications. Another interesting application lies in the use of an aperiodic substrate as template for other structures. In particular the adhesion of molecules on quasicrystalline surfaces is interesting for possible applications in catalysis. See, for example, Tsai and Yoshimura (2001), McGrath *et al.* (2002), Hoeft *et al.* (2006), and Belin-Ferré *et al.* (2006). For a general overview over applications of quasicrystals related to their surface properties, see Dubois (2005).

7.3 Magnetic quasiperiodic systems

Before the discovery of the aperiodic structure of γ-Na$_2$CO$_3$, incommensurability was known in magnetic structures. An example is thulium (Koehler *et al.*, 1962). It has an hexagonal close-packed structure. In neutron scattering, magnetic satellite reflections show up below 56 K besides the nuclear reflections. These are due to a modulated anti-ferromagnetic spin ordering with commensurate wave vector $2\mathbf{c}^*/7$. Below 38 K higher-order satellites are seen also, indicating a squaring of the modulation function. The periodicity of the magnetic ordering is approximately 7. Actually the wave vector depends on temperature (Brun *et al.*, 1970) and drops from 2/7 at 35 K to 0.27 at the Néel temperature 56 K. Therefore, the structure below 56 K is incommensurate, and there is a lock-in transition at 38 K. In the *incommensurate phase* the spin ordering is incommensurate with respect to the lattice.

The ground state of some metals may be a spin-density wave. Its origin is comparable to a charge-density wave and is a consequence of the electron-electron interaction (Grüner, 1994), the second term in the *Hubbard–Hamiltonian*

$$H = \sum_{k,\sigma} \epsilon_k a^\dagger_{k,\sigma} a_{k,\sigma} + \frac{U}{N} \sum_{k,k',q} a^\dagger_{k,\sigma} a_{k+q,\sigma} a^\dagger_{k',-\sigma} a_{k'-q,-\sigma}.$$

The spin σ in this one-dimensional model is along the applied magnetic field

$$H(x) = \sum_q H_q \exp(2\pi i q x).$$

In this model the susceptibility peaks at $q = 2k_F$, and in a molecular field approximation the spin is given by

$$\langle S(x) \rangle = 2|S| \cos(4\pi k_F x + \phi).$$

In general, the Fermi wave vector is incommensurate with the reciprocal lattice, and the spin-density wave is a magnetic structure which is incommensurate with respect to the underlying structure of nuclei. If there is a spin-lattice interaction the incommensurate spin-density wave will induce an incommensurate displacement wave in the crystal as well.

An example is chromium. Cr is paramagnetic above 112 K with a body-centred cubic structure. Below 112 K there are several spin-density wave phases (Tsunoda et al., 1974; Eagen and Werner, 1975; Pynn et al., 1976). The spin-density wave is incommensurate with the lattice. Therefore, the latter is (slightly) modulated. Its basic structure is Immm between 122 K and 312 K, and I4/mmm below 122 K. The magnetic structure is nearly anti-ferromagnetic: the spin wave vector is close to half a reciprocal lattice vector along the c-direction: $\mathbf{q} = (1 - \epsilon)\mathbf{c}^*$ (notice that the lattice is I centred). The nuclear structure becomes modulated with the wave vector $2\mathbf{q}$. The space groups are, in this case, magnetic space groups. The elements of the four-dimensional superspace group may be combined with the time-reversal operator (indicated by a prime on the symbol) or not. The magnetic space groups for the basic structure are Im'm'm and I4/mm'm', respectively (Janner and Janssen, 1980a). They are the basic groups for the magnetic superspace groups

$$P \begin{array}{cccc} I_p & m' & m' & m \\ b & 1 & 1 & \bar{1} \end{array}, \quad P \begin{array}{ccccc} I_p & 4/ & m & m' & m' \\ b & 1 & \bar{1} & 1 & 1 \end{array}$$

in the two-line notation. In the one-line notation, they are Im'm'm(00γ) and I4/mm'm'(00γ), respectively. For the notation of magnetic superspace groups, see Chapter 2, Section 2.7.3.

In the family of Eu-chalcogenides (Eu-M, M=O, S, Se, Te) it has been observed that the first two are ferromagnetic, the last one is anti-ferromagnetic, and the third compound has several phase transitions (Griessen et al., 1971). At the phase transitions the spin ordering and the lattice constants change, which shows the spin-lattice coupling. The phase diagram has been interpreted as consisting of commensurate phases, but it is more likely that these phases are incommensurate, with a displacive modulation induced by the incommensurate spin structure. The spin structure becomes incommensurate because of the frustration between exchange and dipolar interactions. The phase diagram can be satisfactorily explained with a generalization of the Elliott model (Janssen, 1972). Later, incommensurate phases were studied with the so-called ANNNI model Selke (1988) (see also p. 416).

Complex phase diagrams with incommensurate magnetic structures occur in compounds rare earth elements. Examples are DyGe$_3$ and TbGe$_3$ (Schobinger-Papamantellos et al., 1992, 1995). In these materials there is, in a certain temperature range, a spin-density wave. For TbGe$_3$ the basic structure has symmetry Cmcm. There are two spin waves as can be seen from the diffraction pattern with wave vectors

$$\mathbf{H} = h\mathbf{a}^* + k\mathbf{b}^* + \ell\mathbf{c}^* + m_1 \mathbf{q}_1 + m_2 \mathbf{q}_2,$$

with $\mathbf{q}_1 = (\alpha, 0, 0)$ and $\mathbf{q}_2 = (\beta, 0, \gamma)$, where α, β, and γ are temperature dependent. The spin wave is given as $\mathbf{S}_j(\mathbf{n})$ for the spin of atom j in unit cell $\mathbf{n}$ by

$$\mathbf{S}_j(\mathbf{n}) = \sum_{\mathbf{q}} \mathbf{S}(\mathbf{q}) \exp(2\pi i \mathbf{q}.\mathbf{n}),$$

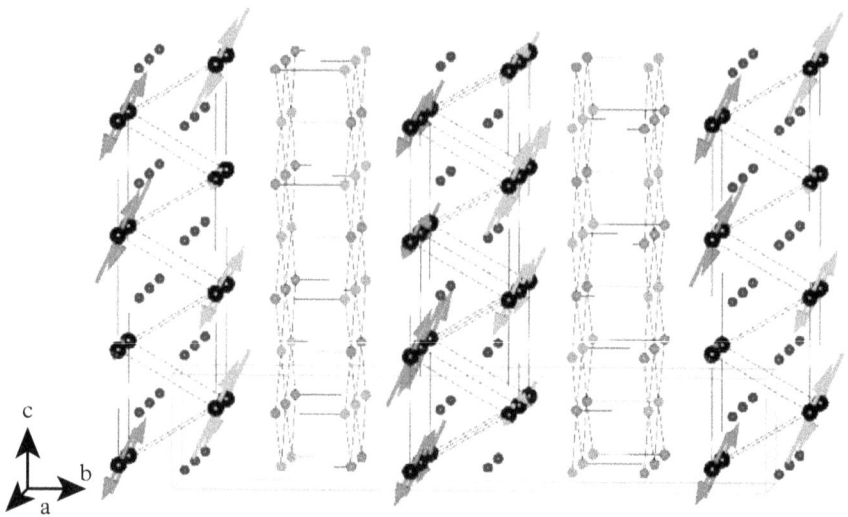

Figure 7.12 *Incommensurate magnetic structure of $HoGe_3$ (Schobinger-Papamantellos et al., 2008). The magnetic moments of Ho form a collinear incommensurate structure with basic vectors $(\frac{1}{2}\beta_1 0)$ and $(\frac{1}{2}\beta_2 \frac{1}{3})$.*

with i-th component of the wave

$$S_j(\mathbf{n})_i = A_{ji} \cos\left(2\pi \mathbf{q}_1.\mathbf{n} + \phi_{ji}^{(1)}\right) + B_{ji} \cos\left(2\pi \mathbf{q}_2.\mathbf{n} + \phi_{ji}^{(2)}\right).$$

Other *rare earth compounds* with such complex magnetic phase diagrams occur in the family of pseudo-binary intermetallics $RGe_{1-x}Si_x$, with R= Ce, Tb, Ho, or Er, as has been found by neutron powder diffraction (Schobinger-Papamantellos et al., 1994, 2008) (see Fig. 7.12). The reason for these incommensurate magnetic structures is probably the competition between the long-range oscillatory exchange interaction between the localized 4f moments. In these materials the quasiperiodic structure stems from the frustration in the magnetic interactions. Due to magneto-elasticity this may also lead to a quasiperiodic structure of the nuclei.

If the nuclei form a quasiperiodic structure, the presence of magnetic moments may lead to a *magnetic quasiperiodic structure* as well. Magnetism due to d-states may be found in compounds with transition metal elements. If the local magnetic moments are non-zero in a quasiperiodic structure, in principle a quasiperiodic magnetic structure may be found as well. Candidates include Mn (e.g. in AlPdMn), Fe (as in AlCuFe), and Co (as in the decagonal AlCoNi). For the occurrence of magnetism in a quasiperiodic structure two conditions have to be satisfied. First, enough localized magnetic moments should be present, and the frustration which may have its origin in the many different local environments has to be solved in an ordered fashion. The occurrence of localized magnetic moments in quasicrystals has been discussed in Trambly de Laissardière and Mayou (2002). From susceptibility and NMR measurements it has been shown that there are localized moments, but the fraction of magnetic atoms in a quasicrystal is much lower than in the corresponding liquid. However, the Curie–Weiss law, which the susceptibility follows in a number of cases,

$$\chi \sim 1/(T-\theta)$$

has a negative value of θ, which means that there is no magnetic phase transition. Moreover, the magnetic properties depend strongly on the precise composition. (Trambly de Laissardière and Mayou, 2000; Rau et al., 2003, 2004; Kashimoto et al., 2002; Jagličić et al., 2004; Dolinšek et al., 2001; Sato et al., 2000; Chernikov et al., 1993). Like other properties, the magnetic susceptibility of decagonal quasicrystals is anisotropic, in contrast to that in icosahedral systems. There are indications of a spin-glass transition.

Compared to the nuclear structure of quasiperiodic crystals, the magnetic structure looks more complex, and has more in common with disordered systems.

The symmetry of a magnetic structure may be given in several ways. The spin wave is given by its value at atom j of cell $\mathbf{n}$ as

$$\mathbf{S}_j(\mathbf{n}) = \sum_\mathbf{q} \hat{\mathbf{S}}_j(\mathbf{q}) \exp\left(2\pi i R \mathbf{q}.(\mathbf{n}+\mathbf{r}_j)\right). \quad (7.4)$$

Under a space group element $\{R|\mathbf{v}\}$ it transforms to

$$\mathbf{S}'_j(\mathbf{n}) = \det(R) \sum_\mathbf{q} R\hat{\mathbf{S}}_j(\mathbf{q}) \exp\left(2\pi i R \mathbf{q}.(\mathbf{n}+\mathbf{r}_j-\mathbf{v})\right). \quad (7.5)$$

If for every pair $\mathbf{n}$, $\mathbf{r}_j$ the relation $\mathbf{S}'_j(\mathbf{n}) = \mathbf{S}_{j'}(\mathbf{n}')$ for some pair $\mathbf{n}'$, $\mathbf{r}_{j'}$ the space group element leaves the spin wave invariant. For periodic crystals there was a controversy between advocates of magnetic space and point groups and representations of the non-magnetic groups. Now, we know that these are almost the same. Similarly, one can use representation theory, superspace groups, and magnetic superspace groups for aperiodic magnetic crystals. Take, as an example, an orthorhombic structure with transversal spin wave with wave vector $\mathbf{q} = \gamma \mathbf{c}^*$. If γ is irrational there is no lattice periodicity. Consider a spin wave given by

$$\mathbf{S}(\mathbf{n}) = S\left(\cos(2\pi\gamma n_3), 0, 0\right).$$

(a) This structure transforms with a representation of the space group Pmmm. The representation belongs to the wave vector $\mathbf{q}$ from the star ($\pm\mathbf{q}$). The group of the vector $\mathbf{q}$ has point group mm2. Because the spin configuration is invariant under m_x and changes sign under m_y the character of the representation for the elements $e, m_x, m_y, 2_z$ is $(1, 1, -1, -1)$. This means that the spin configuration transforms according to an irreducible representation of Pmmm with star $\pm \mathbf{q}$ and the irreducible representation Γ_2 (which has the given character).

(b) Because the translations in the c-direction get an irrational phase factor, there is no periodicity in this direction. However, one can restore periodicity by embedding the function $\mathbf{S}$ in four dimensions: $\mathbf{S}(\mathbf{n}, t) = S_0 (\cos(2\pi(\gamma n_3 + t), 0, 0)$. This spin function in four dimensions is invariant under the translations $(\mathbf{a}, 0)$, $(\mathbf{b}, 0)$, $(\mathbf{c}, -\gamma)$, and $(\mathbf{0}, 1)$. Moreover, it is invariant under the four-dimensional orthogonal transformations $(m_x, 1)$, $(m_y, 1)$ and $(m_z, -1)$, provided the latter two are combined with a phase shift of 1/2 in the internal space. Hence, the superspace group is Pmmm$(00\gamma)0ss$.

(c) A third possibility is the use of a magnetic space group. Because of the aperiodicity this should be a four-dimensional superspace group, and certain elements are combined with the time reversal operator T, indicated by a prime on the elements which are combinations with T. In this case it is the group with symbol Pmm$'$m$'(00\gamma)$.

For this example, the three descriptions are equivalent. Differences may appear if there are, for example, two-dimensional representations, as is the case in canted fan-shaped magnetic structures. The description in terms of representations of space groups breaks down if there is not a lattice periodic basic structure. This is the case for incommensurate composites and quasicrystals. This is not the place to treat these technicalities in detail.

A more detailed discussion of the use of magnetic superpace groups is given in Perez-Mato et al. (2012).

7.4 Incommensurate multiferroics

Multiferroics are materials with at least two coupled order parameters, e.g. a ferroelectric and a (anti-)ferromagnetic or ferroelastic one. An early example was $BiFeO_3$ Kubel and Schmid (1990); Schmid (1994). In this compound, there is a ferroelectric phase transition and, at lower temperatures, a ferromagnetic phase transition. Later (Khomskii, 2006), this was called a type I multiferroic. Type II multiferroics have two phase transitions at (approximately) the same temperature. Examples of this type are $TbMnO_3$ and $TbMn_2O_5$. The coupling in this family, and also in comparable two-dimensional systems, is much larger, which is the reason for considering them as materials with possible applications: by using an external (magnetic) field, one may switch to the other (ferroelectric) order parameter.

This coupling between the order parameters may also be present in an aperiodic crystal. We consider here the coupling between magnetic and displacive order parameters. In this case the magnetic moments and displacements are given by

$$\mathbf{M}(\mathbf{n}, j) = M_o \mathbf{S}(\mathbf{n} + \mathbf{r}_j), \quad \mathbf{U}(\mathbf{n}, j) = u_o \mathbf{U}(\mathbf{n} + \mathbf{r}_j), \quad (7.6)$$

where $\mathbf{S}$ and $\mathbf{U}$ are quasiperiodic functions. Just as for position and magnetic modulated structures these can be embedded in superspace as lattice periodic functions:

$$\mathbf{M}(\mathbf{r}, \mathbf{r}_I) \text{ and } \mathbf{U}(\mathbf{r}, \mathbf{r}_I). \quad (7.7)$$

Materials that may be described in this way are RMn_2O_5 and related structures.

The lattice periodic structure in superspace is invariant under a magnetic superspace group with elements g and $g\theta$, where θ is the time-reversal operator. Under these elements the functions $\mathbf{M}$ and $\mathbf{U}$ transform as follows:

$$\begin{aligned} T_g \mathbf{M}(\mathbf{r}, \mathbf{r}_I) &= \text{Det}(R) R \mathbf{M}(g^{-1}(\mathbf{r}, \mathbf{r}_I)) \\ T_g \theta \mathbf{M}(\mathbf{r}, \mathbf{r}_I) &= -\text{Det}(R) R \mathbf{M}(g^{-1}(\mathbf{r}, \mathbf{r}_I)) \\ T_g \mathbf{U}(\mathbf{r}, \mathbf{r}_I) &= R \mathbf{U}(g^{-1}(\mathbf{r}, \mathbf{r}_I)) \\ T_g \theta \mathbf{U}(\mathbf{r}, \mathbf{r}_I) &= R \mathbf{U}(g^{-1}(\mathbf{r}, \mathbf{r}_I)), \end{aligned} \quad (7.8)$$

where R is the external (three-dimensional) component of the point group element (R, R_I). The group of all operators g and $g\theta$ that satisfy these equations is the magnetic superspace group of the system.

The coupled order parameters correspond to coupling terms in the Hamiltonian. A simple one-dimensional model is, for example, given by the Hamiltonian

$$H = \sum_n \left(\frac{p_n^2}{2m} + \frac{a}{2} u_n^2 + \frac{1}{4} u_n^4 + b u_n u_{n-1} + c S_n S_{n-1} + d S_n S_{n-2} \right. \\ \left. + (-1)^n g u_n^2 S_{n+1} S_{n-1} \right). \quad (7.9)$$

If the first- and second-neighbour interaction are in competition, the spin system may get an incommensurate structure, and via the spin-lattice interaction this may lead to an incommensurate modulation (Janssen, 1994). Then for decreasing temperature one may have a magnetic phase transition and afterwards a displacive phase transition towards incommensurate structures. At high temperatures, the structure is non-magnetic and not modulated. It has the one-dimensional space group $p\bar{1}1'$. Below the first transition, the spin structure is incommensurate with a magnetic superspace group $p_d\bar{1}(\alpha)$ and wave vector $\mathbf{q}$. Below the second transition the chain is displacively modulated with a wave vector $2\mathbf{q}$. (Fig. 7.14) The three symmetry groups are $p\bar{1}1'$, $p_d\bar{1}(\alpha)$, and $p\bar{1}'(\alpha)$, respectively. An element of the second group is the anti-translation $(0, \frac{1}{2})'$. This is indicated by the subscript d in the symbol for the magnetic superspace group.

In a phenomenological formulation, the free energy may be written to third order as

$$F(E,H) = F_0 + \mathbf{P}.\mathbf{E} + \mathbf{M}.\mathbf{H} + \sum_{ij} \epsilon_{ij} E_i E_j + \sum_{ij} \mu_{ij} H_i H_j \qquad (7.10)$$

$$+ \sum_{ij} \alpha_{ij} E_i H_j + \sum_{ijk} \beta_{ijk} E_i H_j H_k + \sum_{ijk} \gamma_{ijk} H_i E_j E_k.$$

This expresses the coupling between the (anti-)ferromagnetic and the ferroelectric order parameters. The coupling parameters β and γ are usually quite small, but for the newer materials big enough to expect possible applications.

Such incommensurate structures are found in the family of RMn_2O_5, where R is a rare earth element (Cheong and Mostovoy, 2007; Khomskii, 2009). They show several phase transitions in both the magnetic structure and the nuclear structure, as exemplified in the one-dimensional model above. At the phase transitions the symmetry group, the (magnetic) superspace group, changes. Starting from the high-symmetry phase an order parameter develops that belongs to a representation of the high-symmetry group. The low-symmetry group is the subgroup for which the representation is the identical representation. This is, in the case of a second-order phase transition, a subgroup of the high-symmetry magnetic superspace group. An example is given by the one-dimensional model of Eq. 7.9. In many of the compounds one observes a phase transition to an incommensurate ordering of the R spins, another to a commensurate spin system with simultaneously a ferroelectric phase transition, and then a re-entrance to the incommensurate phase and finally to an ordering of the R spins. In $TbMn_2O_5$ the spins of Mn^{4+} and Mn^{3+} are helicoidally ordered with an incommensurate wave vector, and coupled to the Tb^{3+} ions. These R spins have an incommensurate non-collinear structure. Several experiments have been reported concerning the compound $TbMnO_3$. Here the Mn^{3+} spins order to a helimagnetic structure with an incommensurate wave vector. Below 41 K, there is also ferroelectric order. The high-temperature magnetic superspace group is Pbnm1'. The magnetic superspace groups for the structure at lower temperatures are subgroups $Pbnm(0\beta0)x_1x_2x_3$ (where x_i is 0 or s), with primed elements when they involve the time reversal θ. For a detailed discussion see (Perez-Mato et al., 2012). There the sequence Pbnm1' $\to$ Pbnm1'$(0\beta0)s00s$ $\to$ $Pbn2_11'(0\beta0)s00s$ is derived. (Notice that $1'(0\beta0)s$ is the same operation as $(0, 1)'$ in Fig. 7.14. It is another choice of basis.)

Another example of an incommensurate multiferroic material is $RbFe(MoO_4)_2$ (Kenzelmann et al., 2007). It is a multiferroic structure on a triangular lattice with a displacive modulation with the wave vector $(\frac{1}{3}, \frac{1}{3}, \gamma)$ (see Fig. 7.13). A trilinear

Incommensurate multiferroics 469

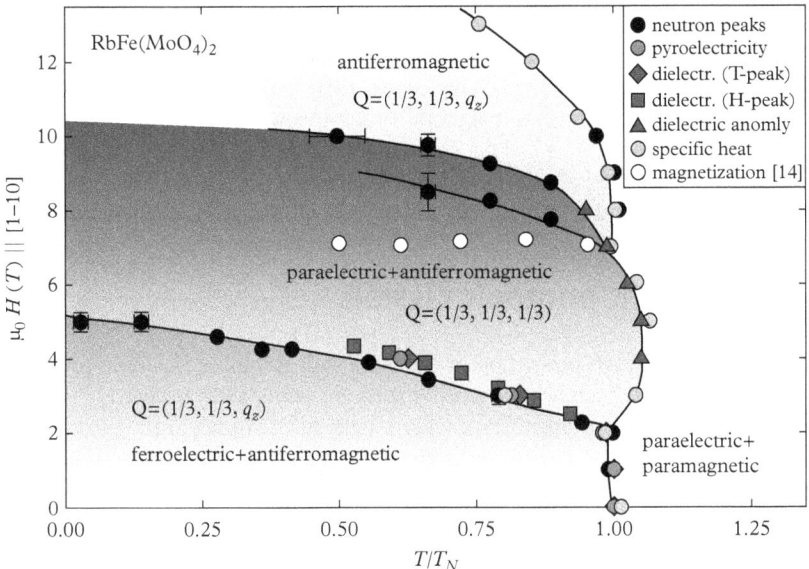

Figure 7.13 *H–T phase diagram of RbFe(MoO$_4$)$_2$. (Kenzelmann et al., 2007)*

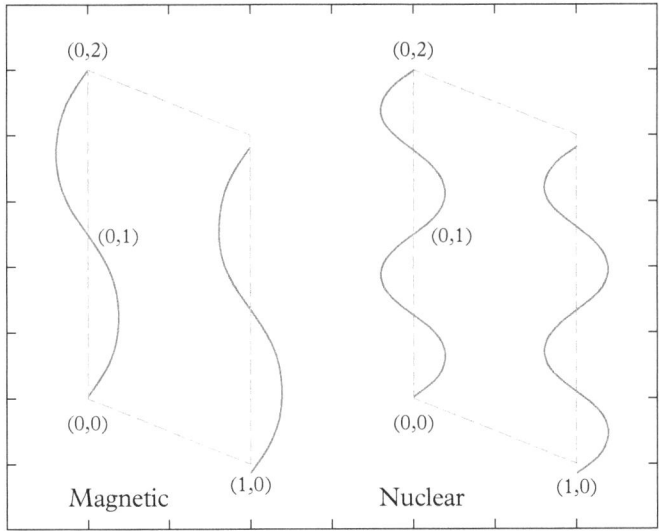

Figure 7.14 *Embedding of a linear chain with incommensurate modulation and spin waves. Both magnetic and nuclear structures are invariant under the magnetic translation $(0, 1)'$.*

phenomenological theory for the coupling between the magnetic and the ferroelectric order explains quite well the measurements. In this case there is simultaneity of the ferroelectric and the anti-ferromagnetic phase transitions at H = 0.

Several other such systems have since then been found. An example is $CaCu_xMn_{7-x}O_{12}$, which has magnetic order below 250 K and an additional nuclear order below 90 K. The superspace group is $R\bar{3}(00\gamma)$ with $\gamma \approx 0.9$ (Sławinski et al., 2009). Another is CuO (Kimura et al., 2008). The spiral spin configuration breaks the inversion symmetry, so allowing ferroelectric order.

7.5 Aperiodic photonic crystals

Nanostructure systems consisting of domains with different dielectric properties are called photonic crystals when the domains are periodically ordered in one, two, or three directions. For light with a wave length in the order of the periodicities, band gaps may occur comparable to those for electrons. The study of photonic crystals has become an important field of investigation, with possible applications. Electrons behave differently in aperiodic crystals than in lattice periodic systems (see Chapter 5, Section 5.6). Therefore, aperiodic photonic crystals may be of interest as well.

On a mesoscopic scale one may construct artificially aperiodic structures. One method is to construct a series of layers of two different types. Choosing the order according to a quasiperiodic series, like the Fibonacci chain, one gets a model system for an aperiodic crystal. Using other sequences, e.g. the Thue–Morse series (see p. 167), one may construct more complicated systems, which do not have to be quasiperiodic. When the layers are dielectrics with different dielectric constants one may get photonic crystals. These have been of interest for their applications. With a suitable thickness of the layers one may create materials with a photonic band gap, such that light with a frequency within this gap cannot propagate (Jin et al., 1999).

Such artificial aperiodic structures may even show non-crystallographic symmetries, like quasicrystals. By means of a set of five laser bundles, one may create a two-dimensional electromagnetic field with five-fold symmetry (Jurdik et al., 2004). Aperiodic photonic crystals with non-crystallographic symmetry, called photonic quasicrystals, can be constructed (Jin et al., 2000). Equivalent structures with a gap in the elastic wave spectrum, and also with non-crystallographic symmetry, are called phononic quasicrystals. Because of the possibility to influence the propagation of light or acoustic waves, these systems are also interesting for possible applications (Nitomi, 2011).

Photonic quasicrystals may show larger photonic band gaps than lattice periodic crystals (Nitomi, 2010). This has been shown for two-dimensional photonic crystals with the structure of a Penrose tiling (DellaVilla et al., 2005). Therefore, such aperiodic photonics could have better possibilities for applications. Moreover, they are isotropic.

7.6 Mesoscopic quasicrystals

Solid-state systems that are close to crystal structures are liquid crystals. Nematic, smectic, and chiral liquid crystals show an increasing positional ordering of the molecules. Molecules in a smectic phase have periodicity in one direction but are disordered in the planes perpendicular to this axis. In smectic-C the molecules in a layer are tilted, and the tilt direction is the same in each layer. In chiral liquid crystals the tilt direction changes periodically along the axis. If this periodicity is irrational with respect to the layer periodicity, the structure is called an incommensurate C⋆-type *liquid crystal*. An example of an incommensurate chiral smectic is MHPOBC (for 4-(1-methylheptyloxycarbonyl)phenyl 4′-octyloxybiphenyl-4-carboxylate). However, because of the positional disorder in the planes these structures are not crystals in the sense of the definition of the International Union of Crystallography.

There are several spontaneously formed systems with non-crystallographic symmetry. The size of the constituent particles may vary considerably, between 1 and 50 nm. When they have a non-crystallographic symmetry (i.e. are incompatible with 3D lattice periodicity) they are called *mesoscopic quasicrystals* or *soft quasicrystals* (for a review, see Dotera (2011)).

A first example is a system of dendrimers (Zeng *et al.*, 2004; Ungar and Zeng, 2005; Zeng, 2005). These are tree-like molecules that are much larger than the atoms of metallic alloys. Therefore, the scale is much larger than that of metallic quasicrystals, and this makes them candidates for applications with light (photonic quasicrystals). On changing the temperature a dodecagonal structure was found: periodic in one direction and with 12-fold symmetry in the plane perpendicular to this axis. This is a quasicrystal on a mesoscopic scale, analogous to the dodecagonal phases in intermetallics (see p. 290 and Fig. 4.93). When the system is heated, there is a phase transition to a lattice periodic structure with space group $Pm\bar{3}m$ or $P4_2/mnm$. The periodicity of the dodecagonal phase (along the 12-fold axis) then remains the same. The stabilizing mechanism of these phases has been presented in Lifshitz and Diamant (2007) and Barkan *et al.* (2011). It considers spherical particles with an interaction potential that is strong for smaller distances, weak up to a maximum distance, and zero beyond; the three-body term interaction is of crucial importance in this approach. An approximant to a dodecagonal tiling was found in a numerical simulation as an Archimedean tiling in Dotera (2006). The same author reported a simulation of a dodecagonal polymeric quasicrystal in Dotera (2012). Such simulations of (two-dimensional) quasicrystals are based either on a multi-component system with simple interactions or on a single-component system with a two-length scale potentials (hard core–soft shell) as in Dotera *et al.* (2014).

Also other structures, with more than one component, were found showing self-organization with a point group symmetry of a type found before in quasicrystals, usually dodecagonal. Such a system is one with polymers (Hayashida *et al.*, 2007).

It is a three-component polymer system with polyisoprene, polystyrene, and poly(2-vinylpyridine).

To another type of mesoscopic quasicrystalline structures belong systems with a mixture of different nanoparticles (Talapin *et al.*, 2009). An example is a mixture of 5 nm-sized gold particles and 13.4 nm-sized Fe_2O_3 particles.

Also in a soft quasicrystal one has discovered an 18-fold symmetry (Fischer *et al.*, 2011). This 2D system has a rank 6, because the Euler function gives $\Phi(18) = 6$. A model system producing such an 18-fold structure was studied in Bekku *et al.* (2017). It reports on Monte Carlo simulations on a system of 2D discs with an interaction with a hard core and a square shoulder potential.

7.7 Defects

As for three-dimensional lattice periodic crystals, aperiodic crystals described with a superspace are generally not perfectly ordered. There may be defects in aperiodic crystals as well. For quasicrystals, this was already discussed in Levine *et al.* (1985) and Socolar *et al.* (1986). The defects may be described using hydrodynamic variables. These are displacements from the ideal positions. For rank 3 crystals, these are just three-dimensional vectors. For aperiodic crystals, there are two types of deviations from the ideal positions. For the first type the displacements are in physical space V_E and are denoted by u_E (or $u_\|$), for the second type they are in internal space and denoted by u_I (or $u_\perp$). A dislocation is characterized by integrating u_E and u_I along a closed path around the dislocation core. The result is analogous to the *Burgers vector* in lattice periodic crystals, and is also called Burgers vector. It is a vector of the lattice in superspace and has a number of dimensions equal to the dimension of superspace, i.e. the rank of the Fourier module. Together, the components form the n-dimensional Burgers vector $b_s = (b_E, b_I)$. The term b_I describes displacements in V_I, similar to phason displacements. Therefore, b_I is called the phason contribution, and then b_E is called the phonon contribution to the Burgers vector. This is the same terminology as used for the tensorial properties in Chapter 5, Section 5.2. Dislocations at quasicrystal surfaces may be seen from HREM pictures, where some lines of points have an endpoint.

The determination of the Burgers vector follows the analogue of a Volterra process. For lattice periodic crystals, the dimension of the lattice is 3. One takes out a two-dimensional half-plane that ends on the one-dimensional dislocation core. Integrating the displacement along a loop around the core one obtains the Burgers vector. If it is parallel to the core, one has a screw dislocation, if it is perpendicular to the core, it is an edge dislocation. In the n-dimensional lattice in superspace one removes an $(n-1)$-dimensional hyperplane, ending in an $(n-2)$-dimensional core. The Burgers vector is the result of an integration of the displacements $u_E(\mathbf{r})$ and $u_I(\mathbf{r})$ around this kernel. Here $\mathbf{r}$ is in the physical space. If the core does not intersect V_E, the result is zero. Otherwise, it is an n-dimensional vector of the lattice

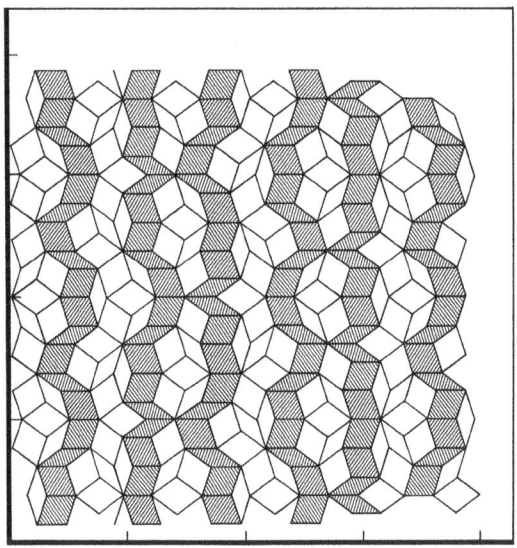

Figure 7.15 *In a Penrose tiling one may distinguish 'worms', series of rhombi with parallel edges.*

in superspace. The analogon to a screw dislocation is found by shifting particles on one side of the removed half-hyperplane over a lattice constant in this hyperplane.

A simple example is given by a Penrose tilling with fat and skinny rhombi. The tiling shows '*worms*' (Fig. 7.15). Cutting the worm at what can be called the dislocation core, and gluing the opposite edges together gives an example of a defect. For example, for the Penrose tiling and a scar perpendicular to the x-axis, the Burgers vector is the lattice vector e_{s1}, the first basis vector in superspace, which has components in both V_E and V_I.

Deviations from strict quasiperiodicity occur also in modulated structures. The discommensurations are usually not perpendicular to the modulation wave vector, but show meandering. See, for example, Fig. 6.8 in Chapter 6. Nevertheless. the diffraction pattern shows sharp Bragg peaks.

An extensive discussion of defects in quasicrystals and their determination from high-resolution electron images is given in Feuerbacher (2012).

Appendix A
Space groups in arbitrary dimensions

A.1 Crystallographic operations in *n* dimensions

Physical systems are zero-, one-, two-, or three-dimensional (a point, a line, a surface, bulk). Furthermore, as we have seen, aperiodic crystals of finite rank may be embedded as periodic structures in still higher dimensions. This makes it convenient to consider the crystallography from a more general point of view. All positive dimensions have crystallographic properties in common. Of course, in lower dimensions more details and complete classifications are known than in higher dimensions, but the general concepts are to a large extent, the same. Therefore, we want to give a brief overview of crystallography in spaces of arbitrary dimension. An introduction to crystallographic groups and their application to physical properties is Janssen (1973).

Crystallographic operations are elements of the *Euclidean group* in n dimensions, the group of all distance-preserving motions in Euclidean space. All *Euclidean transformations* are pairs of an orthogonal transformation and a translation. What types of orthogonal transformations may one encounter?

A real orthogonal transformation in n dimensions is a linear transformation leaving all distances invariant. It may be decomposed as the direct sum of two-dimensional rotations, and one-dimensional identity (+1) or inversion (−1) operations. Its determinant is ±1. If the determinant of the transformation is +1, the number of inverses is even ($2p$), and these can be viewed as p two-fold rotations in a plane. If the determinant is −1, there is one unpaired inversion. As example, consider the four-fold rotation in three dimensions, and its related roto-inversion $\bar{4}$. They can be written in block form:

$$4 = \begin{pmatrix} \cos(\pi/2) & -\sin(\pi/2) & 0 \\ \sin(\pi/2) & \cos(\pi/2) & 0 \\ 0 & 0 & 1 \end{pmatrix}; \quad \bar{4} = \begin{pmatrix} \cos(3\pi/2) & -\sin(3\pi/2) & 0 \\ \sin(3\pi/2) & \cos(3\pi/2) & 0 \\ 0 & 0 & -1 \end{pmatrix}.$$

The first is the sum of a four-fold rotation and a one-dimensional identity, and the second a four-fold rotation and a one-dimensional inversion. Each N-fold rotation in the plane is a rotation over $2\pi m/N$, where $0 < m < N$, and N and m are co-prime. Such a planar rotation then may be denoted by N^m, and an arbitrary rotation in n dimensions can be denoted by a string $N_1^{m_1} N_2^{m_2} \ldots$, whereas an orthogonal transformation which is not a rotation (if its determinant is −1) has a similar string with a $\bar{1}$ at the end. Two examples of this are as follows: the four-fold rotation in three dimensions becomes the string 41, and the roto-inversion $\bar{4}$ becomes $4^3 \bar{1}$. The inverse of the latter is $4\bar{1}$. The eigenvalues of the two-dimensional rotation N^m are $\exp(\pm 2\pi i m/N)$. Because the rotations are real, an eigenvalue $\exp(2\pi i m/N)$ comes along with an eigenvalue $\exp(-2\pi i m)/N)$. In the notation the values of m can therefore be limited to the range $0 < m \leq N/2$.

Not all these rotations leave a lattice invariant. A theorem from number theory leads to the statement that such an invariant lattice exists if and only if the rotation N is accompanied by its conjugates (the rotations N^m with m co-prime to N), and the set of all two-dimensional rotations may be divided into complete sets of conjugates. The number of conjugates for the eigenvalue $\exp(2\pi i/N)$ is the *Euler function* of N: $\Phi(N)$ is the number of positive integers smaller than N which are co-prime with N. The eigenvalues $\exp(2\pi im/N)$ and $\exp(2\pi i(N-m)/N)$ then combine to give the rotation N^m. Examples of the value of the Euler function are $\Phi(4) = 2$ (1, 3), $\Phi(5) = 4$ (1, 2, 3, 4), $\Phi(6) = 2$ (1, 5), $\Phi(7) = 6$, etc. Therefore, a four-fold rotation is crystallographic (leaves a lattice invariant) in two dimensions, a five-fold rotation in four dimensions, a three- or six-fold rotation in two dimensions, and a seven-fold rotation in six dimensions. That is the reason why one has to go to four dimensions to get a crystallographic embedding of the Penrose tiling.

Because in crystallography the crystallographic rotations are relevant, the set of conjugated rotations is indicated by [N]. If $N = 8$ the symbol [8] stands for the direct sum of the rotations 8 and 8^3. It is a four-dimensional rotation. Crystallographic transformations therefore are indicated by strings of symbols [N], N, $\bar{1}$, and 1, where the numbers in square brackets correspond to four- or higher-dimensional rotations, numbers without brackets to three-, four-, or six-fold rotations in two dimensions ($N = 3, 4,$ or 6), 2's to 2-fold rotations and $\bar{1}$ appears when the transformation has the determinant -1. As an example, [8]62$\bar{1}$ is a transformation with the determinant -1 in nine dimensions: an eight-fold rotation in four dimensions, a six-fold rotation in two dimensions, a two-fold rotation in two dimensions and an inversion in one dimension.

For aperiodic crystals, there is an embedding in an n-dimensional space with an invariant D-dimensional physical space ($D = 1, 2, 3$). The invariant physical space consists of one or more of the invariant subspaces of the orthogonal transformation. The action of the transformation in the physical space may be given separately, and the action in the internal space then is put in parentheses. A symmetry operation for a three-dimensional octagonal aperiodic structure may act as a rotation 8_1 in physical space and, because 8 needs its conjugate 8^3, as 8^3 in internal space. The operator [8]1 in five dimensions is then denoted as $8_1(8^3 1)$ for the aperiodic crystal. Very often, one denotes this transformation by 8_1, because for a crystallographic transformation the component 8^3 follows necessarily. Strictly speaking, 8_1 is a non-crystallographic eight-fold rotation in three dimensions, but the shorter symbol is certainly less heavy. In the context of the crystallography of quasiperiodic crystals, there will be no confusion, and then the shorter notation is more convenient.

A.2   Lattices

A *lattice translation group* in n dimensions is a group of translations spanned by n linearly independent basis vectors. That means that the lattice translations span the whole space. If the translations do not span the whole space, or if the basis vectors are not linearly independent, it is better to use the expression **Z**-*module* or *Bravais module*. In both cases any translation may be written as a sum of basis translations:

$$\mathbf{a} = \sum_{j=1}^{n} n_j \mathbf{a}_j, \quad \text{(integer } n_j\text{)}.$$

If n is larger than the dimension of the space, or if the basis vectors $\mathbf{a}_j$ do not span the whole space, the vectors form a **Z**-module, not a lattice. As a Euclidean transformation a translation is written as $\{E|\mathbf{a}\}$. The lattice is characterized by its *metric tensor*, a symmetric tensor with elements

$$g_{ij} = \mathbf{a}_i.\mathbf{a}_j.$$

This implies only the mutual relations between the basis vectors, but not their absolute orientation. Rotating a lattice does not change its metric tensor.

The dual of a lattice is the *reciprocal lattice* with basis vectors $\mathbf{a}_j^*$ defined by

$$\mathbf{a}_i.\mathbf{a}_j^* = \delta_{ij}.$$

Its metric tensor is g^* with elements

$$g_{ij}^* = \mathbf{a}_i^*.\mathbf{a}_j^* = (g^{\mathrm{T}})_{ij}^{-1}: \quad g^* = (g^{\mathrm{T}})^{-1}.$$

The *holohedry* of a lattice is the group of all orthogonal transformations leaving the lattice invariant. With respect to a lattice basis the (arithmetic) holohedry is a group of integer matrices. It is the group of all integer matrices S that satisfy

$$S = S^{\mathrm{T}} g S, \tag{A.1}$$

as one may verify easily.

Two lattices are equivalent if their arithmetic holohedries are arithmetically equivalent. This means that there are bases for the two lattices such that the holohedries are the same groups of matrices. All mutually equivalent lattices form the equivalence class, called a *Bravais class*. In three dimensions there are fourteen Bravais classes.

A crystallographic point group leaves a lattice invariant, and actually a whole set of lattices. The maximal point group that leaves all the lattices invariant that are invariant under the point group K is called the *system group* of K.

A crystallographic point group K that leaves a lattice with metric tensor g invariant corresponds to a group of integer matrices $D(K)$ after the choice of a lattice basis. Then

$$D(R)^{\mathrm{T}} g D(R) = g, \quad \text{for all } R \in K. \tag{A.2}$$

All tensors g satisfying Eq. A.2 form a set G. A group that includes the arithmetic group $D(K)$ is the group of matrices $D(S)$ satisfying

$$D(S)^{\mathrm{T}} g D(S) = g, \quad \text{for all } S \in G$$

Clearly the elements of $D(K)$ belong to this group, but it may be larger. It is called the *Bravais group* $B(D(K))$ of $D(K)$.

For example, in three dimensions the point group 3 leaves a family of lattices invariant that belong either to the rhombohedral Bravais class or to the hexagonal Bravais class. The holohedries are $\bar{3}$m and 6/mmm, respectively. The first is a subgroup of the latter. So the system group of 3 is the point group $\bar{3}$m. The arithmetic holohedries of the two Bravais classes are $\bar{3}$mR and 6/mmmP, respectively. With respect to a lattice basis the point group 3 is either of the type 3R or of the type 3P. The Bravais group for 3R is the arithmetic holohedry $\bar{3}$mR, that for 3P is 6/mmmP.

A.3 Crystal classes

Point groups are subgroups of the orthogonal group $O(n)$ in n dimensions. On an orthogonal basis they are represented by orthogonal matrices, for which the product with their transpose is the identity matrix. If the point group leaves a lattice invariant, it is called a crystallographic point group. On the basis of an invariant lattice the point group is represented by integer matrices. This implies that the trace of a crystallographic point group transformation is an integer. This is called the *crystallographic condition* For example, in three dimensions the trace of an orthogonal transformation with determinant +1 is $1 + 2\cos(\phi)$, and this is only an integer if $\phi/2\pi$ is 1/2, 1/3, 1/4, 1/6, or 0. This is the reason why there are 'forbidden' symmetries in two dimensions and three dimensions. In higher dimensions the crystallographic condition means that an N-fold rotation in n dimensions is non-crystallographic if n is smaller than the Euler function $\Phi(N)$.

A point group may leave a subspace invariant. Because it is a group of orthogonal transformations, the complement of the invariant subspace is also an invariant subspace. For example, the tetragonal group 4/mmm leaves both the unique axis and the plane perpendicular to it invariant. This means that there is a basis with respect to which the matrices of the point group are the direct sum of matrices corresponding to the action in the subspace and its complement. In general, these matrices are real and non-integer. That means there is, in general, not a basis for an invariant lattice with vectors only in the two invariant subspaces. If there is an invariant subspace, the point group is said to be *R-reducible*; otherwise, it is R-irreducible. Here R stands for the real numbers. If there is an invariant lattice having a basis with vectors in the invariant space and its complement, the point group is *Z-reducible*; otherwise, it is Z-irreducible (Z is the group of integers). If there is a sublattice with vectors in the two subspaces, the point group is *Q-reducible* (Q for the rational numbers). This means that there is a sublattice for which the invariant lattice is a centring.

If the point group is R-reducible with an n_1-dimensional and an n_2-dimensional invariant subspace, the n-dimensional matrices are the direct sum of an n_1-dimensional and an n_2-dimensional matrix. The former form a point group K_1 in n_1 dimensions, and the latter a point group K_2 in n_2 dimensions. If K_1 and K_1 are both subgroups of the point group K, then one says that K is the external product of the two subgroups. This is denoted as $K = K_1 \perp K_2$. For example, the tetragonal group 4/mmm has two complementary invariant subspaces. In the one-dimensional subspace along the unique axis it acts as K_2 = m; in the two-dimensional perpendicular subspace it acts as K_1 = 4mm. Then 4/mmm = 4mm $\perp$ m. If K_1 and K_2 are not subgroups, the group K is a subdirect product of the two groups. Then the group K consists of pairs (R_1, R_2), the R_1s forming K_1, and the R_2s forming K_2. An example is the group 422. The unique axis and the plane perpendicular to it are, also here, invariant subspaces. The elements of 422 act as 4mm in the two-dimensional plane, and as the group m in the complementary one-dimensional subspace. The generators now are pairs 4_z = (4, 1) and 2_y = (m_x, m_z); 422 is the subdirect product of 4mm and m.

Two point groups are considered to be the same if there is an orthogonal transformation transforming one into the other: $K' = SKS^{-1}$ for some orthogonal transformation S. Equivalently, one may say that there are two bases such that K with respect to one gives the same matrices as K' with respect to the other. All point groups equivalent to each other according to this definition form an equivalence class, called a *geometric crystal class*. All point groups belonging to one geometric crystal class can be seen as the same group of transformations, but with a different orientation. In three dimensions there are 32 geometric crystal classes of crystallographic point groups.

With respect to a basis of an an invariant lattice, a crystallographic point group corresponds to a group of integer matrices $M(K)$. With respect to another basis, obtained from the former by a basis transformation S, the matrices form the group $SM(K)S^{-1}$. Two such groups of integer matrices are called arithmetically equivalent. The equivalence classes, the sets of all groups of integer matrices arithmetically equivalent to each other, form an *arithmetic crystal class*. Two arithmetically equivalent point groups are also geometrically equivalent, but the inverse is not true. In three dimensions there are 73 arithmetic crystal classes. Each of the 32 geometric crystal classes is composed of one or more complete arithmetic crystal classes.

If two groups of integer matrices $D_1(K)$ and $D_2(K)$ are geometrically equivalent, but not arithmetically equivalent, then there is a matrix S with rational coefficients such that

$$S^{-1}D_1(K)S = D_2(K).$$

The matrix S (or its inverse) describes the transformation of a lattice basis on which the point group K corresponds to the matrix group $D_1(K)$ to the basis of a sublattice. Then in the unit cell of this sublattice there are lattice points of the other lattice. One lattice is a so-called *centring* of the other.

For example, in two dimensions the point group mm leaves two types of lattices invariant, a primitive rectangular lattice with as basis vectors $(a, 0)$ and $(0, b)$, and a centred lattice with basis vectors $(a/2, b/2)$ and $(-a/2, b/2)$. The point groups $D_1(mm)$ and $D_2(mm)$ are related by a rational matrix S:

$$D_1(m_x) = \begin{pmatrix} -1 & 0 \\ 0 & 1 \end{pmatrix} = S^{-1}\begin{pmatrix} 0 & 1 \\ 1 & 0 \end{pmatrix} S = S^{-1}D_2(m_x)S, \quad S = \begin{pmatrix} \frac{1}{2} & \frac{1}{2} \\ -\frac{1}{2} & \frac{1}{2} \end{pmatrix}.$$

In the notation of the *International Tables for Crystallography* geometrically equivalent but arithmetically non-equivalent point groups are distinguished by a symbol that indicates the centring of the lattice. For example, the geometric crystal class 432 contains the three arithmetic crystal classes 432P, 432I, and 432F, corresponding to the three types of centring of the cubic lattice. The symbol for an arithmetic crystal class is the symbol for the geometric crystal class with a postfix indicating the centring.

A.4 Space groups

Space groups are groups of Euclidean transformations that contain a translation subgroup, the intersection with the translation group, which is generated by n independent lattice translations. The elements are denoted, in the *Seitz notation*, by $\{R|\mathbf{t}\}$, where R is an orthogonal transformation in n dimensions. The product of two elements is given by

$$\{R_1|\mathbf{t}_1\}\{R_2|\mathbf{t}_2\} = \{R_1R_2|\mathbf{t}_1 + R_1\mathbf{t}_2\}.$$

The translations are all elements $\{E|\mathbf{a}\}$, with the identity transformation E. They form an invariant subgroup, because

$$\{R|\mathbf{t}\}\{E|\mathbf{a}\}\{R|\mathbf{t}\}^{-1} = \{E|R\mathbf{a}\}.$$

Then each element of the space group can be decomposed into the product of a lattice translation and a representative of the coset to which the element belongs:

$$\{R|\mathbf{t}\} = \{R|\mathbf{t}_R\}\{E|\mathbf{a}\},$$

where the translation $\mathbf{t}_R$ is determined up to a lattice vector. To each element R of the point group one may choose a translation $\mathbf{t}_R$. The elements $\mathbf{t}_R$ form the so-called *system of non-primitive translations* or the *vector system*. The translations satisfy

$$\mathbf{t}_R + R\mathbf{t}_S \equiv \mathbf{t}_{RS} \quad \text{(modulo lattice translations)}.$$

These are called the *Frobenius congruences*.

Euclidean transformations are the product of an orthogonal transformation with a centre O and a translation. Choosing another origin for the orthogonal transformations, the associated translations $\mathbf{t}$ change to $(1-R)\mathbf{t}$. Therefore, the vector system $\mathbf{t}_R$ is determined up to a lattice translation and an origin shift: $\mathbf{t}_R$ and $\mathbf{t}_R + \mathbf{a} + (1-R)\mathbf{v}$ are equivalent for every lattice vector $\mathbf{a}$ and *origin shift* $\mathbf{v}$.

The point groups may be generated by a (finite) number of generators R_j, which satisfy a number of '*defining relations*'. These relations fix the isomorphism class of the group. For example, the point group mm2 has two generators, $R_1 = \mathrm{m}_x$ and $R_2 = \mathrm{m}_y$, which satisfy $R_1^2 = R_2^2 = R_1 R_2 R_1 R_2 = E$. The defining relations are of the form of words (or strings) consisting of generators, which are equal to the identity: $W_j(R_1,\ldots,R_m) = E$. For a space group, with a vector system $\mathbf{t}_R$, the defining relations of the point group imply that

$$W_j(\{R_1|\mathbf{t}_{R_1}\},\ldots,\{R_m|\mathbf{t}_{R_m}\}) = \{E|\mathbf{b}_j\}$$

for lattice translations $\mathbf{b}_j$. Any solution of these equations gives a vector system that determines a space group. For example, for the point group generated by m_x and m_y the vector system $\mathbf{t}_1 = \left(\frac{1}{2},\frac{1}{2}\right), \mathbf{t}_2 = \left(\frac{1}{2},\frac{1}{2}\right)$ satisfies

$$\{\mathrm{m}_x|\mathbf{t}_1\}^2 = \{E|(0,1)\}$$
$$\{\mathrm{m}_y|\mathbf{t}_2\}^2 = \{E|(1,0)\}$$
$$\{\mathrm{m}_x|\mathbf{t}_1\}\{\mathrm{m}_y|\mathbf{t}_2\}\{\mathrm{m}_x|\mathbf{t}_1\}\{\mathrm{m}_y|\mathbf{t}_2\} = \{E|(0,0)\}.$$

Therefore, $\mathbf{t}_R$ is a solution to the equations. The space group here is pgg. For known point group matrices, this is a way to determine all space groups. However, in this way one finds an infinite number of space groups. We shall have to say when space groups can be considered to be the same. This will be discussed in Section A.5.

A space group then is specified by the point group K, and the vector system $\mathbf{t}_R$ ($R \in K$). For a basis of the lattice, the group K corresponds to a group of matrices $D(K)$ with integer coefficients. On the same basis the vectors $\mathbf{t}_R$ can be specified. A non-primitive translation $\mathbf{t}_R$ may be decomposed in the sum of a vector that may be transformed to 0 by an origin shift, and an intrinsic part that cannot be transformed away. The symbol for a symmorphic space group then consists of the symbol for the point group K, and a prefix indicating the centring (together with symbol for K this fixes the arithmetic crystal class). For a non-symmorphic space group a variation on the symbol for the symmorphic space group is used. The variation consists in a change of the symbol for a generator of K to a symbol that indicates the intrinsic part of the associated non-primitive translation. The non-primitive translations in two dimensions are translations along a mirror line, a glide. In three dimensions they are translations in a mirror plane or a translation along the axis of a rotation. The combinations are called glide and screw, respectively.

For example, the arithmetic crystal class 2/mP in three dimensions has a symmorphic space group with symbol P2/m. If the two-fold rotation has a non-primitive translation along the rotation axis, the combined operation is a screw operation. If this is the only non-primitive translation the non-symmorphic space group has symbol P2$_1$/m, indicating that the intrinsic part of the non-primitive translation is half the lattice constant in the direction of the axis. If there is a glide plane, instead of a mirror, with intrinsic part of the non-primitive translation along the a-axis, the symbol of the non-symmorphic space group is P2/a.

Another way to fix a space group type or isomorphism class is by using a set of generators and defining relations. Suppose that the point group K has m generators $(R_1, \ldots, R_m)$ and k defining relations $W_i(R_1, \ldots, R_m) = E$ $(i = 1, \ldots, k)$. Furthermore, the translation group has a basis with n vectors $\mathbf{a}_j$ $(j = 1, \ldots, n)$. Then the defining relations of the space group are

(1) relations between the generating translations

$$\{E|\mathbf{a}_i\}\{E|\mathbf{a}_j\} = \{E|\mathbf{a}_j\}\{E|\mathbf{a}_i\},$$

(2) relations between the elements $\{R_j|\mathbf{t}_j\}$

$$W_i(\{R_1|\mathbf{t}_1\}, \ldots, \{R_m|\mathbf{t}_m\}) = \{E|\mathbf{b}_i\},$$

and (3) relations between these elements and the translations

$$\{R_i|\mathbf{t}_i\}\{E|\mathbf{a}_p\}\{R_i|\mathbf{t}_i\}^{-1} = \{E|R_i\mathbf{a}_p\} = \{E|\sum_{q=1}^{n} D(R_i)_{qp}\mathbf{a}_q\},$$

with $i = 1, \ldots, m$ and $p = 1, \ldots, n$. The translation group is abelian, which is seen from the first relation. And, the action of the point group on a lattice basis is given by the matrices $D(R_i)$. Notice that in these relations only the integer matrices $D(R_i)$ and the lattice translations $\mathbf{b}_i$ occur. These data can all be given by integers, in contrast to the other approach where, generally, non-primitive translations appear. The reason is that the structure of a space group bears some similarity to that of a direct product of two groups. Here we have a combination of the point group K and the translation group A. The group A is an invariant subgroup of the space group G, and the factor group G/A is isomorphic to K. In mathematical terms G is a *group extension* of K by A. And the direct product $A \times K$ is the simplest example of such an extension. However, space groups are not the direct product of the translation group and the point group. Symmorphic space groups are semi-direct products of A and K, but the structure of a non-symmorphic space group is a more general *group extension*.

A.5 Classification

The answer to the question of which space groups should be considered as being the same depends on what one wants to do with it. For use in physical applications one may require that two groups are equivalent if they are connected by a change in coordinate system. In mathematical terms this is called affine conjugated. *Bieberbach* has shown, in 1911, that this requirement is equivalent to the requirement of isomorphism between the two. In two dimensions one finds in this way 17, and in three dimensions 219 different space groups. Equivalent groups belong to one equivalence class. So there are 219 *equivalence classes* in three dimensions. An equivalence class here is also called a space group type. For some physical applications it is useful to have a stronger equivalence relation. Then the coordinate transformation between two equivalent space groups should be orientation preserving. This is stronger, because there is an additional requirement. With this definition of equivalence, there are 230 space group types in three dimensions. In higher dimensions physical properties do not play an immediately obvious role. Therefore, one might use isomorphism as equivalence relation. However, for incommensurate modulated phases, for example, it makes sense to use a stronger equivalence relation. It is obvious that the number of equivalence classes depends on the equivalence relation. A stronger relation leads to a larger number of classes. For incommensurate phases this will be discussed in Sections A.6 and A.8. The role of enantiomorphism in superspace is unclear, as long as no experiments are performed where mirror operations in internal space play a role.

A space group has a lattice of translations and a point group. The operations of the point group with respect to a basis of the translations give a group of $n \times n$ integer matrices, the arithmetic point group. All space groups of one space group type have arithmetic point groups in the same arithmetic crystal class. So, one can get all space group types by taking one representative of each arithmetic crystal class, determining all solutions to the equations above, and eliminating the equivalent ones.

The *translation subgroup* of a space group, which is the intersection of the space group with the translation subgroup, is a group isomorphic with the group $\mathbf{Z}^n$, the abelian group of n-tuples of integers, vectors $(n_1, \ldots, n_n)$. Each n-tuple corresponds to a lattice point. Moreover, the basis vectors span the whole n-dimensional space.

♯ If one wants to be more precise, one should distinguish the translation group with translations $\{E|\mathbf{a}\}$ and the lattice points one obtains by the action of the translation group on an origin. The latter is a set of vectors $\mathbf{a}$ in the n-dimensional space. In fact, the collection of points is the *orbit* of a point (the origin) under the translation group. We shall not make this distinction here. Then one may say that for a space group the translations span the whole n-dimensional space if the coefficients are real. With integer coefficients one obtains the discrete point lattice. ♭

The point group of all orthogonal transformations leaving the lattice invariant is the holohedry of the lattice. It is a finite subgroup of $O(n)$. On the basis of the invariant lattice the holohedry gives a group of integer matrices, the arithmetic holohedry. For another basis the holohedry gives an arithmetic holohedry in the same arithmetic crystal class. All lattices with arithmetic holohedries in the same arithmetic crystal class belong to the same Bravais class. *Bravais classes* are equivalence classes for lattices.

Because a space group determines uniquely an arithmetic crystal class (that of its point group on the basis of the lattice) and a geometric crystal class, one may assign a space group to an arithmetic crystal class and a geometric crystal class. Each geometric crystal class contains complete arithmetic crystal classes, and the latter complete space group types.

If the *metric tensor* of the lattice is g, defined by

$$g_{ij} = \mathbf{a}_i.\mathbf{a}_j,$$

the elements of the arithmetic point group $D(K)$ with respect to the same basis satisfy

$$D(R)^T g D(R) = g, \text{ for all } R \in K.$$

All non-singular integer matrices S satisfying $S^T g S = g$ then form the Bravais group of $D(K)$. This is an arithmetic holohedry. Therefore, a space group may also be assigned to a Bravais class. On the other hand, the subgroup of $O(n)$ that leaves all lattices invariant that are left invariant by K forms the system group of K. All lattices with a holohedry in the same geometric class as this *system group* form a system. So, each space group may be assigned to an arithmetic point group, which may be assigned to a Bravais class. Similarly, each space group may be assigned to a geometrical crystal class and a point group system. Finally, these two lines may be combined in still larger classes, the families. A family is the smallest union of arithmetic crystal and geometric crystal classes such that with each crystallographic point group all point groups belonging to the same system or the same Bravais class are contained in the *family*. Schematically this can be presented as in Fig. A.1.

As an example, consider the hexagonal family in three dimensions. It contains two systems, the hexagonal system with holohedry P6/mmm, and the rhombohedral system with holohedry $\bar{3}$m. The first contains geometric crystal classes 6, 6mm, $\bar{6}$, 622, $\bar{6}$2m, and 6/mmm. The second contains 3, 3m, 32, $\bar{3}$, and $\bar{3}$m. The geometric crystal classes of the first series contain only a single arithmetic crystal class, those of the second series contain each two arithmetic crystal classes, one in the Bravais class 6/mmmP, and one in the rhombohedral Bravais class $\bar{3}$mR. Then each arithmetic crystal class contains one or more space group types. Strictly speaking, space group types are equivalence classes of space groups, (arithmetic and geometric) crystal classes are equivalence classes of point groups, and Bravais classes and systems are classes of lattices. But, because each space

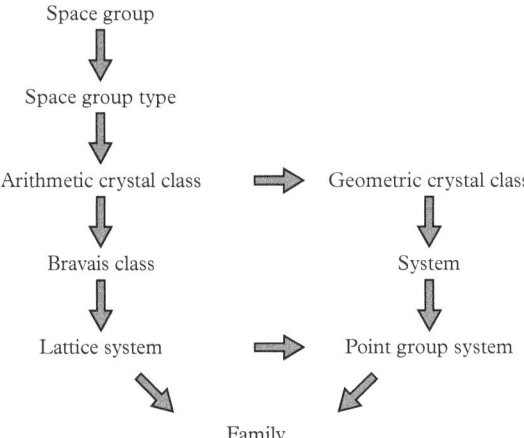

Figure A.1 *Relation between the various crystallographic notions.*

group uniquely determines the other classes, one may consider them also as equivalence classes of space groups.

The number of space group types and that of crystal classes increase rapidly with increasing dimension. For the lower dimensions these numbers are given in the following table (Janssen et al., 1999).

Dimension	1	2	3	4	5	6
Families	1	4	6	23	32	91
Bravais classes	1	5	14	64	189	841
Geometric classes	2	10	32	227	955	7,104
Arithmetic classes	2	13	73	710	6,079	85,311
Space group types	2	17	219	4,783	222,018	28,927,922

♯ Space groups may be calculated starting from representatives of the arithmetic crystal classes. They are what is known as *group extensions* of a point group K with a translation group A. The first is a finite subgroup of $O(n)$ that leaves the n-dimensional lattice invariant. The latter is isomorphic to the group of n-tuples of integers ($\mathbf{Z}^n$) corresponding to the lattice nodes. The representatives of the cosets of A in G are $r(R)$, the elements of A are a. Then an arbitrary element may be written either as a pair (R, a) or as a product $a r(R)$. One introduces an action of K on A by $Ra = r(R) \, a \, r(R)^{-1}$. Finally, the product of $r(R)$ and $r(S)$ is, generally, not $r(RS)$ but it is in the same coset. So, there is a translation vector $\omega(R, S)$ such that $r(R)r(S) = \omega(R, S)r(RS)$. Then the product of two elements of G is

$$a \, r(R) \, b \, r(S) = a \, r(R) \, b \, r(R)^{-1} r(R) r(S) = a \, Rb \, \omega(R, S) \, r(RS).$$

The elements $\omega(R, S)$ are lattice translations, elements of $\mathbf{Z}^n$. So the product may be written as

$$(R, a)(S, b) = (R, a + Rb + \omega(R, S)). \tag{A.3}$$

The pairs (R, a) consist of an orthogonal transformation R and a lattice translation a. In the Seitz notation the elements are $\{R|\mathbf{t}\}$, pairs of R, and a not necessarily primitive translation $\mathbf{t}$. The relation is $\mathbf{t} = \mathbf{t}_R + \mathbf{a}$, with $r(R) = \{R|\mathbf{t}_R\}$. The mapping $\omega(R_1, R_2)$ from pairs of elements of K to a lattice translation is called a two-cocycle in cohomology theory. These satisfy certain relations, and every solution of these equations gives an extension. The extensions correspond to what is called the second cohomology group $H^2_\phi(A, \mathbf{Z}^n)$, where ϕ is the mapping from K to the automorphisms of $\mathbf{Z}^n$ described by an integer matrix group.

The vector system $\mathbf{t}_R$ in the space group elements $\{R|\mathbf{t}_R\}$ forms a one-cocycle, for which we have the relations

$$\mathbf{t}_R + R\mathbf{t}_S = \mathbf{t}_{RS} \bmod \mathbf{Z}^n.$$

The various extensions now correspond to the first cohomology group $H^1_\phi(A, \mathbf{Z}^n)$. But for finite point groups these two groups are isomorphic. Therefore, for space groups in Euclidean space, where the point groups are finite, both approaches are equivalent. Derivation of the list of non-equivalent space groups in n dimensions is based on the calculation of the first or second cohomology group (Burckhardt, 1947; Zassenhaus, 1947; Fast and Janssen, 1971; Brown et al., 1978).

A.6 Space groups for aperiodic crystals

Superspace groups for aperiodic crystals are space groups in n dimensions, but not every n-dimensional space group occurs as such. In superspace there is a physical subspace (of $D = 1$, 2, or 3 dimensions) which is invariant under the point group. Therefore, the point groups of superspace groups are R-reducible into a D- and an $(n - D)$-dimensional component. This eliminates a large number of n-dimensional space groups as candidates for aperiodic crystal symmetries. But the remaining space groups do not all allow an invariant space without periodicity. For example, in $n = 2$ the point group mm2 has two invariant subspaces, but both contain translations from an invariant lattice. However, when one wants to treat commensurate phases on the same footing as the incommensurate, there is a lattice in physical space, and it makes sense to call such an n-dimensional space group a *superspace group* if its point group leaves a D-dimensional subspace invariant. The condition that the point group of a superspace group leaves a subspace invariant restricts the number of space groups which are superspace groups.

The equivalence relation has to be reconsidered. Because of the special role of the physical space two superspace groups are called equivalent if they are conjugated in the n-dimensional affine group by an affine transformation $\{S|\mathbf{a}\}$ such that the physical space is left invariant by S. The additional constraint may be the reason why two groups that are *equivalent as space groups* are non-equivalent as superspace groups. Therefore, on the one hand, the number of space groups that are superspace groups is limited by the condition that the point group leaves a D-dimensional subspace invariant; on the other hand, the stronger equivalence condition (the affine transformation leaves the subspace invariant) increases this number. Therefore, the number of superspace groups is, generally, different than the number of space groups in the same dimension. In addition, similar to three dimensions, a finer classification may be introduced, by limiting the affine transformations to those for which the restriction to the physical subspace has a linear part with positive determinant. Of course, a definition of equivalence is important. Without such a definition a statement about the number of (super)space groups or equivalence classes is meaningless.

Let us consider a number of examples.

1. The hypercubic group in four dimensions is R-irreducible. Therefore, the corresponding symmorphic space group is not a superspace group.

2. The superspace groups P2(1) and $P\bar{1}(\bar{1})$ are isomorphic as four-dimensional space groups, but the connecting affine transformation interchanges the rotation axis in the physical space with the one-dimensional internal space. Therefore, they are non-equivalent as superspace groups.

3. The five primitive icosahedral six-dimensional superspace groups with the generators (besides three- and two-fold elements)

$$(x, y, z, u, v, w) \to (x + p/5, z, u + p/5, v, w, y - p/5) \quad (p = 0, 1, \ldots, 4)$$

fall into two equivalence classes. The first is symmorphic ($p=0$). The groups with $p=1$ and $p=4$, and those with $p=2$ and $p=3$ are equivalent by a mirror, just as the screw operations 4_1 and 4_3 in three dimensions. The groups with $p=1$ and $p=2$ are equivalent because they are related by a centralizer element of the point group, which leaves the physical space invariant. Therefore, they are equivalent as superspace groups, and, a fortiori, also as six-dimensional space groups. This means that there are two non-equivalent superspace groups for this arithmetic crystal class.

As for the general space groups, the number of superspace groups grows rapidly with the dimension, but generally there are fewer superspace groups than space groups, because of the reducibility condition. The number of superspace groups for modulated phases is smaller than (or equal to) the number of general superspace groups of the same dimension, because the point groups for the former should be isomorphic to a D-dimensional crystallographic group, because the physical component of the point group should leave the basis structure invariant. For example, the (2 + 2)-dimensional space group $p8m(8^3m)$ does not occur as symmetry group for a modulated structure, but it is the symmetry group of the Ammann–Beenker tiling. For lower dimensions the numbers are given in the following table. For higher dimensions there are only partial lists.

Dimension	2 + 1 (modulated)	2 + 2 (modulated)	2 + 2 (general)	3 + 1 (modulated)
Bravais classes	4	17	20	24
Geometric crystal classes	5	13	21	30
Superspace group types	22	73	83	777

A.7 Notation

The notation for superspace groups for quasiperiodic crystals is based on the notation for two-dimensional and three-dimensional plane and space groups as used in the *International Tables for Crystallography* and for higher-dimensional groups as recommended in Janssen *et al.* (1999, 2002). The extension to higher dimensions requires new symbols, and the fact that the physical space plays a special role in the space of the superspace group leads to additional information one should give in the symbol. The symbol is a compromise between just a number, such as the number of a space group in the *International Tables for Crystallography*, and full specification. The symbol contains symbols for the geometric crystal class of the space group, the arithmetic crystal class, and the vector system of non-primitive translations.

A.7.1 Superspace groups for incommensurate phases

For incommensurate modulated phases, the fact that here one may distinguish main reflections and satellites in the diffraction pattern is taken into account. In this case one may always choose a basis of the n-dimensional reciprocal space with D basis vectors in the physical space. These are the basis vectors of the reciprocal lattice of the basis structure. Then the basis of the direct lattice has $n-D$ basis vectors in the internal space. With respect to such a basis the point group elements $D(R)$ may be written as

$$D(R) = \begin{pmatrix} D(R)_E & 0 \\ D(R)_M & D(R)_I \end{pmatrix}.$$

The elements $\{R_E|\mathbf{t}_E\}$ form in this case a D-dimensional space group, for which a symbol can be given according to the *International Tables of Crystallography*. The internal components of the point group elements are determined by the action on the basic satellites. For example, if the physical part of the superspace group is the three-dimensional space group Pcmn, and the basic satellite is $(0, 0, \gamma) = \gamma \mathbf{c}^*$, then the point group generators are $(m_x, 1)$, $(m_y, 1)$, and $(m_z, \bar{1})$. Therefore, indicating the physical part of the point group and the basic satellites fixes the higher-dimensional point group. A second, alternative, possibility is indicating the groups K_E and K_I. This is the more general approach, which is also applicable for quasiperiodic structures that are not modulated phases. So, the space group Pmmm$(0, 0, \gamma)$ can also be denoted by Pmmm$(11\bar{1})$. It is a matter of convenience what to choose.

The notation for the non-primitive translations (the vector system) may be differentiated as well. A non-primitive translation consists of an intrinsic (or rational) and a non-intrinsic (or irrational) part:

$$\mathbf{t} = \mathbf{t}_r + \mathbf{t}_i.$$

The latter may be changed, or even eliminated, by another choice of origin, and corresponds to the components of the general solutions which have continuous solutions. For example, in two dimensions, the mirror m_y in a rectangular lattice has as general solutions in the case of mm2P: $\mathbf{t} = (\alpha, \beta)$ with $\beta = 0$ or $\frac{1}{2}$. The intrinsic (or rational) part is here $(0, \beta)$. The intrinsic part has rational components with respect to a lattice basis. The irrational, not intrinsic, part is $(\alpha, 0)$. For modulated phases this means that

$$\mathbf{t}_r = \left(\sum_{i=1}^{D} f_i \mathbf{a}_{Ei}, \sum_{i=1}^{D} f_i \mathbf{a}_{Ii} + \sum_{j=D+1}^{n} f_j \mathbf{a}_{Ij} \right) = (\mathbf{t}_E, \ -\Delta \mathbf{t}_E + \mathbf{t}_I),$$

where $-\Delta \mathbf{t}_E$ is the internal component associated with $\mathbf{t}_E$:

$$\sum_{i=1}^{D} \xi_i \mathbf{a}_i = \sum_{i=1}^{D} \xi_i (\mathbf{a}_{Ei}, \mathbf{a}_{Ii}) = \sum_{i=1}^{D} \xi_i (\mathbf{a}_{Ei}, -\Delta \mathbf{a}_{Ei}).$$

So the intrinsic part of the non-primitive translation has an internal component stemming from the component in physical space and the rational components f_j for $j = D+1, \ldots, n$.

In the case that $n-D = 1$ ('co-dimension one') the fractions f_4 have denominators equal to 1, 2, 3, 4, or 6, because the maximal order for a point group element which is a symmetry for a modulated structure is six, the point group being crystallographic in D dimensions. These values are denoted by 1, s, t, q, and h, respectively. Then the symbol Pcmn$(00\gamma)1ss$ means that the mirror m_x has internal component 1 and non-primitive translation $\left(\frac{1}{2}\mathbf{c}, -\Delta\frac{1}{2}\mathbf{c}\right) = \left(\frac{1}{2}\mathbf{c}, -\gamma/2\right)$. The mirror m_y has also internal component 1 but a non-primitive translation

$\frac{1}{2}\mathbf{a}_4 = \left(0, \frac{1}{2}\right)$, and the mirror m_z has an internal component -1 (because $m_z(00\gamma) = (00 - \gamma)$), and non-primitive translation $\frac{1}{2}(\mathbf{a}_1 + \mathbf{a}_2) + \frac{1}{2}\mathbf{a}_4 = \left(\frac{1}{2}(\mathbf{a} + \mathbf{b}), \frac{1}{2}\right)$. Notice that $\Delta \mathbf{a} = \Delta \mathbf{b} = 0$, because $\Delta \mathbf{a} = \mathbf{q}.\mathbf{a} = (0, 0, \gamma).(1, 0, 0) = 0$. Here $\mathbf{q}$ is the modulation wave vector in the symbol Pcmn$(00\gamma)1ss$. The point group is mmm$(11\bar{1})$.

For higher co-dimension the principle remains the same. The internal component of a non-primitive translation is the sum of the mapping Δ applied to the physical component and the component along the last $n - D$ basis vectors. This means that the same letters $(1, s, t, q,$ and $h)$ may be used to indicate the last $n-D$ components. Now, for each generator of the point group one has to give an m-tuple of the letters ($m = n - D$). As an example, the two-dimensional system of rank four with point group mm2(mm2) and two modulation wave vectors $(\alpha, 0)$ and $(0, \beta)$, has several superspace groups. One of them has symbol pgg$(\alpha, 0; 0\beta)m_d m_c$. It has a point group with two generators

$$A = \begin{pmatrix} -1 & 0 & 0 & 0 \\ 0 & 1 & 0 & 0 \\ 0 & 0 & -1 & 0 \\ 0 & 0 & 0 & 1 \end{pmatrix}, \quad B = \begin{pmatrix} 1 & 0 & 0 & 0 \\ 0 & -1 & 0 & 0 \\ 0 & 0 & 1 & 0 \\ 0 & 0 & 0 & -1 \end{pmatrix},$$

and corresponding superspace group generators $\{(m_x, m_z)|\frac{1}{2}(\mathbf{a}_2 + \mathbf{a}_4)\}$, and $\{(m_y, m_u)|\frac{1}{2}(\mathbf{a}_1 + \mathbf{a}_3)\}$. Here x, y, z, u are the Cartesian coordinates in superspace. The physical space has the coordinates x and y, the internal space z and u. In six dimensions the Cartesian coordinates are usually x, y, z, u, v, and w. The intrinsic non-primitive translations are translations in the irreducible subspaces of dimension 1. For example, in 3D the point group mmm has three such spaces, and one has to indicate to which subspace intrinsic non-primitive translations belong. That leads to symbols like Pcmn, with non-primitive translations in the x-direction and in the xy-plane. For the tetragonal group 4, there is only one invariant subspace (that of the axis) and it is sufficient to indicate the nominator of the intrinsic translation; one can then have symbols 4, 4_1, and 4_2.

The symbol for a superspace group for a modulated phase consists of a symbol for the space group in V_E, the basis reflections, and symbols for the non-primitive translations in internal space, such as P$4_1\left(\frac{1}{2}\frac{1}{2}\gamma\right)q$. Such symbols are called *one-line symbols*. Earlier symbols for the superspace groups for modulated phases used *two-line symbols*. For instance, Pcmn$(00\gamma)0s0$ and P$\frac{\text{Pcmn}}{0\,s\,\bar{1}}$ denote the same group. This convention has been abandoned for typographical reasons. They are explained in the *International Tables for Crystallography*,C.

A.7.2 General space groups

For composites and quasicrystals there is no obvious reciprocal lattice of main reflections. In the case of composites it is still possible to choose the first three reciprocal basis vectors in the physical space, although there is not always an obvious choice. In general, all basis vectors of the n-dimensional reciprocal lattice have internal components. In this case the notation may be based on the property that still holds, namely that there is a distinguished physical subspace in the n-dimensional superspace. This means that the physical space is invariant under the point group. This implies that the point group K may be written as (K_E, K_I) with elements (R_E, R_I). The non-primitive translations may be indicated as subindices to the generator symbols. Because the symbol for a point group element now has two components, the symbols for the non-primitive translations may be attached, as subindex, to the external part.

For example, the four-dimensional group Pcmn$(00\gamma)1ss$ according to this principle gets the symbol Pcm$_d$n$_d$$(11\bar{1})$. One keeps the notation for the physical components. So, now the first generator $(m_x, 1)$ has non-primitive translation $\left(\frac{1}{2}\mathbf{c}, -\gamma/2\right)$, the second generator $(m_y, 1)$ has non-primitive translation $\left(0, \frac{1}{2}\right)$, and the third generator $(m_z, \bar{1})$ has non-primitive translation $\left(\frac{1}{2}(\mathbf{a} + \mathbf{b}), \frac{1}{2}\right)$.

The four-dimensional group pgg$(\alpha, 0; 0, \beta)$m$_d$m$_c$ would get, in this context, the notation pb$_d$a$_c$(mm).

A number of examples for quasicrystals with icosahedral point group are given in Table A.8.

A.8 Equivalence of (super)space groups

The number of space groups and superspace groups is infinite. However, it is useful to divide this set into equivalence classes by considering when they are essentially the same. We discussed this point before in Section A.5. Here we go more in detail. Often, lists of symmetry groups are given without specifying the equivalence relation. For the usual plane and space groups, two groups are called *equivalent*, if they are isomorphic. Bieberbach (Bieberbach, 1910, 1912) showed that this means that there is an affine transformation from one group to the other. With this definition, there are 219 space groups in three dimensions. However, one may require that the linear component of the affine transformation has a positive determinant. One may see the difference by means of optical activity. Including this requirement in the definition leads to 230 non-equivalent space groups in three dimensions.

Superspace groups for modulated structures require a finer classification, because one may distinguish main reflections from satellites. This means that two superspace groups are equivalent if there is an affine transformation mapping one group to the other, and the linear part of the affine transformation should be of the form

$$\Gamma = \begin{pmatrix} \Gamma_E & 0 \\ \Gamma_M & \Gamma_I \end{pmatrix}. \tag{A.4}$$

A finer classification is obtained by the requirement that the determinant of Γ_E be positive. Then, each of the 230 three-dimensional space groups has its own family of superspace groups. For example, the groups P3$_1(00\gamma)$ and P3$_2(00\gamma)$ then are non-equivalent, but P3$(00\gamma)t$ and P3$(00\gamma)\bar{t}$ are equivalent.

For incommensurate composites, the subsystems are, in principle, on the same footing, and symmetry elements can transform particles from one subsystem to particles of another subsystem. Moreover, the choice of the host lattice, which determines the space group of the basic structure, is usually arbitrary (cf. Exercise 2.5). The additional requirement that main reflections are mapped onto main reflections, as formulated in Eq. A.4, no longer holds. For such composites superspace groups are simply equivalent if they are conjugated by an affine transformation. However, sometimes there is a clear host system and guest system. Then it is logical to choose the symmetry of host basic symmetry as the basis for the symbol. This is just another setting, but the group is equivalent to one for another choice of basis.

Anyhow, the same symbols as for modulated phases can be used for composites, but the number of equivalence classes is smaller.

For quasicrystals, there is no basic structure. Consequently, the affine transformation does not take necessarily the form of Eq. A.4. The equivalence relation here is, again, isomorphism. The symbols for the superspace group are the same as for other superspace groups: the symbol for the point group in external space, in parentheses the corresponding symbol for the point group in internal space, and indices for the non-primitive translations.

In principle, the equivalence of superspace groups may be refined, just as one does for 3D space groups. This, however, only makes sense if there is a difference between the structures that has consequences for physical properties. Because left- and right-rotating structures sometimes may be distinguished by optical effects, it makes sense to discriminate structures with symmetry groups that are not related by a transformation via Eq. A.4 with $\text{Det}(\Gamma_E) = +1$. Something similar may happen in internal space. If there is an optical effect due to left- or right-turning modulation, which is quite conceivable, one should make this clear in the notation as well. Then one should have a difference between, for example, the superspace groups $P3(00\gamma)t$ and $P3(00\gamma)\bar{t}$. To our knowledge such an effect has not yet been observed. But one may define equivalence via the matrix of Eq. A.4 with the additional requirement that the determinant of Γ_I also be positive. Then the number of equivalence classes is higher. Notice that this is not a matter of taste, but of experimental meaningfulness.

A.9 Extinction rules

Intrinsic non-primitive translations in elements of a non-symmorphic space group lead to extinction rules. If a vector **H** of the Fourier module is invariant under the homogeneous part R of the superspace group element $\{(R, R_I)|(\mathbf{v}, \mathbf{v}_I)\}$, then the intensity of the diffraction at **H** is zero if $\exp(2\pi i(\mathbf{H}\cdot\mathbf{v} + \mathbf{H}_I\cdot\mathbf{v}_I)) \neq 1$. If the indices of **H** with respect to a basis of the Fourier module are integers $h_1, \ldots, h_n$ and the coordinates of $(\mathbf{v}, \mathbf{v}_I)$ are $\xi_1, \ldots, \xi_n$, then the intensity can only be non-zero if

$$\sum_{i=1}^{n} h_i \xi_i = 0 \mod \mathbf{Z}. \quad (A.5)$$

These are the *reflection conditions*. The *extinction rules* are the complement of this: the intensity is zero if $\sum_i h_i \xi_i \neq 0$ modulo integers. The non-primitive translations $(\mathbf{v}, \mathbf{v}_I)$ can be changed by an origin shift, but, of course, this does not affect the extinction rules. Therefore, one may use the intrinsic part of the non-primitive translation in Eq. A.5.

Extinction rules (or their complement, the reflections conditions) also occur because of centring. If the translations $(\mathbf{v}, \mathbf{v}_I)$ are given with respect to the basis of a conventional cell, a unit cell of a sublattice, there are lattice translations with non-integer, but rational, coordinates. This means that the non-zero intensities are found on positions in reciprocal space which form a superlattice of the conventional lattice (the lattice with the conventional cell as unit cell). Therefore, there are additional extinction rules. For example, in two dimensions the rectangular lattice may be primitive or centred. The primitive cell has basis $(a, 0)$ and $(0, b)$, and the centred cell has basis $(a/2, b/2)$ and $(-a/2, b/2)$. The reciprocal basis for the p-lattice is $(1/a, 0)$ and $(0, 1/b)$. Then the points of the reciprocal c-lattice are $((h_1 - h_2)/a, ((h_1 + h_2)/b)$. With respect to the reciprocal p-lattice the coordinates are $(h_1 - h_2)$ and $(h_1 + h_2)$ which are both even or both odd: their sum is even. Then the reflection condition given in terms of the indices with respect to the p-lattice H_1 and H_2 is $H_1 + H_2$ = even.

A.10 Tables

A.10.1 Introduction

In this section a number of examples are given of point and space groups, and the related character tables for quasiperiodic crystals. Their total number is very large. Therefore, it does not make sense to incorporate them here. As illustration, and as an aid to what has been said in the chapters of this book, we just give examples here. We start with the representations of point groups. Notice that the n-dimensional crystallographic point groups for superspace groups for aperiodic crystals are isomorphic to three-dimensional point groups, in general non-crystallographic. We can analyse the point groups in terms of irreducible representations.

For the two-dimensional group 5, one reads from Table A.1 that the matrices of the point group form a two-dimensional reducible representation with two one-dimensional irreducible complex components, D_2 and D_5. The characters of these two representations are complex conjugates of each other, and their sum is R-irreducible. The character of this representation is $\chi(A^n) = 2\cos(2n\pi/5)$. It is not a crystallographic group in two dimensions, but it becomes crystallographic in four dimensions. Then the four-dimensional point group is the sum of the four non-trivial irreducible representations. For this sum all characters are integers: $\chi(E) = 4$, and $\chi(A^n) = -1$ $(n \neq 0)$.

The two-dimensional group 5m forms the irreducible representation D_3, which is not crystallographic. The group becomes crystallographic in four dimensions, and then has two irreducible components: D_3 and D_4, with integer character. In three dimensions the point groups 5m and 52 are reducible representations (see Table A.2); A is the five-fold rotation, and B the mirror or the two-fold rotation, respectively. The first has components D_3 and D_1, and the second has D_3 and D_2.

In three dimensions there are four geometric crystal classes with point groups isomorphic to the dihedral group of order 20. They are 10mm, 10 22, $\overline{10}$2m, and $\bar{5}$m. Their decomposition into irreducible components (cf. Table A.3) shows the difference:

$$10mm = D_5 + D_1$$
$$10\ 22 = D_5 + D_2$$
$$\overline{10}2m = D_8 + D_4$$
$$\bar{5}m = D_7 + D_3$$

A similar situation occurs for the cyclic groups of order 8 and 12, and for the dihedral groups of order 16 and 24.

The group 10mm is isomorphic to the direct product of the group 5m and the group with two elements. The irreducible representations of such a direct product may simply be obtained from those of the subgroup of index 2 (5m, in this case). The irreducible representations of the direct product K × $\mathbb{Z}_2$ are obtained from the irreps $D_1,\ldots,D_m$ as

$$D_{j\pm}(R,E) = D_j(R), \quad D_{j\pm}(R,A) = \pm D_j(R),$$

for all elements R in K, when the group of order two consists of E and A. The character table of K × $\mathbb{Z}_2$ then looks like

	(R,E)	(R,A)
D_{j+}	$\chi_j(R)$	$\chi_j(R)$
D_{j-}	$\chi_j(R)$	$-\chi_j(R)$

For the group 10mm both notations are given.

The three-dimensional icosahedral group is an irreducible representation of the isomorphism class. It corresponds to representation D_3 (Table A.7). In three dimensions it is a non-crystallographic point group. The corresponding Z-irreducible representation is $D_3 + D_4$. It is a six-dimensional Z-irreducible integer representation. There are other integer representations, for example, in four dimensions, but these are not R-reducible. There is no invariant three-dimensional space that can be identified with the physical space.

In this subsection we have presented examples of character tables for some of the point groups which are relevant for quasicrystals. A full list may be found in the *International Tables for Crystallography, Volume D*.

A.10.2 Tables for irreducible representations of point groups: Point groups of 5-, 8-, 10-, 12-fold, or icosahedral symmetry

Table A.1 *Character table for the point group 5.*

Class:	E	A	A^2	A^3	A^4
Order:	1	5	5	5	5
No.:	1	1	1	1	1
D_1	1	1	1	1	1
D_2	1	$\exp(2\pi i/5)$	$\exp(4\pi i/5)$	$\exp(6\pi i/5)$	$\exp(8\pi i/5)$
D_3	1	$\exp(4\pi i/5)$	$\exp(8\pi i/5)$	$\exp(2\pi i/5)$	$\exp(6\pi i/5)$
D_4	1	$\exp(6\pi i/5)$	$\exp(2\pi i/5)$	$\exp(8\pi i/5)$	$\exp(4\pi i/5)$
D_5	1	$\exp(8\pi i/5)$	$\exp(6\pi i/5)$	$\exp(4\pi i/5)$	$\exp(2\pi i/5)$

Table A.2 *Character table for the point groups 5m and 52.*

Class:	E	A	A^2	B
Order:	1	5	5	2
No.:	1	2	2	5
D_1	1	1	1	1
D_2	1	1	1	−1
D_3	2	Φ	$-\tau$	0
D_4	2	$-\tau$	Φ	0

$\tau = (\sqrt{5}+1)/2, \quad \Phi = (\sqrt{5}-1)/2$

Table A.3 *Character table for the point groups 10mm, 10 22, $\bar{5}m$, and $\overline{10}\,2m$; the representations $d_j(a)$ are the representations of Table A.2 with $a = A^2$.*

Class:		E	A	A^2	A^3	A^4	A^5	B	AB
Order:		1	10	5	10	5	2	2	2
No.:		1	2	2	2	2	1	2	2
D_1	d_{1+}	1	1	1	1	1	1	1	1
D_2	d_{2+}	1	1	1	1	1	1	−1	−1
D_3	d_{1-}	1	−1	1	−1	1	−1	1	−1
D_4	d_{2-}	1	−1	1	−1	1	−1	−1	1
D_5	d_{3+}	2	τ	Φ	$-\Phi$	$-\tau$	−2	0	0
D_6	d_{4+}	2	Φ	$-\tau$	$-\tau$	Φ	2	0	0
D_7	d_{4-}	2	$-\Phi$	$-\tau$	τ	Φ	−2	0	0
D_8	d_{3-}	2	$-\tau$	Φ	Φ	$-\tau$	2	0	0

$\tau = (\sqrt{5}+1)/2, \quad \Phi = (\sqrt{5}-1)/2$

A.10.3 Examples of superspace groups for modulated phases

The full list of all superspace groups in $(3+1)$ dimensions may be found in the *International Tables for Crystallography, Volume C*. Here, we give examples mentioned in the text.

Example 1. The (3 + 1)-dimensional superspace groups Pcmn(00γ) and Pcmn(00γ)0s0 have numbers 62.5 and 62.6 in the *International Tables for Crystallography*, Volume C. Their basic space group is Pcmn, number 62 in the *International Tables for Crystallography*, Volume A, where the standard setting is Pnma. The point group has generators m_x, m_y, and m_z, and because $m_z(00\gamma) = (00-\gamma)$ it has the internal component -1. Therefore, the four-dimensional generators are $(m_x, 1)$, $(m_y, 1)$, and $(m_x, -1)$. The three generators of

Table A.4 *Character table for the point group 8.*

Class:	E	A	A^2	A^3	A^4	A^5	A^6	A^7
Order:	1	8	4	8	2	8	4	8
No.:	1	1	1	1	1	1	1	1
D_1	1	1	1	1	1	1	1	1
D_2	1	α	i	β	-1	$-\alpha$	$-i$	$-\beta$
D_3	1	i	-1	$-i$	1	i	-1	$-i$
D_4	1	β	$-i$	α	-1	$-\beta$	i	$-\alpha$
D_5	1	-1	1	-1	1	-1	1	-1
D_6	1	$-\alpha$	i	$-\beta$	-1	α	$-i$	β
D_7	1	$-i$	-1	i	1	$-i$	-1	i
D_8	1	$-\beta$	$-i$	$-\alpha$	-1	β	i	α

$\alpha = \exp(\pi i/4) = \frac{1}{2}\sqrt{2}(1+i), \beta = \exp(3\pi i/4) = \frac{1}{2}\sqrt{2}(1-i)$

Table A.5 *Character table for the point groups 8mm, 822, and $8\bar{2}m$.*

Class:	E	A	A^2	A^3	A^4	B	AB
Order:	1	8	4	8	2	2	2
No.:	1	2	2	2	1	4	4
D_1	1	1	1	1	1	1	1
D_2	1	1	1	1	1	-1	-1
D_3	1	-1	1	-1	1	1	-1
D_4	1	-1	1	-1	1	-1	1
D_5	2	$\sqrt{2}$	0	$-\sqrt{2}$	-2	0	0
D_6	2	0	-2	0	2	0	0
D_7	2	$-\sqrt{2}$	0	$\sqrt{2}$	-2	0	0

Table A.6 *Character tables for the point groups 23 and 432.*

Class:	E	A	A²	B	Class:	E	B	A²	A	AB
Order:	1	3	3	2	Order:	1	3	2	4	2
No.:	1	4	4	3	No.:	1	8	3	6	6
D_1	1	1	1	1	D_1	1	1	1	1	1
D_2	1	ω	ω^2	1	D_2	1	1	1	-1	-1
D_3	1	ω^2	ω	1	D_3	2	-1	2	0	0
D_4	3	0	0	-1	D_4	3	0	-1	1	-1
$\omega = \exp(2\pi i/3)$					D_5	3	0	-1	-1	1

Table A.7 *Character table for the point group 532.*

Class:	E	A	A²	B	AB
Order:	1	5	5	3	2
No.:	1	12	12	20	15
D_1	1	1	1	1	1
D_2	3	$-\Phi$	τ	0	-1
D_3	3	τ	$-\Phi$	0	-1
D_4	4	-1	-1	1	0
D_5	5	0	0	-1	1

$\tau = (\sqrt{5}+1)/2, \quad \Phi = (\sqrt{5}-1)/2$

the superspace group in four dimensions act as follows on a point with lattice coordinates x, y, z, u:

$$\left(-x+\frac{1}{2}, y, z+\frac{1}{2}, u\right), \quad \left(x, -y+\frac{1}{2}, z, u\right), \quad \left(x+\frac{1}{2}, y+\frac{1}{2}, -z+\frac{1}{2}, -u\right);$$

and $\left(-x+\frac{1}{2}, y, z+\frac{1}{2}, u\right), \quad \left(x, -y+\frac{1}{2}, z, u+\frac{1}{2}\right), \quad \left(x+\frac{1}{2}, y+\frac{1}{2}, -z+\frac{1}{2}, -u\right),$

for the two superspace groups. The (special) reflection conditions are, respectively:

Group 1: $0k\ell m: \ell = 2n$ $hk00: h+k = 2n$
Group 2: $0k\ell m: \ell = 2n$ $h0\ell m: m = 2n$ $hk00: h+k = 2n.$

This is immediately seen if one takes into account that, if $[hk\ell m]$ is invariant under a point transformation with the non-primitive translation (x, y, z, u), then the reflection condition is $hx + ky + \ell z + mu = 0$ (mod integers).

Example 2. The $(3 + 1)$-dimensional superspace group $I4_1/acd(00\gamma)s0s0$ has basic space group $I4_1/acd$. This has three generators for the point group, 4_z, m_x, and m_z. The latter changes the sign of the modulation wave vector $\gamma \mathbf{c}^*$. Therefore, the four-dimensional operation $(m_z, -1)$ belongs to the point group. The three-dimensional group is the group 142 in the *International Tables for Crystallography*, Volume A. The action of the three generators in the four-dimensional superspace group transforms (x, y, z, u) into

$$\left(-y, x+\frac{1}{2}, z+\frac{1}{4}, u+\frac{1}{2}\right), \quad \left(-x, y, z+\frac{1}{2}, u+\frac{1}{2}\right), \quad \left(x+\frac{1}{2}, y, -z+\frac{1}{2}, -u\right).$$

There is a centring condition $h + k + \ell = 2n$. Moreover, one has

$$00\ell m : \ell + 2m = 4n, \quad 0k\ell m : \ell + m = 2n, \quad hk00 : h = 2n, \quad hh\ell m : 2h + \ell = 4n,$$

because of the existence of the operations $(4_z, 1)$, $(m_x, 1)$, $(m_z, -1)$, and $(m_{\bar{x}y}, 1)$, together with their non-primitive translations.

Example 3. The group $Abm2(\frac{1}{2}0\gamma)$ has three-dimensional basic space group $Abm2$, nr. 39 in the *International Tables for Crystallography*, Volume A. The centring A causes centring conditions for the main reflections $(hk\ell 0 : k + \ell = 2n)$. In this case, the modulation wave vector has a rational component $\mathbf{a}^*/2$. This implies that there is a further centring in four dimensions. A general reflection is $(h + m/2)\mathbf{a}^* + k\mathbf{b}^* + (\ell + m\gamma)\mathbf{c}^*$. Introducing a conventional cell such that $\mathbf{a}_c^* = \mathbf{a}^*/2$, $\mathbf{b}_c^* = \mathbf{b}^*$, and $\mathbf{c}_c^* = \mathbf{c}^*$, the new indices are $H = 2h + m$, $K = k$, and $L = \ell$, and the modulation wave vector is written as $[H, K, L, m]$. Because $k + \ell = 2n$, one has the reflection conditions $K + L =$ even, and $H + m =$ even. These are the centring conditions. In addition, there is the reflection condition $0KLm : K = 2n$, because of the existence of a non-primitive translation associated with $(m_x, 1)$.

A.10.4 Superspace groups for quasiperiodic structures with 5-, 8-, 10-, 12-fold, or icosahedral symmetry

The list of superspace groups occurring in quasicrystals found so far, is rather limited. However, the list of possibly relevant space groups for the three-dimensional quasicrystals with point groups with 5-, 8-, 10- or 12-fold rotation symmetries (including the icosahedral systems) is not too long. We give here a condensed form of the information concerning these space groups in five and six dimensions.

Again, the generators of the space group are the n basis translations and a number of elements $\{R|\mathbf{v}\}$. A position $\mathbf{r}$ in the unit cell is given by the coordinates with respect to the lattice basis: the first n of the letters x, y, z, u, v, w. Then the lattice translation $\mathbf{a}_1$ is indicated by the transformed point $x + 1, y, z, u, v, \ldots$, and the element $\{R|\mathbf{v}\}$ by the transformed point $\mathbf{r}'$ with coordinates $x', y', z', \ldots$. For non-symmorphic groups or groups with centred lattices there are extinction rules or reflection conditions.

Superspace groups for the pentagonal, octagonal, decagonal, and dodecagonal quasicrystals are listed in Janssen (1988). The symbols given have a shorter version: because the action of the point group in V_I is usually given by that in V_E, one may suppress the former. This leads to a simpler notation, given in the table in column 'Short'.

Table A.8 *Superspace groups: Icosahedral family.*

Symbol	Short	Rank	Generators	Reflection conditions
P532($5^2$32)	P532	6	(x, z, u, v, w, y)	
			$(w, x, u, v, \bar{z}, \bar{u}, y)$	
P$5_1$32($5^2$32)	P$5_1$32	6	$(x+\frac{1}{5}, z, u+\frac{1}{5}, v, w, y-\frac{1}{5})$	$h_2 = h_3 = h_4 = h_5 = h_6$:
			$(w, x, u, v, \bar{z}, \bar{u}, y)$	$h_1 = 5m$
I532($5^2$32)	I532	6	(x, z, u, v, w, y)	
			$(w, x, u, v, \bar{z}, \bar{u}, y)$	$\sum_i h_i$ = even
I$5_1$32($5^2$32)	I$5_1$32	6	$(x+\frac{1}{5}, z+\frac{1}{5}, u-\frac{1}{5}, v+\frac{2}{5}, w-\frac{1}{5}, y-\frac{1}{5})$	$h_2 = h_3 = h_4 = h_5 = h_6$:
			$(w, x, u, v, \bar{z}, \bar{u}, y)$	$h_1 = 5m$
				$\sum_i h_i$ = even
F532($5^2$32)	F532	6	(x, z, u, v, w, y)	
			$(w, x, u, v, \bar{z}, \bar{u}, y)$	$h_i + h_j$ = even (all i, j)
F532($5^2$32)	F532	6	$(x-\frac{3}{10}, z+\frac{1}{10}, u+\frac{1}{5}, v, w-\frac{1}{5}, y-\frac{1}{5})$	
			$(w, x, u, v, \bar{z}, \bar{u}, y)$	$h_i + h_j$ = even (all i, j)

P$\bar{5}$3m(5^23m)	P$\bar{5}$3m	6	(x,z,u,v,w,y)	
			$(w,x,u,v,\bar{z},\bar{u},y)$	
			$(\bar{x},\bar{y},\bar{z},\bar{u},\bar{v},\bar{w})$	
P$\bar{5}$3m$_{34}$($\bar{5}$3m)	P$\bar{5}$3(cd)	6	$(x,z+\frac{1}{2},u+\frac{1}{2},v+\frac{1}{2},w+\frac{1}{2},y)$	$h_5 = -h_2, h_6 = -h_1$:
			$(w,x,u,v,\bar{z},\bar{u},y)$	$h_3 + h_4 =$ even
			$(\bar{x},\bar{y},\bar{z},\bar{u},\bar{v},\bar{w})$	
I$\bar{5}$3m(5^232)	I$\bar{5}$3m	6	(x,z,u,v,w,y)	
			$(w,x,u,v,\bar{z},\bar{u},y)$	
			$(\bar{x},\bar{y},\bar{z},\bar{u},\bar{v},\bar{w})$	$\sum_i h_i =$ even
F$\bar{5}$3m(5^232)	F$\bar{5}$3m	6	(x,z,u,v,w,y)	
			$(w,x,u,v,\bar{z},\bar{u},y)$	
			$(\bar{x},\bar{y},\bar{z},\bar{u},\bar{v},\bar{w})$	$h_i + h_j =$ even (all i,j)
F$\bar{5}$3a(5^232)	F$\bar{5}$3a	6	$(x,z+\frac{1}{4},u,v,w+\frac{1}{4},y+\frac{1}{2})$	$h_6 = h_1, h_5 = h_2, h_4 = -h_3$:
			$(w,x,u,v,\bar{z},\bar{u},y)$	h_1 even
			$(\bar{x},\bar{y},\bar{z},\bar{u},\bar{v},\bar{w})$	$h_i + h_j =$ even (all i,j)

Appendix B
Exercises: Solutions

Exercise 1.1

The point group 4/mmm has 16 elements. The star of (α, β, γ) has 16 points. The points with third component $-\gamma$ are inverses of the corresponding eight points with the third component γ. Among the latter eight, there are five points independent when only integer combinations are allowed, because four pairs with positive γ give the sum 002γ. Consequently, three among them may be expressed as integer combinations of the other. Therefore, the rank of the Z module generated by the 16 vectors of the star is 5.

Exercise 1.2

The basic structure of the chain has unit cell of length 1, and the wave vector of the modulation is q. Then the unit cell of the 2D unit cell is spanned by (1, -q) and (0, 1). The modulation wave is sinusoidal with amplitude a. Then the structure factor is given by

$$F(H) = \int_0^1 \exp(2\pi i(Ha\sin(2\pi t) + H_I t))\, dt.$$

Here $H = h_1 + h_2 q$ and $H_I = h_2$. Defining

$$T_1 = \int_0^1 \cos(2\pi(Ha\sin(2\pi t) + H_I t))\, dt,$$

$$T_2 = \int_0^1 \sin(2\pi(Ha\sin(2\pi t) + H_I t))\, dt$$

leads to $|F(H)| = \sqrt{T_1^2 + T_2^2}$. Some values for $|F(H)|$ are as follows:

h_1	h_2	$F(H)$	h_1	h_2	$F(H)$
0	0	1	1	0	0.90
0	1	0.	1	1	0.30
0	2	0.	1	2	0.05

Exercise 2.1

The two basis vectors of the Fourier module are $1/a$ and $1/\beta$. For the vector $H = h_1/a + h_2/\beta$ the structure factor is

$$F(H) = \int_0^{1/2} \exp(2\pi i(0.1Ha + H_I t))\, dt + \int_{1/2}^1 \exp(2\pi i(-0.1Ha + H_I t))\, dt,$$

where $H_I = h_2$, and

$$|F(H)| = \left(A^2 + B^2\right)^{1/2}/2\pi b$$

with

$$A = -\sin(2\pi c) + \sin(\pi(2c+b)) - \sin(2\pi(c-b)) + \sin(\pi(2c-b)),$$
$$B = \cos(2\pi c) - \cos(\pi(2c+b)) - \cos(2\pi(c-b)) + \cos(\pi(2c-b)),$$

where $c = 0.1Ha$ and $b = H_I$.

Exercise 2.2

(a) The orbit of a spin + at (x, y) contains a spin − at $(x + 1, y)$ and a spin + at $(1 - x, y)$. This implies a mirror m_x. The same holds for m_y: it appears both with and without time reversal. The magnetic space group is, consequently, $p_a mm$.

(b) Generators: For the first magnetic superspace group: $\{1|(\mathbf{a}, -\alpha)\}, \{1|(\mathbf{b}, 0)\}, \{1|(\mathbf{c}, 0)\}$, $\{m_x|(0, 0)\}, \{m_y| (0, 0)\}$ and $\{m_z| (0, 0) \}'$. For the second group: $\{1|(\mathbf{a}, 0)\}, \{1|(\mathbf{b}, 0\}$, $\{1|(\mathbf{c}, 0)\}, \{1| (0, \frac{1}{2})\}'$, $\{m_x|(0, 0)\}, \{m_y|(0, 0)\}$, and $\{m_z|(0, 0)\}$.

Exercise 2.3

Consider a square lattice with lattice constant 1 $(-\frac{1}{2} < x, y < +\frac{1}{2})$ and five atoms at positions $\mathbf{r}_n = d(\cos(2\pi n/5), \sin(2\pi n/5))$ $(n = 0, \ldots, 4)$. The structure factor is given by

$$F(h_1, h_2) = \sum_{n=0}^{4} \exp(2\pi i(h_1 x_n + h_2 y_n)).$$

A number of diffraction spots is shown in Fig. B.1. The lattice has symmetry P4mm, and the motif point group 5m. The intersection plus the total inversion (which appears automatically in diffraction) has point group 2mm.

Exercise 2.4

The basis of the lattice of the embedding is as for the Fibonacci chain: $(1, -\Phi)/(3 - \Phi)$ and $(\Phi, 1)/(3 - \Phi)$. The modulation does not change the rank of the structure, because the modulation changes the atomic surfaces the same way in every unit cell. Therefore, the structure factor is

$$F(H) = \int_{-1}^{\Phi} \exp(2\pi i(HA\sin(2\pi \Phi t) + H_I t)) \, dt.$$

Exercise 2.5

(a) Basis subsystem 1: $(a, 0), (0, b)$; subsystem 2: $(\frac{1}{2}c, b), (-\frac{1}{2}c, b)$ or $(c, 0), (0, 2b)$ with centring. Bases for reciprocal lattices: $(1/a, 0)$ and $(0, 1/b)$, respectively; $(1/c, 1/2b)$ and $(-1/c, 1/2b)$ or $(1/c, 0)$ and $(0, 1/2b)$, plus centring. A basis for the Fourier module is $(1/a, 0), (0, 1/b)$, and $(1/c, 1/2b)$.

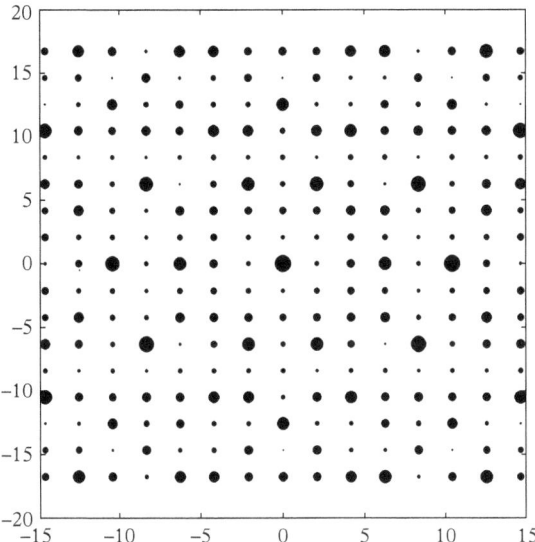

Figure B.1 *Diffraction of a square lattice with a pentagon inside. The radius of the circles corresponds to the intensity of the Bragg peak. There is no sign of a five-fold symmetry. This information is hidden in the intensities. But it is clear that the square symmetry is lost.*

(b) The superspace group is pmm$(\alpha, \frac{1}{2})$ if one takes the first subsystem as the host. Otherwise, it is cmm$(\alpha 0)$. These groups are not equivalent as superspace groups for modulated phases.

(c) The generators of the point group for the two cases are:

$$A_1 = \begin{pmatrix} -1 & 0 & 0 \\ 0 & 1 & 0 \\ 0 & 0 & -1 \end{pmatrix}, A_2 = \begin{pmatrix} 1 & 0 & 0 \\ 0 & -1 & 0 \\ 0 & 0 & 1 \end{pmatrix}; B_1 = \begin{pmatrix} 0 & 1 & 0 \\ 1 & 0 & 0 \\ 0 & 0 & -1 \end{pmatrix}, B_2 = \begin{pmatrix} 0 & -1 & 0 \\ -1 & 0 & 0 \\ 0 & 0 & -1 \end{pmatrix}.$$

Exercise 3.1

(a) The Fourier module of the chain has rank 2 with basis vectors A and $A\Phi$ with $A = (2 - \Phi)^{-1}$ and it is 4 with basis vectors $A(1, 0)$, $A(\Phi, 0)$, $A(0, 1)$, and $A(0, \Phi)$ for the 2D tiling.

(b) The embedding in 4D superspace has the basis vectors $(1, 0, -\Phi, 0)$, $(\Phi, 0, 1, 0)$, $(0, 1, 0, -\Phi)$, and $(0, \Phi, 0, 1)$. The atomic surface is the projection of the unit cell on V_I, which is a square with edge length $1 + \Phi$.

Exercise 3.2

(a) The lattice of the basic structure has two generators: (1, 0) and (0, 1). The reciprocal lattice has as basis also (1, 0) and (0, 1); the modulation wave vector is $(\alpha, 0)$. Therefore, the rank is 3. The basis of the Fourier module (1, 0), (0, 1) and $(\alpha, 0)$ is the projection of the basis (1, 0, 0), (0, 1, 0), and $(\alpha, 0, 1)$ of the three-dimensional reciprocal basis in superspace.

(b) The unit cell of the three-dimensional structure is spanned by the vectors $(1, 0, -\alpha)$, (0, 1, 0), and (0, 0, 1). It contains four atomic surfaces, in this case one-dimensional line segments of length 1/2: from $(-\ell/2, 0, 0)$ till $(-\ell/2, 0, 1/2)$, from $(\ell/2, 0, 0)$ till $(\ell/2, 0, 1/2)$, from $(0, -\ell/2, 1/2)$ till $(0, -\ell/2, 1)$, and from $(0, \ell/2, 1/2)$ till $(0, \ell/2, 1)$. The structure factor is the integral of $\exp(2\pi i(\mathbf{H}.\mathbf{r}+H_I.t))$ over these lines, with $\mathbf{r}$ the projection of the line on V_E, and t running from 0 to 1/2, respectively, from 1/2 to 1.

Exercise 3.3

(a) There are five points in the star: $x = \cos(2\pi n/5)$, and $y = \sin(2\pi n/5)$, with $n = 0, \ldots, 4$. However, the sum of the all the points is zero. Consequently, the basis is given by the points with $n > 0$: $\mathbf{e}_i$ ($i = 1, \ldots, 4$).

(b) The points of the star are obtained by a rotation over $2\pi/5$. However, this rotation does not have an integer trace. Therefore, it has to be accompanied by the associated rotations in internal space over $4\pi/5$. Then the basis for the reciprocal lattice in the four-dimensional superspace is given by the following four vectors $(\mathbf{e}_n, \mathbf{e}_{n \bmod 5})$: $(\mathbf{e}_1, \mathbf{e}_2)$, $(\mathbf{e}_2, \mathbf{e}_4)$, $(\mathbf{e}_3, \mathbf{e}_1)$, and $(\mathbf{e}_4, \mathbf{e}_3)$.

(c) The four-dimensional star is left invariant by the rotation $(5, 5^2)$ and the mirror (m, m) with matrices

$$A = \begin{pmatrix} 0 & 1 & 0 & 0 \\ 0 & 0 & 1 & 0 \\ 0 & 0 & 0 & 1 \\ -1 & -1 & -1 & -1 \end{pmatrix}; \quad B = \begin{pmatrix} 0 & 0 & 0 & 1 \\ 0 & 0 & 1 & 0 \\ 0 & 1 & 0 & 0 \\ 1 & 0 & 0 & 0 \end{pmatrix}.$$

The metric tensor g is invariant under A and B: $A.g.A^T = g$, and $B.g.B^T = g$. This gives the metric tensor

$$g = \begin{pmatrix} 2a & b & -a-b & -a-b \\ b & 2a & b & -a-b \\ -a-b & b & 2a & b \\ -a-b & -a-b & b & 2a \end{pmatrix} \quad (a, b \text{ arbitrary}).$$

The diffraction pattern is invariant under the (non-crystallographic) point group 5m. The holohedry of the four-dimensional structure then is $5m(5^2m)$ with as generators the matrices A and B or on an orthogonal basis

$$A' = \begin{pmatrix} \cos(2\pi/5) & \sin(2\pi/5) & 0 & 0 \\ -\sin(2\pi/5) & \cos(2\pi/5) & 0 & 0 \\ 0 & 0 & \cos(4\pi/5) & \sin(4\pi/5) \\ 0 & 0 & -\sin(4\pi/5) & \cos(4\pi/5) \end{pmatrix}, \quad B' = \begin{pmatrix} -1 & 0 & 0 & 0 \\ 0 & 1 & 0 & 0 \\ 0 & 0 & -1 & 0 \\ 0 & 0 & 0 & 1 \end{pmatrix}.$$

References

Abe, E., Pennycook, S.J., and Tsai, A.P. (2003a). Direct observation of a local thermal vibration anomaly in a quasicrystal. *Nature*, **421**, 347–50.

Abe, E., Yan, Y., and Pennycook, S.J. (2004). Quasicrystals as cluster aggregates. *Nat. Mater.*, **3**, 759–67.

Abe, H., Saitoh, H., Ueno, T., Nakao, H., Matsuo, Y., Ohshima, K., and Matsumoto, H. (2003b). Anomalous Debye-Waller factor associated with an order-disorder transformation in an $Al_{72}Ni_{20}Co_8$ decagonal quasicrystal. *J. Phys. Jpn.*, **72**, 1828–31.

Amemiya, Y. (1995). Imaging plates for use with synchrotron-radiation. *J. Synchrotron Radiat.*, **2**, 13–21.

Angel, R.J, McMullan, R.K., and Prewitt, C.T. (1991). Substructure and superstructure of mullite by neutron diffraction. *Am. Mineral.*, **76**, 332–42.

Arakcheeva, A., Bykov, M., Bykova, E., Dubrovinsky, L., Pattison, P., Dmitriev, V., and Chapuis, G. (2017). Incommensurate atomic density waves in the high-pressure *IVb* phase of barium. *IUCrJ*, **4**(2), 152–7.

Arakcheeva, A. and Chapuis, G. (2005). Atomic clusters and phase transitions in the metastable β-Ta phase between 4.2 and 293 K. *Europhysi. Lett.*, **69**, 378–84.

Arakcheeva, A. and Chapuis, G. (2006). Getting more out of an incommensurately modulated structure: The example of $K_5Yb(MoO_4)_4$. *Acta Cryst. B*, **62**, 52–9.

Arakcheeva, A. and Chapuis, G. (2008). Capabilities and limitations of a (3+d)-dimensional incommensurately modulated structure as a model for the derivation of an extended family of compounds: example of the scheelite-like structures. *Acta Cryst. B*, **64**, 12–25.

Arakcheeva, A., Chapuis, G., Petříček, V., Dušek, M., and Schönleber, A. (2003). The incommensurate structure of $K_3In(PO_4)_2$. *Acta Cryst. B*, **59**, 17–27.

Arakcheeva, A., Logvinovich, D., Chapuis, G., Morozov, V., Eliseeva, S.V., Bunzli, J.C.G., and Pattison, P. (2012). The luminescence of $Na_xEu^{3+}_{(2-x)/3}MoO_4$ scheelites depends on the number of Eu-clusters occurring in their incommensurately modulated structure. *Chem. Sci.*, **3**, 384–90.

Ashraff, J.A., Luck, J-M., and Stinchcombe, R.B. (1990). Dynamical properties of two-dimensional quasicrystals. *Phys. Rev. B*, **41**, 4314–29.

Aslanyan, T.A., Shigenari, T., and Abe, K. (1998). Debye-Waller factors for incommensurate structures. *J. Phys. Cond. Matt.*, **10**, 4565–76.

Aubry, S. (1978). The new concept of transitions by breaking of analyticity in a crystallographic model. In *Solitons and Condensed Matter Physics: Proceedings of the Symposium on Nonlinear (Soliton) Structure and Dynamics in Condensed Matter, Oxford, England, June 27–29, 1978* (ed. A. Bishop and T. Schneider), Berlin, Heidelberg, pp. 264–77. Springer Berlin-Heidelberg.

Aubry, S. and André, G. (1980). Analyticity breaking and Anderson localization in incommensurate lattices. *Ann. Israel Phys. Soc.*, **3**, 133.

Aubry, S. and Le Daeron, P.Y. (1983). The discrete Frenkel-Kontorova model and its extensions I. Exact results for the ground-states. *Physica D*, **8**, 381–422.

Audier, M., Duneau, M., de Boissieu, M., Boudard, M., and Létoublon, A. (1999). Superlattice ordering of cubic symmetry in an icosahedral AlPdMn phase. *Phil. Mag. A*, **79**, 255–70.

Audier, M. and Guyot, P. (1989). Rhombohedral to icosahedral solid state transformation in the $Al_{65}Cu_{20}Fe_{15}$ alloy. In *Quasicrystals and Incommensurate Structures in Condensed Matter, Third International Meeting on quasicrystals* (ed. M. Yacaman, D. Romeu, V. Castano, and A. Gomez), 29 May–2 June 1989, Mexico, pp. 288–99. World Scientific, 1990.

Audier, M., Guyot, P., and Brechet, Y. (1990). High temperature stability and faceting of the icosahedral AlFeCu phase. *Phil. Mag.*, **61**, 55–62.

Axe, J.D. (1980). Debye-Waller factor for incommensurate structures. *Phys. Rev. B*, **21**, 4181–90.

Axe, J.D., Iizumi, M., and Shirane, G. (1980). Lattice dynamics of commensurate and incommensurate K_2SeO_4. *Phys. Rev. B*, **22**, 3408–13.

Axel, F. and Aubry, S. (1981). Spatially modulated phases of a one-dimensional lattice model with competing interactions. *J. Phys. C*, **14**, 5433–51.

Baake, M., Grimm, U., and Moody, V. (2002). What is aperiodic order? *arXiv:math-ph*, **0203252v1**, 1–27.

Baake, M., Klitzing, R., and Schlottmann, M. (1992). Fractally shaped acceptance domains of quasiperiodic square-triangle tilings with dodecagonal symmetry. *Physica A*, **191**, 554–8.

Baake, M., Lenz, D., and van Enter, A. (2015). Dynamical versus diffraction spectrum for structures with finite local complexity. *Ergod. Theory Dyn. Syst.*, **35**, 2017–43.

Baake, M. and Moody, R.V. (2000). *Directions in mathematical quasicrystals*. American Mathematical Society.

Baake, M., Moody, R.V., Richard, C., and Sing, B. (2003). Which distributions of matter diffract? In *Quasicrystals* (ed. H.-R. Trebin), pp. 188–207. Wiley-VCH.

Bachheimer, J.P. (1980). An anomaly in the β-phase near the α-β transition of quartz. *J. Physique Lett.*, **41**, L559–61.

Bak, P. (1985). Symmetry, stability and elastic properties of icosahedral quasicrystals. *Phys. Rev. B*, **32**, 5764–72.

Bak, P. (1986). Icosahedral crystals: Where are the atoms? *Phys. Rev. Lett.*, **56**, 861–4.

Bak, P. and von Boehm, J. (1980). Ising model with solitons, phasons and 'the devil's staircase'. *Phys. Rev. B*, **21**, 5297–308.

Barbier, L., Floc, D. Le, Calvayrac, Y., and Gratias, D. (2002). Identification of the atomic structure of the fivefold surface of an icosahedral Al-Pd-Mn quasicrystal: Helium diffraction and scanning tunneling microscopy studies. *Phys. Rev. Lett.*, **88**, 0855061–4.

Barkan, K., Diamant, H., and Lifshitz, R. (2011). Stability of quasicrystals composed of soft isotropic particles *Phys. Rev. B*, **83**, 172201-1-4.

Bekku, S., Ziher, P., and Dotera, T. (2017). Origin of 18-fold quasicrystal. *J.Phys. Conf. Ser.*, **809**, 012003.

Belin-Ferré, E., Fontaine, M.-F., Thirion, J., Kameoka, S., Tsai, A.P., and Dubois, J.M. (2006). Electronic structure of leached Al-Cu-Fe quasicrystals used as catalysts. *Phil. Mag.*, **86**, 687–92.

Bennema, P., Balzuweit, K., Dam, B., Meekes, H., Verheijen, M.A., and Vogels, L.J.P. (1991). Morphology of modulated crystals and quasicrystals. *J. Phys. D*, **24**, 186–98.

Bergerhoff, G. and Brandenburg, K. (1999). Typical interatomic distances: Inorganic compounds. In *International Table of Crystallography*, Volume C, pp. 770–81. Kluwer.

Bernard, L., Currat, R., Delamoye, P., Zeyen, C.M.E., Hubert, S., and de Kouchkovsky, R. (1983). ThBr$_4$ phasons. *J. Phys. C*, **16**, 433.

Beurskens, P.T., Beurskens, G., Bosman, W.P., de Gelder, R., Garcia-Granda, S., Gould, R.O., Israel, R., and Smits, J.M.M. (1996). *The DIRDIF-96 program system*. Crystallography Laboratory, University of Nijmegen, The Netherlands.

Bieberbach, L. (1910). Bewegungsgruppen Euklidischer Räume I. *Math. Annalen*, **70**, 297–336.

Bieberbach, L. (1912). Bewegungsgruppen Euklidischer Räume II. *Math. Annalen*, **72**, 400–12.

Bindi, L., Bonazzi, P., Dušek, M, Petříček, V., and Chapuis, G. (2001). Five dimensional structure refinement of natural melilite. *Acta Cryst. B*, **47**, 739–46.

Bindi, L., Eller, J., Guan, Y., Hollster, L., MacPherson, G., Steinhardt, P.J., and Yao, N. (2012). Evidence for the extraterrestial origin of a natural quasicrystal. *PNAS*, **109**, 1396–401.

Bindi, L., Lin, C., Ma, C., and Steinhardt, P.J. (2016). Collisions in outer space produced an icosahedral phase in the Khatyrka meteorite never observed previously in the laboratory. *Scient. Rep.*, **6**, 38117.

Bindi, L., Steinhardt, P.J., Yao, N., and Lu, P.J. (2009). Natural quasicrystals. *Science*, **324**, 1306.

Bindi, L., Yao, N., Lin, C., Hollister, L.S., Andronicos, C.L., Distler, V., Eddy, M.P, Kostin, A., Kryachko, V., MacPherson, G.J., Steinhardt, W.M., Yudovskaya, M., and Steinhardt, P.J. (2015). Natural quasicrystal with decagonal symmetry. *Scient. Rep.*, **5**, 9111.

Blinc, R. and Levanyuk, A.P. (ed.) (1986a). *Incommensurate Phases in Dielectrics. 1. Fundamentals*. North-Holland.

Blinc, R. and Levanyuk, A.P. (ed.) (1986b). *Incommensurate Phases in Dielectrics. 2. Materials*. North-Holland.

Böhm, H. (1975). Interpretation of X-ray-scattering patterns due to periodic structural fluctuations. *J. Appl. Cryst.*, **8**, 202.

Bohr, H. (1952). *Collected Mathematical Works. II Almost Periodic Functions*. Dansk Matematisk Forening.

Bombieri, E. and Taylor, J.E. (1986). Which distributions of matter diffract? An initial investigation. In *International Workshop on Aperiodic Crystals* (ed. D. Gratias and L. Michel), pp. C3: 19–28. Les Ulis: Les éditions de physique.

Boudard, M., Bourgeat-Lami, E., de Boissieu, M., Janot, C., Durand-Charre, M., Klein, H., Audier, M., and Hennion, B. (1995). Czokralski growth of AlPdMn quasicrystal. *Phil. Mag. Lett.*, **71**, 11–6.

Boudard, M., de Boissieu, M., Janot, C., Heger, G., Beeli, C., Nissen, H-U., Vincent, H., Ibberson, R., Audier, M., and Dubois, J.-M. (1992). Neutron and X-ray single-crystal study of the Al-Pd-Mn icosahedral phase. *J. Phys. Cond. Matt.*, **4**, 10149–68.

Boudard, M., de Boissieu, M., Kycia, S., Goldman, A.I., Hennion, B., Bellissent, R., Quilichini, M., Currat, R., and Janot, C. (1995). Optic modes in the AlPdMn icosahedral phase. *J. Phys. Cond. Matt.*, **7**, 7299–7308.

Boudard, M., de Boissieu, M., Létoublon, A., Hennion, B., Bellissent, R., and Janot, C. (1996). Phason softening in the AlPdMn icosahedral phase. *Europhys. Lett.*, **33**, 199–204.

Brill, W. and Ehses, K-H. (1985). The incommensurate phase of betaine calcium chloride-dihydrate. *Jap. J. Appl. Phys. Suppl.*, **24**, 826.

Brouns, E., Visser, J.W., and de Wolff, P.M. (1964). An anomaly in the crystal structure of Na_2CO_3. *Acta Cryst.*, **17**, 614.

Brown, H., Bülow, R., Neubüser, J., Wondratschek, H., and Zassenhaus, H. (1978). *Crystallographic Groups of Four-Dimensional Space*. John Wiley.

Brown, I.D., Cutforth, B.D., Davies, C.G., Gillespie, R.J., Ireland, P.R., and Vekris, J.E. (1974). Alchemists' gold $Hg_{2.86}AsF_6$: An X-ray crystallographic study of a novel mercury compound containing metallically bonded infinite cations. *Can. J. Chem.*, **52**, 791–3.

Brown, K.S., Avansov, A.A., and Elser, V. (2000). Solving the crystallographic phase problem in i(AlPdMn). *Phys. Rev. Lett.*, **85**, 4084–7.

Brun, T.O., Sinha, S.K., Wakabayashi, N., Lander, G.H., Edwards, L.R., and Spedding, F.H. (1970). Temperature dependence of the periodicity of the magnetic structure of thullium metal. *Phys. Rev. B*, **1**, 1251–3.

Bungaro, C., Rabe, K.M., and DalCorso, A. (2003). First-principles study of lattice instabilities in ferromagnetic Ni_2MnGa. *Phys. Rev. B*, **68**, 134104.

Burckhardt, J.J. (1947). *Die Bewegungsgruppen der Kristallographie*. Birkhäuser.

Burckhardt, J.J. (1988). *Die Symmetrie der Kristalle*. Birkhäuser.

Burkov, S.E. (1991). Structure model of the Al-Cu-Co decagonal quasicrystal. *Phys. Rev. Lett.*, **67**, 614–617.

Bussien-Gaillard, V., Paciorek, W., Schenk, K., and Chapuis, G. (1996). Hexamethylenetetramine suberate, a strongly anharmonic modulated structure. *Acta Cryst. B*, **52**, 1036–47.

Cahn, J.W., Gratias, D., and Mozer, B. (1988). Six-dimensional Fourier analysis of the icosahedral $Al_{73}Mn_{21}Si_6$ alloy. *Phys. Rev. B*, **38**, 1643–7.

Cahn, J.W., Shechtman, D., and Gratias, D. (1986). Indexing of icosahedral quasiperiodic crystals. *J. Mat. Res.*, **1**, 13–26.

Cailleau, H., Moussa, F., and Mons, J. (1979). Incommensurate phases in biphenyl. *Sol. St. Comm.*, **31**, 521–4.

Canfield, P.C., Caudle, M.L., Ho, C.S., Kreyssig, A., Nandi, S., Kim, M.G., Lin, X., Kracher, A., Dennis, K.W., McCallum, R.W., and Goldman, A.I. (2010). Solution growth of a binary icosahedral quasicrystal of $Sc_{12}Zn_{88}$. *Phys. Rev. B*, **81**, 020201.

Cervellino, A., Haibach, T., and Steurer, W. (2002). Structure solution of the basic decagonal Al-Co-Ni phase by the atomic surfaces modelling method. *Acta Cryst. B*, **58**, 8–33.

Chaikin, P.M. and Lubensky, T.C. (1995). *Principles of Condensed Matter Physics*. Cambridge University Press.

Chen, X., Bansal, D., Sullivan, S., Abernathy, D., Aczel, A., Zhouand, J., Delaire, O., and Shi, Li (2016). Weak coupling of pseudoacoustic phonons and magnon dynamics in the incommensurate spin-ladder compound $Sr_{14}Cu_{24}O_{41}$. *Phys. Rev. B*, **94**, 134309.

Chen, Z.Y. and Walker, M.B. (1991a). Symmetry modes, competing interactions and universal description of modulated phases in the dielectric A_2BX_4 family. *Phys. Rev. B*, **43**, 5634–48.

Chen, Z.Y. and Walker, M.B. (1991b). Theory of the phase transition sequence in betaine calcium chloride dihydrate BCCD. *Phys. Rev. B*, **43**, 760–4.

Cheong, S.W. and Mostovoy, M. (2007). Multiferroics: A magnetic twist for ferroelectricity. *Na. Mate.*, **6**, 13–20.

Chernikov, M.A., Bernasconi, A., Beeli, C., Schilling, A., and Ott, H.R. (1993). Low-temperature magnetism in icosahedral $Al_{70}Mn_9Pd_{21}$. *Phys. Rev. B*, **48**, 3058–65.

Cipriani, F., Castagna, J.C., Lehmann, M.S., and Wilkinson, C. (1995). A large image-plate detector for neutrons. *Physica B*, **213**, 975–7.

Coddens, G., Lyonnard, S., Hennion, B., and Calvayrac, Y. (1999). Correlated simultaneous phason jumps in an icosahedral Al-Pd-Mn quasicrystal. *Phys. Rev. Lett.*, **83**, 3226–9.

Cornier-Quiqandon, M., Quivy, A., Lefebvre, S., Elkaim, E., Heger, G., Katz, A., and Gratias, D. (1991). Neutron-diffraction study of icosahedral Al-Cu-Fe single quasicrystals. *Phys. Rev. B*, **44**, 2071–84.

Curien, H. (1952). Diffusion thermique des rayons-X par des monocristaux de Fer-α et dynamique du réseau cubique centré. *Acta. Cryst.*, **5**, 393.

Currat, R. and Janssen, T. (1988). Excitations in incommensurate crystal phases. In *Solid State Physics* (ed. H. Ehrenreich and D. Turnbull), Volume 41, pp. 201–302. Academic Press.

Currat, R., Kats, E., and Luk'yanchuk, I. (2002). Sound modes in composite incommensurate crystals. *Eur. Phys. J. B*, **26**, 339–47.

Cutforth, B.D., Datars, W.R., Gillespie, R.J., and van Schijndel, A. (1976). $Hg_{2.85}AsF_6$ Novel structure with unusual electrical properties. In *Unusual Properties of Inorganic Complexes*, Volume 150, pp. 56–62. Adv. Chem. Ser.

Dam, B. (1985). In situ observation of a roughening transition of the ($101\bar{2}$) satellite crystal surface of modulated TMA-ZC. *Phys. Rev. Lett.*, **5**, 2806–9.

Dam, B. and Janner, A. (1986). A superspace approach to the structure and morphology of tetramethylammonum tetrachlorozincate. *Acta Cryst. B*, **42**, 69–77.

Daniel, V. and Lipson, H. (1943). An X-ray study of the dissociation of an alloy of copper, iron and nickel. *Proc. Roy. Soc. A*, **181**, 368–78.

Daniel, V. and Lipson, H. (1944). The dissociation of an alloy of copper, iron and nickel: further X-ray work. *Proc. Roy. Soc. A*, **182**, 378–87.

Dattoli, G., Giannessi, L., Mezi, L., and Torre, A. (1990). Theory of generalized Bessel-functions. *Nuov. Cim. B*, **105**, 327–48.

de Boissieu, M., Boudard, M., Bellissent, R., Quilichini, M., Hennion, B., Currat, R., Goldman, A.I., and Janot, C. (1993). Dynamics of the AlPdMn Icosahedral Phase. *J. Phys. Cond. Matt.*, **5**, 4945–66.

de Boissieu, M., Boudard, M., Hennion, B., Bellissent, R., Kycia, S., Goldman, A.I., Janot, C., and Audier, M. (1995). Diffuse scattering and phason elasticity in the AlPdMn icosahedral phase. *Phys. Rev. Lett.*, **75**, 89.

de Boissieu, M., Boudard, M., Ishimasa, T., Lauriat, J.P., Létoublon, A, Audier, M., Duneau, M., and Davroski, A. (1998a). Reversible transformation between an icosahedral AlPdMn phase and a modulated structure of cubic symmetry. *Phil. Mag. A*, **78**, 305–326.

de Boissieu, M., Boudard, M., Ishimasa, T., Lauriat, J.P., Létoublon, A., Audier, M., Duneau, M., and Davroski, A. (1998b). Reversible transformation between an icosahedral AlPdMn phase and a modulated structure of cubic symmetry. *Phil. Mag. A*, **78**, 305–26.

de Boissieu, M., Durand-Charre, M., Bastie, P., Carabelli, A., Boudard, M., Bessiere, M., Lefebvre, S., Janot, C., and Audier, M. (1992). Centimetre-size single grain of the perfect Al-Pd-Mn icosahedral phase. *Phil. Mag. Lett.*, **65**, 147–53.

de Boissieu, M. and Francoual, S. (2005). Diffuse scattering and phason modes in the i-AlPdMn quasicrystalline phases. *Z. Krist.*, **220**, 1043–51.

de Boissieu, M., Francoual, S., Kaneko, Y., and Ishimasa, T. (2005). Diffuse Scattering and phason fluctuations in the Zn-Mg-Sc icosahedral quasicrystal and its Zn-Sc periodic approximant. *Phys. Rev. Lett.*, **95**, 105503.

de Boissieu, M., Francoual, S., Mihalkovic, M., Shibata, K., Baron, A.Q.R., Sidis, Y, Ishimasa, T., Wu, D., Lograsso, T., Regnault, L.P., Gaehler, F., Tsutsui, S., Hennion, B., Bastie, P., Sato, T.J., Takakura, H., Currat, R., and Tsai, A.P. (2007). Lattice dynamics of the ZnMgSc icosahedral quasicrystal and its ZnSc periodic 1/1 approximant. *Nat. Materi.*, **6**, 977–84.

de Boissieu, M., Janot, C., Dubois, J.M., Audier, M., and Dubost, B. (1991). Atomic structure of the icosahedral AL-Li-Cu quasicrystal. *J. Phys. Cond. Matt.*, **3**, 1–25.

de Boissieu, M., Stephens, P., Boudard, M., and Janot, C. (1994a). Is the Al-Pd-Mn icosahedral phase centrosymmetric? *J. Phys. Cond. Matt.*, **6**, 363–73.

de Boissieu, M., Stephens, P., Boudard, M., Janot, C., Chapman, D., and Audier, M. (1994b). Anomalous X-ray diffraction study of the AlPdMn icosahedral phase. *J. Phys. Cond. Matt.*, **6**, 10725–45.

de Boissieu, M., Stephens, P., Boudard, M., Janot, C., Chapman, D., and Audier, M. (1994c). Disorder and complexity in the atomic structure of the perfect icosahedral alloy Al-Pd-Mn. *Phys. Rev. Lett.*, **72**, 3538–41.

de Boissieu, M., Takakura, H., Bletry, M., Guo, J.Q., and Tsai, A.P. (2002). Phason fluctuations in i-AlPdRe and i-CdYb phases. *J. Alloys Comp.*, **342**, 265–70.

de Bruijn, N.G. (1981). Algebraic theory of Penrose's non-periodic tilings of the plane. *Ind. Math.*, **43**, 39–66.

de Lange, C. and Janssen, T. (1981). Incommensurability and recursivity: Lattice dynamics of incommensurate modulated crystals. *J. Phys. C*, **14**, 5269–92.

de Lange, C. and Janssen, T. (1984). Modulated Kronig-Penney model in superspace. *Physica A*, **127**, 125–40.

De Ridder, R., van Tendeloo, G., and Amelinckx, S. (1976). A cluster model for the transition state and its study by means of electron diffraction. *Phys. Stat. Sol. (a)*, **33**, 383.

de Wolff, P.M. (1974). The pseudo-symmetry of modulated crystal structures. *Acta Cryst. A*, **30**, 777–85.

de Wolff, P.M., Janssen, T., and Janner, A. (1981). The superspace groups for incommensurate crystal structures with a one-dimensional modulation. *Acta Cryst. A*, **37**(5), 625–36.

Deguchi, K., Matsukawa, S., Sato, N.K., Hattori, T., Ishida, K., Takakura, H., and Ishimasa, T. (2012). Quantum critical state in a magnetic quasicrystal. *Nat. Mater.*, **11**, 1013–16.

Dehlinger, U. (1929). Zur Theorie der Rekristallisation reiner Metalle. *Ann. Phys.*, **2**, 749–93.

DellaVilla, A., Enoch, S., Tayeb, G., Pierro, V., Galdi, V., and Capolino, F. (2005). Band gap formation and multiple scattering in photonic quasicrystals with a Penrose-type lattice. *Phys. Rev. Lett.*, **94**, 183903.

Dmitriev, S.V., Semagin, D.A., Shigenari, T., Abe, K., Nagamine, M., and Aslanyan, T.A. (2003). Molecular and lattice dynamical study on modulated structures in quartz. *Phys. Rev. B*, **68**, 052101.

Dmitriev, S.V., Yoshikawa, N., Kohyama, M., Tanaka, S., Yang, R., and Kagawa, Y. (2004). Atomistic structure of the Cu(111)α-Al$_2$O$_3$(0001) interface in terms of interatomic potentials fitted to ab initio results. *Act. Mat.*, **52**, 1959–70.

Dolino, G., Bastie, P., Capelle, B., Chamard, V., Härtwig, J., and Guzzo, P.L. (2005). Origin of the opalescence at the α-β transition of quartz. *Phys. Rev. Lett.*, **94**, 155701.

Dolinšek, J., Klanjšek, M., Apih, T., Gavilano, J.L., Gianno, K., Ott, H.R., Dubois, J.M., and Urban, K. (2001). Mn magnetism in icosahedral quasicrystalline Al$_{72.4}$Pd$_{20.5}$Mn$_{7.1}$. *Phys. Rev. B*, **64**, 024203.

Dotera, T. (2006). Dodecagonal quasicrystal in a polymeric alloy. *Phil. Mag.*, **86**, 1085–91.

Dotera, T. (2011). Quasicrystals in soft mater. *Isr. J. Chem.* **51**, 1197–205.

Dotera, T. (2012). Toward the discovery of new soft quasicrystals: From a numerical study viewpoint. *Polymer Phys.*, **50**, 155–67.

Dotera, T. and Steinhard, P.J. (1994). Ising-like transition and phason unlocking in icosahedral quasicrystals. *Phys. Rev. Lett.*, **72**, 1670.

Dotera, T., Oshiro, T., and Ziherl, P. (2014). Mosaic two-lengthscale quasicrystals. *Nature* **506**, 208–11.

Dubbeldam, G.C. and de Wolff, P.M. (1969). The average crystal structure of γ-Na$_2$CO$_3$. *Acta Cryst.*, **B25**, 2665–67.

Dubois, J.M. (2004). A model of wetting of quasicrystals in ambient air. *J. Non-Cryst. Sol*, **334/5**, 481–4.

Dubois, J.M. (2005). *Useful Quasicrystals*. World Scientific.

Dubost, B., Lang, J.M., Tanaka, M., Sainfort, P., and Audier, M. (1986). Large AlCuLi single quasicrystals with triacontahedral solidification morphology. *Nature*, **324**, 48–50.

Dugain, F., de Boissieu, M., Shibata, K., Currat, R., Sato, T.J., Kortan, A.R., Suck, J.-B., Hradil, K., Frey, F., and Tsai, A.P. (1999). Inelastic neutron scattering study of the dynamics of the AlNiCo phase. *Eur. Phys. J. B*, **7**, 513–516.

Duneau, M. and Katz, A. (1985). Quasiperiodic patterns. *Phys. Rev. Lett.*, **54**, 2688–91.

Durand, D., Papoular, R., Currat, R., Lambert, M., Legrand, J.F., and Mezei, F. (1990). NaNO$_2$ relaxation rate. *Phys. Rev. B*, **43**, 10690.

Dušek, M., Chapuis, G., Meyer, M., and Petříček, V. (2003). Sodium carbonate revisited. *Acta Cryst. B*, **59**, 337–52.

Dušek, M., Chapuis, G., Schobinger-Papamantellos, P., Wilkinson, C., Petříček, V., Tung, L.D., and Buschow, K.H.J. (2000). Modulated structure of La$_2$Co$_{1.7}$ from neutron and X-ray diffraction data. *Acta Cryst. B*, **56**, 959–71.

Eagen, C.F. and Werner, S.A. (1975). Neutron-diffraction investigation of the sinusoidal displacement-wave in pure Cr. *Sol. State Comm.*, **16**, 1113–16.

Ebert, Ph., Yurechko, M., Kluge, F., Horn, K., and Urban, K. (2003). Cleavage surfaces of quasicrystals. In *Quasicrystals* (ed. H. Trebin), pp. 572–97. EVCH-Wiley.

Edagawa, K., Suzuki, K., and Takeuchi, S. (2000). High resolution transmission electron microscopy observation of thermally fluctuating phasons in decagonal Al-Cu-Co. *Phys. Rev. Lett.*, **85**, 1674–7.

Eijt, S.W.H., Currat, R., Lorenzo, J.E., Saint-Grégoire, P., Katano, S., Janssen, T., Hennion, B., and Vysochanskii, Yu.M. (1998). Soft modes and phonon interactions in Sn$_2$P$_2$Se$_6$ studied by inelastic neutron scattering. *J. Phys. CM*, **10**, 4811–44.

Eikenberry, E.F., Bronnimann, C., Hulsen, G., Toyokawa, H., Horisberger, R., Schmitt, B., Schulze-Briese, C., and Tomizaki, T. (2003). Pilatus: A two-dimensional x-ray detector for macromolecular crystallography. *Nucl. Instr. Meth. Phys. Res. A*, **501**, 260–6.

Elcoro, L., Perez-Mato, J.M., Darriet, J., and El Abed, A. (2003). Superspace description of trigonal and orthorhombic $A_{1+x}A'_xB_{1-x}O_3$ compounds as modulated layered structures: Application to the refinement of trigonal $Sr_6Rh_5O_{15}$. *Acta Cryst. B*, **59**, 217–33.

Elcoro, L., Perez-Mato, J.M., and Withers, R. (2000). Intergrowth polytypoids as modulated structures: The example of the cation deficient oxides $LaTi_{1-x}O_3$. *Z. Krist.*, **215**, 727–39.

Elhor, H., Mihailkovic, M., Roujiaa, M., Scheffer, M., and Suck, J.-B. (2003). Dynamical properties of quasicrystalline alloys investigated by neutron inelastic scattering and computer simulations based on realistic potentials. In *Quasicrystals* (ed. H. Trebin), pp. 382–413. Wiley-VCH.

Elliott, R.J. (1961). Phenomenological discussion of magnetic ordering in the heavy rare-earth metals. *Phys. Rev.*, **124**, 346–53.

Elser, V. (1985). Indexing problems in quasicrystal diffraction. *Phys. Rev. B*, **32**, 4892–8.

Elser, V. (1999). X-ray phase determination by the principle of minimum charge. *Acta Cryst. A*, **55**, 489–99.

Elser, V. and Henley, C.L. (1985). Crystal and quasicrystal structures in Al-Mn-Si alloys. *Phys. Rev. Lett.*, **55**, 2883–6

Engel, M., Damasceno, P.F., Phillips, C.L., and Glotzer, Sharon C. (2015). Computational self-assembly of a one-component icosahedral quasicrystal. *Nat. Mater.*, **14**, 109–15.

Engel, M. and Trebin, H.-R. (2007). Self-assembly of monatomic complex crystals and quasicrystals with a double-well interaction potential. *Phys. Rev. Lett.*, **98**, 225505.

Estermann, M.A. and Steurer, W. (1998). Diffuse scattering data acquisition techniques. *Phase Trans.*, **67**, 165–95.

Etrillard, J., Bourgeois, L., Bourges, P., Liang, B., Lin, C.T., and Keimer, B. (2004). Low-frequency structural dynamics in the incommensurate composite crystal $Bi_2Sr_2CuO_{6+\delta}$. *Eur. Phys. Lett.*, **66**, 246–52.

Etrillard, J., Bourges, Ph., He, H.F., Keimer, B., Liang, B., and Lin, C.T. (2001). Acoustic phonons in the aperiodic layered crystal of $Bi_2Sr_2CaCu_2O_{8+\delta}$. *Eur. Phys. Lett.*, **55**, 201–7.

Etrillard, J., Toudic, B., Bertault, M., Even, J., Gourdji, M., Peneau, A., and Guibe, L. (1993). ^{35}Cl NQR and calorimetric reinvestigation of the incommensurate phase of bis(4-chlorophenyl) sulfone: Evidence for no lock-in transition. *J. Phys. I*, **3**, 2437–49.

Etxebarria, I., Perez-Mato, J. M., and Madariaga, G. (1992). Lattice dynamics, structural stability, and phase transitions in incommensurate and commensurate A_2BX_4 materials. *Phys. Rev. B*, **46**, 2764–74.

Euchner, H., Pailhes, S., Nguyen, L.T.K., Assmus, W., Ritter, F., Haghighirad, A., Grin, Y., Paschen, S., and de Boissieu, M. (2012). Phononic filter effect of rattling phonons in the thermoelectric clathrate $Ba_8Ge_{40+x}Ni_{6-x}$. *Phys. Rev. B*, **86**, 2243031–9.

Euchner, H., Yamada, T., Rols, S., Ishimasa, T., Kaneko, Y., Ollivier, J., Schober, H., Mihalkovic, M., and de Boissieu, M. (2013). Tetrahedron dynamics in the icosahedral quasicrystals i-ZnMgSc and i-ZnAgSc and the cubic 1/1-approximant Zn_6Sc. *J. Phys. Cond. Matt.*, **25**, 115405.

Euchner, H., Yamada, T., Schober, H., Rols, S., Mihalkovic, M., Tamura, R., Ishimasa, T, and de Boissieu, M. (2012). Ordering and dynamics of the central tetrahedron in the 1/1 Zn_6Sc periodic approximant to quasicrystal. *J. Phys. Cond. Matt.*, **24**, 415403.

Fan, H.F. (1999). Direct methods in electron crystallography: Image processing and solving incommensurate structures. *Microsc. Res. Tech.*, **46**, 104–16.

Fan, H.F., van Smaalen, S., Lam, E.J.W., and Beurskens, P.T. (1993). Direct methods for incommensurate intergrowth compounds. I. Determination of the modulation. *Acta Cryst. A*, **49**, 704–8.

Fast, G. and Janssen, T. (1971). Determination of n-dimensional space groups by means of an electronic computer. *J. Comp. Phys.*, **7**, 1–11.

Feuerbacher, M. (2012). Dislocations in icosahedral quasicrystals. *Chem. Soc. Rev.*, **41**, 6745–59.

Fischer, S., Exner, A., Zielske, K., Perlich, J., Deloudi, S., Steurer, W., Lindner, P., and Förster, S. (2011). Colloidal quasicrystals with 12-fold and 18-fold diffraction symmetry. *PNAS*, **106**, 1810–14.

Fisher, M.E. and Selke, W. (1980). Infinite many commensurate phases in a simple Ising mode. *PRL*, **44**, 1503–5.

Foerster, S., Meinel, K., Hammer, R., Trautmann, M., and Widdra, W. (2013). Quasicrystalline structure formation in a classical crystalline thin-film system. *Nature*, **502**, 215–18.

Foerster, S., Trautmann, M., Roy, S., Adeagbo, W. A, Zollner, E.M, Hammer, R., Schumann, F.O, Meinel, K., Nayak, S.K, Mohseni, K., Hergert, W., Meyerheim, H.L, and Widdra, W. (2016). Observation and structure determination of an oxide quasicrystal approximant. *Phys. Rev. Lett.*, **117**, 095501.

Francoual, S. (2006). Phonons et phasons dans les quasicristaux de symétrie icosaédrique et dans leurs approximants 1/1 périodiques. Ph.D. thesis, Laboratoire de Thermodynamique et Physicochimie Metallurgiques, Grenoble.

Francoual, S., Livet, F., de Boissieu, M., Yakhou, F., Bley, F., Létoublon, A., Caudron, R., and Gastaldi, J. (2003). Dynamics of phason fluctuations in the i-AlPdMn quasicrystal. *Phys. Rev. Lett.*, **91**, 225501.

Francoual, S., Shibata, K., de Boissieu, M., Baron, A.Q.R., Tsutsui, S., Currat, R., Takakura, H., Tsai, A. P., Lograsso, T., and Ross, A.R. (2004). Experimental study of phonon dynamics in the i-CdYb phase and its 1/1 periodic approximant. *Ferroelectrics*, **305**, 235–8.

Frank, F.C. and der Merwe, J.H. Van (1949). One-dimensional dislocations. i. Static Case. *Proc. Roy. Soc. London*, **A198**, 205–16.

Frenkel, Y.I. and Kontorova, T. (1938). In Russian. *Zh. Eksp. Teor. Fiz.*, **8**, 89–95, 1340–9, 1349–59.

Fujita, N. (2017). Quasiperiodic canonical-cell tiling with pseudo-icosahedral symmetry. *Ann. Phys.*, **385**, 225–86.

Fujiwara, T. (1999). Theory of electronic structure in quasicrystals. In *Physical Properties of Quasicrystals* (ed. Z. Stadnik), pp. 169–208. Springer.

Fujiwara, T., Mitsui, T., and Yamamoto, S. (1996). Scaling properties of wave functions and transport properties in quasicrystals. *Phys. Rev. B*, **53**, R2910(R).

Garcia, A., Perez-Mato, J.M., and Madariaga, G. (1989). Dynamics of incommensurate structures and inelastic neutron scattering. *Phys. Rev. B*, **39**, 2476–83.

Gastaldi, J., Agliozzo, S., Létoublon, A., Wang, J., Mancini, L., Klein, H., Hartwig, J., Baruchel, J., Fisher, I.R., Sato, T., Tsai, A.P., and de Boissieu, M. (2003). Degree of structural perfection of icosahedral quasicrystalline grains investigated by synchrotron X-ray diffractometry and imaging techniques. *Phil. Mag. B*, **83**, 1–29.

Gastaldi, J., Reiner, E., Jourdan, C., Grange, G., Quivy, A., and Boudard, M. (1995). Loop- and band-shaped defects observed in quasicrystals by X-ray topography. *Phil. Mag. Lett.*, **72**, 311–21.

Gesi, K. (1965). Phenomenological theory of the phase transitions in $NaNO_2$. *J. Phys. Soc. Jpn.*, **20**, 1764–1772.

Goldman, A., Kong, T., Kreyssig, A., Jesche, A., Ramazanoglu, M., Dennis, K.W., Budko, S.L., and Canfield, P. (2013). A family of binary magnetic icosahedral quasicrystals based on rare earths and cadmium. *Nat. Mater.*, **12**, 714–18.

Goldman, A.I., Stassis, C., Bellissent, R., Mouden, H., Pyka, N., and Gayle, F.W. (1991). Inelastic neutron scattering measurements of phonons in icosahedral AlLiCu. *Phys. Rev. B*, **43**, 8763.

Goldman, A.I., Stassis, C., de Boissieu, M., Currat, R., Janot, C., Bellissent, R., Mouden, H., and Gayle, F.W. (1992). Phonons in icosahedral and cubic Al-Li-Cu. *Phys. Rev. B*, **45**, 10280–91.

Goldschmidt, V., Palache, Ch., and Peacock, M. (1931). Ueber Calaverit. *N. Jahrb. f. Min., Beil. Abt. A*, **63**, 1–58.

Gomez, C.P. and Lidin, S. (2001). Structure of $Ca_{13}Cd_{76}$ A novel approximant to the $MCd_{5.7}$ quasicrystals (M. Ca, Yb). *Angew. Chem. Int.*, **40**, 4037–9.

Gomez, C.P. and Lidin, S. (2003). Comparative structural study of the disordered MCd_6 quasicrystal approximants. *Phys. Rev. B*, **68**, 024203.

Gommonai, A.V., Grabar, A.A., Vysochanskii, Yu.M., Belyaev, A.D., Maluchin, V.F., Gurzan, M.I., and Slivka, V.Yu. (1981). In Russian. *Fiz. Tverd. Tela*, **23**, 3602.

Gratias, D., Puyraimond, F., and Quiqandon, M. (2001). Atomic clusters in icosahedral F-type quasicrystals. *Phys. Rev. B*, **63**, 024202.

Griessen, R., Landolt, M., and Ott, H.R. (1971). New antiferromagnetic phase in EuSe below 1.8 K. *Sol. State Comm.*, **9**, 2219–23.

Grimm, U. and Baake, M. (2013). *Aperiodic Order*. Cambridge University Press.

Grimm, U. and Joseph, D. (2002). Modeling quasicrystal growth. In *Quasicrystals* (ed. J. Suck, M. Schreiber, and P. Häussler), pp. 199–218. Springer Berlin-Heidelberg.

Grimm, U. and Schreiber, M. (2003). Energy spectra and eigenstates of quasiperiodic tight-binding Hamiltonians. In *Quasicrystals* (ed. H. Trebin), pp. 210–35. Wiley-VCH.

Grünbaum, B. and Shephard, G.C. (1987). *Tilings and Patterns*. W.H. Freeman.

Grüner, G. (1994). *Density Waves in Solids*. Addison-Wesley.

Gummelt, P. (1996). Penrose tilings as coverings of congruent decagons. *Geom. Ded.*, **62**, 1–17.

Guyot, P. and Audier, M. (1985). A quasicrystal structure model for Al-Mn. *Phil. Mag. B*, **52**, L15–L19

Hafner, J. and Krajči, M. (1992). Electronic structure and stability of quasicrystals: Quasiperiodic dispersion relations and pseudogaps. *Phys. Rev. Lett.*, **68**, 2321–24.

Hafner, J. and Krajči, M. (1999). Elementary excitations and physical properties. In *Physical Properties of Quasicrystals* (ed. Z. Stadnik), pp. 209–56. Springer.

Hargreaves, M.E. (1951). Modulated structures in some Cu-Ni-Fe alloys. *Acta Cryst. A*, **4**, 301–9.

Hartman, P. and Perdok, W.G. (1955). On the relations between structure and morphology of crystals. III. *Acta Cryst.*, **8**, 525–9.

Hatsui, T. and Graafsma, H. (2015). X-ray imaging detectors for synchrotron and XFEL sources. *IUCrJ*, **2**, 371–83.

Haussermann, U., Soderberg, K., and Norrestam, R. (2002). Comparative study of the high-pressure behavior of As, Sb, and Bi. *J. Am. Chem. Soc.*, **124**, 15359–67.

Hayashida, K., Dotera, T., Takano, A., and Matsushita, Y. (2007). Polymeric quasicrystal: Mesoscopic quasicrystalline tiling in abc star polymers. *Phys. Rev. Lett.*, **98**, 195502.

Heilmann, I.U, Axe, J.D., Hastings, J.M., and Shirane, G. (1979). Neutron investigation of the dynamical properties of the mercury-chain compound $Hg_{3-\delta}AsF_6$. *Phys. Rev. B*, **20**, 751–62.

Henley, C.L. (1986). Sphere packings and local environments in Penrose tilings. *Phys. Rev. B*, **34**, 797–816.

Henley, C.L. (1991a). Cell geometry for cluster-based quasicrystal models. *Phys. Rev. B*, **43**, 993–1020.

Henley, C.L. (1991b). Random tiling models. In *Quasicrystals: The State of the Art* (ed. D. DiVicenzo and P. Steinhardt), pp. 429–524. World Scientific.

Hoeft, J.T., Ledieu, J., Haq, S., Lograsso, T.A., Ross, A.R., and McGrath, R. (2006). Adsorption of benzene on the five-fold surface of the i-$Al_{70}Pd_{21}Mn_9$ quasicrystal. *Phil. Mag.*, **86**, 869–74.

Hof, A. (1997a). Diffraction by aperiodic structures. In *The Mathematics of Long-Range Aperiodic Order* (ed. R. Moody), Mathematical and Physical Sciences 489, pp. 239–68. Kluwer.

Hof, A. (1997b). On scaling in relation to singular spectra. *Comm. Math. Phys.*, **184**, 567–77.

Hofstadter, D.R. (1976). Energy levels and wave functions of Bloch electrons in rational and irrational magnetic fields. *Phys. Rev. B*, **14**, 2239–49.

Ingersent, K. (1991). Matching rules for quasicrystalline tilings. In *Quasicrystals : The State of the Art* (ed. D. Divincenzo and P. Steinhardt), Volume 11 of *Directions in Condensed Matter Physics*, pp. 185–212. World Scientific.

Ishibashi, Y. and Dvorak, V. (1978). Incommensurate phase transitions under the existence of the Lifshitz invariants. *J. Phys. Soc. Jpn.*, **44**, 32–9.

Ishibashi, Y. and Shuba, H. (1978). Succesive phase transitions in ferroelectric $NaNO_2$ and $SC(NH_2)_2$. *J. Phys. Soc. Jp.*, **45**, 409–13.

Ishii, Y. (1989). Mode locking in quasicrystals. *Phys. Rev. B*, **39**, 11862–71.

Ishii, Y. (1990). Soft phason modes inducing rhombohedral-icosahedral transformation. *Phil. Mag. Lett.*, **62**, 393–7.

Ishii, Y. (1992). Phason softening and structural transitions in icosahedral quasicrystals. *Phys. Rev. B*, **45**, 5228–39.

Ishii, Y. (2000). Anisotropic phasonic diffuse scattering from decagonal quasicrystals. *Mat. Sc. Eng. A*, **294–6**, 377–80.

Ishii, Y. and Fujiwara, T. (2001). Hybridization mechanism for cohesion of Cd-based quasicrystals. *Phys. Rev. Lett.*, **87**, 206408/1–4.

Ishimasa, T. (1995). Superlattice ordering in the low-temperature icosahedral phase of AlPdMn. *Phil. Mag. Lett.*, **71**, 65–73.

Ishimasa, T., Iwami, S., Sakaguchi, N., Oota, R., and Mihalkovič, M. (2015). Phason space analysis and structure modelling of 100 -scale dodecagonal quasicrystal in Mn-based alloy. *Phil. Mag.*, **95**, 3745–67.

Ishimasa, T., Kasano, Y., Tachibana, A., Kashimoto, S., and Osaka, K. (2007). Low temperature phase of the Zn-Sc approximant. *Phil. Mag.*, **87**, 2887–97.

Iwami, S. and Ishimasa, T. (2015). Dodecagonal quasicrystal in Mn-based quaternary alloys containing Cr, Ni and Si. *Phil. Mag. Lett.*, **95**, 229–36.

Jach, T., Zhang, Y., Colella, R., de Boissieu, M., Boudard, M., Goldman, A.I., Lograsso, T.A., Delaney, D.W., and Kycia, S. (1999). Dynamical diffraction and X-ray standing waves from atomic planes normal to a twofold symmetry axis of the quasicrystal AlPdMn. *Phys. Rev. Lett.*, **82**, 2904–7.

Jagličić, Z., Dolinšek, J., and Trontelj, Z. (2004). Magnetic properties of TbMgZn and TbMgCd quasicrystals in comparison with canonical spin glasses. *J. Magn. Magn Mat.*, **272–6**, 597–8.

Jagodzinski, H. and Korekawa, M. (1965). Supersatelliten im Beugungsbild des Labradorits $(Ca, Na)(Si, Al)_2O_8$. *Naturwisenschaften*, **52**, 640.

Jagodzinski, H. and Korekawa, M. (1972). X-ray investigations of lunar plagioclases and pyroxenes. *Geochim. Cosmochim. Acta*, **3**, 555–68.

Jamieson, P.B., de Fontaine, D., and Abrahams, S.C. (1969). Determination of atomic ordering arrangements from a study of satellite reflections. *J. Appl. Cryst.*, **2**, 24–30.

Janner, A. and Ascher, E. (1970). Algebraic aspects of crystallography. II. *Helv. Phys. Acta*, **43**, 296.

Janner, A. and Dam, B. (1989). The Morphology of calaverite ($AuTe_2$) from data of 1931: Solution of an old problem of rational indexes. *Acta Cryst. A*, **45**, 115–23.

Janner, A. and Janssen, T. (1971). Electromagnetic compensating gauge transformations. *Physica*, **53**, 1–27.

Janner, A. and Janssen, T. (1972). A charged particle in the field of a transverse electromagnetic plane wave. *Physica*, **60**, 292–321.

Janner, A. and Janssen, T. (1977). Symmetry of periodically distorted crystals. *Phys. Rev. B*, **15**, 643–58.

Janner, A. and Janssen, T. (1980a). Symmetry of incommensurate crystal phases I. Commensurate basis structure. *Acta Cryst. A*, **36**, 399–408.

Janner, A. and Janssen, T. (1980b). Symmetry of incommensurate crystal phases II. Incommensurate basis structure. *Acta Cryst. A*, **36**, 409–15.

Janner, A., Rasing, Th., Bennema, P., and van der Linden, W.H. (1980). Identification of satellite faces on single crystals of the incommensurate structures Rb_2ZnBr_4 and Rb_2ZnCl_4. *Phys. Rev. Lett.*, **45**, 1700–2.

Janot, C. (2012). *Quasicrystal: A Primer*, 2e. Oxford University Press.

Janot, C. and de Boissieu, M. (1994). Quasicrystals as a hierarchy of clusters. *Phys. Rev. Lett.*, **72**, 1674–7.

Janssen, T. (1969). Crystallographic groups in space and time. III Four-dimensional Euclidean crystal classes. *Physica*, **42**, 71–99.

Janssen, T. (1972). The magnetic structure of an f.c.c. spin lattice system with exchange and dipolar interactions. *Phys. Kond. Mat.*, **15**, 142–57.

Janssen, T. (1973). *Crystallographic Groups*. North Holland.

Janssen, T. (1986). On the application of a frustration model to the phase diagram of TMA-chlorometallates and other A_2BX_4 compounds. *Ferroelectrics*, **66**, 203–16.

Janssen, T. (1988). Aperiodic crystals: A contradictio in terminis? *Physics Repts.*, **168**, 55–113.

Janssen, T. (1991). The symmetry of quasiperiodic systems. *Acta Cryst. A*, **47**, 243–55.

Janssen, T. (1992). Models for incommensurate phases in crystals with Pcmn symmetry. *Z. Phys. B*, **86**, 277–83.

Janssen, T. (1994). Dynamics of (anti)ferromagnetic/ferroelectric domain walls. *Ferroelectrics*, **162**, 265–73.

Janssen, T. (2005). Nonlinear dynamics in aperiodic crystals. In *Collective Dynamics of Nonlinear and Disordered Systems* (ed. G. Radons, W. Just, and P. Häussler), pp. 237–66. Springer.

Janssen, T., Birman, J.L., Denoyer, F., Koptsik, V.A., Verger-Gaugry, J.L., Weigel, D., Yamamoto, A., Abrahams, S.C., and Kopsky, V. (2002). Report of a subcommittee on

the nomenclature of n-dimensional crystallography: II Symbols for arithmetic crystal classes, Bravais classes and space groups. *Acta Cryst. A*, **58**, 605–21.
Janssen, T., Birman, J., Koptsik, V.A., Senechal, M., Yamamoto, A., Abrahams, S.C., and Hahn, Th. (1999). Report of a subcommittee on the nomenclature of n-dimensional crystallography: Symbols for point group transformations, families, systems and geometric crystal classes. *Acta Cryst. A*, **55**, 761–82.
Janssen, T. and Janner, A. (1987). Incommensurability in crystals. *Adv. Phys.*, **36**, 519–624.
Janssen, T., Janner, A., and Ascher, E. (1969). Crystallographic groups in space and time. I. General definitions and basic properties. *Physica*, **41**, 541–65.
Janssen, T., Janner, A., and Bennema, P. (1989). On the morphology of quasicrystals. *Phil. Mag. B*, **59**, 233–42.
Janssen, T., Janner, A., Looijenga-Vos, A., and de Wolff, P.M. (1992). Incommensurate and commensurate modulated structures. In *International Table of Crystallography*, Volume C, pp. 899–937. Kluwer.
Janssen, T., Radulescu, O., and Rubtsov, A.N. (2002). Phasons, sliding modes and friction. *Eur. Phys. J. B*, **29**, 85–95.
Janssen, T. and Tjon, J.A. (1981). One-dimensional model for a crystal with displacive modulation. *Phys. Rev. B*, **24**, 2245–48.
Janssen, T. and Tjon, J.A. (1982). Microscopic model for incommensurate crystal phases. *Phys. Rev.*, **25**, 3767–85.
Janssen, T. and Tjon, J.A. (1983). Incommensurate crystal phases in mean-field approximation. *J. Phys. C*, **16**, 4789–810.
Jarič, M.V. and Nelson, D.R. (1988). Diffuse scattering from quasicrystals. *Phys. Rev. B*, **37**, 4458–72.
Jarič, M.V. and Qiu, S.Y. (1993). On the solution of the phase problem in quasiperiodic crystals. *Acta Cryst. A*, **49**, 576–85.
Jaszczak, J. A. (1994). Quasicrystals: Novel forms of solid matter. *Mineralog. Rec.*, **25**, 85–93.
Jehanno, G. and Perio, P. (1962). Structure of AuCu II. *J. Phys. Radium*, **23**, 854.
Jeong, H.-C. and Steinhardt, P.J. (1997). Constructing Penrose-like tilings from a single prototile and the implications for quasicrystals. *Phys. Rev. B*, **55**, 3520–32.
Jin, C., Cheng, B., Man, B., Li, Z., and Zhang, D. (2000). Two-dimensional dodecagonal and decagonal quasiperiodic photonic crystals in the microwave region. *Phys. Rev. B*, **61**, 10762–7.
Jin, C., Cheng, B., Man, B., Li, Z., Zhang, D., Ban, S., and Sun, B. (1999). Band gap and wave guiding effect in a quasiperiodic photonic crystal. *Appl. Phys. Lett.*, **75**, 1848–50.
Johnson, C.K. and Watson, C.R. (1976). Superstructure and modulation wave analysis for the unidimensional conductor hepta-(tetrathiafulvalene) pentaiodide. *J. Chem. Phys.*, **64**, 2271–86.
Jurdik, E., Myszkiewicz, G., Hohlfeld, J., Tsukamoto, A., Toonen, A.J., van Etteger, A.F., Gerritsen, J., Hermsen, J., Goldbach-Aschemann, S., Meerts, W.L., van Kempen, H., and Rasing, Th. (2004). Quasiperiodic structures via atom-optical nanofabrication. *Phys. Rev. B*, **69**, 201102.
Kalning, M., Dorna, V., Press, W., Kek, S., and Boysen, H. (1997). Profile analysis of the superlattice reflections in labradorite. *Z. Krist.*, **212**, 545–9.
Kalugin, P.A., Kitaev, A.Y., and Levitov, L.S. (1985). $Al_{0.86}Mn_{0.14}$: a six-dimensional crystal. *JETP Lett.*, **41**, 145–9.
Kashimoto, S., Motomura, S., Nakano, H., Kaneko, Y., Ishimasa, T., and Matsuo, S. (2002). Magnetic property of a ZnMgSc icosahedral quasicrystal. *J. Alloys Comp.*, **342**, 384–8.

Katz, A. and Gratias, D. (1994). Tilings and quasicrystals. In *Lectures on Quasicrystals* (ed. F. Hippert and D. Gratias), pp. 187–264. Les Ulis: Les éditions de physique.

Keen, D.A. and Goodwin, A.L. (2015). The crystallography of correlated disorder. *Nature*, **521**, 14455.

Kellendonk, J., Lenz, D., and Savien, J. (2015). *Mathematics of Aperiodic Order*. Birkhäuser.

Kenzelmann, M., Lawes, G., Harris, A.B., Gasparovic, G., Broholm, C., Ramirez, A.P., Jorge, G.A., Jaime, M., Park, S., Huang, Q., Shapiro, A.Ya., and Demianets, L.A. (2007). Direct transition from a disordered to a multiferroic phase on a triangular lattice. *Phys. Rev. Lett.*, **98**, 267205.

Keys, A.S. and Glotzer, S.C. (2007). How do quasicrystals grow? *Phys. Rev. Lett.*, **99**, 235503.

Khomskii, D. (2009). Classifying multiferroics: mechanisms and effects. *Physics*, **2**, 20.

Kimura, T., Sekio, Y., Nakmura, H., Siegrist, T., and Ramirez, A.P. (2008). Cupric oxide as an induced multiferroic with high Tc. *Na. Mater.*, **291-4**, 20.

Kittel, Ch. (1996). *Introduction to Solid State Physics, 7th edition*. John Wiley.

Kobas, M., Weber, T., and Steurer, W. (2005). Structural disorder in decagonal Al-Ni-Co. Part II: Modelling. *Phys. Rev. B*, **71**, 224206.

Kocinski, J. (1985). *Theory of Symmetry Changes in Phase Transitions*. Elsevier.

Koehler, W.C., Cable, J.W., Wollan, E.O., and Wilkinson, M.K. (1962). Neutron diffraction study of magnetic ordering in Thulium. *J. Appl. Phys.*, **33**, 1124–5.

Kohmoto, M., Kadanoff, L.P., and Tang, C. (1983). Localization problem in one dimension: Mapping and escape. *Phys. Rev. Lett.*, **50**, 1870–2.

Korekawa, M. (1967). *Theorie der Satellitenreflexe*. Habilitationsschrift, Ludwig-Maximilians-Universität.

Koschella, U., Gähler, F., Roth, J., and Trebin, H.-R. (2002). Phason elastic constants of a binary tiling quasicrystal. *J. Alloys Comp.*, **342**, 287–90.

Krajčí, M. and Hafner, J. (2002). Phonons and electrons in quasicrystals. In *Quasicrystals* (ed. J.-B. Suck, M. Schreiber, and P. Häussler), pp. 393–422. Springer.

Krajčí, M., Hafner, J., and Mihalkovič, M. (2000). Atomic and electronic structure of decagonal Al-Ni-Co alloys and approximant phases. *Phys. Rev. B*, **62**, 243–55.

Kramer, P. (1982). Non-periodic central space filling with icosahedral symmetry using copies of seven elementary cells. *Acta Cryst. A*, **38**, 257–64.

Kramer, P. and Neri, R. (1984). On periodic and non-periodic space fillings of E^m obtained by projection. *Acta Cryst. A*, **40**, 580–7.

Kremers, M., Meekes, H., Bennema, P., Balzuweit, K., and Verheijen, M.A. (1994). A superspace description for the morphology of modulated crystals: An explanation for the occurrence of faces $(hklm)$. *Phil. Mag. B*, **69**, 69–82.

Krutzen, B. and Inglesfield, J. (1990). First-principles electronic structure calculations for incommensurately modulated sylvanite. *J. Phys. Cond. Mat.*, **2**, 4829–48.

Kubel, F. and Schmid, H. (1990). Structure of a ferroelectric and ferroelastic monodomain crystal of the perovskite $BiFeO_3$. *Acta Cryst. B*, **46**, 698–702.

Kycia, S.W., Goldman, A.I., Lograsso, T.A., Delaney, D.W., Black, D., Sutton, M., Dufresne, E., Bruning, R., and Rodricks, B. (1993). Dynamical X-ray diffraction from an icosahedral quasicrystal. *Phys. Rev. B*, **48**, 3544–7.

Lagarias, J.C. (2000). Mathematical quasicrystals and the problem of diffraction. In *Directions in Mathematical Quasicrystals* (ed. M. Baake and R. Moody), pp. 61–93. American Mathematical Society.

Landau, L.D. and Lifshitz, E.M. (1959). *Statistical Physics*. Pergamon Press.

Laval, J. (1938). Diffusion of X-rays by a crystal. *C. R. Acad. Sci.*, **207**, 169.
Lei, T. and Henley, C.L. (1991). Equilibrium faceting shape of quasicrystals at low temperature: Cluster model. *Phil. Mag. B*, **63**, 677–85.
Létoublon, A., de Boissieu, M., Boudard, M., Mancini, L., Gastaldi, J., Hennion, B., Caudron, R., and Bellissent, R. (2001). Phason elastic constants of the icosahedral Al-Pd-Mn phase derived from diffuse scattering measurements. *Phil. Mag. Lett.*, **81**, 273–83.
Létoublon, A., Fisher, I.R., Sato, T.J., de Boissieu, M., Boudard, M., Agliozzo, S., Mancini, L., Gastaldi, J., Canfield, P.C., Goldman, A.I., and Tsai, A.P. (2000a). Phason strain and structural perfection in the Zn-Mg-rare-earth icosahedral phases. *Mat. Sci. Eng. A*, **A294**, 127–30.
Létoublon, A., Ishimasa, T., de Boissieu, M., Boudard, M., Hennion, B., and Mori, M. (2000b). Stability of the F2-(Al-Pd-Mn) phase. *Phil. Mag. Lett.*, **80**, 205–13.
Létoublon, A., Yakhou, F., Livet, F., Bley, F., de Boissieu, M., Mancini, L., Caudron, R., Vettier, C., and Gastaldi, J. (2001). Coherent X-ray diffraction and phason fluctuations in quasicrystals. *Europhy. Lett.*, **54**, 753–9.
Levanyuk, A.P. and Sannikov, D.G. (1975). Theory of structural phase transitions in boracites. *Sov. Phys. Sol. St.*, **17**, 245–8 and 1122–5.
Levine, D., Lubensky, T.C., Ostlund, S., Rawaswamy, S., Steinhardt, P.J., and Toner, J. (1985). Elasticity and dislocations in pentagonal and icosahedral quasicrystals. *Phys. Rev. Lett.*, **54**, 1520.
Levine, D. and Steinhardt, P.J. (1984). Quasicrystals: A new class of ordered structures. *Phys. Rev. Lett*, **53**, 2477–80.
Li, F. and Franzen, H.F. (1996). Incommensuration, and phase transitions in pyrrhotite: Part ii: A high-temperature x-ray powder diffraction and thermomagnetic study. *J. Sol. St. Chem.*, **126**, 108–20.
Lifshitz, R. (1997). Theory of color symmetry for periodic and quasiperiodic crystals. *Rev. Mod. Phys.*, **69**, 1181.
Lifshitz, R. and Diamant, H. (2007). Soft quasicrystals: Why are they stable? *Phil. Mag.*, **87**, 3021–30.
Livet, F., Bley, F., Mainville, J., Caudron, R., Mochrie, S.G., Geissler, E., Dolino, G., D., Abernathy, Grebel, G., and Sutton, M. (2000). Using direct illumination CCDs as high-resolution area detectors for X-ray scattering. *Nuc. Instr. Meth.*, **451**, 596.
Loa, I., Nelmes, R.J., Lundegaard, L.F., and McMahon, M.I. (2012). Extraordinarily complex crystal structure with mesoscopic patterning in barium at high pressure. *Nat. Mater.*, **11**, 627–32.
Los, J., Janssen, T., and Gähler, F. (1993a). The phonon spectra of the octagonal tiling. *Int. J. Mod. Phys. B*, **7**, 1505–25.
Los, J., Janssen, T., and Gähler, F. (1993b). Scaling properties of vibrational spectra and eigenstates for tiling models of icosahedral quasicrystals. *J. Physique*, **3**, 107–34.
Lubensky, T.C (1988). Symmetry, elasticity, and hydrodynamics in quasiperiodic structures. In *Aperiodicity and Order: Introduction to Quasicrystals* (ed. M. Jarič), Volume 1, pp. 199–277. Academic Press.
Lubensky, T.C., Socolar, J.E.S., Steinhardt, P.J., Bancel, P.A., and Heiney, P. (1986). Distortion and peak broadening in quasicrystal diffraction patterns. *Phys. Rev. Lett.*, **57**, 1440–3.
Lyons, K.B., Bhatt, R.N., Negram, T.J., and Guggenheim, H.J. (1982). Incommensurate structural phase transition in $BaMnF_4$. *Phys. Rev. B*, **25**, 1791–804.

Mackay, A.L. (1962). A dense non-crystallographic packing of equal spheres. *Acta Cryst.*, **15**, 916–8.
Mackay, A.L. (1981). De Nive Quinquangula: On the pentagonal snowflake. *Kristallografia*, **26**, 910–19.
Mackay, A.L. (1982). Crystallography and the Penrose pattern. *Physica*, **114A**, 609–613.
Maple, J.R., Hwang, M.J., Stockfisch, T.P., Dinur, U., Waldman, M., Ewig, C.S., and Hagler, A.T. (1994). Derivation of class-ii force-fields.1. Methodology and quantum force-field for the alkyl functional-group and alkane molecules. *J. Comp. Chem.*, **15**, 162–82.
Mariette, C., Guérin, L., Rabiller, Ph., Chen, YS., Bosak, A., Popov, A., M.D., Hollingsworth, and Toudic, B. (2015). The creation of modulated monoclinic aperiodic composites in n-alkane/urea compounds. *Z. Krist.*, **230**, 5.
Martin, P.C., Parodi, O., and Pershan, P.S. (1972). Unified hydrodynamic theory for crystals, liquid crystals, and normal fluids. *Phys. Rev. A*, **6**, 2401–20.
Matsukawa, S., Deguchi, K., Imura, K., Ishimasa, T., and Sato, N.K. (2016). Pressure-driven quantum criticality and T/H scaling in the icosahedral Au-Al-Yb approximant. *J. Phys. Soc. Jp*, **85**, 063706.
McConnell, J.D.C. (1978). Intermediate plagioclase feldspars: Example of a structural resonance. *Z. Krist.*, **147**, 45–62.
McGrath, R., Ledieu, J., Cox, E.J., and Diehl, R.D. (2002). Quasicrystal surfaces: structure and potential as template. *J. Phys. Cond. Mat.*, **14**, R119–44.
McMahon, M. and Nelmes, R. (2004). Incommensurate crystal structures in the elements at high pressure. *Z. Krist.*, **219**, 742–8.
McMillan, W.L. (1976). Theory of discommensurations and the commensurate-incommensurate charge-density-wave phase transition. *Phys. Rev. B*, **14**, 1496–502.
Menguy, N., Audier, M., Guyot, P., and Vacher, M. (1993a). Pentagonal phases as a transient state of the reversible icosahedral rhombohedral transformation in Al-Fe-Cu. *Phil. Mag. B*, **68**, 595–606.
Menguy, N., de Boissieu, M., Guyot, P., Audier, M., Elkaim, E., and Lauriat, J.P. (1993b). Single crystal X-ray study of a modulated icosahedral AlCuFe phase. *J. Phys. I France*, **3**, 1953–68.
Mermin, N.D. (1992). Copernican crystallography. *Phys. Rev. Lett.*, **68**, 1172–5.
Mihalkovic, M., Al-Lehyani, I., Cockayne, E., Henley, C.L., Moghadam, N., Moriarty, J.A., Wang, Y., and Widom, M. (2002). Total-energy-based prediction of a quasicrystal structure. *Phys. Rev. B*, **65**, 104205/1–6.
Mihalkovič, M., Francoual, S., Shibata, K., de Boissieu, M., Baron, A.Q.R., Sidis, Y., Ishimasa, T., Wu, D., Lograsso, T., Regnault, L.P., Gähler, F., Tsutsui, S., Hennion, B., Bastie, P., Sato, T.J., Takakura, H., Currat, R., and Tsai, A.P. (2008). Atomic dynamics of i-ScZnMg and its 1/1 approximant phase: Experiment and simulation. *Phil. Mag.*, **88**, 2311–18.
Mihalkovič, M. and Henley, C.L. (2004). Temperature dependent phason elasticity in a random tiling quasicrystal. *Phys. Rev. B*, **70**, 09202.
Mihalkovič, M. and Henley, C.L. (2012). Empirical oscillating potentials for alloys from ab initio fits and the prediction of quasicrystal-related structures in the Al-Cu-Sc system. *Phys. Rev. B*, **85**, 092102.
Mihalkovič, M. and Mrafko, P. (1993). Tiling of canonical cell: Large Pa3 approximants. *Europhys. Lett.*, **21**, 463–7.

Mihalkovič, M. and Widom, M. (2004). First-principles prediction of a decagonal quasicrystal containing boron. *Phys. Rev. Lett.*, **93**, 95507.

Mihalkovič, M. and Widom, M. (2006). Canonical cell model of cadmium-based icosahedral alloys. *Phil. Mag.*, **86**, 519–27.

Minchau, B., Szeto, K.Y., and Villain, J. (1987). Growth of hard-sphere models with two different sizes: Can a quasicrystal result? *Phys. Lett.*, **58**, 1960.

Mischo, P., Decker, F., Häcker, U., Holzer, K.-P., and Petersson, J. (1997). Low-frequency phason and amplitudon dynamics in the incommensurate phase of Rb_2ZnCl_4. *Phys. Rev. Lett.*, **78**, 2152–5.

Moody, R.V. (1997). *The Mathematics of Long-Range Aperiodic Order*. Kluwer.

Morozov, V. A., Arakcheeva, A. V., Chapuis, G., Guiblin, N., Rossell, M. D., and Van Tendeloo, G. (2006). $KNd(MoO_4)_2$: A new incommensurate modulated structure in the scheelite family. *Chem. Mater.*, **18**, 4075–82.

Moudden, A.H., Denoyer, F., Benoit, J.P., and Fitzgerald, W. (1978). Inelastic neutron-scattering study of commensurate-incommensurate phase-transition in thiourea. *Sol. St. Comm.*, **28**, 575.

Nagao, K., Inuzuka, T., Nishimoto, K., and Edagawa, K. (2015). Experimental observation of quasicrystal growth. *Phys. Rev. Lett.*, **115**, 075501.

Nayak, J., Maniraj, M., Gloskovskii, A., Krajči, M., Sebastian, S., Fisher, I.R., Horn, K., and Barman, S.R. (2015). Bulk electronic structure of Zn-Mg-Y and Zn-Mg-Dy icosahedral quasicrystals. *Phys. Rev. B*, **91**, 235116.

Nayak, J., Maniraj, M., Rai, A., Singh, S., Rajput, P., Gloskovskii, A., Zegenhagen, J., Schlagel, D.L., Lograsso, T.A., Horn, K., and Barman1, S.R. (2012). A bulk electronic structure of quasicrystals. *Phys. Rev. Lett.*, **109**, 216403.

Nelmes, R.J., Allan, D.R., McMahon, M.I., and Belmonte, S.A. (1999). Self-hosting incommensurate structure of barium IV. *Phys. Rev. Lett.*, **83**, 4081–4.

Niizeki, K. (1989). A classification of special points of icosahedral quasilattices. *J. Phys. A*, **22**, 4295–302.

Niizeki, K. (1990a). Icosahedral black-and-white Bravais quasilattices and order-disorder transformations of icosahedral quasicrystals. *J. Phys. A*, **23**, L1069–72.

Niizeki, K. (1990b). Three-dimensional black-and-white Bravais quasilattices with $(2 + 1)$-reducible point groups. *J. Phys. A*, **23**, 5011–16.

Niizeki, K. and Akamatsu, T. (1990). Special points in the reciprocal space of an icosahedral quasicrystal and the quasidispersion relation of electrons. *J. Phys. Cond. Mat.*, **2**, 2759–71.

Nitomi, M. (2010). Manipulating light with strongly modulated photonic crystals. *Rep. Prog. Phys.*, **73**, 096501.

Nitomi, M. (2011). Strong light confinement with periodicity. *Proc. IEEE*, **99**, 1768–79.

Ollivier, J., Etrillard, J., Toudic, B., Ecolivet, C., Bourges, P., and Levanyuk, A.P. (1998). Direct observation of a phason gap in an incommensurate molecular compound. *Phys. Rev. Lett.*, **81**, 3667–70.

Olmer, Ph (1948). Dispersion des ondes acoustiques dans l'aluminium. *Acta Cryst.*, **1**, 57.

Orlov, I. and Chapuis, G. (2005a). New tool based on the superspace concept to discover structure relations. *Acta Cryst. A*, **61**, C105.

Orlov, I. and Chapuis, G. (2005b). Superspace tools; http://superspace.epfl.ch/.

Oszlányi, G. and Sütő, A. (2004). Ab initio structure solution by charge-flipping. *Acta Cryst. A*, **60**, 134–41.

Oszlányi, G. and Sütő, A. (2005). Ab initio structure solution by charge flipping. II. Use of weak reflections. *Acta Cryst. A*, **61**, 147–52.

Overhauser, A.W. (1971). Observability of charge-density waves by neutron diffraction. *Phys. Rev. B*, **3**, 373–82.

Paciorek, W. and Chapuis, G. (1992). A new algorithm for incommensurate structure refinement. *J. Appl. Cryst.*, **25**, 317–22.

Paciorek, W. and Chapuis, G. (1994). Generalized Bessel functions in incommensurate structure analysis. *Acta Cryst. A*, **50**, 194–203.

Paciorek, W., Bussien Gaillard, V., Schenk, K., and Chapuis, G. (1996). Molecular geometry of incommensurate structures. *Acta Cryst. A*, **52**, 349–64.

Paciorek, W. and Kucharczyk, D. (1985). Structure factor calculations in refinement of a modulated crystal structure. *Acta Cryst. A*, **41**, 462–6.

Pailhes, S., Euchner, H., Giordano, V.M., Debord, R., Assy, A., Gomes, S., Bosak, A., Machon, D., Paschen, S., and de Boissieu, M. (2014). Localization of propagative phonons in a perfectly crystalline solid. *Phys. Rev. Lett.*, **113**, 25506.

Palatinus, L. (2004). Ab initio determination of incommensurately modulated structures by charge flipping in superspace. *Acta Cryst. A*, **60**, 604–10.

Palatinus, L. and Chapuis, G. (2006). Superflip, a program for the solution of crystal structures in arbitrary dimension using the charge-flipping algorithm; http://superspace.epfl.ch/superflip.

Palatinus, L. and Chapuis, G. (2007). Superflip: A computer program for the solution of crystal structures by charge flipping in arbitrary dimensions. *J. Appl. Cryst.*, **40**, 786–90.

Pan, Y.S., Brown, D., and Chapuis, G. (2003). Molecular dynamics simulation of hexamine and suberic acid. *Mol. Sim.*, **29**, 509–18.

Papadopolos, Z., Kasner, G., Ledieu, J., Cox, E.J., Richardson, N.V., Chen, Q., Diehl, R.D., Lograsso, T.A., Ross, A.R., and McGrath, R. (2002). Bulk termination of the quasicrystalline fivefold surface of $Al_{70}Pd_{21}Mn_9$. *Phys. Rev. B*, **66**, 184207–1/13.

Papadopolos, Z., Pleasants, P., Kasner, G., Fournee, V., Jenks, C.J., Ledieu, J., and McGrath, R. (2004). Maximum density rule for bulk terminations of quasicrystals. *Phys. Rev. B*, **69**, 224201–1–7.

Park, J.Y., Ogletree, D.F., Salmeron, M., Jenks, C.J., and Thiel, P.A. (2004). Friction and adhesion properties of clean and oxidized AlNiCo quasicrystals. *Tribol. Lett.*, **17**, 629–36.

Park, J.Y., Ogletree, D.F., Salmeron, M., Ribeiro, R.A., Canfield, P.C., Jenks, C.J., and Thiel, P.A. (2005). High frictional anisotropy of periodic and aperiodic directions on a quasicrystal surface. *Science*, **309**, 1354–6.

Parlinski, K. and Chapuis, G. (1993). Mechanisms of phase transitions in a hexagonal model with 1q and 3q incommensurate phases. *Phys. Rev. B*, **47**, 13983–91.

Parlinski, K. and Chapuis, G. (1994). Phase-transition mechanisms between hexagonal commensurate and incommensurate structures. *Phys. Rev. B*, **49**, 11643–51.

Parlinski, K., Kwiecinski, S., and Urbanski, A. (1992). Phase-diagram of a hexagonal model with incommensurate phases. *Phys. Rev. B*, **46**, 5110–15.

Penfield, S.L. and Ford, W.E. (1902). Ueber den Calaverit. *Z. Kristallogr. Mineral.*, **35**, 430–51.

Penrose, R. (1979). Pentaplexity. A class of non-periodic tilings of the plane. *Math. Intell.*, **2**, 32–37.

Perez-Mato, J.M., Madariaga, G., and Elcoro, L. (1991). Influence of phason dynamics of atomic Debye-Waller factors of incommensurate modulated structures and quasicrystals. *Sol. St. Comm.*, **78**, 33–7.

Perez-Mato, J.M., Ribeiro, J.L., Petříček, V., and Aroyo, M. (2012). Magnetic superspace groups and symmetry constraints in incommensurate magnetic phases. *J. Phys. Cond. Matter*, **24**, 63201.

Perez-Mato, J.M., Zakhour-Nakl, M., and Darriet, J. (1999). Structure of composites $A_{1+x}(A'_xB_{1-x})O_3$ related to the 2H hexagonal perovskite: relation between composition and modulation. *J. Chem. Mater.*, **9**, 2795–808.

Peterkova, J., Dušek, M., Petříček, V., and Loub, J. (1998). Structures of fluoroarsenates $KAsF_{6-n}(OH)_n$, n = 0, 1, 2: Application of the heavy-atom method for modulated structures. *Acta Cryst. B*, **54**, 809–18.

Petříček, V., Coppens, P., and Becker, P. (1985). Structure-analysis of displacively modulated molecular-crystals. *Acta Cryst. A*, **41**, 478–83.

Petříček, V. and Dušek, M. (2000). Jana2000, Crystallographic computing system. Technical report, Institute of Physics, Praha, Czech Republic.

Petříček, V., Dušek, M., and Palatinus, L. (2014). Crystallographic Computing System JANA2006: General features. *Z. Krist.*, **229**, 345–52.

Petříček, V., Gao, Y., Lee, P., and Coppens, P. (1990). X-ray-analysis of the incommensurate modulation in the 2-2-1-2 Bi-Sr-Ca-Cu-O superconductor including the oxygen atoms. *Phys. Rev. B*, **42**, 387–92.

Petříček, V., van der Lee, A., and Evain, M. (1995). On the use of crenel functions for occupationally modulated structures. *Acta Cryst. A*, **51**, 529–35.

Pouget, J.P., Shirane, G., Hastings, J.M., Heeger, A.J., Miro, N.D., and McDiarmid, A.G. (1978). Elastic neutron scattering of the 'phase ordering' phase transition in $Hg_{3-\delta}AsF_6$. *Phys. Rev. B*, **18**, 3646–56.

Prince, E. (1994). *Mathematical Techniques in Crystallography and Material Science*. Springer Verlag.

Pryde, A. and Dove, M. (1998). On the sequence of phase transitions in tridymite. *Phys. Chem. Min.*, **26**, 171–9.

Pynn, R., Press, W., and Shapiro, S.M. (1976). Second and third harmonics of the spin density wave in chromium metal. *Phys. Rev. B*, **13**, 295–8.

Queffélec, M. (1987). *Substitution Dynamical Systems*, Volume 1294, of Lecture Notes in Mathematics. Springer.

Quilichini, M., Heger, G., Hennion, B., Lefebvre, S., and Quivy, A. (1990). Inelastic neutron scattering study of acoustic modes in a monodomain AlCuFe quasicrystal. *J. Phys. France*, **51**, 1785–90.

Quilichini, M., Hennion, B., Heger, G., Lefebvre, S., and Quivy, A. (1992). Inelastic neutron scattering study of icosahedral AlFeCu quasicrystal. *J. Phys. II France*, **2**, 125–30.

Quilichini, M. and Janssen, T. (1997). Phonon excitations in quasicrystals. *Rev. Mod. Phys.*, **69**, 277–314.

Quiquandon, M. and Gratias, D. (2006). An unique ideal structure model for AlPdMn and AlCuFe icosahedral phases. *Phys. Rev. B*, **74**, 214205.

Radulescu, O., Janssen, T., and Etrillard, J. (2002). Dynamics of modulated and composite aperiodic crystals: The signature of the inner polarization in the neutron inelastic scattering. *Eur. Phys. J. B*, **29**, 385–98.

Rasing, Th., Wyder, P., Janner, A., and Janssen, T. (1982). Far-infrared and Raman studies of the incommensurate structure Rb_2ZnBr_4. *Phys. Rev. B*, **25**, 7504–19.
Rau, D., Gavilano, J.L., Mushkolaj, Sh., Beeli, C., and Ott, H.R. (2003). Magnetism of decagonal $Al_{69.8}Pd_{12.1}Mn_{18.1}$. *Physica B*, **329–33**, 1103–4.
Rau, D., Gavilano, J.L., Mushkolaj, Sh., Beeli, C., and Ott, H.R. (2004). Inhomogeneous Mn-magnetism in AlPdMn quasicrystals. *J. Magn. Magn. Mat.*, **272–6**, 1330–1.
Robertson, J.L., Misenheimer, M.E., Moss, S.C., and Bendersky, L.A. (1986). X-ray and electron metallographic study of quasi-crystalline Al-Mn-Si alloys. *Acta Metal.*, **34**, 2177–89.
Rubtsov, A.N. and Janssen, T. (2001). Numerical study of the paraelectric-incommensurate-ferroelectric transition in the DIFFFOUR model. *Europhys. Lett.*, **53**, 216–20.
Sato, H. and Toth, R.S. (1961). Effect of additional elements on the period of CuAu II and the origin of the long-period superlattice. *Phys. Rev.*, **124**, 1833.
Sato, T.J., Takakura, H., Tsai, A.P., Shibata, K., Ohoyama, K., and Andersen, K.H. (2000). Antiferromagnetic spin correlations in the Zn-Mg-Ho icosahedral quasicrystal. *Phys. Rev. B*, **61**, 476–86.
Savkin, V.V., Rubtsov, A.N., and Janssen, T. (2003). Monte Carlo simulations of the classical two-dimensional DIFFFOUR model. *Eur. Phys. J. B*, **31**, 525–31.
Schellnhuber, H.J. and Urbschat, H. (1987). Analyticity breaking of wave functions and fractal phase diagram for simple incommensurate systems. *Phys. Stat. Sol. (b)*, **140**, 509–19.
Schmid, H. (1994). Multi-ferroic magnetoelectrics. *Ferroelectrics*, **162**, 317–38.
Schobinger-Papamantellos, P., Buschow, K.H.J., and Janssen, T. (1994). Neutron powder diffraction study of 2d and 3d incommensurately modulated magnetic structures. *Mat. Sci. For.*, **166–9**, 479–86.
Schobinger-Papamantellos, P., de Mooij, D.B., and Buschow, K.H.J. (1992). Crystal structure of the compound $DyGe_3$. *J. Alloys Comp.*, **183**, 181.
Schobinger-Papamantellos, P., Rodriguez-Carvajal, J., Janssen, T., and Buschow, K.H.J. (1995). Structure and magnetic phase transitions of the novel compound $TbGe_3$. In *Aperiodic '94*, pp. 302–6.
Schobinger-Papamantellos, P., Rodriguez-Carvajal, J., Tung, L., Ritter, C., and Buschow, K. (2008). Competing multiple-q magnetic structures in $HoGe_3$: II. Magnetic structures observed in $HoGe_3$. *J. Phys. Cond. Mat.*, **20**, 195202.
Schönleber, A. (2002). The Superspace Approach Applied to Modulated Crystals. Ph.D. thesis, Lausanne.
Schönleber, A. and Chapuis, G. (2004a). Quininium (R)-mandelate, a structure with large Z' described as an incommensurately modulated structure in (3+1)-dimensional superspace. *Acta Cryst. B*, **60**, 108–20.
Schönleber, A. and Chapuis, G. (2004b). The structure of diaqua (15-crown-5) copper (II) dinitrate described in (3+1)-dimensional superspace. *Ferroelectrics*, **305**, 99Ð102.
Schönleber, A., Meyer, M., and Chapuis, G. (2001). NADA: A computer program for the simultaneous refinement of orientation matrix and modulation vector(s). *J. Appl. Cryst.*, **34**, 777–9.
Schönleber, A., Pattison, P., and Chapuis, G. (2003). The (3 + 1)-dimensional superspace description of the commensurately modulated structure of p-chlorobenzamide (α-form) and its relation to the γ-form. *Z. Krist.*, **218**, 507–13.

Schultz, A.J., Williams, J.M., Miro, N.D., MacDiarmid, A.G., and Heeger, A.J. (1978). A neutron diffraction investigation of the crystal and molecular structure of the anisotropic superconductor Hg_3AsF_6. *Inorg. Chem.*, **17**, 646–9.

Schutte, W.J. and de Boer, J.L. (1988). Valence fluctuations in the incommensurately modulated structure of calaverite $AuTe_2$. *Acta Cryst. B*, **44**, 486–94.

Selke, W. (1988). The ANNNI model: Theoretical analysis and experimental application. *Phys. Rep.*, **170**, 213–64.

Selke, W. and Duxbury, P.M. (1984). The mean-field theory of the 3-dimensional ANNNI model. *Z. Physik B*, **57**, 49–58.

Shapiro, S.M (2010). Lattice dynamics and structural phase transitions. *Appl. Phys. A*, **99**, 543–8.

Shechtman, D., Blech, I., Gratias, D., and Cahn, J.W. (1984). Metallic phase with longe-range orientational order and no translational symmetry. *Phys. Rev. Lett.*, **53**, 1951–3.

Shibata, K., Currat, R., de Boissieu, M., Sato, T.J., Takakura, H., and Tsai, A.P. (2002). Dynamics of the ZnMgY icosahedral phase. *J. Phys. Cond. Matt.*, **14**, 1847–63.

Shimoda, M., Sharma, H.R., and Tsai, A.P. (2005). Scanning tunneling microscopy study of the five-fold surface of icosahedral Al-Cu-Ru quasicrystal. *Sur. Sc.*, **598**, 88–95.

Sławinski, W., Przeniosło, R., Sosnowska, I., Bieringer, M., Margiolakic, I., and Suard, E. (2009). Modulation of atomic positions in $CaCu_xMn_{7-x}O_{12}$ ($x \leq 0.1$). *Acta Cryst. B*, **65**, 535–42.

Slot, J.J.M. and Janssen, T. (1988). Dynamics of kinks in modulated crystals. *Physica D*, **32**, 27–71.

Smith, H. (1903). Ueber das bemerkenswerthe Problem der Entwickelung der Krystallfor-men des Calaverit. *Z. Kristallogr. Mineral.*, **37**, 209–34.

Snoeck, E., Roucau, C., and Saint-Grégoire, P. (1986). Electron-microscopy study of the modulated phases in berlinite $AlPO_4$ and quartz. *J. Physique*, **47**, 2041–53.

Socolar, J.E.S., Lubensky, T.C., and Steinhardt, P.J. (1986). Phonons, phasons and disloca-tions in quasicrystals. *Phys. Rev. B*, **34**, 3345.

Solbrig, H. and Landauro, C.V. (2003). Electronic transport parameters and spectral fine structure: From approximants to quasicrystals. In *Quasicrystals* (ed. H. Trebin), pp. 254–71. Wiley-VCH.

Stadnik, Z.M. (1999). Physical Properties of Quasicrystals. Springer.

Steinhardt, P.J., Jeong, H.-C., Saitoh, K., Tanaka, M., Abe, E., and Tsai, A.P. (1998). Exper-imental verification of the quasi-unit-cell model of quasicrystal structure. *Nature*, **365**, 55–7.

Steurer, W. (2005). Structural phase transitions from and to the quasicrystalline state. *Acta Cryst. A*, **61**, 28–38.

Steurer, W. and Cervellino, A. (2001). Quasiperiodicity in decagonal phases forced by inclined net planes? *Acta Cryst. A*, **57**, 333–40.

Steurer, W. and Deloudi, S. (2009). Crystallography of Quasicrystals. Springer.

Steurer, W. and Haibach, T. (1999). The periodic average structure of particular quasicrys-tals. *Acta Cryst. A*, **55**, 48–57.

Steurer, W., Haibach, T., Zhang, B., Kek, S., and Luck, R. (1993). The structure of decagonal $Al_{70}Ni_{15}Co_{15}$. *Acta Cryst. B*, **49**, 661–75.

Stokes, H.T., Hatch, D.M., and Campbell, B.J. (2011). ISOTROPY Software Suite: http://iso.byu.edu/iso/isotropy.php.

Suck, J.-B. (2002). Vibrational density of states of stable and metastable quasicrystalline solids. In *Quasicrystals* (ed. J.-B. Suck, M. Schreiber, and P. Häussler), pp. 454–71. Springer.

Sueno, S., Kimata, M., and Ohmasa, M. (1979). Te atom splitting and modulated structure in calaverite. In *Modulated Structures-1979*, AIP Conference Proceedings No. 53, pp. 333. American Institute of Physics.

Sutton, M. (2001). Coherent X-ray diffraction. In *Third-Generation Hard X-ray Synchrotron Radiation Sources: Source Properties, Optics, and Experimental Techniques* (ed. M. Mills, D.), pp. 101–23. John Wiley.

Swainson, I.P., Dove, M.T., and Harris, M.J. (1995). Neutron powder diffraction study of the ferroelastic phase-transition and lattice melting in sodium-carbonate, Na_2CO_3. *J. Phys. Cond. Matt.*, **7**, 4395–417.

Takakura, H., Shiono, M., Sato, T.J., Yamamoto, A., and Tsai, A.P. (2001a). Ab initio structure determination of icosahedral Zn-Mg-Ho quasicrystals by density modification method. *Phys. Rev. Lett.*, **86**, 236–9.

Takakura, H., Yamamoto, A., de Boissieu, M., Sato, T.J., and Tsai, A.P. (2004b). Ab initio structure determination of quasicrystals via single crystal X-ray diffraction. *Mat. Res. Soc. Symp. Proc.*, **805**, LL2.2.1–10.

Takakura, H., Yamamoto, A., de Boissieu, M., and Tsai, A.P. (2004a). Ab initio structure solution of icosahedral Cd-Yb quasicrystals by a density modification method. *Ferroelectrics*, **305**, 209–12.

Takakura, H., Yamamoto, A., Gomez, C.P., de Boissieu, M., and Tsai, A.P. (2007). Atomic structure of the binary icosahedral Yb-Cd quasicrystal. *Nat. Mater.*, **6**, 58.

Takakura, H., Yamamoto, A., and Tsai, A.P. (2001b). The structure of a decagonal $Al_{72}Ni_{20}Co_8$. *Acta Cryst. A*, **57**, 576–85.

Takeuchi, T. and Mizutani, U. (1995). Electronic structure, electronic transport properties, and relative stability of quasicrystals and their 1/1 and 2/1 approximants in the Al-Zn-Mg alloy system. *Phys. Rev. B*, **52**, 9300–9.

Talapin, D.V., Shevchenko, E.V., Bodnarchuk, M.I., Ye, Xingchen, Chen, Jun, and Murray, C.B. (2009). Quasicrystalline order in self-assembles binary nanoparticle superlattices. *Nature*, **461**, 964–7.

Tamura, R., Edagawa, K., Aoki, C., Takeuchi, S., and Suzuki, K. (2003). Low-temperature structural phase transition in a Cd_6Y 1/1 approximant. *Phys. Rev. B*, **68**, 174105.

Tamura, R., Edagawa, K., Murao, Y., Takeuchi, S., Suzuki, K., Ichihara, M., Isobe, M., and Ueda, Y. (2004). Order-disorder transition in cubic Cd_6Yb and Cd_6Ca. *J. Non-Cryst. Solids*, **334–5**, 173–6.

Tamura, R., Murao, Y., Takeuchi, S., Ichihara, M., Isobe, M., and Ueda, Y. (2002). A low-temperature order-disorder transition in a cubic Cd_6Yb crystalline approximant. *Jpn. J. Appl. Phys.*, **41**, L524–6.

Tang, L.H. (1990). Random tiling quasicrystals in three dimensions. *Phys. Rev. Lett.*, **64**, 2390–3.

Taniguchi, T., Nakata, K., Takaki, Y., and Sakurai, K. (1978). The crystal structures of the α form of p-chlorobenzamide at room temperature and at 120 °C. *Acta Cryst. B*, **34**, 2574–8.

Tanisaki, S. (1961). Microdomain structure in paraelectric phase of $NaNO_2$. *J. Phys. Soc. Jpn.*, **16**, 579.

Tate, M.W., Eikenberry, E.F., and Gruner, S.M. (2001). CCD detectors. In *International Tables of Crystallography*, Volume F, pp. 148–53. Kluwer.
Theis, W., Rotenberg, E., Franke, K.J., Gille, P., and Horn, K. (2003a). Electronic valence bands in decagonal Al-Ni-Co. *Phys. Rev. B*, **68**, 104205.
Theis, W., Sharma, H.R., Franke, K.J., and Rieder, K.H. (2003b). Surface phonons and quasicrystalline monolayers. In *Quasicrystals* (ed. H.-R. Trebin), pp. 622–8. VCH-Wiley.
Thiel, P. (2004). Structure and oxidation at quasicrystal surfaces. *Prog. Surf. Sc.*, **75**, 191–204.
Toledano, J.C. and Toledano, P. (1987). *The Landau Theory of Phase Transitions*. World Scientific.
Toudic, B., Lefort, R., Ecolivet, C., Guérin, L., Currat, R., Bourges, P., and Breczewski, T. (2011a). Mixed acoustic phonons and phase modes in an aperiodic composite crystal. *Phys. Rev. Lett.*, **107**, 205502.
Toudic, B., Rabiller, P., Bourgeois, L., Huard, M., Ecolivet, C., McIntyre, G. J., Bourges, P., Breczewski, T., and Janssen, T. (2011b). Temperature-pressure phase diagram of an aperiodic host guest compound. *Europhys. Lett.*, **93**, 16003.
Trambly de Laissardière, G. and Mayou, D. (2000). Magnetism in Al(Si)-Mn quasicrystals and related phases. *Phys. Rev. Lett.*, **85**, 3273–6.
Trambly de Laissardière, G. and Mayou, D. (2002). Magnetic properties of quasicrystals and approximants. In *Quasicrystals* (ed. J.-B. Suck, M. Schreiber, and P. Häussler), pp. 487–506. Springer.
Trebin, H-R. (ed.) (2003). *Quasicrystals*. Wiley-VCH.
Tsai, A.P., Guo, J.Q., Abe, E., Takakura, H., and Sato, T.J. (2000). A stable binary quasicrtystal. *Nature*, **408**, 537–8.
Tsai, A.P., Inoue, A., Yokoyama, Y., and Masumoto, T. (1990). Stable icosahedral AlPdMn and AlPdRe alloys. *Mat. Trans., JIM*, **31**, 98–103.
Tsai, A.P. and Yoshimura, M. (2001). Highly active quasicrystalline Al-Cu-Fe catalyst for steam reforming of methanol. *Appl. Catal. A*, **214**, 237–41.
Tsai, A.-P., Inoue, A., and Masumoto, T. (1987). A stable quasicrystal in Al-Cu-Fe system. *Japan. J. Appl. Phys.*, **26**, L1505–7.
Tsunoda, Y., Mori, M., Kunitomi, N., Teraoka, Y., and Kanamori, J. (1974). Strain wave in pure chromium. *Sol. State Comm.*, **14**, 287–9.
Ungar, G. and Zeng, X. (2005). Frank-Kasper, quasicrystalline and related phases in liquid crystals. *Soft Matt.*, **1**, 95–106.
van Aalst, W., den Holander, J., Peterse, W.J.A.M., and de Wolff, P.M. (1976). The modulated structure of γ-Na_2CO_3 in a harmonic approximation. *Acta Cryst. B*, **32**, 47–58.
van Erp, T., Fasolino, A., Radulescu, O., and Janssen, T. (1999). Pinning and phonon localization in Frenkel-Kontorova models on quasiperiodic substrates. *Phys. Rev. B*, **60**, 6522–8.
van Landuyt, J., van Tendeloo, G., and Amelinckx, S. (1974). Electron diffraction patterns of distortion modulated structures. *Phys. Stat. Sol. (a)*, **26**, K9–11.
van Smaalen, S. (1991). Symmetry of composite crystals. *Phys. Rev. B*, **43**, 11330–41.
van Smaalen, S. (1993). Theory of incommensurate crystal facets. *Phys. Rev. Lett.*, **70**, 2419–22.

van Smaalen, S. (1999). The morphology of quasicrystals, incommensurate composite crystals and modulated crystals derived from the broken-bond model. *Acta Cryst. A*, **55**, 401–12.

van Smaalen, S. (2012). *Incommensurate Crystallography*. Oxford University Press.

van Smaalen, S. and George, T. F. (1987). Determination of the incommensurately modulated structure of α-uranium below 37 K. *Phys. Rev. B*, **35**, 7939–51.

van Smaalen, S., Palatinus, L., and Schneider, M. (2003). The maximum-entropy method in superspace. *Acta Cryst. A*, **59**, 459–69.

van Tendeloo, G., Amelinckx, S., Darriet, B., Bontchev, R., Darriet, J., and Weill, F. (1994). Structural considerations and high-resolution electron-microscopy observations on $La_n Ti_{n-\delta} O_{3n}$ ($n \geq 4\delta$). *J. Sol. State Chem.*, **108**, 314–35.

van Tendeloo, G., Gregoriades, P., and Amelinckx, S. (1983). Electron-microscopy studies of modulated structures in (Au, Ag)Te_2. 1. Calaverite $AuTe_2$. *J. Solid. State Chem.*, **50**, 321–34.

Vanossi, A., Röder, J., Bishop, A.R., and Bortolani, V. (2000). Driven, underdamped Frenkel-Kontorova model on a quasiperiodic substrate. *Phys. Rev. E*, **63**, 017203.

Walker, C.B. (1956). X-ray study of lattice vibrations in aluminium. *Phys. Rev.*, **B**, 547–57.

Watanabe, S. and Miyake, K. (2016). Origin of the quantum criticality in Yb-Al-Au approximant crystal and quasicrystal. *J. Phys. Soc. Jap.*, **85**, 063703.

Welberry, T.R. (2004). *Diffuse X-Ray Scattering and Models of Disorder*. Oxford University Press.

Widom, M. (1991). Elastic stability and diffuse scattering in icosahedral quasicrystals. *Phil. Mag. Lett.*, **64**, 297–305.

Withers, R. (2015). A modulation wave approach to the order hidden in disorder. *IUCrJ*, **2**, 74–84.

Yamada, T., Euchner, H., Gómez, C.P., Takakura, H., Tamura, R., and de Boissieu, M. (2013). Short- and long-range ordering during the phase transition of the $Zn_6 Sc$ 1/1 cubic approximant. *J. Phys. Cond. Matt.*, **25**, 205405.

Yamada, T., Takakura, H., Euchner, H., Gómez, C.P., Bosak, A., Fertey, P., and de Boissieu, M. (2016a). Atomic structure and phason modes of the Sc-Zn icosahedral quasicrystal. *IUCrJ*, **3**, 247–58.

Yamada, T., Takakura, H., Kong, T., Das, P., Jayasekara, W.T., Kreyssig, A., Beutier, G., Canfield, P.C., de Boissieu, M., and Goldman, A.I. (2016b). Atomic structure of the i-R-Cd quasicrystals and consequences for magnetism. *Phys. Rev. B*, **94**, 060103.

Yamamoto, A. (1982a). Modulated structure of CuAu II (one-dimensional modulation). *Acta Cryst. B*, **38**, 1446–51.

Yamamoto, A. (1982b). Structure factor of modulated crystal structures. *Acta Cryst. A*, **38**, 87–92.

Yamamoto, A. (1985). Displacive modulation in the sinusoidal antiferroelectric phase of $NaNO_2$. *Phys. Rev. B*, **31**, 5941.

Yamamoto, A. (1992). Unified setting and symbols of superspace groups for composite crystals. *Acta Cryst. A*, **48**, 476–83.

Yamamoto, A. (1993). Determination of composite crystal-structures and superspace groups. *Acta Cryst. A*, **49**, 831–46.

Yamamoto, A. (1999). Superspace groups for one-, two and three-dimensionally modulated structures: http://www.nims.go.jp/aperiodic/yamamoto/spgr.new.html.

Yamamoto, A. (2004). Section method for projected structures of icosahedral quasicrystals and its application to electron-microscopy-image and surface analysis. *Phys. Rev. Lett.*, **93**, 195505.

Yamamoto, A., Takakura, H., and Tsai, A.P. (2003). Six-dimensional model of icosahedral Al-Pd-Mn quasicrystals. *Phys. Rev. B*, **68**, 94201-1–13.

Yan, Y., Pennycook, S.J., and Tsai, A.P. (1998). Direct imaging of local chemical disorder and columnar vacancies in ideal decagonal Al-Ni-Co quasicrystals. *Phys. Rev. Lett.*, **81**, 5145–8.

Yokoyama, Y., Tsai, A., Inoue, A., and Masumoto, T. (1991). Production of quasicristalline AlPdMn alloys with large single domain size. *Mater. Trans., JIM*, **32**, 1089–97.

Yoshimori, A. (1959). A new type of antiferromagnetic structure in the rutile type crystal. *J. Phys. Soc. Jpn.*, **14**, 807–21.

Zassenhaus, H. (1947). Über einen Algorithmus zur Bestimmung der Raumgruppen. *Comm. Math. Helv.*, **21**, 117–41.

Zeng, X. (2005). Liquid quasicrystals. *Curr. Op. Coll. Inter. Sci.*, **9**, 384–9.

Zeng, X., Ungar, G., Liu, Y., Percec, V., Dulcey, A.E., and Hobbs, J.K. (2004). Supramolecular dendritic liquid quasicrystals. *Nature*, **428**, 157–60.

Zeyher, R. and Finger, W. (1982). Phason dynamics of incommensurate crystals. *Phys. Rev. Lett.*, **49**, 1833–7.

Zijlstra, E.S., Fasolino, A., and Janssen, T. (1999). Existence and localization of surface states on Fibonacci quasicrystals: A tight-binding study. *Phys. Rev. B*, **59**, 302–7.

Zijlstra, E.S. and Janssen, T. (2000a). Density of states and localization of electrons in the Penrose tiling. *Phys. Rev. B*, **61**, 3377–83.

Zijlstra, E.S. and Janssen, T. (2000b). Non-spiky density of states of an icosahedral quasicrystal. *Europhys. Lett.*, **52**, 578–83.

Index

2p-norm 348

ab initio calculations 350
acceptance domain 126
almost periodic 34
almost-Mathieu
    equation 359
Ammann tiling 81, 139
Ammann–Beenker
    tiling 124
amplitudon 328
ANNNI model 416
anti-ferroic 409, 420
aperiodic crystals 15
aperiodic phase 409
aperiodic tight-binding
    model 359
aperiodic tiling 115
approximant 29, 91
    L/N 143
    icosahedral tiling 142
    octagonal tiling 142
area detectors 167
arithmetic crystal class 479
atomic density 153
atomic surface 59, 62, 126
    fractal surface 74, 119
    non-flat surface 74
autocorrelation function 150
average structure 20, 180, 452

basic structure 17, 20, 35
Bessel functions 174, 178
    generalized 174, 175, 179
Bieberbach Theorem 482
Bienenstock–Ewald
    formulation 102
binary icosahedral
    quasicrystal 269
Bloch function 8
Bragg peaks 148
Bragg's law 148
Bravais class 477, 482
Bravais group 477
Bravais module 115, 476
Bravais–Friedel law 458
    generalization 459

breaking of analyticity 324, 422
Brillouin zone 5, 8
broken bonds 455, 459
Burgers vector 472

Cahn–Gratias
    convention 165, 238
calaverite 444
canonical cell tiling 146
canonical tiling 126, 345
CCD, *see* charge-coupled
    devices
centralizer 98
centred i-lattices 140
centring of a lattice 479
charge-coupled devices 168
charge-density-wave
    system 440
charge-flipping 184
chemical composition 81
chemistry 215, 221
choice of basis vectors
    modulated phase 162
    quasicrystals 163
closeness condition 64, 71,
    81, 200, 255, 285
clusters 29
    Bergman 29
    Mackay 29
    Tsai 29
co-dimension 34
commensurate
    approximation 180
commensurate structure 191
commensurately modulated
    structures 191
compensating gauge
    transformation 47, 94
composite structure 189, 198
compounds
    α-Na$_2$CO$_3$ 193
    α-U 208
    β-Na$_2$CO$_3$ 193
    β-Ta 207
    β-U 207
    δ-Na$_2$CO$_3$ 191

γ-Na$_2$CO$_3$ 13
γ-Na$_2$CO$_3$ 193
[(C$_6$H$_5$)$_4$P]$_2$[TeBr$_6$
    (Se$_2$Br$_2$)$_2$] 186
[(Pb, Bi)S]$_x$
    [(Pb,Bi)$_2$S$_3$] 189
[PbS]$_x$[VS$_2$] 189
p-chlorobenzamide 209
A$_{1+x}$A$'_x$B$_{1-x}$O$_3$ 212
A$_2$BX$_4$ 193
Al$_{71}$Co$_7$Ni$_{22}$ 452
AlCoNi 172
AlCuFe 450
AlCuLi 450
AlMnSi 450
AlPdMn 234
alkane–urea 202, 337
Au$_{1-p}$Ag$_p$Te$_2$ 444
Aurivillius phases 212
Ba^{II} 204
Ba^{IV} 204
BCCD 412
BCPS 368
Bi^{III} 207
Bi$_2$Sr$_2$CaCu$_2$)$_{8+\delta}$ 368
Bi$_2$Sr$_2$CuO$_{8+\delta}$ 336
Bi$_{2m}$A$_{n-m}$B$_n$O$_{3(n+m)}$ 212
C$_4$H$_{12}$N$_2$ ZnCl$_4$ 448
C$_7$H$_6$ClNO 209
calaverite 444
cannizzarite 189
CdYb 269
CrFe 207
CuAuII 11
ferrites 217
Frank–Kasper σ-phase 207
hexamine suberate 434
Hg$_{3-\delta}$AsF$_6$ 13, 336
K$_3$In[PO$_4$]$_2$ 216
KSm(MoO$_4$)$_2$ 215
La$_2$Co$_{1.7}$ 170, 189, 198
La$_4$Ti$_3$O$_{12}$ 212, 214
LaTi$_{1-x}$O$_3$ 212
LaTiO$_3$ 212
NaNO$_2$ 12, 412
palmierites 217
quartz 425

530 Index

compounds (cont.)
  quininium
    (R)-mandelate 181, 183
  Rb^{IV} 207
  Sb^{II} 207
  $Sn_2P_2(Se_{1-x}S_x)_2$ 412
  Sr^V 207
  $TbMnO_3$ 468
  Te^{III} 208
  tetramethyammonium
    tetrachlorozincate 448
  $ThBr_4$ 367
  thiourea 412
  $TTF_7I_{5-\delta}$ 14
constant amplitude
  approximation 410
contrast variation 229
covering 143
crenel function 63, 176, 183
critical mode 320
critical wave function 357
crystallographic condition 4, 64, 71, 478
crystallographic
  operation 475
cut-and-project method 75, 125

damping 349
Dan Shechtman 14, 15, 25
Debye–Waller factor 149, 162, 333, 351, 383
decorated tiling 76
defects 472
defining relations 480
definition of crystal 15
definition of quasicrystal 16, 43
Delaunay cell 112, 127
Delaunay set 111
delta map 62
dense atomic planes 251
density function 185
deperiodization loop 427
devil's staircase 415, 422
difference Fourier
  synthesis 182
DIFFFOUR model 420
diffraction 148, 165
diffraction pattern as
  projection 166
diffuse scattering 351, 383
Dirichlet domain 112, 114
discommensuration 411, 421

discommensuration
  columns 427
discommensuration
  planes 426
discommensuration
  transition 422
displacements parameters
  anisotropic 173
displacive limit 419
domain structures 433
double chain model 334, 437
dynamic structure
  factor 319, 332

edge 113
equivalence
  space groups 482
  superspace groups 485, 489
Euclidean group 475
Euclidean transformation 2
Euler function 136, 476
extended mode 320
external space 56
extinction rule 94, 490
  centring 96
  general 95, 96

family 483
Farey numbers 322
Farey tree 417
ferroic 408, 420
FFT algorithm 185
Fibonacci chain 27, 28, 71, 114, 117, 119
  dynamics 337
Fibonacci tight-binding
  model 359
finite local complexity 166
floating phase 424
force field 433
  consistent 433
Fourier module 16
Fourier series truncation 187
Fourier transform and
  superspace 166
Frank–Van der Merwe
  model 323, 413
Frenkel–Kontorova
  model 323, 413
Frobenius congruences 480

$G_\mathbf{k}$, the group of $\mathbf{k}$ 8
gauge transformation 102

Gaussian integration
  method 177
geometric crystal class 478
glue atoms 29, 144
Goldstone mode 313, 314
grid method 122
group extension 481, 484
group representation 7
GVDOS 366

harmonic terms 180, 182
  and satellites 187
Harper's equation 359
Hartman–Perdok 455, 459
helical magnetism 11
Hermann theorem 309
Hofstadter butterfly 323
holohedry 477
  arithmetic 477
host–guest structures 14
host–guest system 41
Hubbard Hamiltonian 462
Hume–Rothery
  mechanism 440
hydrodynamic mode 313
hyperspace 56

icosahedral basis 26, 81, 137
icosahedral lattice
  body-centred 99
  face-centred 99
IDOS 323
image plate 169
incommensurate
  composite 13, 22, 40
  liquid crystals 471
  magnetic structure 462
  minerals 46
  modulation 12
  multiferroics 466
  photonic crystals 470
indefinite metric 98
indexing 224
indistinguishability 103
inelastic neutron scattering
  quasicrystals 368
inflation–deflation 97, 120
integrated density of
  states 323
internal space 56
*International Tables for
  Crystallography
  Volume C* 179
inverse Mathiessen rule 363

Index 531

irreducible representation 8, 40, 101

Jacobi–Anger expansion 175
k-hedral 114
Katz–Gratias model 254
kinematic approximation 148
klassengleich 106
Kronig–Penney model 358

lattice 476
lattice coordinates 5
lattice translation group 476
Laue diffractogram 170
Laue symmetry 16
Lifshitz invariant 410
Lifshitz point 412, 420
Lifshitz term 409
line broadening 339
local isomorphism 79, 114
localized mode 320
lock-in
   phase 198
lock-in transition 19, 411, 420
   partial 19

magnetic modulated structure 467
magnetic quasicrystals 465
magnetic space group
   klassengleich 106
   non-trivial magnetic group 105
   translationengleich 106
   trivial magnetic group 105
magnetic structure
   chromium 463
   modulated magnetism 462
   quasicrystals 465
   rare-earth elements 463
magnetic symmetry 465
main reflections 19
matching rules 116, 441
maximum entropy 183
Mermin formulation 103
mesoscopic quasicrystals 471
metric tensor 130, 306, 477
Meyer set 113
Miller indices 454
minerals
   incommensurate 46
MLD 114
mode counting 313

model set 112
modulated Kronig–Penney model 359
modulated phase 17
modulated quasicrystal 76
modulation
   atomic 173
   composition 62
   density 39
   displacive 12, 39
   function 19
   occupation 39, 62
   substitution 173
   vector 19
monohedral 114
monolithic active-pixel sensors 169
morphological importance 458
morphology 30, 444
multiferroics 466
mutual local derivability 114

natural
   incommensurate minerals 46
   quasicrystals 46
nesting properties 440
net plane 5, 123, 452, 454
Nobel Prize 2011 15
non-crystallographic symmetry 43
non-magnetic group 105
non-symmorphic space group 3
non-trivial magnetic group 105
normalizer 98
notation
   arithmetic crystal classes 479
   crystallographic operation 475
   orthogonal transformation in n dimensions 475
   point group operation 477
   space groups 479
   space groups modulated phases 486
   superspace groups 486
Nowotny phases 9

octonacci chain 118
one-line symbols 488

orbit 482
order parameter 313
order-disorder system 419
origin shift 480
orthogonal transformation 2

parallel space 56
paraphase 408, 420
participation ratio 348
particle density 81
Patterson analysis 226
Patterson function 149, 151
Patterson map 182
Pauling 15
Peierls mechanism 439
Penrose tiling 28, 116
perfect matching rules 116
periodic average structure 452
perpendicular (perp) space 56
phase transition 21
   1q–2q 424, 431
   incommensurate 19, 38
phason 306, 326, 328
   degrees of freedom 306
   jump 145, 306, 315
   mode 314, 383
photonic quasicrystals 470
physical space 56
Pim de Wolff 12
Pisot property 118
Pisot–Vijayaraghavan number 118
pixel detectors 169
plane group 4, 10
point group 2
projector on V_E, V_I 55
prototiles 113
pseudo-Brillouin zone 318
pseudo-gap 364, 440
pseudo-Mackay cluster 251

quasicrystal 14, 43
   n-dimensional 46
   growth 441
   natural 46
quasiperiodicity 15, 34

random tiling model 345
rank
   Fourier module 16, 34
   tensor 306
rare-earth elements 464
reciprocal basis 5

Index

reciprocal lattice 5, 32, 477
reciprocal space 5
reciprocal superspace 55
reducible
  Q- 478
  R- 478
  Z- 478
reflection condition 490
relatively dense 34, 140
resolution methods
  ab initio 184
  charge-flipping 7, 184, 226, 248, 283
roughening transition 458

$S(\mathbf{Q}, \omega)$ 332
Samson β-phase 10
satellites 19
sawtooth function 177, 183, 199, 209
scale-space group 100
scaling factor
  icosahedral case 99
  octagonal case 99
scaling symmetry 79, 96, 97
scanning electron microscopy 149
scattering function 332, 366
  static 149
section method 122
Seitz notation 479
self-similarity 323
setting 162
sliding mode 335
soft mode 328, 367, 408
soft quasicrystals 471
software
  DIRDIF 181
  JANA 178, 180
  MSR 179
  NADA 179
  REMOS 178
soliton 411
soliton density 411
space group 2, 479
space group type 482
space-time group 48
  generalized magnetic 49
spectrum
  absolute continuous spectrum 165, 166, 349
  discrete spectrum 166, 349
  point 166
  singular continuous spectrum 165, 166, 349

spin-density wave 440, 463
split basis 56
stability of facets 455, 460
star map 56
STM 149
stripes 427
stripples 421, 424, 427
  anti-stripples 429
structure
  host–guest 204
  modular 193, 211
  self-hosting 204
structure factor 150
  dynamic 319
  incommensurate structures 173, 175
  modulated phase 151, 154
  quasicrystal 160
structure refinement 182
subdirect product 104
substitution chain 117
subsystems 40
superspace 56, 172
superspace group 485
superstructure 35, 180, 191, 408
p-chlorobenzamide 209
surface
  clusters 458
  properties 462
  quasicrystals 458
  steps 457
  terraces 456
symmetry equivalent 114
symmorphic space group 3
system 483
system group 477, 483
system of non-primitive translations 480
systematic absence 194

t-plots 216
tensor 306
  contravariant tensor 306
  covariant tensor 306
  dielectric tensor 306
  elasticity tensor 311
  metric tensor 309
  piezo-electric tensor 311
  rank of a tensor 306
  strain tensor 310
thermal factor 162
Thue–Morse chain 167
tight-binding model 358

tiling 113
3DPT 138
Ammann tiling 139
Ammann–Beenker tiling 124, 125
decorated tiling 76
dodecagonal tiling 132
icosahedral tiling 138
kite-and-dart tiling 115
Penrose tiling 116, 133
rhombic Penrose tiling 116
Robinson tiling 116
square-triangle tiling 133
Tübingen Triangle Tiling 121
trace map 339, 341
transition
  displacive 12, 381
  order–disorder 12, 381
transition metal elements 464
translation
  non-primitive 3
  primitive 3
translation subgroup 2, 482
translationengleich 106
triacontahedron 450
trivial magnetic group 104
two-line symbols 488
Type I incommensurate phase 411
Type II incommensurate phase 411

unit cell 4, 113
unit cell nD 72, 81

VDOS 323
vector representation 307
vector system 480
vertex 113
vibrational density of states 323
Voronoi cell 112

Wigner–Seitz cell 4, 112
window 126
worms 473
Wyckoff positions 4

Yamamoto model 260

Z-module 476

The manufacturer's authorised representative in the EU for product safety is Oxford University Press España S.A. of el Parque Empresarial San Fernando de Henares, Avenida de Castilla, 2 – 28830 Madrid (www.oup.es/en or product. safety@oup.com). OUP España S.A. also acts as importer into Spain of products made by the manufacturer.